ADVANCES IN HEMATOPOIETIC STEM CELL RESEARCH

Edited by **Rosana Pelayo**

Advances in Hematopoietic Stem Cell Research
Edited by Rosana Pelayo

Published by InTech
Janeza Trdine 9, 51000 Rijeka, Croatia

Publishing Process Manager Maja Bozicevic

Technical Editor Teodora Smiljanic

Cover Designer InTech Design Team

Additional hard copies can be obtained from orders@intechopen.com

Advances in Hematopoietic Stem Cell Research, Edited by Rosana Pelayo
p. cm.
978-953-307-930-1

For EU product safety concerns:
IN TECH d.o.o., Prolaz Marije Krucifikse Kozulić 3, 51000 Rijeka, Croatia,
info@intechopen.com or visit our website at intechopen.com.

Meet the editor

Rosana Pelayo obtained her Bachelors, Masters and Doctorate degrees from the National Autonomous University of Mexico and completed her postdoctoral fellowship at the Oklahoma Medical Research Foundation, in the Immunobiology and Cancer Program. Dr Pelayo is currently Senior Research Scientist and Principal Investigator in the Oncology Research Unit of the Mexican Institute for Social Security, in Mexico City.

Contents

Preface

The prospective isolation of primitive blood-forming cells along with depicting of transcriptional networks that control early cell fate decisions, and characterization of microenvironmental signals influencing differentiation pathways during normal hematopoiesis, have been critical to the construction of a hierarchical model for the hematopoietic development, that has served as a paradigm for a number of systems within vertebrate development. Hematopoietic stem cell research has been helpful to elucidate mechanisms that govern tissue regeneration, to give an insight into perspectives that may allow protection of the system during disease, to design life-saving therapies, and to discover novel drug activities. This promising field is being accelerated by significant contributions in genomics, molecular biology, and technologies including fluorescent activated cell sorting and mouse engineering, that give us a more integrated view of the nature of stem cells .

This book, *Advances in Hematopoietic Stem Cell Research,* is devoted to current and in-progress scientific knowledge on basic aspects of these seminal cells and their therapeutic applications. The text consists of 20 chapters grouped into four sections: 1) Hematopoietic stem cell properties, 2) Regulation of hematopoietic stem cells, 3) Hematopoietic stem cells in aging and disease, and 4) Hematopoietic stem cell therapy.

The first section provides four comprehensive chapters on the functional characteristics and biological properties that make distinctive the conspicuous population of hematopoietic stem cells, including multipotency, self-renewal, quiescence and novel surface markers. Five chapters in the second section contain powerful information about intrinsic and extrinsic factors that determine cell fates in early development. The authors have discussed cellular and molecular aspects of the hematopoietic microenvironment within the bone marrow, and progress in searching of procedures to make the expansion of truly stem cells possible . Behavior of the stem and progenitor cells in aging and during disease is analyzed in the third section of this text, which highlights recent achievements in unraveling the role of primitive cells in the pathogenesis of hematological malignancies like leukemia.

Finally, to provide a comprehensive overview of the advancements in therapeutic applications of hematopoietic stem cells, a fourth section with five chapters addresses a number of diseases for which stem cell transplantation is the indicated therapy. Of special interest, has been the evaluation of quality and potency of stem and progenitor cells for therapy purposes. Substantial efforts to assemble the bridge between basic research and clinical applications are currently being recorded worldwide, and their review in this section may increase the interest for the book.

It is hoped that *Advances in Hematopoietic Stem Cell Research* will prove to be an enjoyable read and that it contributes to this area of Modern Medicine. This book is in effect a compilation of the latest advances resulted from the participation of distinguished and dedicated authors, experts in the field, to whom I am extremely thankful. I would like to acknowledge the terrific contribution of Maja Bozicevic in the professional editing of the book.

Rosana Pelayo
Oncology Research Unit, Oncology Hospital,
Mexican Institute for Social Security, Mexico City,
Mexico

Part 1

Hematopoietic Stem Cell Properties

1

Networks Establishing Hematopoietic Stem Cell Multipotency and Self-Renewal

Eliana Abdelhay, Luciana Pizzatti and Renata Binato
Instituto Nacional de Câncer, Rio de Janeiro, Brazil

1. Introduction

Hematopoiesis is a tightly regulated process maintained by a small pool of hematopoietic stem cells (HSC) capable of undergoing self-renewal and generating mature progeny of all of the hematopoietic cell lineages. To sustain the proper levels of blood cells, HSCs must continuously monitor and regulate the balance between self-renewal and lineage differentiation. To produce all hematopoietic cells, hematopoiesis proceeds in a step-wise manner from the primordial long-term (LT)-HSCs. LT-HSCs possess the ability to self-renew and the capacity for long-term reconstitution of lethally irradiated hosts. After a first step of differentiation, LT-HSCs lose their capacity for self-renewal and give rise to a population of short-term (ST)-HSCs. The ST-HSCs has a limited ability to self-renew and reconstitute lethally irradiated hosts, but differentiate into a multipotent progenitor (MPP) population. The MPPs lack the capacity to undergo self-renewal, but retain multipotency. From these multipotent progenitors develops a series of intermediate progenitors that give rise to the assorted hematopoietic lineages. In the classical pathway of hematopoiesis, these intermediates include the common lymphoid progenitors (CLPs) that differentiate into lymphoid, but not myeloid progeny, and the common myeloid progenitors (CMPs), which retain full erythromyeloid potential. The CMPs further differentiate to form the granulocyte/macrophage progenitors (GMPs) that differentiate to the myelomonocytic lineage and the megakaryocytic/erythrocyte progenitors (MEPs) that eventually differentiate to form red blood cells and platelets. All these blood cells produced daily in high numbers (1×10^{12} cells/day) are derived from a relatively small but rare fraction of multipotent cells, the HSCs (Weissman, 2000).

Transcriptional regulation is a key mechanism controlling HSC homeostasis, development, and lineage commitment.

2. Transcription factors in hematopoietic development

Hematopoiesis is regulated at the level of pluripotent HSCs and committed progenitors through growth and/or differentiation inducing factors (like EPO, G-CSF, GM-CSF, IL-1, IL-3) that interact with receptors and initiate signal transduction processes that culminate in the activation of new genetic programs. These external stimuli trigger intrinsic determinants of cell fate, the transcription factors which contribute to the reprogramming of HSCs into cell-lineage restricted pathways of maturation (Zon, 2008). Although the transcription factors involved in hematopoietic development belong to all classes of DNA-binding proteins some of them are involved in the regulation of self-renewal function primarily on HSCs while the others act on MPPs and/or early committed progenitors entering the cell-lineage restricted pathways of differentiation. While transcription factors as MLL, RUNX1, TEL/ETV6, SCL/TAL1 and LMO2 are required for HSC formation and function, others are necessary as key lineage restricted factors acting at the level of early pre-committed progenitors, using key partners that act synergistically or competing to restrict cell-lineage hematopoietic differentiation. GATA-1 and PU-1, for example, physically interact and antagonize with each other to promote either myeloid or erythroid maturation (Rekhtman et al., 1999; Zhang et al., 1999), which means that suppression of GATA-1 expression favors myeloid differentiation while inhibition of PU-1 promotes erythroid maturation. Additional, antagonistic interactions with other transcription factors have also been reported as C/EBPα that antagonizes FOG-1 in eosinophilic differentiation, EKLF antagonizes Fli-1 for erythroid versus megakaryocytic differentiation. Finally, repression of the *Pax-5* gene prevented Pro-B cell maturation to B cells, while promoting multi-potentiality into macrophage, T-NKs and dendritic cells (Huntly & Gilliland, 2005).

Transcription factors also interact with other proteins associated with chromatin modification and form active or repressive transcriptional complexes. Knockout of *Scl/Tal1* or *Lmo2* abrogates hematopoietic development. The precise mechanism through which such transcription activator or repressor complexes regulate the expression of several genes is critical, since the gene expression pattern regulates cell fate decision via cell-lineage restricted maturation. A critical point, however, for all these transcriptional complexes is the concentration of the transcription factor itself and its affinity to other interactive proteins.

Under normal hematopoiesis, several groups of hematopoietic and mature blood cells are generated. Hematopoiesis occurs unidirectionally and commitment from one step to the next occurs irreversibly, suggesting that transcription factors regulate cell fate along the specific cell-lineage pathways irreversibly. This occurs in such a way because intrinsic transcription factor network is coordinated with inputs resulting from external stimuli initiated within the hematopoietic cell niche. The question, however, whether one cell type of progenitors can be reprogrammed into another phenotype at the level of manipulation of transcription factor activation, is a potentially interesting one. Indeed, evidence now indicates that transfection of *Gata-1* into CMPs and/or CLPs redirects their commitment to another cell-lineage restricted pathway as megakaryocytic/erythroid. Similarly, pre-T cells can be reprogrammed to myeloid dendritic cells upon *PU-1* overexpression (Laiosa et al., 2006; Orkin & Zon, 2008).

3. Ontogeny of HSCs

In vertebrates, the production of blood stem cells is accomplished by the allocation and specification of distinct embryonic cells in a variety of sites that change during development. In mammals, the sequential sites of hematopoiesis include the yolk sac; an area surrounding the dorsal aorta termed the aorta-gonad mesonephros (AGM) region, the fetal liver, and finally the bone marrow. Recently, the placenta has been recognized as an additional site that participates during the AGM to fetal liver period. The properties of HSCs in each site differ, presumably reflecting diverse niches that support HSC expansion and/or differentiation and intrinsic characteristics of HSCs at each stage. For instance, HSCs present in the fetal liver are in cycle, whereas adult bone marrow HSCs are largely quiescent.

The initial wave of blood production in the mammalian yolk sac is termed "primitive." The primary function for primitive hematopoiesis is the production of red blood cells that facilitate tissue oxygenation as the embryo undergoes rapid growth. The hallmark of primitive erythroid cells is expression of embryonic globin proteins. The primitive hematopoietic system is transient and rapidly replaced by adult-type hematopoiesis that is termed "definitive". In mammals, the next site of hematopoietic potential is the AGM region. Hematopoietic cells were first detected in the aorta of the developing pig more than 80 years ago. Morphological examination revealed that a sheet of lateral mesoderm migrates medially, touches endoderm, and then forms a single aorta tube. Clusters of hematopoietic cells subsequently appear in the ventral wall. Similarly, an intraembryonic source of adult HSCs in mice capable of long-term reconstitution of irradiated hosts resides in the AGM region (Muller et al., 1994). At embryonic day 10.5, little HSC activity is detectable, whereas by day 11 engrafting activity is present. Additional hematopoietic activity in the mouse embryo was detected subsequently in other sites, including the umbilical arteries and the allantois in which hematopoietic and endothelial cells are co-localized (Inman & Downs, 2007). Umbilical veins lack hematopoietic potential, suggesting that a hierarchy exists during definitive hematopoiesis in which HSCs arise predominantly during artery specification. In addition, significant numbers of HSCs are found in the mouse placenta (Gekas et al., 2005; Ottersbach & Dzierzak, 2005), nearly coincident with the appearance of HSCs in the AGM region and for several days thereafter. Placental HSCs could arise through de novo generation or colonization upon circulation, or both. The relative contribution of each of the above sites to the final pool of adult HSCs remains largely unknown.

Subsequent definitive hematopoiesis involves the colonization of the fetal liver, thymus, spleen, and ultimately the bone marrow. It is believed that none of these sites is accompanied by de novo HSC generation. Rather, their niches support expansion of populations of HSCs that migrate to these new sites. However, until very recently, there has been no evidence by fate mapping or direct visualization that HSCs from one site colonize subsequent sites.

4. Pathways involved in the emergence of HSCs

The AGM has been characterized largely by morphology and functional assays, but the pathways involved in HSC generation remain incompletely defined. Studies of chick embryos demonstrate that endoderm has a prominent role and secretes inducing factors. Somitic mesoderm also contributes to the dorsal aspect of the aorta, and the addition of factors such as VEGF, TGF-β, and FGF to the somitic mesoderm leads to induction of hematopoietic tissue. In contrast, TGF-α and EGF suppressed formation of hematopoietic cells (Pardanaud & Dieterlen-Lievre, 1999).

Signaling pathways that regulate the induction of the AGM have been uncovered in mouse and zebrafish, *Notch 1* is required for artery identity and aortic HSC production (Kumano et al., 2003). The fate decisions imposed on mesodermal progenitors within the AGM are clearly influenced by the *Notch* pathway (Burns et al., 2005). For instance, mice deficient in *RBPj* (a downstream component of the *Notch* pathway) show expanded *VE-Cadherin* and *CD31/PECAM* endothelial cell expression with concomitant loss of definitive HSCs (Robert-Moreno et al., 2005). Ablation of the COUP-TFII transcription factor in endothelial cells enabled veins to acquire arterial characteristics, including the expression of *Notch1* and the formation of ectopic HSCs (You et al., 2005). This result would favor *Notch* acting to induce HSCs from a hemogenic endothelial cell. The model in which the *Notch* pathway regulates arterial and HSC fate choice either from distinct mesodermal populations or over different developmental windows since each decision can be uncoupled *in vivo* is very attractive. The finding that both aorta and vein express HSC markers in the *Notch*-activated state with minimal change in *ephrinB2a* expression indicates that *Notch* independently regulates mesoderm–HSC and artery–vein cell fate decisions.Lateral inhibition has been proposed in the central nervous system whereby *Notch* signaling promotes non-neural fates while inhibiting neural development (Lewis, 1998). HSC fate may be established by a similar mechanism whereby *Notch* activation in an endothelial or mesenchymal cell causes down-regulation of ligand production. Consequently, a cell that produces more ligand will force its neighbor to produce less, thus generating a salt-and-pepper pattern of cells containing elevated *Notch* activity. In this model, cells containing high levels of Notch Intra Cytoplasmatic Domain (NICD) would become HSCs, while those with low NICD activity would remain endothelial or mesenchymal.

5. Hematopoietic niches

Stem cells depend on their microenvironment, the niche, for regulation of self-renewal and differentiation. As the site of hematopoiesis changes during vertebrate development, the nature of the stem cell niche must also change. Mutant mice in which the *BMP* pathway is disrupted have increased numbers of osteoblasts and HSCs (Calvi et al., 2003; Zhang et al., 2003). These findings suggest that osteoblasts may represent a critical component of the bone marrow niche for HSCs. Microscopical examination revealed that HSCs appear to reside in the periosteal area of calvarium marrow, where osteoblasts represent an essential component of the bone marrow niche (Papadimitriou et al., 1994). Most recent live animal tracking experiments by using real-time imaging of individual HSCs have indicated that endosteum forms a special zone where HSCs reside (Lo Celso et al., 2009; Xie et al., 2009). The bone marrow HSC niche is constituted of mesenchymal cells type osteoblasts, extracellular matrix components and minerals (high density calcium salts), all of which contribute to the unique micro-environment (niche) (Moore & Lemischka, 2006; Wilson & Trumpp, 2006). At least two distinct hematopoietic progenitor cell supportive niches in bone marrow have been identified thus far: the osteoblastic, which is regulated by *BMP*, *osteopontin*, *angiopoietin-1*, *notch* and maybe others (Adams & Scadden, 2006; Wilson & Trumpp, 2006) and the other one, the vascular niche. The vascular niche is thought to be the site where actively dividing stem or progenitor cells is located, and osteoblastic niche is an environment promoting maintenance of quiescent HSCs (Calvi et al., 2003). Currently, how these two different niches communicate with each other is largely unknown.

The number of HSCs in the bone marrow niche is highly controlled through physical interactions among different cell types, in a way that maintains stem cell state. HSCs remain in a quiescent state through close interaction with osteoblasts where this interaction is not only crucial to attach HSCs to niche osteoblasts, but is also essential to maintain HSC dormancy and function. Many factors, including ligands for Notch receptors and N-cadherin, are liberated by osteoblasts, although the contribution of these to adult hematopoiesis remains to be established. The role of N-cadherin as a mediator of interactions with osteoblasts (Zhang et al., 2003), as well as the prominence of osteoblasts for HSC adherence, has been challenged (Kiel et al., 2007). Recent findings suggest that HSCs are maintained in a quiescent state through interaction with thrombopoietin-producing osteoblasts (Yoshihara et al., 2007). Thrombopoietin (TPO) is the primary cytokine that regulates megakaryocyte and platelet development. Thrombopoietin and its receptor Mpl also exert profound effects on primitive hematopoietic cells. All HSCs express Mpl; *TPO−/−* or *Mpl−/−* mice have a decreased number of repopulating HSCs (Solar et al., 1998). *In vitro* culture studies (Matsunaga et al., 1998) also indicate a role of TPO in promoting the survival of repopulating HSCs. Through study of AGM and fetal liver *Mpl−/−* HSCs, Petit-Cocault et al. (2007) showed that TPO contributes to both generation and expansion of HSCs during definitive hematopoiesis. An intracellular adaptor, Lnk, induces a negative signaling pathway downstream of TPO in HSCs (Buza-Vidas et al., 2006; Seita et al., 2007). Another study (Tong et al., 2007) on mice that express *Mpl* lacking the C-terminal 60 amino acids revealed a pivotal role of an unknown signal emanating from the membrane proximal region of the Mpl receptor or from JAK2 that is critical for maintenance of HSC activity.

The association of HSCs with osteoblasts is countered by other studies that place HSCs adjacent to vascular cells. The chemokine CXCL12 regulates the migration of HSCs to the vascular cells (Kiel & Morrison, 2006). Taken together, these findings suggest that HSCs reside in various sites within the marrow and that their function might depend on their precise localization. Much of the existing debate may be semantic, however, if the osteoblastic and vascular niches are intertwined and not physically separate. Alternatively, HSCs may truly reside in distinct sub-regions, which may endow them with different activities. Cellular dynamics within the niche are relevant to clinical marrow transplantation. For example, recent findings suggest that antibody-mediated clearance of host HSCs facilitates occupancy of the niche and transplantation by exogenous HSCs (Czechowicz et al., 2007).

The physical interactions between individual HSCs and osteoblasts may be effective in determining the stem cell number by facilitating asymmetric or symmetric divisions, which in turn enable HSCs to either self-renew themselves or give birth to early progenitors for blood cells production (Moore & Lemischka, 2006; Wilson and Trumpp, 2006). HSCs are not of static nature, but exist in a dynamic state, since they migrate from the bone marrow into the peripheral blood (frequent trafficking). Whether, or not, HSCs contribute into the repair of the vascular system, is still not known (Janzen & Scadden, 2006).

5.1 Signaling in the niche

Many cell culture experiments have shown that HSCs respond to multiple cytokines and that the fate of a HSC self renewal, apoptosis, mobilization from the niche, formation of differentiated progeny cells depends on multiple cytokines, adhesion proteins, and other signals produced by stromal cells and likely other cells in the body. Since osteoblast (a cell derived from mesenchymal stem cells) is a key component in the HSC niche for the regulation of HSC number via self-renewal (Adams & Scadden, 2006; Huang et al., 2007), modifications of osteoblast functions in co-orchestration with other niche components, would be pivotal for HSCs survival, self-renewal, differentiation and apoptosis under certain circumstances.

HSCs fate decisions is activated by external environmental stimuli and coordinated by intrinsic factors. External stimuli include hematopoietic growth factors such as SCF, BMP/TGF-β, FGF, TPO, WNT proteins (WNT3A), Angiopoietin-1, IL-3, IL-6, Flt3-ligand, as well as Ca2+, hypoxia, PGE2 and retinoic acid (Wilson & Trumpp, 2006) while intrinsic factors are essentially genes controlling cell cycle, apoptosis and chromatin remodeling.

5.2 How some extrinsic factors act in the niche

Stem cell factor receptor, also known as *c-kit* and its ligand *SCF* play a central role in hematopoiesis, melanogenesis and gametogenesis (Edling & Hallberg, 2007; Kent et al., 2008). *C-kit* is a member of the type-III subfamily of receptor tyrosine kinases that also includes the receptor for *M-CSF*, *Flt-3* and *PDGF*. It is expressed in HSCs (LT-HSCs, ST-HSCs and MPPs) (Zayas et al., 2008), normal B- and T-cell progenitors, mast cells, germ cells, melanocytes, neurons, glial cells, placenta, kidney, lung and gut cells. Deficiency and/or deregulation in *SCF* or *c-kit* produce defects in hematopoiesis leading to Acute Myeloid Leukemia (AML) (Scholl et al., 2008). Sporadic mutations of *c-kit* and autocrine/paracrine activation pathways of the *SCF/c-kit* pathway have been implicated in a variety of malignancies. Gain of function mutations of *c-kit* are associated with malignancies such AML, gastrointestinal tumors and mastocytomas. Moreover, expression of a defective *c-kit* leads to a decrease in repopulating HSCs (Ikuta & Weissman, 1992).

Binding of SCF to c-kit promotes dimerization and activation of protein kinase that auto-phosphorylates the receptor. Although SCF may not be essential for the generation of HSCs, numerous studies have shown that it prevents HSC apoptosis. Almost all cytokine combinations used to date for culturing HSCs include SCF. SCF potentiates the greater ability of fetal liver HSCs than adult HSCs to undergo symmetric self-renewal in culture this activity likely needs the cooperation of other factors. The membrane-bound form of SCF is also an adhesive molecule for HSCs to the bone marrow environment (Heissig et al., 2002) as interruption of the interaction between the membrane-bound stem cell factor on osteoblasts with the c-kit receptor on HSCs by blocking antibodies has demonstrated that *c-kit* signaling is essential to maintain HSC dormancy and function (Suzuki et al., 2006), and an increased number of osteoclasts was associated with HSC mobilization. Receptor activator of nuclear factor (NF)-κB (RANK) ligand and cathepsin K mediate the cleavage of membrane-bound SCF; this decreases the abundance of SCF and, therefore, increases HSC mobilization (Kollet et al., 2006). The involvement of SCF in survival, mobility and possibly self-renewal of HSCs in culture and in the HSC niche likely reflects the complex relationship of different cell fates of HSCs.

Transforming growth factor (TGF)-β potently inhibits HSC activity *in vitro* (Blank et al., 2008). However, a *TGF-β* signaling deficiency *in vivo* does not affect proliferation of HSCs. TGF-β and BMP are secreted ligands that are recognized by different receptors that dimerizes and activates downstream cytosolic targets, culminating with the translocation of these activated transcription factors to the nucleus. BMPs, members of the TGF-β superfamily, play important roles in HSC specification during development. A negative role of *BMP* signaling in maintenance of mouse HSCs was shown by its control of the size of the HSC endosteal niche (Ross & Li, 2006). BMP4 supports HSC expansion in culture and partially mediates the effects of Sonic hedgehog on cultured human HSCs (Bhardwaj et al., 2001). Recently, expression characterization of TGF-β superfamily ligands, receptors, and Smads in mouse HSCs was published; primary HSCs and the Lhx2-HPC cell line express most of the proteins required to transmit signals from several TGF-β family ligands (Utsugisawa et al., 2006). In addition, Pimanda et al. (2007) demonstrated the integration of *BMP4/Smad* pathway and *Scl* and *Runx1* activity in HSC development.

All long-term repopulating bone marrow HSCs express a fibroblast growth factor (FGF) receptor (Yeoh et al., 2006); both FGF-1 and FGF-2 support HSC expansion when unfractionated mouse bone marrow cells are cultured in serum-free medium. Crcareva et al. (2005) confirmed that FGF-1 stimulates *ex vivo* expansion of HSCs and showed that the expanded cells were efficiently transduced by retrovirus vectors. Conditional derivatives of FGF receptor-1 have also been used to support short-term HSC expansion and long-term HSC survival in culture (Schiedlmeier et al, 2007). However, the role of the *FGF* pathway in regulating adult HSCs or embryonic hematopoietic development is controversial as the same authors showed that the treatment of purified mouse HSCs that ectopically express *HoxB4* with the fibroblast growth factor receptor (FGFR) inhibitor SU5402 enhanced HSC repopulating activity. Similar results were obtained using primitive hematopoietic colonies derived from embryonic stem cells. These inconsistent results were obtained from different starting cell populations and under different culture conditions, suggesting that the crosstalk of FGF signaling with other pathways is complex.

The WNT protein binds to a receptor complex consisting of a member of the Frizzled family of seven transmembrane proteins and the LDL receptor-related proteins LRP5 or LRP6 (Clevers, 2006). In the canonical *Wnt* pathway, receptor activation leads to stabilization of *β-catenin*, which accumulates and translocates to the nucleus where it activates target gene expression in concert with transcription factors such as TCF and LEF. Fleming et al. (2008) analyzed the role of *Wnt* signaling on HSC activity, including its effects on cell-cycle quiescence and the capacity of HSCs to reconstitute the hematopoietic system of recipient mice (whose bone marrow has been ablated by radiation). In contrast to previous studies that genetically manipulated the HSCs themselves, they analyzed the effects of blocking *Wnt* signaling in the mouse bone marrow microenvironment by overexpression of dickkopf1 (*Dkk1*), an antagonist of *Wnt/ β-catenin* signaling. Dkk1 is a soluble secreted protein that interacts with the Wnt co-receptors LRP5 and LRP6 (Kawano & Kypta, 2003). It is known that the number of osteoblasts directly affects the number of long-term repopulating HSCs (Calvi et al., 2003; Zhang et al., 2003). The overexpression of *Dkk1* in the osteoblastic lineage under the control of a 2.3 kb fragment of the *collagen 1α* promoter reduced activation of the *Tcf/Lef* transcription factors in HSCs in a non-cell-autonomous manner.

The transgenic mice showed no significant alteration in the proportion of HSCs and common lymphoid progenitor cells under steady-state conditions. Although HSCs from the *Dkk1* transgenic mice could reconstitute the hematopoietic system of irradiated recipient mice, they lost their reconstituting capacity after repeated bone marrow transplantation, indicating that the inhibition of *Wnt* signaling in the niche results in the premature loss of self-renewal activity. These findings show that *Wnt/β-catenin* activity is crucial for the maintenance of HSC quiescence in the bone marrow niche.

The angiopoietin (*Ang*) family of growth factors is composed of four members that bind to the Tie-2 tyrosine kinase receptor; Ang growth factors are important modulators of angiogenesis. Members of the angiopoietin family of proteins contain an N-terminal coiled–coil domain that mediates homo-oligomerization and a C-terminal fibrinogen-like domain that binds Tie-2. To identify the HSCs in situ, Arai et al. (2004) analyzed the receptor tyrosine kinase Tie-2 expression in bone marrow and found that 5-FU-resistant Tie-2 expressing HSCs adhere to osteoblasts at the endosteal surface, in agreement with previous findings of Calvi et al. (2003) and Zhang et al (2003). They also demonstrated that angiopoietin-1 (*Ang-1*), a Tie-2 receptor ligand, is produced primarily by osteoblasts, indicating that Tie-2 and Ang-1 are expressed complementarily in the niche. Tie-2 together with Tie-1 was also found required for homing of HSCs to bone marrow. Taken together, Tie receptors seem one group of the likely candidates for localizing stem cells to the stem cell niche.

Mineral content of bone contributes to compose a unique extracellular matrix in bone marrow and distinguishes it from other mesenchymal tissues. The extracellular calcium concentrations are recognized by the seven-transmembrane calcium-sensing receptors and therefore can initiate an intracellular G protein–coupled response. Those receptors are found on hematopoietic cells and have also been identified on the surface of HSCs (Adams et al., 2006). Local calcium gradient is involved in retaining HSCs in close physical proximity to the endosteal surface of bone. Extracellular calcium ion concentrations in the endosteum are likely higher than in the central marrow region (Silver et al., 1988). In receptor deficient mice models, HSCs were found not to engraft in the bone marrow (Adams et al., 2006) suggesting that the ability of stem cells to sense and respond to the increased calcium concentrations at the endosteal surface participates in creating the unique stem cell-niche interaction that enables bone marrow hematopoiesis.

Most slow-cycling hematopoietic cells are found in the hypoxic zones close to bone surface and distant from capillaries (Kubota et al., 2008), raising the possibility that these hypoxic niches are important for diminished HSC proliferation. Evidence for quiescent HSCs situated in a hypoxic environment has lately been confirmed by analyzing bone marrow cells from mice injected with a Hoechst dye. Transplantation results showed that the bone marrow fraction with the lowest Hoechst-dye uptake, inferred to be hypoxic, had the highest amount of long-term repopulating cells (Parmar et al., 2007). Consistently, HSCs were found to be the most positive for binding of the hypoxic probe pimonidazole. The molecular mechanisms involve the hypoxia-inducible factor-1a regulated gene expressions in stromal cells, such as c-Kit, stromal cell derived factor−1, and others (Ceradini et al., 2004).

Other molecules were recently identified to have role in signaling pathways inside the niche. DNA array experiments showed that, among other proteins, IGF-2 is specifically expressed in cells that do support HSC expansion in culture. Moreover, it was showed that all fetal liver and bone marrow HSCs express receptors for IGF-2. The inclusion of IGF-2 with SCF, TPO, and FGF-1 supports an eight-fold increase of highly enriched HSCs in culture (Zhang & Lodish, 2004). Whether IGF-2 acts on self-renewal, apoptosis, differentiation, or homing of HSCs is unclear. Interestingly, IGF-2 was found to bind and stimulate self-renewal of human embryonic stem cells (Bendall et al., 2007). Angiopoietin-like proteins (Angptls) were also implicated in HSC expansion. Angptls are a family of seven secreted glycoproteins that share sequence homology with the angiopoietins (Morisada et al., 2006). Similar to the angiopoietins, each Angptl contains an N-terminal coiledcoil domain and a C-terminal fibrinogen-like domain. However, unlike angiopoietins, Angptls do not bind to Tie-2 or Tie-1 and their receptors are unknown. This suggests that Angptls may have different functions from the angiopoietins. Angptl7 was suggested to be a target of the *Wnt/β-catenin* signaling pathway. However, most of the physiological activities of the Angptls remain unknown. Recently Angptl2 and Angptl3 were identified as growth factors that stimulate *ex vivo* expansion of bone marrow HSCs. Other analogues, including Angptl5, Angptl7, and Mfap4, also support *ex vivo* expansion of HSCs.

5.3 Signaling through cell adhesion molecules

In addition to signaling pathways as described above, extracellular matrix components of the niche have also been shown to play role in regulating the HSC dynamics. A matrix glycoprotein, osteopontin (OPN), as a constraining factor on HSCs within the bone marrow microenvironment is produced by osteoblasts in response to stimulation (Stier et al., 2005). Using studies that combine OPN-deficient mice and exogenous OPN, Stier et al. (2005) demonstrated that OPN modifies primitive hematopoietic cell number and function in a stem cell non-autonomous manner. The OPN-null microenvironment is sufficient to increase the number of stem cells associated with increased stromal *Jagged-1* and *Ang-1* expression and reduced primitive hematopoietic cell apoptosis. The activation of the stem cell microenvironment with PTH was shown to induce a super-physiologic increase in stem cells in the absence of OPN. Therefore, OPN seems to be a negative regulatory element of the stem cell niche that limits the size of the stem cell pool and may provide a mechanism for restricting excess stem cell expansion under conditions of niche stimulation.

The production of OPN by osteoblasts is likely to be an essential requirement as shown by Karahuseyinoglu et al. (2007). Osteogenically induced umbilical cord stromal cells express OPN during the first week of induction followed by a third week expression of another matricellular protein, bone sialoprotein-2 (BSP-2). In the following weeks, in conditioned media differentiating osteoblasts express osteonectin and osteocalcin that led us to suggest that all those proteins have roles in autocrine regulation of osteoblast maturation and thus might serve to determine the conditional status of the partner cell(s) in hematopoietic niche microenvironment.

Previous studies showed that cell adhesion molecules, such as cadherins and integrins, are crucial for the interactions between HSCs and the osteoblastic niche. N-cadherin−mediated adhesion mediates slowing cell cycling of HSCs and may keep HSCs quiescent. Some studies showed that specialized spindle-shaped N-cadherin$^+$ osteoblasts are a key component of the bone marrow stem cell niche. HSCs are thought to be anchored to spindle-shaped N-cadherin$^+$ osteoblast cells via a homotypic N-cadherin interaction. Also, N-cadherin and β1-integrin are identified as the downstream targets in *Tie-2/Ang-1* signaling and *TPO/MPL* signaling (Yoshihara et al., 2007) in HSCs, respectively, suggesting a link between adhesion molecules and cell-cycle regulators in modulating the HSC−niche interaction. These data suggest cell-adhesion molecules not only contribute to the anchoring of HSCs to the niche, but also regulate cell-cycle quiescence of HSCs in the niche. However, the studies by conditional deletion of N-cadherin fail to support the effects of N-cadherin on hematopoiesis (Kiel et al., 2007).

The members of the *Notch* family are developmental morphogens shown to be expressed in self-renewing tissues, enhance the self-renewal capacity of HSCs and promote T-cell differentiation. *Notch* signaling is initiated by the involvement of the extracellular portion of Notch with its ligands Jagged/Delta. Activation of the *Notch* signaling pathway has been shown to potentiate self-renewal of HSCs. It is initiated by the binding of Jagged ligand to Notch protein followed by metalloproteinase (γ-secretase) cleavage in the extracellular receptor portion leading to the intracellular release of Notch (NICD). Then Notch translocate into the nucleus, where it forms a multimeric transcriptional complex with other transcription factors (Huntly & Gilliland, 2005). Inhibitors of γ-secretase abrogate the Notch signaling activation (Rizzo et al., 2008; Shih & Wang, 2007).

Calvi et al. (2003) and Duncan et al. (2005) demonstrated that the *Notch* signaling pathway plays a role in the osteoblast bone marrow HSCs niche. Notch ligands have positive effects on *ex vivo* expansion of HSCs: activated *Notch* is able to immortalize primitive mouse hematopoietic progenitors and Notch ligands support HSC expansion in culture (Chiba, 2006). Recently, by culturing human cord blood cells in serum-free medium supplemented with SCF, TPO, Flt3L, IL-3, IL-6/sIL-6R, and Delta 1, Suzuki et al. (Suzuki et al., 2006) reported an approximate six-fold increase in SCID-repopulating cell (SRC) number. It is noteworthy that there exists a dose effect for Notch ligands in HSC culture. Whereas a low amount of Delta 1 supports human cord blood SRC expansion, high amounts of the cytokine induce apoptosis (Chiba, 2006).

This emphasizes the complicated relationship among the different fates of HSCs. As conditional knockouts of *Notch1* and *Jagged1* have normal *in vivo* HSC activities (Mancini et al., 2005), there likely is functional redundancy of different *Notch* isoforms and their ligands.

Endothelial cells in the vascular niche environment contacting HSCs also provide maintenance signals on the HSC behaviour (Coultas et al., 2005 ; Li & Li, 2006). The main components of vascular niche - hematopoietic cells and endothelial cells - are closely related during development since they are both derived from haemangioblasts (Kopp et al., 2005). Previous studies have suggested that the vascular niche is the place for HSC differentiation and mobilization (Avencilla et al., 2004). Endothelial cells expressing vascular cell-adhesion molecule-1 (VCAM-1) associate closely with megakaryocytes and their progenitors through VLA-4 in response to chemotactic factors, stromal cell-derived factor- 1 (SDF1) and fibroblast growth factor-4 (FGF4), and thus provide a niche for megakaryocyte maturation and platelet production. The immediate juxtaposition of HSCs to endothelial cells also facilitates their rapid mobilization and entry into circulation in response to stress and, in the case of megakaryocytes, release of platelets directly into the blood. Endothelial cells promote survival of HSCs in culture, but this seems to be limited to certain populations of endothelial cells (Li et al., 2004). Fractions of HSCs in both adult bone marrow and spleen were found in close association with endothelial sinusoids (Kiel et al., 2005), suggesting that endothelial cells provide support to HSCs *in vivo*. Depending on these data, it is now plausible to note that while the osteoblastic niche provides a quiescent environment for HSC maintenance, the vascular niche offers an alternative niche for mobilized stem cells and promotes proliferation and further differentiation or maturation into the circulatory system. It would be interesting to further define the respective contributions of endothelial and endosteal niches to HSC behaviour.

5.4 Cell intrinsic responses

Recent studies have shown that Polycomb group (PcG) proteins and their interaction are important in the regulation of HSC self-renewal and lineage restriction. In particular, members of the PRC1 (Polycomb repression complex 1), such as *Bmi1*, *Mel18* and *Rae28*, have been implicated. *Bmi1* plays an important role in regulating the proliferative activity of stem and progenitor cells. It is required for the self-renewal of both adult HSCs and neural stem cells (Molofsky et al., 2005; Park et al, 2003). *Bmi1* enhances symmetrical expansion of the stem cell pool through self-renewal, induces a marked *ex vivo* expansion of multipotent progenitors, and increases the ability of HSCs to repopulate bone marrow *in vivo* (Iwama et al., 2004). Leukemic cells lacking *Bmi1* undergo proliferation arrest, differentiation and apoptosis, leading to failure of leukemia in a mouse transplant model (Lessard & Sauvageau, 2003). In *Bmi1*-deficient bone marrow there is an up-regulation of cell cycle inhibitors *p16* and *p19*, and the *p53*-induced gene *Wig1*, and a down-regulation of the apoptosis inhibitor *AI-6*. This suggests that a mechanism exists whereby *Bmi1* functions by modulating proliferation and preventing apoptosis (Park et al., 2004). *Bmi1* has also been shown to regulate the expression of *Hox* genes that are required for differentiation during hematopoiesis (van der Lugt et al., 1996).Loss or knockdown of another Polycomb gene, *Mel18*, leads to increased expression of *Hoxb4* (Kajiume et al., 2004), and transplanted *Mel18*-deficient bone marrow showed an increase in overall HSC numbers but a decrease in their activity owing to arrest in G_0 phase of the cell cycle. *Rae28*-deficient HSCs were defective in their long-term repopulating ability in serial transplantation experiments (Kim et al., 2004; Ohta et al., 2002). Taken together, these studies show the importance of the Polycomb proteins in HSC self-renewal and maintenance of the blood system.

Transcriptional repression by PcG proteins is essential for maintenance of HSC identity. Part of the mechanism by which it functions is by repression of genes that promote lineage specification, cell death and cell cycle arrest. More recently, PcG complexes have been shown to be essential for maintenance of the undifferentiated state in murine embryonic stem (ES) cells and human ES cells by directly repressing a large number of developmental regulators (Boyer et al., 2006; Lee et al., 2006). PcG complexes bind to and presumably repress the expression of a subset of these genes linked to differentiation. This represents a dynamic repression of genes required for differentiation, and a scenario in which PcG proteins act as transcription repressors by cooperating with a specific set of transcription factors in stem cells. Some target genes include *Hox* family members important for induction of differentiation. Expression of *Hox* genes that are involved in differentiation is repressed in the ES cells by PcG proteins. Thus, PcG complex repression is also necessary for ES cell identity. Taken together, these studies suggest that differentiation is the default state during stem cell replication, and self-renewal requires active repression of transcription factors that prevent self-renewal

The transcription factor *Tel* (Translocation Ets leukemia; also known as Etv6 [Ets variant gene 6]), an *Ets* (E-26 transforming-specific)-related transcriptional repressor, is also required for HSC maintenance. Conditional inactivation of *Tel/Etv6* in HSCs rapidly leads to the depletion of *Tel/Etv6*-deficient bone marrow. However, *Tel/Etv6* is not required for the maintenance of committed precursors. When it is conditionally inactivated in most hematopoietic lineages, it does not affect their differentiation or survival (Hock et al., 2004). At the moment, the mechanism by which *Tel/Etv6* modulates adult HSCs renewal is not known. Study of the downstream targets it represses should shed light on other players essential for HSC maintenance.

Pbx1 (pre–B-cell acute lymphoblastic leukemia) is a TALE class homeodomain transcription factor that critically regulates numerous embryonic processes, including hematopoiesis. *Pbx1* is preferentially expressed in LT-HSCs compared to more mature short-term HSCs and multipotent progenitor cells (Ficara et al., 2008). By using *Pbx1*-conditional knockout mice, it was revealed that *Pbx1* positively regulates HSC quiescence. Transcriptional profiling showed that a significant proportion of *Pbx1*-dependent genes are associated with the *TGF-β* pathway.

The homeobox (*Hox*) genes encode transcription factors that regulate embryonic body patterning and organogenesis. They play a role in the regulation of hematopoiesis. Overexpression of *HoxB4* in bone marrow leads to expansion of HSCs *in vivo* and *in vitro*, therefore appearing to be a positive regulator of HSC self-renewal (Antonchuck et al., 2002; Krosl et al., 2003; Miyake et al., 2006; Sauvageau et al., 1995). It therefore came as a surprise when *HoxB4*-deficient mice had normal hematopoietic development but exhibited only mild proliferative HSC defects (Brun et al., 2004). In an attempt to determine if this was due to compensatory mechanisms, the entire *HoxB* cluster was deleted. However, this did not lead to major defects in hematopoiesis (Bijl et al., 2006), possibly owing to compensation by *HoxA4* and/or *HoxC4*.

Gfi1 (Growth factor independence 1), a zinc-finger repressor, has been recently implicated as a regulator of HSC self-renewal. Two groups working independently determined that *Gfi1* controls self-renewal of HSCs by restraining their proliferative potential (Hock et al 2004; Zeng et al., 2004). They showed that *Gfi1*-deficient HSCs display increased proliferation rates and are also functionally compromised in competitive repopulation and serial transplantation assays. *Gfi1* might exert its effects on HSC proliferation by regulating the cell cycle inhibitor p21.

Gfi1 is originally recognized for its role in T-cell differentiation and lymphoma. *Gfi1* gene knockout is one of the first targeted mutants to exhibit the combination of an increase in cycling HSCs at the expense of HSC function. Both *Gfi1* knockout models displayed an increase in cycling cells within the HSC pool, a large decrease in HSC function in transplantation experiments. Profoundly reduced expression of *p21*, the cyclin-dependent kinase inhibitor, in *Gfi1* null HSCs may account for the mechanism. Thus, under normal homeostasis, *Gfi1* is thought to suppress the proliferation of HSCs, thereby keeping HSCs in quiescence.

Numerous studies have identified roles for *p53* in the proliferation, differentiation, apoptosis, and aging of hematopoietic cells. LT-HSCs express high levels of *p53* transcripts, which is an indication of roles of *p53* in HSC physiology (Dumble et al., 2007). Recently, *p53* has been identified as a positive regulator of HSC quiescence through analysis of *p53*$^{-/-}$ mice (Liu et al., 2009). Furthermore, in the same study, it was demonstrated that the increased quiescence of HSCs from MEF null mice, in which both *p53* and *p21* are up-regulated, is dependent on *p53*, but not *p21*, further confirming the positive role of *p53* in maintaining HSC in quiescence. *Gfi1* was identified as *p53* target gene, which is both shown important in regulating HSC quiescence by up-regulation or knockdown experiments.

Stem cell leukemia/T-cell acute lymphoblastic leukemia 1 (*SCL/TAL1*) plays a key role in controlling development of primitive and definitive hematopoiesis during mouse development. In adult HSCs, it is highly expressed in LT-HSCs compared with short-term HSCs and progenitors (Lacombe et al., 2010). SCL impedes G_0-G_1 transition in HSCs. The function of HSCs from *Scl* $^{+/-}$ mice or with decreased dosage of SCL protein by *in vitro* interference was shown decreased in various transplantation assays. At the molecular level, SCL maintains HSC quiescence by regulating gene expression of *Cdkn1a* and *Id1*.

Recently, many other transcriptional factors, such as interferon regulatory factor−2, a transcriptional suppressor of interferon signaling (Sato et al., 2009); *Nurr1*, a nuclear receptor transcription factor (Sirin et al., 2010); and thioredoxin-interacting protein, a transcriptional repressor (Shao et al., 2010), have been identified as positive regulators of HSC quiescence. Loss of HSC quiescence was observed in mice with deletion of each of these factors.

Individual member of Retinoblastoma (*Rb*) tumor suppressor gene family serves critical roles in the control of cellular proliferation and differentiation with functional redundancy for each other. The mice with conditional triple knockout of *Rb* family genes including *Rb*, *p107*, and *p130* display a cell-intrinsic myeloproliferation that originates from hyperproliferative early hematopoietic progenitors due to the loss of quiescence, and the mutant HSCs show strong short-term repopulation capacity but impaired long-term repopulation ability on transplantation. Thus, *Rb* family members collectively maintain HSC quiescence (Viatour et al., 2008).

It has been shown that the conditional inactivation of *c-Myc* induces excessive expression of *integrins* and *N-cadherin* in HSCs, leading to the enhanced HSC interaction with the niche, which subsequently enable *Myc*-deficient HSCs stay in quiescence. Conversely, enforced *c-Myc* expression in HSCs downregulates *N-cadherin* and *integrins*, leading to a loss of HSC function (Wilson et al., 2004).

p21 mRNA expression levels are dramatically lower in the *Gfi1*-deficient HSCs. *p21* itself has been implicated in the regulation of HSCs (Cheng et al., 2000). In its absence, HSCs have an impaired serial transplantation capacity. Another cell cycle inhibitor, *p18* , has also been shown to affect HSC self-renewal. The absence of *p18* leads to increased HSC self-renewal (Yuan et al., 2004; Yu et al., 2006). Therefore, intricate control of the cell cycle and proliferation machinery is required for self-renewal regulation.

In contrast to *p21*, little is known about the role of *p57* in adult stem cell populations. Using primary human hematopoietic cells and microarray analysis, Scandura et al. (2004) identified *p57* as the only cyclin-dependent kinase inhibitor induced by TGF-β. Up-regulation of *p57* is essential for TGF-β−induced cell-cycle arrest in these cells, which may represent the mechanisms by which TGF-β affects cell-cycle arrest and stem cell quiescence.

Bone marrow is a very low oxygen tension environment that would protect cells from exposure to oxidative stress. Various intrinsic factors have also been identified to function in maintaining low oxidant levels in HSCs. ATM, a cell-cycle checkpoint regulator activated after DNA damage, is shown to regulate oxidant levels in HSCs (Ito et al., 2006). ATM deficiency-induced ROS elevation in HSCs specifically activates the p38 mitogen-activated protein kinase (MAPK) pathway, a signaling pathway responding to diverse cellular stresses, leading to a defect in the maintenance of HSC quiescence (Ito et al., 2004). *ATM*$^{-/-}$ mice over the age of 24 weeks show progressive bone marrow failure due to a defect in HSC function associated with elevated levels of ROS. Treatment with anti-oxidative reagents, N-acetyl cysteine or with a MAPK inhibitor restores reconstitutive capacity and quiescence of *ATM*$^{-/-}$ HSCs.

Members of the FoxO subfamily of forkhead transcription factors have been shown to protect HSCs from oxidative stress by up-regulating genes involved in their detoxification. Triple knockout mice of *FoxO1*, *FoxO3*, and *FoxO4* exhibited defective long-term repopulating activity of HSCs, which correlated with increased cycling and apoptosis of HSCs, as well as increased levels of ROS in HSCs (Tothova et al., 2007). Similarly, the HSC compartment in *FoxO3a* null mice suffers from augmented levels of ROS and subsequent bone marrow failure (Miyamoto et al., 2007). The HSC defect resulting from loss of *FoxOs* could also be rescued by administration of the antioxidant N-acetyl cysteine.

It is conceivable that both the hypoxic environment in which the HSCs reside and the intrinsic factors in HSCs serve to protect HSCs from oxygen radicals, keeping HSCs' quiescent status.

The *JAK-STAT* (Janus family kinase–signal transducer and activator of transcription) pathway is a common downstream pathway of cytokine signaling that promotes hematopoiesis. Constitutive activation of the transcription factors of the *Stat* family, particularly *Stat3* and *Stat5*, are frequently detected in leukemias, lymphomas and solid tumors. In order to evaluate their role in HSCs, constitutively active *Stat* mutants were used to activate signaling in HSCs. Activation of *Stat5* in HSCs led to the dramatic expansion of multipotent progenitors and promoted HSC self-renewal *ex vivo* (Kato et al., 2005). Deletion of *Stat5* resulted in profound defects in hematopoiesis and markedly reduced ability of the mutant cells to repopulate the bone marrow of lethally irradiated mice (Snow et al., 2002). In a mouse model of myeloproliferative disease (MPD), sustained *Stat5* activation in HSCs and not multipotent progenitors induced fatal MPD, suggesting that the capacity of *Stat5* to promote self-renewal of hematopoietic stem cells is crucial for MPD development. Another group showed that transduction of adult mouse bone marrow cells with a constitutively activated form of *Stat3* increased their regenerative activity in lethally irradiated recipients, whereas the transduction of these cells with a dominant negative form of *Stat3* suppressed their regenerative activity (Chung et al., 2006). These studies suggest that Stat proteins play a role in HSC self-renewal and potentially in other tissues; owing to the wide range of solid tissue and blood malignancies that harbor constitutively activated Stats.

Studies using transgenic mice constitutively expressing *BCL2* (*B-cell lymphoma 2*) in all hematopoietic tissues provide evidence directly supporting this theory. The forced expression of the oncogene *Bcl2* resulted in increased numbers of transgenic HSCs *in vivo* and gave these cells a competitive edge over wild type HSCs in competitive reconstitution experiments (Domen et al., 1998; Domen et al., 2000) suggesting that cell death plays a role in regulating the homeostasis of HSCs. Recently, *Mcl1* (Myeloid cell leukemia 1), another anti-apoptotic *Bcl2* family member, has been shown to be required for HSC survival (Opferman et al., 2005).

6. Quiescence or self-renewal

In order to both maintain a supply of mature blood cells and not exhaust HSCs throughout the lifetime of an individual, under steady state, most HSCs remain quiescent and only a small number enter the cell cycle. However, in response to hematopoietic stress such as blood loss, HSCs exit quiescence and rapidly expand and differentiate to repopulate the peripheral hematopoietic compartments. When quiescence is disrupted, HSCs displayed defective maintenance in G_0 phase of cell cycle, leading to premature exhaustion of the stem cell pool under conditions of hematopoietic stress, impaired self-renewal, and loss of competitive repopulating capacity, eventually causing hematological failure.

Quiescence of HSCs is not only critical for protecting the stem cell compartment and sustaining stem cell pools over long periods, but it is also critical for protecting stem cells by minimizing their accumulation of replication-associated mutations. The balance between quiescence and proliferation is tightly controlled by both HSC-intrinsic and -extrinsic mechanisms. Understanding quiescence regulation in HSC is of great importance not only for understanding the physiological foundation of HSCs, but also for understanding the pathophysiological origins of many related disorders.

In steady state conditions HSCs are in a slowly dividing state, termed relative quiescence, with a cell division cycle in the mouse in the range of 2–4 weeks, localized in close contact with stromal cells, including osteoblasts (Calvi et al., 2003; Zhang et al., 2003). This is in contrast to the rapidly cycling hematopoietic progenitor cells, which are more committed to differentiation than HSCs. The balance between quiescent and cycling stem cells was proposed to rely on the amount of soluble cytokines, which result in HSCs relocating from the osteoblastic to the vascular niche (Heissig et al., 2002). However new results indicate that it depends on a complex network of signals.

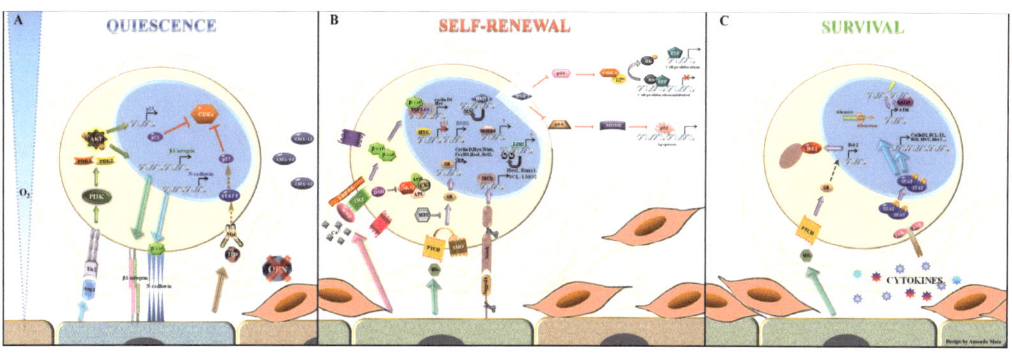

Fig. 1. Networks interaction model for: A) quiescence, B) self-renewal and C) survival.

In part, the dramatic contrast in cell cycle status between stem and progenitor cells has led to the hypothesis that cell cycle regulation plays a fundamentally important role in stem cell fate determination. This hypothesis is supported by recent data demonstrating a slower rate of division in Hoxb4hiPbx1lo cells, which extensively self-renewal *in vitro*, compared to control cells (Cellot et al., 2007). It is essential for an HSC to undergo cell division if it is to self-renew, but how the cell division cycle is integrated into the process of self-renewal is unclear. It is also unknown as to whether cell cycle regulation represents an intrinsic or extrinsic modifier of HSC fate.

6.1 How HSC maintain quiescence

Negative regulators of both *Cdk2* and *Cdk4/6* activity, and therefore *Rb* function, have been demonstrated to have roles in regulating HSCs (Cheng et al., 2000; Janzen et al., 2006; Stepanova & Sorrentino, 2005; Van Os et al., 2007; Walkley et al., 2005). For the most part however these phenotypes have been relatively subtle, particularly when compared to hematopoietic phenotypes apparent after disruption of transcription factors such as C/EBPα (Hock et al., 2004) and Tel (Zhang et al., 2004) amongst others, and are often apparent only after serial transplantation. The "Rb pathway" has also been implicated in phenotypes observed in both the *Bmi1-/-* and *ATM-/-* HSCs (Ito et al., 2004; Lessard & Sauvageau, 2003; Park et al., 2003). The interaction of cell cycle regulators with other factors such as Hoxb4 or telomerase deficiency has produced much more striking phenotypes than that observed for the cell cycle mutants in isolation (Choudhury et al., 2007; Miyake et al., 2006). While clearly demonstrating that cell cycle modifiers have roles in regulating stem cells, particularly HSCs, the aforementioned studies have not been able to clearly discriminate between intrinsic or extrinsic contributions to HSC fate as all studies to date had utilized non-hematopoietic restricted mutant alleles. A study demonstrating that the *p27Kip1-/-* microenvironment mediates the myelo-lymphoid expansion observed in the *p27Kip1-/-* animals raises the possibility that the HSC expansion observed in *p27Kip1-/-* bone marrow is extrinsic in nature (Chien et al., 2006; Walkley et al, 2005). This result suggested that cell cycle regulators might play a role in regulating the competence of the hematopoietic niche, in addition to having potential intrinsic roles in HSC fate determination. Moreover Daria et al (2007) observed a requirement for *Rb* in the stress response of HSCs and this has also previously been suggested in the context of the role of Rb in erythropoiesis (Spike et al., 2004; Spike & Macleod, 2005).

Also of note is that the cell division dynamics of HSCs change during development, from rapidly cycling and dividing cells during the fetal liver and early stages of life to relatively quiescent and more slowly cycling in the adult context (Bowie et al., 2007; Bradford et al., 1997; Ito et al, 2000; Kiel et al., 2007; Sato et al., 1999). Thus the role for *Rb* may be context dependent, both in terms of stress response and developmentally in the regulation of HSC fate.

One important point that is becoming clearer recently is how some HSCs are maintained quiescent while others enter self-renewal program. Although bone-lining cells in the endosteal surface are often described as osteoblasts in the literature, they are heterogeneous in their degree of differentiation, and only a minority of these cells are actually bone synthesizing osteoblasts. So a good hypothesis is that in the endosteal niche some cells are in contact with true osteoblasts that expresses the necessary factors to maintain quiescence while others are not receiving the same signalization so will follow other fate.

Li (2008) proposed that HSC quiescence is maintained through several signaling pathways including positive and negative regulators from extrinsic and intrinsic factors already described. In this context the *Tie-2/Ang-1* signaling pathway plays a critical role in the maintenance of HSCs in a quiescent state in the bone marrow niche (Adams et al., 2006). HSCs express the receptor tyrosine kinase Tie-2 and osteoblasts are the source of the Ang-1 ligand for Tie-2. *Tie-2/Ang-1* signaling activates its key downstream targets, *β1-integrin* and *N-cadherin* in lineage-negative, *Sca-1*, *C-kit* double-positive (LSK), and *Tie-2*-positive cells, and promotes HSC interactions with extracellular matrix and cellular components of the niche. This interaction is sufficient to maintain the quiescence and enhanced survival of HSCs by preventing cell division (Arai et al., 2007). *Ang-1/Tie-2* signaling also activates the phosphatidylinositol 3-kinase/Akt signaling pathway (Visnjic et al., 2004). Phosphatidylinositol 3-kinase/Akt signaling regulates several cell-cycle regulators, such as the CDK inhibitor, p21, which in turn leads to HSC quiescence.

Other pathway that has been proven to enhance quiescence is TPO/MPL. TPO is secreted by osteoblast while MPL is expressed in the membrane of HSCs. Interaction of these two proteins maintain HSCs attached to osteoblasts by activation of a pathway that results in the expression of their adhesion molecules targets and at the same time activate genetic programs which will control entry in cell cycle and survival of HSC. How these positive regulators interact with other positive and negative regulators is not completely understood. Moreover, which signaling pathways are being activated and which genes have their expression changed waits to be clarified.

A recent study by Wang et al. (2009) recently identified STAT5, a downstream target of MPL, as a positive regulator of HSC quiescence by analyzing *STAT5*$^{-/-}$ mice. Expression of quiescence regulators including *Tie-2* and *p57* are decreased in *STAT5*$^{-/-}$ HSCs. This study demonstrated that *STAT5* might mediate *MPL* effects in maintaining HSC quiescence during steady state hematopoiesis and that the same pathway directly or indirectly regulates *Tie-2* and *p57*. Interestingly, up-regulation of *p57* is essential for TGF-β–induced cell-cycle arrest. How these pathways are connected awaits more investigation.

Two other signaling pathways that act as positive and negative regulators of quiescence deserve more discussion, the Hypoxia induced factor (HIF) and Osteopontina signaling pathway. Hypoxia microenvironment seems to be important for maintaining HSC quiescence The molecular mechanisms for this involve the hypoxia-inducible factor-1a (HIF-1a) regulated gene expressions in stromal cells. Two genes known to be targets of the HIF pathway are *c-Kit* and stromal cell derived factor−1 (*SDF-1* or *CXCL12*) that both have proven to be important to HSC maintenance.

Osteopontina (*OPN*) is a negative regulator of HSC quiescence as an OPN-null microenvironment is sufficient to increase the number of stem cells associated with increased stromal *Jagged-1* and *Ang-1* expression and reduced primitive hematopoietic cell apoptosis. OPN seems to function by preventing HSC cycling. It is interesting to note that the release of this inhibition occurs in parallel with the possible activation of the *Notch* pathway.

As we can see there is many connections between intrinsic factors and extrinsic cues and between different intrinsic factors or different extrinsic factors. Some intrinsic factors function through affecting extrinsic factors, such as *c-Myc*, which negatively regulates HSC quiescence by controlling *N-cadherin* expression level, reducing the *N-cadherin*–mediated interaction between HSCs and niche. Some extrinsic cues function through certain intrinsic factors, such as *STAT5*, which may serve as a component of *MPL*–induced signaling pathway, mediating MPL's effects in maintaining HSC quiescence.

Interestingly, among those HSC quiescence regulators that have been identified so far, the majority are positive regulators; few are negative for the maintenance of HSCs quiescence. This is consistent with the idea showing that the bias toward reduced gene expression that actively maintains HSC quiescence is an important mechanism of HSC proliferation, suggesting that various positive regulators of HSC quiescence are actively restricting proliferation of HSCs, and that there may exist signals in the environment to promote HSC proliferation.

6.2 Deciding for self-renewal

Many signaling pathways are thought to contribute to stem cell self-renewal in the marrow niche including *Notch* (Maillard et al., 2003), Wnt (Duncan et al. 2005; Reya et al., 2003; Willert et al., 2003) and *Hedgehog* (Baron, 2001 ; Bhardwaj et al., 2001; Gering & Patient 2005).

Activated *Notch* expands the stem and progenitor cell compartment by either influencing undifferentiated cells to adopt a HSC fate or by causing a G_0 HSC population to up-regulate *runx1*-dependent gene expression. Findings that the stem cell markers *runx1*, *scl*, and *lmo2* were transcriptionally increased in response to NICD (Notch Intra-Cytoplamatic Domain) indicate that stem and progenitor cells were expanded in the adult marrow, possibly by increasing stem cell self-renewal. A conditional allele of *runx1* was generated in the mouse to study the loss of *Runx1* function during adult hematopoiesis (Growney et al., 2005; Ichikawa et al., 2004). In transplantation studies, *Runx1*-excised marrow cells showed a reduced competitive repopulating ability in long-term engraftment assays (Growney et al., 2005), demonstrating that Runx1 is essential for normal stem cell function.

The *Wnt/β-catenin* signaling pathway also plays a crucial role during self-renewal of HSCs (Nemeth & Bodine, 2007). Deregulation of this pathway has been implicated in the formation of solid tumors, like lung epidermal adenocarcinomas, breast carcinomas and intestinal colorectal tumors just to mention a few (Reya & Clevers, 2005). Although several *Wnt* genes are expressed in bone marrow, the precise role of *Wnt* signaling pathway in HSCs and its mechanism(s) of action remained unclear until very recently.

There is a multitude of *Wnt* signaling cascades some of them regulating one another. Using different receptors, Wnt proteins can trigger at least three intracellular signaling pathways: the canonical b-catenin pathway, the non-canonical calcium pathway and the c-Jun N-terminal kinase pathway (Zeng et al., 2004). Several components of the *Wnt* signaling machinery have been shown to play a role in HSC self-renewal. Both canonical as well as non-canonical pathways seem to be involved, since the canonical ligand *Wnt3a* intrinsically promotes self-renewal (Luis et al., 2009). On the other hand, the non-canonical ligand *Wnt5a* has been shown to extrinsically promote self-renewal by inhibiting canonical signaling (Murdoch et al., 2003). The mechanistic basis for the balance between canonical and non-canonical pathways is not fully understood. It is likely that numerous *Wnt* inhibitors or antagonists are modulating *Wnt* signaling.

Taken together, the existing studies suggest that canonical Wnt signaling may not be strictly required for HSC function, but that canonical Wnt signaling may affect self-renewal and differentiation of HSCs depending on the extent of canonical Wnt signaling and on the context of expression of additional genes. Non-canonical Wnt signaling and/or other signaling pathways may also compensate for the absence of canonical Wnt signaling in maintaining the self-renewal of HSCs (Huang, 2007).

To exactly control the fine tune of *Wnt* in HSC it is likely that the numerous Wnt-signaling inhibitors (Dickkopf homolog (Dkk), Wnt inhibitory factor (Wif) or secreted frizzled-related protein (Sfrp), or other Wnt antagonists, such as Kremen, Ctgf, Cyr61, Sost and Sostdc1) have to be the correctly expressed. Interestingly, some of these molecules also directly stimulate certain Fzds independent of Wnt factors. For example, Sfrp1 directly activates Fzd2, as well as Fzd4, and Fzd7 but can also interact with Wnt5a (Rodriguez et al., 2005; Dufourc et al, 2008; Matsuyama et al, 2009 & Kirstetter et al., 2006). This balance and feedback mechanisms between canonical and non-canonical Wnt signaling, suggests that *β-catenin* is the primary regulatory target of Wnt signaling. However, overexpression or stabilization of *β-catenin* results in expansion of the HSC pool, but, at the same time, the loss of myelopoiesis is due to a differentiation block (Renstrom et al., 2010), suggesting that b-catenin promotes self-renewal and/or inhibits differentiation.

Conversely, Wnt signaling also induces increased expression of *HOXB4, Bmi1* and targets of *Notch-1*, genes that are implicated in self-renewal of HSCs . Transcription factors of homeodomain family (HOX family) have been found to regulate HSC self-renewal and downregulate differentiation. Disruption of HOX genes in mice led to abnormalities in multiple hematopoietic cell lineages. Moreover, overexpression of HOX genes (like HOXB4) has been associated with HSCs ex vivo expansion and HOX gene mutation with acute leukemia. *Bmi1*,a polycomb gene, seems to have a repressor role over *p16* inhibiting apoptosis of HSCs and thus contributing to its maintenance . So the correct Wnt signaling seems to be essential to integrate the intracellular response in the decision to self-renew or differentiate (Reya et al., 2003).

The investigation of the interactions between Bmi1 and Hoxb4, showed that Bmi1 is not required for the in vivo expansion of fetal HSCs but is essential for the long-term maintenance of adult HSCs. Moreover, Hoxb4 overexpression induces an expansion of Bmi1-/- STR-HSCs leading to a rescue of their repopulation defect. Together, these results support the emerging concept that fate and sustainability of this fate are two critical components of self-renewal in adult stem cells such as HSCs.

Moreover Polycomb group (PcG) proteins play a role in the transcriptional repression of genes through histone modifications. Recent studies have clearly demonstrated that PcG proteins are required for the maintenance of embryonic as well as a broad range of adult stem cells, including hematopoietic stem cells (HSCs). PcG proteins maintain the self-renewal capacity of HSCs by repressing tumor suppressor genes and keep differentiation programs poised for activation in HSCs by repressing a cohort of hematopoietic developmental regulator genes via bivalent chromatin domains. Enforced expression of one of the PcG genes, Bmi1, augments the self-renewal capacity of HSCs. PcG proteins also maintain redox homeostasis to prevent premature loss of HSCs. These findings established PcG proteins as essential regulators of HSCs and underscored epigenetics as a new field of HSC research (Li et al., 2010; Komuna, 2010).

Recently we demonstrated that another polycomb group member, Suz12 gene, is activated by the non canonical Wnt pathway and may epigenetically inhibit genes involved in hematopoietic differentiation. These data pointed to cell cycle changes, deregulation of early differentiation genes and regulation of PRC2 polycomb complex genes, due to Suz12 role in CML blast crisis. This observation indicates that the cross talk between Wnt and Polycomb pathways may promotes hematopoietic differentiation. (Pizzatti et al., 2010).

Taken together all these data fits in a model were HSCs fated to self renew are in contact in the endosteal niches with osteoblasts expressing *Notch* legands (*Jagged*) so the pathway that will be induced is *Notch* pathway.

The *Hedgehog* (*Hh*) is a ligand that binds and represses the *Patched* receptor and thereby releases the latent activity of the multipass membrane protein Smoothened, which is essential for transducing the *Hh* signal. Using *Patched*$^{+/-}$ mouse with increased *Hh* signaling activity, it was demonstrated that constitutive activation of the *Hh* signaling pathway results in the steady-state accumulation of phenotypically defined HSCs and an increase in the proportion of cycling cells within this population (Kuhn et al., 1995). However, HSC activity on secondary transplantation is reduced 3-fold, indicating the functional exhaustion of the HSC pool in this mutant. In vivo treatment with an inhibitor of the *Hh* pathway rescues these transcriptional and functional defects in HSCs. This study establishes *Hh* signaling as a negative regulator of the HSC quiescence. In contrast to the germline *Patched*$^{+/-}$ mode, the mode of conditional deletion of Smoothened in the adult hematopoietic compartment was used in other two studies. However, the negative effects of the Hh pathway on HSC quiescence were shown in one study (Walkley & Orkin, 2006) and not in another (Stead et al., 2002). The discrepancy is possibly due to a distinct mode of deletion. How this pathway collaborate with the two others is not clear although interaction through GSK3 have been already proposed.

The outcome of *Hh* signaling varies according to the receiving cell type. GLI, the cytoplasmic effector of *Hh* signaling activates the transcription of several target genes as *CyclinD1 and D2,N-Myc,Wnts, FoxM1,Hes 1,Bcl2,Osteopontin* and others. If these genes are activated in HSCs has not yet been defined but if they are, a clear interconnection between several important signaling pathways is visualized.

One important point when talking about self-renewal is how to prevent exhaustion of the HSC pool.

7. Role of telomerase in hematopoietic stem cell

Stem cells self renewal capacity is believed to be closely associated with tissue degeneration during aging. Studies of human genetic diseases and gene-targeted animal models have provided evidence that functional decline of telomeres and deregulation of cell cycle checkpoints contribute to the aging process of tissue stem cells. Telomere dysfunction can induce DNA damage response via key cell cycle checkpoints, leading to cellular senescence or apoptosis depending on the tissue type and developmental stage of a specific stem cell compartment (Ju Z et al., 2011).

Studies in hematopoietic stem cell (HSC) biology are often focused on "self-renewal" and differentiation. Implicit in the word self-renewal is that the two daughter cells generated by a self-renewal division are identical to the parental cell. Strictly speaking, this is not possible because DNA is continuously damaged and repaired by DNA-repair mechanisms that are not 100% efficient (Lansdorp et al., 2005).

It is important to note that the efficiency of DNA repair varies greatly among different stem cell types. For example, embryonic stem cells are quite resistant to DNA damage and maintain the length of telomere repeats on serial passage, whereas HSCs are quite sensitive to DNA damage and less able to maintain telomere length. This idea has given rise to the notion that many aspects of normal aging could primarily reflect limitations in DNA repair and telomere-maintenance pathways in the (stem) cells of the soma (Lansdorp et al., 2005).

The loss of telomere repeats in adult hematopoietic cells (including purified "candidate" HSCs) relative to fetal hematopoietic cells also fits a model that postulates a finite and limited replicative potential of HSCs (Vaziri et al., 1994; Lansdorp et al., 1995; Lansdorp et al., 1997). How this collaborates with the model of LT-HSCs given rise to ST-HSCs has not still been addressed.

Eukaryotic chromosomes are capped by special structures called telomeres, which are guanine-rich, simple repeat sequences. Telomeres act to guarantee chromosome integrity by preventing illegitimate recombination, degradation, and end fusions (Blackburn et al., 1991; Stain et al ., 2004).

Synthesis and maintenance of telomeric repeats are accomplished by a specialized ribonucleoprotein complex known as telomerase. Telomerase consists of an essential RNA template and protein components, one of which appears to resemble reverse transcriptase. In the absence of telomerase, the failure of DNA polymerase to fully synthesize DNA termini leads to chromosome shortening (Stain et al., 2004; Lee et al., 1998).

In contrast to mice were short telomeres maintain cell survival for some generations, a modest two fold reduction in telomerase levels in humans (resulting from haploinsufficiency for the telomerase RNA template gene) typically results in premature death from complications of aplastic anemia or immune deficiency. Recent studies indicate that short telomeres and eventual marrow failure may also result from haploinsufficiency for the telomerase reverse transcriptase(hTERT) gene (Yamaguchi et al., 2005).

Moreover the large number of HSCs typically used in clinical transplant settings may effectively prevent their replicative exhaustion. Variations in telomere length between cells and individuals have even made it difficult to reproducibly document a decline in telomere length following transplantation. Nevertheless, a significant shortening of telomeres was observed in the first year after allogeneic bone marrow transplantation (Landsdorp et al., 2005). Furthermore, marrow failure with pronounced telomere shortening has been described in a few long-term survivors of HSC transplants. Although there is little evidence to suggest that telomere shortening will result in an epidemic of marrow failure in HSC transplant recipients, caution remains warranted when the cell number available for transplantation is limited or when the telomere length in HSCs for transplantation is short, as in cells from old donors or patients with telomerase deficiencies. It is tempting to speculate that some of the advantages of cord blood HSC transplants are related to the longer telomeres in individual cord blood HSCs (Awaya et al ., 2002).

It has been proposed that telomeres can switch between an open state (in principle allowing elongation by telomerase) and a closed state (inaccessible to telomerase) with the likelihood of the open state inversely related to the length of the repeat track (Blackburn et al., 2001). In most human cells, telomerase appears to be present at limiting levels, allowing elongation of only a limited number of critically short telomeres. Accumulation of short telomeres before replicative senescence has been observed and replicative senescence or apoptosis could result when the number of critically short telomeres exceeds the telomere repair capacity in a cell (Ju Z et al., 2011). In this context the ability of HSCs to modulate telomerase activity may be crucial in maintaining the self-renewal process. In human BM cells, low telomerase activity levels were demonstrated in multipotent HSCs, whereas significant upregulation of enzyme activity was apparent in the presence of proliferation-inducing cytokines (Samper et al., 2002; Stein et al., 2004). So cytokines and JAK-STAT signaling pathway may contribute to self renewal by maintaining telomerase activity.

The role of DNA repair pathways and telomeres/telomerase in the biology of normal and malignant human HSCs cells as well as the biology of aging clearly needs further study. New insight in the role of telomerase in HSCs has been provided by recent studies of patients with inborn errors in telomerase activity. Therefore, a further understanding of the molecular mechanisms underlying HSC aging may help identity new therapeutic targets for stem cell-based regenerative medicine.

8. Asymmetric cell division – A mechanism to generate progenitors while maintaining HSC

In both invertebrates like the insect *Drosophila*, and mammals, the major characteristic of stem cells is their ability to self-renew. Using various modes of proliferation, stem cells maintain or expand the available stem cell pool, but they can also generate more specialized progeny that constitute the majority of cells in an adult individual. In multi-cellular organisms, totipotent zygotes generate pluripotent stem cells, which become increasingly restricted in their lineage potential during development, and subsequently give rise to mature tissue-specific, multipotent stem cells. Stem cells show either 'proliferative' symmetric divisions or 'differentiative' asymmetric divisions to regulate a balance between the maintenance of stem cell pool and the supply of mature cells. It is critical for stem cells to tightly control this balance between the two different modes of division, both during development and adulthood, because, failure in maintaining cellular homeostasis may lead to incomplete tissue or organ development, whereas uncontrolled proliferation can lead to tumorigenesis.

Symmetric cell divisions commonly occur during development of invertebrates and vertebrates, phenomena that can also be observed during wound healing and regeneration of tissues. This mode of division is defined by the generation of two daughter cells that acquire the same fate, thereby expanding the pool of stem cells required or generating two differentiating daughter cells. Asymmetric cell divisions play a key role in generating cellular diversity during development by generating two daughter cells that are committed to different fates in a single division, simultaneously self-renewing to generate a daughter cell with stem cell properties, as well as to give rise to a more differentiated progeny.

Asymmetric stem cell divisions can be controlled by intrinsic mechanisms or the asymmetric exposure to extrinsic cues. Intrinsic mechanisms use apical-basal or planar polarity along the mitotic spindle to asymmetrically segregate cell fate determinants into only one daughter cell. Extrinsic mechanisms rely on contact with the so called stem cell niche, a cellular microenvironment that provides external cues (Doe, 2008; Li & Xie, 2005; Morrison & Spradling, 2008). Orientation of its mitotic spindle perpendicular to the niche surface allows the asymmetric segregation of cell fate determinants relative to the external stimuli to maintain self-renewal potential.

Much progress has been made in understanding intercellular mechanisms, especially the identification of niches for various types of tissue stem cells and elucidation of the role of the niche in regulating asymmetric stem cell division.

Although the role of niche in the asymmetric division of mammalian stem cells has not been clearly illustrated, Fuchs (2008) have shown that embryonic basal epidermal cells use their polarity to divide asymmetrically with respect to the underlying basal lamina, generating a committed suprabasal cell and a proliferative basal cell. Because skin stem cells are a subpopulation of mitotically active basal epidermal cells, it is conceivable that these stem cells divide in an asymmetric fashion to self-renew and to produce differentiated keratinocytes. Moreover, integrins and cadherins in the basal lamina are essential for the proper localization of apical complexes containing atypical PKC (aPKC), the Par3 - LGN - Inscuteable protein, and NuMA (nuclear mitotic apparatus protein) - dynactin. This asymmetric localization may be functionally important because similar complexes in *Drosophila* neuroblasts are essential for asymmetric division (Chia et al., 2008).

In addition to basal epidermal cells, mouse neuroepithelial stem cells and hematopoietic precursor cells undergo both asymmetric and symmetric divisions. In the mammalian central nervous system, embryonic neuroepithelial cells first undergo symmetric division to expand their population and then switch to asymmetric divisions for neurogenesis. This switch involves a change in cleavage plane orientation from perpendicular to parallel to the plane of the apical lamina, leading to an asymmetric distribution to the daughter cells of the apical plasma membrane, which constitutes only a minute fraction (1 - 2%) of the entire neuroepithelial cell plasma membrane (Kosodo et al., 2004). Somewhat similarly, mouse hematopoietic progenitor cells are capable of both symmetric and asymmetric divisions in cultures supported by stromal cells (Wu et al., 2007). A pro-differentiation stromal cell line increased the frequency of asymmetric division, whereas a pro-proliferation stromal cell line promoted symmetric division. These observations indicate that niche signaling can also control the asymmetry of stem cell division at a populational level.

Although niche induction accounts for asymmetric division in some types of stem cells, it may not play a role in all types of stem cells. In some stem cells there is an intrinsic polarity where molecules are segregated along an axis and serves as determinant of cell fate after cell division. In these cases the orientation of mitotic spindle has to be coordinated with the asymmetric localization of the cell fate regulators. This is the case, for example of *Numb* homologue during hematopoietic precursor cell division. So looking for hematopoietic niche and signaling between microenviroment cells and the stem cell we could " envisage"(imagine; idealize, construct) a model for stem cell to decide between symmetric or asymmetric division.

A particularly exciting development in basic stem cell research in the past few years is the discovery of novel functions of cell cycle regulators in controlling the asymmetry of stem cell division, as timely reviewed by Chia et al (2008). For example, the *cdc2/cdk1* level controls whether a neural or muscle progenitor undergoes symmetric or asymmetric division. In neuroblasts, high levels of CDK1 during mitosis are required for the asymmetric localization of apical and basal protein complexes. In addition, Aurora and Polo kinases act as tumor suppressors in neuroblasts by preventing excess self-renewal, implicating the function of asymmetric division in restricting over-proliferation. The mutations of these two kinase genes affect the asymmetric localization of aPKC, Numb, Partner of Numb, and Notch, causing symmetric division to generate two daughter neuroblasts. In addition, anaphase-promoting complex/cyclosome is also required for the localization of Miranda and its cargo proteins (Prospero, Brain Tumor, and Staufen). More surprisingly, even cyclin E, a G1 cyclin, is involved in asymmetric neuroblast division.

Interestingly in epidermal progenitors the decision of choosing between symmetric cell division or asymmetric cell decision is tightly regulated. Two control points have been indentified: expression of Inscutable and recruitment of NuMA to the apical cell cortex. Moreover in embryonic lung distal epithelium Eya1 protein regulates cell polarity, spindle orientation and the localization of Numb, which inhibits Notch signaling with the participation of NuMa protein (El-Hashad et al , 2011).

9. Conclusion

In the last ten years a body of evidence has accumulated on the hematopoietic stem cell niches and the mechanisms by which they regulate HSCs homeostasis. However many questions remain to be addressed. What we can speculate with today data is that in the endosteal niche dissimilar stages of differentiating osteoblasts provides diverse signals. These signals induce a quiescent or a self-renewal fate. However after leaving quiescence HSCs are prone to accumulate mutations that will lead to senescence. To prevent HSCs exhaustion mechanisms to enhance survival are also induced. Several of these additional signals come from the osteoblasts but also from other cells from microenvironment as stromal cells and endothelial cells. This is specially visualized in the decision of a symmetric or asymmetric division. From the control of all these interconnected pathways depends a normal hematopoiesis.

10. Acknowledgements

We thank Amanda Maia and André Mencalha for the excellent figures design.

11. References

Adams, G.B. & Scadden, D.T.(2006). The hematopoietic stem cell in its place. *Nat Immunol.* Apr;7(4):333-7.

Antonchuk, J.; Sauvageau, G. & Humphries, R.K.(2002). HOXB4-induced expansion of adult hematopoietic stem cells ex vivo. *Cell.* Apr 5;109(1):39-45.

Arai, F.; Hirao, A.; Ohmura, M.; Sato, H.; Matsuoka, S.; Takubo, K.; Ito, K.; Koh, G.Y. & Suda,T.(2004). Tie2/angiopoietin-1 signaling regulates hematopoietic stem cell quiescence in the bone marrow niche. *Cell.* Jul 23;118(2):149-61.

Arai, F.& Suda T.(2007). Maintenance of quiescent hematopoietic stem cells in the osteoblastic niche. *Ann N Y Acad Sci.* Jun;1106:41-53.

Avecilla, S.T.; Hattori, K.; Heissig, B.; Tejada, R.; Liao, F.; Shido, K.; Jin, D.K.; Dias, S.; Zhang, F.; Hartman. T.E.; Hackett, N.R.; Crystal, R.G.; Witte, L.; Hicklin, D.J.; Bohlen, P.; Eaton, D.; Lyden, D.; de Sauvage, F.& Rafii, S.(2004). Chemokine-mediated interaction of hematopoietic progenitors with the bone marrow vascular niche is required for thrombopoiesis. *Nat Med.* Jan;10(1):64-71.

Awaya, N.; Baerlocher, G.M.; Manley, T.J.; Sanders, J.E.; Mielcarek ,M.; Torok-Storb, B. & Lansdorp, P.M. (2002). Telomere shortening in hematopoietic stem cell transplantation: a potential mechanism for late graft failure? *Biol Blood Marrow Transplant.* 8(11):597-600.

Baron, M.(2001). Induction of embryonic hematopoietic and endothelial stem/progenitor cells by hedgehog-mediated signals. *Differentiation.* Oct;68(4-5):175-85.

Bendall, S.C.; Stewart, M.H.; Menendez, P. George,D.; Vijayaragavan, K.; Werbowetski-Ogilvie, T.; Ramos-Mejia, V.; Rouleau, A.; Yang, J.; Bossé, M.; Lajoie, G.& Bhatia, M.(2007). IGF and FGF cooperatively establish the regulatory stem cell niche of pluripotent human cells in vitro. *Nature.* Aug 30; 448 (7157):1015-21.

Bhardwaj, G.; Murdoch, B.; Wu, D.; Baker, D.P.; Williams, K.P.; Chadwick, K.; Ling, L.E.; Karanu, F.N. & Bhatia, M.(2001). Sonic hedgehog induces the proliferation of primitive human hematopoietic cells via BMP regulation. *Nat Immunol.* Feb;2(2):172-80.

Bijl, J.; Thompson, A.; Ramirez-Solis, R.; Krosl, J.; Grier, D.G.; Lawrence, H.J. & Sauvageau,G.(2006). Analysis of HSC activity and compensatory Hox gene expression profile in Hoxb cluster mutant fetal liver cells. *Blood.* Jul 1;108(1):116-22.

Blank, U.; Karlsson, G. & Karlsson, S.(2008). Signaling pathways governing stem-cell fate. *Blood.* Jan 15;111(2):492-503.

Blackburn, E.H.(1991). Structure and function of telomeres. *Nature.* 350:569–573.

Blackburn, E.H. (2001). Switching and signaling at the telomere. *Cell* 106: 661–673.

Bowie, M.B.; McKnight, K.D.; Kent, D.G.; McCaffrey, L.; Hoodless, P.A. & Eaves, C.J.(2006).Hematopoietic stem cells proliferate until after birth and show a reversible phase-specific engraftment defect. *J Clin Invest.* Oct;116(10):2808-16.

Boyer, L.A.; Plath, K.; Zeitlinger, J.; Brambrink, T.; Medeiros, L.A.; Lee, T.I.; Levine, S.S.; Wernig, M.; Tajonar, A.; Ray, M.K.; Bell, G.W.; Otte, A.P.; Vidal, M.; Gifford, D.K.; Young, R.A.& Jaenisch, R.(2006). Polycomb complexes repress developmental regulators in murine embryonic stem cells. *Nature.* May 18;441(7091):349-53.

Bradford, G.B.; Williams, B.; Rossi, R. & Bertoncello, I.(1997). Quiescence, cycling, and turnover in the primitive hematopoietic stem cell compartment. *Exp Hematol.* May;25(5):445-53.

Brun, A.C.; Björnsson, J.M.; Magnusson, M.; Larsson, N.; Leveén, P.; Ehinger, M.; Nilsson, E. & Karlsson, S.(2004). Hoxb4-deficient mice undergo normal hematopoietic development but exhibit a mild proliferation defect in hematopoietic stem cells. *Blood.* Jun1;103(11):4126-33.

Burns, C.E.; Traver, D.; Mayhall, E.; Shepard, J.L. & Zon, L.I.(2005). Hematopoietic stem cell fate is established by the Notch-Runx pathway. *Genes Dev.* Oct 1;19(19):2331-42.

Buza-Vidas, N.; Antonchuk, J.; Qian, H.; Månsson, R.; Luc, S,; Zandi, S.; Anderson, K.; Takaki, S.; Nygren, J.M.; Jensen, C.T. & Jacobsen, S.E.(2006). Cytokines regulate postnatal hematopoietic stem cell expansion: opposing roles of thrombopoietin and LNK. *Genes Dev.*Aug 1;20(15):2018-23.

Calvi, L.M.; Adams, G.B.; Weibrecht, K.W.; Weber, J.M.; Olson, D.P.; Knight, M.C.; Martin, R.P.;Schipani, E.; Divieti, P.; Bringhurst, F.R.; Milner, L.A.; Kronenberg, H.M. & Scadden, D.T.(2003).Osteoblastic cells regulate the haematopoietic stem cell niche. *Nature.* Oct23;425(6960):841-6.

Cellot, S.; Krosl, J.; Chagraoui, J.; Meloche, S.; Humphries, R.K. & Sauvageau, G.(2007). Sustained in vitro trigger of self-renewal divisions in Hoxb4hiPbx1(10)hematopoietic stem cells. *Exp Hematol.* May;35(5):802-16.

Ceradini, D.J.; Kulkarni, A.R.; Callaghan, M.J.; Tepper, O.M.; Bastidas, N.; Kleinman, M.E.; Capla, J.M.; Galiano, R.D.; Levine, J.P.& Gurtner, G.C.(2004). Progenitor cell trafficking is regulated by hypoxic gradients through HIF-1 induction of SDF-1. *Nat Med.* Aug;10(8):858-64.

Cheng, T.; Rodrigues, N.; Shen, H.; Yang, Y.; Dombkowski, D.; Sykes, M.& Scadden, D.T.(2000). Hematopoietic stem cell quiescence maintained by p21cip1/waf1. *Science.* Mar 10;287(5459):1804-8.

Chia, W.; Somers, W.G. & Wang, H.(2008). Drosophila neuroblast asymmetric divisions: cell cycle regulators, asymmetric protein localization, and tumorigenesis. *J Cell Biol.* Jan 28;180(2):267-72.

Chiba, S.(2006). Notch signaling in stem cell systems. *Stem Cells.* Nov;24(11):2437-47.

Chien, W.M.; Rabin, S.; Macias, E.; Miliani de Marval, P.L.; Garrison, K.; Orthel, J.; Rodriguez-Puebla, M. & Fero, M.L.(2006). Genetic mosaics reveal both cell-autonomous and cell-nonautonomous function of murine p27Kip1. *Proc Natl Acad Sci U S A.* Mar 14;103(11):4122-7.

Choudhury, A.R.; Ju, Z.; Djojosubroto, M.W.; Schienke, A.; Lechel, A.; Schaetzlein, S.; Jiang, H.; Stepczynska, A.; Wang, C. ; Buer, J.; Lee, H.W.; von Zglinicki, T.; Ganser, A.; Schirmacher, P.; Nakauchi, H. & Rudolph, K.L.(2007). Cdkn1a deletion improves stem cell function and lifespan of mice with dysfunctional telomeres without accelerating cancer formation. *Nat Genet.* Jan;39(1):99-105.

Chung,Y.J.; Park, B.B.; Kang, Y.J.; Kim, T.M.; Eaves, C.J.& Oh, I.H.(2006). Unique effects of Stat3 on the early phase of hematopoietic stem cell regeneration. *Blood.* Aug 15;108(4):1208-15.

Clevers, H.(2006) Wnt/beta-catenin signaling in development and disease. *Cell.* Nov 3;127(3):469-80.

Coultas, L.; Chawengsaksophak, K. & Rossant, J.(2005). Endothelial cells and VEGF in vascular development. *Nature.* Dec 15;438(7070):937-45.

Crcareva, A.; Saito, T.; Kunisato, A.; Kumano, K.; Suzuki, T.; Sakata-Yanagimoto, M.;Kawazu, M.; Stojanovic, A.; Kurokawa, M.; Ogawa, S.; Hirai, H. & Chiba, S.(2005). Hematopoietic stem cells expanded by fibroblast growth factor-1 are excellent targets for retrovirus-mediated gene delivery. *Exp Hematol.*Dec;33(12):1459-69.

Czechowicz, A.; Kraft, D.; Weissman, I.L. & Bhattacharya, D.(2007). Efficient transplantation via antibody-based clearance of hematopoietic stem cell niches. *Science.* Nov 23;318(5854):1296-9.

Daria, D.; Filippi, M.D.; Knudsen, E.S.; Faccio, R.; Li, Z.; Kalfa, T. & Geiger, H.(2008).The retinoblastoma tumor suppressor is a critical intrinsic regulator for hematopoietic stem and progenitor cells under stress. *Blood.* Feb 15;111(4):1894-902.

Doe, C.Q.(2008). Neural stem cells: balancing self-renewal with differentiation. *Development.* May;135(9):1575-87.

Domen, J.; Gandy, K.L.& Weissman, I.L.(1998). Systemic overexpression of BCL-2 in the hematopoietic system protects transgenic mice from the consequences of lethal irradiation. *Blood.* Apr 1;91(7):2272-82.

Domen, J.; Cheshier, S.H.& Weissman, I.L.(2000). The role of apoptosis in the regulation of hematopoietic stem cells: Overexpression of Bcl-2 increases both their number and repopulation potential. *J Exp Med.* Jan 17;191(2):253-64.

Dufourcq, P.; Descamps, B.; Tojais, N.F.; Leroux, L.; Oses, P.; Daret, D.; Moreau, C.; Lamaziere, J.M.; Couffinhal, T. & Duplaa, D. (2008). Secreted frizzled-related protein-1 enhances mesenchymal stem cell function in angiogenesis and contributes to neovessel maturation. *Stem Cells* 26 (11) 2991–3001.

Dumble, M.; Moore, L.; Chambers, S.M.; Geiger, H.; Van Zant, G.; Goodell, M.A.& Donehower, L.A.(2007). The impact of altered p53 dosage on hematopoietic stem cell dynamics during aging. *Blood.* Feb 15;109(4):1736-42.

Duncan, A.W.; Rattis, F.M.; DiMascio, L.N.; Congdon, K.L.; Pazianos, G.; Zhao, C.; Yoon, K.; Cook, J.M.; Willert, K.; Gaiano, N. & Reya, T.(2005). Integration of Notch and Wnt signaling in hematopoietic stem cell maintenance. *Nat Immunol.* Mar;6(3):314-22.

Edling, C.E. & Hallberg, B.(2007). c-Kit--a hematopoietic cell essential receptor tyrosine kinase. *Int J Biochem Cell Biol.* 39(11):1995-8.

El-Hashash, A.H.; Turcatel, G.; Al Alam, D.; Buckley, S.; Tokumitsu, H.; Bellusci, S. & Warburton, D. (2011). Eya1 controls cell polarity, spindle orientation, cell fate and Notch signaling in distal embryonic lung epithelium. *Development.* Apr;138(7):1395-407.

Ficara, F.; Murphy, M.J.; Lin, M.& Cleary, M.L.(2008). Pbx1 regulates self-renewal of long-term hematopoietic stem cells by maintaining their quiescence. *Cell Stem Cell.* May 8;2(5):484-96.

Fleming, H.E.; Janzen, V.; Lo Celso, C.; Guo, J.; Leahy, K.M.; Kronenberg, H.M. & Scadden, D.T.(2008).Wnt signaling in the niche enforces hematopoietic stem cell quiescence and is necessary to preserve self-renewal in vivo. *Cell Stem Cell.* Mar 6;2(3):274-83.

Fuchs, E.(2008). Skin stem cells: rising to the surface. *J Cell Biol.* Jan 28;180(2):273-84.

Gekas, C.; Dieterlen-Lièvre, F.; Orkin, S.H. & Mikkola, H.K.(2005). The placenta is a niche for hematopoietic stem cells. *Dev Cell.* Mar;8(3):365-75.

Gering, M. & Patient R.(2005). Hedgehog signaling is required for adult blood stem cell formation in zebrafish embryos. *Dev Cell.* Mar;8(3):389-400.

Heissig, B.; Hattori, K.; Dias, S.; Friedrich, M.; Ferris, B; Hackett, N.R.; Crystal, R.G.; Besmer, P.; Lyden, D.; Moore, M.A.; Werb, Z. & Rafii, S.(2002). Recruitment of stem and progenitor cells from the bone marrow niche requires MMP-9 mediated release of kit-ligand.*Cell.* May 31;109(5):625-37.

Hock, H.; Meade, E.; Medeiros, S.; Schindler, J.W.; Valk, P.J.; Fujiwara, Y.& Orkin, S.H.(2004).Tel/Etv6 is an essential and selective regulator of adult hematopoietic stem cellsurvival. *Genes Dev*. Oct 1;18(19):2336-41.

Huang, X.; Cho, S. & Spangrude, G.J.(2007) Hematopoietic stem cells: generation and self-renewal. *Cell Death Differ*. Nov;14(11):1851-9.

Huntly, B.J. & Gilliland, D.G.(2005) Cancer biology: summing up cancer stem cells. *Nature*. Jun 30;435(7046):1169-70.

Ichikawa, M.; Asai, T.; Chiba, S.; Kurokawa, M. & Ogawa S.(2004) Runx1/AML-1 ranks as a master regulator of adult hematopoiesis. *Cell Cycle*. Jun;3(6):722-4.

Ikuta, K.& Weissman, I.L.(1992). Evidence that hematopoietic stem cells express mouse c-kit but do not depend on steel factor for their generation. *Proc Natl Acad Sci U S A*. Feb 15;89(4):1502-6.

Inman, K.E. & Downs, K.M.(2007). The murine allantois: emerging paradigms in development of the mammalian umbilical cord and its relation to the fetus. *Genesis*. May;45(5):237-58.

Ito, T.; Tajima, F.& Ogawa, M.(2000). Developmental changes of CD34 expression by murine hematopoietic stem cells. *Exp Hematol*. Nov;28(11):1269-73.

Ito, K.; Hirao, A.; Arai, F.; Matsuoka, S.; Takubo, K.; Hamaguchi, I.; Nomiyama, K.; Hosokawa, K.; Sakurada, K.; Nakagata, N.; Ikeda, Y.; Mak, T.W. & Suda, T.(2004). Regulation of oxidative stress by ATM is required for self-renewal of haematopoietic stem cells. *Nature*. Oct 21;431(7011):997-1002.

Ito, K.; Hirao, A.; Arai, F.; Takubo, K.; Matsuoka, S.; Miyamoto, K.; Ohmura, M.; Naka, K.;Hosokawa, K.; Ikeda, Y. & Suda T. (2006). Reactive oxygen species act through p38 MAPK to limit the lifespan of hematopoietic stem cells. *Nat Med*. Apr;12(4):446-51.

Iwama, A.; Oguro, H.; Negishi, M.; Kato, Y.; Morita, Y.; Tsukui, H.; Ema, H.; Kamijo, T.; Katoh-Fukui, Y.; Koseki, H.; van Lohuizen, M.& Nakauchi, H. (2004). Enhanced self-renewal of hematopoietic stem cells mediated by the polycomb gene product Bmi-1. *Immunity*. Dec;21(6):843-51.

Janzen, V. & Scadden, D.T.(2006). Stem cells: good, bad and reformable. *Nature*. May 25;441(7092):418-9.

Janzen, V.; Forkert, R.; Fleming, H.E.; Saito, Y.; Waring, M.T.; Dombkowski, D.M.; Cheng, T.; DePinho, R.A.; Sharpless, N.E. & Scadden, D.T.(2006). Stem-cell ageing modified by the cyclin-dependent kinase inhibitor p16INK4a. *Nature*. 2006 Sep 28;443(7110):421-6.

Ju, Z.; Zhang, J.; Gao, Y. & Cheng, T.(2011). Telomere dysfunction and cell cycle checkpoints in hematopoietic stem cell aging. *Int J Hematol*. Jul;94(1):33-43.

Kajiume, T.; Ninomiya, Y.; Ishihara, H.; Kanno, R.& Kanno, M.(2004). Polycomb group gene mel-18 modulates the self-renewal activity and cell cycle status of hematopoietic stem cells. *Exp Hematol*. Jun;32(6):571-8.

Karahuseyinoglu, S.; Cinar, O.; Kilic, E.; Kara, F.; Akay, G.G.; Demiralp, D.O.; Tukun, A.; Uckan, D.& Can, A.(2007).Biology of stem cells in human umbilical cord stroma: in situ and in vitro surveys. *Stem Cells*. Feb;25(2):319-31.

Kato, Y.; Iwama, A.; Tadokoro, Y.; Shimoda, K.; Minoguchi, M.; Akira, S.; Tanaka, M.; Miyajima, A.; Kitamura, T. & Nakauchi, H.(2005). Selective activation of STAT5 unveils its role in stem cell self-renewal in normal and leukemic hematopoiesis. *J Exp Med.* Jul 4;202(1):169-79.

Kawano, Y. & Kypta, R.(2003). Secreted antagonists of the Wnt signaling pathway. *J Cell Sci.* Jul 1;116(Pt 13):2627-34.

Kent, D.; Copley, M.; Benz, C.; Dykstra, B.; Bowie, M.& Eaves, C.(2008).Regulation of hematopoietic stem cells by the steel factor/KIT signaling pathway. *Clin Cancer Res.*Apr 1;14(7):1926-30.

Kiel, M.J.; Yilmaz, O.H.; Iwashita, T.; Yilmaz, O.H.; Terhorst, C. & Morrison, S.J.(2005). SLAM family receptors distinguish hematopoietic stem and progenitor cells and reveal endothelial niches for stem cells. *Cell.* Jul 1;121(7):1109-21.

Kiel, M.J.& Morrison, S.J.(2006). Maintaining hematopoietic stem cells in the vascular niche. *Immunity.* Dec;25(6):862-4.

Kiel, M.J.; He, S.; Ashkenazi, R.; Gentry, S.N.; Teta, M.; Kushner, J.A.; Jackson, T.L. & Morrison, S.J.(2007). Haematopoietic stem cells do not asymmetrically segregate chromosomes or retain BrdU. *Nature.* Sep 13;449(7159):238-42.

Kiel, M.J.; Radice, G.L. & Morrison, S.J.(2007). Lack of evidence that hematopoietic stem cells depend on N-cadherin-mediated adhesion to osteoblasts for their maintenance. *Cell Stem Cell.* Aug 16;1(2):204-17.

Kim, J.Y.; Sawada, A.; Tokimasa, S.; Endo, H.; Ozono, K.; Hara, J. & Takihara, Y.(2004). Defective long-term repopulating ability in hematopoietic stem cells lacking the Polycomb-group gene rae28. *Eur J Haematol.* Aug;73(2):75-84.

Kollet, O.; Dar, A.; Shivtiel, S.; Kalinkovich, A.; Lapid, K.; Sztainberg, Y.; Tesio, M.; Samstein, R.M.; Goichberg, P.; Spiegel, A.; Elson, A. & Lapidot T. (2006). Osteoclasts degrade endosteal components and promote mobilization of hematopoietic progenitor cells. *Nat Med.* Jun;12(6):657-64.

Kopp, H.G.; Avecilla, S.T.; Hooper, A.T. & Rafii, S.(2005). The bone marrow vascular niche: home of HSC differentiation and mobilization. *Physiology (Bethesda).* Oct;20:349-56.

Kosodo, Y.; Röper, K.; Haubensak, W.; Marzesco, A.M.; Corbeil, D. & Huttner, W.B.(2004). Asymmetric distribution of the apical plasma membrane during neurogenic divisions of mammalian neuroepithelial cells. *EMBO J.* Jun 2;23(11):2314-24.

Konuma, T.; Oguro, H. & Iwama, A.(2010). Role of the polycomb group proteins in hematopoietic stem cells. *Dev Growth Differ.* Aug;52(6):505-16.

Krosl, J.; Austin, P.; Beslu, N.; Kroon, E.; Humphries, R.K. & Sauvageau, G.(2003). In vitro expansion of hematopoietic stem cells by recombinant TAT-HOXB4 protein. *Nat Med.*Nov;9(11):1428-32.

Kubota, Y.; Takubo, K. & Suda, T.(2008). Bone marrow long label-retaining cells reside in the sinusoidal hypoxic niche. *Biochem Biophys Res Commun.* 2008 Feb 8;366(2):335-9.

Kühn, R.; Schwenk, F.; Aguet, M. & Rajewsky, K.(1995). Inducible gene targeting in mice. *Science.* Sep 8;269(5229):1427-9.

Kumano, K.; Chiba, S.; Kunisato, A.; Sata, M.; Saito, T.; Nakagami-Yamaguchi, E.; Yamaguchi, T.; Masuda, S.; Shimizu, K.; Takahashi, T.; Ogawa, S.; Hamada, Y. & Hirai, H.(2003). Notch1 but not Notch2 is essential for generating hematopoietic stem cells from endothelial cells. *Immunity.* May;18(5):699-711.

Lacombe, J.; Herblot, S.; Rojas-Sutterlin, S.; Haman, A.; Barakat, S.; Iscove, N.N.; Sauvageau, G.& Hoang, T.(2010). Scl regulates the quiescence and the long-term competence of hematopoietic stem cells. *Blood.* Jan 28;115(4):792-803.

Laiosa, C.V.; Stadtfeld, M.; Xie, H.; de Andres-Aguayo, L. & Graf, T.(2006).Reprogramming of committed T cell progenitors to macrophages and dendritic cells by C/EBP alphaand PU.1 transcription factors. *Immunity*. 2006 Nov;25(5):731-44.

Lansdorp, P.M. (1995). Telomere length and proliferation potential of hematopoietic stem cells. *J. Cell Sci*. 108: 1–6.

Lansdorp, P.M. (1997). Self-renewal of stem cells. *Biol. Blood Marrow Transplant* 3: 171–178.

Lansdorp, P.M. (2005). Role of telomerase in Hematopoietic Stem Cells. *Ann. N.Y.Acad.Sci*. 220-227.

Li, J.(2011). Quiescence regulators for hematopoietic stem cell. *Exp Hematol*. May;39(5):511-20.

Lee, H.W.; Blasco, M.A.; Gottlieb, G.J.; Horner, J.W2nd.; Greider, C.W. & Depinho, R.A. (1998). Essential role of mouse telomerase in highly proliferative organs. *Nature*. 392:569–574.

Lee, T.I.; Jenner, R.G.; Boyer, L.A.; Guenther, M.G.; Levine, S.S.; Kumar, R.M.; Chevalier, B.; Johnstone, S.E.; Cole, M.F.; Isono, K.; Koseki, H.; Fuchikami, T.; Abe, K.; Murray, H.L.; Zucker, J.P.; Yuan, B.; Bell, G.W.; Herbolsheimer, E.; Hannett, N.M.; Sun, K.; Odom, D.T.; Otte, A.P.; Volkert, T.L.; Bartel, D.P.; Melton, D.A.; Gifford, D.K.; Jaenisch, R. & Young, R.A. (2006). Control of developmental regulators by Polycomb in human embryonic stem cells. *Cell*. Apr 21;125(2):301-13.

Lessard, J.& Sauvageau, G.(2003). Bmi-1 determines the proliferative capacity of normal and leukaemic stem cells. *Nature*. May 15;423(6937):255-60.

Lewis, J.(1998). Notch signaling and the control of cell fate choices in vertebrates. *Semin Cell Dev Biol*. Dec;9(6):583-9.

Li, J. (2011). Quiescence regulators for hematopoietic stem cell. *Exp Hematol*. May; 39(5):511-20.

Li, L. & Xie, T.(2005). Stem cell niche: structure and function. *Annu Rev Cell Dev Biol*. 21:605-31.

Li, Z. & Li, L.(2006). Understanding hematopoietic stem-cell microenvironments. *Trends Biochem Sci*. Oct;31(10):589-95.

Li, W.; Johnson, S.A.; Shelley, W.C. & Yoder, M.C.(2004). Hematopoietic stem cell repopulating ability can be maintained in vitro by some primary endothelial cells. *Exp Hematol*. Dec;32(12):1226-37.

Li, X.; Han, Y. & Xi, R. (2010). Polycomb group genes Psc and Su(z)2 restrict follicle stem cell self-renewal and extrusion by controlling canonical and noncanonical Wnt signaling. *Genes Dev*. May;24(9):933-46.

Liu, Y.; Elf, S.E.; Miyata, Y.; Sashida, G.; Liu, Y.; Huang, G.; Di Giandomenico, S.; Lee, J.M.; Deblasio, A.; Menendez, S.; Antipin, J.; Reva, B.; Koff, A. & Nimer, S.D.(2009). p53 regulates hematopoietic stem cell quiescence. *Cell Stem Cell*. Jan 9;4(1):37-48.

Lo Celso, C.; Fleming, H.E.; Wu, J.W.; Zhao, C.X.; Miake-Lye, S.; Fujisaki, J.; Côté, D.; Rowe,D.W.; Lin, C.P. & Scadden, D.T.(2009) Live-animal tracking of individual haematopoietic stem/progenitor cells in their niche. *Nature*. Jan 1;457(7225):92-6.

Luis, T.C.; Weerkamp, F.; Naber, B.A.; Baert, M.R.; de Haas, E.F.; Nikolic, T.; Heuvelmans, S.; De Krijger, R.R.; van Dongen, J.J.& Staal, F.J.(2009). Wnt3a deficiency irreversibly impairs hematopoietic stem cell self-renewal and leads to defects in progenitor cell differentiation. *Blood*. Jan 15;113(3):546-54.

Maillard, I.; He, Y. & Pear, W.S.(2003). From the yolk sac to the spleen: New roles for Notch in regulating hematopoiesis. *Immunity*. May;18(5):587-9.

Mancini, S.J.; Mantei, N.; Dumortier, A.; Suter, U.; MacDonald, H.R. & Radtke, F.(2005). Jagged1-dependent Notch signaling is dispensable for hematopoietic stem cell self-renewal and differentiation. *Blood*. Mar 15;105(6):2340-2.

Matsunaga, T.; Kato, T.; Miyazaki, H.& Ogawa, M.(1998). Thrombopoietin promotes the survival of murine hematopoietic long-term reconstituting cells: comparison with the effects of FLT3/FLK-2 ligand and interleukin-6. *Blood*. Jul 15;92(2):452-61.

Matsuyama, M.; Aizawa, S. & Shimono, A. (2009). Sfrp controls apicobasal polarity and oriented cell division in developing gut epithelium. *PLoS Genet*. 5 (3) e1000427.

Miyake, N.; Brun, A.C.; Magnusson, M.; Miyake, K.; Scadden, D.T.& Karlsson, S.(2006).HOXB4-induced self-renewal of hematopoietic stem cells is significantly enhanced by p21 deficiency. *Stem Cells*.Mar;24(3):653-61.

Miyamoto, K.; Araki, K.Y.; Naka, K.; Arai, F.; Takubo, K.; Yamazaki, S.; Matsuoka, S.; Miyamoto, T.; Ito, K.; Ohmura, M.; Chen, C.; Hosokawa, K.; Nakauchi, H.; Nakayama, K.; Nakayama, K.I.; Harada, M.; Motoyama, N.; Suda, T. & Hirao, A. (2007). Foxo3a is essential for maintenance of the hematopoietic stem cell pool. *Cell Stem Cell*. Jun 7;1(1):101-12.

Molofsky, A.V.; Pardal, R.; Iwashita, T.; Park, I.K.; Clarke, M.F. & Morrison, S.J.(2005). Bmi-1 dependence distinguishes neural stem cell self-renewal from progenitor proliferation. *Nature*. Oct 30;425(6961):962-7.

Moore, K.A. & Lemischka, I.R.(2006). Stem cells and their niches. *Science*. Mar 31;311(5769):1880-5.

Morisada, T.; Kubota, Y.; Urano, T.; Suda, T. & Oike Y.(2006). Angiopoietins and angiopoietin-like proteins in angiogenesis. *Endothelium*. Mar-Apr;13(2):71-9.

Morrison, S.J.& Spradling, A.C.(2008). Stem cells and niches: mechanisms that promote stem cell maintenance throughout life. *Cell*. Feb 22;132(4):598-611.

Müller, A.M.; Medvinsky, A.; Strouboulis, J.; Grosveld, F. & Dzierzak, E.(1994)Development of hematopoietic stem cell activity in the mouse embryo. *Immunity*.Jul;1(4):291-301.

Murdoch, B.; Chadwick, K.; Martin, M.; Shojaei, F.; Shah, K.V.; Gallacher, L.; Moon, R.T.& Bhatia, M.(2003). Wnt-5A augments repopulating capacity and primitive hematopoietic development of human blood stem cells in vivo. *Proc Natl Acad Sci U S A*. Mar 18;100(6):3422-7.

Nemeth, M.J.& Bodine, D.M.(2007). Regulation of hematopoiesis and the hematopoietic stem cell niche by Wnt signaling pathways. *Cell Res*. Sep;17(9):746-58.

Ohta, H.; Sawada, A.; Kim, J.Y.; Tokimasa, S.; Nishiguchi, S.; Humphries, R.K.; Hara, J. & Takihara, Y. (2002). Polycomb group gene rae28 is required for sustaining activity of hematopoietic stem cells. *J Exp Med*. Mar 18;195(6):759-70.

Opferman, J.T.; Iwasaki, H.; Ong, C.C.; Suh, H.; Mizuno, S.; Akashi, K.& Korsmeyer, S.J.(2005).Obligate role of anti-apoptotic MCL-1 in the survival of hematopoietic stem cells. *Science*. Feb 18;307(5712):1101-4.

Orkin, S.H. & Zon, L.I.(2008) Hematopoiesis: an evolving paradigm for stem cell biology.*Cell*. Feb 2;132(4):631-44.

Ottersbach, K. & Dzierzak, E. (2005). The murine placenta contains hematopoietic stem cells within the vascular labyrinth region. *Dev Cell*. Mar;8(3):377-87.

Papadimitriou, J.C.; Drachenberg, C.B.; Shin, M.L. & Trump, B.F.(1994). Ultrastructural studies of complement mediated cell death: a biological reaction model to plasma membrane injury. *Virchows Arch.* 424(6):677-85.

Pardanaud, L. & Dieterlen-Lièvre, F.(1999). Manipulation of the angiopoietic/hemangiopoietic commitment in the avian embryo. *Development.* Feb;126(4):617-27.

Park, I.K.; Morrison, S.J. & Clarke, M.F.(2004). Bmi1, stem cells, and senescence regulation. *J Clin Invest.* Jan;113(2):175-9.

Park, I.K.; Qian, D.; Kiel, M.; Becker, M.W.; Pihalja, M.; Weissman, I.L.; Morrison, S.J. & Clarke, M.F.(2003). Bmi-1 is required for maintenance of adult self-renewing haematopoietic stem cells. *Nature.* May 15;423(6937):302-5.

Parmar, K.; Mauch, P.; Vergílio, J.A.; Sackstein, R. & Down, J.D.(2007). Distribution of hematopoietic stem cells in the bone marrow according to regional hypoxia. *Proc Natl Acad Sci U S A.* Mar 27;104(13):5431-6.

Petit-Cocault, L.; Volle-Challier, C.; Fleury, M.; Péault, B.& Souyri M.(2007). Dual role of Mpl receptor during the establishment of definitive hematopoiesis. *Development.* Aug;134(16):3031-40.

Pimanda, J.E.; Donaldson, I.J.; de Bruijn, M.F.; Kinston, S.; Knezevic, K.; Huckle, L.; Piltz, S.; Landry, J.R.; Green, A.R.; Tannahill, D. & Göttgens, B.(2007). The SCL transcriptional network and BMP signaling pathway interact to regulate RUNX1 activity. *Proc Natl Acad Sci U S A.* Jan 16;104(3):840-5.

Pizzatti, L.; Binato, R.; Cofre, J.; Gomes, B.E.; Dobbin, J.; Haussmann, M.E.; D'Azambuja, D.; Bouzas, L.F. & Abdelhay, E.(2010). SUZ12 is a candidate target of the non-canonical WNT pathway in the progression of chronic myeloid leukemia. *Genes Chromosomes Cancer.* Feb;49(2):107-18.

Rekhtman, N.; Radparvar, .F; Evans, T. & Skoultchi, A.I.(1999). Direct interaction of hematopoietic transcription factors PU.1 and GATA-1: functional antagonism in erythroid cells. *Genes Dev.* Jun 1;13(11):1398-411.

Renstrom, J.; Kroger, M.; Peschel, C. & Robert, A.J.(2010). Oostendorp. How the niche regulates hematopoietic stem cells. *Chemico-Biological Interactions.* 184: 7-15.

Reya, T.& Clevers, H.(2005). Wnt signaling in stem cells and cancer. *Nature.* Apr 14;434(7035):843-50.

Reya, T.; Duncan, A.W.; Ailles, L.; Domen, J.; Scherer, D.C.; Willert, K.; Hintz, L.; Nusse, R. & Weissman, I.L.(2003). A role for Wnt signaling in self-renewal of haematopoietic stem cells. *Nature.* May 22;423(6938):409-14.

Rizzo, P.; Osipo, C.; Foreman, K.; Golde, T.; Osborne, B. & Miele, L.(2008). Rational targeting of Notch signaling in cancer. *Oncogene.* Sep 1;27(38):5124-31.

Robert-Moreno, A.; Espinosa, L.; de la Pompa, J.L. & Bigas, A.(2005). RBPjkappa-dependent Notch function regulates Gata2 and is essential for the formation of intra-embryonic hematopoietic cells. *Development.* Mar; 132(5):1117-26.

Rodriguez, J.; Esteve, P. ; Weinl, C. ; Ruiz, J.M.; Fermin, Y.; Trousse, F.; Dwivedy, A.; Holt, C. & Bovolenta, P. (2005). SFRP1 regulates the growth of retinal ganglion cell axons through the Fz2 receptor, *Nat. Neurosci.* 8 (10) 1301–1309.

Ross, J. & Li, L.(2006). Recent advances in understanding extrinsic control of hematopoietic stem cell fate. *Curr Opin Hematol.* ul;13(4):237-42.

Samper, E.; Fernández, P.; Eguía, R.; Martín-Rivera, L.; Bernad, A.; Blasco, M.A.& Aracil, M. (2002). Long-term repopulating ability of telomerase-deficient murine hematopoietic stem cells. *Blood.* Apr 15;99(8):2767-75.

Sato, T.; Laver,J.H. & Ogawa, M.(1999). Reversible expression of CD34 by murine hematopoietic stem cells. *Blood.* Oct 15;94(8):2548-54.

Sato, T.; Onai, N.; Yoshihara, H.; Arai, F.; Suda, T. & Ohteki, T. (2009). Interferon regulatory factor-2 protects quiescent hematopoietic stem cells from type I interferon-dependent exhaustion. *Nat Med.* Jun;15(6):696-700.

Sauvageau, G.; Thorsteinsdottir, U.; Eaves, C.J.; Lawrence, H.J.; Largman, C.; Lansdorp,P.M. & Humphries, R.K.(1995). Overexpression of HOXB4 in hematopoietic cells causes the selective expansion of more primitive populations in vitro and in vivo. *Genes Dev.* Jul 15;9(14):1753-65.

Scandura, J.M.; Boccuni, P.; Massagué, J. & Nimer, S.D.(2004). Transforming growth factor beta-induced cell cycle arrest of human hematopoietic cells requires p57KIP2 up-regulation. *Proc Natl Acad Sci U S A.* Oct 19;101(42):15231-6.

Schiedlmeier, B.; Santos, A.C.; Ribeiro, A.; Moncaut, N.; Lesinski, D.; Auer, H.; Kornacker,K.; Ostertag, W.; Baum, C.; Mallo, M. & Klump, H.(2007). HOXB4's road map to stem cell expansion. *Proc Natl Acad Sci U S A.* Oct 23;104(43):16952-7.

Scholl, C.; Gilliland, D.G. & Fröhling, S.(2008). Deregulation of signaling pathways in acute myeloid leukemia. *Semin Oncol.* Aug;35(4):336-45.

Seita, J.; Ema, H.; Ooehara, J.; Yamazaki, S.; Tadokoro, Y.; Yamasaki, A.; Eto, K.; Takaki,S.; Takatsu, K.& Nakauchi H.(2007). Lnk negatively regulates self-renewal of hematopoietic stem cells by modifying thrombopoietin-mediated signal transduction. *Proc Natl Acad Sci U S A.* Feb 13;104(7):2349-54.

Shao, Y.; Kim, S.Y.; Shin, D.; Kim, M.S.; Suh, H.W.; Piao, Z.H.; Jeong, M.; Lee, S.H.; Yoon, S.R.; Lim,B.H.; Kim, W.H.; Ahn, J.K. & Choi, I.(2010). TXNIP regulates germinal center generation by suppressing BCL-6 expression. *Immunol Lett.* Apr 8;129(2):78-84.

Shih, Ie.M. & Wang, T.L.(2007). Notch signaling, gamma-secretase inhibitors, and cancer therapy. *Cancer Res.* Mar 1;67(5):1879-82.

Silver, I.A.; Murrills, R.J.& Etherington, D.J.(1988). Microelectrode studies on the acid microenvironment beneath adherent macrophages and osteoclasts. *Exp Cell Res.*Apr;175(2):266-76.

Sirin, O.; Lukov, G.L.; Mao, R.; Conneely, O.M. & Goodell, M.A.(2010). The orphan nuclear receptor Nurr1 restricts the proliferation of haematopoietic stem cells. *Nat Cell Biol.* Dec;12(12):1213-9.

Snow, J.W.; Abraham, N.; Ma, M.C.; Abbey, N.W.; Herndier, B.& Goldsmith, M.A. (2002). STAT5 promotes multilineage hematolymphoid development in vivo through effects on early hematopoietic progenitor cells. *Blood.* Jan 1;99(1):95-101.

Solar, G.P.; Kerr, W.G.; Zeigler, F.C.; Hess, D.; Donahue, C.; de Sauvage, F.J. & Eaton, D.L.(1998).Role of c-mpl in early hematopoiesis. *Blood.* Jul 1;92(1):4-10.

Spike, B.T.; Dirlam, A.; Dibling, B.C.; Marvin, J.; Williams, B.O.; Jacks, T.& Macleod, K.F.(2004).The Rb tumor suppressor is required for stress erythropoiesis. *EMBO J.* Oct 27;23(21):4319-29.

Stein, M.I.; Zhu, J. & Emerson, S.G. (2004). Molecular pathways regulation the self-renewal of hematopoietic stem cells. *Exp Hematology.* 32- 1129-1136.

Stead, E.; White, J.; Faast, R.; Conn, S.; Goldstone, S.; Rathjen, J.; Dhingra, U.; Rathjen,P.; Walker, D. & Dalton, S.(2002). Pluripotent cell division cycles are driven by ectopic Cdk2, cyclin A/E and E2F activities. *Oncogene*. Nov 28;21(54):8320-33.

Stepanova, L.& Sorrentino, B.P.(2005). A limited role for p16Ink4a and p19Arf in the loss of hematopoietic stem cells during proliferative stress. *Blood*. Aug 1;106(3):827-32.

Stier, S.; Ko, Y.; Forkert, R.; Lutz, C.; Neuhaus, T.; Grünewald, E.; Cheng, T.; Dombkowski,D.; Calvi, L.M.; Rittling, S.R. & Scadden, D.T.(2005). Osteopontin is a hematopoietic stem cell niche component that negatively regulates stem cell pool size. *J Exp Med*. Jun 6;201(11):1781-91.

Suzuki, T.; Yokoyama, Y.; Kumano, K.; Takanashi, M.; Kozuma, S.; Takato, T.; Nakahata, T.;Nishikawa, M.; Sakano, S.; Kurokawa, M.; Ogawa, S.& Chiba S. (2006). Highly efficient ex vivo expansion of human hematopoietic stem cells using Delta1-Fc chimeric protein.*Stem Cells*. Nov;24(11):2456-65.

Tong, W.; Ibarra, Y.M. & Lodish, H.F.(2007). Signals emanating from the membrane proximal region of the thrombopoietin receptor (mpl) support hematopoietic stem cell self-renewal. *Exp Hematol*. Sep;35(9):1447-55.

Tothova, Z.; Kollipara, R.; Huntly, B.J.; Lee, B.H.; Castrillon, D.H.; Cullen, D.E.; McDowell, E.P.; Lazo-Kallanian, S.; Williams, I.R.; Sears, C.; Armstrong, S.A.; Passegué, E.; DePinho, R.A. & Gilliland, D.G.(2007). FoxOs are critical mediators of hematopoietic stem cell resistance to physiologic oxidative stress. *Cell*. Jan 26;128(2):325-39.

Utsugisawa, T.; Moody, J.L.; Aspling, M.; Nilsson, E.; Carlsson, L.& Karlsson, S.(2006). A road map toward defining the role of Smad signaling in hematopoietic stem cells. *Stem Cells*. Apr;24(4):1128-36.

Van der Lugt, N.M.; Alkema, M.; Berns, A. & Deschamps, J.(1996). The Polycomb-group homolog Bmi-1 is a regulator of murine Hox gene expression. *Mech Dev*. Aug;58(1-2):153-64.

Van Os, R.; Kamminga, L.M.; Ausema, A.; Bystrykh, L.V.; Draijer, D.P.; van Pelt, K.; Dontje, B. & de Haan. G. (2007). A Limited role for p21Cip1/Waf1 in maintaining normal hematopoietic stem cell functioning. *Stem Cells*. Apr;25(4):836-43.

Viatour, P.; Somervaille, T.C.; Venkatasubrahmanyam, S.; Kogan, S.; McLaughlin, M.E.; Weissman, I.L.; Butte, A.J.; Passegué, E.& Sage, J.(2008). Hematopoietic stem cell quiescence is maintained by compound contributions of the retinoblastoma gene family. *Cell Stem Cell*. Oct 9;3(4):416-28.

Vaziri, H.; Dragowska, W.; Allsopp, R.C.; Thomas, T.E.; Harley, C.B. & Lansdorp, P.M. (1994). Evidence for a mitotic clock in human hematopoietic stem cells: loss of telomeric DNA with age. *Proc Natl Acad Sci U S A*. Oct 11;91(21):9857-60.

Visnjic, D.; Kalajzic, Z.; Rowe, D.W.; Katavic, V.; Lorenzo, J. & Aguila, H.L.(2004). Hematopoiesis is severely altered in mice with an induced osteoblast deficiency. *Blood*. May 1;103(9):3258-64.

Walkley, C.R. & Orkin, S.H.(2006). Rb is dispensable for self-renewal and multilineage differentiation of adult hematopoietic stem cells. *Proc Natl Acad Sci U S A*. Jun 13;103(24):9057-62.

Walkley, C.R.; Fero, M.L.; Chien, W.M.; Purton, L.E.& McArthur, G.A.(2005). Negative cell-cycle regulators cooperatively control self-renewal and differentiation of haematopoietic stem cells. *Nat Cell Biol*. Feb;7(2):172-8.

Wang, Z.; Li, G.; Tse, W. & Bunting, K.D.(2009). Conditional deletion of STAT5 in adult mouse hematopoietic stem cells causes loss of quiescence and permits efficient nonablative stem cell replacement. *Blood.* May 14;113(20):4856-65.

Weissman, I.L. (2000).Stem cells: units of development, units of regeneration, and units in evolution. *Cell* Jan 7;100(1):157-68.

Willert, K.; Brown, J.D.; Danenberg, E.; Duncan, A.W.; Weissman, I.L.; Reya, T.; Yates, J.R3rd. & Nusse, R.(2003). Wnt proteins are lipid-modified and can act as stem cell growth factors. *Nature.* May 22;423(6938):448-52.

Wilson, A.& Trumpp, A.(2006). Bone-marrow haematopoietic-stem-cell niches. *Nat Rev Immunol.*Feb;6(2):93-106.

Wilson, A.; Murphy, M.J.; Oskarsson, T.; Kaloulis, K.; Bettess, M.D.; Oser, G.M.; Pasche, A.C.; Knabenhans, C.; Macdonald, H.R. & Trumpp, A.(2004). c-Myc controls the balance between hematopoietic stem cell self-renewal and differentiation. *Genes Dev.* Nov 15;18(22):2747-63.

Wu, M.; Kwon, H.Y.; Rattis, F.; Blum, J.; Zhao, C.; Ashkenazi, R.; Jackson, T.L.; Gaiano, N.; Oliver, T. & Reya, T.(2007). Imaging hematopoietic precursor division in real time. *Cell Stem Cell.*Nov;1(5):541-54.

Xie, Y.; Yin, T.; Wiegraebe, W.; He, X.C.; Miller, D.; Stark, D.; Perko, K.; Alexander, R.;Schwartz, J.; Grindley, J.C.; Park, J.; Haug, J.S.; Wunderlich, J.P.; Li, H.; Zhang, S.; Johnson,T.; Feldman, R.A.& Li, L.(2009). Detection of functional haematopoietic stem cell niche using real-time imaging. *Nature.* Jan 1;457(7225):97-101.

Yamashita, Y.M.& Fuller, M.T.(2008). Asymmetric centrosome behavior and the mechanisms of stem cell division. *J Cell Biol.* Jan 28;180(2):261-6.

Yeoh, J.S.; van Os, R.; Weersing, E.; Ausema, A.; Dontje, B.; Vellenga, E. & de Haan, G.(2006).Fibroblast growth factor-1 and -2 preserve long-term repopulating ability of hematopoietic stem cells in serum-free cultures. *Stem Cells.* Jun;24(6):1564-72.

Yoshihara, H.; Arai, F.; Hosokawa, K.; Hagiwara, T.; Takubo, K.; Nakamura, Y.; Gomei, Y.;Iwasaki, H.; Matsuoka, S.; Miyamoto, K.; Miyazaki, H.; Takahashi, T. & Suda, T.(2007).Thrombopoietin/MPL signaling regulates hematopoietic stem cell quiescence and interaction with the osteoblastic niche. *Cell Stem Cell.* Dec 13;1(6):685-97.

You, L.R.; Lin, F.J.; Lee, C.T.; DeMayo, F.J.; Tsai, M.J. & Tsai, S.Y.(2005). Suppression of Notch signaling by the COUP-TFII transcription factor regulates vein identity. *Nature.* May 5;435(7038):98-104.

Yu, H.; Yuan, Y.; Shen, H. & Cheng, T.(2006). Hematopoietic stem cell exhaustion impacted by p18 INK4C and p21 Cip1/Waf1 in opposite manners. *Blood.* Feb 1;107(3):1200-6.

Yuan, Y.; Shen, H.; Franklin, D.S.; Scadden, D.T. & Cheng, T.(2004). In vivo self-renewing divisions of haematopoietic stem cells are increased in the absence of the early G1-phase inhibitor, p18INK4C. *Nat Cell Biol.* May;6(5):436-42.

Zayas, J.; Spassov, D.S.; Nachtman, R.G. & Jurecic, R.(2008). Murine hematopoietic stem cells and multipotent progenitors express truncated intracellular form of c-kit receptor. *Stem Cells Dev.* Apr;17(2):343-53.

Zeng, H.; Yücel, R.; Kosan, C.; Klein-Hitpass, L. & Möröy T. (2004).Transcription factor Gfi1 regulates self-renewal and engraftment of hematopoietic stem cells. *EMBO J.*Oct 13;23(20):4116-25.

Zhang, C.C. & Lodish, H.F.(2004). Insulin-like growth factor 2 expressed in a novel fetal liver cell population is a growth factor for hematopoietic stem cells. *Blood.* Apr 1;103(7):2513-21.

Zhang, J.; Niu, C.; Ye, L.; Huang, H.; He, X.; Tong, W.G.; Ross, J.; Haug, J.; Johnson, T.; Feng,J.Q.; Harris, S.; Wiedemann, L.M.; Mishina, Y. & Li, L.(2003). Identification of the haematopoietic stem cell niche and control of the niche size. *Nature.* Oct 23;425(6960):836-41.

Zhang, P.; Behre, G.; Pan, J.; Iwama, A.; Wara-Aswapati, N.; Radomska, H.S.; Auron, P.E.; Tenen, D.G. & Sun, Z.(1999) Negative cross-talk between hematopoietic regulators: GATA proteins repress PU.1. *Proc Natl Acad Sci U S A.* Jul 20;96(15):8705-10.

Zhang, P.; Iwasaki-Arai, J.; Iwasaki, H.; Fenyus, M.L.; Dayaram, T.; Owens, B.M.; Shigematsu, H.; Levantini, E.; Huettner, C.S.; Lekstrom-Himes, J.A.; Akashi, K. & Tenen, D.G.(2004). Enhancement of hematopoietic stem cell repopulating capacity and self-renewal in the absence of the transcription factor C/EBP alpha. *Immunity.* Dec;21(6):853-63.

Zon, L.I.(2008). Self-renewal and differentiation at Cell Stem Cell. *Cell Stem Cell.* Jun 5;2(6):510.

2

Regulation of Hematopoietic Stem Cell Fate: Self-Renewal, Quiescence and Survival

Yasushi Kubota[1,2] and Shinya Kimura[1]
[1]*Division of Hematology, Respiratory Medicine and Oncology, Department of Internal Medicine, Faculty of Medicine, Saga University*
[2]*Department of Transfusion Medicine, Saga University Hospital*
Japan

1. Introduction

Hematopoietic stem cells (HSCs) are probably the most extensively characterized somatic stem cells and are the only stem cells that have been clinically used to treat diseases such as leukemia, germ cell tumors, and congenital immunodeficiencies. Because of their capacity for self-renewal and their ability to differentiate into different lineages, HSCs are able to continually replenish the cells that make up the hematopoietic system (Kondo et al., 2003). Decades of intensive study using multicolor cell sorting techniques have allowed investigators to identify these cells within a small population in the mouse bone marrow (BM) (i.e., $CD34^{low/-}$, Kit^{+} $Sca\text{-}1^{+}$ lineage marker-negative cells: $CD34^{low/-}$ KSL) and thereby allow the prospective isolation of nearly-homogenous HSC populations for further characterization (Osawa et al., 1996).

Under steady-state conditions, the majority of HSCs are maintained in a quiescent state in which they divide infrequently to produce proliferative progenitors that eventually give rise to the mature hematopoietic cells that sustain blood homeostasis (Cheshier et al., 1999). However, in response to external stresses such as bleeding, myeloablative chemotherapy and total body irradiation, HSCs proliferate extensively to produce very high numbers of primitive progenitor cells, thereby enabling rapid hematological regeneration (Randall et al., 1997). Once recovery from myelosuppression has been achieved, the activated HSCs return to a quiescent state via a number of negative feedback mechanisms (Venezia et al., 2004). The cell fate decisions (including life and death, self-renewal and differentiation) of HSCs are important processes that regulate the number and lifespan of the HSC pool within a host. Defects in these processes may contribute to hematopoietic failures and to the development of hematologic malignancies.

Understanding the molecular mechanisms underlying HSC regulation is of great importance to basic stem cell biology and for the development of HSCs for use in various clinical applications. Information regarding the regulation of HSC fate has been gained using conventional experimental approaches such as gene deletion, gene overexpression, and the direct stimulation of HSCs with cytokines. Although many studies have elucidated the factors controlling HSC fate using these methods, they can occasionally be misleading because they lack physiological relevance and do not identify phenomena such as genetic redundancy. For example, family genes or alternative pathways can compensate functionally for deleted genes in gene-ablated mouse models in a manner that masks the true physiology. One approach to identifying the individual components involved in the molecular pathways underlying HSC regulation is to define the molecular signature of the HSCs by comparative transcriptional profiling of distinct subsets of hematopoietic cells. Over the past decade, several attempts have been made by independent investigators, including ourselves, to define the molecular signature of HSCs (Park et al., 2002; Ramalho-Santos et al., 2002; Ivanova et al., 2002; Akashi et al., 2003; Venezia et al., 2004; Zhong et al., 2005; Forsberg et al., 2005; Ramos et al., 2006; Chambers et al., 2007; Kubota et al., 2009). A list of gene expression profiling studies using purified mouse HSCs performed to date is shown in Table 1. Although this information has, more or less, clarified the molecular makeup of HSCs and several critical factors have been identified based on the data reported in these studies, it is still extremely time-consuming to elucidate the physiological function of each individual gene involved in HSC regulation. The transcriptional regulation of stem cell fate, particularly by factors that have specific functions in HSCs, is only beginning to be understood.

In this chapter, we briefly review the recent advances in our knowledge of cell-intrinsic regulators of HSC self-renewal, differentiation, quiescence, cycling, and survival.

	Year	HSC phenotype	Compared population	References
Park et al.	2002	Rho^{low} KSL	Rho^{high} KSL	*Blood* 99(2):488-498.
Ramalho-Santos et al.	2002	$CD34^{-/low}$KSL-SP	MP	*Science* 298(5593):597-600.
Ivanova et al.	2002	Rho^{low} KSL	Rho^{high} KSL, LCP, MBC	*Science* 298(5593):601-604.
Akashi et al.	2003	Rho^{low} Thy-1.1^{low}KSL	MPP, CLP, CMP	*Blood* 101(2):383-389.
Venezia et al.	2004	Sca-1^{+}-SP	5-FU treated SP	*PLoS Biol* 2(10):e301.
Zhong et al.	2005	$CD34^{-}CD38^{+}$ KSL	$CD38^{+}$ or $CD38^{-}CD34^{+}$ KSL	*PNAS* 102(7):2448-2453.
Kiel et al.	2005	$Thy1.1^{low}$ KSL	Thy-1.1^{lo}Sca-1^{+}Mac-$1^{lo}CD4^{lo}B220^{-}$	*Cell* 121(7):1109-1121.
Forsberg et al.	2005	$Flk2^{-}$ $Thy1.1^{low}$ KSL	$Thy1.1^{low}$, $Thy1.1^{-}$ Flk2+ KSL	*PLoS Genet* 1(3):e28.
Ramos et al.	2006	Sca-$1^{+}Gr1^{-}$-SP	$CD8^{+}$ Tcell	*PLoS Genet* 2(9):e159.
Chambers et al.	2007	KSL-SP	Erythrocyte Granulocyte Native T Activated T B-cell Monocyte NK	*Cell Stem Cell* 1(5):578-591.
Kubota et al.	2009	$CD34^{-/low}$ KSL	$CD34^{+}$ KSL	*Blood* 114(20):4383-4392.

Rho, rhodamine; SP, side population; LCP, lineage-committed progenitor; MBC, mature blood cell;
MPP, multipotent progenitor; CLP, common lymphoid progenitor; CMP, common myeloid progenitor

Table 1. Gene expression profiling analyses of adult HSCs

2. Regulators of HSC fate

2.1 Regulation of HSC self-renewal and quiescence

The outstanding feature of adult stem cells is their relative quiescence (Orford et al., 2008; Wilson et al., 2008). Quiescence is critical for the maintenance and self-renewal of HSCs. Unscheduled HSC proliferation results in the loss of self-renewal or stem cell exhaustion (Orford et al., 2008; Wilson et al., 2009; Trumpp et al., 2010). Identification of the molecules that regulate adult HSCs has largely been achieved through the use of gene-targeted mouse models. Increasing or decreasing HSC cell-cycling results in the accelerated production of more committed progenitors at the expense of self renewal, or the insufficient production of progeny cells, which eventually results in BM failure.

2.1.1 Positive regulation

2.1.1.1 GATA-2

GATA-2 is highly expressed in immature progenitors within hematopoietic lineages (Tsai & Orkin, 1997; Akashi et al., 2000). The haploinsufficient *GATA-2*$^{+/-}$ mouse model shows mildly increased quiescence of both HSCs and progenitor cells (Rodrigues et al., 2005). However, Tipping et al. recently showed that enforced expression of GATA-2 in a murine cell line (Ba/F3), or human cord blood HSCs ($CD34^+CD38^-$) and progenitors ($CD34^+CD38^+$), increases quiescence and inhibits proliferation (Tipping, et al, 2009).

2.1.1.2 Bmi1

Bmi1 belongs to the polycomb group (PcG) of proteins, which play a role in the transcriptional repression of genes via histone modification (Rajasekhar et al., 2007). Bmi1 is highly expressed in HSCs. The expression of Bmi1 is maintained at high levels in lymphoid lineage cells but is downregulated during myeloid differentiation (Iwama et al., 2004). Although *Bmi1*$^{-/-}$ mice show normal fetal liver hematopoiesis, progressive pancytopenia emerges in postnatal *Bmi1*$^{-/-}$ mice. This hematopoietic defect can be attributed to impaired HSC self-renewal. Transplanted fetal liver and bone marrow cells from *Bmi1*$^{-/-}$ mice cannot contribute to long-term hematopoiesis, although they do maintain the ability to repopulate in the short-term (Park et al., 2003; Iwama et al., 2004). Conversely, enforced expression of Bmi1 promotes HSC self-renewal (Iwama et al., 2004). Thus, Bmi1 is essential for the maintenance of HSC self-renewal.

The activity of Bmi1 in HSCs largely depends on the silencing of its target, the *Ink4a* locus (Jacobs et al., 1999). The expression of $p16^{INK4a}$ and $p19^{ARF}$ (both cell-cycle inhibitors encoded by the *Ink4a* locus) is markedly upregulated in hematopoietic cells in *Bmi1*-deficient mice, and the overexpression of $p16^{INK4a}$ and $p19^{ARF}$ in HSCs induces cell-cycle arrest and p53-dependent apoptosis (Park et al., 2003). On the contrary, the deletion of both $p16^{INK4a}$ and $p19^{ARF}$ restores the self-renewal ability of *Bmi1*$^{-/-}$ HSCs (Oguro et al., 2006). Thus, Bmi1 prevents the premature loss of HSCs by repressing the $p16^{INK4a}$- and $p19^{ARF}$-dependent senescence pathways.

2.1.1.3 Gfi-1

Gfi1 is a SNAG-domain–containing zinc-finger transcriptional repressor, which plays a role in T cell proliferation and the development of lymphoid tumors (Gilks et al., 1993). It is suggested that Gfi-1 restricts proliferation and preserves functional integrity of hematopoietic stem cells. Gfi-1-null HSCs show excessive cell cycling and a decreased capacity for self-renewal in competitive repopulation assays (Hock et al., 2004; Zeng et al., 2004).

2.1.1.4 Pbx1

Pbx1 is a TALE class homeodomain transcription factor that critically regulates numerous embryonic processes, including hematopoiesis (DiMartino et al., 2001). Although a potential role was suggested by the observation that Pbx1 is preferentially expressed in long-term repopulating HSCs (LT-HSCs) compared with more mature progenitor cells (Forsberg et al., 2005), its functional analysis in adult HSCs has been hampered because Pbx1 mutant mice are embryonic lethal. Therefore, Pbx1-conditional knockout (KO) mice have been used to study the role of Pbx1 in the adult mouse hematopoietic system (Ficara et al., 2008). Conditional inactivation of Pbx1 in hematopoietic cells results in the loss of HSCs, which is associated with decreased quiescence. This leads to a defect in the maintenance of self-renewal in serial transplantation assays. Global gene expression profiling analyses show that a significant proportion (~8%) of the downregulated genes in Pbx1-deficient HSCs belong to the TGF-β signaling pathway, which has been implicated in maintaining HSC quiescence (Yamazaki et al., 2009). Also, in contrast to WT LT-HSCs, Pbx1-mutant LT-HSCs do not upregulate the expression of several downstream transcripts in response to TGF-β stimulation *in vitro*. These results suggest that Pbx1 regulates HSC self-renewal and quiescence, at least in part by affecting the response to TGF-β.

2.1.1.5 Evi-1

The ecotropic viral integration site-1 (Evi-1) was first identified in murine model systems as the integration site for the ecotropic retrovirus that causes myeloid leukemia (Morishita et al., 1988; Mucenski et al., 1988). Several studies using gene-targeting mice show that Evi-1 is required for HSC regulation. Yuasa et al. showed that Evi-1 is preferentially expressed in HSCs in embryos and adult BM. Evi-1–deficient embryonic HSCs are severely decreased in number, and show defective repopulating capacity. In addition, the expression of GATA-2 mRNA is markedly reduced in HSCs from Evi-1–null embryos. GATA-2 promoter analysis revealed that Evi-1 directly binds to the GATA-2 promoter and acts as an enhancer (Yuasa et al., 2005). Another study using conditional Evi-1 knockout mice showed that Evi-1 also regulates adult HSC proliferation in a dose-dependent manner. Evi-1–deficient BM HSCs did not maintain definitive hematopoiesis and lost their ability to reconstitute the cell population. Mutant mice heterozygous for Evi-1 exhibited an intermediate phenotype in terms of HSC activity (Goyama et al., 2008). Furthermore, gene expression profiling of Evi-1–deleted HSCs and leukemic cells identified Pbx1 as a downstream target for Evi-1 in HSCs (Shimabe et al., 2009).

2.1.1.6 JunB

The AP-1 transcription factor, JunB, is a transcriptional regulator of myelopoiesis and a potential tumor suppressor gene in mice (Passegue et al., 2001). Compared with normal HSCs, JunB-deficient LT-HSCs showed an average 2-fold increase in the percentage of cycling cells, suggesting that JunB functions to limit cell-cycle entry. Gene expression analyses revealed that JunB-deficient LT-HSCs show increased expression of cyclins and decreased expression of cyclin-dependent kinase inhibitors (Santaguida et al., 2009). These results suggest that the absence of JunB induces quiescent cells to enter the cell cycle.

2.1.1.7 p53

The p53 tumor suppressor protein functions as a transcription factor, regulating the transcription of genes that induce cell-cycle arrest, senescence, and apoptosis. LT-HSCs express high levels of p53 (Dumble et al., 2007). Although p53-deficient mice show almost normal hematopoiesis (Lotem & Suchs., 1993), a number of studies have identified a role for p53 in the proliferation, differentiation, apoptosis, and aging of HSCs (Kastan et al., 1991; Shounan et al., 1996; Park et al., 2003; Dumble et al., 2007). Recent detailed analyses of p53-null mice have unraveled other important functions of p53 in HSCs. Liu et al. found that p53 promotes HSC quiescence, and that p53-deficient HSCs enter the cell cycle more easily (Liu et al., 2009). Competitive BM repopulation assays revealed that p53-null cells out-compete wild-type cells (TeKippe et al., 2003; Chen et al., 2008; Liu et al., 2009), indicating that p53 is a negative regulator of HSC self-renewal. In addition, Liu et al. also identified Gfi-1 and necdin as p53 target genes by performing comparative transcriptional profiling of HSCs isolated from wild-type and p53-deficient mice. The results of *in vitro* overexpression and knockdown experiments identified a role for necdin in the maintenance of HSC quiescence and self-renewal. However, necdin appears to have a modest functional role in HSCs *in vivo* (Kubota et al., 2009), and necdin overexpression does not result in enhanced HSC quiescence (Sirin et al., 2010).

2.1.1.8 Nurr1

Gene expression profiling analyses identified Nurr1 (also known as Nr4a2), an orphan nuclear receptor, as a candidate molecule that may play a functional role in HSC quiescence (Venezia et al., 2004; Chambers et al., 2007). Overexpression of Nurr1 resulted in HSC quiescence. On the other hand, loss of one Nurr1 allele resulted in enhanced cycling and sensitivity to the chemotherapeutic agent 5-fluorouracil (5-FU). Molecular analysis showed that Nurr1 overexpression is positively correlated with the upregulation of the cell-cycle inhibitor $p18^{INK4C}$, suggesting a mechanism by which Nurr1 may regulate HSC quiescence (Sirin et al., 2010).

2.1.1.9 Reactive oxygen species, FoxOs

Reactive oxygen species (ROS) play an important role in the regulation of HSC quiescence. The forkhead O (FoxO) family of transcription factors (FoxO1, FoxO3, FoxO4, and FoxO6) participates in various cellular processes, including the induction of cell-cycle arrest, stress resistance, apoptosis, differentiation, and metabolism (Greer & Brunet., 2005). Two groups reported that FoxOs play a regulatory role in a number of physiologic processes that influence HSC numbers and function. Both aged germline FoxO3-deficient mice and conditional triple knockout (FoxO1, 3, 4) mice show a reduction in HSC numbers with a deficient repopulating capacity in competitive reconstitution assays and serial competitive transplantation assays (Tothova et al., 2007; Miyamoto et al., 2007). These phenotypes correlate with increased cell-cycling and apoptosis of HSCs, caused by increased levels of ROS. Furthermore, treatment with the antioxidant, N-acetyl-L-cysteine (NAC), rescues the FoxO-deficient HSC phenotype.

2.1.1.10 Fbxw7

Fbxw7 is the F-box protein subunit of an SCF-type ubiquitin ligase complex that targets positive regulators of the cell-cycle, including Notch, c-Myc, cyclin E, and c-Jun. Two independent groups investigated the functions of Fbxw7 in HSCs using conditional Fbxw7 knockout mice (Matsuoka et al., 2008; Thompson et al., 2008). Conditional ablation of Fbxw7 rapidly and severely affects hematopoietic progenitor maintenance within the BM. *Fbxw7*$^{-/-}$ HSCs show increased cycling and defective long-term repopulation capacity in competitive transplantation assays. As Fbxw7 is able to ubiquitinate several target proteins, studies were conducted to examine the protein expression of Notch1, c-Myc, and cyclin E. The results showed that c-Myc protein was substantially overexpressed in *Fbxw7*$^{-/-}$ HSCs, suggesting that the activation of the cell-cycle in Fbxw7-null HSCs induced by excess c-Myc causes the premature exhaustion of HSCs.

2.1.1.11 HIF-1α

Leukemic stem cells (LSCs) reside in the niches near epiphysis of the bone (Ishikawa et al., 2007) and oxygen concentration of this area is quite low. Thus, it may be very important for leukemic cells, especially for LSCs to survive and adapt to hypoxia (Takeuchi et al., 2010). Cellular responses to hypoxia are mediated by hypoxia-inducible factors (HIFs), which regulate gene expression to facilitate adaptation to hypoxic conditions (Kaelin & Ratcliffe., 2008). Hypoxia inducible factor-1α (HIF-1α) is stabilized under low-oxygen conditions, such as those present in the BM. Recently, two groups investigated the importance of hypoxia and its related signaling pathways in HSC function using different approaches (Simsek et al., 2010; Takubo et al., 2010). HIF-1α levels are elevated in adult HSCs and its transcription is regulated by the homeodomain protein Meis1, which is essential for hematopoiesis (Hisa et al., 2004; Simsek et al., 2010). HIF-1α conditional knockout mice show that HIF-1α–deficient HSCs have an increased cell cycling rate and show progressive loss of long-term repopulation ability in serial transplantation assays (Takubo et al., 2010). Taken together, these data indicate that the precise regulation of HIF-1α levels is required to maintain HSC quiescence.

2.1.1.12 Lkb1

The control of energy metabolism within HSCs is poorly understood, although they are highly sensitive to oxidative stress. Recently, several groups examined the role of the protein, Lkb1, in the metabolic regulation of HSCs (Nakada et al., 2010; Gurumurthy et al., 2010; Gan et al., 2010). Lkb1 is a kinase enzyme that regulates the activity of AMP-activated protein kinase (AMPK). Conditional inactivation of Lkb1 (*Mx1-Cre; LKB1*$^{fl/fl}$ or *RosaCreERT2; LKB1*$^{L/L}$) in adult mice causes the loss of HSC quiescence, rapid HSC depletion, and pancytopenia. Interestingly, Lkb1 seems to regulate HSC homeostasis primarily through pathways that are independent of its downstream effectors, AMPK and mTORC1.

2.1.1.13 Cyclin-dependent kinase inhibitors

p21$^{cip1/waf1}$ (hereafter referred to as p21) is a mammalian member of the CIP/KIP family and was the first cyclin-dependent kinase inhibitor to be identified (Serrano et al., 1993; Harper et al., 1993; Stier et al., 2003). Serial transplantation assays using p21-deficient cells showed premature HSC exhaustion; also, p21-null mice were more sensitive to 5-FU (Cheng et al., 2000). These results suggest that p21 restricts HSC entry into the cell cycle and regulates the size of the HSC pool under conditions of stress. However, a later study demonstrated that p21 plays a minor role in regulating HSC quiescence under conditions of steady-state hematopoiesis (van Os et al., 2007).

Although p57^{kip2} (hereafter referred to as p57) is highly expressed in HSCs (Table 2) (Kubota et al., 2009; Umemoto et al., 2005), little is known about its functional role. Microarray analysis studies of human CD34+ HSC/progenitor cells identified p57 as the only cyclin-dependent kinase inhibitor induced by TGFβ (Scandura et al., 2004). Knockdown of p57 in hematopoietic cell lines using small interfering RNA (siRNA) results in more rapid proliferation of hematopoietic cells in the absence of TGF-β. These results suggest that p57 is required for the TGF-β–mediated cell cycle entry of hematopoietic cells and for repressing the proliferation of these cells.

Gene Name	Gene Symbol
Apoptosis	
serine (or cysteine) peptidase inhibitor, clade A, member 3G	Serpina3g
Cell surface	
adhesion molecule with Ig like domain 2	Amigo2
claudin 5	Cldn5
junction adhesion molecule 2	Jam2
vascular cell adhesion molecule 1	Vcam1
Cell Cycle Regulation	
cyclin-dependent kinase inhibitor 1C (P57)	Cdkn1c
Cell Signaling	
frizzled homolog 4 (Drosophila)	Fzd4
insulin-like growth factor 1	Igf1
interferon inducible GTPase 1	Iigp
multiple PDZ domain protein	Mpdz
nik related kinase	Nrk
regulator of G-protein signaling 4	Rgs4
ras homolog gene family, member J	Rhoj
suppressor of cytokine signaling 2	Socs2
Cellular Metabolism	
cytochrome P450, family 4, subfamily b, polypeptide 1	Cyp4b1
fatty acid binding protein 4, adipocyte	Fabp4
RIKEN cDNA 4432416J03 gene	4432416J03Rik
Endocytosis	
intersectin 1 (SH3 domain protein 1A)	Itsn
Extracellular	
bone morphogenetic protein 2	Bmp2
connective tissue growth factor	Ctgf
nidogen 1	Nid1
tissue factor pathway inhibitor	Tfpi
tissue inhibitor of metalloproteinase 3	Timp3
Transcription Factor	
forkhead box A3	Foxa3
kruppel-like factor 9	Klf9
myeloid/lymphoid or mixed lineage-leukemia translocation to 3 homolog (Drosophila)	Mllt3
necdin	Ndn
nuclear protein 1 (p8)	Nupr1
retinoid X receptor gamma	Rxrg
Unknown	
tripartite motif-containing 47	Trim47
RIKEN cDNA 2310051E17 gene	2310051E17Rik
RIKEN cDNA 2810432L12 gene	2810432L12Rik

Table 2. Genes expressed at higher levels in HSCs than in other subsets.

Genes showing at least 2-fold higher expression in CD34$^{-/low}$ KSL cells than in CD34^{+} KSL cells were selected by microarray analysis. The selected genes were then evaluated by Q-PCR, and genes whose transcripts were expressed at ≥ 2-foltd higher levels in CD34$^{-/low}$ KSL cells than all other samples are listed.

2.1.2 Negative regulation

2.1.2.1 E3 ubiquitin ligase

The E3 ubiquitin ligase, c-Cbl, is a member of the RING finger-type ubiquitin ligase Cbl (casitas B-cell lymphoma) family. The c-Cbl protein is thought to implement the degradation of various cellular proteins, receptors, and signaling molecules including Notch1, STAT5, and c-Kit (Jehn et al., 2002; Goh et al., 2002; Zeng et al., 2005). c-Cbl–deficient mice were used to study the role of c-Cbl in HSCs (Rathinam et al., 2008). The number of HSCs and progenitors was significantly higher in the BM of c-Cbl-null mice due to increased proliferation. Interestingly, detailed analyses revealed augmented STAT5 phosphorylation in *c-Cbl*$^{-/-}$ HSCs in response to TPO/c-MPL signaling which is crucial for the proliferation and self-renewal of HSCs (Kimura et al., 1998), and this led to enhanced c-Myc expression. C-Cbl–deficient HSCs also showed an increased repopulating ability in competitive reconstitution assays, including serial transplantation. These results suggest that c-Cbl acts as a negative regulator of both the size of the HSC pool and self-renewal (Rathinam et al., 2008).

Recently, Itch, another E3 ligase belonging to the HECT family (Bernassola et al., 2008), was also identified as a negative regulator of HSC homeostasis and function. The phenotype of *Itch*$^{-/-}$ HSCs was similar to that of *c-Cbl*$^{-/-}$ HSCs. However, unlike c-Cbl, Itch-deficient HSCs showed augmented Notch1 signaling. Furthermore, knockdown of Notch1 in Itch-null HSCs resulted in the reversion of the phenotype (Rathinam et al., 2011). Taken together, these studies underscore the pivotal roles of E3 ubiquitin ligases and the importance of post-translational modification of HSCs in the molecular control of HSC self-renewal.

2.1.2.2 Egr1

Egr1 is a member of the immediate early response gene family (Gashler et al., 1995). Egr1 is highly expressed in LT-HSCs under steady-state conditions and is downregulated upon proliferative stimulation and migration in response to pharmacological mobilization (Min et al., 2008). Egr1-deficient mice show a significant increase in the frequency of cycling HSCs. This phenomenon results in a slightly higher frequency of HSCs in the BM of *Egr1*$^{-/-}$ mice. Interestingly, loss of Egr1 results in a striking increase (up to 10-fold) in the number of circulating HSCs. Importantly, HSCs isolated from both the BM and peripheral blood of *Egr1*$^{-/-}$ mice show a greater degree of long-term multi-lineage repopulation after transplantation, although their life span is slightly reduced. Quantitative RT-PCR analysis shows that Bmi1 is upregulated in *Egr1*$^{-/-}$ HSCs. In addition, *Egr1*$^{-/-}$ HSCs also show the downregulation of $p21^{CIP1/WAF1}$ and increased expression of cyclin-dependent kinase 4 (cdk4), which is consistent with their increased cell-cycling status (Min et al., 2008). Taken together, the deletion of Egr1 causes an increase in the number of cycling HSCs but does not lead to stem cell exhaustion. This may be due to Bmi1 upregulation.

2.1.2.3 Lnk

Lnk is a member of an adaptor protein family that possesses a number of protein-protein interaction domains: a proline-rich amino-terminus, a pleckstrin homology (PH) domain, a Src homology 2 (SH2) domain, and many potential tyrosine phosphorylation motifs (Rudd., 2001). Studies using Lnk-deficient mice show that Lnk-null HSCs are expanded during post-natal development (Ema et al., 2005; Buza-Vidas et al., 2006). The *Lnk*$^{-/-}$ HSC population contains an increased proportion of quiescent cells and shows decelerated cell cycle kinetics and enhanced resistance to repeat treatment with 5-FU *in vivo* compared with wild-type HSCs. Genetic evidence demonstrates that Lnk controls HSC self-renewal and quiescence, predominantly through c-Mpl. Furthermore, Lnk-deficient HSCs show higher levels of symmetric proliferation in response to thrombopoietin (TPO) in *ex vivo* culture than wild-type HSCs (Seita et al., 2007). Biochemical analyses revealed that Lnk directly binds to phosphorylated tyrosine residues in JAK2 after TPO stimulation (Bersenev et al., 2008). Therefore, Lnk is a physiologic negative regulator of JAK2 in HSCs, and TPO/c-Mpl/JAK2/Lnk constitute a major regulatory pathway controlling HSC quiescence and self-renewal.

2.1.2.4 Myc

Human c-MYC was the second proto-oncogene to be identified and encodes a basic helix-loop-helix leucine zipper transcription factor (c-Myc) (Sheiness et al., 1978). Overexpression of one of the three family members has been detected in numerous human cancers including Burkitt's lymphoma (c-MYC), neuroblastoma (N-MYC), and small cell lung cancer (L-MYC) (Nesbit et al., 1999). Conditional deletion of c-Myc in the BM results in cytopenia and the accumulation of functionally defective HSCs. In the absence of c-Myc, HSC differentiation into more committed progenitors is inhibited because they upregulate a number of adhesion molecules, such as N-cadherin, that anchor them in the niche. Conversely, enforced c-Myc expression in HSCs causes marked repression of N-cadherin and integrin expression leading to the loss of self-renewal ability at the expense of differentiation (Wilson et al., 2004). These results suggest that c-Myc activity controls the first differentiation step of LT-HSCs *in vivo*. Unexpectedly, conditional ablation of both c-myc and N-myc results in pancytopenia and rapid lethality due to HSC apoptosis via the accumulation of the cytotoxic molecule, Granzyme B (Laurenti et al., 2008). Thus, Myc activity controls important aspects of HSC function such as proliferation, survival and differentiation.

2.1.2.5 MEF/ELF4

MEF (also known as ELF4), an Ets transcription factor, was identified as a novel component of the transcriptional circuit that dynamically regulates HSC quiescence (Lacorazza et al., 2006). Mef-deficient HSCs grow more slowly than wild-type HSCs in response to cytokine stimulation Pyronin Y staining and BrdU incorporation show increased quiescence. Enhanced HSC quiescence in Mef-null mice also increases HSC resistance to cytotoxic agents that target dividing cells and allows more rapid hematological recovery after chemotherapy or irradiation. These findings suggest that Mef normally functions to induce or facilitate the entry of quiescent HSCs into the cell cycle and imply that Mef expression and/or activity may be dynamically regulated in HSCs. To explain this, Lacorazza et al. proposed a model in which Mef acts at an earlier stage than p18 and antagonizes p21.

2.2 Survival of HSCs

HSC self-renewal and apoptosis represent major factors that determine the size of the HSC mass. The number of HSCs is also controlled by their capacity to survive during homeostasis or under conditions of stress.

2.2.1 Bcl-2 family

Accumulating evidence suggests that the suppression of apoptosis is required for HSC survival. Forced expression of Bcl-2 increases the number of HSCs and provides them with enhanced competitive repopulation ability (Domen et al., 1998, 2000), suggesting that cell death plays a role in regulating HSC homeostasis.

Mcl-1, another anti-apoptotic Bcl-2 family member, is also an essential regulator of HSC survival. Mcl-1 is highly expressed in LT-HSCs, and conditional deletion of MCl-1 results in the loss of the early BM progenitor population, including HSCs, leading to fatal hematopoietic failure (Opferman et al., 2005). Recently, it was reported that Mcl-1 is an indispensable regulator of self-renewal in human stem cells and that functional dependence on Mcl-1 defines the human stem cell hierarchy (Campbell et al., 2010).

2.2.2 Scl, Lyl1

Scl/Tal1 is a basic helix-loop-helix (bHLH) transcription factor that is essential for the development of HSCs in the embryo (Robb et al., 1995; Shivdasani et al., 1995). During adult hematopoiesis, Scl/Tal1 is highly expressed in LT-HSCs compared with short-term HSCs and progenitor cells (Lacombe et al., 2010). However, a study using conditional Scl/Tal1 knockout mice revealed that Scl/Tal1 is required for the generation of, but not the maintenance of, adult HSCs (Mikkola et al., 2003). Another group showed that conditional deletion of Scl/Tal1 in adult HSCs has a relatively mild effect: Scl-null HSCs show impaired short-term repopulating ability, but no defect in long-term repopulating capacity (Curtis et al., 2004). Redundant activity caused by the expression of Lyl1, a related bHLH transcription factor, in adult HSCs may provide an explanation for these "mild" phenotypes. While adult HSCs in single-knockout mice show no or only a mild phenotype, Lyl1;Scl-conditional double-knockout mice show a gene dosage defect on HSC survival, as HSCs and progenitor cells are immediately lost due to apoptosis (Souroullas et al., 2009).

Recently, Lacombe et al. demonstrated that Scl/Tal1 is required for the maintenance of the quiescent stem cell pool (Lacombe et al., 2010). Cell-cycle analyses revealed that Scl/Tal1 negatively regulates the G0-G1 transit of LT-HSCs; however, these phenomena were specific to adult HSCs and were not observed in perinatal HSCs. The reconstituting ability of $Scl^{+/-}$ HSCs or HSCs with decreased Scl protein expression induced by RNA interference was impaired in various transplantation assays. Furthermore, gene expression analysis and chromatin immunoprecipitation experiments revealed that the Cdkn1a and Id1 genes are direct SCL targets.

2.2.3 Tel/Etv6

The transcription factor Tel (also known as Etv6), an Ets-related transcriptional repressor, is a frequent target of the diverse chromosomal translocations observed in leukemias (Golub et al., 1994). Tel/ETV6 is also required for HSC survival in adult hematopoiesis. Following conditional inactivation of Tel/Etv6, HSCs are rapidly lost from the adult BM. However, Tel/Etv6 is not required for the maintenance of lineage-committed progenitors. Conditional deletion of Tel/Etv6 after lineage commitment does not affect the differentiation or survival of these progenitors, although it does impair the maturation of megakaryocytes (Hock et al., 2004).

2.2.4 Zfx

Zfx is a zinc finger protein belonging to the Zfx/ZFy family. Mammalian Zfx is encoded on the X chromosome and contains an acidic transcriptional activation domain, a nuclear localization sequence, and a DNA binding protein domain consisting of 13 C2H2-type zinc fingers (Schneider-Gadicke et al., 1989). Zfx is highly expressed in both HSCs and undifferentiated embryonic stem cells (ESCs). Using conditional gene targeting, Zfx was identified as an essential transcriptional regulator of HSC function (Galan-Caridad et al., 2007). Constitutive or inducible deletion of Zfx in HSCs (using *Tie2-Cre* and *Mx1-Cre* deletion strains, respectively) impairs self-renewal, resulting in increased apoptosis and the upregulation of stress-inducible genes.

2.2.5 ADAR1

ADAR (adenosine deaminase acting on RNA) catalyzes the deamination of adenosine to inosine in double-stranded RNA. Conventional *Adar*$^{-/-}$ mice die around embryonic day 11.5–12 because of widespread apoptosis and defective hematopoiesis (Hartner et al., 2004; Wang et al., 2004). Conditional deletion of Adar in HSCs shows that ADAR1 is essential for the maintenance of both fetal and adult HSCs, and leads to global upregulation of type I and II interferon-inducible transcripts and rapid apoptosis (Hartner et al., 2009). Interferon regulatory factor-2 (Irf2), a transcriptional suppressor of type I interferon signaling, is a positive regulator of HSC quiescence (Sato et al., 2009). Irf2-deficient HSCs are unable to restore hematopoiesis in irradiated mice, but the reconstituting capacity of *Irf2*$^{-/-}$ HSCs can be restored in these cells by disabling type I IFN signaling.

2.3 Response to hematopoietic emergency

Various external stresses, such as myelosuppressive chemotherapy, bleeding, infection, and total body irradiation, put HSCs under stress, as they must proliferate to produce large numbers of primitive progenitor cells, thereby enabling rapid hematologic regeneration. Although this property has long been recognized, the molecular basis underlying the reaction of HSCs to hematologic emergency remains enigmatic. However, some key players have been identified.

2.3.1 Heme oxygenase-1

Heme promotes the proliferation and differentiation of hematopoietic progenitor cells (HPCs) (Chertkov et al., 1991) and stimulates hematopoiesis (Porter et al., 1979; Abraham, 1991). The degradation of heme is catalyzed by heme oxygenase (HO). HO-1, encoded by the *Hmox1* gene, is the stress-inducible isozyme of HO and is highly expressed in the spleen and BM (Abraham, 1991). Heterozygous HO-1–deficient mice (*HO-1*$^{+/-}$) show accelerated hematologic recovery from myelotoxic injury induced by 5-FU treatment, and mice transplanted with *HO-1*$^{+/-}$ BM cells show more rapid hematopoietic repopulation than those transplanted with *Ho-1*$^{+/+}$ BM cells. However, *HO-1*$^{+/-}$ HSCs show a reduced capacity to rescue lethally irradiated mice and to serially repopulate irradiated recipients (Cao et al., 2008). These results suggest that HO-1 limits the proliferation and differentiation of HPCs under stressful conditions, and that the failure of this mechanism can lead to the premature exhaustion of the HSC pool.

2.3.2 Necdin

Necdin is a member of the melanoma antigen family of molecules, whose physiological roles have not been well characterized (Xiao et al., 2004). Necdin acts as a cell cycle regulator in post-mitotic neurons (Yoshikawa, 2000). Intriguingly, recent genetic analyses show that aberrant genomic imprinting of *NDN* on the human 15q11-q13 chromosomal region is, at least in part, responsible for the pathogenesis of Prader-Willi syndrome (MacDonald & Wevrick, 1997; Nakada et al., 1998; Barker et al., 2002), a disorder associated with a mildly increased risk of myeloid leukemia (Davies et al., 2003). Necdin interacts with multiple cell-cycle related proteins, such as SV-40 large T antigen, adenovirus E1A, E2F1, and p53 (Taniura et al., 1998, 1999, 2005; Hu et al., 2003). As shown in Table 2, necdin is one of 32 genes that show higher expression in HSCs than in differentiated hematopoietic cells (Kubota et al., 2009). Other groups also found that necdin is highly expressed in HSCs (Forsberg et al., 2005; Liu et al., 2009). Necdin-deficient mice show accelerated recovery of hematopoietic systems after myelosuppressive stress, such as 5-FU treatment and BM transplantation, whereas no overt abnormality is seen under conditions of steady-state hematopoiesis. Considering necdin as a potential negative cell-cycle regulator, it was reasoned that the enhanced hematologic recovery in necdin-null mice could be the result of an increased number of proliferating HSCs and progenitor cells. As expected, after 5-FU treatment, necdin-deficient mice had an increased number of HSCs, but this was only transiently observed during the recovery phase (Kubota et al., 2009). These data suggest that the repression of necdin function in HSCs may present a novel strategy for accelerating hematopoietic recovery, thus providing therapeutic benefits after clinical myelosuppressive treatments (e.g., cytoablative chemotherapy or HSC transplantation).

2.3.3 Slug

Slug belongs to the highly conserved Slug/Snail family of zinc-finger transcriptional repressors found in diverse species ranging from *C. elegans* to humans. SLUG is a target gene for the E2A-HLF chimeric oncoprotein in pro-B cell acute leukemia (Inukai et al., 1999). Slug-deficient mice show normal peripheral blood counts, but they are very sensitive to γ-irradiation (Inoue et al., 2002). Slug is induced by p53 and protects primitive hematopoietic cells from apoptosis triggered by DNA damage. Slug exerts this function by repressing Puma, a proapoptotic target of p53 (Wu et al., 2005). Sun et al. recently showed that Slug negatively regulates the repopulating ability of HSCs under conditions of stress. Slug deficiency increases HSC proliferation and reconstitution potential *in vivo* after myelosuppressive treatment, and accelerates HSC expansion during *in vitro* culture (Sun et al., 2010).

3. Cancer stem cells

Accumulating evidence strongly suggests that tumors are organized into cellular hierarchies initiated and maintained by a small pool of self-renewing cancer stem cells (CSCs) (Dick, 2008; Reya et al., 2001). CSCs are thought to be resistant to various cancer treatments because of their relative quiescence (Komarova & Wodarz., 2007). Cancer relapses may occur because the dormancy of CSCs protects them from elimination by various cancer therapies (Dick, 2008). In an acute myelogenous leukemia (AML) xenograft model, AML leukemic stem cells (LSCs) localized in the endosteal region of the BM show cellular quiescence and resistance to chemotherapy (Ishikawa et al., 2007; Saito et al., 2010). In patients with chronic myelogenous leukemia (CML), $CD34^+$ progenitor cells contain dormant cells that are resistant to BCR/ABL tyrosine kinase inhibitors (Bhatia et al., 2003).

It is well documented that regulators of HSC maintenance are also involved in the development of leukemias (Rizo et al., 2006). A number of cancer-related proteins, such as Bmi1, c-Myc, p53, Gfi-1, and PTEN, are key participants in HSC regulation, demonstrating the close relationship between normal HSCs and CSCs. Therefore, further understanding the mechanisms regulating HSC fate is needed if we are to develop new strategies for targeting CSCs and successfully treat cancer.

4. Conclusions

In this review, we have briefly summarized a number of critical regulators involved in the control of HSC self-renewal, quiescence, survival, and responses to external insults. Recent evidence strongly suggests that the BM niche also plays an integral role by providing critical signals that maintain HSCs in a stat of hibernation, thus preventing them from exhausting themselves. However, HSCs are critical for the maintenance and regeneration of an organism after injury/illness. This process must be tightly regulated and coordinated. Intensive studies have uncovered the molecular signatures and key molecules regulating HSC behavior. Moreover, new systems approaches, such as microRNA expression profiling and protein expression profiling, are expected to provide further useful information about HSC biology in the future. However, the overall picture of the molecular mechanisms that govern HSC fate is still unclear. Further understanding of the systems that regulate HSCs will enable the manipulation of stem cells for use in tissue engineering and cell-based therapies.

5. Acknowledgments

This work was supported by a Grant-in-Aid for Young Scientists to Y.K. (no. 23791083) from the Ministry of Education, Culture, Sports, Science and Technology (MEXT), Japan.

6. References

Abraham, N.G. (1991) Molecular regulation--biological role of heme in hematopoiesis. *Blood Rev* 5(1):19-28.

Akashi, K., Traver, D., Miyamoto, T., & Weissman, I.L. (2000) A clonogenic common myeloid progenitor that gives rise to all myeloid lineages. *Nature* 404(6774):193-197.

Akashi, K., He, X., Chen, J., Iwasaki, H., Niu, C., Steenhard, B., Zhang, J., Haug, J., & Li, L. (2003) Transcriptional accessibility for genes of multiple tissues and hematopoietic lineages is hierarchically controlled during early hematopoiesis. *Blood* 101(2):383-389.

Barker, P.A., & Salehi, A. (2002) The MAGE proteins: emerging roles in cell cycle progression, apoptosis, and neurogenetic disease. *J Neurosci Res* 67(6):705-712.

Bernassola, F., Karin, M., Ciechanover, A., & Melino, G. (2008) The HECT family of E3 ubiquitin ligases: multiple players in cancer development. *Cancer Cell* 14(1):10-21.

Bersenev, A., Wu, C., Balcerek, J., & Tong, W. (2008) Lnk controls mouse hematopoietic stem cell self-renewal and quiescence through direct interactions with JAK2. *J Clin Invest* 118(8):2832-2844.

Bhatia, R., Holtz, M., Niu, N., Gray, R., Snyder, D.S., Sawyers, C.L., Arber, D.A., Slovak, M.L., & Forman, S.J. (2003) Persistence of malignant hematopoietic progenitors in chronic myelogenous leukemia patients in complete cytogenetic remission following imatinib mesylate treatment. *Blood* 101(12):4701-4707.

Buza-Vidas, N., Antonchuk, J., Qian, H., Månsson, R.,Luc, S., Zandi, S., Anderson, K., Takaki, S., Nygren, J.M., Jensen, C.T., & Jacobsen S.E. (2006) Cytokines regulate postnatal hematopoietic stem cell expansion: opposing roles of thrombopoietin and LNK. *Genes Dev* 20(15):2018-2023.

Campbell, C.J.V., Lee, J.B., Levadoux-Martin, M., Wynder, T., Xenocostas, A., Leber, B., & Bhatia, M. (2010) The human stem cell hierarchy is defined by a functional dependence on Mcl-1 for self-renewal capacity. *Blood* 116(9):1433-1442.

Cao, Y.A., Wagers, A.J., Karsunky, H., Zhao, H., Reeves, R., Wong, R.J., Stevenson, D.K., Weissman, I.L., & Contag, C.H. (2008) Heme oxygenase-1 deficiency leads to disrupted response to acute stress in stem cells and progenitors. *Blood* 112(12):4494-4502.

Chambers, S.M., Boles, N.C., Lin, K.Y., Tierney, M.P., Bowman, T.V., Bradfute, S.B., Chen, A.J., Merchant, A.A., Sirin, O., Weksberg, D.C., Merchant, M.G., Fisk, C.J., Shaw, C.A., & Goodell, M.A. (2007) Hematopoietic fingerprints: an expression database of stem cells and their progeny. *Cell Stem Cell* 1(5): 578-591.

Chen, J., Ellison, F.M., Keyvanfar, K., Omokaro, S.O., Desierto, M.J., Eckhaus, M.A., & Young, N.S. (2008) Enrichment of hematopoietic stem cells with SLAM and LSK markers for the detection of hematopoietic stem cell function in normal and Trp53 null mice. *Exp Hematol* 36(10):1236-1243.

Cheng, T., Rodrigues, N., Shen, H., Yang, Y., Dombkowski, D., Sykes, M., & Scadden D.T. (2000) Hematopoietic stem cell quiescence maintained by $p21^{cip1/waf1}$. *Science* 287(5459):1804-1808.

Chertkov, J.L., Jiang, S., Lutton, J.D., Levere, R.D., & Abraham, N.G. (1991) Hemin stimulation of hemopoiesis in murine long-term bone marrow culture. *Exp Hematol* 19(9):905-909.

Cheshier, S.H., Morrison, S.J., Liao, X., & Weissman, I.L. In vivo proliferation and cell cycle kinetics of long-term self-renewing hematopoietic stem cells. (1999) *Proc Natl Acad Sci U S A* 96(6):3120-3125.

Curtis, D.J., Hall, M.A., Van Stekelenberg, L.J., Robb, L., Jane, S.M., & Begley, C.G. (2004) SCL is required for normal function of short-term repopulating hematopoietic stem cells. *Blood* 103(9):3342-3348.

Davies, H.D., Leusink, G.L., McConnell, A., Deyell, M., Cassidy, S.B., Fick, G.H., & Coppes, M.J. (2003) Myeloid leukemia in Prader-Willi syndrome. *J Pediatr* 142(2):174-178.

Dick, J.E. (2008) Stem cell concepts renew cancer research. *Blood* 112(13):4793-4807.

DiMartino, J.F., Selleri, L., Traver, D., Firpo, M.T., Rhee, J., Warnke, R., O'Gorman, S., Weissman, I.L., & Cleary, M.L. (2001) The Hox cofactor and protooncogene Pbx1 is

required for maintenance of definitive hematopoiesis in the fetal liver. *Blood* 98:618-626.
Domen, J., Gandy, K.L., & Weissman, I.L. (1998) Systemic overexpression of BCL-2 in the hematopoietic system protects transgenic mice from the consequences of lethal irradiation. *Blood* 91(7):2272-2282.
Domen, J., Cheshier, S.H., & Weissman, I.L. (2000) The role of apoptosis in the regulation of hematopoietic stem cells: Overexpression of Bcl-2 increases both their number and repopulation potential. *J Exp Med* 191(2):253-264.
Dumble, M., Moore, L., Chambers, S.M., Geiger, H., Zant, G.V., Goodell, M.A. & Donehower, L.A. (2007) The impact of altered p53 dosage on hematopoieitic stem cell dynamics during aging. *Blood* 109(4):1736-1742.
Ema, H., Sudo, K., Seita, J., Matsubara, A., Morita, Y., Osawa, M., Takatsu, K., Takaki, S., & Nakauchi, H. (2005) Quantification of self-renewal capacity in single hematopoietic stem cells from normal and Lnk-deficient mice. *Dev Cell* 8(6):907-914.
Ficara, F., Murphy, M.J., Lin, M., & Cleary, M.L. (2008) Pbx1 regulates self-renewal of long-term hematopoietic stem cells by maintaining their quiescence. *Cell Stem Cell* 2(5):484-496.
Forsberg, E.C., Prohaska, S.S., Katzman, S., Heffner, G.C., Stuart, J.M., & Weissman, I.L. (2005) Differential expression of novel potential regulators in hematopoietic stem cells. *PLoS Genet* 1(3):e28.
Galan-Caridad, J.M., Harel, S., Arenzana, T.L., Hou, Z.E., Doetsch, F.K., Mirny, L.A., & Reizis, B. (2007) Zfx controls the self-renewal of embryonic and hematopoietic stem cells. *Cell* 129(2):345-357.
Gan, B., Hu, J., Jiang, S., Liu, Y., Sahin, E., Zhuang, L., Fletcher-Sananikone, E., Colla, S., Wang, Y.A., Chin, L., & Depinho, R.A. (2010) Lkb1 regulates quiescence and metabolic homeostasis of haematopoietic stem cells. *Nature* 468(7324):701-704.
Gashler, A., & Sukhatme, V.P. (1995) Early growth response protein 1 (Egr-1): prototype of a zinc-finger family of transcription factors. *Prog Nucleic Acid Res Mol Biol* 50:191-224.
Gilks, C.B., Bear, S.E., Grimes, H.L., & Tsichlis, P.N. (1993) Progression of interleukin-2 (IL-2)-dependent rat T cell lymphoma lines to IL-2-independent growth following activation of a gene (Gfi-1) encoding a novel zinc finger protein. *Mol Cell Biol* 13(3): 1759-1768.
Golub,T.R., Barker, G.F., Lovett, M., & Gilliland, D.G. (1994) Fusion of PDGF receptor β to a novel ets-like gene, tel, in chronic myelomonocytic leukemia with t(5;12) chromosomal translocation. *Cell* 77(2): 307-316.
Goh, E.L., Zhu, T., Leong, W.Y., & Lobie, P.E. (2002) c-Cbl is a negative regulator of GH-stimulated STAT5-mediated transcription. *Endocrinology* 143(9):3590-3603.
Goyama, S., Yamamoto, G., Shimabe, M., Sato, T., Ichikawa, M., Ogawa, S., Chiba, S., & Kurokawa, M. (2008) Evi-1 is a critical regulator for hematopoietic stem cells and transformed leukemic cells. *Cell Stem Cell* 3(2):207-220.
Greer, E.L., & Brunet, A. (2005) FOXO transcription factors at the interface between longevity and tumor suppression. *Oncogene* 24(50):7410-7425.
Gurumurthy, S., Xie, S.Z., Alagesan, B., Kim, J., Yusuf, R.Z., Saez, B., Tzatsos, A., Ozsolak, F., Milos, P., Ferrari, F., Park, P.J., Shirihai, O.S., Scadden, D.T., & Bardeesy, N. (2010) The Lkb1 metabolic sensor maintains haematopoietic stem cell survival. *Nature* 468(7324):659-663.

Harper, J.W., Adami, G.R., Wei, N., Keyomarsi, K., & Elledge, S.J. (1993) The p21 Cdk-interacting protein Cip1 is a potent inhibitor of G1 cyclin-dependent kinases. *Cell* 75(4):805-816.

Hartner, J.C., Schmittwolf, C., Kispert, A., Müller, A.M., Higuchi, M., & Seeburg, P.H. (2004) Liver disintegration in the mouse embryo caused by deficiency in the RNA-editing enzyme ADAR1. *J Biol Chem* 279(6):4894-4902.

Hartner,J.C., Walkley, C.R., Lu, J., & Orkin, S.H. (2009) ADAR1 is essential for the maintenance of hematopoiesis and suppression of interferon signaling. *Nat Immunol* 10(1):109-115.

Hisa, T., Spence, S.E., Rachel, R.A., Fujita, M., Nakamura, T., Ward, J.M., Devor-Henneman, D.E., Saiki, Y., Kutsuna, H., Tessarollo, L., Jenkins, N.A., & Copeland, N.G. (2004) Hematopoietic, angiogenic and eye defects in Meis1 mutant animals. *EMBO J* 23(2):450-459.

Hock, H., Meade, E., Medeiros, S., Schindler, J.W., Valk, P.J.M., Fujiwara, Y., & Orkin, S.H. (2004) Tel/Etv6 is an essential and selective regulator of adult hematopoietic stem cell survival. *Genes Dev* 18(19):2336-2341.

Hock, H., Hamblen, M.J., Rooke, H.M., Schindler, J.W., Saleque, S., Fujiwara, Y., & Orkin, S.H. (2004) Gfi-1 restricts proliferation and preserves functional integrity of haematopoietic stem cells. *Nature* 431(7011):1002-1007.

Hu, B., Wang, S., Zhang, Y., Feghali, C.A., Dingman, J.R., & Wright, T.M. (2003) A nuclear target for interleukin-1alpha: interaction with the growth suppressor necdin modulates proliferation and collagen expression. *Proc Natl Acad Sci U S A* 100(17):10008-10013.

Inoue, A., Seidel, M.G., Wu, W., Kamizono, S., Ferrando, A.A., Bronson, R.T., Iwasaki, H., Akashi, K., Morimoto, A., Hitzler, J.K., Pestina, T.I., Jackson, C.W., Tanaka, R., Chong, M.J., McKinnon, P.J., Inukai, T., Grosveld, G.C., & Look, A.T. (2002) Slug, a highly conserved zinc finger transcriptional repressor, protects hematopoietic progenitor cells from radiation-induced apoptosis in vivo. *Cancer Cell* 2(4):279-288.

Inukai, T., Inoue, A., Kurosawa, H., Goi, K., Shinjyo,T., Ozawa, K., Mao, M., Inaba, T., & Look, A.T. (1999) SLUG, a ces-1-related zinc finger transcription factor gene with antiapoptotic activity, is a downstream target of the E2A-HLF oncoprotein. *Mol Cell* 4(3):343-352.

Ishikawa, F., Yoshida, S., Saito, Y., Hijikata, A., Kitamura, H., Tanaka, S., Nakamura, R., Tanaka, T., Tomiyama, H., Saito, N., Fukata, M., Miyamoto, T., Lyons, B., Ohshima, K., Uchida, N., Taniguchi, S., Ohara, O., Akashi, K., Harada, M., & Shultz, L.D. (2007) Chemotherapy-resistant human AML stem cells home to and engraft within the bone-marrow endosteal region. *Nat Biotechnol* 25(11):1315-1321.

Ivanova, N.B., Dimos, J.T., Schaniel, C., Hackney, J.A., Moore, K.A., & Lemischka, I.R. (2002) A stem cell molecular signature. *Science* 298(5593):601-604.

Iwama, A., Oguro, H., Negishi, M., Kato, Y., Morita, Y., Tsukui, H., Ema, H., Kamijo, T., Katoh-Fukui, Y., Koseki, H., van Lohuizen, M., & Nakauchi, H. (2004) Enhanced self-renewal of hematopoietic stem cells mediated by the polycomb gene product Bmi-1. *Immunity* 21(6):843-851.

Jacobs, J.J.L., Kieboom, K., Marino, S., DePinho, R.A., & van Lohuizen, M. (1999) The oncogene and Polycomb-group gene bmi1 regulates proliferation and senescence through the ink4a locus. *Nature* 397(6715):164-168.

Jehn, B.M., Dittert, I., Beyer, S., von der Mark, K., & Bielke, W. (2002) c-Cbl binding and ubiquitin-dependent lysosomal degradation of membrane-associated Notch1. *J Biol Chem* 277(10):8033-8040.

Kaelin, W.G., Jr., & Ratcliffe, P.J. (2008) Oxygen sensing by metazoans: the central role of the HIF hydroxylase pathway. *Mol Cell* 30(4):393-402.

Kastan, M.B., Radin, A.I., Kuerbitz, S.J., Onyekwere, O., Wolkow, C.A., Civin, C.I., Stone, K.D., Woo, T., Ravindranath, Y., & Craig, R.W. (1991) Levels of p53 protein increase with maturation in human hematopoietic cells. *Cancer Res* 51(16):4279-4286.

Kiel, M.J., Yilmaz, O.H., Iwashita, T., Yilmaz, O.H., Terhorst, C., & Morrison, S.J. (2005) SLAM family receptors distinguish hematopoietic stem and progenitor cells and reveal endothelial niches for stem cells. *Cell* 121(7):1109-1121.

Kimura, S., Roberts, A.W., Metcalf, D., & Alexander, W.S. (1998) Hematopoietic stem cell deficiencies in mice lacking c-Mpl, the receptor for thrombopoietin. *Proc Natl Acad Sci USA* 95(3):1195-1200.

Komarova, N.L., & Wodarz, D. (2007) Effect of cellular quiescence on the success of targeted CML therapy. *PLoS One* 2(10):e990.

Kondo, M., Wagers, A.J., Manz, M.G., Prohaska, S.S., Scherer, D.C., Beilhack, G.F., Shizuru, J.A., & Weissman, I.L. (2003) Biology of hematopoietic stem cells and progenitors: implications for clinical application. *Annu Rev Immunol* 21:759-806.

Kubota, Y., Osawa, M., Jakt, L.M., Yoshikawa, K., & Nishikawa, S-I. (2009) Necdin restricts proliferation of hematopoietic stem cells during hematopoietic regeneration. *Blood* 114(20):4383-4392.

Lacombe, J., Herblot, S., Rojas-Sutterlin, S., Haman, A., Barakat, S., Iscove, N.N., Sauvageau, G., & Hoang, T. (2010) Scl regulates the quiescence and the long-term competence of hematopoietic stem cells. *Blood* 115(4):792-803.

Lacorazza, H.D., Yamada, T., Liu, Y., Miyata, Y., Sivina, M., Nunes, J., & Nimer, S.D. (2006) The transcription factor MEF/ELF4 regulates the quiescence of primitive hematopoietic cells. *Cancer Cell* 9(3):175-187.

Laurenti, E., Varnum-Finney, B., Wilson, A., Ferrero, I., Blanco-Bose, W.E., Ehninger, A., Knoepfler, P.S., Cheng, P.F., MacDonald, H.R., Eisenman, R.N., Bernstein, I.D., & Trumpp, A. (2008) Hematopoietic stem cell function and survival depend on c-Myc and N-Myc activity. *Cell Stem Cell* 3(6):611-624.

Liu, Y., Elf, S.E., Miyata, Y., Sashida, G., Liu, Y., Huang, G., Di Giandomenico, S., Lee, J.M., Deblasio, A., Menendez, S., Antipin, J., Reva, B., Koff, A., & Nimer, S.D. (2009) p53 regulates hematopoietic stem cell quiescence. *Cell Stem Cell* 4(1):37-48.

Lotem, J., & Sachs, L. (1993) Hematopoietic cells from mice deficient in wild-type p53 are more resistant to induction of apoptosis by some agents. *Blood* 82(4):1092-1096.

MacDonald, H.R. & Wevrick, R. (1997) The necdin gene is deleted in Prader-Willi syndrome and is imprinted in human and mouse. *Hum Mol Genet* 6(11):1873-1878.

Matsuoka, S., Oike, Y., Onoyama, I., Iwama, A., Arai, F., Takubo, K., Mashimo, Y., Oguro, H., Nitta, E., Ito, K., Miyamoto, K., Yoshiwara, H., Hosokawa, K., Nakamura, Y., Gomei, Y., Iwasaki, H., Hayashi, Y., Matsuzaki, Y., Nakayama, K., Ikeda, Y., Hata, A., Chiba, S., Nakayama, K.I., & Suda, T. (2008) Fbxw7 acts as a critical fail-safe against premature loss of hematopoietic stem cells and development of T-ALL. *Genes Dev* 22(8):986-991.

Mikkola, H.K., Klintman, J., Yang, H., Hock, H., Schlaeger, T.M., Fujiwara, Y., & Orkin, S.H. (2003) Haematopoietic stem cells retain long-term repopulating activity and multipotency in the absence of stem-cell leukaemia SCL/tal-1 gene. *Nature* 421(6922):547-551.

Min, I.M., Pietramaggiori, G., Kim, F.S., Passegué, E., Stevenson, K.E., & Wagers, A,J. (2008) The transcription factor EGR1 controls both the proliferation and localization of hematopoietic stem cells. *Cell Stem Cell* 2(4):380-391.

Miyamoto, K., Araki, K.Y., Naka, K., Arai, F., Takubo, K., Yamazaki, S., Matsuoka, S., Miyamoto, T., Ito, K., Ohmura, M., Chen, C., Hosokawa, K., Nakauchi, H., Nakayama, K., Nakayama, K.I., Harada, M., Motoyama, N., Suda, T., & Hirao, A. (2007) Foxo3a is essential for maintenance of the hematopoietic stem cell pool. *Cell Stem Cell* 1(1):101-112.

Morishita, K., Parker, D.S., Mucenski, M.L., Jenkins, N.A., Copeland, N.G., & Ihle, J.N. (1988) Retroviral activation of a novel gene encoding a zinc finger protein in IL-3-dependent myeloid leukemia cell lines. *Cell* 54(6):831-840.

Mucenski, M.L., Taylor, B.A., Ihle, J.N., Hartley, J.W., Morse, H.C., Jenkins, N.A., Copeland, N.G. (1988) Identification of a common ecotropic viral integration site, Evi-1, in the DNA of AKXD murine myeloid tumors. *Mol Cell Biol* 8(1):310-308.

Nakada, D., Saunders, T.L., & Morrison, S.J. (2010) Lkb1 regulates cell cycle and energy metabolism in haematopoietic stem cells. *Nature* 468(7324):653-658.

Nakada, Y., Taniura, H., Uetsuki, T., Inazawa, J., & Yoshikawa, K. (1998) The human chromosomal gene for necdin, a neuronal growth suppressor, in the Prader-Willi syndrome deletion region. *Gene* 213(1-2):65-72.

Nesbit, C.E., Tersak, J.M., & Prochownik, E.V. (1999) MYC oncogenes and human neoplastic disease. *Oncogene* 18(19):3004-3016.

Oguro, H., Iwama, A., Morita, Y., Kamijo, T., van Lohuizen, M., & Nakauchi, H. (2006) Differential impact of *Ink4a* and *Arf* on hematopoietic stem cells and their bone marrow microenvironment in *Bmi1*-deficient mice. *J Exp Med* 203(10):2247-2253.

Opferman, J.T., Iwasaki, H., Ong, C.C., Suh, H., Mizuno, S., Akashi, K., & Korsmeyer, S.J. (2005) Obligate role of anti-apoptotic MCL-1 in the survival of hematopoietic stem cells. *Science* 307(5712):1101-1104.

Orford, K.W., & Scadden, D.T. (2008) Deconstructing stem cell self-renewal: genetic insights into cell-cycle regulation. *Nat Rev Genet* 9(2):115-128.

Osawa, M., Hanada, K., Hamada, H., & Nakauchi, H. (1996) Long-term lymphohematopoietic reconstitution by a single CD34-low/negative hematopoietic stem cell. *Science* 273(5272):242-245.

Park, I.K., He, Y., Lin, F., Laerum, O.D., Tian, Q., Bumgarner, R., Klug, C.A., Li, K., Kuhr, C., Doyle, M.J., Xie, T., Schummer, M., Sun, Y., Goldsmith, A., Clarke, M.F., Weissman, I.L., Hood, L., & Li, L. (2002) Differential gene expression profiling of adult murine hematopoietic stem cells. *Blood* 99(2):488-498.

Park, I.K., Qian, D., Kiel, M., Becker, M.W., Pihalja, M., Weissman, I.L., Morrison, S.J., & Clarke, M.F. (2003) Bmi-1 is required for maintenance of adult self-renewing haematopoietic stem cells. *Nature* 423(6937):302-305.

Passegué E., Jochum W., Schorpp-Kistner M., Möhle-Steinlein U., & Wagner E.F. (2001) Chronic Myeloid Leukemia with Increased Granulocyte Progenitors in Mice Lacking JunB Expression in the Myeloid Lineage. *Cell* 104(1):21-32.

Porter, P.N., Meints, R.H., & Mesner,K. (1979) Enhancement of erythroid colony growth in culture by hemin. *Exp Hematol* 7(1):11-16.

Rajasekhar, V.K., Begemann, M. (2007) Concise review: roles of polycomb group proteins in development and disease: a stem cell perspective. *Stem Cells* 25(10):2498-2510.

Ramalho-Santos, M., Yoon, S., Matsuzaki, Y., Mulligan, R.C., & Melton, D.A. (2002) "Stemness": transcriptional profiling of embryonic and adult stem cells. *Science* 298(5593):597-600.

Ramos, C.A., Bowman, T.A., Boles, N.C., Merchant, A.A., Zheng, Y., Parra, I., Fuqua, S.A., Shaw, C.A., & Goodell, M.A. (2006) Evidence for diversity in transcriptional profiles of single hematopoietic stem cells. *PLoS Genet* 2(9):e159.

Randall, T.D., & Weissman, I.L. (1997) Phenotypic and functional changes induced at the clonal level in hematopoietic stem cells after 5-fluorouracil treatment. *Blood* 89(10):3596-3606.

Rathinam, C., Thien, C.B., Langdon, W.Y., Gu,H., & Flavell, R.A. (2008) The E3 ubiquitin ligase c-Cbl restricts development and functions of hematopoietic stem cells. *Genes Dev* 22(8):992-997.

Rathinam, C., Matesic, L.E., & Flavell, R.A. (2011) The E3 ligase Itch is a negative regulator of the homeostasis and function of hematopoietic stem cells. *Nat Immunol* 12(5):399-407.

Reya, T., Morrison, S.J., Clarke, M.F., & Weissman, I.L. (2001) Stem cells, cancer, and cancer stem cells. *Nature* 414(6859):105-111.

Rizo, A., Vellenga, E., de Haan, G., & Schuringa, J.J. (2006) Signaling pathways in self-renewing hematopoietic and leukemic stem cells: do all stem cells need a niche? *Hum Mol Genet* 15 Spec No 2:R210-9.

Robb, L., Lyons, I., Li, R., Hartley, L., Köntgen, F., Harvey, R.P., Metcalf, D., & Begley, C.G. (1995) Absence of yolk sac hematopoiesis from mice with a targeted disruption of the scl gene. *Proc Natl Acad Sci U S A* 92(15):7075-7079.

Rodrigues, N.P., Janzen, V., Forkert, R., Dombkowski, D.M., Boyd, A.S., Orkin, S.H., Enver, T., Vyas, P., & Scadden, D.T. (2005) Haploinsufficiency of GATA-2 perturbs adult hematopoietic stem cell homeostasis. *Blood* 106(2):477-484.

Rudd, C.E. (2001) Lnk adaptor: novel negative regulator of B cell lymphopoiesis. *Sci STKE* 2001(85):pe1

Saito, Y., Uchida, N., Tanaka, S., Suzuki, N., Tomizawa-Murasawa, M., Sone, A., Najima, Y., Takagi, S., Aoki, Y., Wake, A., Taniguchi, S., Shultz, L.D., & Ishikawa, F. (2010) Induction of cell cycle entry eliminates human leukemia stem cells in a mouse model of AML. *Nat Biotechnol* 28(3):275-280.

Santaguida, M., Schepers, K., King, B., Sabnis, A.J., Forsberg, E.C., Attema, J.L., Braun, B.S., & Passegué, E. (2009) JunB protects against myeloid malignancies by limiting hematopoietic stem cell proliferation and differentiation without affecting self-renewal. *Cancer Cell* 15(4):341-352.

Sato, T., Onai, N., Yoshihara, H., Arai, F., Suda, T., & Ohteki, T. (2009) Interferon regulatory factor-2 protects quiescent hematopoietic stem cells from type I interferon-dependent exhaustion. *Nat Med* 15(6):696-700.

Scandura, J.M., Boccuni, P., Massagué, J., Nimer, S.D. (2004) Transforming growth factor beta-induced cell cycle arrest of human hematopoietic cells requires p57KIP2 up-regulation. *Proc Natl Acad Sci U S A* 101(42):15231-15236.

Schneider-Gadicke, A., Beer-Romero, P., Brown, L.G., Mardon, G., Luoh, S.W., & Page, D.C. (1989) Putative transcription activator with alternative isoforms encoded by human ZFX gene. *Nature* 342(6250):708-711.

Seita, J., Ema, H., Ooehara, J., Yamazaki, S., Tadokoro, Y., Yamasaki, A., Eto, K., Takaki, S., Takatsu, K., & Nakauchi, H. (2007) Lnk negatively regulates self-renewal of hematopoietic stem cells by modifying thrombopoietin-mediated signal transduction. *Proc Natl Acad Sci U S A* 104(7):2349-2354.

Serrano, M., Hannon, G.j., & Beach, D. (1993) A new regulatory motif in cell-cycle control causing specific inhibition of cyclin D/CDK4. *Nature* 366(6456):740-707.

Sheiness, D., Fanshier, L., & Bishop, J.M. (1978) Identification of nucleotide sequences which may encode the oncogenic capacity of avian retrovirus MC29. *J Virol* 28(2):600-610.

Shimabe, M., Goyama, S., Watanabe-Okochi, N., Yoshimi, A., Ichikawa, M., Imai, Y., & Kurokawa, M. (2009) Pbx1 is a downstream target of Evi-1 in hematopoietic stem/progenitors and leukemic cells. *Oncogene* 28(49):4364-4374.

Shivdasani, R.A., Mayer, E.L., & Orkin, S.H. (1995) Absence of blood formation in mice lacking the T-cell leukaemia oncoprotein tal-1/SCL. *Nature* 373(6513):432-434.

Shounan, Y., Dolnikov, A., MacKenzie, K.L., Miller, M., Chan, Y.Y., & Symonds, G. Retroviral transduction of hematopoietic progenitor cells with mutant p53 promotes survival and proliferation, modifies differentiation potential and inhibits apoptosis. *Leukemia* 10(10):1619-1628.

Simsek, T., Kocabas, F., Zheng, J., Deberardinis, R.J., Mahmoud, A.I., Olson, E.N., Schneider, J.W., Zheng, C.C., & Sadek, H.A. (2010) The distinct metabolic profile of hematopoietic stem cells reflects their location in a hypoxic niche. *Cell Stem Cell* 7(9):380-390.

Sirin, O., Lukov, G.L., Mao, R., Conneely, O.M., & Goodell, M.A. (2010) The orphan nuclear receptor Nurr1 restricts the proliferation of haematopoietic stem cells. *Nat Cell Biol* 12(12):1213-1219.

Souroullas. G.P., Salmon, J.M., Sablitzky, F., Curtis, D.J., & Goodell, M.A. (2009) Adult hematopoietic stem and progenitor cells require either *Lyl1* or *Scl* for survival. *Cell Stem Cell* 4(2):180-186.

Stier S, Cheng T, Forkert R, Lutz C, Dombkowski DM, Zhang JL, Scadden DT. (2003) Ex vivo targeting of p21Cip1/Waf1 permits relative expansion of human hematopoietic stem cells. *Blood* 102(4):1260-1266.

Sun, Y., Shao, L., Bai, H., Wang, Z.Z., & Wu, W.S. (2010) Slug deficiency enhances self-renewal of hematopoietic stem cells during hematopoietic regeneration. *Blood* 115(9):1709-1717.

Takeuchi, M., Kimura, S., Kuroda, J., Ashihara, E., Kawatani, M., Osada, H., Umezawa, K., Yasui, E., Imoto, M., Tsuruo, T., Yokota, A., Tanaka, R., Nagao, R., Nakahara, T., Fujiyama, Y., & Maekawa, T. (2010) Glyoxalase-I is a novel target against Bcr-Abl^{+} leukemic cells acquiring stem-like characteristics in hypoxic environment. *Cell Death Diff* 17(7):1211-1220.

Takubo, K., Goda, N., Yamada, W., Iriuchishima, H., Ikeda, E., Kubota, Y., Shima, H., Johnson, R.S., Hirao, A., Suematsu, M., & Suda, T. (2010) Regulation of the HIF-1α level is essential for hematopoietic stem cells. *Cell Stem Cell* 7(9):391-402.

Taniura, H., Taniguchi, N., Hara, M., & Yoshikawa, K. (1998) Necdin, a postmitotic neuron-specific growth suppressor, interacts with viral transforming proteins and cellular transcription factor E2F1. *J Biol Chem* 273(2):720-728.

Taniura, H., Matsumoto, K., & Yoshikawa, K. (1999) Physical and functional interactions of neuronal growth suppressor necdin with p53. *J Biol Chem* 274(23):16242-16248.

Taniura, H., Kobayashi, M., & Yoshikawa, K. (2005) Functional domains of necdin for protein-protein interaction, nuclear matrix targeting, and cell growth suppression. *J Cell Biochem* 94(4):804-815.

TeKippe, M., Harrison, D.E., & Chen, J. (2003) Expansion of hematopoietic stem cell phenotype and activity in Trp53-null mice. *Exp Hematol* 31(6):521-527.

Thompson, B.J., Jankovic, V., Gao, J., Buonamici, S., Vest, A., Lee, J.M., Zavadil, J., Nimer, S.D., & Aifantis, I. (2008) Control of hematopoietic stem cell quiescence by the E3 ubiquitin ligase Fbw7. *J Exp Med* 205(6):1395-1408.

Tipping, A. J., Pina, C., Caster, A., Hong, D., Rodrigues, N.P., Lazzari, L., May, G.E., Jacobsen, S.E., & Enver, T. (2009) High GATA-2 expression inhibits human hematopoietic stem and progenitor cell function by effects on cell cycle. *Blood* 113(12): 2661-2672.

Tothova, Z., Kollipara, R., Huntly, B.J., Lee, B.H., Castrillon, D.H., Cullen, D.E., McDowell, E.P., Lazo-Kallanian, S., Williams, I.R., Sears, C., Armstrong, S.A., Passegué, E., DePinho, R.A., & Gilliland, D.G. (2007) FoxOs are critical mediators of hematopoietic stem cell resistance to physiologic oxidative stress. *Cell* 128(2):325-339.

Trumpp, A., Essers, M., & Wilson, A. (2010) Awakening dormant haematopoietic stem cells. *Nat Rev Immunol* 10(3):201-209.

Tsai, F. Y. & Orkin, S. H. (1997) Transcription factor GATA-2 is required for proliferation/survival of early hematopoietic cells and mast cell formation, but not for erythroid and myeloid terminal differentiation. *Blood* 89(10): 3636-3643.

Umemoto, T., Yamato, M., Nishida, K., Yang, J., Tano, Y., & Okano, T. (2005) $p57^{Kip2}$ is expressed in quiescent mouse bone marrow side population cells. *Biochem Biophys Res Commun* 337(1):14-21.

van Os, R., Kamminga, L.M., Ausema, A., Bystrykh, L.V., Draijer, D.P., van Pelt, K., Dontje, B., & de Haan, G. (2007) A Limited role for p21Cip1/Waf1 in maintaining normal hematopoietic stem cell functioning. *Stem Cells* 25(4):836-843.

Venezia, T.A., Merchant,A.A., Ramos, C.A., Whitehouse, N.L., Young, A.S., Shaw, C.A., & Goodell, M.A. (2004) Molecular signatures of proliferation and quiescence in hematopoietic stem cells. *PLoS Biol* 2(10): e301.

Wang, Q., Miyakoda, M., Yang, W., Khillan, J., Stachura, D.L., Weiss, M.J., & Nishikura, K. (2004) Stress-induced apoptosis associated with null mutation of ADAR1 RNA editing deaminase gene. *J Biol Chem* 279(6):4952-4961.

Wilson, A., Murphy, M.J., Oskarsson, T., Kaloulis, K., Bettess, M.D., Oser, G.M., Pasche, A-C., Knabenhans, C., MacDonald, H.R., & Trumpp, A. (2004) c-Myc controls the balance between hematopoietic stem cell self-renewal and differentiation. *Genes Dev* 18(22):2747-2763.

Wilson, A., Laurenti, E., Oser, G., van der Wath, R.C., Blanco-Bose, W., Jaworski, M., Offner, S., Dunant, C.F., Eshkind, L., Bockamp, E., Lió, P., Macdonald, H.R., & Trumpp, A.

(2008) Hematopoietic stem cells reversibly switch from dormancy to self-renewal during homeostasis and repair. *Cell* 135(6):1118-1129.

Wilson, A., Laurenti, E., & Trumpp, A. (2009) Balancing dormant and self-renewing hematopoietic stem cells. *Curr Opin Genet Dev* 19(5):461-468.

Wu, W.S., Heinrichs, S., Xu, D., Garrison, S.P., Zambetti, G.P., Adams, J.M., & Look, A.T. (2005) Slug antagonizes p53-mediated apoptosis of hematopoietic progenitors by repressing puma. *Cell* 23(4):641-653.

Xiao, J., & Chen, H.S. (2004) Biological functions of melanoma-associated antigens. *World J Gastroenterol* 10(13):1849-1853.

Yamazaki, S., Iwama, A., Takayanagi, S., Eto, K., Ema, H., & Nakauchi, H. (2009) TGF-beta as a candidate bone marrow niche signal to induce hematopoietic stem cell hibernation. *Blood* 113(6):1250-1256.

Yoshikawa, K. (2000) Cell cycle regulators in neural stem cells and postmitotic neurons. *Neurosci Res* 37:1–14.

Yuasa, H., Oike, Y., Iwama, A., Nishikata, I., Sugiyama, D., Perkins, A., Mucenski, M.L., Suda, T., & Morishita, K. (2005) Oncogenic transcription factor Evi1 regulates hematopoietic stem cell proliferation through GATA-2 expression. *EMBO J* 24(11):1976-1987.

Zeng, H., Yücel, R., Kosan, C., Klein-Hitpass, L., & Möröy, T. (2004) Transcription factor Gfi1 regulates self-renewal and engraftment of hematopoietic stem cells. *EMBO J* 23(20): 4116-4125.

Zeng, S., Xu, Z., Lipkowitz, S., & Longley, J.B. (2005) Regulation of stem cell factor receptor signaling by Cbl family proteins (Cbl-b/c-Cbl). *Blood* 105(1):226-232.

Zhong, J.F., Zhao, Y., Sutton, S., Su, A., Zhan, Y., Zhu, L., Yan, C., Gallaher, T., Johnston, P.B, Anderson, W.F., & Cooke, M.P. (2005) Gene expression profile of murine long-term reconstituting vs. short-term reconstituting hematopoietic stem cells. *Proc Natl Acad Sci U S A* 102(7):2448-2453.

3

Transcriptional Quiescence of Hematopoietic Stem Cells

Rasmus Freter
Ludwig Institute for Cancer Research, University of Oxford
United Kingdom

1. Introduction

Haematopoietic stem cells (HSC) have the exceptional capacity to undergo continuous self-renewal and differentiation into multiple lineages, which is essential for haematopoietic homeostasis and response to injury. To achieve this life long function, these cells have to be protected from cytotoxic and genetic damage. On the other hand, rapid activation of haematopoietic stem cell proliferation in response to stimuli must be ensured. While cellular quiescence is thought to be the key mechanism underlying this paradoxical nature of HSC, the molecular basis of induction and maintenance of quiescence remains unresolved.

Quiescence is commonly defined as a reversible cell cycle exit. Induction and maintenance of stem cell quiescence has been studied at the level of cell cycle regulation (Orford & Scadden, 2008), cellular metabolism (Tothova & Gilliland, 2007) or interaction with the specific niche (Fuchs *et al.*, 2004). Genome-wide association studies have been performed on a variety of quiescent model systems, such as serum starvation of fibroblasts (Coller *et al.*, 2006), primary lymphocytes (Garriga *et al.*, 1998) or yeast in stationary phase (Patturajan *et al.*, 1998, Radonjic *et al.*, 2005). All of these studies revealed a significant decrease of productive mRNA transcription in these model systems. However, if quiescent adult stem cells share this down regulation of mRNA transcription has never been examined.

Due to their relative ease of isolation, cells of the haematopoietic lineage have been extensively studied. Importantly, several assays for hematopoietic stem cell function have been developed, such as colony forming ability and rescue of lethally irradiated mice. These functional tests are lacking in most other adult stem cell models, with the exception of spermatogonia and mammary gland stem cells (Brinster & Nagano, 1998, Shackleton *et al.*, 2006). Functional assays for HSC ability have provided us with the notion that most defined populations of long term repopulating HSC still contain progenitor cells, which can only transiently contribute to repopulation of the haematopoietic system. This heterogeneity is not only evident in defined cell populations, but also in the in vivo niche for HSC, the bone marrow. HSC in the bone marrow are interspersed with transient amplifying cells and differentiated cells, complicating stem cell identification by spatial organization of the tissue. Other stem cell systems, such as spermatogonia, keratinocyte or crypt stem cells have a clearly defined niche architecture, enabling stem cell identification by location only (Fuchs *et al.*, 2004). In this case, resting stem cells and activated progenitors can be separated by location and molecular markers can be easily identified. If all adult stem cells share a repertoire of molecular markers, findings from other adult stem cells can be transferred to HSC and should lead to characterization of haematopoietic stem cell subpopulations.

In our previous work, we found that adult melanocyte stem cells exhibit a 10 to 100fold lower level of housekeeping gene mRNA compared to differentiated cells, suggesting a global repression of mRNA transcription (Osawa *et al.*, 2005). We could then show that the largest subunit of RNA polymerase II (RNApII), which is responsible for all mRNA transcription, exhibits a partly phosphorylated C-terminal domain (CTD), characteristic of initiated, but paused mRNA transcription (Freter *et al.*, 2010). In line with this, we found the RNApII kinase CDK9 absent in adult melanocyte stem cells. Inhibition of CDK9 resulted in cellular resistance to withdrawal of essential growth factors, conferring a stem cell-like phenotype to progenitor cells. Interestingly, various other adult stem cells, including keratinocyte, muscle, spermatogonia and also HSC exhibited a similar partial phosphorylation of RNApII (Freter *et al.*, 2010). We concluded that transcriptional quiescence is an early, specific and conserved marker for adult stem cells. This feature can be used to isolate and characterize pure populations of stem cell-like cells from any tissue, enabling a deeper understanding of stem cell biology and recapitulation of the stem cell niche, in order to expand immature stem cells in vitro.

In this chapter I would like to summarize our findings that HSC exhibit a reduction in productive mRNA transcription. I would like to elaborate on the implications arising from transcriptional quiescence of a subset of HSC, both in development and disease. Technical challenges and resulting applications of identifying and isolating transcriptionally quiescent HSC in vitro will be discussed.

2. The mRNA transcription cycle

Regulation of gene expression is essential for all single- and multi cellular organisms. This fundamental process is executed at the level of mRNA transcription by RNApII, typically in distinct transcription steps. The different stages of mRNA transcription, initiation, promoter clearance, elongation, mRNA processing and release of RNApII from DNA are tightly regulated by modifications of the CTD of RNApII (Sims *et al.*, 2004, Fig1). This characteristic domain is in mammalian cells composed of 52 repeats of the consensus sequence YS_2PTS_5PS. During mRNA transcription several posttranslational modifications of the CTD are occurring, most prominently phosphorylation of Serine 5 (Ser5) and Serine 2 (Ser2). These phosphorylation events are requisite for binding of proteins essential for RNA processing, splicing and polyadenylation. Using antibodies specifically detecting these phosphorylation events enables determination of global mRNA transcription activity in single cells in vivo.

Gene expression in mammalian cells has been comprehensively studied at the transcription initiation step that is controlled by cell-specific transcription factors. In fact, it has been thought for long time that assembly of the preinitiation complex and subsequent recruitment on RNApII is the rate-limiting step for gene transcription. However, early results indicated that RNApII is initiated, but paused at Drosophila heat shock genes (Boehm *et al.*, 2003, Ni *et al.*, 2004). More recently, it was observed using genome-wide association studies that initiated but stalled polymerase is not only present on immediate-response or developmentally regulated genes, but also many non-expressed genes, suggesting transcription elongation as the critical step in gene expression (Guenther *et al.*, 2007, Muse *et al.*, 2007, Zeitlinger *et al.*, 2007).

Transcription initiation requires phosphorylation of Ser5 of the CTD by TFIIH, a heterodimeric kinase consisting of CDK7 and Cyclin H. These phosphorylation events enable binding of the mRNA capping machinery (Ho & Shuman, 1999) and promoter clearance. Typically, a short (~40nt) nascent RNA is then produced by RNApII. However, mRNA transcription is paused at many genes due to the action of negative elongation factor (NELF) and DRB-sensitivity inducing factor (DSIF) (Wu *et al.*, 2003, Yamaguchi *et al.*, 2002).

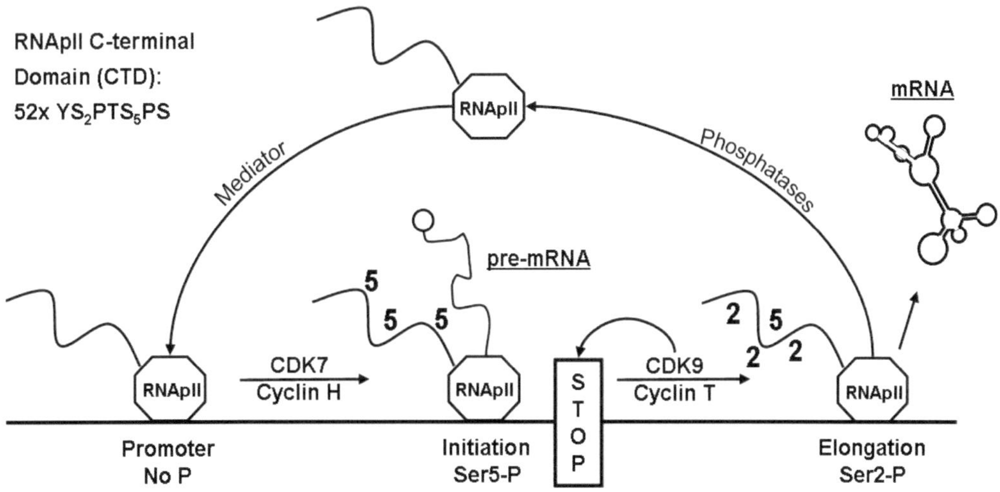

Fig. 1. The mRNA transcription cycle is characterized by phosphorylation of the RNA polymerase (RNApII) CTD. Phosphorylation of Ser5 (5) by CDK7/Cyclin H induces promoter clearance and pre-mRNA capping. RNApII is then halted (STOP), until activation of CDK9/Cyclin T. Phosphorylation of the inhibitory complex and RNA polymerase II CTD Ser2 (2) leads to productive elongation and release of mature mRNA.

Recruitment of positive transcription elongation factor b (P-TEFb), a heterodimeric protein consisting of the kinase CDK9 and one of the regulatory subunits Cyclin T1, T2 or K, to stalled polymerases is required for alleviation of the transcriptional block by NELF and DSIF (Peterlin & Price, 2006, Rahl *et al.*, 2010). P-TEFb phosphorylates RNA recognition motif-containing protein RD, a component of NELF, and Spt5, a subunit of DSIF (Aida *et al.*, 2006, Fujinaga *et al.*, 2004). NELF then dissociates from RNApII, while DSIF remains associated with RNApII and becomes a positive transcription elongation factor (Chen *et al.*, 2009). Importantly, phosphorylation of Ser2 of the CTD by P-TEFb triggers transcription elongation, mRNA processing, and release of mature mRNA (Kohoutek, 2009, Ni *et al.*, 2004).

CDK9 has first been identified as a CDC2-related kinase with a PITALRE motif (Grana *et al.*, 1994). The cyclin partner of CDK9 is Cyclin T1, T2 or K. Unlike other CDK/Cyclin heterodimers, neither P-TEFb levels nor kinase activity is fluctuating during the cell cycle (Garriga *et al.*, 2003, Grana *et al.*, 1994). CDK9 exists in two isoforms, a major 42kD sized peptide and an N-terminal extended peptide, whose transcription starts from an alternative

TATA box upstream of the housekeeping-type promoter of the 42kD isoform (Shore *et al.*, 2003). The expression of both isoforms varies between developmental stages and organs (Shore *et al.*, 2005). For example, the expressed isoform shifts from the longer 55kD form to the shorter 42kD isoform during lymphocyte activation (Liu & Herrmann, 2005). However, the target gene specificity is very similar between the two isoforms (Liu & Herrmann, 2005).

P-TEFb is present in the cell as either a large or small multiprotein complex. The inactive, large complex consists of 7SK non-coding RNA, Hexim1, LARP7 and MEPCE (Li *et al.*, 2005), while the small, active complex is formed by binding of Brd4 to P-TEFb (Yang *et al.*, 2005). Brd4 binds to acetylated histones and may therefore target P-TEFb to actively transcribed genes, when no specific transcription factor is present (Jang *et al.*, 2005) but can also recruit P-TEFb to inducibly acetylated histones (Hargreaves *et al.*, 2009). However, even the large complex contains primarily active P-TEFb, which is sequestered away from the kinase targets. Cellular stresses, such as UV irradiation, cytokines or changes in the microenvironment result in release of active P-TEFb to support quick mRNA transcription to respond to stimuli. Several transcription factors have been shown to interact directly with P-TEFb to stimulate transcription elongation of RNApII. These include NF-kB (Barboric *et al.*, 2001), c-Myc (Eberhardy & Farnham, 2001), CIITA (Kanazawa *et al.*, 2000), GATA1 (Elagib *et al.*, 2008) and Runx1 (Jiang *et al.*, 2005), amongst others. Thus, P-TEFb can be either recruited directly to promoters by specific transcription factors or to acetylated histones by Brd4.

Phosphorylation of Ser2 in the CTD of RNApII and productive transcription elongation is the critical target for eukaryotic gene expression (Bentley, 1995, Chao & Price, 2001). Inhibition of the CTD Ser2 kinase CDK9 by 5,6-Dichloro-1-β-D-ribofuranosylbenzimidazole (DRB) or Flavopiridol results in degradation of most mRNA (Chao & Price, 2001, Sehgal & Darnell, 1976) and induces apoptosis (Chen *et al.*, 2005, Gojo *et al.*, 2002). Similarly, knockdown of CDK9 in vivo results in complete absence of mRNA synthesis and embryonic lethality (Eissenberg *et al.*, 2007, Shim *et al.*, 2002). Cyclin T1 KO mice exhibit minor immunological defects (Oven *et al.*, 2007), while the Cyclin T2 KO mouse is embryonic lethal (Kohoutek *et al.*, 2009) with an extremely early phenotype before implantation of the blastocyst. This difference in phenotype of CDK9 Cyclin partners can partly be explained by a limited overlap in target genes of these isoforms (Ramakrishnan *et al.*, 2011).

On the other hand, ectopic activation of P-TEFb by ablation of Hexim1 results in embryonic lethality as well, due to hypertrophy of the heart (Huang *et al.*, 2004). Similarly, inactivation of the inhibitory large complex member LARP7 results in aberrant splicing and embryonic death in zebrafish, highlighting the essential role of P-TEFb for pre-mRNA splicing (Barboric *et al.*, 2009). However, overexpression of CDK9 from the endogenous Rosa26 promoter did not result in any phenotype in mice, mainly due to a low level of expression of CDK9 from this promoter (Freter *et al.*, 2010). Limiting P-TEFb activity is necessary for development of germ line cells in both D. melanogaster and C. elegance (Batchelder *et al.*, 1999, Hanyu-Nakamura *et al.*, 2008, Zhang *et al.*, 2003). Ectopic activation of P-TEFb by overexpression or knock down of inhibitors results in misexpression of somatic genes in germ line cells and their subsequent degeneration, resulting in sterile offspring. In summary, levels of P-TEFb and thus the global activity of RNApII need to be maintained within a certain limits, not exceeding and not below a basal threshold.

Most cells, including proliferating, terminally differentiated and senescent cells, actively synthesize mRNA. In these cells, RNApII is phosphorylated on CTD Ser2 and Ser5 independent of the cell cycle (Garriga *et al.*, 2003, Marshall *et al.*, 2005). However, some cells do not display this active phosphorylation pattern. For example, deeply dormant cells, such as primary T and B lymphocytes, exhibit an almost complete absence of RNApII phosphorylation (Garriga *et al.*, 1998, Marshall *et al.*, 2005). Activation of these cells by antigen encounter results in upregulation of Ccnt1 both on mRNA and protein level and subsequent phosphorylation of CTD Ser2 (Marshall *et al.*, 2005).

On the other hand, RNApII in yeast cells in the stationary phase (Patturajan *et al.*, 1998, Radonjic *et al.*, 2005) or on Drosophila heat shock genes (Boehm *et al.*, 2003, Ni *et al.*, 2004) is phosphorylated on Ser5, but not on Ser2. Stimulation of these cells, such as the addition of nutrients or heat shock, ensures rapid activation of gene transcription. Thus, analysis of the specific phosphorylated sites in RNApII can distinguish cells featuring phases of productive mRNA elongation or paused mRNA transcription initiation.

3. mRNA transcription in the hematopoietic lineage

Most of our knowledge of P-TEFb function derived from studies involving HIV replication (Barboric & Peterlin, 2005). Human Cyclin T1 is an essential co-factor of the immediate-early HIV gene product Tat, which recruits P-TEFb to the tar RNA located 5′ on HIV genes to activate gene expression (Mancebo *et al.*, 1997, Zhu *et al.*, 1997). In resting lymphocytes, P-TEFb activity is low and thus HIV replication is blocked. Upon stimulation, upregulation of Cyclin T1 results in activation of P-TEFb and transcription of viral genes (Garriga *et al.*, 1998). Human Cyclin T1 (Ccnt1), the major Cyclin associated with CDK9, is characterized by a TAR recognition motif, which is essential for the formation of a ternary complex between tar RNA/Tat and P-TEFb to activate HIV gene transcription in cells (Wei *et al.*, 1998). This motif contains an essential Cysteine, which is required for complex formation with HIV Tat. Mutagenesis of mouse Ccnt1 at this position, which normally contains a Tyrosine in mouse, activates HIV transcription in murine cells (Fujinaga *et al.*, 2002).

P-TEFb is also required for normal hematopoietic development and function. Knock down of CDK9 in zebrafish embryos results in severe defects in definitive erythropoiesis, but no gross developmental defects despite a smaller body size (Meier *et al.*, 2006). Given the ubiquitous requirement of CDK9 for mRNA transcription this surprising observation may be explained by incomplete knockdown using morpholino DNA. Similarly, partial depletion of Ccnt1 in mice results in modest immunological phenotypes, such as appearance of autoimmunity due to impaired negative selection of autoreactive T cells in thymus (Oven *et al.*, 2007). Together, these results suggest that the hematopoietic lineage may be very susceptible for small changes in P-TEFb activity.

Recruitment of P-TEFb by transcription factors to heterochromatin converts this general elongation factor to a repressor of transcription. Runx1 binds the CD4 silencer in thymocytes and leads to active suppression of CD4 transcription during development. Interestingly, despite an engaged RNApII on the CD4 promoter and in the presence of an active CD4 enhancer in these cells, CD4 transcription is silenced (Jiang *et al.*, 2005). This is achieved by binding of Cyclin T1 to Runx1 and sequestering of P-TEFb into inactive chromatin loops (Jiang & Peterlin, 2008). Thus, inactive genes can be loaded with a poised polymerase and induction of chromatin remodelling complexes result in rapid activation of gene transcription by release of active P-TEFb from adjacent loci (Jiang & Peterlin, 2008). In this sense, assembly of the transcriptional machinery on inactive promoters can be seen as a transcription bookmark, to facilitate future expression.

Activation of hematopoietic gene transcription can also be achieved via binding of P-TEFb to actively acetylated chromatin. Studies on LPS-induced inflammatory gene expression in macrophages revealed that primary response genes have a stalled polymerase at their promoters, already phosphorylated at Ser5 of the CTD (Hargreaves *et al.*, 2009). In response to LPS stimulation, acetylation of H4K5/8/12 recruits Brd4, this in turn engages P-TEFb leading to CTD Ser2 phosphorylation and mRNA transcription (Hargreaves *et al.*, 2009).

Hematopoietic lineages are very susceptible for inhibition of P-TEFb activity and require specific co-factors for their respective differentiation. For example, differentiating macrophages and murine erythroleukemic cells down regulate the 42kD isoform, and up regulate the 55 kD isoform of CDK9 (Liu & Herrmann, 2005). Megakaryocyte differentiation depends on activation of P-TEFb and can be blocked by CDK9 inhibitors or dominant negative CDK9 (Elagib *et al.*, 2008). Erythroid differentiation depends on stabilization of a GATA-1/SCL/LMO2 complex on β-globin chromatin, subsequent association of P-TEFb and RNApII Ser2 phosphorylation in the locus by the ubiquitous enhancer facilitator/chromatin factor Ldb1 (Song *et al.*, 2010). Interestingly, deletion of Ldb1 in vivo results in defects in adult haematopoietic stem cell maintenance and diminished long-term reconstitution potential upon transplantation (Li *et al.*, 2011). However, the authors did not examine activity of P-TEFb in their knock-out mice, which may be reduced and thus result in defects in HSC specification. Certainly, many other genes and cells depend on P-TEFb activity during development and differentiation. Using the hematopoietic lineage as a model system for the basal mRNA transcription machinery will shed light onto many aspects of eukaryotic mechanisms of transcription control.

Conversely, some leukemic cancers are characterized by dysregulation of P-TEFb activity. Several fusion genes of the histone methytransferase MLL1 involved in chromosomal rearrangements leading to myeloid and lymphoblastic leukaemia associate with mRNA transcription elongation factor encoded by ELL or P-TEFb (Benedikt *et al.*, 2011, Lin *et al.*, 2010), suggesting that one major mechanism for leukomogenesis is deregulated transcription elongation (Shilatifard *et al.*, 1996). Indeed, targeting P-TEFb with the specific CDK9 inhibitor Flavopiridol induces apoptosis in chronic lymphocytic leukemic cells by suppression of short-lived anti-apoptotic genes, such as Mcl-1 (Chen *et al.*, 2005). Dysregulation of P-TEFb activity is involved in several other cancer types. For example Hexim1, a negative regulator of P-TEFb activity, was found down regulated in invasive breast cancer samples compared to normal breast tissue (Wittmann *et al.*, 2003). Exploiting the susceptibility of the hematopoietic lineage for disturbance of their mRNA transcription may result in novel targets of cancer therapy.

4. Transcriptional quiescence of hematopoietic stem cells

Hematopoietic stem cells (HSC) have been in the focus of basic and applied research since many decades. Definition of subsets of transplantable HSC and their in vitro culture have advanced considerably in recent years. However, so far no reliable marker for the isolation of pure HSC exists and our use of transplants is limited by the inability to expand these cells ex vivo. A sensitive marker of HSC quiescence and activation could be useful to isolate unadulterated long-term repopulating HSC and screen for factors that enable stem cell expansion while maintaining their undifferentiated state. It has been known for three decades that HSC down regulate productive mRNA transcription. Low retention of Pyronin Y, an RNA binding dye, can be used to isolate HSC (Shapiro, 1981), suggesting that global suppression of mRNA transcription is a feature of quiescent HSC. However, this observation has not been followed up by analysis of the global status of RNApII activity, which is responsible for all mRNA transcription.

We have previously observed that adult melanocyte stem cells (MelSC) down regulate many housekeeping genes, including ActB, ActG and GapDH, suggesting a global repression of mRNA synthesis in these cells (Osawa *et al.*, 2005). Melanocytes are pigmented cells in the hair follicle and skin, providing melanin granules to differentiating keratinocytes. The MelSC system in mouse hair follicles can serve as a model system for adult stem cell systems. It has the advantage of spatial separation of stem and differentiated cells and a non-lethal but obvious hair graying phenotype if this system is perturbed (Nishimura *et al.*, 2002). We observed that adult MelSC show a complete absence of RNApII CTD Ser2 phosphorylation, while Ser5 was phosphorylated (Freter *et al.*, 2010), suggesting a global down regulation of productive mRNA transcription. In line with this, CDK9 protein and mRNA was down regulated in MelSC as well. This suggests that RNApII is present at many genes in quiescent MelSC, but P-TEFb levels are not sufficient to induce active transcription elongation. Importantly, inhibition of CDK9 in vitro protected melanocyte precursors from stress-induced apoptosis and converted them to a stem cell-like state (Freter *et al.*, 2010).

We then expanded our observation to other stem cell systems, and found CTD Ser2-negative cells in all stem cell systems tested, including keratinocyte, muscle, spermatogonia and hematopoietic stem cells. This suggests that global suppression of mRNA transcription elongation is a conserved feature of adult stem cells. Interestingly, some stem cell systems showed heterogeneity of CTD Ser2 staining. For example, we observed that spermatogonia stem cells attached to the basal lamina are negative for CTD Ser2 phosphorylation, while those detaching up-regulate Ser2 phosphorylation, even though they are still positive for the spermatogonia stem cell marker CD9 (Freter *et al.*, 2010). Attachment to the basal lamina is often a requirement for stem cell function by directing planes of division or maintenance of the undifferentiated state. Thus, the CTD Ser2 negative population seems to be the more stem cell-like population in CD9 positive spermatogonia.

Similarly, murine CD34- c-Kit+Sca1+Lin- (KSL) long-term repopulating hematopoietic stem cells clearly showed two different populations. One population exhibited CTD Ser2 phosphorylation levels as high as short-term repopulating CD34+ KSL cells, while ~27% of all CD34- KSL cells were negative for CTD Ser2 phosphorylation (Freter *et al.*, 2010). Heterogeneity of the HSC pool has been described previously, with a transplantable fraction of 15-25% of CD34- KSL HSC population, also using additional markers (Ema *et al.*, 2005, Foudi *et al.*, 2009, Wilson *et al.*, 2008). Importantly, analysis of transcriptionally quiescent HSC requires isolation of pure subpopulations of cells. We found that adult MelSC exhibit up to 100fold lower levels of total RNA per cell (Osawa *et al.*, 2005), suggesting that one activated stem cell may be sufficient to mask the RNA signal of 100 quiescent stem cells.

In order to identify the transcriptionally quiescent subpopulation of CD34- KSL cells, we performed sorting of bone marrow cells and antibody staining of sorted populations. Unfortunately, this procedure always includes fixation of cells, so they can not be used for assessment of in vivo repopulation ability. In order to validate stem cell function in vivo, it is necessary to convert the negative CTD Ser2 phosphorylation event into a readout which can be measured in living cells. Kinase activity can be measured using fusion proteins of Cyan and Yellow fluorescent proteins (CFP and YFP respectively), separated by a kinase target, a flexible linker and a phosphoprotein binding domain. Phosphorylation events result in binding of the phosphoacceptor to the kinase target and folding of the fusion protein resulting in Foerster resonance energy transfer (FRET) between CFP and YFP (Sato *et al.*, 2007). However, autofluorescence of cellular organelles in low energetic wavelengths, such as CFP, results in low signal to noise ratios, which is particularly difficult for cell sorting of multiparametric cell suspensions, such as bone marrow cells. Furthermore, even a complete lack of FRET by spatial separation results in FRET signal due to high concentration of expressed fluorescent proteins (Nguyen & Daugherty, 2005). Thus, a FRET-based approach can be useful for single-cell based imaging approaches, such as time lapse imaging of individual cells in vitro, but rather not for FACS sorting of heterogenic cell populations.

Recently, the development of circular permuted green fluorescent proteins (cpGFP) has enabled researchers to measure phosphorylation events in living cells using a single wavelength (Kawai *et al.*, 2004). However, the increase in fluorescence was only around 10-15%, which would be too little for separation by FACS. Novel mutations of circular GFP and ratiometric measurement of absorbance at different wavelengths increased the dynamic range up to 16fold between free and saturated forms of cpGFP, at least for measurement of pH or Ca^{2+} in living cells (Bizzarri *et al.*, 2006, Souslova *et al.*, 2007). Yet, if a kinase reporter can be constructed using these advanced cpGFP mutants has still to be shown.

We could observe a specific down regulation of CDK9, the RNApII CTD Ser2 kinase, in adult melanocyte stem cells both on the mRNA and protein levels (Freter *et al.*, 2010). The CDK9 promoter has many features of a housekeeping gene promoter (Bagella *et al.*, 2000, Liu & Rice, 2000), thus down regulation of this promoter may be due to a similar mechanism as other down regulated housekeeping genes in MelSC. Reporters for promoter function, for example GFP, Luciferase or LacZ have been used extensively to isolate or trace specific cells in vivo. However, isolating CDK9 promoter negative cells as transcriptionally quiescent stem cells could be biased by secondary effects on the reporter, such as silencing of reporter constructs or heterogeneity of expression between cells. One solution would be to label all cells with a constitutive reporter, and isolate constitutive promoter positive, CDK9 promoter reporter negative cells. The constitutive promoter has to be carefully chosen, as for example expression from the CMV promoter highly depends on CDK9 (Peng *et al.*, 1998). Given our observation of low activity of CDK9 in HSC, it is not surprising that HSC show limited CMV promoter activity (Salmon *et al.*, 2000). Constitutive promoters, but also promoters used for overexpression of genes in quiescent stem cells thus need to be validated for activity in vivo.

In vitro culture and expansion of immature hematopoietic stem cells could help to achieve better transplantation response in patients, but has not been achieved yet. So far culture of immature HSC leads to almost immediate differentiation and loss of multi-lineage repopulation ability. It has been shown recently that the xenobiotic Aryl receptor is present on HSC (Singh *et al.*, 2009). Activation by ligands results in nuclear translocation, recruitment of Ccnt1 and activation of hematopoietic gene transcription (Tian *et al.*, 2003). Interestingly, antagonists of the Aryl receptor prevent differentiation of HSC in vitro (Boitano *et al.*, 2010), suggesting that transcriptional quiescence may be beneficial for in vitro stem cell expansion. It would be very interesting to determine if inhibitors of P-TEFb activity have an effect on maintenance of undifferentiated HSC.

Transcriptional quiescence could also be used as a read-out of stem cell function in vitro. Screens of small molecular compounds in in vitro culture of primary HSC for maintenance of the human HSC markers CD34 and CD133 have led to some promising results (Boitano *et al.*, 2010). However, surface proteins may be unstable or unreliable, and may not be an immediate read-out of stem cell function. Transcriptional quiescence could serve as an alternative marker for stem cell identity. The development of fluorescent reporters for this screening is required to evaluate the impact of cytokines or small molecules on maintenance and expansion of HSC in vitro.

We and others have shown that inhibition of P-TEFb activity can be favourable for cell survival during cellular stresses, such as serum or growth factor starvation (Freter *et al.*, 2010, Kanazawa *et al.*, 2003). Down regulation of P-TEFb activity could thus also be advantageous for cancer cell survival during metastasis or therapy. Flavopiridol, a very specific CDK9 inhibitor has been used in BLL with some success (Chen *et al.*, 2005), but failed in most cases as a single agent in cancer chemotherapy (Blagosklonny, 2004). If transcriptionally quiescent cancer cells are present in primary or metastatic tumours, further inhibition of CDK9 activity may not be required. Rather, activation of mRNA transcription in these cells may render them susceptible to therapy and prevent metastasis and relapse. The first step towards this goal would be to identify and isolate transcriptionally quiescent cancer cells from a given tumour using a kinase activity reporter or CDK9 promoter reporter. Next, it would be necessary to determine if transcriptionally quiescent tumour cells have an enhanced tumour forming capacity in vivo or survive treatment with chemotherapeutic agents better than transcriptionally activated cells. Finally, high throughput screens for small molecular compounds which activate transcriptional quiescent cancer cells using the same reporter systems will enable us to activate dormant cancer stem cell-like cells in vivo and improve treatment of metastasis and prevent relapse of cancer in patients.

5. Conclusion

Adult stem cells have the unique capacity to self-renew and give rise to differentiated cells. To fulfil their lifelong function, these cells must be protected from cellular and genetic damage. Most adult stem cells are thought to enter a state of reversible cell cycle quiescence to preserve their role (Orford & Scadden, 2008). Indeed, activation of the cell cycle leads to premature stem cell exhaustion (Cheng *et al.*, 2000, Kippin *et al.*, 2005, Park *et al.*, 2003). However, most cells in the adult body have withdrawn from the cell cycle and can be induced to proliferate again, resulting in their eventual depletion (Bond *et al.*, 2004, Pajalunga *et al.*, 2007). Some somatic cells are able to resume proliferation and even self-renew, for example differentiated T and B lymphocytes (Fearon *et al.*, 2001). It has thus been proposed that adult stem cells are distinguished by other mechanisms rather than cell cycle quiescence (Mikkers & Frisen, 2005). Yet, what kind of defining mechanisms or marker this property integrates, and if it is shared by various adult stem cells, is unclear at the moment.

It has been long known that adult HSC can be isolated by their low retention of Pyronin Y, an RNA binding dye (Shapiro, 1981), suggesting that global mRNA transcriptional quiescence is a hallmark of quiescent stem cells. This observation however, has never been addressed further in terms of the precise molecular mechanism underlying global transcriptional quiescence. Analysis of the distinct phosphorylation patterns of the CTD of RNApII during the different stages of mRNA transcription can reveal initiation, paused transcription or productive elongation. Our analysis of various adult stem cell systems showed that down regulation of productive mRNA transcription elongation is a conserved, specific and early feature of adult stem cells (Freter *et al.*, 2010). In line with this, the CTD kinase CDK9 was absent and its inhibition improved cell survival during cellular stresses, suggesting a beneficial function of transcriptional quiescence for stem cell maintenance and survival. However, we could not induce activation of mRNA transcription by overexpression of CDK9 in vivo. It is thus not clear what the in vivo function of this down regulation is. New animal models with a constitutively active RNApII have to be developed to elucidate if transcriptional quiescence is necessary for adult stem cell maintenance or just a symptom of their quiescence.

Interestingly, we could observe some heterogeneity in terms of mRNA transcription elongation in the HSC pool, where only 27% of cells showed a clear absence of mRNA transcription elongation. Yet, identification and isolation of these transcriptionally quiescent cells depends on the availability of genetically encoded reporters of mRNA transcription. CDK9 activity reporters need to be specific for this kinase, and a large dynamic range has to be provided to clearly separate cell populations. A particular challenge will be to convert the negative observation of CTD Ser2 dephosphorylation into a signal-positive output, in order to avoid false-positive events by untransfected cells.

Taken together, we defined a novel molecular mechanism for adult stem cell quiescence, which may lead to the identification of pure stem cell-like cell populations from various sources, including heterogeneous adult stem cell populations or cancerous tissue. Even though some technical questions and functional tests are still to be answered, transcriptional quiescence is a novel and exciting mechanism to detect, isolate and characterize adult stem cells in an unprecedented purity from various sources.

6. References

Aida, M., Chen, Y., Nakajima, K., Yamaguchi, Y., Wada, T. & Handa, H. (2006) Transcriptional pausing caused by NELF plays a dual role in regulating immediate-early expression of the junB gene. *Mol Cell Biol,* 26, 6094-6104.

Bagella, L., Stiegler, P., De Luca, A., Siracusa, L.D. & Giordano, A. (2000) Genomic organization, promoter analysis, and chromosomal mapping of the mouse gene encoding Cdk9. *J Cell Biochem,* 78, 170-178.

Barboric, M., Lenasi, T., Chen, H., Johansen, E.B., Guo, S. & Peterlin, B.M. (2009) 7SK snRNP/P-TEFb couples transcription elongation with alternative splicing and is essential for vertebrate development. *Proc Natl Acad Sci U S A,* 106, 7798-7803.

Barboric, M., Nissen, R.M., Kanazawa, S., Jabrane-Ferrat, N. & Peterlin, B.M. (2001) NF-kappaB binds P-TEFb to stimulate transcriptional elongation by RNA polymerase II. *Mol Cell,* 8, 327-337.

Barboric, M. & Peterlin, B.M. (2005) A new paradigm in eukaryotic biology: HIV Tat and the control of transcriptional elongation. *PLoS Biol,* 3, e76.

Batchelder, C., Dunn, M.A., Choy, B., Suh, Y., Cassie, C., Shim, E.Y., Shin, T.H., Mello, C., Seydoux, G. & Blackwell, T.K. (1999) Transcriptional repression by the Caenorhabditis elegans germ-line protein PIE-1. *Genes Dev,* 13, 202-212.

Benedikt, A., Baltruschat, S., Scholz, B., Bursen, A., Arrey, T.N., Meyer, B., Varagnolo, L., Muller, A.M., Karas, M., Dingermann, T. & Marschalek, R. (2011) The leukemogenic AF4-MLL fusion protein causes P-TEFb kinase activation and altered epigenetic signatures. *Leukemia,* 25, 135-144.

Bentley, D.L. (1995) Regulation of transcriptional elongation by RNA polymerase II. *Curr Opin Genet Dev,* 5, 210-216.

Bizzarri, R., Arcangeli, C., Arosio, D., Ricci, F., Faraci, P., Cardarelli, F. & Beltram, F. (2006) Development of a novel GFP-based ratiometric excitation and emission pH indicator for intracellular studies. *Biophys J,* 90, 3300-3314.

Blagosklonny, M.V. (2004) Flavopiridol, an inhibitor of transcription: implications, problems and solutions. *Cell Cycle,* 3, 1537-1542.

Boehm, A.K., Saunders, A., Werner, J. & Lis, J.T. (2003) Transcription factor and polymerase recruitment, modification, and movement on dhsp70 in vivo in the minutes following heat shock. *Mol Cell Biol,* 23, 7628-7637.

Boitano, A.E., Wang, J., Romeo, R., Bouchez, L.C., Parker, A.E., Sutton, S.E., Walker, J.R., Flaveny, C.A., Perdew, G.H., Denison, M.S., Schultz, P.G. & Cooke, M.P. (2010) Aryl hydrocarbon receptor antagonists promote the expansion of human hematopoietic stem cells. *Science,* 329, 1345-1348.

Bond, J., Jones, C., Haughton, M., DeMicco, C., Kipling, D. & Wynford-Thomas, D. (2004) Direct evidence from siRNA-directed "knock down" that p16(INK4a) is required for human fibroblast senescence and for limiting ras-induced epithelial cell proliferation. *Exp Cell Res,* 292, 151-156.

Brinster, R.L. & Nagano, M. (1998) Spermatogonial stem cell transplantation, cryopreservation and culture. *Semin Cell Dev Biol,* 9, 401-409.

Chao, S.H. & Price, D.H. (2001) Flavopiridol inactivates P-TEFb and blocks most RNA polymerase II transcription in vivo. *J Biol Chem,* 276, 31793-31799.

Chen, R., Keating, M.J., Gandhi, V. & Plunkett, W. (2005) Transcription inhibition by flavopiridol: mechanism of chronic lymphocytic leukemia cell death. *Blood,* 106, 2513-2519.

Chen, Y., Yamaguchi, Y., Tsugeno, Y., Yamamoto, J., Yamada, T., Nakamura, M., Hisatake, K. & Handa, H. (2009) DSIF, the Paf1 complex, and Tat-SF1 have nonredundant, cooperative roles in RNA polymerase II elongation. *Genes Dev,* 23, 2765-2777.

Cheng, T., Rodrigues, N., Shen, H., Yang, Y., Dombkowski, D., Sykes, M. & Scadden, D.T. (2000) Hematopoietic stem cell quiescence maintained by p21cip1/waf1. *Science,* 287, 1804-1808.

Coller, H.A., Sang, L. & Roberts, J.M. (2006) A new description of cellular quiescence. *PLoS Biol,* 4, e83.

Eberhardy, S.R. & Farnham, P.J. (2001) c-Myc mediates activation of the cad promoter via a post-RNA polymerase II recruitment mechanism. *J Biol Chem,* 276, 48562-48571.

Eissenberg, J.C., Shilatifard, A., Dorokhov, N. & Michener, D.E. (2007) Cdk9 is an essential kinase in Drosophila that is required for heat shock gene expression, histone methylation and elongation factor recruitment. *Mol Genet Genomics,* 277, 101-114.

Elagib, K.E., Mihaylov, I.S., Delehanty, L.L., Bullock, G.C., Ouma, K.D., Caronia, J.F., Gonias, S.L. & Goldfarb, A.N. (2008) Cross-talk of GATA-1 and P-TEFb in megakaryocyte differentiation. *Blood,* 112, 4884-4894.

Ema, H., Sudo, K., Seita, J., Matsubara, A., Morita, Y., Osawa, M., Takatsu, K., Takaki, S. & Nakauchi, H. (2005) Quantification of self-renewal capacity in single hematopoietic stem cells from normal and Lnk-deficient mice. *Dev Cell,* 8, 907-914.

Fearon, D.T., Manders, P. & Wagner, S.D. (2001) Arrested differentiation, the self-renewing memory lymphocyte, and vaccination. *Science,* 293, 248-250.

Foudi, A., Hochedlinger, K., Van Buren, D., Schindler, J.W., Jaenisch, R., Carey, V. & Hock, H. (2009) Analysis of histone 2B-GFP retention reveals slowly cycling hematopoietic stem cells. *Nat Biotechnol,* 27, 84-90.

Freter, R., Osawa, M. & Nishikawa, S. (2010) Adult stem cells exhibit global suppression of RNA polymerase II serine-2 phosphorylation. *Stem Cells,* 28, 1571-1580.

Fuchs, E., Tumbar, T. & Guasch, G. (2004) Socializing with the neighbors: stem cells and their niche. *Cell,* 116, 769-778.

Fujinaga, K., Irwin, D., Huang, Y., Taube, R., Kurosu, T. & Peterlin, B.M. (2004) Dynamics of human immunodeficiency virus transcription: P-TEFb phosphorylates RD and dissociates negative effectors from the transactivation response element. *Mol Cell Biol,* 24, 787-795.

Fujinaga, K., Irwin, D., Taube, R., Zhang, F., Geyer, M. & Peterlin, B.M. (2002) A minimal chimera of human cyclin T1 and tat binds TAR and activates human immunodeficiency virus transcription in murine cells. *J Virol,* 76, 12934-12939.

Garriga, J., Bhattacharya, S., Calbo, J., Marshall, R.M., Truongcao, M., Haines, D.S. & Grana, X. (2003) CDK9 is constitutively expressed throughout the cell cycle, and its steady-state expression is independent of SKP2. *Mol Cell Biol,* 23, 5165-5173.

Garriga, J., Peng, J., Parreno, M., Price, D.H., Henderson, E.E. & Grana, X. (1998) Upregulation of cyclin T1/CDK9 complexes during T cell activation. *Oncogene,* 17, 3093-3102.

Gojo, I., Zhang, B. & Fenton, R.G. (2002) The cyclin-dependent kinase inhibitor flavopiridol induces apoptosis in multiple myeloma cells through transcriptional repression and down-regulation of Mcl-1. *Clin Cancer Res,* 8, 3527-3538.

Grana, X., De Luca, A., Sang, N., Fu, Y., Claudio, P.P., Rosenblatt, J., Morgan, D.O. & Giordano, A. (1994) PITALRE, a nuclear CDC2-related protein kinase that phosphorylates the retinoblastoma protein in vitro. *Proc Natl Acad Sci U S A,* 91, 3834-3838.

Guenther, M.G., Levine, S.S., Boyer, L.A., Jaenisch, R. & Young, R.A. (2007) A chromatin landmark and transcription initiation at most promoters in human cells. *Cell,* 130, 77-88.

Hanyu-Nakamura, K., Sonobe-Nojima, H., Tanigawa, A., Lasko, P. & Nakamura, A. (2008) Drosophila Pgc protein inhibits P-TEFb recruitment to chromatin in primordial germ cells. *Nature,* 451, 730-733.

Hargreaves, D.C., Horng, T. & Medzhitov, R. (2009) Control of inducible gene expression by signal-dependent transcriptional elongation. *Cell,* 138, 129-145.

Ho, C.K. & Shuman, S. (1999) Distinct roles for CTD Ser-2 and Ser-5 phosphorylation in the recruitment and allosteric activation of mammalian mRNA capping enzyme. *Mol Cell,* 3, 405-411.

Huang, F., Wagner, M. & Siddiqui, M.A. (2004) Ablation of the CLP-1 gene leads to down-regulation of the HAND1 gene and abnormality of the left ventricle of the heart and fetal death. *Mech Dev,* 121, 559-572.

Jang, M.K., Mochizuki, K., Zhou, M., Jeong, H.S., Brady, J.N. & Ozato, K. (2005) The bromodomain protein Brd4 is a positive regulatory component of P-TEFb and stimulates RNA polymerase II-dependent transcription. *Mol Cell,* 19, 523-534.

Jiang, H. & Peterlin, B.M. (2008) Differential chromatin looping regulates CD4 expression in immature thymocytes. *Mol Cell Biol,* 28, 907-912.

Jiang, H., Zhang, F., Kurosu, T. & Peterlin, B.M. (2005) Runx1 binds positive transcription elongation factor b and represses transcriptional elongation by RNA polymerase II: possible mechanism of CD4 silencing. *Mol Cell Biol,* 25, 10675-10683.

Kanazawa, S., Okamoto, T. & Peterlin, B.M. (2000) Tat competes with CIITA for the binding to P-TEFb and blocks the expression of MHC class II genes in HIV infection. *Immunity,* 12, 61-70.

Kanazawa, S., Soucek, L., Evan, G., Okamoto, T. & Peterlin, B.M. (2003) c-Myc recruits P-TEFb for transcription, cellular proliferation and apoptosis. *Oncogene,* 22, 5707-5711.

Kawai, Y., Sato, M. & Umezawa, Y. (2004) Single color fluorescent indicators of protein phosphorylation for multicolor imaging of intracellular signal flow dynamics. *Anal Chem,* 76, 6144-6149.

Kippin, T.E., Martens, D.J. & van der Kooy, D. (2005) p21 loss compromises the relative quiescence of forebrain stem cell proliferation leading to exhaustion of their proliferation capacity. *Genes Dev,* 19, 756-767.

Kohoutek, J. (2009) P-TEFb- the final frontier. *Cell Div,* 4, 19.

Kohoutek, J., Li, Q., Blazek, D., Luo, Z., Jiang, H. & Peterlin, B.M. (2009) Cyclin T2 is essential for mouse embryogenesis. *Mol Cell Biol,* 29, 3280-3285.

Li, L., Jothi, R., Cui, K., Lee, J.Y., Cohen, T., Gorivodsky, M., Tzchori, I., Zhao, Y., Hayes, S.M., Bresnick, E.H., Zhao, K., Westphal, H. & Love, P.E. (2011) Nuclear adaptor Ldb1 regulates a transcriptional program essential for the maintenance of hematopoietic stem cells. *Nat Immunol,* 12, 129-136.

Li, Q., Price, J.P., Byers, S.A., Cheng, D., Peng, J. & Price, D.H. (2005) Analysis of the large inactive P-TEFb complex indicates that it contains one 7SK molecule, a dimer of HEXIM1 or HEXIM2, and two P-TEFb molecules containing Cdk9 phosphorylated at threonine 186. *J Biol Chem,* 280, 28819-28826.

Lin, C., Smith, E.R., Takahashi, H., Lai, K.C., Martin-Brown, S., Florens, L., Washburn, M.P., Conaway, J.W., Conaway, R.C. & Shilatifard, A. (2010) AFF4, a component of the ELL/P-TEFb elongation complex and a shared subunit of MLL chimeras, can link transcription elongation to leukemia. *Mol Cell,* 37, 429-437.

Liu, H. & Herrmann, C.H. (2005) Differential localization and expression of the Cdk9 42k and 55k isoforms. *J Cell Physiol,* 203, 251-260.

Liu, H. & Rice, A.P. (2000) Genomic organization and characterization of promoter function of the human CDK9 gene. *Gene,* 252, 51-59.

Mancebo, H.S., Lee, G., Flygare, J., Tomassini, J., Luu, P., Zhu, Y., Peng, J., Blau, C., Hazuda, D., Price, D. & Flores, O. (1997) P-TEFb kinase is required for HIV Tat transcriptional activation in vivo and in vitro. *Genes Dev,* 11, 2633-2644.
Marshall, R.M., Salerno, D., Garriga, J. & Gra√±a, X. (2005) Cyclin T1 expression is regulated by multiple signaling pathways and mechanisms during activation of human peripheral blood lymphocytes. *J Immunol,* 175, 6402-6411.
Meier, N., Krpic, S., Rodriguez, P., Strouboulis, J., Monti, M., Krijgsveld, J., Gering, M., Patient, R., Hostert, A. & Grosveld, F. (2006) Novel binding partners of Ldb1 are required for haematopoietic development. *Development,* 133, 4913-4923.
Mikkers, H. & Frisen, J. (2005) Deconstructing stemness. *EMBO J,* 24, 2715-2719.
Muse, G.W., Gilchrist, D.A., Nechaev, S., Shah, R., Parker, J.S., Grissom, S.F., Zeitlinger, J. & Adelman, K. (2007) RNA polymerase is poised for activation across the genome. *Nat Genet,* 39, 1507-1511.
Nguyen, A.W. & Daugherty, P.S. (2005) Evolutionary optimization of fluorescent proteins for intracellular FRET. *Nat Biotechnol,* 23, 355-360.
Ni, Z., Schwartz, B.E., Werner, J., Suarez, J.R. & Lis, J.T. (2004) Coordination of transcription, RNA processing, and surveillance by P-TEFb kinase on heat shock genes. *Mol Cell,* 13, 55-65.
Nishimura, E.K., Jordan, S.A., Oshima, H., Yoshida, H., Osawa, M., Moriyama, M., Jackson, I.J., Barrandon, Y., Miyachi, Y. & Nishikawa, S. (2002) Dominant role of the niche in melanocyte stem-cell fate determination. *Nature,* 416, 854-860.
Orford, K.W. & Scadden, D.T. (2008) Deconstructing stem cell self-renewal: genetic insights into cell-cycle regulation. *Nat Rev Genet,* 9, 115-128.
Osawa, M., Egawa, G., Mak, S.S., Moriyama, M., Freter, R., Yonetani, S., Beermann, F. & Nishikawa, S. (2005) Molecular characterization of melanocyte stem cells in their niche. *Development,* 132, 5589-5599.
Oven, I., Brdickova, N., Kohoutek, J., Vaupotic, T., Narat, M. & Peterlin, B.M. (2007) AIRE recruits P-TEFb for transcriptional elongation of target genes in medullary thymic epithelial cells. *Mol Cell Biol,* 27, 8815-8823.
Pajalunga, D., Mazzola, A., Salzano, A.M., Biferi, M.G., De Luca, G. & Crescenzi, M. (2007) Critical requirement for cell cycle inhibitors in sustaining nonproliferative states. *J Cell Biol,* 176, 807-818.
Park, I.K., Qian, D., Kiel, M., Becker, M.W., Pihalja, M., Weissman, I.L., Morrison, S.J. & Clarke, M.F. (2003) Bmi-1 is required for maintenance of adult self-renewing haematopoietic stem cells. *Nature,* 423, 302-305.
Patturajan, M., Schulte, R.J., Sefton, B.M., Berezney, R., Vincent, M., Bensaude, O., Warren, S.L. & Corden, J.L. (1998) Growth-related changes in phosphorylation of yeast RNA polymerase II. *J Biol Chem,* 273, 4689-4694.
Peng, J., Zhu, Y., Milton, J.T. & Price, D.H. (1998) Identification of multiple cyclin subunits of human P-TEFb. *Genes Dev,* 12, 755-762.
Peterlin, B.M. & Price, D.H. (2006) Controlling the elongation phase of transcription with P-TEFb. *Mol Cell,* 23, 297-305.
Radonjic, M., Andrau, J.C., Lijnzaad, P., Kemmeren, P., Kockelkorn, T.T., van Leenen, D., van Berkum, N.L. & Holstege, F.C. (2005) Genome-wide analyses reveal RNA polymerase II located upstream of genes poised for rapid response upon S. cerevisiae stationary phase exit. *Mol Cell,* 18, 171-183.

Rahl, P.B., Lin, C.Y., Seila, A.C., Flynn, R.A., McCuine, S., Burge, C.B., Sharp, P.A. & Young, R.A. (2010) c-Myc regulates transcriptional pause release. *Cell,* 141, 432-445.

Ramakrishnan, R., Yu, W. & Rice, A.P. (2011) Limited redundancy in genes regulated by Cyclin T2 and Cyclin T1. *BMC Res Notes,* 4, 260.

Salmon, P., Kindler, V., Ducrey, O., Chapuis, B., Zubler, R.H. & Trono, D. (2000) High-level transgene expression in human hematopoietic progenitors and differentiated blood lineages after transduction with improved lentiviral vectors. *Blood,* 96, 3392-3398.

Sato, M., Kawai, Y. & Umezawa, Y. (2007) Genetically encoded fluorescent indicators to visualize protein phosphorylation by extracellular signal-regulated kinase in single living cells. *Anal Chem,* 79, 2570-2575.

Sehgal, P.B. & Darnell, T. (1976) The inhibition by DRB (5,6-dichloro-1-beta-D-ribofuranosylbenzimidazole) of hnRNA and mRNA production in HeLa cells. *Cell,* 9, 473-480.

Shackleton, M., Vaillant, F., Simpson, K.J., Stingl, J., Smyth, G.K., Asselin-Labat, M.L., Wu, L., Lindeman, G.J. & Visvader, J.E. (2006) Generation of a functional mammary gland from a single stem cell. *Nature,* 439, 84-88.

Shapiro, H.M. (1981) Flow cytometric estimation of DNA and RNA content in intact cells stained with Hoechst 33342 and pyronin Y. *Cytometry,* 2, 143-150.

Shilatifard, A., Lane, W.S., Jackson, K.W., Conaway, R.C. & Conaway, J.W. (1996) An RNA polymerase II elongation factor encoded by the human ELL gene. *Science,* 271, 1873-1876.

Shim, E.Y., Walker, A.K., Shi, Y. & Blackwell, T.K. (2002) CDK-9/cyclin T (P-TEFb) is required in two postinitiation pathways for transcription in the C. elegans embryo. *Genes Dev,* 16, 2135-2146.

Shore, S.M., Byers, S.A., Dent, P. & Price, D.H. (2005) Characterization of Cdk9(55) and differential regulation of two Cdk9 isoforms. *Gene,* 350, 51-58.

Shore, S.M., Byers, S.A., Maury, W. & Price, D.H. (2003) Identification of a novel isoform of Cdk9. *Gene,* 307, 175-182.

Sims, R.J., 3rd, Belotserkovskaya, R. & Reinberg, D. (2004) Elongation by RNA polymerase II: the short and long of it. *Genes Dev,* 18, 2437-2468.

Singh, K.P., Casado, F.L., Opanashuk, L.A. & Gasiewicz, T.A. (2009) The aryl hydrocarbon receptor has a normal function in the regulation of hematopoietic and other stem/progenitor cell populations. *Biochem Pharmacol,* 77, 577-587.

Song, S.H., Kim, A., Ragoczy, T., Bender, M.A., Groudine, M. & Dean, A. (2010) Multiple functions of Ldb1 required for beta-globin activation during erythroid differentiation. *Blood,* 116, 2356-2364.

Souslova, E.A., Belousov, V.V., Lock, J.G., Stromblad, S., Kasparov, S., Bolshakov, A.P., Pinelis, V.G., Labas, Y.A., Lukyanov, S., Mayr, L.M. & Chudakov, D.M. (2007) Single fluorescent protein-based Ca2+ sensors with increased dynamic range. *BMC Biotechnol,* 7, 37.

Tian, Y., Ke, S., Chen, M. & Sheng, T. (2003) Interactions between the aryl hydrocarbon receptor and P-TEFb. Sequential recruitment of transcription factors and differential phosphorylation of C-terminal domain of RNA polymerase II at cyp1a1 promoter. *J Biol Chem,* 278, 44041-44048.

Tothova, Z. & Gilliland, D.G. (2007) FoxO transcription factors and stem cell homeostasis: insights from the hematopoietic system. *Cell Stem Cell,* 1, 140-152.

Wei, P., Garber, M.E., Fang, S.M., Fischer, W.H. & Jones, K.A. (1998) A novel CDK9-associated C-type cyclin interacts directly with HIV-1 Tat and mediates its high-affinity, loop-specific binding to TAR RNA. *Cell,* 92, 451-462.

Wilson, A., Laurenti, E., Oser, G., van der Wath, R.C., Blanco-Bose, W., Jaworski, M., Offner, S., Dunant, C.F., Eshkind, L., Bockamp, E., Li√≥, P., Macdonald, H.R. & Trumpp, A. (2008) Hematopoietic stem cells reversibly switch from dormancy to self-renewal during homeostasis and repair. *Cell,* 135, 1118-1129.

Wittmann, B.M., Wang, N. & Montano, M.M. (2003) Identification of a novel inhibitor of breast cell growth that is down-regulated by estrogens and decreased in breast tumors. *Cancer Res,* 63, 5151-5158.

Wu, C.H., Yamaguchi, Y., Benjamin, L.R., Horvat-Gordon, M., Washinsky, J., Enerly, E., Larsson, J., Lambertsson, A., Handa, H. & Gilmour, D. (2003) NELF and DSIF cause promoter proximal pausing on the hsp70 promoter in Drosophila. *Genes Dev,* 17, 1402-1414.

Yamaguchi, Y., Inukai, N., Narita, T., Wada, T. & Handa, H. (2002) Evidence that negative elongation factor represses transcription elongation through binding to a DRB sensitivity-inducing factor/RNA polymerase II complex and RNA. *Mol Cell Biol,* 22, 2918-2927.

Yang, Z., Yik, J.H., Chen, R., He, N., Jang, M.K., Ozato, K. & Zhou, Q. (2005) Recruitment of P-TEFb for stimulation of transcriptional elongation by the bromodomain protein Brd4. *Mol Cell,* 19, 535-545.

Zeitlinger, J., Stark, A., Kellis, M., Hong, J.W., Nechaev, S., Adelman, K., Levine, M. & Young, R.A. (2007) RNA polymerase stalling at developmental control genes in the Drosophila melanogaster embryo. *Nat Genet,* 39, 1512-1516.

Zhang, F., Barboric, M., Blackwell, T.K. & Peterlin, B.M. (2003) A model of repression: CTD analogs and PIE-1 inhibit transcriptional elongation by P-TEFb. *Genes Dev,* 17, 748-758.

Zhu, Y., Pe'ery, T., Peng, J., Ramanathan, Y., Marshall, N., Marshall, T., Amendt, B., Mathews, M.B. & Price, D.H. (1997) Transcription elongation factor P-TEFb is required for HIV-1 tat transactivation in vitro. *Genes Dev,* 11, 2622-2632.

4

Markers for Hematopoietic Stem Cells: Histories and Recent Achievements

Takafumi Yokota[1], Kenji Oritani[1], Stefan Butz[2], Stephan Ewers[2], Dietmar Vestweber[2] and Yuzuru Kanakura[1]
[1]*Department of Hematology and Oncology, Osaka University Graduate School of Medicine, Suita*
[2]*Department of Vascular Cell Biology, Max-Planck-Institute for Molecular Biomedicine, Münster,*
[1]*Japan*
[2]*Germany*

1. Introduction

Hematopoietic stem cells (HSC) are characterized with the capacity for self-renewal as well as multi-lineage differentiation, maintaining the immune system and blood cell formation throughout life. Although studies for the HSC biology have been in the forefront of the stem cell research field, many questions still remain with regard to the origin, development, and aging of HSC. Furthermore, needless to say, HSC are very useful for clinical medicine, particularly in the transplantation and/or regeneration therapy for hematological malignancies. Success of those therapies depends on how effectively HSC are purified and transplanted to the patients. In order to address those important issues in both basic and clinical science, information of cell surface molecules that selectively mark HSC is essential.

Since the frequency of HSC in bone marrow or peripheral blood is extremely low, many studies have attempted to identify unique markers associated with those rare cells. As a result, it is now possible to purify long-term reconstituting HSC from mouse bone marrow with very high efficiency. However, many of these parameters change dramatically during ontogeny or inflammation, and what is worse still, they differ between mouse and man. Efficient HSC-based therapies and the emerging field of tissue-regenerative medicine will benefit from more precise information about what defines HSC.

In this chapter, we summarize a large body of information with respect to the HSC-related markers and introduce Endothelial cell-selective adhesion molecule (ESAM) as a novel marker for HSC (Yokota et al., 2009). Indeed, ESAM is expressed throughout the ontogeny in mouse and can be used as a gating parameter for sorting long-term repopulating HSC. In addition, the marker appears to be useful for the purification of human HSC.

2. Development of methodology for HSC purification from mouse bone marrow

In 1988, Spangrude et al tried to find a set of cell surface proteins that were associated with multi-lineage reconstitution ability, and succeeded to enrich such multipotential progenitors in the Lineage marker (Lin; generally including TER119, Mac1, Gr1, CD45R/B220, CD3, CD4, CD8)$^{-}$ Thy-1^{Low} Sca-1^{+} fraction of mouse bone marrow (Spangrude et al, 1988). Indeed, they showed that only 30 Lin^{-} Thy-1^{Low} Sca-1^{+} cells injected via a tail vein could rescue 50% of lethally irradiated mice. Three years later, in 1991, Ogawa et al reported that hematopoietic progenitor activity of mouse bone marrow was excusive to the cells expressing c-kit, which is a receptor for stem cell factor (Ogawa et al., 1991). Since then, Lin^{-} Sca-1^{+} c-kit^{+} (LSK) has been generally used as a canonical marker set for HSC enrichment.

It has been gradually recognized that the LSK fraction is heterogeneous, including long-term self-renewing HSC, short-term non-self-renewing HSC and lineage-committed progenitors. In 1996, Osawa et al reported that long-term HSC in adult bone marrow exist in the CD34 low to negative fraction among LSK cells (Osawa et al., 1996). Injection of a single CD34$^{-/Low}$ LSK cell resulted in multi-lineage long-term reconstitution in 21% of lethally irradiated mice whereas CD34^{+} LSK cells revealed early but only short-term reconstitution. Transplantation of graded numbers of CD34$^{-/Low}$ LSK cells showed that the CD34$^{-/Low}$ LSK fraction contains long-term HSC at the frequency of 1 out of 5 cells. In 2001, Christensen and Weissman also showed that the LSK fraction is heterogeneous and long-term HSC are highly enriched in the Flk2/Flt3 receptor tyrosine kinase negative cells (Christensen & Weissman, 2001).

In addition to the cell surface markers, another approach has been developed to enrich long-term HSC activity by focusing on their high efflux activity. Using the fluorescent DNA-binding dye Hoechst33342, in 1996, Goodell et al found that cells in a small Hoechst^{low}-stained population (termed "Side population") can protect recipients from lethal irradiation at low cell doses (Goodell et al., 1996). A following study by Matsuzaki et al showed that, in combination with the CD34$^{-/Low}$ LSK phenotype, the strongest Hoechst33342 efflux activity (Tip-side population) can purify long-term multi-lineage HSC with almost absolute efficiency (Matsuzaki et al., 2004).

Recently, Morrison and colleagues reported an alternative method for HSC purification based on the expression pattern of the signaling lymphocytic activation molecule (SLAM) family proteins, i.e. CD150, CD244, and CD48 (Kiel et al., 2005). They showed that CD150^{+} CD48^{-} cells were uniformly CD244^{-} and a simple gating for CD150^{+} CD48^{-} could enrich long-term HSC at approximately 1 in 5 cells. Moreover, combined with the canonical HSC marker LSK, the SLAM code could purify the HSC at 1 in 2 cells (Kiel et al., 2005).

Representative achievements during these 2 decades are summarized in Table 1. With surface markers, we can now purify the long-term multi-lineage HSC from adult mouse bone marrow with extremely high efficiency as Lin^{-} Sca-1^{+} c-kit^{+} Thy1^{Low} CD34$^{-/low}$ CD150^{+} CD48^{-} cells. In fact, recent studies have demonstrated that the Lin^{-} Sca-1^{+} c-kit^{+} CD34^{-} CD150^{+} CD48^{-} fraction in adult mouse bone marrow contains truly dormant HSC, which divide only 5-6 times during the life span (Wilson et al., 2008; Foudi et al., 2009).

Markers	references
Lin^- $Thy\text{-}1^{Low}$ $Sca\text{-}1^+$	Spangrude et al., 1988
$CD34^{-/Low}$ Lin^- $Sca\text{-}1^+$ $c\text{-}kit^+$	Osawa et al., 1996
Side Population (high Hoechst-efflux ability)	Goodell et al., 1996
*Tip-SP Lin^- $Sca\text{-}1^+$ $c\text{-}kit^+$	Matsuzaki et al., 2004
$CD150^+$ CD244 $CD48^-$	Kiel et al., 2005
BrdU or Histone 2B-GFP-retaining, $CD150^+$ $CD48^-$ $CD34^-$ Lin^- $Sca\text{-}1^+$ $c\text{-}kit^+$	Wilson et al., 2008 Foudi et al., 2009

*Tip-SP: The highest Hoechst-efflux fraction in the Side Population

Table 1. Markers for hematopoietic stem cells in adult mouse bone marrow

3. Fickleness of HSC surface markers

It is important to stress here that none of the surface markers shown above is entirely specific to the long-term HSC. In addition, many of these parameters differ between strains of mice and change dramatically during developmental age. For example, Sca1, which has been a center in the canonical HSC marker "LSK", is not detectable on the emerging HSC in the aorta-gonad-mesonephros (AGM) area and only appears after day 11.5 of gestation on HSC in fetal liver (Matsubara et al., 2005; Our unpublished observation). Furthermore, the expression level of Sca1 on HSC differs between strains and is not very effective to enrich HSC from Balb/c mice (Spangrude & Brooks, 1993). Likewise, the SLAM family CD150 is not useful for the emerging and developing HSC in embryos (McKinney-Freeman et al., 2009). On the contrary, CD41, CD11b/Mac1, vascular endothelial (VE)-cadherin, and CD34 are known to mark the emerging and developing HSC during the fetal period, but gradually disappear along the ontogeny (Mikkola & Orkin, 2006).

It is also a well-recognized fact that cell surface markers on HSC in adult bone marrow do fluctuate according to the cell-cycle status and the differentiating behavior, which change depending on the physiological requirement. Bone marrow suppression by irradiation and/or chemotherapy revives several disappeared markers including CD11b/Mac1 and CD34 whereas it significantly down-regulates the expression level of c-kit on long-term HSC (Randall & Weissman, 1997; Ogawa 2002). Molecular crosstalk between HSC and bone marrow microenvironment is thought to control the status of HSC and influence their surface phenotypes, but precise mechanisms remain largely unknown. Therefore, researchers in the HSC field need to carefully choose an appropriate marker set and a sorting gate depending on the HSC characteristics, otherwise they would miss important target cells even in the lineage depletion step.

4. Difference between mouse and man

Another very critical issue on the topic of the HSC markers is their diversity between species. Although essential difference has not been observed between mouse and man regarding either the organs producing HSC or the transcription factors regulating their differentiation, completely different markers have been used to sort HSC in the two species. Human HSC do not express Sca1 or the SLAM family CD150 (Larochelle et al., 2011). While the CD34$^+$ CD38$^-$ phenotype has been regarded as the canonical marker set for human HSC, it has been repeatedly reported that murine adult HSC locate in the CD34$^-$ CD38$^+$ fraction (Randall et al., 1996; Matsuoka et al., 2001; Tajima et al., 2001). There is no reasonable explanation so far for the change along evolution, and such phenotypic differences between murine and human HSC have been an obstacle to apply achievement in mouse studies to human.

Early studies by Berenson et al demonstrated that autologous CD34$^+$ cells enriched from bone marrow effectively radioprotected baboons and promoted hematopoietic recovery in human patients after marrow ablative therapy (Berenson et al, 1988, 1991). Over the past 2 decades, the use of CD34 as a marker for hematopoietic stem/progenitor cells has been a strong tool in the field of clinical hematology. Since the CD34$^+$ fraction of human bone marrow contains lineage-committed progenitors as well as long-term multi-lineage HSC, many laboratories have sought additional markers to further enrich the CD34$^+$ population for long-term HSC. CD90/Thy1, Tie, CD117/c-kit, and CD133/AC133 have been found as positive markers to enrich long-term-HSC whereas several negative markers including CD38 have been reported (Baum et al., 1992; Hasiyama et al., 1996; Gunji et al., 1993; Yin et al., 1997; Terstappen et al., 1991).

Recent advances of xenotransplantation models and techniques have enabled the assessment of pluripotency as well as self-renewal of human hematopoietic progenitors in vivo (Shultz et al., 2007). A series of studies by John Dick's laboratory have successfully enriched human long-term HSC within the Lin$^-$ CD34$^+$ CD38$^-$ population (McKenzie et al., 2007; Doulatov et al., 2010). In a very recent report, they have purified human HSC from cord blood with a maker set of Lin$^-$ CD34$^+$ CD38$^-$ CD45RA$^-$ CD90/Thy1$^+$ Rhodamin123^{Low} CD49f$^+$. Indeed, those cells were capable of long-term multilineage engraftment in NOD/SCID/IL2 receptor common-γ chain null mice at a single-cell level (Notta et al., 2011). The information regarding human HSC markers is summarized in Table 2.

While CD34 has been playing an important role as a reliable marker for human hematopoietic stem/progenitor cells in the practical medicine, several studies have demonstrated that long-term reconstituting activity is also detectable in the CD34$^-$ Lin$^-$ population (Bhatia et al., 1998; Gallacher et al., 2000; Wang et al., 2003). A prior study using Hoechst 33342 by Goodell et al also identified CD34$^-$ cells in the side-population of human and rhesus bone marrow, and actually rhesus CD34$^-$ side-population cells acquired the ability to form hematopoietic colonies after long-term cultivation on bone marrow stromal cells (Goodell et al., 1997). It should be interesting to examine molecular signatures associated with those CD34$^-$ HSC in primates, and compare their features with murine CD34$^-$ LSK cells.

Markers	references
$CD34^+$	Berenson et al., 1988, 1991
$CD34^+$ $CD38^-$	Terstappen et al., 1991
$CD34^+$ Lin^- $Thy1^+$	Baum et al., 1992
$CD34^+$ c-kit^+	Gunji et al., 1993
$CD34^+$ Tie^+	Hashiyama et al., 1996
$CD34^+$ $CD133/AC133^+$	Yin et al., 1997
$CD34^-$ Lin^- $CD133/AC133^+$ $CD7^-$	Gallacher et al., 2000
$CD34^+$ $CD38^-$ Lin^- $Rhodamine123^{Low}$	McKenzie et al., 2007
$CD34^+$ $CD38^-$ Lin^- $CD45RA^-$ $Rhodamine123^{Low}$ $CD49f^+$	Notta et al., 2011

Table 2. Markers for human hematopoietic stem cells

5. Endothelial-related markers

Hematopoietic cells are thought to originate from the hemangioblast and/or the hemogenic endothelium, which can produce hematopoietic cells and endothelial cells. Therefore, it seems quite natural that HSC share some surface molecules with the endothelial lineage. CD34, PECAM-1/CD31, endoglin, Tie2 and VE-cadherin are well-known endothelial antigens that also mark HSC particularly at early developmental stages (Mikkola & Orkin 2006; Takakura et al., 1998; Yokota et al, 2006). In addition, recent studies have identified endomucin, endothelial protein-C receptor/CD201, and junctional adhesion molecule-A that are common to HSC and endothelial cells (Matsubara et al., 2005; Balazs et al., 2006; Sugano et al., 2008). Although, as discussed above, the expression level of some of these antigens declines or even diminishes at later stages of development (Mikkola & Orkin 2006), each of these advances offered the promise of learning more about how HSC arise de novo and function throughout life. It is crucial to define the means to identify the authentic HSC at all developmental stages so that we can ultimately understand the precise molecular mechanisms of the HSC development.

6. Identification of ESAM as a novel HSC marker

We previously reported that Rag1/GFP$^-$ Lin$^-$ c-kit^{High} Sca1$^+$ cells derived from bone marrow or fetal liver of the Rag1/GFP reporter mice reconstituted lympho-hematopoiesis in lethally irradiated recipients, while Rag1/GFP$^+$ Lin$^-$ c-kit^{High} Sca1$^+$ cells only transiently contributed to T and B lymphopoiesis (Igarashi et al., 2002; Yokota et al., 2003). Those data demonstrated that Rag1 expression is useful to distinguish early lymphoid progenitors (ELP) from the long-term HSC. To learn more about the first step of HSC differentiation to the lymphoid lineage, microarray analyses were conducted to search for genes that characterize the initial transition of HSC to ELP. The search brought us a large body of information about genes potentially related to early lymphopoiesis whereas it also identified genes whose expression seemed to correlate with HSC. Among the HSC-related genes, ESAM strongly drew our attention because of its conspicuous expression in the HSC fraction and sharp down-regulation on differentiation to ELP.

ESAM was originally identified as an endothelial cell-specific protein (Hirata et al., 2001; Nasdala et al., 2002). Flow cytometry analyses with anti-ESAM antibodies showed that the HSC-enriched Rag1$^-$ c-kit^{High} Sca1$^+$ fraction of E14.5 fetal liver could be subdivided into two on the basis of ESAM level (Figure 1A). The subpopulation with the high density of ESAM was enriched for c-kit^{High} Sca1^{High} cells, while ones with negative or low levels of ESAM were found in the c-kit^{High} Sca1^{Low} subset. In addition, ESAM expression well correlated with hematopoietic stem/progenitor activity (Figure 1B). Cells in the ESAM^{High} Rag1$^-$ c-kit^{High} Sca1$^+$ fraction formed more and larger colonies than those in the ESAM$^{-/Low}$ Rag1$^-$ c-kit^{High} Sca1$^+$ fraction. Particularly, majority of CFU-Mix, multi-potent primitive progenitors, were found in the ESAM^{High} fraction (Figure 1B and 1C). In limiting dilution stromal cell co-cultures, we found that 1 in 2.1 ESAM^{High} Rag1$^-$ c-kit^{High} Sca1$^+$ cells and 1 in 3.5 ESAM$^{-/Low}$ Rag1$^-$ c-kit^{High} Sca1$^+$ cells gave rise to blood cells. However, 1 in 8 ESAM^{High} Rag1$^-$ c-kit^{High} Sca1$^+$ cells produced CD19$^+$ B lineage cells whereas only 1 in 125 ESAM$^{-/Low}$ Rag1$^-$ c-kit^{High} Sca1$^+$ cells were lymphopoietic under these conditions. Furthermore, in long-term reconstituting assays, ESAM^{High} Rag1$^-$ c-kit^{High} Sca1$^+$ cells contributed highly to the multi-lineage recovery of lympho-hematopoiesis in recipients, but no chimerism was detected in mice transplanted with ESAM$^{-/Low}$ Rag1$^-$ c-kit^{High} Sca1$^+$ cells. These results suggested that the long-term multi-lineage HSC in E14.5 fetal liver are exclusively present in the ESAM^{High} fraction.

7. ESAM marks HSC in different developmental stages and in different species

Hematopoietic cells arise from mesoderm precursors at different sites and stages of development (de Bruijn et al., 2000; Oberlin et al., 2002). We previously determined that, while myelo-erythroid progenitors emerge from the yolk sac, hematopoietic progenitors with lymphopoietic potential first develop in the paraaortic splanchnopleura (pSp) /AGM region (Yokota et al., 2006). ESAM$^+$ cells in the AGM were found to co-express c-kit and endothelial antigens, Tie2, CD34 and CD31/PECAM-1 that are known as a marker set for emerging HSC. However, the earlier hematopoietic progenitors in the yolk sac that have limited life span and little lymphopoietic activity were harbored in the ESAM^{Low} Tie2^{Low} c-kit^{High} fraction (Figure 2).

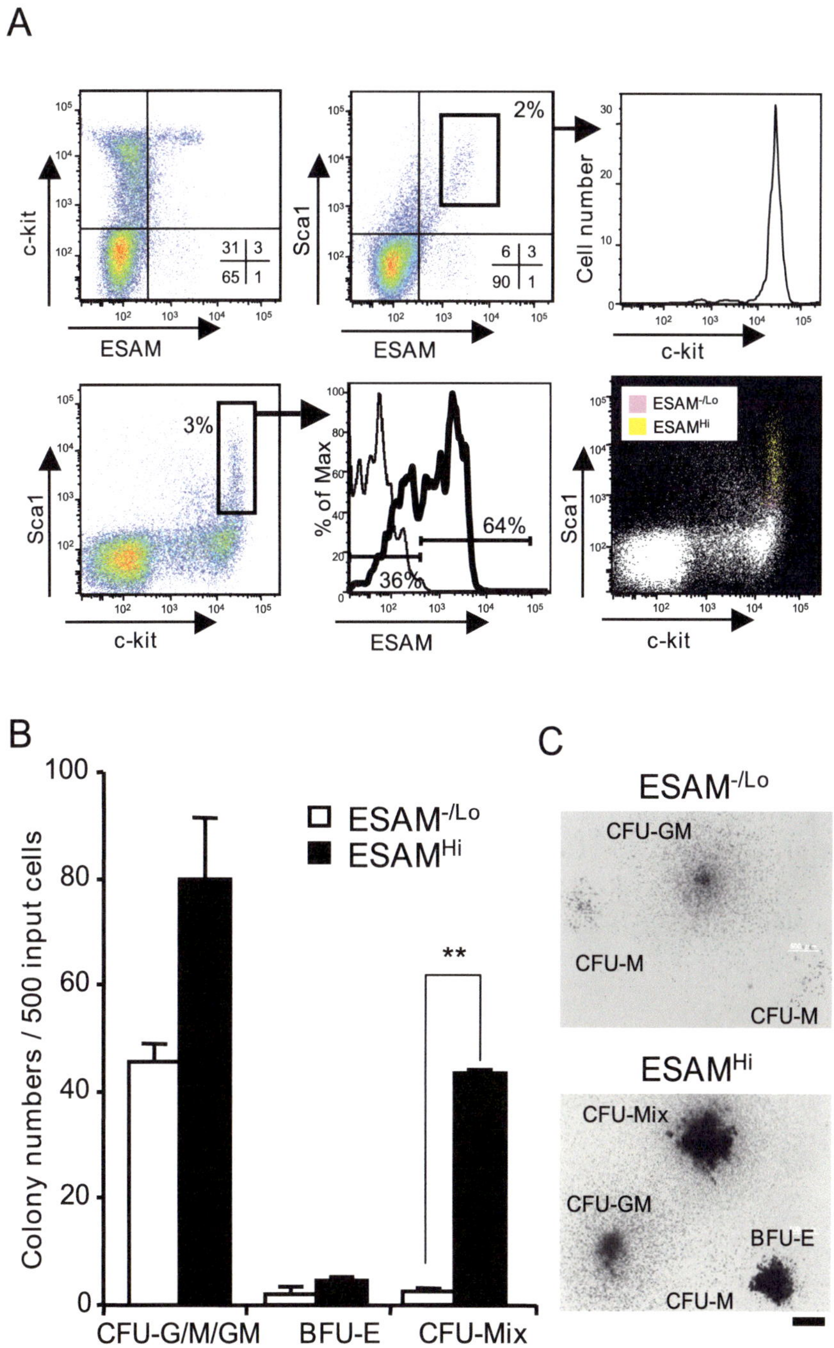

(A) Flow cytometry analysis was performed for mouse E14.5 fetal liver cells using anti-c-kit, anti- Sca1, and anti-ESAM Abs. ESAM$^{-/Lo}$ or ESAM^{Hi} cells of the Rag1/GFP^{-} ckit^{Hi} Sca1^{+} fraction were sorted and subjected to methylcellulose colony formation assay. Numbers of CFUs (B) and morphology of the colonies (C) are shown. (Modified from reference Yokota et al., 2009)

Fig. 1. ESAM expression on the HSC-enriched population of mouse fetal liver

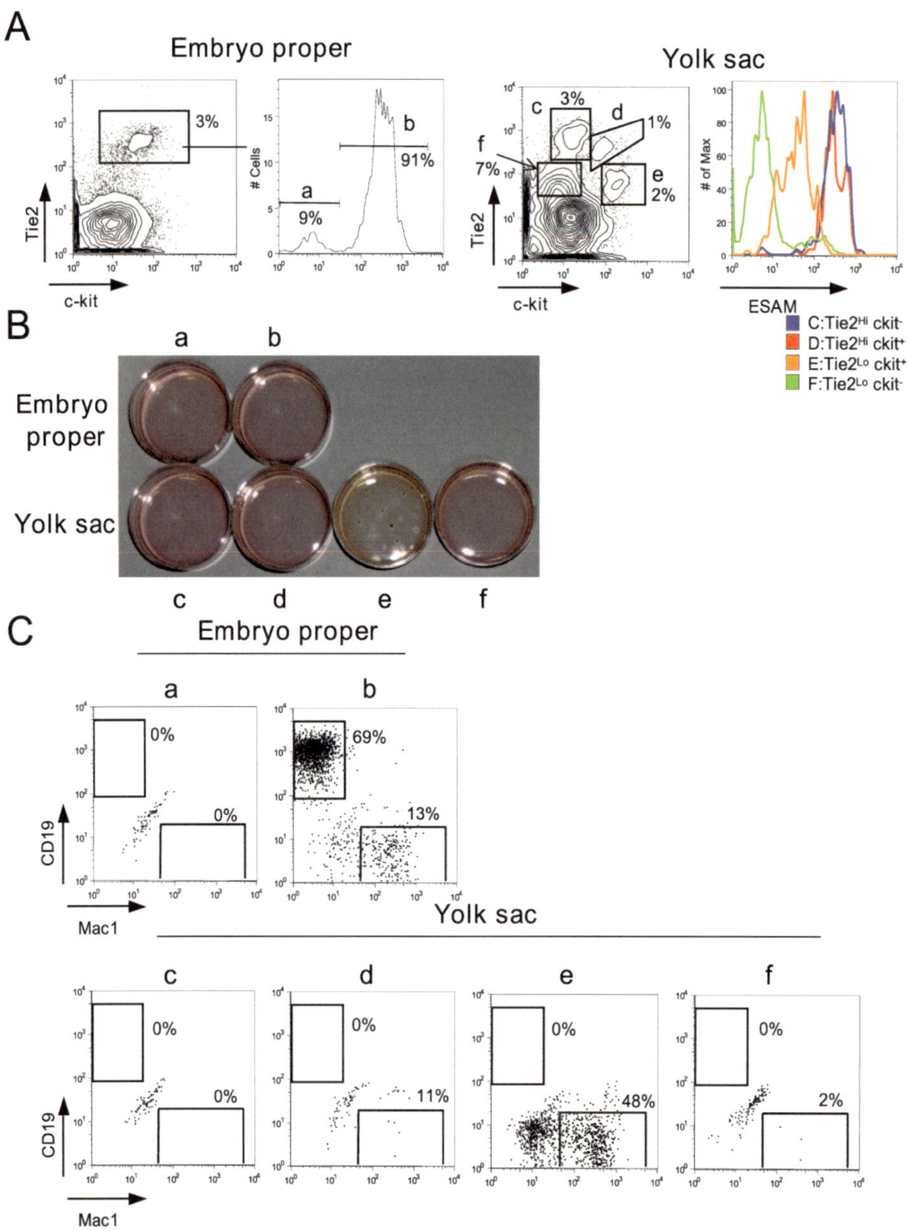

Yolk sac or the caudal half of embryo proper were obtained and pooled from E9.5 embryos of wild type C57B6 mice. The obtained cells were stained with the anti-ESAM Ab followed by goat anti-rat IgG-FITC, anti-c-kit-APC, anti-Tie2-PE, and 7AAD. (A) The profiles of Tie2 and c-kit expression are shown in the left panels. In the right panels, ESAM expression in each gate is shown in histograms. The sorted fractions were labeled with "a" to "f". The sorted cells were subjected to methylcellulose colony formation assay (B) and tested in the MS5 co-culture system (C). (Modified from reference Yokota et al., 2009)

Fig. 2. Yolk sac hematopoietic cells differ from those in the embryo proper with respect to ESAM expression and lymphopoietic activity.

ESAM expression was also detected on HSC within the Lin- c-kit^{High} Sca1^{+} fraction in adult bone marrow. Interestingly, while the expression level was slightly decreased in the adolescent period, it was up-regulated again in aged mice. In addition, Ooi et al showed that the ESAM^{+} Lin^{-} Sca1^{+} gating could more effectively enrich adult bone marrow for the long-term reconstituting HSC than the conventional LSK gating, and that ESAM expression in HSC is conserved between different mouse strains (Ooi et al., 2009). Based on these observations, we conclude that ESAM serves as an effective and durable marker for HSC throughout life in mice.

The importance of ESAM as a HSC marker has been further enhanced by the findings that its expression in HSC is conserved between mouse and man. Ooi et al detected abundant ESAM transcripts in human cord blood CD34^{+} CD38^{-} Lin^{-} Thy1/CD90^{+} cells (Ooi et al., 2009). Furthermore, by using a rabbit anti-human polyclonal ESAM antibody and flow cytometry, we also detected ESAM expression on human cord blood CD34^{+} cells (Figure 3). The intensity of ESAM expression, however, was similar between CD34^{+} CD38^{-} and CD34^{+} CD38^{+} cells, suggesting that the ESAM^{+} gate covers committed as well as non-committed hematopoietic progenitors. ESAM expression might serve as an alternative marker to CD34 for the selection of hematopoietic stem/progenitor cells in human. It is noteworthy that, although majority of human cord blood CD34^{-} CD38^{+} fraction were negative for the ESAM staining, the fraction contains a small ESAM^{+} population. Further study is necessary to characterize those ESAM^{+} CD34^{-} cells.

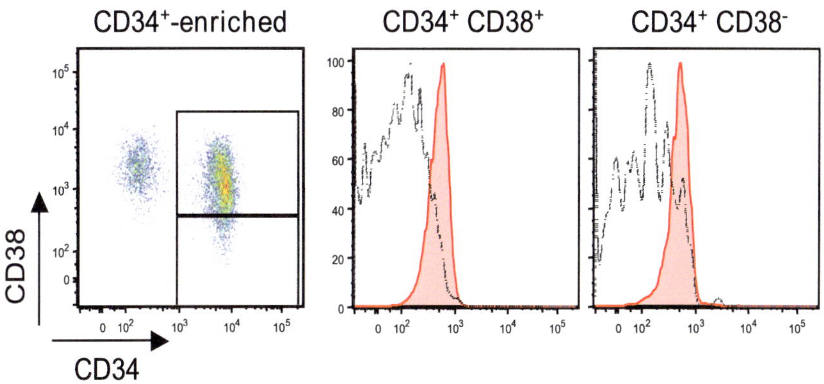

CD34^{+} cells were firstly enriched from cord blood mononuclear cells by magnetic beads conjugated with an anti-human CD34 antibody, and then stained with anti-CD34, anti-CD38, and anti-ESAM antibodies. The left panel shows CD34 and CD38 expression profile of the CD34^{+} enriched population. The middle and right panels indicate ESAM expression (red tinted) on CD34^{+} CD38^{+} or CD34^{+} CD38^{-} cells, respectively. Dot lines show background staining levels with control IgG for an anti-ESAM antibody.

Fig. 3. ESAM expression on human cord blood CD34^{+} cells

8. Conclusion

In this chapter, we summarized 2 decades achievements for the identification of HSC and introduced our recent discovery of human ESAM as a new HSC marker. Although it is possible in mouse to purify the long-term multi-lineage HSC with high efficiency, characterization of human HSC has lagged behind partly due to insufficient information about their cell surface antigens. As a new tool, ESAM expression might contribute to improve the purification strategy of human HSC, not only from human hematopoietic tissues but also from cultures of embryonic stem cells or induced-pluripotent stem cells, for therapeutic application. In addition, our findings will be beneficial to basic studies of HSC themselves. Since ESAM is expressed on mesoderm precursors for HSC in the pSp/AGM region as well as expanding HSC in fetal liver, its expression will be potentially useful to trace developing HSC back to their origin and localize them in distinct hematopoietic organs. In addition, up-regulation of ESAM expression on aged HSC may provide some insights regarding molecular mechanisms of HSC senescence. We are expecting that a new HSC marker ESAM could help future studies regarding HSC in many directions.

9. References

Balazs AB, Fabian AJ, Esmon CT, & Mulligan RC. (2006). Endothelial protein C receptor (CD201) explicitly identifies hematopoietic stem cells in murine bone marrow. *Blood* Vol.107, pp 317-2321.

Baum CM, Weissman IL, Tsukamoto AS, Buckle AM, & Peault B. (1992). Isolation of a candidate human hematopoietic stem-cell population. *Proc Natl Acad Sci USA.* Vol. 89, pp 2804-2808.

Berenson RJ, Andrews RG, Bensinger WI, Kalamasz D, Knitter G, Buckner CD, & Bernstein ID. (1988). Antigen CD34+ marrow cells engraft lethally irradiated baboons. *J Clin Invest.* Vol. 81, pp 951-955.

Berenson RJ, Bensinger WI, Hill RS, Andrews RG, Garcia-Lopez J, Kalamasz DF, Still BJ, Spitzer G, Buckner CD, Bernstein ID, & Thomas ED. (1991). Engraftment after infusion of CD34+ marrow cells in patients with breast cancer or neuroblastoma. *Blood* Vol. 77, pp 1717-1722.

Bhatia M, Bonnet D, Murdoch B, Gan OI, & Dick JE. (1998). A newly discovered class of human hematopoietic cells with SCID-repopulating activity. *Nat Med.* Vol. 4, pp 1038-1045.

Christensen JL, & Weissman IL. (2001). Flk-2 is a marker in hematopoietic stem cell differentiation: a simple method to isolate long-term stem cells. *Proc Natl Acad Sci USA.* Vol. 98, pp 14541-14546.

de Bruijn MF, Speck NA, Peeters MC, & Dzierzak E. (2000). Definitive hematopoietic stem cells first develop within the major arterial regions of the mouse embryo. *EMBO J.* Vol. 19. pp 2465-2474.

Doulatov S, Notta F, Eppert K, Nguyen LT, Ohashi PS, & Dick JE. (2010). Revised map of the human progenitor hierarchy shows the origin of macrophages and dendritic cells in early lymphoid development. *Nat Immunol.* Vol. 11, pp 585-593.

Foudi A, Hochedlinger K, Van Buren D, Schindler JW, Jaenisch R, Carey V, & Hock H. (2009). Analysis of histone 2B-GFP retention reveals slowly cycling hematopoietic stem cells. *Nat Biotechnol.* Vol. 27, pp 84-90.

Gallacher L, Murdoch B, Wu DM, Karanu FN, Keeney M, & Bhatia M. (2000). Isolation and characterization of human CD34-Lin- and CD34+Lin- hematopoietic stem cells using cell surface markers AC133 and CD7. *Blood* Vol. 95, pp 2813-2820.

Goodell MA, Brose K, Paradis G, Conner AS, & Mulligan RC. (1996). Isolation and functional properties of murine hematopoietic stem cells that are replicating in vivo. *J Exp Med.* Vol. 183, pp 1797-1806.

Goodell MA, Rosenzweig M, Kim H, Marks DF, DeMaria M, Paradis G, Grupp SA, Sieff CA, Mulligan RC, & Johnson RP. (1997). Dye efflux studies suggest that hematopoietic stem cells expressing low or undetectable levels of CD34 antigen exist in multiple species. *Nat Med.* Vol. 3, pp 1337-1345.

Gunji Y, Nakamura M, Osawa H, Nagayoshi K, Nakauchi H, Miura Y, Yanagisawa M, & Suda T. (1993). Human primitive hematopoietic progenitor cells are more enriched in KIT^{low} cells than in KIT^{high} cells. *Blood* Vol. 82, pp 3283-3289.

Hashiyama M, Iwama A, Ohshiro K, Kurozumi K, Yasunaga K, Shimizu Y, Masuho Y, Matsuda I, Yamaguchi N, & Suda T. (1996). Predominant expression of a receptor tyrosine kinase, TIE, in hematopoietic stem cells and B cells. *Blood* Vol. 87, pp 93-101.

Hirata K, Ishida T, Penta K, Rezaee M, Yang E, Wohlgemuth J, & Quertermous T. (2001). Cloning of an immunoglobulin family adhesion molecule selectively expressed by endothelial cells. *J Biol Chem.* Vol. 276, pp 16223-16231.

Igarashi H, Gregory SC, Yokota T, Sakaguchi N, & Kincade PW. (2002). Transcription from the RAG1 locus marks the earliest lymphocyte progenitors in bone marrow. *Immunity* Vol. 17, pp 117-130.

Kiel MJ, Yilmaz OH, Iwashita T, Yilmaz OH, Terhorst C, & Morrison SJ. (2005). SLAM family receptors distinguish hematopoietic stem and progenitor cells and reveal endothelial niches for stem cells. *Cell* Vol. 121, pp 1109-1121.

Larochelle A, Savona M, Wiggins M, Anderson S, Ichwan B, Keyvanfar K, Morrison SJ, & Dunbar CE. (2011). Human and rhesus macaque hematopoietic stem cells cannot be purified based only on SLAM family markers. *Blood* Vol. 117, pp1550-1554.

Matsubara A, Iwama A, Yamazaki S, Furuta C, Hirasawa R, Morita Y, Osawa M, Motohashi T, Eto K, Ema H, Kitamura T, Vestweber D, & Nakauchi H. (2005). Endomucin, a CD34-like sialomucin, marks hematopoietic stem cells throughout development. *J Exp Med.* Vol. 202, pp 1483-1492.

Matsuoka S, Ebihara Y, Xu M, Ishii T, Sugiyama D, Yoshino H, Ueda T, Manabe A, Tanaka R, Ikeda Y, Nakahata T, & Tsuji K. (2001). CD34 expression on long-term repopulating hematopoietic stem cells changes during developmental stages. *Blood* Vol. 97, pp 419-425.

Matsuzaki Y, Kinjo K, Mulligan RC, & Okano H. (2004). Unexpectedly efficient homing capacity of purified murine hematopoietic stem cells. *Immunity* Vol. 20, pp 87-93.

McKenzie JL, Takenaka K, Gan OI, Doedens M, & Dick JE. (2007). Low rhodamine 123 retention identifies long-term human hematopoietic stem cells within the $Lin^-CD34^+CD38^-$ population. *Blood* Vol. 109, pp 543-545.

McKinney-Freeman SL, Naveiras O, Yates F, Loewer S, Philitas M, Curran M, Park PJ, & Daley GQ. (2009). Surface antigen phenotypes of hematopoietic stem cells from embryos and murine embryonic stem cells. *Blood* Vol. 114, pp 268-278

Mikkola HKA, & Orkin SH. (2006). The journey of developing hematopoietic stem cells. *Development* Vol. 133, pp 3733-3744.

Nasdala I, Wolburg-Buchholz K, Wolburg H, Kuhn A, Ebnet K, Brachtendorf G, Samulowitz U, Kuster B, Engelhardt B, Vestweber D, & Butz S. (2002). A transmembrane tight junction protein selectively expressed on endothelial cells and platelets. *J Biol Chem.* Vol. 277, pp 16294-16303.

Notta F, Doulatov S, Laurenti E, Poeppl A, Jurisica I, & Dick JE. (2011). Isolation of single human hematopoietic stem cells capable of long-term multilineage engraftment. *Science* Vol. 333, pp 218-221.

Oberlin E., Tavian M., Blazsek I., & Péault B. (2002). Blood-forming potential of vascular endothelium in the human embryo. *Development* Vol. 129, pp 4147-4157.

Ogawa M, Matsuzaki Y, Nishikawa S, Hayashi S, Kunisada T, Sudo T, Kina T, Nakauchi H, & Nishikawa S-I. (1991). Expression and function of c-kit in hemopoietic progenitor cells. *J Exp Med.* Vol. 174, pp 63-71.

Ogawa M. (2002). Changing phenotypes of hematopoietic stem cells. *Exp Hematol.* Vol. 30, pp 3-6.

Ooi AG, Karsunky H, Majeti R, Butz S, Vestweber D, Ishida T, Quertermous T, Weissman IL, & Forsberg EC. (2009). The adhesion molecule esam1 is a novel hematopoietic stem cell marker. *Stem Cells* Vol. 27, pp 653-661.

Osawa M, Hanada K, Hamada H, & Nakauchi H. (1996). Long-term lymphohematopoietic reconstitution by a single CD34-low/negative hematopoietic stem cell. *Science* Vol. 273, pp 242-245.

Randall TD, Lund FE, Howard MC, & Weissman IL. (1996). Expression of murine CD38 defines a population of long-term reconstituting hematopoietic stem cells. *Blood* Vol. 87, pp 4057-4067.

Randall TD, & Weissman IL. (1997). Phenotypic and functional changes induced at the clonal level in hematopoietic stem cells after 5-fluorouracil treatment. *Blood* Vol. 89, pp 3596-3606.

Shultz LD, Ishikawa F, & Greiner DL. (2007). Humanized mice in translational biomedical research. *Nat Rev Immunol.* Vol. 7, pp 118-130.

Spangrude GJ, Heimfeld S, & Weissman IL. (1988). Purification and characterization of mouse hematopoietic stem cells. *Science* Vol. 241, pp 58-62.

Spangrude GJ, & Brooks DM. (1993). Mouse strain variability in the expression of the hematopoietic stem cell antigen Ly-6A/E by bone marrow cells. *Blood* Vol. 82, pp 3327-3332.

Sugano Y, Takeuchi M, Hirata A, Matsushita H, Kitamura T, Tanaka M, & Miyajima A. (2008). Junctional adhesion molecule-A, JAM-A, is a novel cell-surface marker for long-term repopulating hematopoietic stem cells. *Blood* Vol. 111, pp 1167-1172.

Tajima F, Deguchi T, Laver JH, Zeng H, & Ogawa M. (2001). Reciprocal expression of CD38 and CD34 by adult murine hematopoietic stem cells. *Blood* Vol. 97, pp 2618-2624.

Takakura N, Huang XL, Naruse T, Hamaguchi I, Dumont DJ, Yancopoulos GD, & Suda T. (1998). Critical role of the TIE2 endothelial cell receptor in the development of definitive hematopoiesis. *Immunity* Vol. 9, pp 677-686.

Terstappen LW, Huang S, Safford M, Lansdorp PM, & Loken MR. (1991). Sequential generations of hematopoietic colonies derived from single nonlineage-committed $CD34^{+}CD38^{-}$ progenitor cells. *Blood* Vol. 77, pp 1218-1227.

Wang J, Kimura T, Asada R, Harada S, Yokota S, Kawamoto Y, Fujimura Y, Tsuji T, Ikehara S, & Sonoda Y. (2003). SCID-repopulating cell activity of human cord blood-derived $CD34^{-}$ cells assured by intra-bone marrow injection. *Blood* Vol. 101, pp 2924-2931.

Wilson A, Laurenti E, Oser G, van der Wath RC, Blanco-Bose W, Jaworski M, Offner S, Dunant CF, Eshkind L, Bockamp E, Lió P, Macdonald HR, & Trumpp A. (2008). Hematopoietic stem cells reversibly switch from dormancy to self-renewal during homeostasis and repair. *Cell* Vol. 135, pp1118-1129.

Yin AH, Miraglia S, Zanjani ED, Almeida-Porada G, Ogawa M, Leary AG, Olweus J, Kearney J, & Buck DW. (1997). AC133, a novel marker for human hematopoietic stem and progenitor cells. *Blood* Vol. 90, pp 5002-5012.

Yokota T, Kouro T, Hirose J, Igarashi H, Garrett KP, Gregory SC, Sakaguchi N, Owen JJ, & Kincade PW. (2003). Unique properties of fetal lymphoid progenitors identified according to RAG1 gene expression. *Immunity* Vol. 19, pp 365-375.

Yokota T, Huang J, Tavian M, Nagai Y, Hirose J, Zúñiga-Pflücker JC, Péault B, & Kincade PW. (2006). Tracing the first waves of lymphopoiesis in mice. *Development* Vol. 133, pp 2041-2051.

Yokota T, Oritani K, Butz S, Kokame K, Kincade PW, Miyata T, Vestweber D, & Kanakura Y. (2009). The endothelial antigen ESAM marks primitive hematopoietic progenitors throughout life in mice. *Blood* Vol. 113, pp 2914-2923.

Part 2

Regulation of Hematopoietic Stem Cells

5

Interferon Regulatory Factor-2 Regulates Hematopoietic Stem Cells in Mouse Bone Marrow

Atsuko Masumi[1], Shoichiro Miyatake[2],
Tomoko Kohno[3] and Toshifumi Matsuyama[3]
[1]*Department of Safety Research on Blood and Biological Products, National Institute of Infectious Diseases, Tokyo,*
[2]*Laboratory of Self Defense Gene Regulation, Tokyo Metropolitan Institute of Medical Science, Tokyo,*
[3]*Department of Molecular Microbiology and Immunology, Nagasaki University Graduate School of Biomedical Sciences, Nagasaki,*
Japan

1. Introduction

Hematopoiesis is regulated by intrinsic gene-regulatory networks, ensuring the rapid production of differentiated blood cells for the immediate needs of embryos and generation of definitive hematopoietic stem cells (HSCs) that are required for life-long hematopoiesis. Homeostasis in bone marrow is dependent on the ability of HSCs to faithfully self-renew and to generate progenitor cells that undergo limited proliferation and give rise to terminally differentiated cells in the peripheral blood. HSCs are specialized to give rise to all elements of the blood system throughout life (Orkin & Zon, 2002, 2008a, 2008b) and are capable of self-renewal and differentiation into various lineages of the hematopoietic system to form all types of blood cells. Self-renewal is a tightly controlled process through which stem cells divide and generate daughter stem cells with properties identical to those of the mother cells. However, under certain conditions, HSCs differentiate into progenitor cells with less ability to self-renew. Since the discovery of stem cells, intense research aimed at understanding the genetic and molecular bases of self-renewal has identified candidate regulatory factors involved in the process of HSC self-renewal. These include cell-intrinsic regulators, such as transcription factors, signal transducers, cell-cycle inhibitors and surface receptors and cell-extrinsic regulators, such as the bone marrow niche and cytokines (He et al., 2009).

Interferon (IFN) is produced by cells of the immune system in response to challenges by agents, such as viruses, bacteria and tumor cells. IFNs suppress viral replication, have immunomodulatory activities and are used clinically to treat viral diseases and malignancies, such as chronic myeloid leukemia (CML)(Stark, 1998). Type I IFNs are induced by the genomes of many RNA viruses, and this induction can be mimicked by the double-stranded RNA mimetic polyinosinic-polycytidylic acid (poly [I:C]) (Darnell et al.

1994, Pichlmair et al. 2007). However, under steady-state conditions in the absence of infection, small amount of IFN are produced constitutively (Taniguchi & Takaoka, 2001). Recently, Essers et al demonstrated that chronic activation of IFN-α pathway impairs function of HSCs and acute IFN-α treatment promotes the proliferation of dormant HSCs in vivo, and the possibility for new application of type I IFN to target cancer stem cells is expected (Essers et al. 2009) .

Interferon regulatory factors (IRFs) constitute a family of transcription factors involved in regulating the development and functions of the immune system (Honda et al., 2006; Taniguchi et al., 2001). Interferon regulatory factor-2 (IRF-2) is a transcriptional repressor in the interferon system and is thought to function by competing with IRF-1. While IRF-2 acts as a repressor for interferon production, IRF-2 exists ubiquitously and is a positive regulator for H4, vascular adhesion molecule-1 (VCAM-1), CIITA, gp91 phox, Fas ligand, TPO receptor (Vaughan et al.1995, Jesse et al. 1998 , Xi et al. 1999, Luo & Skalnik 1996, Chow et al. 2000, Stellacci et al. 2004, Masumi et al, 2001). Previously, we demonstrated that IRF-2 expression into mouse bone marrow hematopoietic stem/progenitor cells induced megakaryopoiesis through CD41 promoter activation in an inflammatory states (Masumi et al. 2009). IRF-2 regulates cell growth and differentiation through the target gene promoters.

There are several studies of hematopoietic approaches using IRF-2-/- mice. The physiological role in lymphoid and hematopoietic development has been investigated in IRF-2-/-mice, in which a general bone marrow suppression of hematopoiesis and B lymphopoiesis has been reported (Matsuyama et al. 1993). Recently, a marked reduction of hematopoietic stem cells in IRF-2-/- mice involving a type I interferon-dependent mechanism was reported (Sato et al. 2009). The population of bone marrow Lin-c-Kit+Sca-1+ (KSL) cells is increased in IRF-2-/- mice because of the general enhancement of Sca-1-positive cells.

Herein, we show that an enhanced population of Sca-1-positive cells and reduced HSC activity in the Lin-Sca-1+c-kit+ fraction were detected in IRF-2-/-mouse bone marrow cells. HSC abnormalities in IRF-2-/- mice have been demonstrated to be due to elevated type I IFN signaling (Sato et al. 2009). IFN signaling enhances the Sca-1 expression and cell cycle progression of HSCs. It was shown that chronic IFN signaling enhances cell cycle progression of HSCs in IRF-2-/-mice, resulting in the loss of quiescent HSCs. However, our results reveal unknown HSC markers in bone marrow from IRF-2-/- mice. Our present findings demonstrate that IRF-2 acts on long-term (LT)-HSCs, not only through protective type I IFN responses, but also by directly regulating HSC cell-surface molecules.

2. Hematopoietic stem cells in bone marrow derived from interferon regulatory factor-2-deficient mice

To analyze the expression of IRFs in mouse bone marrow cells, Gr1/Mac1-positive, B220-positive, Ter119-positive/lineage (Lin)-negative and KSL (c-kit+Sca-1+Lin-) cells were isolated from mouse bone marrow cells by flow cytometry. Real-time polymerase chain reaction (PCR) analysis and in situ hybridization showed that IRF-2 was present in especially high levels in the CD34-KSL fraction compared with fractions from other lineages (Masumi et al., 2009). The CD34-KSL cells from mouse bone marrow were stained with anti-IRF-2 antibody and DAPI (Fig. 1).

Fig. 1. CD34-KSL cells were stained with anti-IRF-2 as the primary antibody and then with anti-rabbit Alexa594 as the secondary antibody. Concurrently, cells were stained with DAPI.

2.1 Isolation and characterization of KSL cells

We showed that IRF-2 was highly localized in the mouse HSCs (CD34-Lin-c-kit+Sca-1+) in Fig. 1. To examine the role of IRF-2 in mouse hematopoietic stem cells, we isolated Lin-c-kit+sca-1+ (KSL) cells from the bone marrow of IRF-2-/-mice (Fig. 2A). The IRF-2-/- mice had a larger population of KSL cells than did wild-type mice because of enhanced expression Sca-1, which is downstream of interferon-α-receptor (IFNAR)-STAT1 signaling. Enhanced type I IFN signaling in IRF-2-/- mice induces Sca-1. Sca-1 cell-surface glycoprotein is used routinely as a marker of adult HSCs, allowing a >100-fold enrichment of these rare cells from the bone marrow of adult mice. The Sca-1 protein is encoded by the Ly-6A/E gene. This protein is highly inducible by IFNs -α, –β, and -γ, tumor necrosis factor (TNF) and interleukin-1 (IL-1) (Khan et al., 1990, 1993). The presence of a consensus sequence for IFN-γ-responsive elements has been reported to be localized to the Ly-6A/E genes promoter (Ma et al., 2001). Ito et al investigated that competitive repopulation assay using HSC from Sca-1-deficient mice and colony formation assay for Sca-1-deficient bone marrow. They demonstrated that Sca-1 is required for regulating HSC self-renewal and development of committed progenitor cells, megakaryocytes, and platelets (Ito et al. 2003). Bradfute et al investigated the effect of Sca-1 on HSC function and demonstrated that Sca-1 affects c-kit expression, the lineage fate of peripheral blood cells after transplantation, and may be dispensable for HSC self-renewal (Bradfute et al., 2005).We also observed higher populations of KSL cells in the spleens of IRF-2-/- mice (data not shown).

HSCs are contained within the CD150+CD48- population of KSL cells. There were far fewer CD150+CD48- cells in the KSL fraction in IRF-2-/- mice than in wild-type mice (0.97% versus 0.001%). In contrast, within the CD150+ CD48-Lin- subset, the fraction of KSL cells in wild-type mice and IRF-2-/- mice was 31.58% and 0.001%, respectively. Thus, the population of CD150+CD48-KSL cells was very low in IRF-2-/- mice (Fig. 2B). IRF-2 is known to be a transcription factor that attenuates type I IFN (IFN-α/IFN-β) signaling as indicated by the up-regulation of IFN-inducible genes in IRF-2-/-mice. IFN-α/IFN-β produced by plasmacytoid dendritic cells (DCs) in IRF-2-/- mice may have stimulated HSC proliferation, which resulted in loss of stem cells (Fig. 2C).

The number of KSL side-population (SP) cells was much lower in bone marrow cells from IRF-2-/- mice compared to those from wild-type mice (Fig. 3). We observed an increased cell number in the KSL fraction and a great reduction in the number of HSCs in the KSL fraction among the bone marrow cells of IRF-2-/- mice.

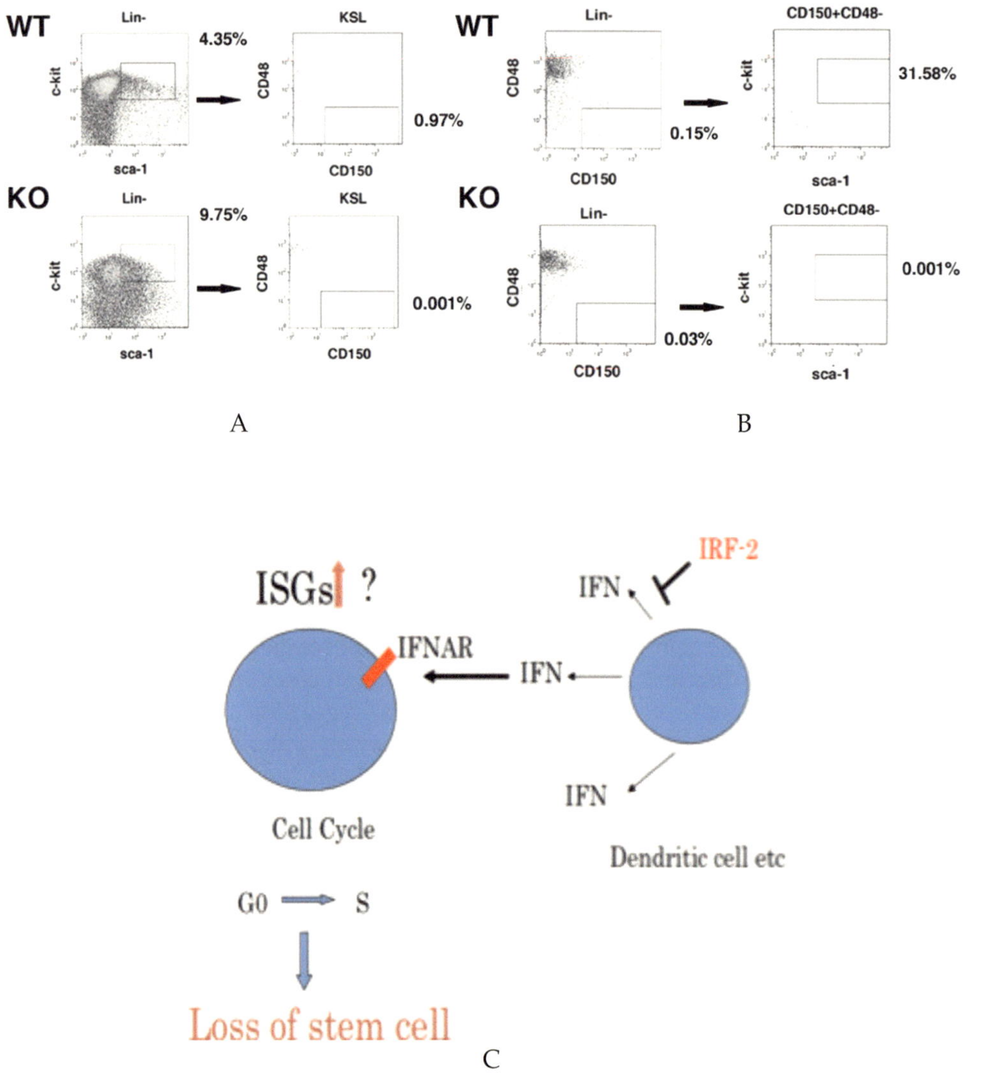

Fig. 2. (A) Lin-c-kit+sca-1+ cells were isolated from wild-type (WT) and IRF-2-/- bone marrow cells (KO), and then Lin-CD48-CD150+ cells were isolated from the KSL fraction. (B) Lin-c-kit+sca-1+ cells were isolated from Lin-CD48-CD150+ cells derived from the bone marrow cells of IRF-2-/- mice. (C) Chronic IFN stress model in KSL cells of IRF-2-/- mice. ISGs : Interferon-stimulated genes

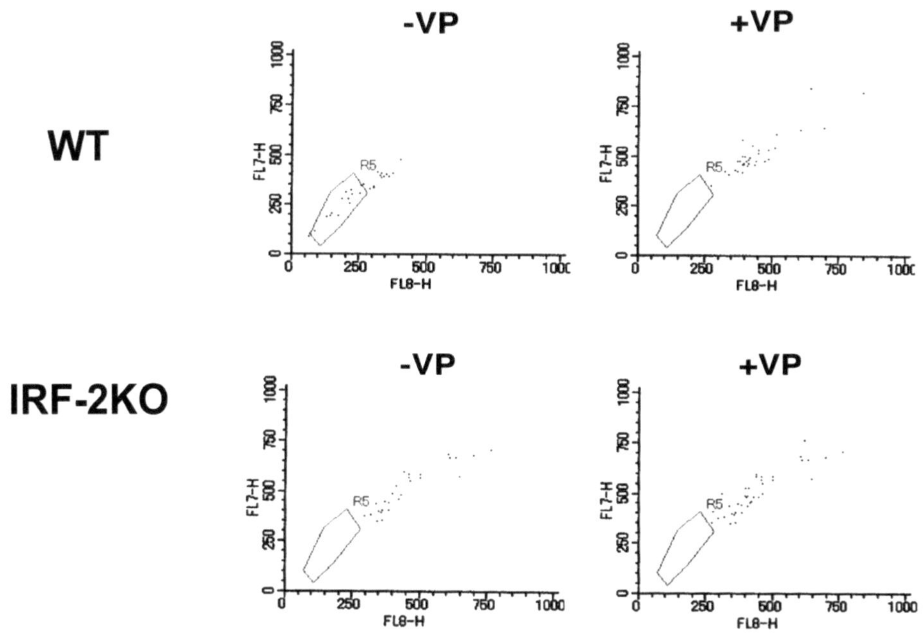

Fig. 3. Reduction in side population of KSL cells in the bone marrow cells from wild-type (WT) and IRF-2-/- mice (IRF-2KO). VP (verapamil) treatment eliminated the side population.

2.1.1 In vitro differentiation of KSL

To analyze the population of the KSL fraction in bone marrow from IRF-2-/- mice, an in vitro colony-forming assay was performed. The assay showed enhanced granulocyte/macrophage progenitor activity and reduced numbers of megakaryocyte progenitors in KSL derived from bone marrow of IRF-2-/- mice (Fig. 4). We did not see any difference for erythrocyte progenitors in the bone marrow derived-KSL cells in either wild-type or IRF-2-/-mice.

To compare ex vivo expansion of HSC between wild type and IRF-2-/-mice, KSL cells were plated at a density of 1 cell/well in Terasaki plates in 20 μL serum-free medium. The Terasaki single colony assay showed the significant colony formation activity in KSL cells from IRF-2-/- mice although its activity is less than that of wild-type mice (Fig. 5). In vitro culture assay, KSL cells from IRF-2-/- mice make colonies in the presence of cytokines, despite of near complete reduction of HSC population (Fig.2).

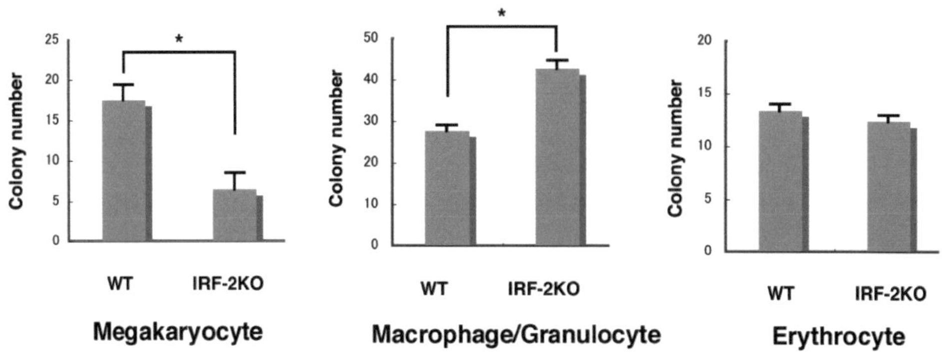

Fig. 4. Clonogenic progenitor assay. Two hundred KSL cells were subjected to a colony-forming unit-granulocyte/macrophage CFU-GM assay in methylcellulose medium M3231 (Stem Cell Technologies, Vancouver, BC, Canada) consisting of 1% methylcellulose, 30% fetal calf serum (FCS), 1% bovine serum albumin (BSA), 10 ng/mL stem cell factor (SCF), 25 ng/mL Flt ligand, 25 ng/mL thrombopoietin (TPO), 5 ng/mL IL-3 and 25 ng/mL granulocyte colony-stimulating factor (G-CSF). For the burst-forming units-erythroid (BFU-E) assay, 200 KSL cells were cultured in M3231 consisting of 1% methylcellulose, 30% FCS, 1% BSA, 50 ng/mL SCF, 50 ng/mL TPO and 5U/mL erythropoietin (EPO) for 7 days. To perform the mouse CFU-megakaryocytic (Mk) assays, $4x10^3$ KSL cells were mixed with Megacult-C 04900 together with 1.1 mg/mL collagen, 50 ng/mL TPO and 10 ng/mL IL-3 in 0.75 mL and added to the wells of chamber slides (177429, LAB-TEK Brand Products). Cells were cultured at 37°C in an incubator with an atmosphere of 5% CO_2 and >95% humidity for 7 days. The chamber slides were placed in acetone solution to fix the cells and dried, and the dried slides were then stained with acetylthiocholiniodide solution (Sigma, St Louis, MO). After they were stained with hematoxylin, megakaryocyte colonies were counted. $*P<0.05$ (Student *t* test). Data are representative of two independent experiments (mean ±SD).

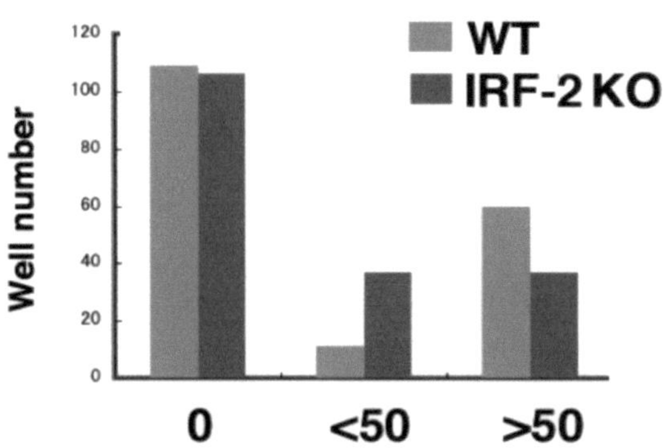

Fig. 5. Terasaki single colony assay. The KSL fraction from wild-type and IRF-2-/- bone marrow cells was fractionated to single cells in Terasaki plates by cell sorter (JSAN). Single cells were cultured with cytokines (hTPO, mSCF, mIL-3 and mFlt-3 ligand) containing 10% BSA and 2-mercaptethanol (0.01 M) in X-VIVO medium for 10 days. Colonies in each well were analyzed. <50 indicates wells that contain under 50 colonies in one well. >50 indicates wells that contain over 50 colonies. 0 indicates the wells containing no colony.

HSCs in wild-type mice were predominantly in a quiescent, intracellular Ki67-negative (icKi67 Hoechst low) G_0 phase. Quiescent cells were observed more frequently in KSL cells derived from IRF-2-/- mice, although the population of HSCs was much smaller than that of wild-type mice (Fig. 6).

2.1.2 Gene expression of KSL cells

Next, we investigated the KSL-specific gene expression in KSL cells derived from bone marrow cells of IRF-2-/-mice. Expression levels of GATA-2 and Tie2 in IRF-2-/- mice were similar to those of wild-type, but p57 expression was much lower than that of wild-type (Fig. 7A). Reduced p57 gene expression is associated with decreased numbers of HSCs in the KSL population, and the increased number of cells in G_0 phase may be associated with an increased frequency of quiescent KSL cells in bone marrow cells from IRF-2-/- mice (Fig. 6). When types I and II IFN were analyzed, expression of IFN-γ, but not of IFN-α and -β was decreased in the KSL fraction of IRF-2-/- mice under no stimulation (Fig. 7BC). Expression of Sca-1 was enhanced in the KSL fraction and in whole bone marrow in the IRF-2-/- mice. We examined PKR, TNF-a, adenosine deaminase 1 (ADAR1) expression which is known to be a suppressor of interferon signaling (Hartner et al. 2009) and Bmi1, which is down regulated in IRF-2 deficient HSC (Sato et al 2009). Expressions of ADAR1, PKR, Bmi1 and TNF-α were comparable between wild-type and IRF-2-deficient mice (Fig. 7C).

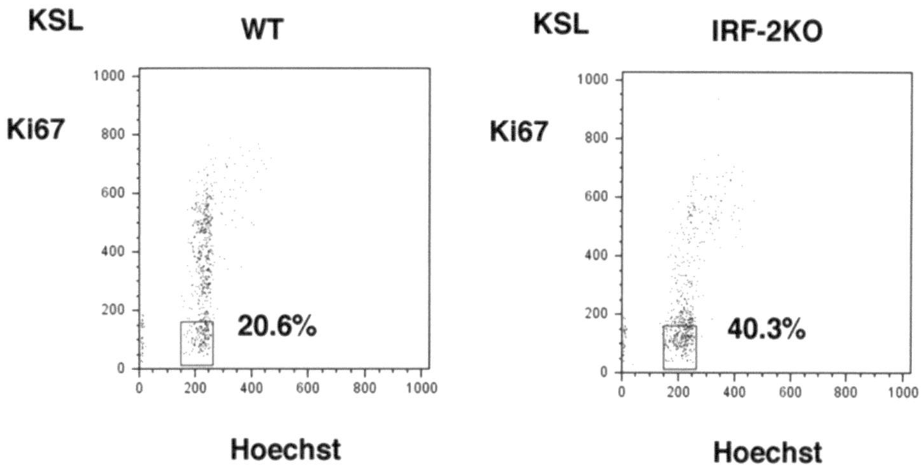

Fig. 6. Cell cycle analysis. Populations of KSL cells were isolated from bone marrow cells of IRF-2-/-mice and stained with Ki67 and Hoechst. Numbers indicate the percentage of cells in G_0 phase. Data are representative of two independent experiments.

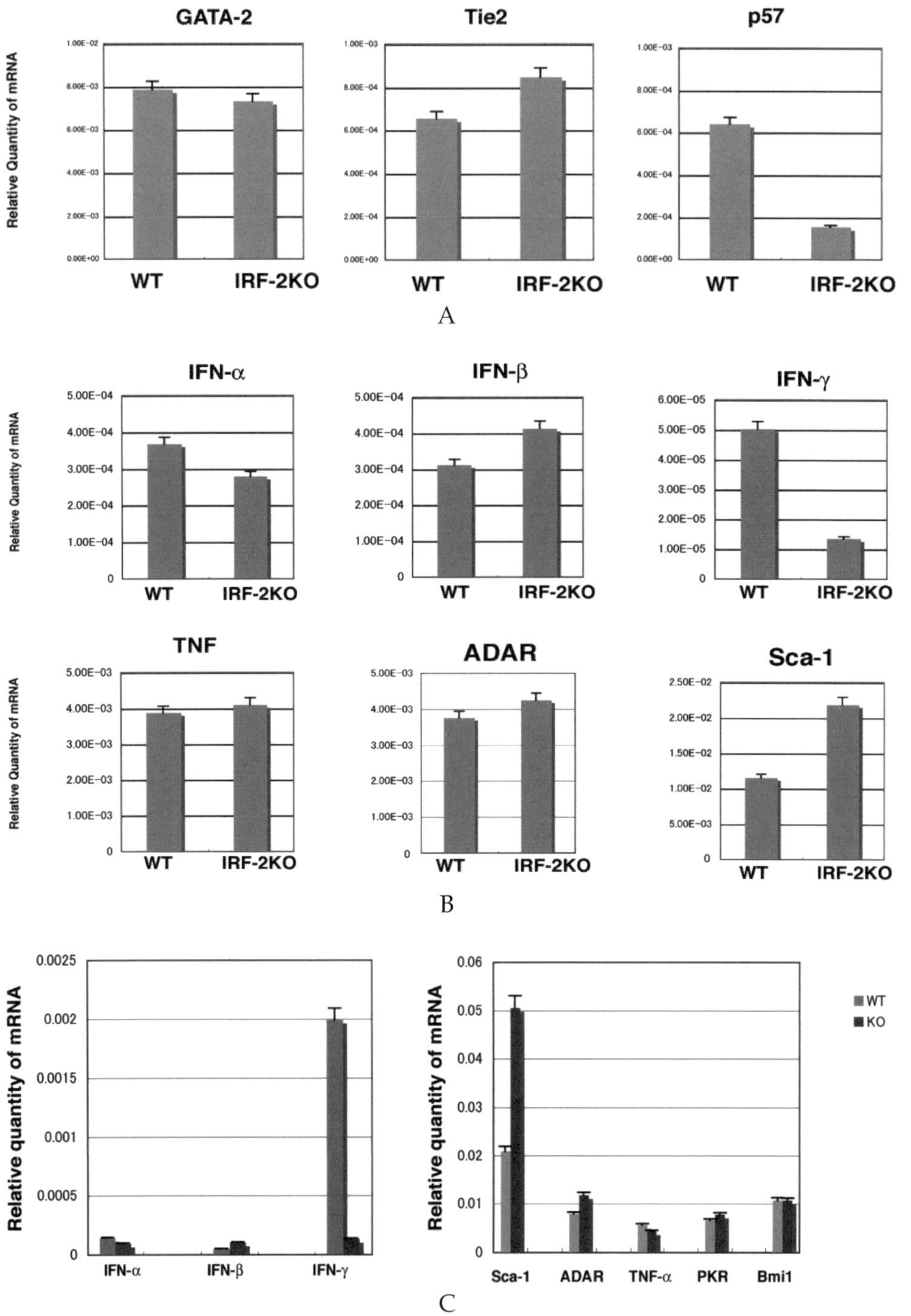

Fig. 7. Gene expression in KSL (A and B) and bone marrow cells (C) from wild-type and IRF-2-/- mice. Data represent the mean ± SD of triplicate reactions and are representative of two independent experiments.

2.1.3 Transplantation

To examine the functional properties of IRF-2-/- KSL cells, transplantation analysis was performed. In competitive repopulation assays, a constant number ($1x10^5$) of wild-type competitor cells was mixed with 1,500 KSL cells from IRF-2-/- mice and injected into lethally irradiated Ly5.1 mice. Engraftment analysis by peripheral blood chimerism (CD45.2 versus CD45.1 x CD45.2) at 4, 8 and 12 weeks after transplantation showed a profound deficit in wild-type recipient marrow, as peripheral blood elements derived from IRF-2-/- HSCs were progressively lost in favor of wild-type cells (Fig. 8A). Many more KSL cells could be engrafted into recipient mice if HSCs existed in the KSL fraction. However, a 24-fold larger number of cells also failed to rescue the recipients (Fig. 8B). When cells from the lineage-CD48 fraction were injected into recipients, no engraftment was shown in the recipients injected with cells from IRF-2-/- mice (Fig. 8C). The proportions of cells in the lineage-CD48 fraction isolated from both wild-type and IRF-2-/- mice were very similar (3.4% versus 2.9%). However, the number of CD150-positive cells was much lower in the IRF-2-/- mice compared to the wild-type mice (Fig.1).

To examine whether HSCs existed in fractions other than the KSL fraction of bone marrow cells in IRF-2-/- mice, whole bone marrow cells were injected into recipient mice and noncompetitive transplant assays were also performed. Transplantation of $1x10^5$ wild-type bone marrow cells is normally sufficient to rescue and fully repopulate the hematopoietic systems of all lethally irradiated recipients. This dose of cells from KO donors failed to rescue any recipients from lethal irradiation, indicating impaired self-renewal of HSCs in whole bone marrow cells derived from IRF-2-/- cells. A dose of 2×10^6 cells from IRF-2-/- mice could rescue recipients from lethal irradiation, although the engraftment efficiency was poorer than that of wild-type at 1 month after transplantation (Fig. 8D).

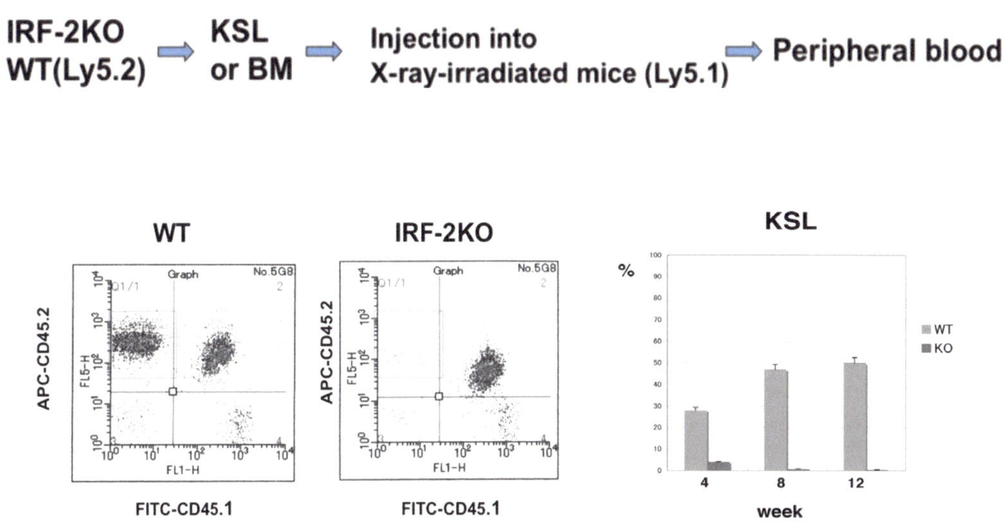

A

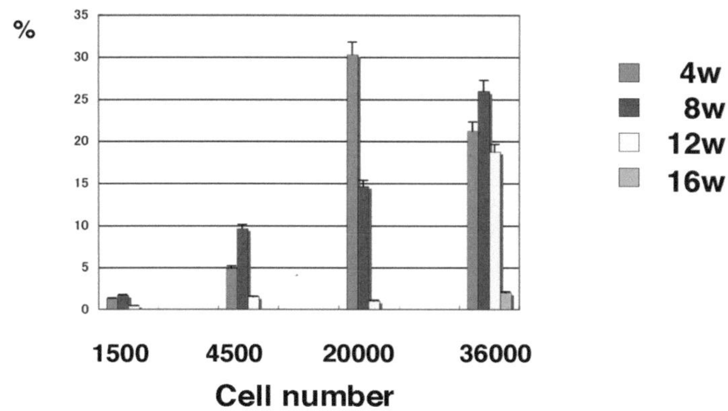

B

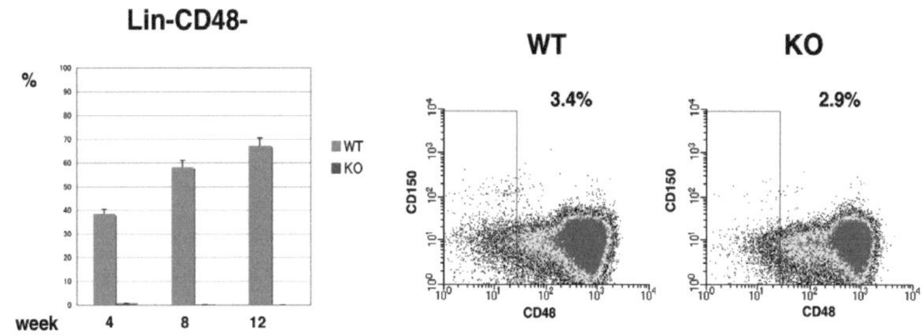

C

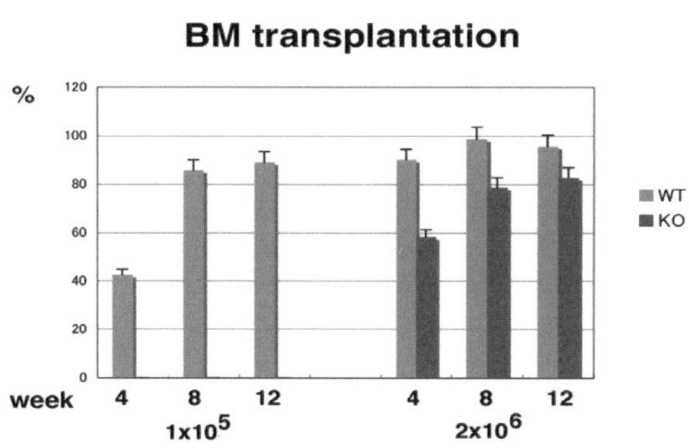

D

Fig. 8. Transplantation analysis. Cells in the KSL population were isolated from IRF-2-/- mice (Ly5.2) and injected into X-irradiated mice. Peripheral blood cells were analyzed 4, 8 and 12 weeks after transplantation. (A) Percentage of donor-derived KSL cells (1500-cell injection) in the blood. (B) Percentage of KSL derived from IRF-2-/- mice (injection of indicated cell numbers) in the blood. (C) Percentage of Lin-CD48- cells from bone marrow of wild-type and IRF-2-/- mice in the blood. (D) Percentage of donor-derived whole bone marrow cells in the blood. Cells were grown in one of three mice injected with $2x10^6$ whole bone marrow cells derived from IRF-2-/- mice. Representative data are shown from three independent experiments.

An IRF-2-expressing retrovirus (Masumi et al. 2009) was transduced into the KSL fraction of IRF-2-deficient mouse bone marrow cells. When these IRF-2-expressing KSL cells were injected into lethally irradiated recipients, no sufficient rescue of engraftment was observed (data not shown). KSL cells from IRF-2-deficient mice may be distinct from those of wild-type mice. IRF-2 expression does not contribute to rescue the HSC function in KSL cells from IRF-2-/- mice in vivo.

IRF-2-/- mice were previously reported to be more sensitive to 5-fluorouracil (5-FU) than IRF-2+/- mice because of a progressive decrease in functional HSCs in IRF-2-/- mice (Sato et al., 2009). We treated both IRF-2-/- and wild-type mice weekly with 5-FU. Four of the IRF-2-/- mice died after the initial injection, but one mouse lived after three injections. We conclude that quiescent HSCs are present in whole bone marrow from IRF-2-/- mice, although the IRF-2-/- mice are more sensitive than wild-type (Fig. 9). As seen in Fig. 8D and Fig. 9, HSC-like cells, which may be isolated using cell surface markers distinct from KSL cells, are thought to be present in IRF-2-/-mice bone marrow cells (Fig. 10).

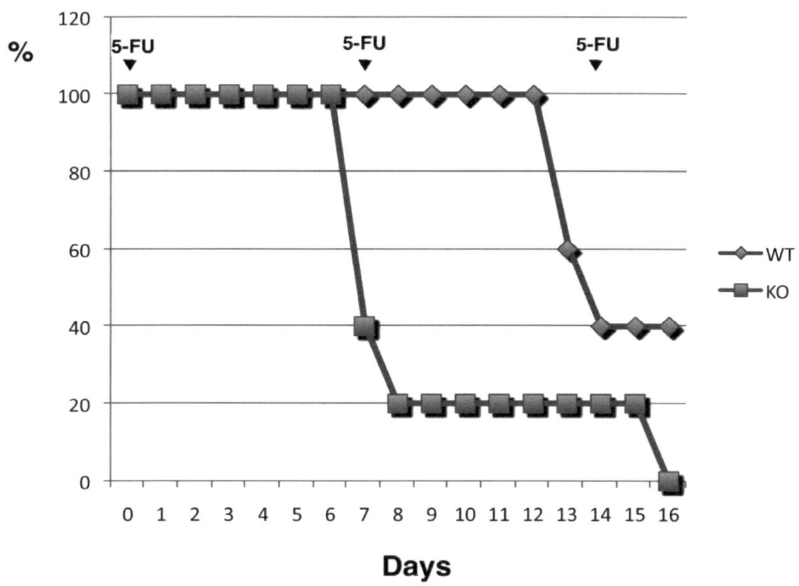

Fig. 9. Rate of survival (%) of wild-type and IRF-2-/- mice that were injected weekly with 150 mg/kg body weight of 5-FU (Sigma Chemical Co.); n= 5 for each group.

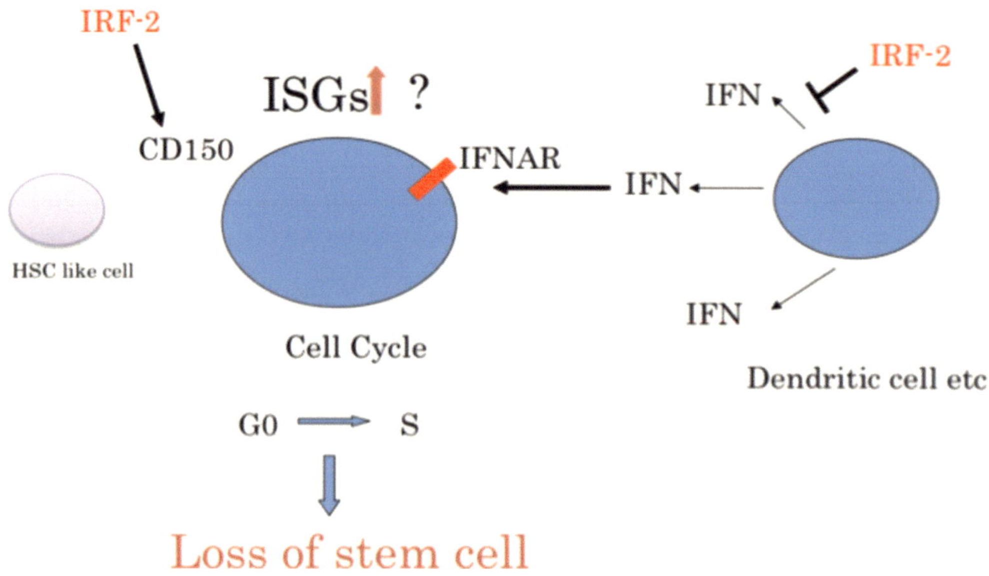

Fig. 10. HSC-like cells will be appeared in IRF-2-/- mice bone marrow cells (Refer to Fig.2C). IRF-2 may inhibit the down-regulation of CD150 gene expression by type I IFN.

2.2 Progenitors in bone marrow cells from IRF-2-/- mice

Next we analyzed the progenitor population in bone marrow cells from IRF-2-/- mice. Akashi et al proposed the model of major hematopoietic maturation pathways from HSCs (Akashi et al. 2000). According to his proposal, granulocyte/macrophage lineage progenitor (GMP), megakaryocyte/erythrocyte lineage progenitor (MEP), and common myeloid progenitor (CMP) and common lymphoid progenitor (CLP) were isolated from wild-type and IRF-2-deficient mice bone marrow. The frequencies of MEPs (Lin-ckit^{lo}sca-1-FcRg^{lo}CD34-) and CMPs (Lin-ckit^{lo}sca-1-FcRg^{lo}CD34+) were slightly decreased in IRF-2-deficient mice compared to wild-type mice. By contrast, GMPs (Lin-ckit^{lo}sca-1-FcRg^{high}CD34+) were slightly increased in IRF-2-deficient mice (Fig. 11A). The frequency of the CLP compartment in IRF-2-/- mouse bone marrow cells was lower than that from wild-type bone marrow (Fig. 11B).

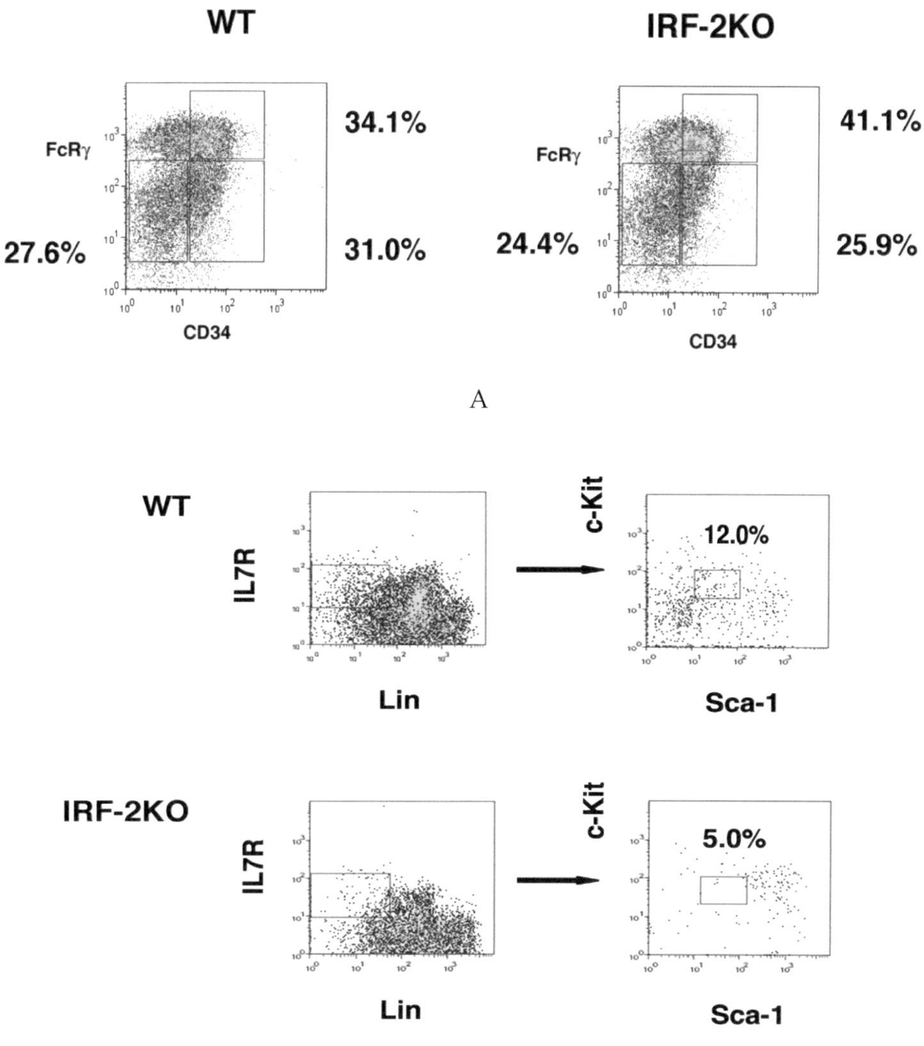

Fig. 11. (A) Lineage relationships among the myeloid progenitor subsets. MEPs (Lin-ckit^{lo}sca-1-FcRg^{lo}CD34-), CMPs (Lin-ckit^{lo}sca-1-FcRg^{lo}CD34+), and GMPs (Lin-ckit^{lo}sca-1-FcRg^{high}CD34+) are indicated. (B) Common lymphoid progenitors in bone marrow from IRF-2-/- mice.

2.2.1 Mouse colony-forming cell (CFC) assays with bone marrow derived from IRF-2-/- mice

The size of the GMP population from IRF-2-/- mice was higher than that from wild-type. We analyzed bone marrow and spleen cells from IRF-2-/- mice with CFU-GM assay. Colony numbers were higher in the bone marrow and spleen cells from IRF-2-/- mice, likely to the KSL population (Fig. 12 and Fig. 4).

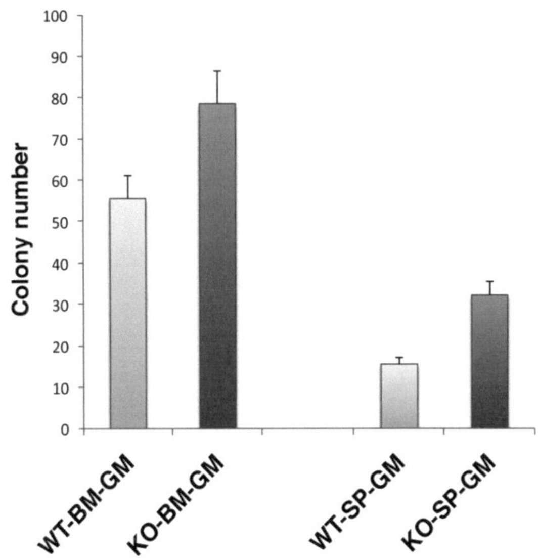

Fig. 12. Clonogenic progenitor assay with bone marrow and spleen cells from wild-type and IRF-2-/- mice. Assays performed as described in methods for Fig. 4. BM: bone marrow; SP: spleen

2.3 IRF-2 is required for bone marrow lymphopoiesis

To confirm the direct effect of IRF-2 deficiency in KSL cells, complementary DNA microarray analysis was performed on sorted Lin-c-Kit+Sca-1+ cells from bone marrow of wild-type and IRF-2-/- mice. This analysis showed that the up-regulated genes included IFN-inducible genes, such as Ly6s and Ifits, and the pre-B lymphocyte gene family (data not shown). As shown in Fig.13, when bone marrow B cell progenitors are analyzed, severe reduction in the frequency of mature IgM+ B cells was detected in the IRF-2-/- mice; and an enhanced frequency of pre-pro-B cells was detected in young IRF-2-deficient mice. These defects were correlated with the KSL array data (not shown). These data indicate a requirement for IRF-2 in maintaining bone marrow B homeostasis and B cell differentiation.

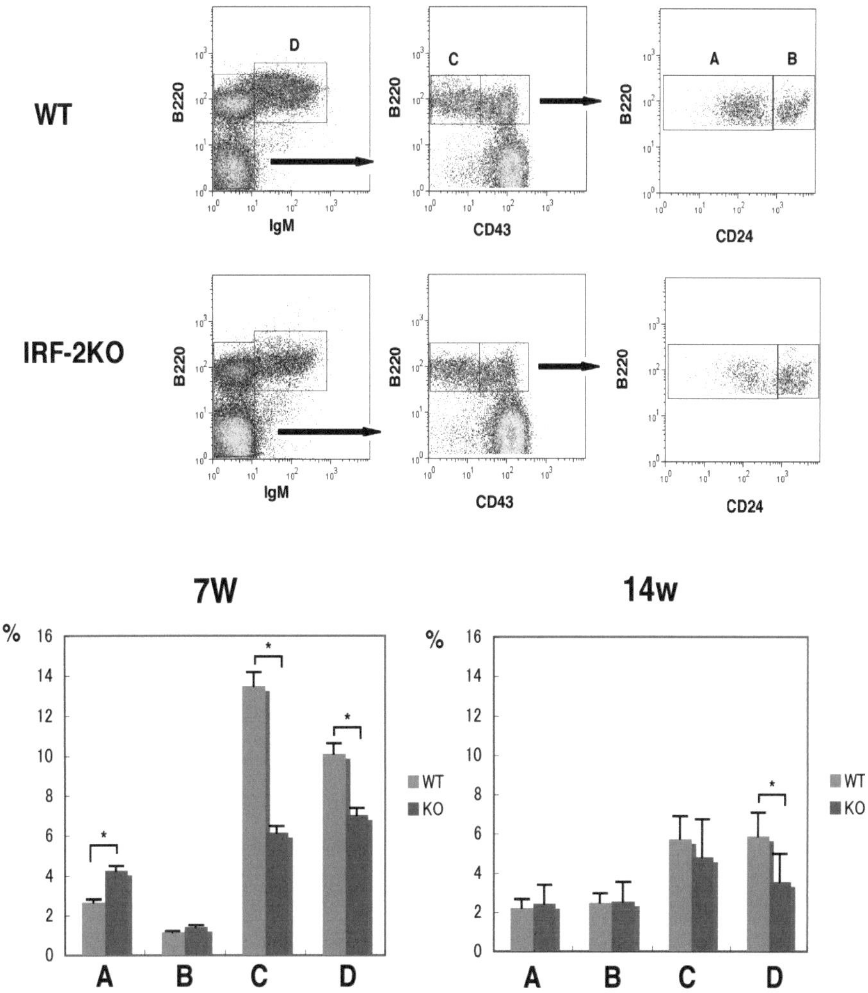

A: pre pro-B, B: pro-B, C: pre-B, D: Mature B

Fig. 13. Loss of IRF-2 in hematopoietic cells results in impaired B cell homeostasis. Top panel: FACS analysis of frequencies of B-cell subsets in bone marrow of IRF-2-/- and age-matched littermate control mice, for expression of B220, IgM, CD43 and CD24. Bottom panel: Four to five (each) young (7W) and old (14W) mice were analyzed. The frequency of mature B220+IgM+B cells (D) was significantly reduced in both young and old IRF-2-/- mice. Young, but not old, IRF-2-/- mice exhibited significant reductions in pre-B fractions (C) and increases in pre pro-B fractions (A).

Fluorescence-activated cell sorting (FACS) analysis of peripheral blood indicated that any significant differences in lineage between wild-type and IRF-2-/- mice were not observed (data not shown). However, an increase in Gr1+Mac1+ neutrophils and a decrease in B220+ cells were detected in the bone marrow and spleens IRF-2-/- mice (Fig.14). These lineage populations almost reflect to that of progenitors in IRF-2-/-mice (Fig.11).

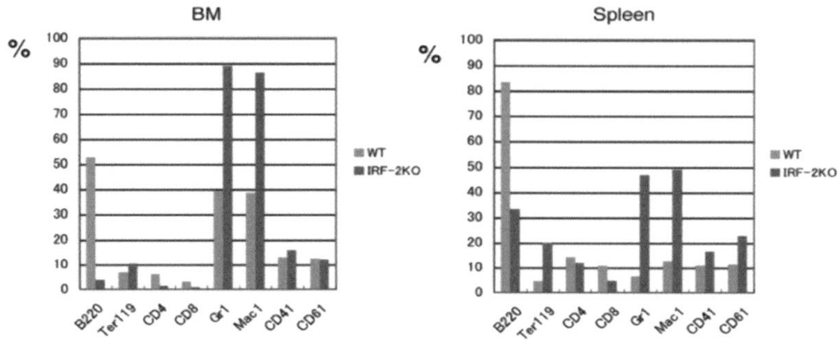

Fig. 14. Lineage populations in IRF-2-/- mouse bone marrow and spleens. Mononuclear cells from mouse bone marrow (BM) or spleen cells were stained with each antibody conjugated to PE and analyzed by FACS. Percentage indicates relative counts per whole mononuclear cells in each surface marker analysis.

2.4 Effect of type I IFN in IRF-2-/- mice

To assess the role of IRF-2 in the regulation of type I IFN signaling, we analyzed gene expression in IRF-2-/- mice compared to IRF-2-/- IFNAR-/- dKO mice, which do not respond to type I IFN. Sca-1 gene expression was enhanced in the bone marrow of IRF-2-/- mice, but not in IRF-2-/-/IFNAR-/- dKO mice. However, reduced expression of IFN-γ in IRF-2-/- mice was not rescued in IRF-2-/-IFNAR-/- dKO mice. Sca-1 expression is regulated by IRF-2 and the type I IFN response. However, IFN-γ may be regulated by IRF-2, independent of the type I IFN response. Arakura et al. reported the up-regulation of IFN-γ resulting from aberrant IFN-α/IFN-β responses in abdominal skin from IRF-2-/- mice (Arakura et al., 2007). In the bone marrow and KSL cells of our IRF-2-/- mice, IFN-γ expression was extremely low and the defect in IFN-α/IFN-β signaling did not rescue the expression. IFN-α/IFN-β expression was not enhanced in bone marrow and KSL cells in IRF-2-/- mice in the absence of stimulation (Fig. 8).

IFN-γ expression decreased in the bone marrow of both IRF-2-/- mice and IRF-2-/- IFNAR-/-dKO mice (Fig. 8B and Fig. 15). IFN-γ reduction is independent of type I IFN signaling in bone marrow cells from IRF-2-/- mice. IRF-2 may regulate IFN-γ gene expression through its promoter or other factors.

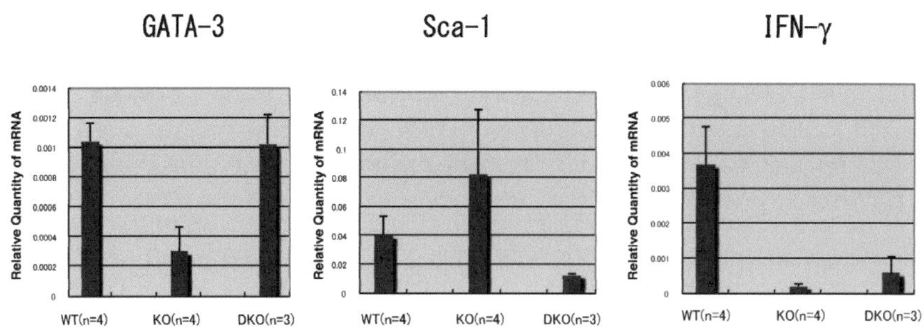

Fig. 15. Real-time PCR analysis for Sca-1, IFN-γ and GATA-3 gene expression in bone marrow cells from wild-type (WT), IRF-2-/- (KO)and IRF-2-/- IFNAR-/- dKO (DKO)mice.

A FACS analysis revealed the reduction of the CD150 surface marker in IRF-2-/- mice. Using real-time RT-PCR for expression of the CD150 gene in IRF-2-deficent mouse bone marrow, we demonstrated a profound decrease of CD150 gene expression (Fig. 16). However, the CD150 expression level in IRF-2-/- IFNAR-/- dKO bone marrow was comparable to wild-type mice. These results indicate that CD150 expression is regulated by the type I IFN response and is transcriptionally regulated in IRF-2-/-mice (Fig. 16). Sato et al. reported that the HSC population, including CD150-positive cells, was reduced through the induction of HSC proliferation by type I IFN signaling. However, we revealed another mechanism in which IRF-2 or type I IFN signaling directly mediated CD150 gene expression. In contrast, more depressed expression of Sca-1 was detected in IRF-2-/- IFNAR-dKO mice compared to that in wild type, supporting that Sca-1 expression was regulated by the type I IFN system (Fig. 16).

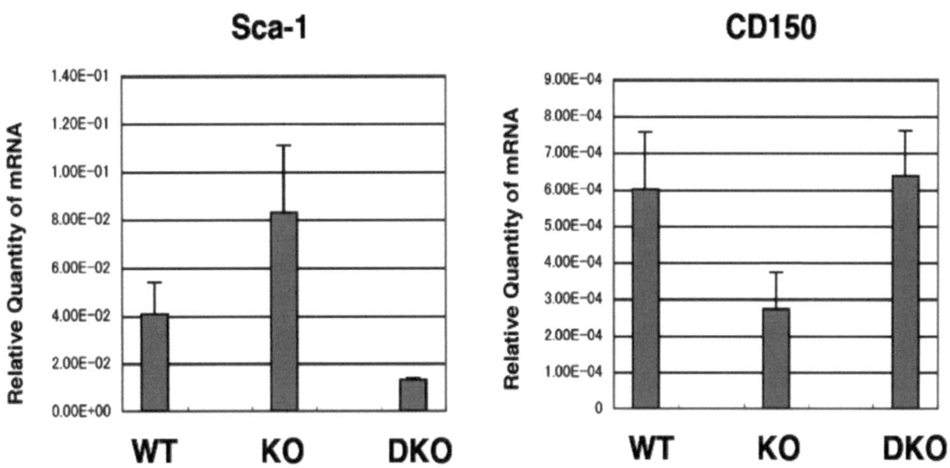

Fig. 16. Real-time PCR analysis for CD150 in bone marrow cells from wild-type (WT), IRF-2-/- (KO) and IRF-2-/- IFNAR-/- dKO (DKO) mice.

2.5 IRF-2 interaction with transcription factors in KSL

We investigated IRF-2-interacting transcription factors that are associated with hematopoiesis. TF(Transcriptional factor)-TF analysis indicates that IRF-2 associates GATA-1/2. However, we did not observe any difference of GATA-1 (data not shown) and GATA-2 (Fig.7A) expressions between wild-type and IRF-2-/- mice mice by KSL array data and real-time PCR analysis. We performed in vitro protein interaction analysis using 293T culture cells. IRF-2 interacted with GATA-2, but not GATA-1, when Flag-tagged IRF-2 and HA-tagged GATAs were transfected into 293T cells. To examine which region of IRF-2 interacts with GATA-2, IRF-2 DNA binding domain (DBD) and IRF-2 without DBD were incubated with several deletion constructs of GATA-2. The IRF-2 DNA-binding domain associated with GATA-2, specifically the N-terminal transcription activation domain in 293T cells (Fig. 17).

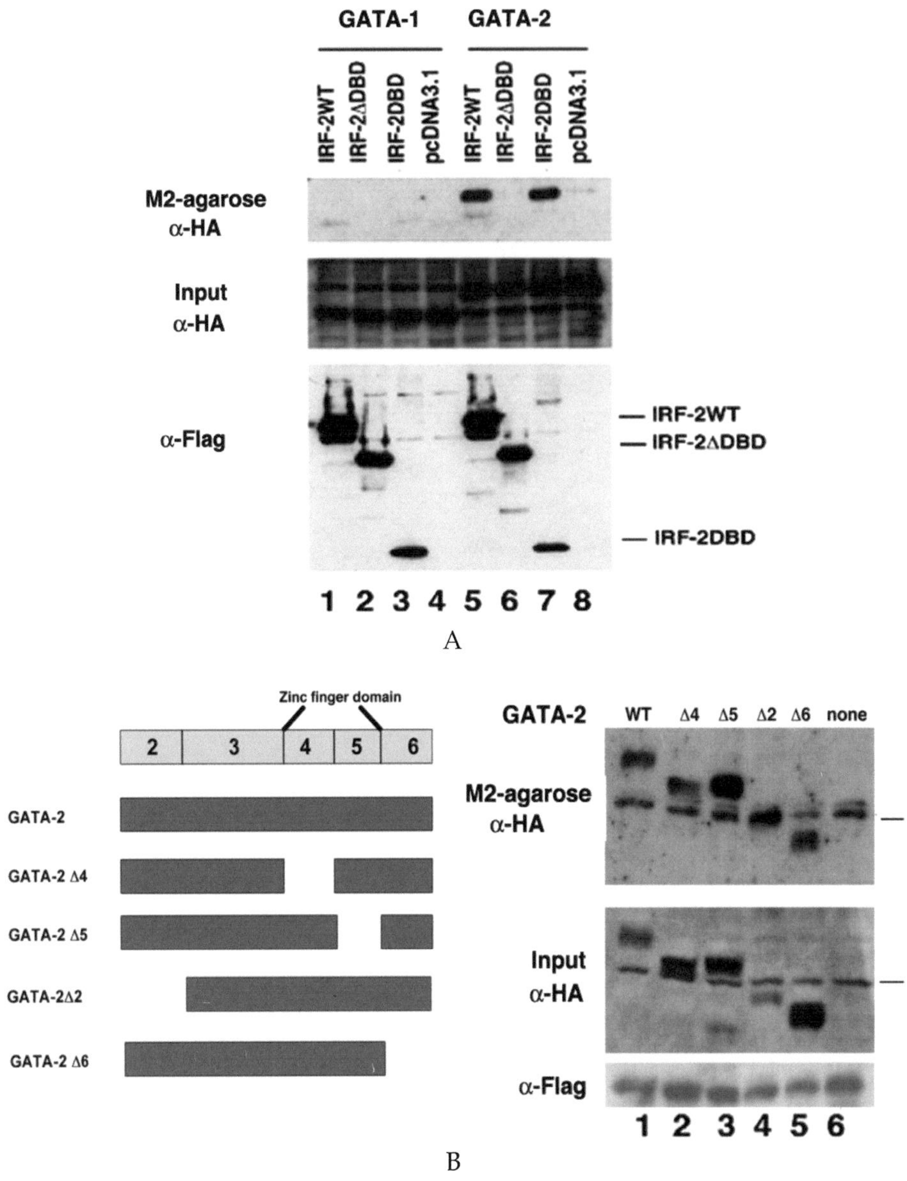

Fig. 17. IRF-2 interacts with GATA-2, but not GATA-1. (A) Flag-tagged IRF-2, Flag-tagged-IRF-2 DNA binding domain (IRF-2DBD) or Flag-tagged IRF-2 without DBD (IRF-2ΔDBD) and HA-tagged GATA-1 or HA-tagged GATA-2 were transfected to 293T cells. Cell lysate were incubated with M2-agarose, and agarose were washed and eluted with Flag peptide solution. Eluted fraction were electrophoresed and Western blot analysis was performed. (B) Protein structure of mouse GATA-2 and its mutants (GATA-2Δ4, GATA-2Δ5, GATA-2Δ2 and GATA-2Δ6) were shown (left) and numbers indicate exons of GATA-2. Each exon deletion mutants tagged with HA was incubated with Flag-tagged IRF-2. Exon2 (transcriptional activation domain) in GATA-2 has high affinity for binding with IRF-2 (right). - indicates non-specific bands.

We found that GATA-3 gene expression was decreased in IRF-2-deficient bone marrow and KSL cells by KSL array data and real-time PCR analysis (Fig. 15). GATA-3 gene expression was comparable to that in IRF-2-/- IFNAR-/- dKO mice, suggesting that the type I IFN response affects GATA-3 expression in IRF-2-/- mice. To examine GATA-3 interacts with IRF-2, Flag-tagged IRF-2 and myc-GATA-3 expression vectors were transfected to 293T cells. We show that GATA-3 interacted with IRF-2 in 293T cells (Fig. 18).

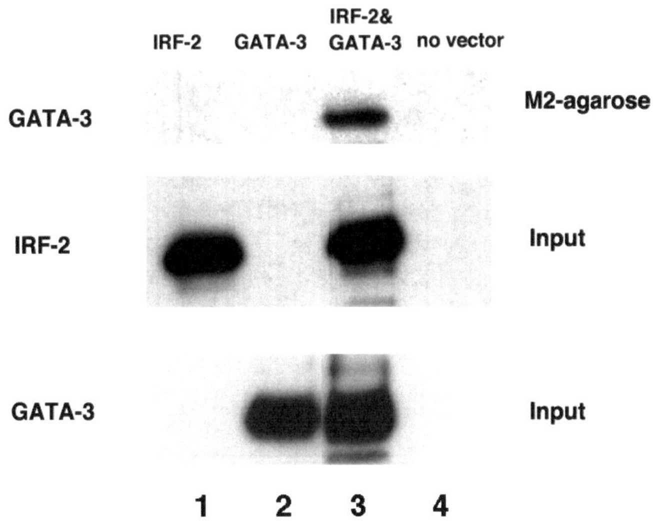

Fig. 18. IRF-2 interacts with GATA-3. Flag-tagged IRF-2 and Myc-tagged GATA-3 were transfected to 293T cells. Western blot analysis was performed as described in Fig.17.

In our investigation, GATA-2 and GATA-3 associated with IRF-2 in the in vitro cell culture system. GATA-2 is expressed abundantly in the mouse HSC population and is necessary for hematopoietic differentiation (Kitajima et al. 2006). GATA-1 is expressed in megakaryocyte/erythrocyte precursors (MEP) and their progenitors, and GATA-3 is expressed in common lymphoid precursors (CLPs) and T cells. There are very few reports regarding GATA-3 expression in HSCs, although we found that the GATA-3 expression level changed in IRF-2-deficient bone marrow and KSL in microarray and real time PCR analysis. Previous reports have shown that forced GATA-3 expression into mouse HSCs induces differentiation toward erythrocytes and megakaryocytes (Chen and Zhang 2001). As indicated by our investigation, GATA-3 may be important for the maintenance of HSC cooperation with IRF-2 or IFN signaling. These findings indicate that the interactions between IRF-2 and the GATAs are required to maintain HSC function in mouse bone marrow cells. Interactions between GATA-2 and GATA-3 with IRF-2 in HSCs should be clarified in future experiments.

3. Conclusion

IRF-2 exists high in the mouse bone marrow HSC population and helps maintain the protective immune response, which responds to viral or bacterial infection and inflammation, resulting in IFN producing system. HSCs are essential for the production of immune cells, such as myeloid or lymphoid cells. Recently, interferon treatment has been reported to be a target for cancer stem cells. Under chronic interferon stimulation, such as in an IRF-2-deficient condition, not only HSCs, but also cancer stem cells can be activated. Blocking the IRF-2 function may induce to eliminate of cancer stem cell through IFN signaling. However, the possibility of the presence of HSC-like cells in IRF-2 deficient mice cell population necessitates further investigation because IRF-2 in part regulates HSC populations independent of the type I IFN system in an another possible mechanism.

4. Acknowledgments

We thank Dr. I. Hamaguchi, Dr. T. Mizukami, Dr. H. Momose, Dr. M. Kuramitsu, Dr. K. Takizawa, and Dr. K. Yamaguchi for their experimental support and useful discussions. Ms. K. Furuhata for cell sorting and Ms. M. Tsuruhara for molecular technical assistance. This work was supported in part the Japan Society for the Promotion of Science and the Ministry of Education, Science, Sports and culture of Japan.

5. References

Akashi, K., Traver, D., Miyamoto, T. & Weissman I (2000). A clonogenic common myeloid progenitor that gives rise to all myeloid lineages. *Nature* Vol.404, 6774, (Mar 2000), pp. 193-197.

Arakura, F., Hida, S., Ichikawa, E., Yajima, C., Nakajima, S., Saida T. & Taki, S. (2007). Genetic control directed toward spontaneous IFN-alpha/IFN-beta responses and downstream IFN-gamma expression influences the pathogenesis of a murine psoriasis-like skin disease. *J Immunol* Vol.179, No.5, (Sep 2007), pp. 3249-3257.

Bradfute, S.B., Graubert, T.A. & Goodell, M.A. (2005). Roles of Sca-1 in hematopoietic stem/progenitor cell function. *Exp Hematol* Vol.33, No.7, (Jul 2005), pp. 836-843.

Chen, D. & Zhang, G. (2001). Enforced expression of the GATA-3 transcription factor affects cell fate decisions in hematopoiesis. *Exp Hematol* Vol.29, No.29, (Aug 2001), pp. 971-980.

Chow, W., Fang, J. & Yee, J. (2000). The IFN regulatory factor family participates in regulation of Fas ligand gene expression in T cells. *J Immunol* Vol. 164, No.7, (Apr 2000), pp. 3512-3518.

Darnell, JE Jr., Kerr, IM. & Stark, G.R. (1994). Jak-STAT pathways and transcriptional activation in response to IFNs and other extracellular signaling proteins. *Science* Vol. 264, 5164, (Jun 1994), pp. 1415-1421.

Essers, M., Offner, S., Blanco-Bose, W., Waibler, Z., Kalinke, U., Duchosal, M. & Trumpp, A. (2009). IFNa activates dormant hematopoietic stem cells in vivo. *Nature* Vol. 458, 7240, (Apr 2009), pp. 904-908.

Hartner, J., Walkley, C., Lu, J.& Orkin, S. (2009). ADAR1 is essential for the maintenance of hematopoiesis and suppression of interferon signaling. *Nat Immunol* Vol.10, No.1, (Jan 2009), pp. 109-115.

He, S., Nakada, D. & Morrison, S.J. (2009). Mechanisms of stem cell self-renewal. *Annu Rev Cell Dev Biol* Vol.25, (2009), pp. 377-406.

Honda, K., Takaoka, A. & Taniguchi, T. (2006). Type I interferon gene induction by the interferon regulatory factor family of transcription factors. *Immunity* Vol.25, No.3, (Sep 2006), pp. 349-360.

Ito, C.Y., Li, C.Y., Bernstein, A., Dick, J.E. & Stanford, W.L. (2003). Hematopoietic stem cell and progenitor defects in Sca-1/Ly-6A-null mice. *Blood* Vol.101, No.2, (Jan 2003), pp. 517-523.

Jesse, T.L, LaChance, R., Iademarco, M.F. & Dean, D.C. (1998). Interferon regulatory factor-2 is a transcriptional activator in muscle where it regulates expression of vascular cell adhesion molecule-1. *J Cell Biol* Vol.140, No.5, (Mar 1998), pp. 1265-1276.

Khan, K.D., Lindwall, G., Maher, S.E. & Bothwell, A.L. (1990). Characterization of promoter elements of an interferon-inducible Ly-6E/A differentiation antigen, which is expressed on activated T cells and hematopoietic stem cells. *Mol Cell Biol* Vol.10, No.10, (Oct 1990), pp. 5150-5159.

Khan, K.D., Shuai, K., Lindwall, G., Maher, S.E., Darnell, J.E., Jr., & Bothwell, A.L. (1993). Induction of the Ly-6A/E gene by interferon alpha/beta and gamma requires a DNA element to which a tyrosine-phosphorylated 91-kDa protein binds. *Proc Natl Acad Sci U S A* Vol.90, No.14, (July 1993), pp. 6806-6810.

Kitajima, K., Tanaka, M., Zheng, J., Yen, H., Sato, A., Sugiyama D., Umehara, H., Sakai, E. & Nakano, T. (2006). Redirecting differentiation of hematopoietic progenitors by a transcription factor, GATA-2. *Blood* Vol.107, No.5, (Mar 2006), pp. 1857-1863.

Luo, W. & Skalnik D (1996). Interferon regulatory factor-2 directs transcription from the gp91phox promoter. *J Biol Chem* Vol.271, No.38, (September 1996), pp. 23445-23451.

Ma, X., Ling, K.W. & Dzierzak, E. (2001). Cloning of the Ly-6A (Sca-1) gene locus and identification of a 3' distal fragment responsible for high-level gamma-interferon-induced expression in vitro. *Br J Haematol* Vol.114, No.3, (Sep 2001), pp. 724-730.

Masumi, A. & Ozato, K. (2001). Coactivator p300 acetylates the interferon regulatory factor-2 in U937 cells following phorbol ester treatment. *J Biol Chem* Vol.276, No.24, (Jun 2001), pp. 20973-20980.

Masumi, A., Hamaguchi, I., Kuramitsu, M., Mizukami, T., Takizawa, K., Momose, H., Naito, S. & Yamaguchi, K. (2009). Interferon regulatory factor-2 induces megakaryopoiesis in mouse bone marrow hematopoietic cells. *FEBS Lett* Vol.583, No.21, (Nov 2009), pp. 3493-3500.

Matsuyama, T., Kimura, T., Kitagawa, M., Pfeffer, K., Kawakami, T., Watanabe, N., Kundig, T. M., Amakawa, R., Kishihara, K., Wakeham, A., Potter, J., Fulonger, C.L., Narendan, A., Suzuki, H., Ohashi, P.S., Paige, C.J., Taniguchi, T. & Mak, T.W. (1993). Targeted disruption of IRF-1 or IRF-2 results in abnormal type I IFN gene induction and aberrant lymphocyte development. *Cell* Vol.75, No.1, (Oct 1993), pp. 83-97.

Orkin, S.H., & Zon, L.I. (2002). Hematopoiesis and stem cells: plasticity versus developmental heterogeneity. *Nat Immunol* Vol.3, No.4, (Apr 2002), pp. 323-328.

Orkin, S.H., & Zon, L.I. (2008a). SnapShot: hematopoiesis. *Cell* Vol. 132, No. 4, (Feb 2008), pp. 712.

Orkin, S.H. & Zon, L.I. (2008b). Hematopoiesis: an evolving paradigm for stem cell biology. *Cell* Vol.132, No.4, (Feb 2008), pp. 631-644.

Pichlmair A & Reis e Sousa C (2007). Innate recognition of viruses. *Immunity* Vol. 27, No.3, (Sep 2007), pp. 370-383.

Sato, T., Onai, N., Yoshihara, H., Arai, F., Suda, T. & Ohteki, T. (2009). Inteferon regulatory factor-2 protects quiescent hematopoietic stem cells from type 1 interferon-dependent exhaustion. *Nat Med* Vol.15, No.6, (Jun 2009), pp. 696-700.

Stark, G.R., Kerr, I.M., Williams B.R, Silverman, RH. & Schreiber, R.D. (1998). How cells respond to interferons. *Annu Rev Biochem* Vol. 67, pp. 227-264.

Stellacci, E., Testa, U., Petrucci, E., Benedetti, E., Orsatti, R., Feccia, T., Stafsnes, M., Marziali, G. & Battistini, A. (2004). Interferon regulatory factor-2 drives megakaryocytic differentiation. *Biochem J* Vol. 377, Pt2, (Jan 2004), pp. 367-378.

Taniguchi, T., Ogasawara, K., Takaoka, A. & Tanaka N. (2001). IRF family of transcription factors as regulators of host defense. *Annu Rev Immunol* Vol.19, pp. 623-655.

Taniguchi, T. & Takaoka, A. (2001). A weak signal for strong responses: interferon-α/β revisited. *Nat. Rev. Mol. Cell Biol* Vol. 2, No.5, (May 2001), pp. 378-386.

Vaughan, P., Aziz, F., van Wijnen, A., Wu, S., Harada, H., Taniguchi ,T., Soprano, K.J., Stein, J.L. & Stein, G.S. (1995). Activation of a cell-cycle-regulated histone gene by the oncogenic transcription factor IRF-2. *Nature* Vol.377, 6547, (Sep 1995), pp. 362-365.

Xi, H., Eason, D., Ghosh, D., Dovhey, S., Wright, K. & Blanck, G. (1999). Co-occupancy of the interferon regulatory element of the class II transactivator (CIITA) type IV promoter by interferon regulatory factors 1 and 2. *Oncogene* Vol.18, No.43, (Oct 1999), pp. 5889-5903.

6

Regulation of Tyrosine Kinase Signaling by Cbl in Hematopoietic Stem Cells

Mayumi Naramura
University of Nebraska Medical Center
USA

1. Introduction

Phosphorylation of tyrosine residues is an essential biochemical reaction in many higher eukaryotes. One of the most important and well-studied functions of tyrosine phosphorylation is to convey extracellular signals to the cytoplasm and ultimately to the nucleus in order to control various cell functions such as proliferation, differentiation, migration and survival.

The first tyrosine kinase (TK) was discovered as the tumor-inducing activity from Rous sarcoma virus, which is now known as v-Src (Rous, 1911). Later studies revealed that the oncogenic properties of v-Src was due to the loss of the regulatory mechanisms to control its kinase activity (Martin, 2001). These findings clearly highlight the critical importance of precise regulation of TK activities in order to avoid detrimental consequences to the homeostasis of the organisms.

Various mechanisms are employed to control TK activities. In the Src family non-receptor tyrosine kinase (non-RTK), phosphorylation status of the tyrosine residue in the C-terminal regulatory region alters intramolecular interactions and therefore serves as a way to modulate kinase activity. The kinases and the phosphatases involved in this regulatory mechanism are, in turn, themselves under additional layers of regulation, thus, creating an intricate network of signal mediators to fine-tune cellular responses (Sen & Johnson, 2011). Incidentally, activity of a typical receptor tyrosine kinase (RTK) is regulated by ligand binding; RTKs alter conformation upon ligand binding and dimerize, which leads to transphosphorylation of critical tyrosine residues in the activation loop of the neighboring kinase in the cytoplasmic domain (He & Hristova). This initiates a cascade of biochemical reactions that activates downstream signaling pathways.

Subcellular localization of TKs is another important determinant of their activity (Murphy et al., 2009). There are now abundant evidence indicating that TKs can generate different signals dependent on their intracellular locations. Therefore, molecules that regulate protein trafficking and localization constitute a critical component of signal regulatory mechanism.

Covalent attachment of small proteins such as ubiquitin and small ubiquitin-like modifier (SUMO) to target proteins serves as a signal for various biological processing, including alteration of its localization and promotion of degradation (Schulman & Harper, 2009; van Wijk & Timmers, 2010). This reaction is mediated by a series of biochemical reactions involving the E1 or activating enzyme, the E2 or conjugating enzyme and the E3 ligase. Human genome encodes for two E1s, thirty E2 and over one thousand E3s for the ubiquitin system. This pathway architecture immediately implies that the substrate specificity of the ubiquitin system must be achieved largely at the level of E3s.

The Casitas B-lineage lymphoma (Cbl) family proteins are RING finger (RF)-containing multi-domain adaptors that function as E3 ubiquitin ligase primarily towards activated TKs (Thien & Langdon, 2001; Duan et al., 2004; Schmidt & Dikic, 2005). Using genetically-engineered mouse models, we and others showed that loss of Cbl, either singly or in combination with another family member Cbl-b, led to the enlargement of the hematopoietic stem cell (HSC) compartment (Naramura et al., 2011a). Additionally, mutations in the *CBL* gene have been identified in a small but significant number of hematological malignancies in human, and experimental evidence proved the oncogenicity of mutant *CBL* products (Naramura et al., 2011b). All together, these observations strongly support that the Cbl family proteins are critical regulators of hematopoietic homeostasis.

Here, we review functions of the Cbl family proteins and some of the candidate Cbl targets in the HSC compartment and discuss potential mechanisms of their regulation.

2. The Cbl family proteins

The Cbl family proteins are evolutionarily conserved signal regulators present through *C. elegans* to human (Figure 1). In mammals, this family includes Cbl (also known as c-Cbl, encoded by the *CBL* gene in human), Cbl-b (*CBLB* gene in human) and Cbl-c (also know as Cbl-3 or Cbl-SL, *CBLC* gene in human). Cbl was originally identified as a cellular homolog of a viral oncogene *v-Cbl* which caused leukemia and lymphoma in mice (Langdon et al., 1989). Cbl's involvement in signal transduction was suggested because it became prominently tyrosine-phosphorylated upon stimulation through various cell surface receptors (Donovan et al., 1994; Galisteo et al., 1995). But it was not until genetic studies in *C. elegans* identified the *sli-1* gene product as a Cbl homolog that Cbl was established as a negative regulator of RTK signaling.

2.1 Structure and biochemical functions of the Cbl family proteins

All Cbl family proteins share a high degree of homology in their N-terminal regions. These include the tyrosine kinase binding (TKB) domain, the RF domain and the short intervening linker region. X-ray crystallography studies revealed that the TKB domain comprised a four-helix bundle (4H), a calcium-binding EF hand and a variant Src homology region 2 (SH2) domain (Meng et al., 1999). The TKB domain mediates specific binding to cognate phosphotyrosine-containing motifs in activated TKs and select non-TK signal mediators (Lupher et al., 1996). The RF domain and the linker region together bind to E2 ubiquitin-conjugating enzymes and both of these motifs are essential for the E3 ubiquitin ligase activity of the Cbl family proteins (Joazeiro et al., 1999; Levkowitz et al., 1999; Yokouchi et al., 1999; Zheng et al., 2000).

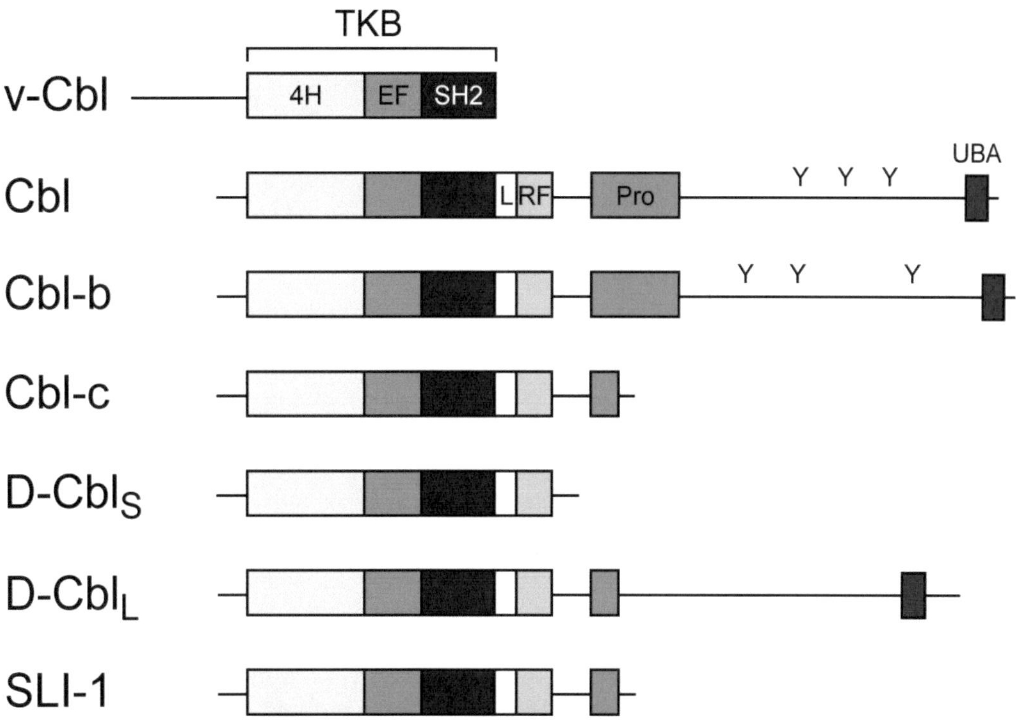

Fig. 1. Structure of the Cbl family proteins. The original oncogenic form of Cbl (v-Cbl), the three mammalian Cbl family proteins (Cbl, Cbl-b and Cbl-c), the short and long forms of *Drosophila* Cbl (D-Cbl$_S$ and D-Cbl$_L$) and the *C. elegans* homolog (SLI-1) are shown. TKB, tyrosine kinase binding; 4H, four-helix bundle; EF, EF hand; SH2, Src homology region 2; L, linker; RF, RING finger; Y, tyrosine; UBA, ubiquitin-associated.

Band and colleagues originally described that the Cbl TKB domain specifically recognized the phosphotyrosine-containing motif D(N/D)XpY, which was later refined as (N/D)XpY(S/T)XXP, found in several TKs such as ZAP70, epidermal growth factor receptor (EGFR), and Src (Lupher et al., 1997). Additional binding motifs, RA(V/I)XNQpY(S/T) and DpYR, were proposed in the adaptor protein APS (Hu & Hubbard, 2005) and the RTK c-Met (also known as hepatocyte growth factor receptor; Peschard et al., 2004), respectively. A recent comprehensive structural study showed that phosphopeptides with diverse sequences bound TKB at the same site, albeit in two different orientations (Ng et al., 2008). These studies collectively revealed the unique binding strategy for the specialized and biologically vital function of the Cbl family proteins and provided means to identify potential Cbl targets based on the amino acid sequences.

The C-terminal half of the Cbl family proteins are more divergent. A proline-rich region follows the RF domain in all mammalian Cbl family proteins, but this domain is more prominent in Cbl and Cbl-b than in Cbl-c. Biochemical studies have demonstrated that Cbl interacted with SH3-domain containing proteins such as Grb2 and Nck through the proline-rich region (Rivero-Lezcano et al., 1994; Fukazawa et al., 1995).

In addition to being a TK regulator, Cbl itself is subject to tyrosine phosphorylation. Phosphorylation at tyrosine residues 700, 731 and 774 have been extensively characterized; residues 700 and 774 provide docking sites for the SH2 domain-containing adaptor protein CrkL (Andoniou et al., 1996). Tyrosine 700 also mediates an interaction with the guanine nucleotide exchange factor Vav (Marengère et al., 1997). Tyrosine 731 provides a docking site of the p85 regulatory subunit of phosphatidylinositol 3-kinase (PI3K) (Hunter et al., 1999). Based on sequence homology and experimental data, tyrosine residues in the C-terminal domain of Cbl-b are thought to share many of the same functions as those in Cbl. Cbl-c does not possess comparable tyrosines.

The C-termini of Cbl and Cbl-b, but not Cbl-c, contain a conserved domain known as a ubiquitin-associated (UBA) domain, which is present in a variety of proteins involved in ubiquitin-mediated processes. Structural studies indicate that this domain is capable of binding ubiquitin and involved in dimerization (Kozlov et al., 2007; Peschard et al., 2007).

2.2 Insights from genetic models

The first critical clue into Cbl's functions came from genetic studies in *C. elegans* (Yoon et al., 1995). The vulval development in *C. elegans* is regulated by signals through the EGFR pathway. A reduction-of-function mutation in *let-23* (encodes the EGFR homolog) leads to death of most worms, and the vulval development is incomplete in surviving worms. However, when loss-of-function mutations in *sli-1* were introduced to this genetic background, worms survived and vulval development was restored. The sequence analysis of the *sli-1* gene revealed that it encoded a protein with a high similarity to Cbl, thus establishing Cbl as a negative regulator of the EGFR pathway.

Genetic studies in gene targeted mouse models provided further insights into the physiological roles of the Cbl family proteins in mammals. Cbl-deficient mice are viable, but they show recognizable changes in the hematopoietic, lymphoid, metabolic and reproductive systems. In contrast, effects of Cbl-b loss is mostly limited to the peripheral immune functions. Cbl-c expression appears to be restricted to the epithelial tissues, but no significant phenotypes were reported in mice deficient in Cbl-c (Table 1).

While mice deficient in either one of the Cbl family members are viable, simultaneous loss of Cbl and Cbl-b is not compatible with the survival of the organism and double-deficient mice do not survive beyond embryonic day 10 (Naramura et al., 2002). This indicates that Cbl and Cbl-b play redundant and overlapping functions in critical organ systems during fetal development. Using the Cre-loxP-mediated conditional gene deletion approach, effects of Cbl, Cbl-b loss have been analyzed in the T, B and HSC compartments (Naramura et al., 2002; Huang et al., 2006; Kitaura et al., 2007; Naramura et al., 2010). These studies demonstrated that, in the adaptive immune system, the Cbl family proteins are required to establish appropriate threshold for selection of T and B cells, and disruption of this process leads to autoimmune-like phenotypes in mice.

In the hematopoietic compartment, Cbl-deficiency leads to moderate splenomegaly and enhanced extramedullary hematopoiesis (Murphy et al., 1998). In the bone marrow, the lineage-negative, Sca-1-positive, c-Kit-positive (LSK) compartment, which is highly enriched for HSCs, is enlarged and Cbl-deficient HSCs showed enhanced capacity to reconstitute myeloabrated recipient's hematopoietic system (Rathinam et al., 2008). However, mice were outwardly normal and had a normal lifespan. When both Cbl and Cbl-b are deleted in the HSC, however, mice succumbed to aggressive myeloproliferative disease-like leukemia within two to three months after birth (Naramura et al., 2010).

Gene	Phenotype	Reference
Cbl	Altered T cell antigen receptor expression Increased tyrosine phosphorylation Enhanced thymic selection Splenomegaly and extramedullary hematopoiesis Decreased fertility Altered metabolism	(Murphy et al., 1998; Naramura et al., 1998; Thien et al., 1999; Molero et al., 2004; El Chami et al., 2005; Rathinam et al., 2008)
Cblb	Co-stiumlation-independent activation of peripheral T cells Predisposition to autoimmune diseases and inflammatory injury Resistance to spontaneous and transplanted tumors	(Bachmaier et al., 2000; Chiang et al., 2000; Krawczyk et al., 2000; Chiang et al., 2007; Loeser et al., 2007; Bachmaier et al., 2007)
Cblc	No apparent phenotypes	(Griffiths et al., 2003)

Table 1. Phenotypes of mice deficient in the Cbl family members

2.3 Cbl and hematological malignancies

Because of the involvement of various RTKs in cancer, it has long been speculated that the Cbl family proteins may play critical roles in the initiation and/or progression of cancer. Oncogenic mutations in RTKs that abrogate interaction with Cbl have been reported (Peschard & Park, 2003), but the direct evidence supporting Cbl's roles in cancer was not established until 2007.

The vast majority of *CBL* mutations reported so far are associated with myeloid disorders. Although the first human *CBL* mutations were described in acute myeloid leukemia (AML) samples (Sargin et al., 2007; Caligiuri et al., 2007; Abbas et al., 2008), later studies documented a significant number of cases in myelodysplastic syndromes-myeloproliferative neoplasms (MDS/MPN), a heterogeneous group of myeloid disorders including the chronic myelomonocytic leukemia (CMML), atypical chronic myeloid leukemia (aCML) and juvenile myelomonocytic leukemia (JMML) (Dunbar et al., 2008; Reindl et al., 2009; Grand et al., 2009; Loh et al., 2009; Sanada et al., 2009; Makishima et al., 2009; Muramatsu et al., 2010; Fernandes et al., 2010; Niemeyer et al., 2010). The association of *CBL* mutations with JMML is particularly thought-provoking because the pathogenesis of this rare pediatric hematological malignancy is closely associated with the activation of the Ras-MAPK signaling pathway (Loh, 2011). Among JMML patients, the activating mutations of *PTPN11*, *NRAS* and *KRAS*, and the loss of *NF1*, a gene encoding for a Ras GTPase-activator account for approximately 75 % of the total cases. Roughly half of the remainder of the cases are now attributed to *CBL* mutations. While *in vitro* experimental data indicate that the loss of the Cbl family proteins lead to prolonged Erk activation, it was never formally demonstrated whether Cbl can regulate Ras activity directly.

It is of note that most *CBL* mutations are either point mutation or internal deletion involving the linker and/or the RF regions rather than complete deletion at the *CBL* locus. As expected from domain-function analysis results, these mutant Cbl proteins lack E3 ubiquitin ligase activity. Interestingly, in patient samples with *CBL* mutations, the wild-type allele is often lost and replaced with the mutant allele by acquired uniparental isodisomy (aUPD). The *CBLB*, *CBLC* alleles are usually unaffected in these patients although mutations in these genes have been reported (Makishima et al., 2009; Makishima et al., 2011). All together, these clinical observations suggest that the presence of one wild-type copy of *CBL* is generally sufficient to maintain the functions of Cbl in the presence of normal Cbl-b and Cbl-c. These findings are consistent with the data in mice expressing a RF-mutant Cbl from the endogenous promoter on a *Cblb*, *Cblc* wild-type background (Thien et al., 2005; Rathinam et al., 2010); homozygous mutant mice are perinatally lethal, but hemizygous mutants over the wild-type *Cbl* allele develop normally. However, when the hemizygous mutant is expressed over the Cbl-null background, mice develop myeloproliferative disease-like leukemia within a year. This is a striking contrast when compared to the rapid progression and fatality of the HSC-specific Cbl, Cbl-b double-deficient mice (Naramura et al., 2010). These differences may reflect that the RF mutant and patient-derived oncogenic mutant Cbl proteins function as gain-of-function mutants rather than as dominant-negative inhibitors of Cbl-b (Cbl-c expression is minimal in the hematopoietic system). While these mutants lack E3 ubiquitin ligase activity and thus defective in promoting target degradation, they possess intact TKB and C-terminal protein-protein interaction motifs, which may enable them to form aberrant but stable multi-protein super-signaling complexes and activate unconventional signaling pathways.

2.4 Potential Cbl targets in the hematopoietic system

What, then, are the target of Cbl-dependent regulation in the HSC compartment? Because Cbl becomes phosphorylated upon stimulation with various cell surface receptors, it is conceivable that Cbl is involved in the regulation of signal transduction downstream of such pathways (Table 2). Among this diverse group of cell surface receptors, Kit and Flt3 are of particular interest because both of them are RTKs expressed in HSC and known to perform critical functions in the HSC compartment (Masson & Rönnstrand, 2009). Colony stimulating factor 1 receptor (CSF1R) is known to interact with Cbl (Lee et al., 1999), but it is expressed primarily in more differentiated myeloid/phagocytic cells than in HSCs. Endothelial-specific receptor tyrosine kinase (Tek, also known as Tie2) is another RTK expressed in the HSC compartment (Arai et al., 2004) and therefore may interact with Cbl. Thrombopoietin (TPO) is indispensable for the maintenance of HSC quiescence (Yoshihara et al., 2007; Qian et al., 2007). Although its receptor (TPO-R, also known as Mpl or c-Mpl) is not an RTK, stimulation with TPO induce phosphorylation of Cbl (Sasaki et al., 1995), Mpl have been shown to be ubiquitinylated (Saur et al., 2010) and Cbl loss alters the signal transduction downstream of TPO (Rathinam et al., 2008; Naramura et al., 2010). Therefore, Mpl may interact with Cbl indirectly. Other potential (direct as well as indirect) Cbl targets in the HSC compartment include the chemokine and integrin pathways.

In following sections, I will discuss how these pathways may be regulated by the Cbl family proteins in the HSCs.

Antigen and other immunological receptors

- T cell antigen receptor complex (Donovan et al., 1994; Meisner et al., 1995; Fukazawa et al., 1995)
- B cell antigen receptor complex (Cory et al., 1995; Tezuka et al., 1996; Panchamoorthy et al., 1996)
- Fcγ receptor (Marcilla et al., 1995)
- Fcε receptor (Matsuo et al., 1996; Suzuki et al., 1997)

RTKs

- Epidermal growth factor receptor (Galisteo et al., 1995)
- Insulin receptor (Ribon & Saltiel, 1997)
- Platelet-derived growth factor receptor (Bonita et al., 1997)
- Kit (Wisniewski et al., 1996; Brizzi et al., 1996)
- Flt3 (Lavagna-Sévenier et al., 1998)
- Fibroblast growth factor receptor (Wong et al., 2002)
- Colony stimulating factor 1 receptor (Wang et al., 1996)
- Met (Fixman et al., 1997; Garcia-Guzman et al., 2000)
- TrkB (McCarty & Feinstein, 1999)
- Tie2 (Wehrle et al., 2009)

Cytokine receptors

- Interleukin 2 receptor (Gesbert et al., 1998)
- Interleukin 3 receptor (Barber et al., 1997)
- Interleukin 4 receptor (Ueno et al., 1998)
- Erythropoietin receptor (Odai et al., 1995; Barber et al., 1997)
- Mpl (Sasaki et al., 1995; Brizzi et al., 1996)
- GM-CSF receptor (Odai et al., 1995)
- Prolactin receptor (Hunter et al., 1997)

Chemokine receptors (Chernock et al., 2001)

Integrins (Ojaniemi et al., 1997; Manié et al., 1997; Meng & Lowell, 1998)

Table 2. Partial list of potential upstream receptors for Cbl

3. Kit

The mouse dominant spotting mutation at the *W* locus was first described in the early 1900s (Durham, 1908). Mutations at this locus were studied extensively not only because they produced visible coat color changes, but also because mutant mice showed defects in hematopoiesis, mast cell development and gametogenesis (Russell, 1979). However, it was not until 1988 that the gene product at the *W* locus was found to encode for the cellular homolog of the *kit* oncogene which had been molecularly identified a few years earlier (Besmer et al., 1986; Chabot et al., 1988; Geissler et al., 1988).

Kit is a type III RTK that shares structural similarities with platelet-derived growth factor receptors (PDGFRs) α and β, Flt3 (also known as Flk-2, discussed below) and CSF1R. They are characterized by an extracellular domain with five immunoglobulin-like domains, a single transmembrane domain and an intracellular tyrosine kinase domain that is split into two by an intervening sequence.

HSCs are functionally defined as rare cells with the capacity to self-renew and give rise to all cell types of the hematopoietic lineage, including erythrocytes, granulocytes, monocytes, megakaryocytes and lymphocytes. No single marker specific for HSCs is known today. However, it is widely accepted that, in mice, most HSCs reside in a population of cells that express Kit and another cell surface protein Sca-1 and lack the expression of committed lineage markers (Ikuta & Weissman, 1992). Thus, Kit expression is intimately tied to HSCs.

The ligand for Kit is called stem cell factor (SCF) and encoded by the Steel (*Sl*) locus. The phenotypes of the *Sl* mutant mice are in most cases similar to those of *W* mutant mice, affecting hematopoiesis, mast cell development, fertility and coat colors (Galli et al., 1993). Collectively, these observations firmly established the essential roles of the SCF-Kit axis in these biological processes.

In the mouse embryo, hematopoietic cells are found in the blood islands in the yolk sac starting around embryonic day 7. Subsequently, at day 10-11 of gestation, HSCs migrate to the fetal liver and then to the spleen and the bone marrow, the primary hematopoietic organs in adult. The hematopoietic defect in *W* mice is detected throughout the course of development. Syngeneic transplantation experiments demonstrated that the defect exerted by *W* mutations was intrinsic to hematopoietic cells. The hematopoietic microenvironment in these animals are not affected and able to support hematopoiesis of normal donor-derived cells (Russell, 1979).

Kit activity is regulated at various levels. Ligand-receptor engagement of Kit initiates receptor dimerization and subsequent activation of its TK activity. Extensive biochemical studies have mapped intracellular phosphorylated tyrosine residues and their interacting proteins. These include Src family TKs, phosphatases such as SHP1 and SHP2, phospholipase Cγ, p85 subunit of phosphoinositide-3 kinase, (p85(PI3K)) and adaptor proteins such as Grb2 and APS (Lennartsson et al., 2005).

Cbl becomes phosphorylated when Kit-expressing cells are stimulated with SCF (Wisniewski et al., 1996). Earlier studies suggested that Cbl interacted with Kit indirectly through Grb2 (Brizzi et al., 1996), CrkL and p85(PI3K) (Sattler et al., 1997), and APS (Wollberg et al., 2003). More recent data suggest that Cbl binds to Kit directly at tyrosine 568, which is located in the juxtamembrane domain, and tyrosine 936, which is located in the carboxyterminal tail, ubiquitinylate Kit and target them for degradation (Masson et al., 2006). Both Cbl and Cbl-b function similarly towards Kit (Zeng et al., 2005). Hematopoietic cells deficient in Cbl functions are hypersensitive to stimulations through Kit (Naramura et al., 2010; Rathinam et al., 2010). These data all together strongly support that Kit may be one of the physiological targets of Cbl proteins in the HSC compartment.

Structurally, sequences surrounding tyrosine 568 partially conform to the canonical Cbl(TKB) recognition sequence while those around tyrosine 936 do not. Further analyses into the mechanisms of binding between Cbl and Kit may reveal novel molecular interactions that remained unknown so far.

4. Flt3

Flt3, another member of the type III RTKs, was originally identified by two separate groups through homology screening for TKs (Matthews et al., 1991; Rosnet et al., 1991). Its expression is detected in placenta, gonads, brain and hematopoietic cells, but its role outside of the hematopoietic system is not clear at present. Ligand for Flt3 (Flt3 ligand; FL) was identified a few years later and its transcript is expressed in wide range of both fetal and adult tissues (Lyman et al., 1993; Hannum et al., 1994).

Roles of Flt3 in HSCs appear to vary among species and also dependent upon developmental stages. The most primitive self-renewing HSCs with long-term reconstituting potential (LT-HSCs) are not found within $Flt3^+$ LSK cells in adult mouse bone marrow while the same biological activity was detected in both $Flt3^+$ and $Flt3^-$ populations in fetal liver (Adolfsson et al., 2001; Christensen & Weissman, 2001). Notably, human HSCs with multi-lineage reconstituting activity are $Flt3^+$ (Sitnicka et al., 2003). Flt3 deficient mice are viable and fertile, but show defects in B lymphocyte progenitors and dendritic cell generation (Mackarehtschian et al., 1995). The role of the FL-Flt3 axis in HSC maintenance and expansion remains controversial. Mackarehtschian et al. originally reported that Flt3-deficient bone marrow cells showed defects in lymphoid and myeloid reconstitution upon transplanting into myeloablated hosts (Mackarehtschian et al., 1995), while a more recent report by Buza-Vidas et al. concluded that Flt3 and FL were dispensable for maintenance and posttransplantation expansion of mouse HSCs (Buza-Vidas et al., 2009). Partly based on the phenotypes of Flt3 and FL deficient mice, models were proposed that Flt3 might function in the lineage restriction process from HSCs to lymphoid progenitors (Luc et al., 2007). However, in human, activating mutations in the *FLT3* gene, either in the form of internal tandem duplication (ITD) mutation in the juxtamembrane domain or point mutations in the kinase domain, are more frequently associated with myeloid malignancies rather than with lymphoid malignancies (Stirewalt & Radich, 2003). Clearly, the roles of Flt3 in the normal and pathological hematopoiesis need to be further delineated.

Cbl becomes tyrosine phosphorylated upon Flt3 engagement (Lavagna-Sévenier et al., 1998). It has been shown to physically interact with Flt3, and overexpression of an E3 ligase-defective mutant Cbl inhibited FL-induced Flt3 ubiquitylation and internalization, indicating involvement of Cbl in Flt3 signaling regulation (Sargin et al., 2007). Mice expressing a RF mutant Cbl from its endogenous locus are hypersensitive to FL stimulation (Rathinam et al., 2010), and we confirmed a similar phenotype in mouse bone marrow cells deficient in both Cbl and Cbl-b (Naramura, manuscript in preparation). Furthermore, deletion of FL blocks leukemia development in Cbl RING finger mutant mice (Rathinam et al., 2010).

Nevertheless, the mode of interaction between Cbl and Flt3 has not been clarified. Direct binding between Cbl and Flt3 has not been demonstrated. The sequences surrounding tyrosine 589 partially conform to the canonical Cbl(TKB) recognition sequence, and this region shares a very high homology to the sequences surrounding tyrosine 568 (a candidate Cbl binding site) in Kit. Notably, this is also the region frequently affected by ITD mutations. Alternatively, or in addition to the direct binding, because Flt3 is known to interact with Grb2 (Dosil et al., 1993; Zhang et al., 1999), a Cbl-binding adaptor protein, Cbl-Flt3 interaction may be mediated through this adaptor protein.

5. Other potential targets

As is clear from the list of potential Cbl upstream receptors, Kit and Flt3 may not be the only targets of Cbl-dependent regulation in HSCs. Although pathways other than Kit or Flt3 have not been as carefully examined in relation to Cbl, existing evidences suggest that following pathways may be regulated by the Cbl family proteins either directly or indirectly.

5.1 Tek

Tie2, encoded by the *TEK* gene, is an RTK expressed predominantly on endothelial cells, but they also provide crucial functions in the maintenance of quiescence and self-renewal capacity of the HSCs (Arai et al., 2004). The interaction between Tie2 and angiopoietin-1 (Ang-1), its ligand, has been shown to promote ubiquinylation of Tie2 by Cbl and receptor internalization (Wehrle et al., 2009). Structurally, the cytoplasmic domain of Tie2 does not contain any tyrosine residues that match the canonical Cbl(TKB) recognition sequence. However, activated Tie2 is known to bind Grb2 (Huang et al., 1995), thus may interact with Cbl indirectly through this adaptor.

5.2 Cytokine receptors

Cytokines such as hematopoietic growth factors and interleukins play essential roles in hematopoiesis. Cbl becomes tyrosine phosphorylated upon stimulation through various cytokine receptors (Table 2), and hematopoietic cells deficient in Cbl activity show enhanced sensitivity to cytokines (Rathinam et al., 2008; Sanada et al., 2009; Naramura et al., 2010). Receptors for these factors do not possess cytoplasmic tyrosine kinases but they activate the Janus kinase/Signal Transducers and Activators of Transcription (JAK/STAT) pathway (Yoshimura, 2009). Ligand binding induces receptor oligomerization, which activate associated JAK kinases and they, in turn, phosphorylate the receptor cytoplasmic domains and create binding sites for SH2-containing proteins.

There is no solid experimental evidence supporting the direct interaction between the JAK/STAT pathway and Cbl. Activation of the JAK/STAT pathway induces the expression of Suppressor of Cytokine Signaling (SOCS) family proteins, which function as E3 ubiquitin ligases for this pathway.

In addition to the JAK/STAT pathway, ligand binding to cytokine receptors activate the Ras-MAPK pathway through adaptor proteins such as APS and Grb2. Activation of this pathway is required for cell proliferation. As discussed above, these adaptor proteins are know to interact with Cbl, providing a potential link between the cytokine pathway and the Cbl family proteins.

6. Conclusion

In spite of its original identification as a cellular homolog of a viral oncogene, pathophysiological roles of the Cbl family proteins remained unclear for some time. Genetic studies in model organisms as well as identification of *CBL* mutations in patient-derived specimen played crucial roles in deciphering their essential functions as regulators of HSC homeostasis. Combined with molecular/biochemical information gathered over the last two decades, we now appreciate the complexity of the regulatory pathways surrounding the Cbl family proteins. While the primary focus of studies in the last ten years has been on Cbl's E3 ubiquitin ligase functions towards phosphotyrosine motif-containing targets, observations in cells expressing mutant Cbl proteins began to challenge this relatively-simplistic viewpoint. Further studies into this multifaceted protein family may uncover opportunities for novel diagnostics and therapeutics.

7. Acknowledgment

Works in the author's laboratory is supported by grants from the US Department of Defense Breast Cancer Research Program (W81XWH-10-1-0740) and the Nebraska Department of Health and Human Services (Stem Cell 2011-06).

8. References

Abbas, S., Rotmans, G., Löwenberg, B., & Valk, P. J. M. (2008). Exon 8 splice site mutations in the gene encoding the E3-ligase CBL are associated with core binding factor acute myeloid leukemias. *Haematologica*, Vol. 93, pp. 1595–1597.

Adolfsson, J., Borge, O. J., Bryder, D., Theilgaard-Mönch, K., Astrand-Grundström, I., Sitnicka, E., Sasaki, Y., & Jacobsen, S. E. (2001). Upregulation of Flt3 expression within the bone marrow Lin(-)Sca1(+)c-kit(+) stem cell compartment is accompanied by loss of self-renewal capacity. *Immunity*, Vol. 15, pp. 659–669.

Andoniou, C. E., Thien, C. B., & Langdon, W. Y. (1996). The two major sites of cbl tyrosine phosphorylation in abl-transformed cells select the crkL SH2 domain. *Oncogene*, Vol. 12, pp. 1981–1989.

Arai, F., Hirao, A., Ohmura, M., Sato, H., Matsuoka, S., Takubo, K., Ito, K., Koh, G. Y., & Suda, T. (2004). Tie2/angiopoietin-1 signaling regulates hematopoietic stem cell quiescence in the bone marrow niche. *Cell*, Vol. 118, pp. 149–161.

Bachmaier, K., Krawczyk, C., Kozieradzki, I., Kong, Y. Y., Sasaki, T., Oliveira-dos-Santos, A., Mariathasan, S., Bouchard, D., Wakeham, A., Itie, A., et al. (2000). Negative regulation of lymphocyte activation and autoimmunity by the molecular adaptor Cbl-b. *Nature*, Vol. 403, pp. 211–216.

Bachmaier, K., Toya, S., Gao, X., Triantafillou, T., Garrean, S., Park, G. Y., Frey, R. S., Vogel, S., Minshall, R., Christman, J. W., et al. (2007). E3 ubiquitin ligase Cblb regulates the acute inflammatory response underlying lung injury. *Nat. Med*, Vol. 13, pp. 920–926.

Barber, D. L., Mason, J. M., Fukazawa, T., Reedquist, K. A., Druker, B. J., Band, H., & D'Andrea, A. D. (1997). Erythropoietin and interleukin-3 activate tyrosine phosphorylation of CBL and association with CRK adaptor proteins. *Blood*, Vol. 89, pp. 3166–3174.

Besmer, P., Murphy, J. E., George, P. C., Qiu, F., Bergold, P. J., Lederman, L., Snyder, H. W., Brodeur, D., Zuckerman, E. E., & Hardy, W. D. (1986). A new acute transforming feline retrovirus and relationship of its oncogene v-kit with the protein kinase gene family. *Nature*, Vol. 320, pp. 415–421.

Bonita, D. P., Miyake, S., Lupher, M. L., Langdon, W. Y., & Band, H. (1997). Phosphotyrosine binding domain-dependent upregulation of the platelet-derived growth factor receptor α signaling cascade by transforming mutants of Cbl: implications for Cbl's function and oncogenicity. *Mol. Cell. Biol*, Vol. 17, pp. 4597–4610.

Brizzi, M. F., Dentelli, P., Lanfrancone, L., Rosso, A., Pelicci, P. G., & Pegoraro, L. (1996). Discrete protein interactions with the Grb2/c-Cbl complex in SCF- and TPO-mediated myeloid cell proliferation. *Oncogene*, Vol. 13, pp. 2067–2076.

Buza-Vidas, N., Cheng, M., Duarte, S., Nozad Charoudeh, H., Jacobsen, S. E. W., & Sitnicka, E. (2009). FLT3 receptor and ligand are dispensable for maintenance and posttransplantation expansion of mouse hematopoietic stem cells. *Blood*, Vol. 113, pp. 3453–3460.

Caligiuri, M. A., Briesewitz, R., Yu, J., Wang, L., Wei, M., Arnoczky, K. J., Marburger, T. B., Wen, J., Perrotti, D., Bloomfield, C. D., et al. (2007). Novel c-CBL and CBL-b ubiquitin ligase mutations in human acute myeloid leukemia. *Blood*, Vol. 110, pp. 1022–1024.

Chabot, B., Stephenson, D. A., Chapman, V. M., Besmer, P., & Bernstein, A. (1988). The proto-oncogene c-kit encoding a transmembrane tyrosine kinase receptor maps to the mouse W locus. *Nature*, Vol. 335, pp. 88–89.

Chernock, R. D., Cherla, R. P., & Ganju, R. K. (2001). SHP2 and cbl participate in α-chemokine receptor CXCR4-mediated signaling pathways. *Blood*, Vol. 97, pp. 608–615.

Chiang, J. Y., Jang, I. K., Hodes, R., & Gu, H. (2007). Ablation of Cbl-b provides protection against transplanted and spontaneous tumors. *J. Clin. Invest*, Vol. 117, pp. 1029–1036.

Chiang, Y. J., Kole, H. K., Brown, K., Naramura, M., Fukuhara, S., Hu, R. J., Jang, I. K., Gutkind, J. S., Shevach, E., & Gu, H. (2000). Cbl-b regulates the CD28 dependence of T-cell activation. *Nature*, Vol. 403, pp. 216–220.

Christensen, J. L., & Weissman, I. L. (2001). Flk-2 is a marker in hematopoietic stem cell differentiation: a simple method to isolate long-term stem cells. *Proc. Natl. Acad. Sci. U.S.A*, Vol. 98, pp. 14541–14546.

Cory, G. O., Lovering, R. C., Hinshelwood, S., MacCarthy-Morrogh, L., Levinsky, R. J., & Kinnon, C. (1995). The protein product of the c-cbl protooncogene is phosphorylated after B cell receptor stimulation and binds the SH3 domain of Bruton's tyrosine kinase. *J. Exp. Med.*, Vol. 182, pp. 611–615.

Donovan, J. A., Wange, R. L., Langdon, W. Y., & Samelson, L. E. (1994). The protein product of the c-cbl protooncogene is the 120-kDa tyrosine-phosphorylated protein in Jurkat cells activated via the T cell antigen receptor. *J. Biol. Chem*, Vol. 269, pp. 22921–22924.

Dosil, M., Wang, S., & Lemischka, I. R. (1993). Mitogenic signalling and substrate specificity of the Flk2/Flt3 receptor tyrosine kinase in fibroblasts and interleukin 3-dependent hematopoietic cells. *Mol. Cell. Biol.*, Vol. 13, pp. 6572–6585.

Duan, L., Reddi, A. L., Ghosh, A., Dimri, M., & Band, H. (2004). The Cbl family and other ubiquitin ligases: destructive forces in control of antigen receptor signaling. *Immunity*, Vol. 21, pp. 7-17.

Dunbar, A. J., Gondek, L. P., O'Keefe, C. L., Makishima, H., Rataul, M. S., Szpurka, H., Sekeres, M. A., Wang, X. F., McDevitt, M. A., & Maciejewski, J. P. (2008). 250K single nucleotide polymorphism array karyotyping identifies acquired uniparental disomy and homozygous mutations, including novel missense substitutions of c-Cbl, in myeloid malignancies. *Cancer Res*, Vol. 68, pp. 10349–10357.

Durham, F. M. (1908). A preliminary account of the inheritance of coat colour in mice. *Reports to the Evolution Committee of the Royal Society*, Vol. 4, pp. 41-53.

El Chami, N., Ikhlef, F., Kaszas, K., Yakoub, S., Tabone, E., Siddeek, B., Cunha, S., Beaudoin, C., Morel, L., Benahmed, M., et al. (2005). Androgen-dependent apoptosis in male germ cells is regulated through the proto-oncoprotein Cbl. *J. Cell Biol*, Vol. 171, pp. 651-661.

Fernandes, M. S., Reddy, M. M., Croteau, N. J., Walz, C., Weisbach, H., Podar, K., Band, H., Carroll, M., Reiter, A., Larson, R. A., et al. (2010). Novel oncogenic mutations of CBL in human acute myeloid leukemia that activate growth and survival pathways depend on increased metabolism. *J. Biol. Chem*, Vol. 285, pp. 32596-32605.

Fixman, E. D., Holgado-Madruga, M., Nguyen, L., Kamikura, D. M., Fournier, T. M., Wong, A. J., & Park, M. (1997). Efficient cellular transformation by the Met oncoprotein requires a functional Grb2 binding site and correlates with phosphorylation of the Grb2-associated proteins, Cbl and Gab1. *J. Biol. Chem*, Vol. 272, pp. 20167-20172.

Fukazawa, T., Reedquist, K. A., Trub, T., Soltoff, S., Panchamoorthy, G., Druker, B., Cantley, L., Shoelson, S. E., & Band, H. (1995). The SH3 domain-binding T cell tyrosyl phosphoprotein p120. Demonstration of its identity with the c-cbl protooncogene product and in vivo complexes with Fyn, Grb2, and phosphatidylinositol 3-kinase. *J. Biol. Chem*, Vol. 270, pp. 19141-19150.

Galisteo, M. L., Dikic, I., Batzer, A. G., Langdon, W. Y., & Schlessinger, J. (1995). Tyrosine phosphorylation of the c-cbl proto-oncogene protein product and association with epidermal growth factor (EGF) receptor upon EGF stimulation. *J. Biol. Chem*, Vol. 270, pp. 20242-20245.

Galli, S. J., Tsai, M., & Wershil, B. K. (1993). The c-kit receptor, stem cell factor, and mast cells. What each is teaching us about the others. *Am. J. Pathol.*, Vol. 142, pp. 965-974.

Garcia-Guzman, M., Larsen, E., & Vuori, K. (2000). The proto-oncogene c-Cbl is a positive regulator of Met-induced MAP kinase activation: a role for the adaptor protein Crk. *Oncogene*, Vol. 19, pp. 4058-4065.

Geissler, E. N., Ryan, M. A., & Housman, D. E. (1988). The dominant-white spotting (W) locus of the mouse encodes the c-kit proto-oncogene. *Cell*, Vol. 55, pp. 185-192.

Gesbert, F., Garbay, C., & Bertoglio, J. (1998). Interleukin-2 stimulation induces tyrosine phosphorylation of p120-Cbl and CrkL and formation of multimolecular signaling complexes in T lymphocytes and natural killer cells. *J. Biol. Chem.*, Vol. 273, pp. 3986-3993.

Grand, F. H., Hidalgo-Curtis, C. E., Ernst, T., Zoi, K., Zoi, C., McGuire, C., Kreil, S., Jones, A., Score, J., Metzgeroth, G., et al. (2009). Frequent CBL mutations associated with 11q acquired uniparental disomy in myeloproliferative neoplasms. *Blood*, Vol. 113, pp. 6182-6192.

Griffiths, E. K., Sanchez, O., Mill, P., Krawczyk, C., Hojilla, C. V., Rubin, E., Nau, M. M., Khokha, R., Lipkowitz, S., Hui, C.-C., et al. (2003). Cbl-3-deficient mice exhibit normal epithelial development. *Mol. Cell. Biol*, Vol. 23, pp. 7708-7718.

Hannum, C., Culpepper, J., Campbell, D., McClanahan, T., Zurawski, S., Kastelein, R., Bazan, J. F., Hudak, S., Wagner, J., Mattson, J., et al. (1994). Ligand for FLT3/FLK2 receptor tyrosine kinase regulates growth of haematopoietic stem cells and is encoded by variant RNAs. *Nature*, Vol. 368, pp. 643-648.

He, L., & Hristova, K. Physical-chemical principles underlying RTK activation, and their implications for human disease. *Biochimica et Biophysica Acta (BBA) - Biomembranes*, Vol. In Press, Uncorrected Proof. Available at: http://www.sciencedirect.com/science/article/pii/S0005273611002495 [Accessed September 1, 2011].

Hu, J., & Hubbard, S. R. (2005). Structural Characterization of a Novel Cbl Phosphotyrosine Recognition Motif in the APS Family of Adapter Proteins. *Journal of Biological Chemistry*, Vol. 280, pp. 18943–18949.

Huang, F., Kitaura, Y., Jang, I., Naramura, M., Kole, H. H., Liu, L., Qin, H., Schlissel, M. S., & Gu, H. (2006). Establishment of the major compatibility complex-dependent development of CD4+ and CD8+ T cells by the Cbl family proteins. *Immunity*, Vol. 25, pp. 571–581.

Huang, L., Turck, C. W., Rao, P., & Peters, K. G. (1995). GRB2 and SH-PTP2: potentially important endothelial signaling molecules downstream of the TEK/TIE2 receptor tyrosine kinase. *Oncogene*, Vol. 11, pp. 2097–2103.

Hunter, S., Burton, E. A., Wu, S. C., & Anderson, S. M. (1999). Fyn Associates with Cbl and Phosphorylates Tyrosine 731 in Cbl, A Binding Site for Phosphatidylinositol 3-Kinase. *Journal of Biological Chemistry*, Vol. 274, pp. 2097–2106.

Hunter, S., Koch, B. L., & Anderson, S. M. (1997). Phosphorylation of cbl after stimulation of Nb2 cells with prolactin and its association with phosphatidylinositol 3-kinase. *Mol. Endocrinol*, Vol. 11, pp. 1213–1222.

Ikuta, K., & Weissman, I. L. (1992). Evidence that hematopoietic stem cells express mouse c-kit but do not depend on steel factor for their generation. *Proceedings of the National Academy of Sciences*, Vol. 89, pp. 1502–1506.

Joazeiro, C. A., Wing, S. S., Huang, H., Leverson, J. D., Hunter, T., & Liu, Y. C. (1999). The tyrosine kinase negative regulator c-Cbl as a RING-type, E2-dependent ubiquitin-protein ligase. *Science*, Vol. 286, pp. 309-312.

Kitaura, Y., Jang, I. K., Wang, Y., Han, Y.-C., Inazu, T., Cadera, E. J., Schlissel, M., Hardy, R. R., & Gu, H. (2007). Control of the B cell-intrinsic tolerance programs by ubiquitin ligases Cbl and Cbl-b. *Immunity*, Vol. 26, pp. 567–578.

Kozlov, G., Peschard, P., Zimmerman, B., Lin, T., Moldoveanu, T., Mansur-Azzam, N., Gehring, K., & Park, M. (2007). Structural basis for UBA-mediated dimerization of c-Cbl ubiquitin ligase. *J. Biol. Chem*, Vol. 282, pp. 27547–27555.

Krawczyk, C., Bachmaier, K., Sasaki, T., Jones, R. G., Snapper, S. B., Bouchard, D., Kozieradzki, I., Ohashi, P. S., Alt, F. W., & Penninger, J. M. (2000). Cbl-b is a negative regulator of receptor clustering and raft aggregation in T cells. *Immunity*, Vol. 13, pp. 463–473.

Langdon, W. Y., Hartley, J. W., Klinken, S. P., Ruscetti, S. K., & Morse, H. C. (1989). v-cbl, an oncogene from a dual-recombinant murine retrovirus that induces early B-lineage lymphomas. *Proc. Natl. Acad. Sci. U.S.A*, Vol. 86, pp. 1168–1172.

Lavagna-Sévenier, C., Marchetto, S., Birnbaum, D., & Rosnet, O. (1998). FLT3 signaling in hematopoietic cells involves CBL, SHC and an unknown P115 as prominent tyrosine-phosphorylated substrates. *Leukemia*, Vol. 12, pp. 301–310.

Lee, P. S., Wang, Y., Dominguez, M. G., Yeung, Y. G., Murphy, M. A., Bowtell, D. D., & Stanley, E. R. (1999). The Cbl protooncoprotein stimulates CSF-1 receptor multiubiquitination and endocytosis, and attenuates macrophage proliferation. *EMBO J*, Vol. 18, pp. 3616–3628.

Lennartsson, J., Jelacic, T., Linnekin, D., & Shivakrupa, R. (2005). Normal and Oncogenic Forms of the Receptor Tyrosine Kinase Kit. *STEM CELLS*, Vol. 23, pp. 16–43.

Levkowitz, G., Waterman, H., Ettenberg, S. A., Katz, M., Tsygankov, A. Y., Alroy, I., Lavi, S., Iwai, K., Reiss, Y., Ciechanover, A., et al. (1999). Ubiquitin ligase activity and tyrosine phosphorylation underlie suppression of growth factor signaling by c-Cbl/Sli-1. *Mol. Cell*, Vol. 4, pp. 1029–1040.

Loeser, S., Loser, K., Bijker, M. S., Rangachari, M., van der Burg, S. H., Wada, T., Beissert, S., Melief, C. J. M., & Penninger, J. M. (2007). Spontaneous tumor rejection by cbl-b-deficient CD8+ T cells. *J. Exp. Med*, Vol. 204, pp. 879–891.

Loh, M. L. (2011). Recent advances in the pathogenesis and treatment of juvenile myelomonocytic leukaemia. *Br. J. Haematol.*, Vol. 152, pp. 677–687.

Loh, M. L., Sakai, D. S., Flotho, C., Kang, M., Fliegauf, M., Archambeault, S., Mullighan, C. G., Chen, L., Bergstraesser, E., Bueso-Ramos, C. E., et al. (2009). Mutations in CBL occur frequently in juvenile myelomonocytic leukemia. *Blood*, Vol. 114, pp. 1859–1863.

Luc, S., Buza-Vidas, N., & Jacobsen, S. E. W. (2007). Biological and molecular evidence for existence of lymphoid-primed multipotent progenitors. *Ann. N. Y. Acad. Sci.*, Vol. 1106, pp. 89–94.

Lupher, M. L., Reedquist, K. A., Miyake, S., Langdon, W. Y., & Band, H. (1996). A novel phosphotyrosine-binding domain in the N-terminal transforming region of Cbl interacts directly and selectively with ZAP-70 in T cells. *J. Biol. Chem*, Vol. 271, pp. 24063–24068.

Lupher, M. L., Songyang, Z., Shoelson, S. E., Cantley, L. C., & Band, H. (1997). The Cbl phosphotyrosine-binding domain selects a D(N/D)XpY motif and binds to the Tyr292 negative regulatory phosphorylation site of ZAP-70. *J. Biol. Chem*, Vol. 272, pp. 33140–33144.

Lyman, S. D., James, L., Vanden Bos, T., de Vries, P., Brasel, K., Gliniak, B., Hollingsworth, L. T., Picha, K. S., McKenna, H. J., & Splett, R. R. (1993). Molecular cloning of a ligand for the flt3/flk-2 tyrosine kinase receptor: a proliferative factor for primitive hematopoietic cells. *Cell*, Vol. 75, pp. 1157–1167.

Mackarehtschian, K., Hardin, J. D., Moore, K. A., Boast, S., Goff, S. P., & Lemischka, I. R. (1995). Targeted disruption of the flk2/flt3 gene leads to deficiencies in primitive hematopoietic progenitors. *Immunity*, Vol. 3, pp. 147–161.

Makishima, H., Cazzolli, H., Szpurka, H., Dunbar, A., Tiu, R., Huh, J., Muramatsu, H., O'Keefe, C., Hsi, E., Paquette, R. L., et al. (2009). Mutations of e3 ubiquitin ligase cbl family members constitute a novel common pathogenic lesion in myeloid malignancies. *J. Clin. Oncol*, Vol. 27, pp. 6109–6116.

Makishima, H., Jankowska, A. M., McDevitt, M. A., O'Keefe, C., Dujardin, S., Cazzolli, H., Przychodzen, B., Prince, C., Nicoll, J., Siddaiah, H., et al. (2011). CBL, CBLB, TET2, ASXL1, and IDH1/2 mutations and additional chromosomal aberrations constitute molecular events in chronic myelogenous leukemia. *Blood*, Vol. 117, pp. e198–e206.

Manié, S. N., Sattler, M., Astier, A., Phifer, J. S., Canty, T., Morimoto, C., Druker, B. J., Salgia, R., Griffin, J. D., & Freedman, A. S. (1997). Tyrosine phosphorylation of the product of the c-cbl protooncogene is [corrected] induced after integrin stimulation. *Exp. Hematol.*, Vol. 25, pp. 45-50.

Marcilla, A., Rivero-Lezcano, O. M., Agarwal, A., & Robbins, K. C. (1995). Identification of the major tyrosine kinase substrate in signaling complexes formed after engagement of Fcγ receptors. *J. Biol. Chem.*, Vol. 270, pp. 9115-9120.

Marengère, L. E., Mirtsos, C., Kozieradzki, I., Veillette, A., Mak, T. W., & Penninger, J. M. (1997). Proto-oncoprotein Vav interacts with c-Cbl in activated thymocytes and peripheral T cells. *J. Immunol*, Vol. 159, pp. 70-76.

Martin, G. S. (2001). The hunting of the Src. *Nat Rev Mol Cell Biol*, Vol. 2, pp. 467-475.

Masson, K., Heiss, E., Band, H., & Rönnstrand, L. (2006). Direct binding of Cbl to Tyr568 and Tyr936 of the stem cell factor receptor/c-Kit is required for ligand-induced ubiquitination, internalization and degradation. *Biochem. J.*, Vol. 399, pp. 59.

Masson, K., & Rönnstrand, L. (2009). Oncogenic signaling from the hematopoietic growth factor receptors c-Kit and Flt3. *Cellular Signalling*, Vol. 21, pp. 1717-1726.

Matsuo, T., Hazeki, K., Hazeki, O., Katada, T., & Ui, M. (1996). Specific association of phosphatidylinositol 3-kinase with the protooncogene product Cbl in Fcγ receptor signaling. *FEBS Lett.*, Vol. 382, pp. 11-14.

Matthews, W., Jordan, C. T., Wiegand, G. W., Pardoll, D., & Lemischka, I. R. (1991). A receptor tyrosine kinase specific to hematopoietic stem and progenitor cell-enriched populations. *Cell*, Vol. 65, pp. 1143-1152.

McCarty, J. H., & Feinstein, S. C. (1999). The TrkB receptor tyrosine kinase regulates cellular proliferation via signal transduction pathways involving SHC, PLCγ, and CBL. *J. Recept. Signal Transduct. Res.*, Vol. 19, pp. 953-974.

Meisner, H., Conway, B. R., Hartley, D., & Czech, M. P. (1995). Interactions of Cbl with Grb2 and phosphatidylinositol 3'-kinase in activated Jurkat cells. *Mol. Cell. Biol.*, Vol. 15, pp. 3571-3578.

Meng, F., & Lowell, C. A. (1998). A β1 integrin signaling pathway involving Src-family kinases, Cbl and PI-3 kinase is required for macrophage spreading and migration. *EMBO J.*, Vol. 17, pp. 4391-4403.

Meng, W., Sawasdikosol, S., Burakoff, S. J., & Eck, M. J. (1999). Structure of the amino-terminal domain of Cbl complexed to its binding site on ZAP-70 kinase. *Nature*, Vol. 398, pp. 84-90.

Molero, J. C., Jensen, T. E., Withers, P. C., Couzens, M., Herzog, H., Thien, C. B. F., Langdon, W. Y., Walder, K., Murphy, M. A., Bowtell, D. D. L., et al. (2004). c-Cbl-deficient mice have reduced adiposity, higher energy expenditure, and improved peripheral insulin action. *J. Clin. Invest*, Vol. 114, pp. 1326-1333.

Muramatsu, H., Makishima, H., Jankowska, A. M., Cazzolli, H., O'Keefe, C., Yoshida, N., Xu, Y., Nishio, N., Hama, A., Yagasaki, H., et al. (2010). Mutations of an E3 ubiquitin ligase c-Cbl but not TET2 mutations are pathogenic in juvenile myelomonocytic leukemia. *Blood*, Vol. 115, pp. 1969-1975.

Murphy, J. E., Padilla, B. E., Hasdemir, B., Cottrell, G. S., & Bunnett, N. W. (2009). Endosomes: A legitimate platform for the signaling train. *Proceedings of the National Academy of Sciences*, Vol. 106, pp. 17615-17622.

Murphy, M. A., Schnall, R. G., Venter, D. J., Barnett, L., Bertoncello, I., Thien, C. B., Langdon, W. Y., & Bowtell, D. D. (1998). Tissue hyperplasia and enhanced T-cell signalling via ZAP-70 in c-Cbl-deficient mice. *Mol. Cell. Biol*, Vol. 18, pp. 4872-4882.

Naramura, M., Band, V., & Band, H. (2011a). Indispensable roles of mammalian Cbl family proteins as negative regulators of protein tyrosine kinase signaling: Insights from in vivo models. *Commun Integr Biol*, Vol. 4, pp. 159-162.

Naramura, M., Jang, I.-K., Kole, H., Huang, F., Haines, D., & Gu, H. (2002). c-Cbl and Cbl-b regulate T cell responsiveness by promoting ligand-induced TCR down-modulation. *Nat. Immunol*, Vol. 3, pp. 1192-1199.

Naramura, M., Kole, H. K., Hu, R. J., & Gu, H. (1998). Altered thymic positive selection and intracellular signals in Cbl-deficient mice. *Proc. Natl. Acad. Sci. U.S.A*, Vol. 95, pp. 15547-15552.

Naramura, M., Nadeau, S., Mohapatra, B., Ahmad, G., Mukhopadhyay, C., Sattler, M., Raja, S. M., Natarajan, A., Band, V., & Band, H. (2011b). Mutant Cbl proteins as oncogenic drivers in myeloproliferative disorders. *Oncotarget*. Available at: http://www.ncbi.nlm.nih.gov/pubmed/21422499 [Accessed March 27, 2011].

Naramura, M., Nandwani, N., Gu, H., Band, V., & Band, H. (2010). Rapidly fatal myeloproliferative disorders in mice with deletion of Casitas B-cell lymphoma (Cbl) and Cbl-b in hematopoietic stem cells. *Proc. Natl. Acad. Sci. U.S.A*, Vol. 107, pp. 16274-16279.

Ng, C., Jackson, R. A., Buschdorf, J. P., Sun, Q., Guy, G. R., & Sivaraman, J. (2008). Structural basis for a novel intrapeptidyl H-bond and reverse binding of c-Cbl-TKB domain substrates. *EMBO J*, Vol. 27, pp. 804-816.

Niemeyer, C. M., Kang, M. W., Shin, D. H., Furlan, I., Erlacher, M., Bunin, N. J., Bunda, S., Finklestein, J. Z., Sakamoto, K. M., Gorr, T. A., et al. (2010). Germline CBL mutations cause developmental abnormalities and predispose to juvenile myelomonocytic leukemia. *Nat. Genet*, Vol. 42, pp. 794-800.

Odai, H., Sasaki, K., Iwamatsu, A., Hanazono, Y., Tanaka, T., Mitani, K., Yazaki, Y., & Hirai, H. (1995). The proto-oncogene product c-Cbl becomes tyrosine phosphorylated by stimulation with GM-CSF or Epo and constitutively binds to the SH3 domain of Grb2/Ash in human hematopoietic cells. *J. Biol. Chem.*, Vol. 270, pp. 10800-10805.

Ojaniemi, M., Martin, S. S., Dolfi, F., Olefsky, J. M., & Vuori, K. (1997). The proto-oncogene product p120(cbl) links c-Src and phosphatidylinositol 3'-kinase to the integrin signaling pathway. *J. Biol. Chem*, Vol. 272, pp. 3780-3787.

Panchamoorthy, G., Fukazawa, T., Miyake, S., Soltoff, S., Reedquist, K., Druker, B., Shoelson, S., Cantley, L., & Band, H. (1996). p120cbl is a major substrate of tyrosine phosphorylation upon B cell antigen receptor stimulation and interacts in vivo with Fyn and Syk tyrosine kinases, Grb2 and Shc adaptors, and the p85 subunit of phosphatidylinositol 3-kinase. *J. Biol. Chem*, Vol. 271, pp. 3187-3194.

Peschard, P., Ishiyama, N., Lin, T., Lipkowitz, S., & Park, M. (2004). A conserved DpYR motif in the juxtamembrane domain of the Met receptor family forms an atypical c-Cbl/Cbl-b tyrosine kinase binding domain binding site required for suppression of oncogenic activation. *J. Biol. Chem*, Vol. 279, pp. 29565-29571.

Peschard, P., Kozlov, G., Lin, T., Mirza, I. A., Berghuis, A. M., Lipkowitz, S., Park, M., & Gehring, K. (2007). Structural basis for ubiquitin-mediated dimerization and activation of the ubiquitin protein ligase Cbl-b. *Mol. Cell*, Vol. 27, pp. 474-485.

Peschard, P., & Park, M. (2003). Escape from Cbl-mediated downregulation: a recurrent theme for oncogenic deregulation of receptor tyrosine kinases. *Cancer Cell*, Vol. 3, pp. 519-523.

Qian, H., Buza-Vidas, N., Hyland, C. D., Jensen, C. T., Antonchuk, J., Månsson, R., Thoren, L. A., Ekblom, M., Alexander, W. S., & Jacobsen, S. E. W. (2007). Critical role of thrombopoietin in maintaining adult quiescent hematopoietic stem cells. *Cell Stem Cell*, Vol. 1, pp. 671–684.

Rathinam, C., Thien, C. B. F., Flavell, R. A., & Langdon, W. Y. (2010). Myeloid leukemia development in c-Cbl RING finger mutant mice is dependent on FLT3 signaling. *Cancer Cell*, Vol. 18, pp. 341–352.

Rathinam, C., Thien, C. B. F., Langdon, W. Y., Gu, H., & Flavell, R. A. (2008). The E3 ubiquitin ligase c-Cbl restricts development and functions of hematopoietic stem cells. *Genes Dev*, Vol. 22, pp. 992–997.

Reindl, C., Quentmeier, H., Petropoulos, K., Greif, P. A., Benthaus, T., Argiropoulos, B., Mellert, G., Vempati, S., Duyster, J., Buske, C., et al. (2009). CBL exon 8/9 mutants activate the FLT3 pathway and cluster in core binding factor/11q deletion acute myeloid leukemia/myelodysplastic syndrome subtypes. *Clin. Cancer Res*, Vol. 15, pp. 2238–2247.

Ribon, V., & Saltiel, A. R. (1997). Insulin stimulates tyrosine phosphorylation of the proto-oncogene product of c-Cbl in 3T3-L1 adipocytes. *Biochem. J.*, Vol. 324 (Pt 3), pp. 839–845.

Rivero-Lezcano, O. M., Sameshima, J. H., Marcilla, A., & Robbins, K. C. (1994). Physical association between Src homology 3 elements and the protein product of the c-cbl proto-oncogene. *J. Biol. Chem.*, Vol. 269, pp. 17363–17366.

Rosnet, O., Marchetto, S., deLapeyriere, O., & Birnbaum, D. (1991). Murine Flt3, a gene encoding a novel tyrosine kinase receptor of the PDGFR/CSF1R family. *Oncogene*, Vol. 6, pp. 1641–1650.

Rous, P. (1911). A SARCOMA OF THE FOWL TRANSMISSIBLE BY AN AGENT SEPARABLE FROM THE TUMOR CELLS. *The Journal of Experimental Medicine*, Vol. 13, pp. 397–411.

Russell, E. S. (1979). Hereditary anemias of the mouse: a review for geneticists. *Adv. Genet.*, Vol. 20, pp. 357–459.

Sanada, M., Suzuki, T., Shih, L.-Y., Otsu, M., Kato, M., Yamazaki, S., Tamura, A., Honda, H., Sakata-Yanagimoto, M., Kumano, K., et al. (2009). Gain-of-function of mutated C-CBL tumour suppressor in myeloid neoplasms. *Nature*, Vol. 460, pp. 904–908.

Sargin, B., Choudhary, C., Crosetto, N., Schmidt, M. H. H., Grundler, R., Rensinghoff, M., Thiessen, C., Tickenbrock, L., Schwäble, J., Brandts, C., et al. (2007). Flt3-dependent transformation by inactivating c-Cbl mutations in AML. *Blood*, Vol. 110, pp. 1004–1012.

Sasaki, K., Odai, H., Hanazono, Y., Ueno, H., Ogawa, S., Langdon, W. Y., Tanaka, T., Miyagawa, K., Mitani, K., & Yazaki, Y. (1995). TPO/c-mpl ligand induces tyrosine phosphorylation of multiple cellular proteins including proto-oncogene products, Vav and c-Cbl, and Ras signaling molecules. *Biochem. Biophys. Res. Commun*, Vol. 216, pp. 338–347.

Sattler, M., Salgia, R., Shrikhande, G., Verma, S., Pisick, E., Prasad, K. V., & Griffin, J. D. (1997). Steel factor induces tyrosine phosphorylation of CRKL and binding of CRKL to a complex containing c-kit, phosphatidylinositol 3-kinase, and p120(CBL). *J. Biol. Chem*, Vol. 272, pp. 10248–10253.

Saur, S. J., Sangkhae, V., Geddis, A. E., Kaushansky, K., & Hitchcock, I. S. (2010). Ubiquitination and degradation of the thrombopoietin receptor c-Mpl. *Blood*, Vol. 115, pp. 1254–1263.

Schmidt, M. H. H., & Dikic, I. (2005). The Cbl interactome and its functions. *Nat. Rev. Mol. Cell Biol*, Vol. 6, pp. 907–918.

Schulman, B. A., & Harper, J. W. (2009). Ubiquitin-like protein activation by E1 enzymes: the apex for downstream signalling pathways. *Nat Rev Mol Cell Biol*, Vol. 10, pp. 319-331.

Sen, B., & Johnson, F. M. (2011). Regulation of Src Family Kinases in Human Cancers. *J Signal Transduct*, Vol. 2011.

Sitnicka, E., Buza-Vidas, N., Larsson, S., Nygren, J. M., Liuba, K., & Jacobsen, S. E. W. (2003). Human CD34+ hematopoietic stem cells capable of multilineage engrafting NOD/SCID mice express flt3: distinct flt3 and c-kit expression and response patterns on mouse and candidate human hematopoietic stem cells. *Blood*, Vol. 102, pp. 881–886.

Stirewalt, D. L., & Radich, J. P. (2003). The role of FLT3 in haematopoietic malignancies. *Nat. Rev. Cancer*, Vol. 3, pp. 650–665.

Suzuki, H., Takei, M., Yanagida, M., Nakahata, T., Kawakami, T., & Fukamachi, H. (1997). Early and late events in Fcε RI signal transduction in human cultured mast cells. *J. Immunol.*, Vol. 159, pp. 5881–5888.

Tezuka, T., Umemori, H., Fusaki, N., Yagi, T., Takata, M., Kurosaki, T., & Yamamoto, T. (1996). Physical and functional association of the cbl protooncogen product with an src-family protein tyrosine kinase, p53/56lyn, in the B cell antigen receptor-mediated signaling. *J. Exp. Med.*, Vol. 183, pp. 675–680.

Thien, C. B. F., Blystad, F. D., Zhan, Y., Lew, A. M., Voigt, V., Andoniou, C. E., & Langdon, W. Y. (2005). Loss of c-Cbl RING finger function results in high-intensity TCR signaling and thymic deletion. *EMBO J*, Vol. 24, pp. 3807–3819.

Thien, C. B., Bowtell, D. D., & Langdon, W. Y. (1999). Perturbed regulation of ZAP-70 and sustained tyrosine phosphorylation of LAT and SLP-76 in c-Cbl-deficient thymocytes. *J. Immunol*, Vol. 162, pp. 7133–7139.

Thien, C. B., & Langdon, W. Y. (2001). Cbl: many adaptations to regulate protein tyrosine kinases. *Nat. Rev. Mol. Cell Biol*, Vol. 2, pp. 294–307.

Ueno, H., Sasaki, K., Honda, H., Nakamoto, T., Yamagata, T., Miyagawa, K., Mitani, K., Yazaki, Y., & Hirai, H. (1998). c-Cbl is tyrosine-phosphorylated by interleukin-4 and enhances mitogenic and survival signals of interleukin-4 receptor by linking with the phosphatidylinositol 3'-kinase pathway. *Blood*, Vol. 91, pp. 46–53.

Wang, Y., Yeung, Y. G., Langdon, W. Y., & Stanley, E. R. (1996). c-Cbl is transiently tyrosine-phosphorylated, ubiquitinated, and membrane-targeted following CSF-1 stimulation of macrophages. *J. Biol. Chem*, Vol. 271, pp. 17–20.

Wehrle, C., Van Slyke, P., & Dumont, D. J. (2009). Angiopoietin-1-induced ubiquitylation of Tie2 by c-Cbl is required for internalization and degradation. *Biochem. J.*, Vol. 423, pp. 375–380.

van Wijk, S. J. L., & Timmers, H. T. M. (2010). The family of ubiquitin-conjugating enzymes (E2s): deciding between life and death of proteins. *The FASEB Journal*, Vol. 24, pp. 981–993.

Wisniewski, D., Strife, A., & Clarkson, B. (1996). c-kit ligand stimulates tyrosine phosphorylation of the c-Cbl protein in human hematopoietic cells. *Leukemia*, Vol. 10, pp. 1436–1442.

Wollberg, P., Lennartsson, J., Gottfridsson, E., Yoshimura, A., & Rönnstrand, L. (2003). The adapter protein APS associates with the multifunctional docking sites Tyr-568 and Tyr-936 in c-Kit. *Biochem. J*, Vol. 370, pp. 1033–1038.

Wong, A., Lamothe, B., Lee, A., Schlessinger, J., Lax, I., & Li, A. (2002). FRS2 α attenuates FGF receptor signaling by Grb2-mediated recruitment of the ubiquitin ligase Cbl. *Proc. Natl. Acad. Sci. U.S.A.*, Vol. 99, pp. 6684–6689.

Yokouchi, M., Kondo, T., Houghton, A., Bartkiewicz, M., Horne, W. C., Zhang, H., Yoshimura, A., & Baron, R. (1999). Ligand-induced ubiquitination of the epidermal growth factor receptor involves the interaction of the c-Cbl RING finger and UbcH7. *J. Biol. Chem*, Vol. 274, pp. 31707–31712.

Yoon, C. H., Lee, J., Jongeward, G. D., & Sternberg, P. W. (1995). Similarity of sli-1, a regulator of vulval development in C. elegans, to the mammalian proto-oncogene c-cbl. *Science*, Vol. 269, pp. 1102–1105.

Yoshihara, H., Arai, F., Hosokawa, K., Hagiwara, T., Takubo, K., Nakamura, Y., Gomei, Y., Iwasaki, H., Matsuoka, S., Miyamoto, K., et al. (2007). Thrombopoietin/MPL signaling regulates hematopoietic stem cell quiescence and interaction with the osteoblastic niche. *Cell Stem Cell*, Vol. 1, pp. 685–697.

Yoshimura, A. (2009). Regulation of cytokine signaling by the SOCS and Spred family proteins. *Keio J Med*, Vol. 58, pp. 73–83.

Zeng, S., Xu, Z., Lipkowitz, S., & Longley, J. B. (2005). Regulation of stem cell factor receptor signaling by Cbl family proteins (Cbl-b/c-Cbl). *Blood*, Vol. 105, pp. 226–232.

Zhang, S., Mantel, C., & Broxmeyer, H. E. (1999). Flt3 signaling involves tyrosyl-phosphorylation of SHP-2 and SHIP and their association with Grb2 and Shc in Baf3/Flt3 cells. *J. Leukoc. Biol.*, Vol. 65, pp. 372–380.

Zheng, N., Wang, P., Jeffrey, P. D., & Pavletich, N. P. (2000). Structure of a c-Cbl-UbcH7 complex: RING domain function in ubiquitin-protein ligases. *Cell*, Vol. 102, pp. 533–539.

7

The Hypoxia Regulatory System in Hematopoietic Stem Cells

Keiyo Takubo
Department of Cell Differentiation, The Sakaguchi Laboratory of Developmental Biology, Keio University School of Medicine, Tokyo, Japan

1. Introduction

Stem cells localize to specific sites called 'niches' in various tissues, where they are preferentially maintained by growth factors from the environment. Mammalian bone marrow (BM) has been shown to be relatively hypoxic compared to other tissues, and primitive hematopoietic cells, including hematopoietic stem cells (HSCs), are thought to localize to the most hypoxic microenvironments in the BM. The hypoxic *ex vivo* culture of BM cells or primitive hematopoietic progenitors results in the maintenance of the primitive phenotype and cell cycle quiescence (Mohyeldin et al., 2010; Suda et al., 2011). *Ex vivo* culture of human HSCs under hypoxia also stabilizes hypoxia-inducible factor-1α (HIF-1α), a master transcriptional regulator of the cellular and systemic hypoxic response, and induces various downstream effectors of HIF-1α (Danet et al., 2003). However, the regulatory mechanisms and functional effects of BM hypoxia on HSCs *in vivo* have not been fully elucidated.

In the stem cell niche, HSCs are quiescent and show slow cell cycling. Various extracellular ligands, including CXCL12 (Sugiyama et al., 2006), angiopoietin-1 (Arai et al., 2004), and/or thrombopoietin (TPO) (Qian et al., 2007; Yoshihara et al., 2007), contribute to the quiescence of HSCs. Quiescent HSCs are maintained at a lower oxidative stress state to avoid their differentiation and exhaustion (Jang & Sharkis 2007). HIF-1α is a bHLH-PAS–type transcription factor (Semenza, 2007, 2009, 2010). Under normoxic conditions, prolyl residues in the HIF-1α oxygen-dependent degradation domain (ODD) are hydroxylated by HIF prolyl hydroxylases (PHDs). The hydroxylated ODD domain of HIF-1α protein is recognized by an E3 ubiquitin ligase, the von Hippel-Lindau protein (VHL). In the autosomal dominant hereditary disorder von Hippel Lindau disease, VHL is mutated, resulting in overstabilized HIF-1α protein by the impaired ubiquitin-proteasome pathway. Under hypoxic conditions, PHDs are inactivated and HIF-1α protein escapes degradation. Several niche factors, such as thrombopoietin (TPO) (Kirito et al., 2005) and stem cell factor (SCF) (Pedersen et al., 2008), also stabilize HIF-1α protein in hematopoietic cells even under normoxic conditions.

Stabilized HIF-1α protein forms a heterodimeric transcriptional complex with the oxygen-independent subunit HIF-1β, translocates to the nucleus, and directly binds hypoxia-responsive elements found in the promoter regions of numerous downstream regulators, thereby activating their transcription. HIF-1β is reportedly required for hematopoietic cell generation during ontogeny. However, a detailed analysis of the contribution of HIF-1α to the maintenance of adult HSCs has not yet been reported.

We analyzed HSCs in HIF-1α– and VHL-deficient mice and found that the cellular pool and cell cycle status of HSCs were regulated by the HIF-1α level (Takubo et al., 2010). Our analysis revealed that the regulation of the HIF-1α dose is critical for HSC maintenance in the hypoxic niche microenvironment of the BM. The critical role for HIF-1α in HSC cell cycle regulation broadens the involvement of oxygen status in the stem cell niche. It also implies a novel strategy for maintaining and expanding HSC resources based on cellular oxygen metabolism reprogramming, including the modulation of HSC quiescence through the oxygenation status of HIF-1α.

2. Quiescence of hematopoietic stem cells

Somatic stem cells contribute to tissue homeostasis throughout life (Suda et al., 2011). Because proliferation will induce senescence, the proper maintenance of stem cells without senescence is mandatory. There are two states for tissue stem cells in terms of the cell cycle. One is the quiescent state. Stem cells in the quiescent state are out of the active cell cycle (S/G2/M phase) and in the G0 phase. The other is the cycling state. Cycling stem cells actively reproduce themselves (self-renewal) to generate progeny. Cycling cells are in the non-G0 phase of the cell cycle. Quiescence is thought to be an effective strategy for stem cells to avoid various forms of cytotoxic damage. If stem cells lost quiescence, they would become susceptible to intrinsic and extrinsic stresses.

Mammalian HSCs are included in the heterogeneous population of lineage marker$^-$, Sca-1$^+$, and c-Kit$^+$ (LSK) cells. LSK cells are a mixture of progenitors and HSCs. Within the LSK population, CD34$^-$CD150^{hi}CD48$^-$CD41$^-$Flt3$^-$ cells, as well as side population (SP) cells (Osawa et al, 1996; Kiel et al, 2005; Goodell et al, 1996), are quiescent.

Measurement of the cell cycle in HSCs by the staining of DNA with Hoechst 33342, DAPI, and/or anti-Ki67 indicates that more than 70% of highly purified HSC (CD34$^-$CD48$^-$CD150^{hi} LSK) are in the G0 phase, whereas less than 10% of CD34$^+$ LSK cells (differentiated progenitors) are in the quiescent phase (Wilson et al, 2008).

Slow-cycling HSCs have long-term (LT) reconstitution activity when they are transplanted into lethally irradiated recipient mice. In contrast, actively cycling HSCs and progenitors exhibit only short-term (ST) reconstitution activity and only maintain hematopoiesis for 3–4 months. Thus, the former are termed "LT-HSC(s)" and the latter are "ST-HSC(s)". LT-HSCs produce ST-HSCs, multipotent progenitors (MPPs), lineage-restricted progenitors, and terminally differentiated hematopoietic cells including erythrocytes, platelets, lymphocytes, granulocytes, and macrophages (Figure 1).

Because slow-cycling stem cells are not in the S or M phase of the cell cycle, they are more resistant to cytotoxic agents such as ultraviolet (UV) light, ionizing radiation and chemicals, compared with actively cycling cells. Recent reports indicate that quiescent stem cell fractions are present in several tissues. For example, in the hair follicle, cell cycle progression of stem cells in bulge regions is suppressed by Wnt inhibitors (Fuchs and Horsley, 2011). In contrast, stem cells in the murine intestinal and gastric epithelia divide every 24 hours

(Snippert and Clevers, 2011). Therefore, quiescence itself is not the only strategy for the long-term maintenance of stem cells.

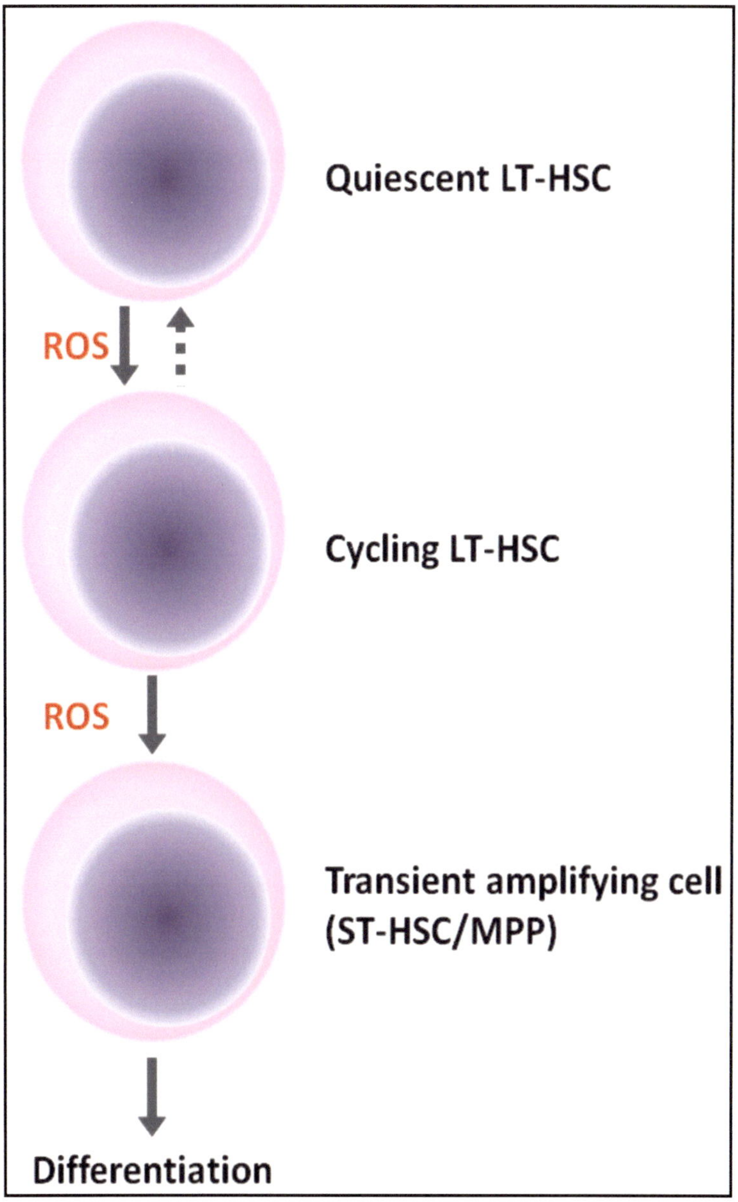

Life-long hematopoiesis is maintained by long-term (LT)-HSCs and their progeny. LT-HSCs have a two cell cycle states: a quiescent state (G0 phase) and a cycling state (non-G0; i.e., G1/S/G2/M phase). LT-HSCs in the former state are resistant to various cytotoxic stresses. Reactive oxygen species (ROS) change the cell cycle state of LT-HSCs from quiescent to cycling. Cycling LT-HSCs are also promoted to differentiate into short-term (ST)-HSCs and multipotent progenitors (MPPs). These differentiated progenitors actively produce various terminally differentiated hematopoietic cells.

Fig. 1. Quiescent and cycling hematopoietic stem cells (HSCs)

One important regulator for the quiescence of HSCs is reactive oxygen species (ROS). ROS are an intrinsic and extrinsic stress for HSCs (Figure 2). Intrinsically, ROS are mainly produced by mitochondria, the energy factory of the cell, as a by-product of the electron transport chain. Because anaerobic energy metabolism in mitochondria utilizes oxygen to generate ATP, oxygen-rich conditions produce intracellular ROS in HSCs. In addition, various immune cells utilize oxygen to generate ROS as an anti-microbial agent. ROS have favourable and unfavourable effects on HSCs. ROS are a signal transducer for essential cytokine signalling in HSCs (Sattler et al., 1999). However, excessive or prolonged ROS exposure is detrimental to HSCs (Naka et al., 2008). Aberrant exposure to ROS induces senescence, apoptosis, or the accumulation of DNA damage in HSCs. These damaged cells are dysfunctional and a potential source for leukemic transformation. Therefore, it is reasonable to hypothesize that HSCs reside in a hypoxic microenvironment.

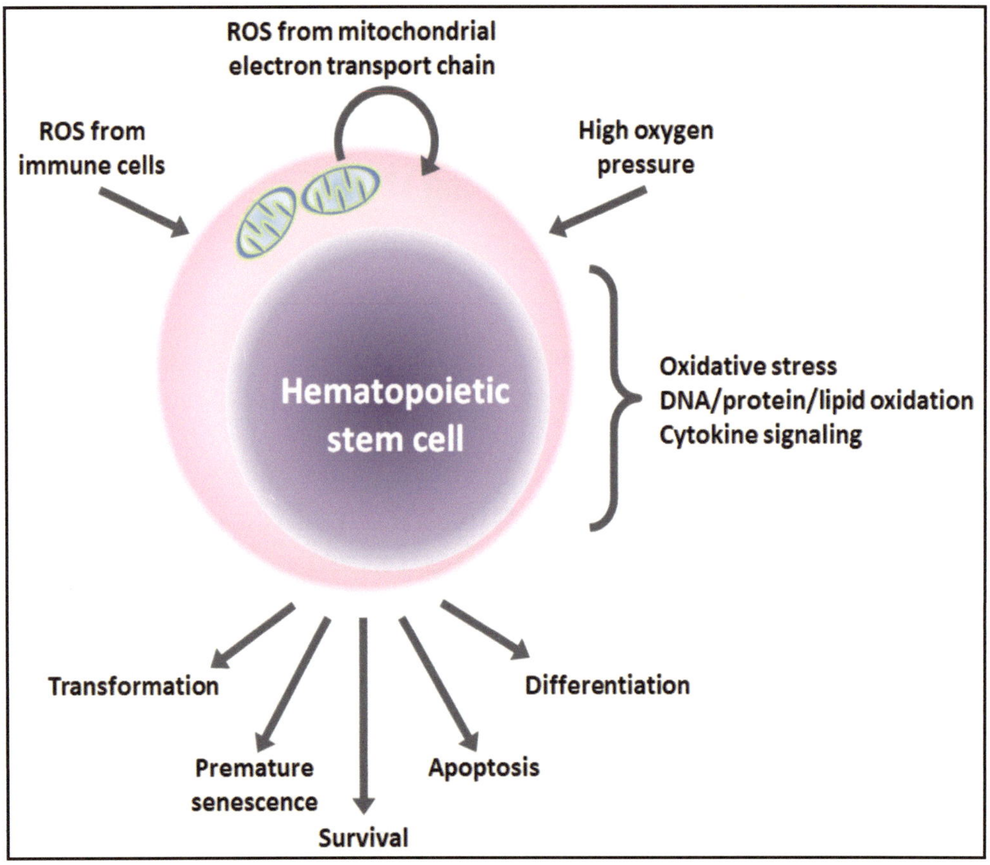

HSCs are exposed to ROS from various sources, including endogenous mitochondria and adjacent immune cells. High O2 pressure in the microenvironment also promotes ROS generation. This ROS burden results in the oxidation of DNA, protein, and lipid in HSCs. Also, the appropriate dose of ROS mediates cytokine signalling in HSCs. These balances determine the fate of HSCs: survival, premature senescence, apoptosis, differentiation, or malignant transformation.

Fig. 2. Intrinsic and extrinsic oxidative stresses and HSCs

3. Hypoxic nature of bone marrow

Although molecular oxygen is critically important for living organisms, HSCs are susceptible to reactive oxygen species or oxidative stresses that are derived from molecular oxygen. To maintain life-long hematopoiesis, it is reasonable for HSCs to avoid high-oxygen conditions. Although classical observations and theoretical studies supported these views, experimental evidence has only been recently provided.

Classically, bone marrow has been thought to be hypoxic. Recently, its exact nature and dynamic regulation were studied. This section will summarize the classical and recent studies related to the functional anatomy of bone marrow oxygenation.

Genetic studies have postulated that LT-HSCs reside primarily in the endosteal zone of the bone marrow (BM) (Calvi et al, 2003; Zhang J, 2003; Arai et al, 2004). Vascular organization around the endosteal zone is unique (Draenert and Draenert, 1980). Nutrient arteries penetrate the cortical bone, enter the medullary canal, and then proceed in a spiral pattern into the metaphyseal region of the bone marrow. The blood in arterial capillaries drains into sinusoids, which are fenestrated and loosely organized.

As a result, hematopoietic cells can easily move across the sinusoidal endothelium. Accordingly, the perfusion of the BM is limited and the partial oxygen pressure (PO2) in the endosteal region is very low.

In addition to hypoperfusion, the BM is tightly packed with blood cells. Oxygen consumption by hematopoietic cells is relatively high, and a simulation of O2 diffusion in the bone marrow suggested that the PO2 is decreased 10-fold at a distance of several cells from the nearest capillary (Chow et al, 2001). The average PO2 in the BM is approximately 55 mmHg and the mean O2 saturation is 87.5% (Harrison et al, 2002). Thus, based on this simulation study, HSCs may well reside in a severely hypoxic environment.

In support of this idea, it has also been reported that murine HSCs live in a hypoxic BM niche. By administering a perfusion tracer into mice, one group found that HSCs accumulated in a hypoperfusion cellular fraction in the BM (Parmar et al, 2007). These hypoperfused cells retained pimonidazole, a probe that selectively binds and forms adducts with protein thiol groups in a hypoxic environment. Administration of a toxin selective for hypoxic cells (tirapazamine) resulted in the depletion of HSCs *in vivo*. It was also shown that LT-HSCs are positive for pimonidazole in mice (Takubo et al, 2010). Moreover, human cord blood stem cells transplanted into super-immunodeficient NOD/scid/IL-2Rγ (NOG) mice homed to the BM niche and became both hypoxic and quiescent after BM transplantation (Shima et al, 2010).

Collectively, these findings suggest the hypoxic nature of HSCs. The hypoxic character of LT-HSCs is potentially determined by their position within the BM. However, in contrast to the simple O2 gradient model for the BM hypoxic niche, immunohistochemical observation of a two-dimensional segment of the murine BM suggests that 60% of LT-HSCs localize closely to BM endothelial cells (Kiel et al, 2005; Sugiyama et al, 2006). These findings do not fit the simple O2 gradient model for the hypoxic status of HSCs in the niche. However, as noted above, the vasculature in the niche near the endosteal zone of the BM may perfuse the bone marrow very poorly. Four-dimensional tracking (time-lapse and three-dimensional observation with a multi-photon microscope) of single LT-HSCs in the BM has shed light on this paradox. Real-time tracking of murine BM revealed that HSCs gradually move away from bone marrow blood vessels and then detach from them and translocate to the osteoblastic zone of the BM after transplantation (Lo Celso et al, 2009; Xie et al, 2009). Based on these observations, it is possible that subpopulations of HSCs residing in different specific locations have different oxygenation statuses.

In parallel with the hypoxic microenvironment for HSCs *in vivo*, hypoxic culture phenotypically and functionally sustained HSCs more effectively than normoxic culture (20% oxygen). Also, hypoxic culture enhances the colony-forming ability (progenitor ability) and transplantation capacity (HSC capacity) of cultured BM cells or isolated HSCs (Cipolleschi et al, 1993; Danet et al, 2003; Ivanovic et al, 2004). Hypoxic treatment also induces cell cycle quiescence in cultured HSC (Hermitte et al, 2006; Shima et al, 2010). Quiescent HSCs are defined by a high amount of efflux of the DNA-binding dye Hoechst 33342 from the cytosol (Goodell et al, 1996). These cells are called "side population (SP)" cells due to their specific staining pattern by flow cytometric analysis. Hypoxic treatment also sustains the SP phenotype in HSC *in vitro* (Krishnamurthy et al, 2004). Exclusion of Hoechst dye from the HSC cytosol is supported by Bcrp1/ABCG2, an ATP-dependent transporter, at the plasma membrane. When HSCs were cultured under hypoxic conditions, mRNA expression of Bcrp1/ABCG2 was significantly increased and the number of SP cells was also increased as compared to HSCs cultured at normoxia. Interestingly, because Bcrp1-/- mice show no significant defect in hematopoiesis (Zhou et al, 2001), the functional role of Bcrp1 in HSCs is still uncharacterized.

4. Hypoxia response system in HSCs

Cells sense, respond, and adapt to hypoxia using hypoxia-responsive regulatory pathways. HSCs utilize the same hypoxia response pathways as a number of other cell types. A central component of these pathways is hypoxia-inducible factor-1 (HIF-1), a transcription factor that is essential for cellular and systemic responses to a low oxygen microenvironment (Semenza, 2010) (Figure 3). HIF-1 is a heterodimeric transcription factor consisting of the oxygen-dependent HIF-1α subunit and an oxygen-independent HIF-1β subunit (Wang and Semenza, 1995). HIF-1α is hydroxylated at proline (Pro) 402 and/or 564 in the oxygen-dependent degradation (ODD) domain under normoxic conditions (Kaelin and Ratcliffe, 2008). HIF-1α is hydroxylated by three prolyl hydroxylases (PHD1–3) which require molecular oxygen, Fe2+, 2-oxoglutarate, and ascorbic acid for their full enzymatic activity (Epstein et al, 2001). Prolyl-hydroxylated HIF-1α protein is recognized by the von Hippel-Lindau (VHL) tumor suppressor protein, which recruits the Elongin C/Elongin B/Cullin2/E3 ubiquitin ligase complex. As a result, prolyl-hydroxylated HIF-1α protein is ubiquitinated and degraded by the proteasome. Under a hypoxic environment, prolyl hydroxylases lose their enzymatic activity. Thus, prolyl hydroxylation of HIF-1α is suppressed, and HIF-1α protein is stabilized without degradation (Kaelin and Ratcliffe, 2008). HIF-1 heterodimers (HIF-1α:HIF-1β) are recruited and bind to hypoxia response elements (HREs) in various target genes and activate transcription programs (Semenza, 2010).

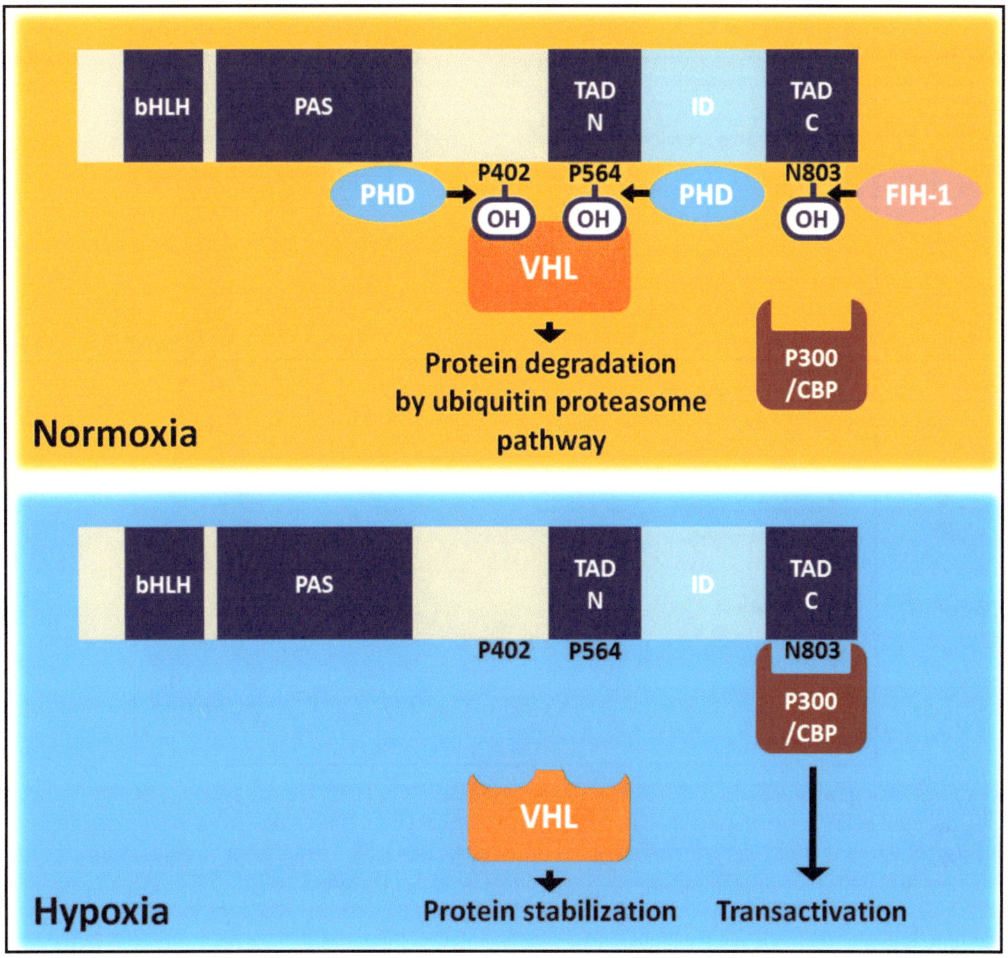

The diagrams represent the regulation of HIF-1α protein and interacting factors under different oxygen conditions. HIF-1α is a substrate for both prolyl and asparaginyl hydroxylases. Under normoxia, proline and asparagine residues are hydroxylated. These modifications regulate the stability and transcriptional activity of HIF-1α. bHLH, basic-helix-loop-helix domain; PAS, Per-ARNT-Sim domain; TAD-N, transactivation domain N-terminal; ID, inhibitory domain; TAD-C, C-terminal transactivation domain; PHD, prolyl hydroxylase domain-containing protein; and FIH-1, factor-inhibiting HIF-1.

Fig. 3. Regulation of hypoxia-inducible factor-1α (HIF-1α)

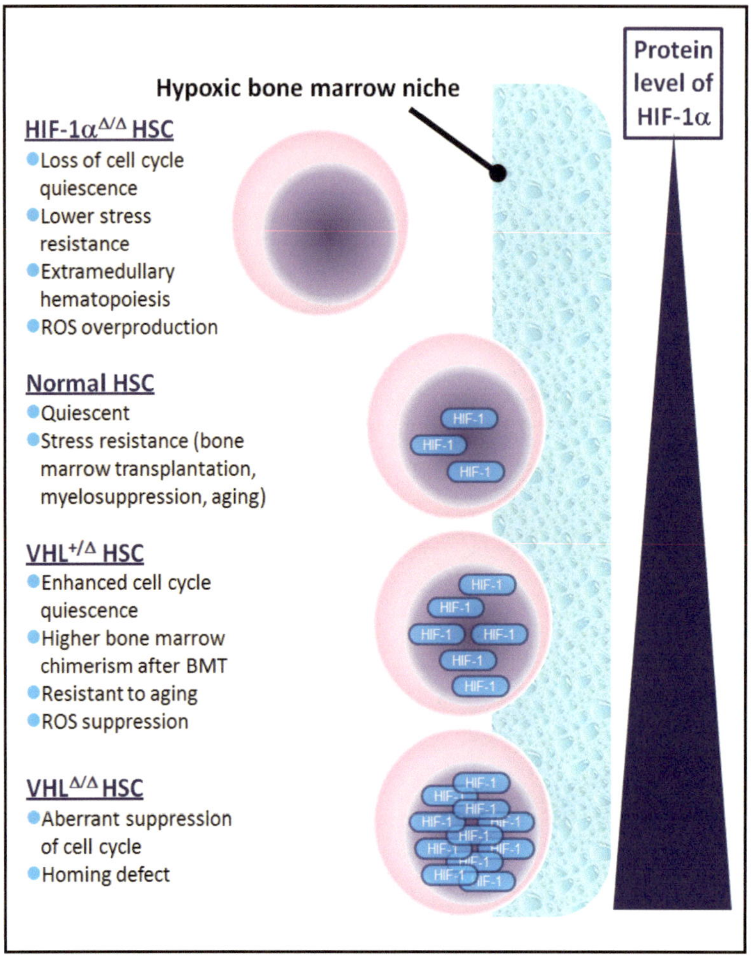

Scheme of biological outcomes of different HIF-1α protein levels in HSCs. This is achieved by HIF-1α or VHL deletion in HSCs using knockout mouse models. Normal HSCs (the second from the top) stabilize HIF-1α, which maintains cell cycle quiescence at the hypoxic bone marrow niche in the endosteum. Preferential stabilization of HIF-1α was observed in HSCs under hypoxia. HIF-1α$^{\Delta/\Delta}$ HSCs (top) lose cell cycle quiescence and stress resistance against transplantation, chemotherapeutic agents, and aging. In addition, HIF-1α$^{\Delta/\Delta}$ HSCs leave the bone marrow niche and drive extramedullary hematopoiesis in the spleen. Production of ROS is accelerated in HIF-1α$^{\Delta/\Delta}$ HSCs. Heterozygous deletion of VHL results in a slight increase in HIF-1α protein. Under these conditions, cell cycle quiescence in HSCs is enhanced. The VHL$^{+/\Delta}$ HSC (the second from the bottom) is resistant to transplantation and aging. ROS production is also suppressed in VHL$^{+/\Delta}$ HSCs. The homozygous VHL mutant (VHL$^{\Delta/\Delta}$) HSC has a maximal dose of HIF-1α protein. In contrast to heterozygous VHL mutant HSCs, VHL$^{\Delta/\Delta}$ HSCs completely lost stem cell capacity potentially due to aberrant suppression of the cell cycle and/or homing capacity to the niche. This defect is HIF-1α–dependent because the co-deletion of HIF-1α in homozygous VHL-deficient hematopoietic cell rescued the defect. Thus, the precise regulation of HIF-1α levels coordinates stem cell proliferation and differentiation. Recently, it has been reported that vascular endothelial growth factor, heat shock proteins, and GRP78 and its ligand Cripto regulate HSC quiescence and maintain HSCs in hypoxia as downstream factors of HIF-1α (Rehn et al, 2011; Miharada et al, 2011).

Fig. 4. Features of HIF-1α or VHL knockout HSCs

HIF-1α mRNA and protein are highly expressed in LT-HSCs (Takubo et al, 2010; Simsek et al, 2010) (Figure 4). HSCs derived from conditional HIF-1α knockout (HIF-1$\alpha^{\Delta/\Delta}$) mice have a defective capacity for marrow reconstitution during serial BM transplantation (Takubo et al., 2010). HIF-1$\alpha^{\Delta/\Delta}$ LT-HSCs lost cell cycle quiescence, entered the cell cycle from G0 phase, proliferated, and showed reduced tolerance to stresses such as 5-fluorouracil administration or aging. These studies suggest that HIF-1α plays an essential role in the regulation of HSC quiescence and stress resistance *in vivo*. In addition to these HIF-1α loss-of-function studies, conditional deletion of the VHL gene in hematopoietic cells was performed as a HIF-1α gain-of-function experiment. Analysis of VHL mutant hematopoietic cells revealed that the functional properties of LT-HSCs and progenitors are differentially influenced by HIF-1α. HIF-1α protein levels are elevated in either biallelic ($VHL^{\Delta/\Delta}$) or monoallelic ($VHL^{+/\Delta}$) conditional knockout hematopoietic cells. For example, only a minor population of normal hematopoietic progenitors ($CD34^+$ LSK cells) are in a quiescent state. In clear contrast, the proportion of $VHL^{+/\Delta}$ hematopoietic progenitors in the quiescent phase is significantly higher. At steady state, HIF-1α protein levels are not high in hematopoietic progenitors, and forced stabilization of HIF-1α protein through monoallelic VHL deletion induces $VHL^{+/\Delta}$ $CD34^+$ LSK progenitors to exit the cell cycle and maintains them in the G0 phase. Severe suppression of cell cycling and transplantation capacity is restored in HIF-1$\alpha^{\Delta/\Delta}$:$VHL^{\Delta/\Delta}$ doubly mutated HSCs. The decreased frequency of LT-HSCs seen in $VHL^{\Delta/\Delta}$ mice is rescued by the co-deletion of the HIF-1α gene *in vivo*. Also, long-term *in vitro* exposure of LT-HSCs to a PHD inhibitor (dimethyloxalylglycine; DMOG), which stabilizes HIF-1α even under normoxic conditions, attenuates stem cell ability especially during BM transplantation (Eliasson P et al, 2010).

Collectively, these results provide evidence that there is an optimal HIF-1α protein level for HSC maintenance. HIF-1α is required for stress resistance and long-term maintenance of HSCs, and within an appropriate range, moderate increases of HIF-1α (to the level caused by VHL heterozygous deletion) are trophic for HSCs through the induction of quiescence. However, aberrantly high HIF-1α levels are also harmful to HSCs and lead to a loss of stem cell capacity and the exhaustion of the HSC pool. Homozygous deletion of VHL results in a severe suppression of the cell cycle and a homing defect during transplantation.

HIF-1α not only acts in the HSC system but also plays an important role in neural stem cells (NSCs) under hypoxic conditions. In this type of cell, HIF-1α induces the activation of the Wnt/β-catenin signalling pathway through the upregulation of β-catenin and the expression of the downstream transcription factors lymphoid enhancer-binding factor 1 and T-cell factor 1 (Mazumdar J et al, 2010). Wnt/β-catenin activity was closely correlated with hypoxic status in the subgranular zone of the hippocampus, which is one of the niches for NSCs. Loss of HIF-1α in NSCs resulted in a defective Wnt-dependent hippocampal neurogenic niche capacity. As a result, NSC proliferation and differentiation, and the production of new neurons, were attenuated. Interestingly, the biological effects of HIF-1α on NSCs (cell cycle promotion) are clearly different from those seen in HSCs (cell cycle quiescence). It will be important to dissect how these different lineage stem cell systems utilize the same protein (HIF-1α) to sustain themselves using different downstream molecular machinery and biological events. It is also of interest to investigate embryonic

HSCs, which actively proliferate in hypoxic conditions, because HIF-1α may support HSC proliferation in that stage. In addition, HIF-1α protein has been reported to inhibit Wnt/β-catenin activity in cancer cells (Kaidi A et al, 2007), suggesting that the interaction of the HIF-1α and Wnt/β-catenin pathway in stem/progenitor cells may differ from that of more differentiated or transformed cell types.

5. Conclusion

In this chapter, I have summarized our current knowledge regarding the hypoxia response and oxygen metabolism in HSCs at the BM niche. These studies open novel fields in stem cell biology. The invisible niche factor, oxygen, is usually essential because mitochondria utilize it for the energy production. However, molecular oxygen is a source of ROS during mitochondrial metabolism. Because an excessive dose of ROS can be damaging to HSC, escape from oxygen (in the hypoxic niche) is a reasonable strategy for the long-term maintenance of HSCs *in vivo*. Adult HSCs are quiescent and contain few mitochondria, whereas hematopoietic progenitor cells actively proliferate and contain many mitochondria. Thus, stem cells and progenitors have distinct metabolic states, and the transition from stem to progenitor cell may correspond to a critical metabolic change, namely from glycolysis to oxidative phosphorylation. Slow cell cycling or long-term quiescence is common in adult tissue stem cells. Dormancy in the cell cycle may be a crucial mechanism for the stress resistance of normal and leukemic stem cells.

Further investigation of oxygen metabolism in tissue stem cells will result in more effective maintenance, expansion, and manipulation of various somatic stem cells *ex vivo* and *in vivo*, maximizing the potential of therapeutic strategies using stem cells in regenerative medicine. Also, an understanding of oxygen homeostasis in HSCs is essential for understanding senescence at the stem cell level as well as therapeutic targeting against leukemic stem cells.

6. Acknowledgments

I would like to thank to Drs. Atsushi Hirao, Makoto Suematsu, Nobuhito Goda, Tomoyoshi Soga, and Randall S. Johnson for providing thoughtful insights and collaborations for this review. Most of our work on hypoxia in HSCs was performed in Dr. Toshio Suda's laboratory at the Keio University School of Medicine, Tokyo, Japan under his careful management. I would like to acknowledge my deep appreciation of fruitful discussions with the previous and current members of the Stem Cell Metabolism group of the Suda lab, especially Dr. Hirono Iriuchishia, Dr. Chiharu Kobayashi, Dr. Hiroshi Kobayashi, Dr. June-Won Cheong, Dr. Ayako Ishizu, and Ms. Wakako Yamada. Also, I would like to thank Ms. Tomoko Muraki and Ms. Takako Hirose for the preparation of this manuscript. K.T. is supported by the Global COE Program for Human Metabolomic Systems Biology and for Stem Cell Medicine of the Japan Society for Promotion of Science, and also in part by a Ministry of Education, Culture, Sports, Science and Technology (MEXT) Grant-in-Aid for Young Scientists (A), a MEXT Grant-in-Aid for Scientific Research (A), and a MEXT Grant-in-Aid for Scientific Research on Innovative Areas. The author dedicates this paper to the memory of Masako Takubo, who passed away October 9, 2011.

7. References

Arai, F., Hirao, A., Ohmura, M., Sato, H., Matsuoka, S., Takubo, K., Ito, K., Koh, G.Y., Suda, T. (2004). Tie2/angiopoietin-1 signaling regulates hematopoietic stem cell quiescence in the bone marrow niche. *Cell.* 118, 149-161.

Calvi, L.M., Adams, G.B., Weibrecht, K.W., Weber, J.M., Olson, D.P., Knight, M.C., Martin, R.P., Schipani, E., Divieti, P., Bringhurst, F.R. et al. (2003). Osteoblastic cells regulate the haematopoietic stem cell niche. *Nature.* 425, 841-846.

Chow, D.C., Wenning, L.A., Miller, W.M., Papoutsakis, E.T. (2001). Modeling pO(2) distributions in the bone marrow hematopoietic compartment. II. Modified Kroghian models. *Biophys J.* 81, 685-696.

Cipolleschi, M.G., Dello, Sbarba P., Olivotto, M. (1993). The role of hypoxia in the maintenance of hematopoietic stem cells. *Blood.* 82, 2031-2037.

Danet, G.H., Pan, Y., Luongo, J.L., Bonnet, D.A., Simon, M.C. (2003). Expansion of human SCID-repopulating cells under hypoxic conditions. *J Clin Invest.* 112, 126-135.

Draenert, K., Draenert, Y. (1980). The vascular system of bone marrow. *Scan Electron Microsc* 113-122.

Eliasson, P., Rehn, M., Hammar, P., Larsson, P., Sirenko, O., Flippin, L.A., Cammenga, J., Jonsson, J.I. (2010). Hypoxia mediates low cell-cycle activity and increases the proportion of long-term-reconstituting hematopoietic stem cells during in vitro culture. *Exp Hematol.* 38, 301-310.

Epstein, A.C., Gleadle, J.M., McNeill, L.A., Hewitson, K.S., O'Rourke, J., Mole, D.R., Mukherji, M., Metzen, E., Wilson, M.I., Dhanda, A. et al. (2001). C. elegans EGL-9 and mammalian homologs define a family of dioxygenases that regulate HIF by prolyl hydroxylation. *Cell.* 107, 43-54.

Fuchs, E., Horsley, V. (2011). Ferreting out stem cells from their niches. *Nat Cell Biol.* 13, 513-518.

Goodell, M.A., Brose, K., Paradis, G., Conner, A.S., Mulligan, R.C. (1996). Isolation and functional properties of murine hematopoietic stem cells that are replicating in vivo. *J Exp Med.* 183, 1797-1806.

Harrison, J.S., Rameshwar, P., Chang, V., Bandari, P. (2002). Oxygen saturation in the bone marrow of healthy volunteers. *Blood.* 99, 394.

Hermitte, F., Brunet, de la Grange P., Belloc, F., Praloran, V., Ivanovic, Z. (2006). Very low O2 concentration (0.1%) favors G0 return of dividing CD34+ cells. *Stem Cells.* 24, 65-73.

Ivanovic, Z., Hermitte, F., Brunet, de la Grange P., Dazey, B., Belloc, F., Lacombe, F., Vezon, G., Praloran, V. (2004). Simultaneous maintenance of human cord blood SCID-repopulating cells and expansion of committed progenitors at low O2 concentration (3%). *Stem Cells.* 22, 716-724.

Jang, Y.Y., Sharkis, S.J. (2007). A low level of reactive oxygen species selects for primitive hematopoietic stem cells that may reside in the low-oxygenic niche. *Blood.* 110, 3056-3063.

Kaelin, W.G. Jr, Ratcliffe, P.J. (2008). Oxygen sensing by metazoans: the central role of the HIF hydroxylase pathway. *Mol Cell.* 30, 393-402.

Kaidi, A., Williams, A.C., Paraskeva, C. (2007). Interaction between beta-catenin and HIF-1 promotes cellular adaptation to hypoxia. *Nat Cell Biol.* 9, 210-217.

Kiel, M.J., Yilmaz, O.H., Iwashita, T., Yilmaz, O.H., Terhorst, C., Morrison, S.J. (2005). SLAM family receptors distinguish hematopoietic stem and progenitor cells and reveal endothelial niches for stem cells. *Cell.* 121, 1109-1121.

Kirito K, Fox N, Komatsu N, Kaushansky K. (2005) Thrombopoietin enhances expression of vascular endothelial growth factor (VEGF) in primitive hematopoietic cells through induction of HIF-1alpha. *Blood.* Jun 1;105(11):4258-63. Epub 2005 Feb 10.

Krishnamurthy, P., Ross, D.D., Nakanishi, T., Bailey-Dell, K., Zhou, S., Mercer, K.E., Sarkadi, B., Sorrentino, B.P., Schuetz, J.D. (2004). The stem cell marker Bcrp/ABCG2 enhances hypoxic cell survival through interactions with heme. *J Biol Chem.* 279, 24218-24225.

Lo, Celso C., Fleming, H.E., Wu, J.W., Zhao, C.X., Miake-Lye, S., Fujisaki, J., Cote, D., Rowe, D.W., Lin, C.P., Scadden, D.T. (2009). Live-animal tracking of individual haematopoietic stem/progenitor cells in their niche. *Nature.* 457, 92-96.

Mazumdar, J., O'Brien, W.T., Johnson, R.S., LaManna, J.C., Chavez, J.C., Klein, P.S., Simon, M.C. (2010). O2 regulates stem cells through Wnt/beta-catenin signalling. *Nat Cell Biol.* 12, 1007-1013.

Miharada K, Karlsson G., Rehn M., Rorby E., Siva K., Cammenga J., Karlsson S. (2011) Cripto regulates hematopoietic stem cells are a hypoxic niche related factor through cell surface receptor GRP78. (2011) *Cell Stem Cell*, Oct 4;9(4):330-44.

Mohyeldin A, Garzón-Muvdi T, Quiñones-Hinojosa A. (2010) Oxygen in stem cell biology: a critical component of the stem cell niche. *Cell Stem Cell.* Aug 6;7(2):150-61.

Naka K, Muraguchi T, Hoshii T, Hirao A. (2008) Regulation of reactive oxygen species and genomic stability in hematopoietic stem cells. *Antioxid Redox Signal.* Nov;10(11):1883-94.

Osawa, M., Hanada, K., Hamada, H., Nakauchi, H. (1996). Long-term lymphohematopoietic reconstitution by a single CD34-low/negative hematopoietic stem cell. *Science.* 273, 242-245.

Parmar, K., Mauch, P., Vergilio, J.A., Sackstein, R., Down, J.D. (2007). Distribution of hematopoietic stem cells in the bone marrow according to regional hypoxia. *Proc Natl Acad Sci U S A.* 104, 5431-5436.

Pedersen M, Löfstedt T, Sun J, Holmquist-Mengelbier L, Påhlman S, Rönnstrand L. (2008) Stem cell factor induces HIF-1alpha at normoxia in hematopoietic cells. *Biochem Biophys Res Commun.* Dec 5;377(1):98-103. Epub 2008 Oct 1.

Qian H, Buza-Vidas N, Hyland CD, Jensen CT, Antonchuk J, Månsson R, Thoren LA, Ekblom M, Alexander WS, Jacobsen SE. (2007) Critical role of thrombopoietin in maintaining adult quiescent hematopoietic stem cells. *Cell Stem Cell.* Dec 13;1(6):671-84. Epub 2007 Nov 20.

Rehn M, Olsson A, Reckzeh K, Diffner E, Carmeliet P, Landberg G, Cammenga J. (2011) Hypoxic induction of vascular endothelial growth factor regulates murine hematopoietic stem cell function in the low-oxygenic niche. *Blood.* 118(6):1534-43. Epub 2011 Jun 13.

Sattler M, Winkler T, Verma S, Byrne CH, Shrikhande G, Salgia R, Griffin JD. (1999) Hematopoietic growth factors signal through the formation of reactive oxygen species. *Blood.* May 1;93(9):2928-35.

Semenza, G.L. (2007). Oxygen-dependent regulation of mitochondrial respiration by hypoxia-inducible factor 1. *Biochem J. 405*, 1-9.

Semenza, G.L. (2009). Regulation of cancer cell metabolism by hypoxia-inducible factor 1. *Semin Cancer Biol. 19*, 12-16.

Semenza, G.L. (2010). Oxygen homeostasis. *Wiley Interdiscip Rev Syst Biol Med. 2*, 336-361.

Shima, H., Takubo, K., Tago, N., Iwasaki, H., Arai, F., Takahashi, T., Suda, T. (2010) Acquisition of G_0 state by CD34-positive cord blood cells after bone marrow transplantation. *Exp Hematol.* 38, 1231-1240.

Simsek, T., Kocabas, F., Zheng, J., Deberardinis, R.J., Mahmoud, A.I., Olson, E.N., Schneider, J.W., Zhang, C.C., Sadek, H.A. (2010). The distinct metabolic profile of hematopoietic stem cells reflects their location in a hypoxic niche. *Cell Stem Cell.* 7, 380-390.

Snippert, H.J., Clevers, H. (2011). Tracking adult stem cells. *EMBO Rep.* 12, 113-122.

Suda T, Takubo K, Semenza GL. (2011) Metabolic regulation of hematopoietic stem cells in the hypoxic niche. *Cell Stem Cell.* Oct 4;9(4):298-310.

Sugiyama, T., Kohara, H., Noda, M., Nagasawa, T. (2006). Maintenance of the hematopoietic stem cell pool by CXCL12-CXCR4 chemokine signaling in bone marrow stromal cell niches. *Immunity.* 25, 977-988.

Takubo, K., Goda, N., Yamada, W., Iriuchishima, H., Ikeda, E., Kubota, Y., Shima, H., Johnson, R.S., Hirao, A., Suematsu, M. et al. (2010). Regulation of the HIF-1alpha level is essential for hematopoietic stem cells. *Cell Stem Cell.* 7, 391-402.

Wang, G.L., Semenza, G.L. (1995). Purification and characterization of hypoxia-inducible factor 1. *J Biol Chem.* 270, 1230-1237.

Wilson, A., Laurenti, E., Oser, G., van, der Wath R.C., Blanco-Bose, W., Jaworski, M., Offner, S., Dunant, C.F., Eshkind, L., Bockamp, E. et al. (2008). Hematopoietic stem cells reversibly switch from dormancy to self-renewal during homeostasis and repair. *Cell.* 135, 1118-1129.

Xie, Y., Yin, T., Wiegraebe, W., He, X.C., Miller, D., Stark, D., Perko, K., Alexander, R., Schwartz, J., Grindley, J.C., Park J, Haug JS, Wunderlich JP, Li H, Zhang S, Johnson T, Feldman RA, Li L. (2009). Detection of functional haematopoietic stem cell niche using real-time imaging. *Nature.* 457, 97-101.

Yoshihara H, Arai F, Hosokawa K, Hagiwara T, Takubo K, Nakamura Y, Gomei Y, Iwasaki H, Matsuoka S, Miyamoto K, Miyazaki H, Takahashi T, Suda T. (2007) Thrombopoietin/MPL signaling regulates hematopoietic stem cell quiescence and interaction with the osteoblastic niche. *Cell Stem Cell.* Dec 13;1(6):685-97.

Zhang, J., Niu, C., Ye, L., Huang, H., He, X., Tong, W.G., Ross, J., Haug, J., Johnson, T., Feng, J.Q. Harris S, Wiedemann LM, Mishina Y, Li L. (2003). Identification of the haematopoietic stem cell niche and control of the niche size. *Nature.* 425, 836-841.

Zhou S, Schuetz JD, Bunting KD, Colapietro AM, Sampath J, Morris JJ, Lagutina I, Grosveld GC, Osawa M, Nakauchi H, Sorrentino BP. (2001) The ABC transporter Bcrp1/ABCG2 is expressed in a wide variety of stem cells and is a molecular determinant of the side-population phenotype. *Nat Med.* Sep;7(9):1028-34.

8

Skeletogenesis and the Hematopoietic Niche

Elizabeth Sweeney and Olena Jacenko
University of Pennsylvania
USA

1. Introduction

The reciprocal regulation of the skeletal and the immune systems has been clinically appreciated for years. In particular, factors produced by immune cells during homeostasis and activation markedly affect the skeleton, which in turn affects the marrow niche environments (as reviewed in (Compston 2002). This relationship also extends to an interdependence between bone and hematopoiesis during immune cell development, however the critical cell types and extracellular matrix components involved in establishing and maintaining hematopoietic niches within the bone marrow are only recently beginning to be defined. Indeed, some immuno-osseous disorders with hematopoietic defects such as bone marrow failure and immune dysfunction, as well as certain cancers, may result from a defective hematopoietic niche (Spranger et al. 1991; Kuijpers et al. 2004; Hermanns et al. 2005; Walkley et al. 2007; Walkley et al. 2007; Raaijmakers et al. 2010). Likewise during aging, a progressive decline in cell replacement and repair manifests in both the skeletal and hematopoietic systems with reduced bone mass and diminished blood cell formation respectively (as reviewed in (Rossi et al. 2008) and (Gruver et al. 2007). Further, this altered hematopoiesis due to aging leads to deficient immune function and increased incidence of malignancies (Rossi et al. 2005; Janzen et al. 2006; Mayack et al. 2010). Thus, the dynamic relationship between skeletal and hematopoietic maintenance throughout life suggests that these clinical outcomes may ensue from cell signaling deficiencies or from defects in the structural environment supporting hematopoiesis. This chapter provides an overview of our current understanding of how hematopoietic niches may be established, how they promote hematopoiesis, and how the skeletal status may modulate niche function.

2. Coordinate skeletal and hematopoietic development

The vertebrate skeleton develops by one of two essential processes, endochondral (EO) and intramembranous (IO) ossification mechanisms (as reviewed in (Chan, D. and Jacenko 1998). The direct differentiation of ectomesenchymal cells to osteoblasts in IO represents the rudimentary mechanism through which many skull bones and all periosteal bones form. The IO-derived bone is referred to as "dense", "compact" or "cortical", and as the names imply, is a solid bone with primary functions relating to weight bearing and protection (**Fig. 1C**). In contrast, EO relies on the generation of a cartilaginous skeletal blueprint that is gradually replaced by a "trabecular", "spongy", "cancellous" bone and a marrow capable of sustaining hematopoiesis (Chan, D. and Jacenko 1998; Mackie et al. 2008) (**Fig. 1C**). This replacement mechanism of EO is responsible for the formation of the vertebrate axial and appendicular skeleton, as well as certain cranial bones (Jacenko et al. 1991; Chan, D. and Jacenko 1998).

As EO initiates during embryogenesis, its distinctive feature is the emergence of hypertrophic cartilage, which is present in all skeletal elements that will develop a marrow cavity, e.g. long bones, hips, vertebrae, ribs, certain skull bones. The eventual replacement of cartilage by bone and marrow via EO relies on the sequential maturation of chondrocytes from resting, to proliferating, to hypertrophic (**Fig. 1A**). Chondrocyte hypertrophy manifests with a dramatic increase in cell size, cessation of proliferation, and synthesis of a new repertoire of differentiation-specific gene products (Godman and Porter 1960; Chan, D. and Jacenko 1998; Alvarez et al. 2001; James et al. 2010). Among these is the matrix protein collagen X, which represents the predominant biosynthetic product of hypertrophic cartilage (Gibson and Flint 1985; Schmid and Linsenmayer 1985). Concomitant with hypertrophy is a transformation from a non-calcified avascular cartilage matrix, to a calcifiable one that is permissive to vascular invasion. Morphometric analysis suggests that before vascular invasion, the terminal hypertrophic chondrocytes undergo either autophagy (Srinivas and Shapiro 2006; Bohensky et al. 2007) or apoptosis (Farnum and Wilsman 1989), the rate of which controls longitudinal growth of the skeletal element, as well as the transition from cartilage to trabecular bone and marrow (Farnum and Wilsman 1989).

Subsequent vascular entry into hypertrophic cartilage is critical to skeleto-hematopoietic development, since it leads to an influx of mesenchymal cells, hematopoietic precursors, and chondro/osteoclasts. This influx of cells, together with growth factors, cytokines and hormones, establishes the primary center of ossification and the marrow environment where hematopoiesis ensues (**Fig. 1**). Specifically, while chondro/osteoclasts degrade hypertrophic cartilage, multipotent stromal cells, including mesenchymal and perivascular reticular cells, form the marrow stroma, a meshwork of non-hematopoietic cells supporting hematopoiesis by providing structural scaffolding and producing hematopoietic factors (Taichman et al. 1996; Bianco et al. 1999). As hypertrophic cartilage continues to be degraded, matrix remnants serve as scaffolds upon which differentiating osteoblasts deposit bone matrix, thus forming trabecular bony spicules with hypertrophic cartilage cores (**Fig. 1B**) (Chan, D. and Jacenko 1998). Of note, the origin of the trabecular bone osteoblasts at the junction between marrow and the hypertrophic cartilage, termed the chondro-osseous junction, is still debated (Roach 1992; Roach et al. 1995; Roach and Erenpreisa 1996; Nakamura et al. 2006; Hilton et al. 2007; Maes et al. 2010). Following the formation of the primary ossification zones in the central or diaphyseal regions of skeletal elements, the establishment of secondary ossification centers at outer epiphyseal ends of bones defines the growth plate regions at the metaphysis (**Fig. 1A**). The growth plates occupy the narrow space that separates the marrow of the primary and secondary ossification centers, and are composed of a gradient of differentiating chondrocytes culminating in a zone of hypertrophic chondrocytes (**Fig. 1A & B**) (as reviewed in (Lefebvre and Smits 2005). The continual replacement of the hypertrophic chondrocytes by trabecular bone and marrow allows for longitudinal skeletal growth, robust hematopoiesis, and the progression of EO without consumption of the skeletal model until maturity, when in most non-rodent vertebrates EO ceases and growth plates close (**Fig. 1C**) (Kilborn et al. 2002). Thus, the end result of EO is a porous network of primary trabecular bone, consisting of a hybrid hyptertrophic cartilage-bone matrix, and engulfed by a hematopoietic marrow (**Fig. 1B & C**). Subsequent bone remodeling gradually leads to a complete replacement of the hybrid primary bone by mature secondary bone, and is coincident with a gradual decline in lymphopoiesis and the onset of immunosenescence (**Fig. 1C**) (as reviewed in (Compston 2002; Gruver et al. 2007).

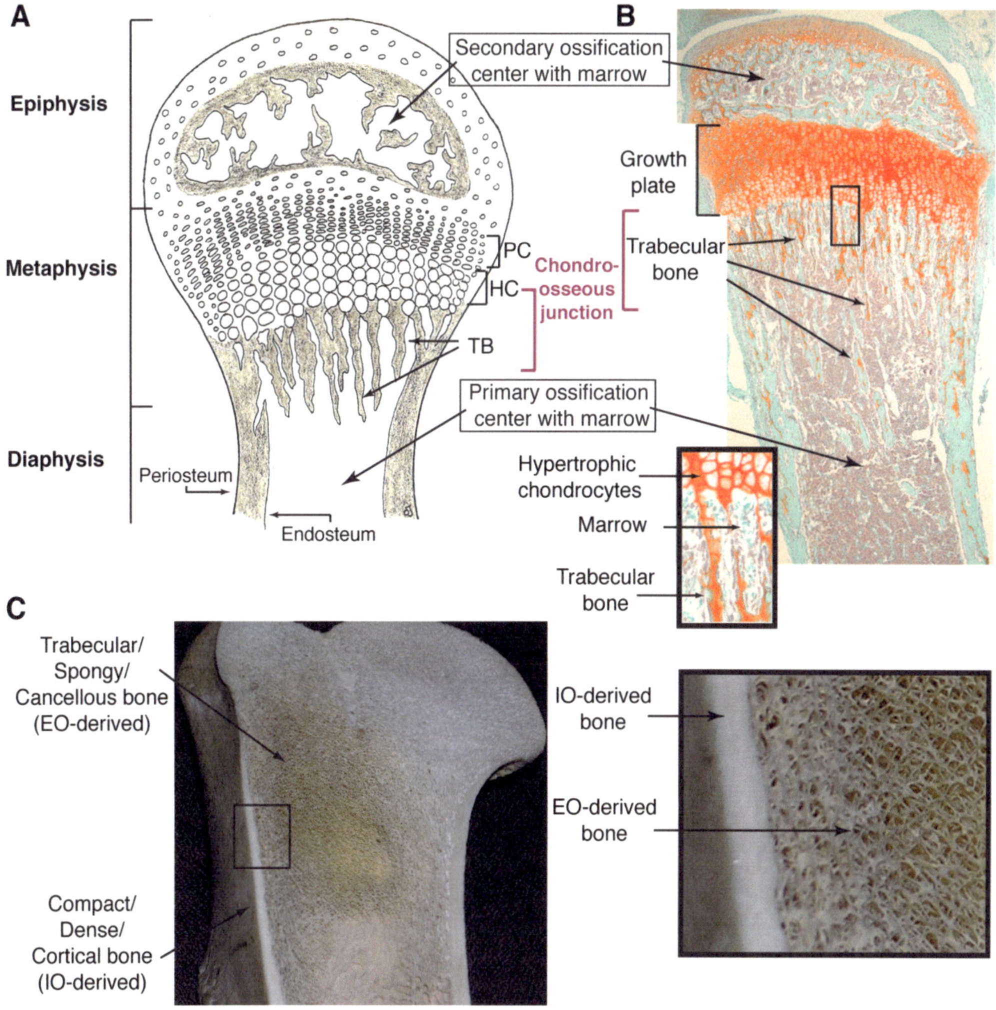

Fig. 1. Architecture of the endochondral bone. A) A schematic of a developing long bone illustrating its architecture. The epiphysis, or the bulbous end, lined by articular cartilage and containing the secondary ossification center with marrow, is supported by the flared metaphysis, which in turn rests upon the slender cylindrical shaft of the diaphysis. The growth plate separates the primary and secondary ossification centers, and consists of a gradient of differentiating chondrocyte zones; the proliferative cartilage (PC) and hypertrophic cartilage (HC) zones are marked, as well as the hybrid trabecular bone (TB) protruding into the marrow. The locations of the two-layered periosteal membrane surrounding the diaphysis and the inner endosteal network are marked. B) A longitudinal tibial section from a week-3 wild type mouse stained with safranin-orange, hematoxylin & eosin (H&E) and counterstained with fast green. Using these stains, the negatively charged cartilaginous matrix appears orange while the bone stains light blue-green; mature erythrocytes stain green, while other marrow elements stain pink-purple with H&E. The boxed inset is a high magnification of the chondro-osseous junction containing hypertrophic chondrocytes, bone and marrow with vascular. The hybrid nature of the trabecular bone can

be appreciated by the orange staining of the cartilaginous core, with green-blue bone matrix deposited on the surface (magnification, 10x). C) The inorganic mineralized matrix of a mature zebra bone illustrates the structural differences between the EO-derived trabecular /spongy/cancellous bone and the IO-derived compact/dense/cortical bone. Boxed is a high magnification of the EO- and IO-derived bone tissues. Note the mesh-like structure of trabecular bone for hematopoietic cell support.

Taken together, the proper differentiation of chondrocytes, vascular invasion and the gradual replacement of the cartilaginous anlagen by trabecular bone and marrow through EO, underscore the intricate orchestration of skeleto-hematopoiteic development. Moreover, the coincident establishment and localization of trabecular bone within the site of active hematopoiesis likely reflects a critical hematopoietic niche in the chondro-osseous region (**Fig. 1B boxed**) (Jacenko et al. 1993; Nilsson et al. 1997; Gress and Jacenko 2000; Nilsson et al. 2001; Jacenko et al. 2002; Yoshimoto et al. 2003; Arai, F. et al. 2004; Balduino et al. 2005; Sweeney et al. 2008; Kohler et al. 2009; Lo Celso et al. 2009; Xie et al. 2009; Sweeney et al. 2010). This skeleto-hematopoietic link is strongly supported by several animal models where alterations in process of EO leads to hematopoietic defects (**Table 1**), including mouse models with altered: collagen X (Jacenko et al. 1993; Gress and Jacenko 2000; Jacenko et al. 2002; Sweeney et al. 2008; Sweeney et al. 2010), parathyroid hormone related protein (PTHrP) receptor in osteoblasts (Calvi et al. 2001; Calvi et al. 2003; Kuznetsov et al. 2004; Wu et al. 2008), osteoblast numbers (Visnjic et al. 2001; Visnjic et al. 2004; Zhu et al. 2007), bone morphogenic protein (BMP) receptor type 1A in marrow cells (Zhang, J. et al. 2003), osteoclast function (Blin-Wakkach et al. 2004; Mansour et al. 2011), retinoic acid receptor gamma (Purton et al. 2006; Walkley et al. 2007), $G_s\alpha$ in ostoblasts (Wu et al. 2008), Dicer in ostoblasts (Raaijmakers et al. 2010), glypican-3 (Viviano et al. 2005), and perlecan (Rodgers et al. 2008). **Table 1** presents a list of mouse models with defects in hematopoiesis due to alterations in a component within the niche environment. Only those mouse models are summarized that were proven, by and large, via bone marrow transplantation experiments to have an aberrant niche environment, since wild type marrow cells could not rescue the disease phenotype of the host.

3. Overview of the hematopoietic niche

Hematopoiesis is the process by which hematopoietic stem cells (HSCs) generate and replenish progenitors that develop into fully mature blood and immune cells, and populate the periphery. During vertebrate ontogeny, hematopoiesis is established sequentially in several different anatomic sites (see Development of Hematopoietic Stem Cells chapter in this book for review). Coincident with the onset of EO (approximately the last third of embryonic development), hematopoiesis shifts from the fetal liver and spleen to the EO-derived marrow, which represents the predominant site of blood cell production after birth (Aguila and Rowe 2005; Cumano and Godin 2007). Therefore, the marrow has become a tissue of study for hematopoietic cell biology post parturition. Additionally, due to the ease of marrow cell isolation in combination with the extensive list of cell markers identifying HSCs at different stages of differentiation (as summarized in (Morrison and Spradling 2008), stem cell niche biology has also utilized the marrow environment for study.

Year	Protein/ cell	Description	Hematopoietic highlights	References
1993	Collagen X	Collagen X is a short chain collagen secreted by hypertrophic chondrocytes in the growth plate of endochondrally developing bones. In these mice, collagen X is either knocked-out or mutated to cause dominant interference.	-Altered chondro-osseous junction -Decreased trabecular bone, -No change in HSPCs, -Decreased B lymphocytes throughout life, -Diminished immunity in vitro and in vivo, -Altered hematopoietic/lymphopoietic cytokines, e.g. decreased SCF, CXCL-12, IL-7, -Disease phenotype retained after transfer of wild type HSPCs into collagen X mice	Jacenko, 1993; Jacenko, 1996; Gress, 2000; Jacenko 2001; Jacenko 2002; Sweeney, 2008; Sweeney, 2010
2001	PTH/PTHrP receptor	Parathyroid hormone (PTH) and the PTH related protein (PTHrP) receptor (PPR) are involved in calcium homeostasis and activation of osteoblasts, thus indirectly osteoclasts. In these mice, a constitutively active form of PPR is expressed in mesenchymal cells under the 2.3kb promotor of collagen I.	-Delayed hematopoiesis due to delay in the bone to marrow transition, -With aging, increased trabecular bone and osteoblast numbers, -Increased HSPCs, -Increase in Notch signaling, -Increase in IL-6, CXCL-12, SCF, -Decrease in B cells at all stages	Calvi, 2001; Calvi, 2003; Kuznetsov, 2004; Wu 2008
2001	Osteoblast deletion	Osteoblasts have been implicated as a niche cell where increased in osteoblasts resulted in increases in HSPCs. These mice were generated with the herpes thymidine kinase gene under the 2.3kb promotor of collagen for deletion of collagen I expressing mesenchymal cells after ganciclovir treatment.	-Loss of bone lining cells and trabecular bone elements, -Decrease in marrow cellularity, -Decreases in HSPCs, -Decreases in lymphoid, erythroid, and myeloid progenitor cells, -Decrease in osteoclasts, -Increased extramedullary hematopoiesis, -Recovery from disease phenotype after ganciclovire removed	Visnjic, 2001; Visnjic, 2004; Zhu, 2007
2003	BMPR1A	The bone morphogenic protein receptor, type 1A (BMPR1A) is a receptor for BMPs, which have been shown to influence hematopoiesis. These mice were generated with a PolyI:C-inducible Mx-1-Cre to delete BMPR1A in hematopoietic and stromal cells.	-Increases in spindle-shaped N-cadherin+ CD45- osteoblasts (SNO), -Increases in HSPCs, -LT-HSC observed attached to SNO cells, -No block in HPC differentiation to the lymphoid or myeloid lineages	Zhang, 2003

Continued on facing page

Table 1. Mouse models with altered hematopoetic niche enviroments.

Year	Protein/ cell	Description	Hematopoietic highlights	References
2004	Tcirg 1	The T cell immune regulator 1 (Tcirg1) is a subunit of the vacuolar protein pump (V-ATPase) involved in osteoclast resorption of bone. In these mice there is a mutation of Tcrig1 rendering osteoclasts ineffective at bone resorption.	-Increases in myelomonocytic differentiation, -Reduced medullary cavity size, -Defective B cell differentiation leading to reduced numbers, -Reduced interferon-γ secretion from T cells, -Decreased IL-7 secretion from bone marrow cells, -Rescue of phenotype with restoration of marrow environment after wild type cell transfer	Blin-Wakkach, 2004; Mansour, 2011
2005	Osteopontin	Osteopontin is an extracellular matrix glycoprotein made by several cell types, including osteoblasts, fibroblasts, chondrocytes, etc. In these mice, osteopontin is knocked-out, and in 1998 Rittling et.al. showed no alterations to bone mophology, but increased osteoclastogenesis.	-Increased number of HSPCs, -No change in lymphoid or myeloid cell production, -Decreased HSPC apoptosis, -Increased Jagged 1 and Angiopoietin 1 stromal expression, -Disease phenotype retained after transfer of wild type HSPCs to Opn-/- mice	Rittling, 1998; Stier, 2005; Nilsson, 2005
2005	p27Kip1 and MAD1	Cyclin-dependent kinase inhibitors MAD1 and p27Kip1 are negative regulators of cell cycle. In these two studies, mice had either p27Kip1 knocked-out (Chien, 2006) or p27Kip1 and MAD1 knocked-out (Walkley, 2005).	-Hyperplasia of hematopoietic organs, -Increased myeloid and erythroid colony forming cells, -Increase in LT-HSCs, -Disease phenotype retained after transfer of wild type HSPCs into p27Kip1-/- mice	Walkley, 2005; Chien, 2006
2005	gp130	Glycoprotein 103 (gp130) is a subunit of the cytokine receptors for the IL-6 family. These mice were generated with the Tcre mouse for excision of gp103 in hematopoietic and endothelial cells.	-Hypocellular marrow, -Impairment in erythro-and thrombopoiesis, -Reduction in T lymphocytes in the thymus, -Reduction of B lymphopoiesis in the marrow, -Extramedullary hematopoiesis, -Disease phenotype retained after transplant of wild type HSPCs into gp130 mice	Yao, 2005

Continued on facing page

Table 1. continued

Year	Protein/ cell	Description	Hematopoietic highlights	References
2006	RAR gamma	Retinoic acid receptors (RAR) are nuclear hormone receptors that act as ligand-dependent transcriptional regulators. In these mice the RARgamma is knocked-out.	-Decrease in HSPCs, -Increased granulopoiesis, -Myloproliferative-like disease with excessive extramedullary hematopoiesis, -Loss of trabecular bone by 12 weeks of age, -Disease phenotype retained after transfer of wild type HSPCs into RARγ-/- mice	Purton, 2006; Walkley, 2007
2007	Rb	Retinoblastoma protein (Rb) is a central regulator of the cell cycle and several downstream regulators have been shown to have affects on hematopoiesis. These mice were generated using an inducible deletion construct in conjunction with Mx-1 Cre for deletion of Rb in hematopoietic and stromal cells.	-Myloproliferative disease phenotype, -Increased HSPC differentiation, -Egress of HSPCs from marrow to extramedullary sites, -Loss of Rb was found to be necessary from both the myeloid-derived cells and the environment for presentation of the disease phenotype	Walkley, 2007
2008	Gsα	Gsa is a heterotrimeric G protein subunit that activates the cAMP-dependent pathway by activating adenylate cyclase and is part of the parathyroid hormone (PTH) and the PTH related protein (PTHrP) receptor. In these mice, Gsa was ablated in cells expressing ostrix, e.g. early osteoprogenitors and chondrocytes.	-Decreases in B cell precursors in the marrow and in the periphery, -No negative affect on other hematopoietic lineages, -Decreases in IL-7 expression from osteoblasts, -Decreases in trabecular bone, -Disease phenotype not transferable to wild type mice with marrow transplant	Wu, 2008
2008	Bis (BAG-3 or CAIR-1)	Bcl-2 interacting cell death suppressor (Bis) is a protein involved in antiapoptotic and antistress pathways. In this mouse, Bis was truncated for loss of function.	-Reduced lymphoid tissues, -Perturbed vasculature with defects in endothelial cells, -Loss of HSPCs, -Defect in B lymphopoiesis, -Decreased splenic hematopoietic cell numbers, -Defects in stromal progenitor cells, -Loss of stromal cells expressing CXCL-12 and IL-7, -Osteoblast lineage unaffected, -Disease phenotype retained after transplant of wild type cells into Bis-/- mice	Youn, 2008; Kwon, 2010

Continued on facing page

Table 1. continued

Year	Protein/ cell	Description	Hematopoietic highlights	References
2008	Hf2/merlin	Neurofibromin 2 (Nf2)/moesin-ezrin-radixin-like (merlin) is a cytoskeletal scaffolding protein involved in cell-cell communications. These mice were generated with a PolyI:C-inducible Mx-1-Cre to delete Nf2/merlin in hematopoietic and stromal cells.	-Egress of HSPCs, -Hypocellular marrow, -Increases in marrow vascularity, -Hematopoietic lineages unaffected, -After time, increases in trabecular bone and osteoblast numbers accompanied by restoration in marrow cellularity, -Disease phenotype retained after transplant of wild type cells into Nf2/merlin mice	Larsson, 2008
2010	Dicer	Dicer is an endoribonuclease that cleaves double-stranded RNA and pre-mircoRNA into small interfering RNA. In these mice, Dicer was ablated in cells expressing ostrix, e.g. early osteoprogenitors and chondrocytes.	-Decreases in leukocytes, platelets and red blood cells, -Decreases in B cell number with increases in myeloid cells, -No change in HSPC number, -Extramedullary hematopoiesis, -Slight decrease in osteoblast number with impaired differentiation, -Bone volume unchanged, but altered bone texture, -Disease phenotype retained after transplant of wild type cells into Dicer mice	Raaijmakers, 2010
2010	CAR cells	CXC chemokine ligand (CXCL) 12 abundant reticular (CAR) cells have been implicated as a hematopoietic niche cell in the marrow. These mice were designed to have inducible selective ablation of CAR cells with the diptheria toxin receptor.	-Reduced cycling of lymphoid and erythroid progenitor cells, -Reduced HSPC numbers, -HSPCs more quiescent, -HSPCs express more myeloid genes, -No changes in osteoblasts or endothelial cells, -Impaired production of SCF and CXCL-12	Omatsu, 2010
2010	Nestin+ cells	Nestin is an intermediate filament protein expressed in nerve cells and reported here in rare non-hematopoietic cells with mesencymal progenitor cell qualities. These mice were generated using the inducible diphtheria toxin with Nes-creERT2 for selective depletion of nestin expressing cells.	-Rapid reduction of HSPCs in marrow and increase in HSPCs in spleen, -Reduction of adrenergic nerve fibers, -Reduction of wild type HSPC homing to marrow of Nestin mice	Mendez-Ferrer, 2010

Continued on facing page

Table 1. continued

Year	Protein/ cell	Description	Hematopoietic highlights	References
2010	Ebf2	Early B cell factor 2 (Ebf2) is a transcription factor expressed in neurons, immature osteoblsts and adipocytes. In these mice, Ebf2 is knocked-out.	-Impaired lymphopoiesis, -Reduced numbers of HSPCs, -Mice were smaller then wild type cohorts, -Disease phenotype rescued after transplant of Ebf2-/- cells into wild type mice	Corradi, 2003; Kieslinger, 2010
2011	Agrin	Agrin is a heparan sulfate proteoglycan matrix protein expressed on MSCs and trabecular osteoblasts in the niche. In these mice, agrin expression is deficient with low levels of muscle specific kinase (MuSK) expression for postnatal survival.	-Marrow hypoplasia with decreases in myeloid and lymphoid cells, -Reduced CD45+ cell numbers in marrow, spleen and marrow, -Decreases in thymus T-lineage populations, -Decreases in ST-HSC in marrow, but normal levels in fetal liver, -Disease phenotype is not transfered to wild type mice after transplant with MuSK-L:Agrin-/- hematopoietic cells	Mazzon, 2011

Table 1. continued

The idea of a unique tissue environment, or stem cell niche, as a tissue setting that can direct progenitor cell behavior, e.g. quiescence, proliferation, differentiation, etc. was proposed over four decades ago (Wolf and Trentin 1968; Trentin 1971; Schofield 1978; Wolf 1979). Hematopoietic niches, or hematopoietic microenvironments (HME), are defined by the association of particular cell types, their secreted matrix products and their soluble hematopoietic factors (Yin and Li 2006; Rodgers et al. 2008). The identity of the cellular, matrix and soluble components that influence HSCs, including the long-term populating (LT-HSC), short-term populating (ST-HSC), or the more differentiated hematopoietic progenitor cells (HPC), as well as the lymphoid and myeloid lineages, remains an active topic of investigation. However, at least two hematopoietic niches have been described, an osteoblastic (or endosteal) niche, ascribed to osteoblasts residing on bone surfaces, and a vascular niche, ascribed to endothelial cells and subendothelial MSCs or pericytes lining marrow sinusoids. Many have argued that more quiescent LT-HSCs and ST-HSCs are located in the osteoblast niche, while differentiating HPCs are located in the vascular niche for mobilization to the periphery (Lord et al. 1975; Shackney et al. 1975; Gong 1978; Nilsson et al. 2001; Heissig et al. 2002; Arai, F. et al. 2004; Balduino et al. 2005; Jang and Sharkis 2007; Bourke et al. 2009). However, these regions in the marrow are so close in proximity (Arai, F. et al. 2004; Kiel et al. 2005; Lo Celso et al. 2009; Xie et al. 2009), that the osteoblast and vascular niches may be the same, or perhaps interchangeable to some degree. Additionally, recent work has identified other cells involved in hematopoiesis that do not fully comply with these proposed niche regions, such as CXC chemokine ligand (CXCL)-12 expressing reticular cells that are scattered throughout the marrow (Tokoyoda et al. 2004; Sugiyama et al. 2006; Omatsu et al. 2010). Below we will discuss the different cellular, matrix and soluble components within the marrow that have been shown to influence hematopoiesis.

3.1 Cells of the niche

Experiments designed to identify the cells of the HME date back over fifty years (Pfeiffer 1948; Tavassoli and Crosby 1968; Tavassoli and Weiss 1971; Meck et al. 1973; Friedenstein et al. 1974; Tavassoli and Khademi 1980; Friedenstein et al. 1982; Patt et al. 1982; Tavassoli 1984; Friedenstein et al. 1987; Gurevitch and Fabian 1993; Kawai et al. 1994; Kuznetsov et al. 1997; Hara et al. 2003; Akintoye et al. 2006; Mankani et al. 2007; Sacchetti et al. 2007; Chan, C.K. et al. 2008; Mankani et al. 2008; Song et al. 2010). In these studies, ectopic bone with a functional HME was generated in host mice using various bone marrow derived osteoprogenitor seed cells. More recently, different osteoprogenitor pools have been isolated that can either generate EO-like bone with active hematopoiesis or compact IO-like bone without an HME (Akintoye et al. 2006; Chan, C.K. et al. 2008). Collectively, these studies have shown that functional HMEs form through the progression of EO with contributions from cartilage, bone, vasculature, and marrow stromal cells. Further, similar conclusions about the necessity of EO-derived components, e.g. cells and matrix molecules, were obtained through analyses of several mouse models with skeleto-hematopoietic defects (**Table 1**), including mice with disrupted collagen X function in the HME (Jacenko et al. 1993; Gress and Jacenko 2000; Jacenko et al. 2002; Sweeney et al. 2008; Sweeney et al. 2010). Discussed below are the data describing which cell type(s), associated matrix and soluble factors are necessary for blood cell development in the marrow, including: multipotent stromal cells (fibroblasts, pericytes, reticular cells, and adipocytes), osteoblasts, chondrocytes, endothelial cells, and cells of hematopoietic origin (hematopoietic stem/progenitor cells (HSPC), osteoclasts and macrophages).

3.1.1 Fibroblasts and perivascular cells

As early as the 1970's, in vitro studies with marrow stromal adherent colonies showed that this pool of cells is able to support hematopoiesis (Friedenstein et al. 1970; Dexter et al. 1973; Friedenstein et al. 1974; Friedenstein et al. 1976; Dexter et al. 1977). These plastic adherent colonies have been thought to contain mesenchymal progenitor cells, and likely mesodermal progenitor cells as well (Petrini et al. 2009). This would account for recent data indicating that marrow derived progenitor cells generate adipocytes, chondrocytes and osteoblasts, traditional mesenchymal cell types, as well as fibroblasts, smooth muscle cells, endothelial cells, and pericytes/subendothelial cells (Bentley and Foidart 1980; Muguruma et al. 2006; Sacchetti et al. 2007; Crisan et al. 2008; Kalajzic et al. 2008; Augello et al. 2010; Mendez-Ferrer et al. 2010). Thus, isolated marrow stromal cells will be referred to as multipotent stromal cells (MSC) throughout (Horwitz et al. 2005). The MSCs isolated for ectopic bone assays have been described as osteoprogenitor cells that generate bone with a HME able to support host-derived hematopoiesis (Kuznetsov et al. 1997; Akintoye et al. 2006; Sacchetti et al. 2007; Chan, C.K. et al. 2008; Morikawa et al. 2009). Of note, many of these osteoprogenitor cells have been shown to have stem-like qualities, such as self-replication and the ability to differentiate into several different cell types. For example, Sacchetti et al. isolated human-derived MCAM/CD146-expressing subendothelial cells, which can self-replicate as well as give rise to osteoblasts, chondrocytes and reticular cells in ectopic HMEs (Sacchetti et al. 2007). More recently, these data were replicated in the mouse by Morikawa et al. who also identified a perivascular cell type that has the ability to self-replicate and give rise to adipocytes, osteoblasts, chondrocytes and endothelial cells (Morikawa et al. 2009). An additional perivascular cell has also been identified as a HME cell type, the nestin-expressing MSCs (Mendez-Ferrer et al. 2010). These, nestin^{+} MSCs have been shown to be spatially associate with HSPCs in the marrow and to express several HSPC maintenance genes, e.g. CXCL-12, stem cell factor (SCF)/kit ligand, angiopoietin (Ang)-1, interlukin (IL)-7, vascular cell adhesion molecule (VCAM)-1, and osteopontin (Mendez-Ferrer et al. 2010). Further, numbers of HSPC were rapidly reduced in the marrow of mice that were selectively depleted of nestin^{+} MSCs (**Table 1**) (Mendez-Ferrer et al. 2010). These mouse models also revealed a necessity for marrow nestin+ MSCs for homing of transferred HSPCs (Mendez-Ferrer et al. 2010). Together, these data suggest that MSCs and associated daughter cells not only make up the physical structure of the HME, but also provide maintenance and differentiation signals to HSPCs.

3.1.2 CXCL-12 abundant reticular cells

Reticular cells are of mesodermal origin and are a type of fibroblast cell localized to the intertrabecular region of the marrow near both the osteoblast and vascular niches (Weiss 1976; Rouleau et al. 1990). Recently, a sub-set of marrow reticular cells has been shown to express high levels of CXCL-12 (or stromal derived factor (SDF)-1) and have been termed CXCL-12 abundant reticular (CAR) cells (Tokoyoda et al. 2004; Sugiyama et al. 2006). CXCL-12 is reportedly involved in several aspects of hematopoiesis, including HSPC homing and maintenance, as well as B cell development (Nagasawa et al. 1994; Nagasawa et al. 1996; Ara et al. 2003; Broxmeyer et al. 2005; Jung et al. 2006; Sugiyama et al. 2006). Using CXCL-12/GFP knock-in mice, HSPCs, early lineage B cells and plasma B cells have been shown to spatially associate with CAR cells (Tokoyoda et al. 2004), suggesting CAR cells are a HME cell type. To address the importance of CXCL-12 expressing cells in the HME, Omatsu et al. designed a mouse model with selective ablation of CAR cells (**Table 1**) (Omatsu et al. 2010). These assays showed no change in the osteoblast or vascular niches, but impaired production of SCF and CXCL-12, combined with marked reduction in cycling lymphoid and erythroid progenitors. Further, the HSPC population in these mice was more quiescent, diminished in numbers and expressed myeloid selector genes. Finally, CAR cells can give rise to adipocytes and osteoblasts (Bianco et al. 1988; Balduino et al. 2005; Sipkins et al. 2005; Omatsu et al. 2010). Thus, these data combined with the ectopic bone assays and the nestin+ reticular cell studies discussed above, raise the possibility that the CAR, osteoprogenitor and nestin+ cells are from a similar cell pool, sharing differentiation capabilities and roles in hematopoiteitc support.

3.1.3 Adipocytes

Within the young marrow there are few adipocytes, however this phenomenon is reversed with aging and after marrow insult, such as post irradiation (Burkhardt et al. 1987; Verma et al. 2002). Although adipocytes have been described as having a positive influence on hematopoiesis via growth factors secretion (Lanotte et al. 1982), other studies have reported that these growth factors are in too low a concentration to influence HSPCs and that adipocytes secrete anti-hematopoietic factors as well (Hotamisligil et al. 1993; Zhang, Y. et al. 1995; Yokota et al. 2000; Corre et al. 2006; Belaid-Choucair et al. 2008; Miharada et al. 2008). Further, an increase in marrow adiposity has been negatively correlated with hematopoiesis in vivo (Touw and Lowenberg 1983; Naveiras et al. 2009). Interestingly,

Naveiras et al. have shown that in the adult mouse spine there is a proximal to distal gradient of marrow adipocity, with thoracic vertebrae being virtually free of adipocytes. This provides an in vivo model to study the affects of adipocytes on hematopoietic cells under homeostatic conditions. These studies showed that the number, frequency and cycling capacity of HSPCs was reduced as the number of adipocytes was increased (Naveiras et al. 2009). In support, after irradiation and marrow transplantation of mice genetically incapable of forming adipocytes, or in wild type mice treated with an inhibitor of adipogenesis, there was enhanced HSPC expansion compared to non-treated wild type cohorts (Naveiras et al. 2009). Of note, in these models of reduced/abrogated adipogenesis, as well as in a model where the fatty marrow is surgically removed, the increase in hematopoiesis is concomitant with an increase in bone formation (Tavassoli et al. 1974; Naveiras et al. 2009). These data suggest that after marrow insult and in the absence of adipocytes, there are signals enhancing osteoblast activity and bone formation, which may contribute to the enhanced hematopoiesis measured, as discussed below. This is supported by the clinical observations that aged patients have an increase in adiposity in the marrow, which is correlated with a decrease in bone formation and decreased hematopoiesis (Verma et al. 2002; Rossi et al. 2005; Mayack et al. 2010).

3.1.4 Osteoblasts

With the identification of osteoblast-like cells in stromal cultures (Friedenstein et al. 1987; Benayahu et al. 1991; Benayahu et al. 1992) and osteoprogenitors within marrow preparations for ectopic bone assays (Kuznetsov et al. 1997; Akintoye et al. 2006; Sacchetti et al. 2007; Chan, C.K. et al. 2008; Morikawa et al. 2009), Emerson and Taichman designed in vitro assays to assess the ability of isolated osteoblasts to support hematopoiesis (Taichman and Emerson 1994; Taichman et al. 1996; Taichman et al. 1997; Jung et al. 2005; Jung et al. 2006; Zhu et al. 2007). These assays showed that osteoblasts can support hematopoiesis through the secretion of pro-hematopoietic cytokines, e.g. granulocyte colony-stimulating factor (G-CSF), granulocyte-macrophage colony-stimulating factor (GM-CSF), CXCL-12, IL-6, and IL-7, and that cell-cell contact is necessary for support via integrin binding (very late antigen (VLA)-4/5 and VCAM/ICAM). Further, these assays confirmed a connection between osteoblasts and B cell development. To assess the contribution of osteoblasts to hematopoiesis in vivo, several different mouse models have been generated that either increase osteoblasts (Calvi et al. 2001; Calvi et al. 2003; Zhang, J. et al. 2003), decrease osteoblasts (Visnjic et al. 2001; Visnjic et al. 2004), disrupt EO-based trabecular bone formation (Jacenko et al. 1993; Gress and Jacenko 2000; Jacenko et al. 2001; Jacenko et al. 2002; Sweeney et al. 2008; Sweeney et al. 2010), alter osteoblast signaling (Wu et al. 2008), or modify osteoblast RNA processing (**Table 1**) (Raaijmakers et al. 2010). These in vivo models have confirmed that osteoblasts can support hematopoiesis, are involved in B lymphopoiesis, make pro-hematopoietic cytokines, make cell-cell contact with HSPCs, and moreover, have implicated osterix expressing osteoprogenitors as mediators of hematopoiesis.

The above studies, as well as many imaging studies of HSPC in bone, support the osteoblast hematopoietic niche theory and suggest that the osteoblast niche may additionally encompass the B lymphopoietic niche (Nilsson et al. 1997; Nilsson et al. 2001; Yoshimoto et al. 2003; Arai, F. et al. 2004; Balduino et al. 2005; Kohler et al. 2009; Lo Celso et al. 2009; Xie et al. 2009). For instance, Xie et al. showed that GFP+ HSPCs home to the trabecular bone surface in the marrow, and others have reported early developing B lymphocytes at the endosteal region of the marrow (Hermans et al. 1989; Jacobsen and Osmond 1990; Osmond 1990; Xie et al. 2009). This zone is the chondro-osseous region where hypertrophic chondrocytes, trabecular osteoblasts and marrow cells are juxtaposed (**Fig. 1**). Hypertrophic chondrocytes and their matrix components are also essential for trabecular bone formation, and are proposed to be part of the osteoblast/lymphopoietic niche (Jacenko et al. 2002; Rodgers et al. 2008; Sweeney et al. 2008; Sweeney et al. 2010). Interestingly, hypertrophic chondrocytes have been described to trans-differentiate into osteoblasts (Roach 1992; Galotto et al. 1994; Roach et al. 1995; Roach and Erenpreisa 1996), and express osteoblast-like markers, e.g. osterix, osteocalcin, osteonectin, osteopontin and collagen I (Roach 1992; Yagi et al. 2003), suggesting similarities in the cells of the chondro-osseous niche. Indeed, it has been suggested that chondro-osteoprogenitor cells that expresses the chondrocyte-like marker collagen II contribute to both the perichondrial and trabecular osteoblast populations (Nakamura et al. 2006; Hilton et al. 2007). In contrast, however Maes et al. do not report any contribution from the collagen II labeled hypertrophic chondrocytes to trabecular bone (Maes et al. 2010). Taken together, these finding are reminiscent of the reports indicating that progenitor cells in the marrow can give rise to different cell types with many similarities, e.g. osteoprogenitors, CAR and nestin+ cells not only share gene expression profiles, but are all located in the chondro-osseous environment (Weiss 1976; Rouleau et al. 1990; Sugiyama et al. 2006). The possible overlap between the cells of the chondro-osseous region can also be appreciated when comparing osteoblasts and reticular cells that can support B lymphopoiesis and both express VCAM-1 and IL-7 (Ryan et al. 1991; Funk et al. 1995; Zhu et al. 2007). Moreover, using an osteoblast lineage tracer mouse generated with an osterix-LacZ construct, cells of the perichondrium, trabecular bone, cartilage and marrow stroma, some intimately associated with blood vessels in a pericyte-like fashion, were all positive for osterix expression (Maes et al. 2010), again confirming an overlap of cell phenotypes in the chondro-osseous HME.

3.1.5 Chondrocytes

As previously discussed, chondrocytes provide the blueprint for future bone with a marrow cavity during EO, and are adjacent to the postulated osteoblast and vascular niches (**Fig. 1B boxed**) (Arai, F. et al. 2004; Kiel et al. 2005; Kohler et al. 2009; Xie et al. 2009). Indeed, growth plate chondrocytes, as well as osteoblasts and vascular cells, express leukemia inhibitory factor (LIF), which can synergize with growth factors to promote the proliferation of HSPCs (Keller et al. 1996; Grimaud et al. 2002). Additionally, Wei et al. recently showed that hypertrophic chondrocytes express CXCR-4, the receptor for CXCL-12 made by stromal cells and osteoblasts (Peled et al. 1999; Kortesidis et al. 2005; Dar et al. 2006; Jung et al. 2006; Sacchetti et al. 2007; Wei et al. 2010). These data begin to reveal the cross talk between the hypertrophic chondrocytes and the cells of the chondro-osseous environment that are players within the hematopoietic niche, e.g. osteoblasts, stromal and hematopoietic cells.

To date, no imaging studies have attempted to localize HSPCs to hypertrophic chondrocytes. However, the contribution of various matrix components, in particular the heparan sulfate proteoglycans (HSPG), in establishing reservoirs of soluble factors for cell signaling and/or retention of HSPCs has been well established (as reviewed in (Rodgers et al. 2008) also see (Gordon et al. 1988; Roberts et al. 1988; Siczkowski et al. 1992; Verfaillie 1993; Allouche and Bikfalvi 1995; Bruno et al. 1995; Klein et al. 1995; Gupta et al. 1996; Gupta et al. 1998; Borghesi et al. 1999; Siebertz et al. 1999; Zweegman et al. 2004; Rodgers et al. 2008; Spiegel et al. 2008). In support, our laboratory has shown altered localization of both hyaluronan and HSPGs within the hypertrophic cartilage zone of the growth plates in the collagen X mouse models that display altered hematopoiesis (Jacenko et al. 2001), which directly links EO and the hypertrophic cartilage matrix to hematopoiesis (Jacenko et al. 1993; Gress and Jacenko 2000; Jacenko et al. 2001; Jacenko et al. 2002; Sweeney et al. 2008; Sweeney et al. 2010). Briefly, collagen X is a short chain, network forming collagen that is the major secreted matrix protein of hypertrophic chondrocytes and is localized to the hypertrophic cartilage/chondro-osseous region (Campbell et al. 2004), where it is proposed to form a hexagonal lattice-like network in the matrix (Jacenko et al. 1991; Chan, D. and Jacenko 1998). Affinity co-electrophoresis studies demonstrated that collagen X and heparin, a structural analog of heparan sulfate, can endogenously bind (Sweeney, unpublished). We thus proposed that the hypertrophic chondrocyte derived matrix, made up of the collagen X network that is likely stabilized by associating with the HSPGs, is enriched with hematopoietic factors and is a vital component of the HME (Jacenko et al. 2001; Rodgers et al. 2008). In accord, mouse models where the function of collagen X was altered via transgenesis or targeted gene knock-out have both an altered HME structure, as well as aberrant hematopoiesis. Specifically, alterations within the EO-derived chondro-osseous junction include aberrant growth plate histomorphometry, collapsed hypertrophic chondrocyte matrix network, diminished and altered localization for HSPG within hypertrophic cartilage and trabecular bone, and decreased trabecular bone. The hematopoietic changes include diminished B lymphopoiesis throughout life, perinatal lethality within a sub-set of mice due to opportunistic infections by the third week of life, decreased responses to concanavalin A by splenocytes from mice at all ages, and the succumbing of all collagen X mice to non-virulent pathogen challenge (**Table 1**) (Jacenko et al. 1993; Rosati et al. 1994; Jacenko et al. 1996; Kwan et al. 1997; Gress and Jacenko 2000; Jacenko et al. 2001; Jacenko et al. 2002; Sweeney et al. 2008; Sweeney et al. 2010). Additionally, altered levels of hematopoietic cytokines have been measured from the collagen X mouse derived hypertrophic chondrocytes and trabecular osteoblasts when compared to wild type cohorts (Sweeney, 2011 Ann N Y Acad. Sci. in press). The altered cytokine availability may negatively affect hematopoiesis, which may be even further amplified by the altered chondro-osseous matrix in the collagen X mice, e.g. loss of functional collagen X and diminished HSPGs at the chondro-ossous HME (Jacenko et al. 2001). In support of the notion that matrix/cytokine signaling can affect cell differentiation in the HME, fate switching was sited as the cause of decreased osteoblast progenitors and bone formation in a knock-out mouse for the critical transforming growth factor (TGF)-β binding proteoglycans, biglycan and decorin (Bi et al. 2005). Overall, the collagen X mouse models have highlighted the contribution of EO-derived cells and matrix components to HMEs and to hematopoietic cell development.

3.1.6 Endothelial cells

Marrow arterioles and capillaries supply the sinusoids, which in turn supply the marrow with cells and nutrients (reviewed in (Kopp et al. 2005). These are the sites of the vascular niche, and examples of active vascular niches can be appreciated during development when hematopoiesis takes place in the yolk sac, aorta-gonad-mesonephros region and placenta perivascularly (Cumano et al. 1996; Medvinsky and Dzierzak 1996; Sanchez et al. 1996; de Bruijn et al. 2002; North et al. 2002; Gekas et al. 2005; Ottersbach and Dzierzak 2005). Additionally, throughout life in some species, such as the zebra fish, hematopoiesis is not coincident with bone (Murayama et al. 2006). Further, the characteristics of the marrow sinusoids, e.g. chemokine and adhesion molecule expression, not only allow them to be conduits for hematopoietic cells to and from the circulation, but also to serve as an area for HSPC differentiation (Rafii et al. 1994; Rafii et al. 1995; Schweitzer et al. 1996; Naiyer et al. 1999; Abkowitz et al. 2003; Avecilla et al. 2004). In agreement, endothelial cells from several sources are able to support HSPC maintenance and differentiation toward lymphoid and myeloid lineages in culture (Rafii et al. 1995; Ohneda et al. 1998; Li et al. 2004; Wittig et al. 2009; Butler et al. 2010). Visualizing the vascular niche in vivo with a pure sub-set of HSCs also provided support for the vascular niche theory. These studies revealed approximately 60% of HSC residing in the chondro-osseous HME, and of that population, approximately 60% were proximal to the vasculature, where as 15% were near bone (Kiel et al. 2005). Moreover, one aspect of the osteoblast niche theory maintains that HSCs are bound to osteoblasts by a N-cadherin-mediated homophilic adhesion, however in a mouse model where HSC specific N-cadherin was depleted, hematopoiesis was fully functional (Zhang, J. et al. 2003; Kiel et al. 2009). More in vivo support is provided by the biglycan deficient mice, which present with decreased trabecular osteoblasts and bone formation, however show no defects in hematopoiesis, HSC frequency or function, and show HSC localization to the vasculature (Kiel et al. 2007). These data suggest an overlap in niche location with perhaps some ability for compensation between the osteoblast and vascular niches, albeit limited. In support, via histochemistry of the long bone, one can appreciate the spatial proximity of osteoblasts to the vasculature (**Fig. 1B boxed**).

3.1.7 Hematopoietic derived cells influence the niche

Discussed above are the data linking non-hematopoietic cells with hematopoiesis, however there are also data supporting reciprocal affects of hematopoietic derived cells influencing non-hematopoietic lineages. For example, HSPCs regulate MSC differentiation toward the osteoblast lineage via expression of BMP-2 and -6, suggesting that the HSPCs can actively maintain the osteoblast niche (Jung et al. 2008). Additionally, macrophages intercalated throughout bone have been described as osteoblast helper cells since they promote osteoblast mineralization in vitro and form a canopy over osteoblasts generating bone in vivo (Chang et al. 2008). An additional player in osteoblastogenesis is the megakaryocyte. In mouse models with increased numbers of megakaryocytes due to maturation arrest, increased osteoblast proliferation and bone mass were measured (Kacena et al. 2004). Additionally, megakaryocytes have been described as niche restoring cells post-irradiation since they migrate to the damaged bone surfaces and increase local concentrations of CXCL-12, platelet-derived growth factor (PDGF)-β and basic fibroblast growth factor (bFGF), which are associated with osteoblast proliferation (Kacena et al. 2006; Dominici et al. 2009).

A reciprocal balance between bone deposition by osteoblasts, bone resorption by osteoclasts and signaling by osteocytes is extensively noted in the literature, and is appreciated clinically. These coupled interactions will not be discussed here other than to acknowledge that the function of one cell type highly depends upon and is affected by the actions of the other (Khosla 2003). Thereby, the continual signaling between cells of the mesenchymal and hematopoietic lineages underlies the tightly coupled process of bone remodeling, and its uncoupling can lead to skeletal disorders such as osteoporosis, osteopetrosis, as well as calcium homeostasis imbalances. Such examples are also seen in mouse models; for example, in one model where osteoclasts are depleted, instead of having increased bone mass due to lack of resorption, there is decreased bone mass compared to wild type cohorts (Kong et al. 1999). Collectively, the intricate cross talk between cells within the HME can both positively and negatively affect the niche environment as well as hematopoiesis, and is a vast area if research that remains to be adequately explored.

3.2 Soluble factors and the extracellular matrix in the niche

3.2.1 Cytokines, chemokines, growth factors, and neurotransmitters

As referred to above, the hematopoietic and non-hematopoietic cells within the HME are surrounded in all dimensions by matrix components and soluble factors. The building blocks of the extracellular matrices found within the chondro-osseous environment can include collagens, proteoglycans (PGs; including the HSPGs) and their glycosaminoglycan (GAG) constituents and glycoproteins. Collagens generally provide structural support for cells in the niche by forming supramolecular aggregates around cells (Jacenko et al. 1991), while the PGs, such as the HSPGs, can trap and store soluble factors for presentation to local cells (as reviewed in (Rodgers et al. 2008). The amount and ratio of these molecules in the matrix also dictates the mechanical properties of the HME, which has recently become a topic of investigation in the stem cell field. For instance, matrix elasticity in the HME can influence fate choices of HSPCs (Holst et al. 2010), a phenomenon also reported with MSCs (as reviewed in (Discher et al. 2009). Thus, the matrix provides structural integrity to the HME, acts as a substrate for cell migration and anchorage, and actively regulates cell morphology, development and metabolic function (Peerani and Zandstra 2010). The cells in the HME receive information for maintenance, development, differentiation, etc. via cell-cell interactions, cell-matrix interactions, and exposure to variable concentrations and combinations of soluble factors, e.g. cytokines, chemokines, hormones, and growth factors. Many soluble factors have been implicated in hematopoiesis, such as CXCL-12, SCF, Fms-related tyrosine kinase 3 ligand (Flt-3L), thrombopoietin (TPO), FGF, G-CSF, GM-CSF, LIF, Wnt, BMP-4, IL-3, 6, 7, 8, 11, 12, 14, 15 (Guba et al. 1992; Heinrich et al. 1993; Verfaillie 1993; Funk et al. 1995; Rafii et al. 1997; Taichman et al. 1997; Peled et al. 1999; Majumdar et al. 2000; Ponomaryov et al. 2000; Petit et al. 2002; Avecilla et al. 2004; Kortesidis et al. 2005; Dar et al. 2006; Jung et al. 2006; Spiegel et al. 2007; Wittig et al. 2009), or in maintenance and quiescence of HSCs, e.g. Ang-1 and TGF (Eaves et al. 1991; Fortunel et al. 2000; Arai, F. et al. 2004) (cytokine functions reviewed in (Zhang, C.C. and Lodish 2008). The HME cell types discussed above are the primary sources of these soluble factors, however lymphocytes have also been shown to stimulate and suppress hematopoiesis through the release of different factors during both homeostasis and immune activation (Nathan et al. 1978; Bacigalupo et al. 1980; Mangan et al. 1982; Harada et al. 1985; Trinchieri et al. 1987; Crawford et al. 2010). Interestingly, many of these soluble factors are sequestered and presented by the matrix, specifically HSPGs, which have been described as key orchestrators of hematopoiesis (Bruno et al. 1995; Gupta et al. 1996; Gupta et al. 1998; Borghesi et al. 1999). This is particularly relevant to the collagen X mouse models that present with a disrupted collagen X network coupled with a decreased HSPG staining intensity at the chondro-osseous HME and diminished cytokine levels, leading to diminished B lymphopoiesis (Gress and Jacenko 2000; Jacenko et al. 2001; Jacenko et al. 2002; Sweeney et al. 2008; Sweeney et al. 2010). These findings imply that the structural defects in the matrix may lead to changes in the cytokine reservoirs, which in turn would negatively affect hematopoietic cell development. By extension, human diseases associated with altered matrix components at the HME may have altered hematopoiesis due to changes in cytokine availability, such as with Simpson-Golabi Behmel syndrome where alterations in the HSPG glypican-3 results in skeletal and hematopoietic abnormalities (Pilia et al. 1996; Viviano et al. 2005).

There is increasing evidence that the nervous system can also affect the immune system through neurotransmitter signaling. The bone and marrow are supplied with autonomic efferent and afferent sensory innervations, specifically at the epiphysis and metaphysis of long bone, which includes the chondro-osseous HME (**Fig. 1A**) (reviewed in (Mignini et al. 2003). Catecholamines, acetylcholine and peptide transmitters of neural and non-neural origin are released in the HME, which contribute to neuro-immune modulations. For example, signaling from the nervous system can regulate HSPC egress and repopulation of the marrow (Katayama et al. 2006; Spiegel et al. 2007), which has been shown to be coupled to the circadian rhythm (Mendez-Ferrer et al. 2008). Indeed, HSPCs have receptors for several neurotransmitters, which can stimulate cell proliferation (Spiegel et al. 2008; Kalinkovich et al. 2009). Of note, beta-adrenergic agonists have been shown to stimulate osteoclast activity (Arai, M. et al. 2003), which could have an effect on the osteoblast niche via two methods, a) physically by decreasing bone lining cells and releasing HSPCs, and b) chemically via the release of soluble factor (Kollet et al. 2006; Mansour et al. 2011). Calcium is an example of one such soluble factor, which serves as an attractant to HSPCs encouraging homing to the osteoblast niche (Adams et al. 2006). Hematopoietic cell egress from the marrow has also been linked to many systemic causes, including exercise, inflammation, bleeding, cytotoxic drugs, and psychological anxiety (reviewed in (Lapid et al. 2008). These data serve as reminders that hematopoiesis and the HME can be influenced by factors outside of the local environment.

3.2.2 Cell and matrix influence

In the HME, cell-cell interactions influence cell fate decisions and mobility/homing; examples of such interactions include: Notch-1/Jagged-1, N-cadherin/N-cadherin and VLA-4/VCAM-1 (reviewed in (Coskun and Hirschi 2010). On the other hand, cell-matrix interactions influence not only cell behavior, but also cell anchorage to the niche. To date, the matrix proteins within the HME include: collagens (Types I, II, III, IV and X), glycoproteins (fibronectin, lamanin, nidogen, tenasin C, thrombospondin, vitronectin), PGs (perlecan, decorin, agrin) and the GAG hyaluronan (Bentley and Foidart 1980; Bentley 1982; Spooncer et al. 1983; Zuckerman and Wicha 1983; Zuckerman et al. 1985; Klein 1995; Ohta et al. 1998; Campbell et al. 2004; Mazzon et al. 2011). These matrix constituents can signal to hematopoietic cells through cell receptors such as: integrins, immunoglobulin-like molecules, cadherins, selectins, and mucins (Teixido et al. 1992; Coulombel et al. 1997; Levesque and Simmons 1999; Zhang, J. et al. 2003; Merzaban et al. 2011). Notably, the HME matrix network is not static, but is continually remodeled by different enzymes including metalloproteinases, neutrophil elastase and hepranase. Matrix turnover can thus assist in the release of hematopoietic cells from the niche (Levesque et al. 2001; Heissig et al. 2002; Petit et al. 2002; Spiegel et al. 2008), as well as liberate bound soluble hematopoietic factors (Heissig et al. 2002; Spiegel et al. 2008). An example of one such cell-matrix interaction in the HME is the VLA-4/fibronectin binding between HSPCs and the matrix, which provides the hematopoietic cell with anchorage as well as proliferation stimuli (Weinstein et al. 1989; Klein et al. 1998; Sagar et al. 2006). Hyaluronan also impacts HSPC maintenance, propagation, homing and homeostasis via CD44 binding (Avigdor et al. 2004; Matrosova et al. 2004; Haylock and Nilsson 2006). Most recently, Mazzon et al. have found the binding between agrin, expressed by trabecular osteoblasts and MSCs in the niche, and HSPCs leads to survival and proliferation signals (Mazzon et al. 2011). Finally, mature plasma B cells homing back to the marrow via CXCL-12 signals are anchored to their marrow niche via matrix-bound ligands produced by local myeloid cells (O'Connor et al. 2004; Crowley et al. 2005; Ingold et al. 2005; Moreaux et al. 2005; Nagasawa 2006; Schwaller et al. 2007; Huard et al. 2008; Moreaux et al. 2009). In fact, it has been shown that this interaction maintains long-lived antibody producing plasma B cells in the marrow by stimulating expression of anti-apoptotic genes in the lymphocytes (O'Connor et al. 2004; Huard et al. 2008). Thus, the cell-matrix interactions in the marrow serve to support hematopoietic maintenance and development, as well as support the persistence of mature hematopoietic cells that have returned to the marrow.

4. Summary and perspectives

All the specialized cells of the blood are generated through hematopoiesis via the directed differentiation of HSCs. The bone marrow, which is the predominant hematopoietic tissue after birth (Aguila and Rowe 2005; Cumano and Godin 2007), is formed through EO, where the cartilage anlage serves as a transient template for trabecular bone, and defines the environment of the marrow stroma. Thereby, either directly or indirectly, the process of EO establishes the hematopoietic niche by providing the niche with both the structure matrix constituents and the cellular components (Jacenko et al. 1993; Taichman and Emerson 1994; Taichman et al. 1996; Taichman et al. 1997; Gress and Jacenko 2000; Calvi et al. 2001; Visnjic et al. 2001; Jacenko et al. 2002; Calvi et al. 2003; Zhang, J. et al. 2003; Visnjic et al. 2004; Jung et al. 2005; Jung et al. 2006; Zhu et al. 2007; Chan, C.K. et al. 2008; Sweeney et al. 2008; Wu et al. 2008; Raaijmakers et al. 2010; Sweeney et al. 2010). Many cell types, matrix components and soluble factors contribute to the HME (**Fig. 2**). Through several different methods, HSPCs have been visualized in the chondro-osseous HME (Nilsson et al. 1997; Nilsson et al. 2001; Yoshimoto et al. 2003; Arai, F. et al. 2004; Balduino et al. 2005; Kiel et al. 2005; Kohler et al. 2009; Xie et al. 2009), which is comprised of vasculature sinusoids, sympathetic nerves, complex and diverse matrix regions, as well as osteoblasts, hypertrophic chondrocytes, endothelial cells, pericytes, CXCL-12 expressing cells, adipocytes, nestin+ cells, MSCs, macrophages intercalated in the endosteum, and cells of the immune system, both developing and recirculating. The cells of the HME express the cytokines, growth factors and chemokines utilized throughout hematopoiesis, as well as the matrix molecules that provides structural support, cell-matrix signaling and reservoirs of soluble factors (**Fig. 2**). We propose that the chondro-oseous HME is not static, but is continuously changing in response to various systemic influences (Lapid et al. 2008), as well as to remodeling of the hybrid trabecular bone-hypertrophic cartilage spicules into mature secondary bone (**Fig. 1C & 2**). During remodeling of the HME, both cells and

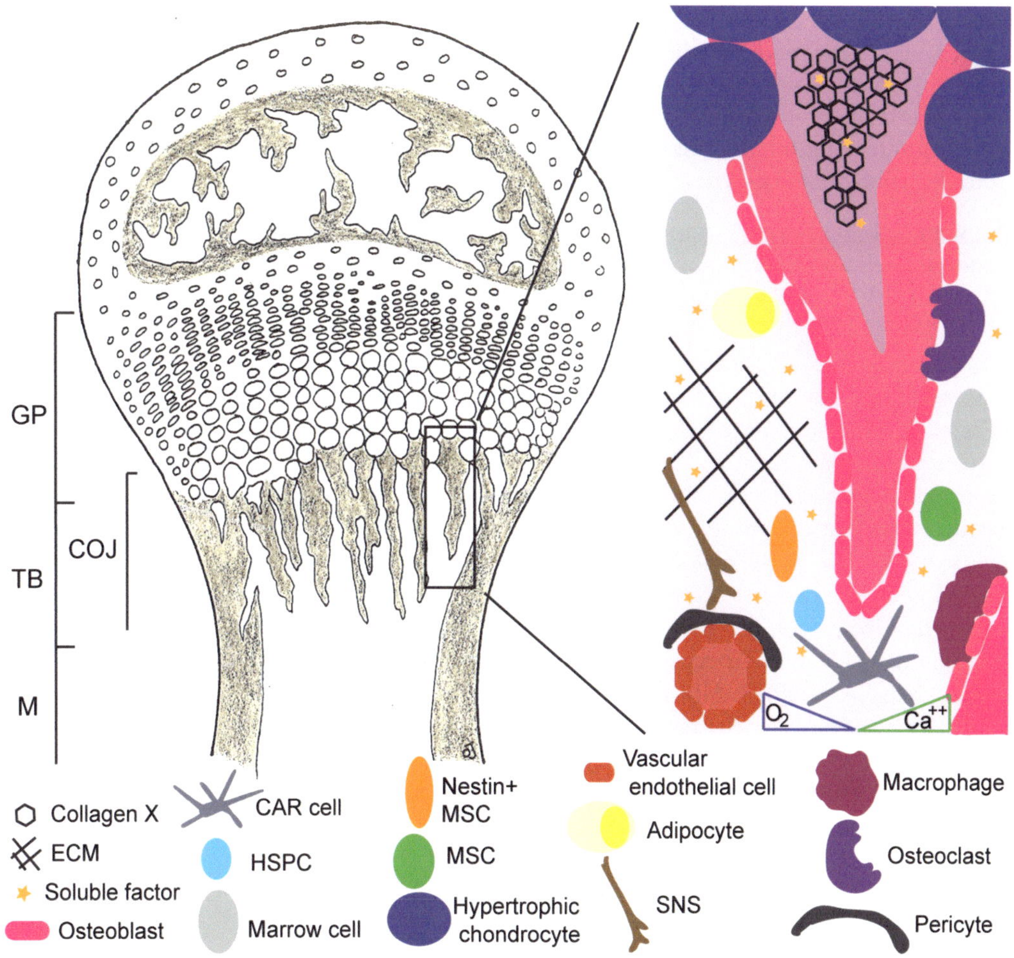

Fig. 2. The cells, matrix components and soluble factors within the chondro-osseous hematopoietic microenvironment. The putative hematopoietic niche has been localized to the chondro-osseous junction (COJ) of endochondrally developing bones, e.g. the juncture of the growth plate (GP) hypertrophic chondrocytes and the trabecular bone (TB) and marrow (M). Boxed, a cartoon in higher magnification represents the chondro-osseous hematopoietic environment, which includes: osteoblasts, hypertrophic chondrocytes, endothelial cells, pericytes, CXCL-12 abundant cells (CAR), adipocytes, endosteal intercalating macrophages, and other marrow cells, such as lymphocytes. Additionally, progenitor cells such as hematopoietic stem/progenitor cells (HSPC), multipotent stromal cells (MSC) and nestin expressing (nestin +) MSC are represented. All of these cells serve as a source of secreted factors, including cytokines, chemokines and growth factors, as well as neurotransmitters from the sympathetic nervous system (SNS). Also note the gradients of oxygen (O_2) and calcium (Ca^{++}) released from the vasculature or remodeled bone respectively. Finally, the extracellular matrix (ECM) in the marrow and in the core of trabecular bone, consisting of hypertrophic cartilage derived lattice-like collagen X and HSPGs, can serve as a substrate for cell anchorage, migration, and/or signaling through cell-ECM binding and as a reservoirs of secreted factors.

soluble factors are released into the milieu; for example matrix degrading enzymes release HSPCs for egress, and the combined activity of osteoclasts and enzymes release soluble factors from bone, including the trabecular bone with a collagen X/HSPG core (**Fig. 2**). Thus, the chondro-osseous HME is a continually active site with intrinsic and systemic signals influencing the cellular cross talk that ensures proper quiescence, maintenance, and differentiation of the HSPCs.

As itemized in **Table 1** and discussed above, various mouse models have revealed important cell and matrix players in the HME, while others have implicated the importance of proper cell cycle (Walkley et al. 2005; Walkley et al. 2007), transcription (Corradi et al. 2003; Purton et al. 2006; Walkley et al. 2007; Kieslinger et al. 2010), cell-cell communication (Larsson et al. 2008), and survival signals (Youn et al. 2008; Kwon et al. 2010) in the HME (**Table 1**). Still, the exact cellular make up and location of the HME is continually under debate, though most agree the cells of the osteoblast and vascular niches are important players in hematopoiesis. The possibility of osteoblast and vascular niches spatially overlapping or having the ability to provide some compensation for each other to some extent, is another intriguing theory. Evidence for this is observed with the biglycan deficient mouse model that presents with reduced bone, yet intact hematopoiesis (Kiel et al. 2007), as well as with the recently identified CAR cells that unite the osteoblast and vascular niches in the chondro-osseous HME by the observation that they are in contact with 90% of HSPCs throughout the trabecular bone region and the marrow sinusoidal region (Sugiyama et al. 2006; Omatsu et al. 2010). Additionally, there is similarity in the expression of hematopoietic soluble factors between the osteoblasts, endothelial cells, and CAR cells, such as CXCL-12 implicated in the homing, growth, development and maintenance of hematopoietic cells (Peled et al. 1999; Ponomaryov et al. 2000; Tokoyoda et al. 2004; Broxmeyer et al. 2005; Kortesidis et al. 2005; Dar et al. 2006; Sugiyama et al. 2006). Moreover, Medici et al. have confirmed an endothelial to osteoblast and chondrocyte transition in vivo, further supporting a cell interchange and overlap theory (Medici et al. 2010). In contrast, there is also evidence that the osteoblast and vascular niches provide different roles in hematopoiesis, e.g. a quiescence and maintenance role verses a differentiation and egress role (Lord et al. 1975; Shackney et al. 1975; Gong 1978; Nilsson et al. 2001; Heissig et al. 2002; Arai, F. et al. 2004; Balduino et al. 2005; Jang and Sharkis 2007; Bourke et al. 2009). Further, although these niches are proximal in the chondro-oseous region, mathematical modeling has predicted a layer of only two myeloid cells is sufficient to deplete most oxygen provided by a near by sinusoid (Chow et al. 2001). Thus, the local environment within each niche may differ significantly in chemical signals, such as with oxygen and calcium (**Fig. 2**).

Overall, the research generated in the hematopoietic niche field is beginning to shed light on many hematologic disorders, such as myelodysplasia, myeloproliferative syndromes and leukemias that seem to be influenced by the quality of the marrow environment (Walkley et al. 2007; Walkley et al. 2007; Raaijmakers et al. 2010). Further information on niche components, specifically the matrix molecules, will assist in generating bio-mimicking composites necessary for in vitro culture and expansion of patient specific hematopoietic tissues, for such clinical applications as autologous marrow transfers. The past sixty years of hematopoietic biology research has increased our understanding of marrow stromal cell types, as well as the three-dimensional regions that provide structure and organization of cell signaling for the maintenance and propagation of HSPCs. The study of the hematopoietic niche will continue to provide details on necessary niche components, which may assist in the understanding of other stem cell niches, including the vascular, skin, hair, and neural niches, and provide therapeutic cues for immuno-osseous diseases that present with skeletal defects and altered hematopoiesis, such as McKusick type metaphyseal chondrodysplasia (cartilage-hair hypoplasia; CHH), Shwachmen-Diamond syndrome, Schimke dysplasia (Spranger et al. 1991; Kuijpers et al. 2004; Hermanns et al. 2005), and others.

5. References

Abkowitz, J. L., A. E. Robinson, S. Kale, M. W. Long & J. Chen (2003). "Mobilization of hematopoietic stem cells during homeostasis and after cytokine exposure." *Blood* 102(4): 1249-1253

Adams, G. B., K. T. Chabner, I. R. Alley, D. P. Olson, Z. M. Szczepiorkowski, M. C. Poznansky, C. H. Kos, M. R. Pollak, E. M. Brown & D. T. Scadden (2006). "Stem cell engraftment at the endosteal niche is specified by the calcium-sensing receptor." *Nature* 439(7076): 599-603

Aguila, H. L. & D. W. Rowe (2005). "Skeletal development, bone remodeling, and hematopoiesis." *Immunol Rev* 208: 7-18

Akintoye, S. O., T. Lam, S. Shi, J. Brahim, M. T. Collins & P. G. Robey (2006). "Skeletal site-specific characterization of orofacial and iliac crest human bone marrow stromal cells in same individuals." *Bone* 38(6): 758-768

Allouche, M. & A. Bikfalvi (1995). "The role of fibroblast growth factor-2 (FGF-2) in hematopoiesis." *Prog Growth Factor Res* 6(1): 35-48

Alvarez, J., M. Balbin, M. Fernandez & J. M. Lopez (2001). "Collagen metabolism is markedly altered in the hypertrophic cartilage of growth plates from rats with growth impairment secondary to chronic renal failure." *J Bone Miner Res* 16(3): 511-524

Ara, T., K. Tokoyoda, T. Sugiyama, T. Egawa, K. Kawabata & T. Nagasawa (2003). "Long-term hematopoietic stem cells require stromal cell-derived factor-1 for colonizing bone marrow during ontogeny." *Immunity* 19(2): 257-267

Arai, F., A. Hirao, M. Ohmura, H. Sato, S. Matsuoka, K. Takubo, K. Ito, G. Y. Koh & T. Suda (2004). "Tie2/angiopoietin-1 signaling regulates hematopoietic stem cell quiescence in the bone marrow niche." *Cell* 118(2): 149-161

Arai, M., T. Nagasawa, Y. Koshihara, S. Yamamoto & A. Togari (2003). "Effects of beta-adrenergic agonists on bone-resorbing activity in human osteoclast-like cells." *Biochim Biophys Acta* 1640(2-3): 137-142

Augello, A., T. B. Kurth & C. De Bari (2010). "Mesenchymal stem cells: a perspective from in vitro cultures to in vivo migration and niches." *Eur Cell Mater* 20: 121-133

Avecilla, S. T., K. Hattori, B. Heissig, R. Tejada, F. Liao, K. Shido, D. K. Jin, S. Dias, F. Zhang, T. E. Hartman, N. R. Hackett, R. G. Crystal, L. Witte, D. J. Hicklin, P. Bohlen, D. Eaton, D. Lyden, F. de Sauvage & S. Rafii (2004). "Chemokine-mediated interaction of hematopoietic progenitors with the bone marrow vascular niche is required for thrombopoiesis." *Nat Med* 10(1): 64-71

Avigdor, A., P. Goichberg, S. Shivtiel, A. Dar, A. Peled, S. Samira, O. Kollet, R. Hershkoviz, R. Alon, I. Hardan, H. Ben-Hur, D. Naor, A. Nagler & T. Lapidot (2004). "CD44 and hyaluronic acid cooperate with SDF-1 in the trafficking of human CD34+ stem/progenitor cells to bone marrow." *Blood* 103(8): 2981-2989

Bacigalupo, A., M. Podesta, M. C. Mingari, L. Moretta, M. T. Van Lint & A. Marmont (1980). "Immune suppression of hematopoiesis in aplastic anemia: activity of T-gamma lymphocytes." *J Immunol* 125(4): 1449-1453

Balduino, A., S. P. Hurtado, P. Frazao, C. M. Takiya, L. M. Alves, L. E. Nasciutti, M. C. El-Cheikh & R. Borojevic (2005). "Bone marrow subendosteal microenvironment harbours functionally distinct haemosupportive stromal cell populations." *Cell Tissue Res* 319(2): 255-266

Belaid-Choucair, Z., Y. Lepelletier, G. Poncin, A. Thiry, C. Humblet, M. Maachi, A. Beaulieu, E. Schneider, A. Briquet, P. Mineur, C. Lambert, D. Mendes-Da-Cruz, M. L. Ahui, V. Asnafi, M. Dy, J. Boniver, B. V. Nusgens, O. Hermine & M. P. Defresne (2008). "Human bone marrow adipocytes block granulopoiesis through neuropilin-1-induced granulocyte colony-stimulating factor inhibition." *Stem Cells* 26(6): 1556-1564

Benayahu, D., A. Fried, D. Zipori & S. Wientroub (1991). "Subpopulations of marrow stromal cells share a variety of osteoblastic markers." *Calcif Tissue Int* 49(3): 202-207

Benayahu, D., M. Horowitz, D. Zipori & S. Wientroub (1992). "Hemopoietic functions of marrow-derived osteogenic cells." *Calcif Tissue Int* 51(3): 195-201

Bentley, S. A. (1982). "Collagen synthesis by bone marrow stromal cells: a quantitative study." *Br J Haematol* 50(3): 491-497

Bentley, S. A. & J. M. Foidart (1980). "Some properties of marrow derived adherent cells in tissue culture." *Blood* 56(6): 1006-1012

Bi, Y., C. H. Stuelten, T. Kilts, S. Wadhwa, R. V. Iozzo, P. G. Robey, X. D. Chen & M. F. Young (2005). "Extracellular matrix proteoglycans control the fate of bone marrow stromal cells." *J Biol Chem* 280(34): 30481-30489

Bianco, P., M. Costantini, L. C. Dearden & E. Bonucci (1988). "Alkaline phosphatase positive precursors of adipocytes in the human bone marrow." *Br J Haematol* 68(4): 401-403

Bianco, P., M. Riminucci, S. Kuznetsov & P. G. Robey (1999). "Multipotential cells in the bone marrow stroma: regulation in the context of organ physiology." *Crit Rev Eukaryot Gene Expr* 9(2): 159-173

Blin-Wakkach, C., A. Wakkach, P. M. Sexton, N. Rochet & G. F. Carle (2004). "Hematological defects in the oc/oc mouse, a model of infantile malignant osteopetrosis." *Leukemia* 18(9): 1505-1511

Bohensky, J., I. M. Shapiro, S. Leshinsky, S. P. Terkhorn, C. S. Adams & V. Srinivas (2007). "HIF-1 regulation of chondrocyte apoptosis: induction of the autophagic pathway." *Autophagy* 3(3): 207-214

Borghesi, L. A., Y. Yamashita & P. W. Kincade (1999). "Heparan sulfate proteoglycans mediate interleukin-7-dependent B lymphopoiesis." *Blood* 93(1): 140-148

Bourke, V. A., C. J. Watchman, J. D. Reith, M. L. Jorgensen, A. Dieudonne & W. E. Bolch (2009). "Spatial gradients of blood vessels and hematopoietic stem and progenitor cells within the marrow cavities of the human skeleton." *Blood* 114(19): 4077-4080

Broxmeyer, H. E., C. M. Orschell, D. W. Clapp, G. Hangoc, S. Cooper, P. A. Plett, W. C. Liles, X. Li, B. Graham-Evans, T. B. Campbell, G. Calandra, G. Bridger, D. C. Dale & E. F. Srour (2005). "Rapid mobilization of murine and human hematopoietic stem and progenitor cells with AMD3100, a CXCR4 antagonist." *J Exp Med* 201(8): 1307-1318

Bruno, E., S. D. Luikart, M. W. Long & R. Hoffman (1995). "Marrow-derived heparan sulfate proteoglycan mediates the adhesion of hematopoietic progenitor cells to cytokines." *Exp Hematol* 23(11): 1212-1217

Burkhardt, R., G. Kettner, W. Bohm, M. Schmidmeier, R. Schlag, B. Frisch, B. Mallmann, W. Eisenmenger & T. Gilg (1987). "Changes in trabecular bone, hematopoiesis and bone marrow vessels in aplastic anemia, primary osteoporosis, and old age: a comparative histomorphometric study." *Bone* 8(3): 157-164

Butler, J. M., D. J. Nolan, E. L. Vertes, B. Varnum-Finney, H. Kobayashi, A. T. Hooper, M. Seandel, K. Shido, I. A. White, M. Kobayashi, L. Witte, C. May, C. Shawber, Y. Kimura, J. Kitajewski, Z. Rosenwaks, I. D. Bernstein & S. Rafii (2010). "Endothelial cells are essential for the self-renewal and repopulation of Notch-dependent hematopoietic stem cells." *Cell Stem Cell* 6(3): 251-264

Calvi, L. M., G. B. Adams, K. W. Weibrecht, J. M. Weber, D. P. Olson, M. C. Knight, R. P. Martin, E. Schipani, P. Divieti, F. R. Bringhurst, L. A. Milner, H. M. Kronenberg & D. T. Scadden (2003). "Osteoblastic cells regulate the haematopoietic stem cell niche." *Nature* 425(6960): 841-846

Calvi, L. M., N. A. Sims, J. L. Hunzelman, M. C. Knight, A. Giovannetti, J. M. Saxton, H. M. Kronenberg, R. Baron & E. Schipani (2001). "Activated parathyroid hormone/parathyroid hormone-related protein receptor in osteoblastic cells differentially affects cortical and trabecular bone." *J Clin Invest* 107(3): 277-286

Campbell, M. R., C. J. Gress, E. H. Appleman & O. Jacenko (2004). "Chicken collagen X regulatory sequences restrict transgene expression to hypertrophic cartilage in mice." *Am J Pathol* 164(2): 487-499

Chan, C. K., C. C. Chen, C. A. Luppen, J. B. Kim, A. T. Deboer, K. Wei, J. A. Helms, C. J. Kuo, D. L. Kraft & I. L. Weissman (2008). "Endochondral ossification is required for haematopoietic stem-cell niche formation." *Nature*

Chan, D. & O. Jacenko (1998). "Phenotypic and biochemical consequences of collagen X mutations in mice and humans." *Matrix Biol* 17(3): 169-184

Chang, M. K., L. J. Raggatt, K. A. Alexander, J. S. Kuliwaba, N. L. Fazzalari, K. Schroder, E. R. Maylin, V. M. Ripoll, D. A. Hume & A. R. Pettit (2008). "Osteal tissue macrophages are intercalated throughout human and mouse bone lining tissues and regulate osteoblast function in vitro and in vivo." *J Immunol* 181(2): 1232-1244

Chow, D. C., L. A. Wenning, W. M. Miller & E. T. Papoutsakis (2001). "Modeling pO(2) distributions in the bone marrow hematopoietic compartment. I. Krogh's model." *Biophys J* 81(2): 675-684

Compston, J. E. (2002). "Bone marrow and bone: a functional unit." *J Endocrinol* 173(3): 387-394

Corradi, A., L. Croci, V. Broccoli, S. Zecchini, S. Previtali, W. Wurst, S. Amadio, R. Maggi, A. Quattrini & G. G. Consalez (2003). "Hypogonadotropic hypogonadism and peripheral neuropathy in Ebf2-null mice." *Development* 130(2): 401-410

Corre, J., C. Barreau, B. Cousin, J. P. Chavoin, D. Caton, G. Fournial, L. Penicaud, L. Casteilla & P. Laharrague (2006). "Human subcutaneous adipose cells support complete differentiation but not self-renewal of hematopoietic progenitors." *J Cell Physiol* 208(2): 282-288

Coskun, S. & K. K. Hirschi (2010). "Establishment and regulation of the HSC niche: Roles of osteoblastic and vascular compartments." *Birth Defects Res C Embryo Today* 90(4): 229-242

Coulombel, L., I. Auffray, M. H. Gaugler & M. Rosemblatt (1997). "Expression and function of integrins on hematopoietic progenitor cells." *Acta Haematol* 97(1-2): 13-21

Crawford, L. J., R. Peake, S. Price, T. C. Morris & A. E. Irvine (2010). "Adiponectin is produced by lymphocytes and is a negative regulator of granulopoiesis." *J Leukoc Biol* 88(4): 807-811

Crisan, M., S. Yap, L. Casteilla, C. W. Chen, M. Corselli, T. S. Park, G. Andriolo, B. Sun, B. Zheng, L. Zhang, C. Norotte, P. N. Teng, J. Traas, R. Schugar, B. M. Deasy, S. Badylak, H. J. Buhring, J. P. Giacobino, L. Lazzari, J. Huard & B. Peault (2008). "A perivascular origin for mesenchymal stem cells in multiple human organs." *Cell Stem Cell* 3(3): 301-313

Crowley, J. E., L. S. Treml, J. E. Stadanlick, E. Carpenter & M. P. Cancro (2005). "Homeostatic niche specification among naive and activated B cells: a growing role for the BLyS family of receptors and ligands." *Semin Immunol* 17(3): 193-199

Cumano, A., F. Dieterlen-Lievre & I. Godin (1996). "Lymphoid potential, probed before circulation in mouse, is restricted to caudal intraembryonic splanchnopleura." *Cell* 86(6): 907-916

Cumano, A. & I. Godin (2007). "Ontogeny of the hematopoietic system." *Annu Rev Immunol* 25: 745-785

Dar, A., O. Kollet & T. Lapidot (2006). "Mutual, reciprocal SDF-1/CXCR4 interactions between hematopoietic and bone marrow stromal cells regulate human stem cell migration and development in NOD/SCID chimeric mice." *Exp Hematol* 34(8): 967-975

de Bruijn, M. F., X. Ma, C. Robin, K. Ottersbach, M. J. Sanchez & E. Dzierzak (2002). "Hematopoietic stem cells localize to the endothelial cell layer in the midgestation mouse aorta." *Immunity* 16(5): 673-683

Dexter, T. M., T. D. Allen & L. G. Lajtha (1977). "Conditions controlling the proliferation of haemopoietic stem cells in vitro." *J Cell Physiol* 91(3): 335-344

Dexter, T. M., T. D. Allen, L. G. Lajtha, R. Schofield & B. I. Lord (1973). "Stimulation of differentiation and proliferation of haemopoietic cells in vitro." *J Cell Physiol* 82(3): 461-473

Discher, D. E., D. J. Mooney & P. W. Zandstra (2009). "Growth factors, matrices, and forces combine and control stem cells." *Science* 324(5935): 1673-1677

Dominici, M., V. Rasini, R. Bussolari, X. Chen, T. J. Hofmann, C. Spano, D. Bernabei, E. Veronesi, F. Bertoni, P. Paolucci, P. Conte & E. M. Horwitz (2009). "Restoration and reversible expansion of the osteoblastic hematopoietic stem cell niche after marrow radioablation." *Blood* 114(11): 2333-2343

Eaves, C. J., J. D. Cashman, R. J. Kay, G. J. Dougherty, T. Otsuka, L. A. Gaboury, D. E. Hogge, P. M. Lansdorp, A. C. Eaves & R. K. Humphries (1991). "Mechanisms that regulate the cell cycle status of very primitive hematopoietic cells in long-term human marrow cultures. II. Analysis of positive and negative regulators produced by stromal cells within the adherent layer." *Blood* 78(1): 110-117

Farnum, C. E. & N. J. Wilsman (1989). "Cellular turnover at the chondro-osseous junction of growth plate cartilage: analysis by serial sections at the light microscopical level." *J Orthop Res* 7(5): 654-666

Fortunel, N., J. Hatzfeld, L. Aoustin, P. Batard, K. Ducos, M. N. Monier, A. Charpentier & A. Hatzfeld (2000). "Specific dose-response effects of TGF-beta1 on developmentally distinct hematopoietic stem/progenitor cells from human umbilical cord blood." *Hematol J* 1(2): 126-135

Friedenstein, A. J., R. K. Chailakhjan & K. S. Lalykina (1970). "The development of fibroblast colonies in monolayer cultures of guinea-pig bone marrow and spleen cells." *Cell Tissue Kinet* 3(4): 393-403

Friedenstein, A. J., R. K. Chailakhyan & U. V. Gerasimov (1987). "Bone marrow osteogenic stem cells: in vitro cultivation and transplantation in diffusion chambers." *Cell Tissue Kinet* 20(3): 263-272

Friedenstein, A. J., R. K. Chailakhyan, N. V. Latsinik, A. F. Panasyuk & I. V. Keiliss-Borok (1974). "Stromal cells responsible for transferring the microenvironment of the hemopoietic tissues. Cloning in vitro and retransplantation in vivo." *Transplantation* 17(4): 331-340

Friedenstein, A. J., J. F. Gorskaja & N. N. Kulagina (1976). "Fibroblast precursors in normal and irradiated mouse hematopoietic organs." *Exp Hematol* 4(5): 267-274

Friedenstein, A. J., N. W. Latzinik, A. G. Grosheva & U. F. Gorskaya (1982). "Marrow microenvironment transfer by heterotopic transplantation of freshly isolated and cultured cells in porous sponges." *Exp Hematol* 10(2): 217-227

Funk, P. E., R. P. Stephan & P. L. Witte (1995). "Vascular cell adhesion molecule 1-positive reticular cells express interleukin-7 and stem cell factor in the bone marrow." *Blood* 86(7): 2661-2671

Galotto, M., G. Campanile, G. Robino, F. D. Cancedda, P. Bianco & R. Cancedda (1994). "Hypertrophic chondrocytes undergo further differentiation to osteoblast-like cells and participate in the initial bone formation in developing chick embryo." *J Bone Miner Res* 9(8): 1239-1249

Gekas, C., F. Dieterlen-Lievre, S. H. Orkin & H. K. Mikkola (2005). "The placenta is a niche for hematopoietic stem cells." *Dev Cell* 8(3): 365-375

Gibson, G. J. & M. H. Flint (1985). "Type X collagen synthesis by chick sternal cartilage and its relationship to endochondral development." *J Cell Biol* 101(1): 277-284

Godman, G. C. & K. R. Porter (1960). "Chondrogenesis, studied with the electron microscope." *J Biophys Biochem Cytol* 8: 719-760

Gong, J. K. (1978). "Endosteal marrow: a rich source of hematopoietic stem cells." *Science* 199(4336): 1443-1445

Gordon, M. Y., G. P. Riley & D. Clarke (1988). "Heparan sulfate is necessary for adhesive interactions between human early hemopoietic progenitor cells and the extracellular matrix of the marrow microenvironment." *Leukemia* 2(12): 804-809

Gress, C. J. & O. Jacenko (2000). "Growth plate compressions and altered hematopoiesis in collagen X null mice." *J Cell Biol* 149(4): 983-993

Grimaud, E., F. Blanchard, C. Charrier, F. Gouin, F. Redini & D. Heymann (2002). "Leukaemia inhibitory factor (lif) is expressed in hypertrophic chondrocytes and vascular sprouts during osteogenesis." *Cytokine* 20(5): 224-230

Gruver, A. L., L. L. Hudson & G. D. Sempowski (2007). "Immunosenescence of ageing." *J Pathol* 211(2): 144-156

Guba, S. C., C. I. Sartor, L. R. Gottschalk, Y. H. Jing, T. Mulligan & S. G. Emerson (1992). "Bone marrow stromal fibroblasts secrete interleukin-6 and granulocyte-macrophage colony-stimulating factor in the absence of inflammatory stimulation: demonstration by serum-free bioassay, enzyme-linked immunosorbent assay, and reverse transcriptase polymerase chain reaction." *Blood* 80(5): 1190-1198

Gupta, P., J. B. McCarthy & C. M. Verfaillie (1996). "Stromal fibroblast heparan sulfate is required for cytokine-mediated ex vivo maintenance of human long-term culture-initiating cells." *Blood* 87(8): 3229-3236

Gupta, P., T. R. Oegema, Jr., J. J. Brazil, A. Z. Dudek, A. Slungaard & C. M. Verfaillie (1998). "Structurally specific heparan sulfates support primitive human hematopoiesis by formation of a multimolecular stem cell niche." *Blood* 92(12): 4641-4651

Gurevitch, O. & I. Fabian (1993). "Ability of the hemopoietic microenvironment in the induced bone to maintain the proliferative potential of early hemopoietic precursors." *Stem Cells* 11(1): 56-61

Hara, H., H. Ohdan, D. Tokita, T. Onoe, W. Zhou & T. Asahara (2003). "Construction of ectopic xenogeneic bone marrow structure associated with persistent multi-lineage mixed chimerism by engraftment of rat bone marrow plugs into mouse kidney capsules." *Xenotransplantation* 10(3): 259-266

Harada, M., K. Odaka, K. Kondo, S. Nakao, M. Ueda, K. Matsue, T. Mori & T. Matsuda (1985). "Effect of activated lymphocytes on the regulation of hematopoiesis: suppression of in vitro granulopoiesis and erythropoiesis by OKT8+ Ia- T cells induced by concanavalin-A stimulation." *Exp Hematol* 13(9): 963-967

Haylock, D. N. & S. K. Nilsson (2006). "The role of hyaluronic acid in hemopoietic stem cell biology." *Regen Med* 1(4): 437-445

Heinrich, M. C., D. C. Dooley, A. C. Freed, L. Band, M. E. Hoatlin, W. W. Keeble, S. T. Peters, K. V. Silvey, F. S. Ey, D. Kabat & et al. (1993). "Constitutive expression of steel factor gene by human stromal cells." *Blood* 82(3): 771-783

Heissig, B., K. Hattori, S. Dias, M. Friedrich, B. Ferris, N. R. Hackett, R. G. Crystal, P. Besmer, D. Lyden, M. A. Moore, Z. Werb & S. Rafii (2002). "Recruitment of stem and progenitor cells from the bone marrow niche requires MMP-9 mediated release of kit-ligand." *Cell* 109(5): 625-637

Hermanns, P., A. A. Bertuch, T. K. Bertin, B. Dawson, M. E. Schmitt, C. Shaw, B. Zabel & B. Lee (2005). "Consequences of mutations in the non-coding RMRP RNA in cartilage-hair hypoplasia." *Hum Mol Genet* 14(23): 3723-3740

Hermans, M. H., H. Hartsuiker & D. Opstelten (1989). "An in situ study of B-lymphocytopoiesis in rat bone marrow. Topographical arrangement of terminal deoxynucleotidyl transferase-positive cells and pre-B cells." *J Immunol* 142(1): 67-73

Hilton, M. J., X. Tu & F. Long (2007). "Tamoxifen-inducible gene deletion reveals a distinct cell type associated with trabecular bone, and direct regulation of PTHrP expression and chondrocyte morphology by Ihh in growth region cartilage." *Dev Biol* 308(1): 93-105

Holst, J., S. Watson, M. S. Lord, S. S. Eamegdool, D. V. Bax, L. B. Nivison-Smith, A. Kondyurin, L. Ma, A. F. Oberhauser, A. S. Weiss & J. E. Rasko (2010). "Substrate elasticity provides mechanical signals for the expansion of hemopoietic stem and progenitor cells." *Nat Biotechnol* 28(10): 1123-1128

Horwitz, E. M., K. Le Blanc, M. Dominici, I. Mueller, I. Slaper-Cortenbach, F. C. Marini, R. J. Deans, D. S. Krause & A. Keating (2005). "Clarification of the nomenclature for MSC: The International Society for Cellular Therapy position statement." *Cytotherapy* 7(5): 393-395

Hotamisligil, G. S., N. S. Shargill & B. M. Spiegelman (1993). "Adipose expression of tumor necrosis factor-alpha: direct role in obesity-linked insulin resistance." *Science* 259(5091): 87-91

Huard, B., T. McKee, C. Bosshard, S. Durual, T. Matthes, S. Myit, O. Donze, C. Frossard, C. Chizzolini, C. Favre, R. Zubler, J. P. Guyot, P. Schneider & E. Roosnek (2008). "APRIL secreted by neutrophils binds to heparan sulfate proteoglycans to create plasma cell niches in human mucosa." *J Clin Invest* 118(8): 2887-2895

Ingold, K., A. Zumsteg, A. Tardivel, B. Huard, Q. G. Steiner, T. G. Cachero, F. Qiang, L. Gorelik, S. L. Kalled, H. Acha-Orbea, P. D. Rennert, J. Tschopp & P. Schneider (2005). "Identification of proteoglycans as the APRIL-specific binding partners." *J Exp Med* 201(9): 1375-1383

Jacenko, O., D. Chan, A. Franklin, S. Ito, C. B. Underhill, J. F. Bateman & M. R. Campbell (2001). "A dominant interference collagen X mutation disrupts hypertrophic chondrocyte pericellular matrix and glycosaminoglycan and proteoglycan distribution in transgenic mice." *Am J Pathol* 159(6): 2257-2269

Jacenko, O., S. Ito & B. R. Olsen (1996). "Skeletal and hematopoietic defects in mice transgenic for collagen X." *Ann N Y Acad Sci* 785: 278-280

Jacenko, O., P. A. LuValle & B. R. Olsen (1993). "Spondylometaphyseal dysplasia in mice carrying a dominant negative mutation in a matrix protein specific for cartilage-to-bone transition." *Nature* 365(6441): 56-61

Jacenko, O., B. R. Olsen & P. LuValle (1991). "Organization and regulation of collagen genes." *Crit Rev Eukaryot Gene Expr* 1(4): 327-353

Jacenko, O., D. W. Roberts, M. R. Campbell, P. M. McManus, C. J. Gress & Z. Tao (2002). "Linking hematopoiesis to endochondral skeletogenesis through analysis of mice transgenic for collagen X." *Am J Pathol* 160(6): 2019-2034

Jacobsen, K. & D. G. Osmond (1990). "Microenvironmental organization and stromal cell associations of B lymphocyte precursor cells in mouse bone marrow." *Eur J Immunol* 20(11): 2395-2404

James, C. G., L. A. Stanton, H. Agoston, V. Ulici, T. M. Underhill & F. Beier (2010). "Genome-wide analyses of gene expression during mouse endochondral ossification." *PLoS One* 5(1): e8693

Jang, Y. Y. & S. J. Sharkis (2007). "A low level of reactive oxygen species selects for primitive hematopoietic stem cells that may reside in the low-oxygenic niche." *Blood* 110(8): 3056-3063

Janzen, V., R. Forkert, H. E. Fleming, Y. Saito, M. T. Waring, D. M. Dombkowski, T. Cheng, R. A. DePinho, N. E. Sharpless & D. T. Scadden (2006). "Stem-cell ageing modified by the cyclin-dependent kinase inhibitor p16INK4a." *Nature* 443(7110): 421-426

Jung, Y., J. Song, Y. Shiozawa, J. Wang, Z. Wang, B. Williams, A. Havens, A. Schneider, C. Ge, R. T. Franceschi, L. K. McCauley, P. H. Krebsbach & R. S. Taichman (2008). "Hematopoietic stem cells regulate mesenchymal stromal cell induction into osteoblasts thereby participating in the formation of the stem cell niche." *Stem Cells* 26(8): 2042-2051

Jung, Y., J. Wang, A. Havens, Y. Sun, J. Wang, T. Jin & R. S. Taichman (2005). "Cell-to-cell contact is critical for the survival of hematopoietic progenitor cells on osteoblasts." *Cytokine* 32(3-4): 155-162

Jung, Y., J. Wang, A. Schneider, Y. X. Sun, A. J. Koh-Paige, N. I. Osman, L. K. McCauley & R. S. Taichman (2006). "Regulation of SDF-1 (CXCL12) production by osteoblasts; a possible mechanism for stem cell homing." *Bone* 38(4): 497-508

Kacena, M. A., C. M. Gundberg & M. C. Horowitz (2006). "A reciprocal regulatory interaction between megakaryocytes, bone cells, and hematopoietic stem cells." *Bone* 39(5): 978-984

Kacena, M. A., R. A. Shivdasani, K. Wilson, Y. Xi, N. Troiano, A. Nazarian, C. M. Gundberg, M. L. Bouxsein, J. A. Lorenzo & M. C. Horowitz (2004). "Megakaryocyte-osteoblast interaction revealed in mice deficient in transcription factors GATA-1 and NF-E2." *J Bone Miner Res* 19(4): 652-660

Kalajzic, Z., H. Li, L. P. Wang, X. Jiang, K. Lamothe, D. J. Adams, H. L. Aguila, D. W. Rowe & I. Kalajzic (2008). "Use of an alpha-smooth muscle actin GFP reporter to identify an osteoprogenitor population." *Bone* 43(3): 501-510

Kalinkovich, A., A. Spiegel, S. Shivtiel, O. Kollet, N. Jordaney, W. Piacibello & T. Lapidot (2009). "Blood-forming stem cells are nervous: direct and indirect regulation of immature human CD34+ cells by the nervous system." *Brain Behav Immun* 23(8): 1059-1065

Katayama, Y., M. Battista, W. M. Kao, A. Hidalgo, A. J. Peired, S. A. Thomas & P. S. Frenette (2006). "Signals from the sympathetic nervous system regulate hematopoietic stem cell egress from bone marrow." *Cell* 124(2): 407-421

Kawai, M., H. Hattori, K. Yasue, H. Mizutani, M. Ueda, T. Kaneda & T. Hoshino (1994). "Development of hemopoietic bone marrow within the ectopic bone induced by bone morphogenetic protein." *Blood Cells* 20(1): 191-199; discussion 200-191

Keller, J. R., J. M. Gooya & F. W. Ruscetti (1996). "Direct synergistic effects of leukemia inhibitory factor on hematopoietic progenitor cell growth: comparison with other hematopoietins that use the gp130 receptor subunit." *Blood* 88(3): 863-869

Khosla, S. (2003). "Parathyroid hormone plus alendronate--a combination that does not add up." *N Engl J Med* 349(13): 1277-1279

Kiel, M. J., M. Acar, G. L. Radice & S. J. Morrison (2009). "Hematopoietic stem cells do not depend on N-cadherin to regulate their maintenance." *Cell Stem Cell* 4(2): 170-179

Kiel, M. J., G. L. Radice & S. J. Morrison (2007). "Lack of evidence that hematopoietic stem cells depend on N-cadherin-mediated adhesion to osteoblasts for their maintenance." *Cell Stem Cell* 1(2): 204-217

Kiel, M. J., O. H. Yilmaz, T. Iwashita, O. H. Yilmaz, C. Terhorst & S. J. Morrison (2005). "SLAM family receptors distinguish hematopoietic stem and progenitor cells and reveal endothelial niches for stem cells." *Cell* 121(7): 1109-1121

Kieslinger, M., S. Hiechinger, G. Dobreva, G. G. Consalez & R. Grosschedl (2010). "Early B cell factor 2 regulates hematopoietic stem cell homeostasis in a cell-nonautonomous manner." *Cell Stem Cell* 7(4): 496-507

Kilborn, S. H., G. Trudel & H. Uhthoff (2002). "Review of growth plate closure compared with age at sexual maturity and lifespan in laboratory animals." *Contemp Top Lab Anim Sci* 41(5): 21-26

Klein, G. (1995). "The extracellular matrix of the hematopoietic microenvironment." *Experientia* 51(9-10): 914-926

Klein, G., S. Conzelmann, S. Beck, R. Timpl & C. A. Muller (1995). "Perlecan in human bone marrow: a growth-factor-presenting, but anti-adhesive, extracellular matrix component for hematopoietic cells." *Matrix Biol* 14(6): 457-465

Klein, G., C. Kibler, F. Schermutzki, J. Brown, C. A. Muller & R. Timpl (1998). "Cell binding properties of collagen type XIV for human hematopoietic cells." *Matrix Biol* 16(6): 307-317

Kohler, A., V. Schmithorst, M. D. Filippi, M. A. Ryan, D. Daria, M. Gunzer & H. Geiger (2009). "Altered cellular dynamics and endosteal location of aged early hematopoietic progenitor cells revealed by time-lapse intravital imaging in long bones." *Blood* 114(2): 290-298

Kollet, O., A. Dar, S. Shivtiel, A. Kalinkovich, K. Lapid, Y. Sztainberg, M. Tesio, R. M. Samstein, P. Goichberg, A. Spiegel, A. Elson & T. Lapidot (2006). "Osteoclasts degrade endosteal components and promote mobilization of hematopoietic progenitor cells." *Nat Med* 12(6): 657-664

Kong, Y. Y., H. Yoshida, I. Sarosi, H. L. Tan, E. Timms, C. Capparelli, S. Morony, A. J. Oliveira-dos-Santos, G. Van, A. Itie, W. Khoo, A. Wakeham, C. R. Dunstan, D. L. Lacey, T. W. Mak, W. J. Boyle & J. M. Penninger (1999). "OPGL is a key regulator of osteoclastogenesis, lymphocyte development and lymph-node organogenesis." *Nature* 397(6717): 315-323

Kopp, H. G., S. T. Avecilla, A. T. Hooper & S. Rafii (2005). "The bone marrow vascular niche: home of HSC differentiation and mobilization." *Physiology (Bethesda)* 20: 349-356

Kortesidis, A., A. Zannettino, S. Isenmann, S. Shi, T. Lapidot & S. Gronthos (2005). "Stromal-derived factor-1 promotes the growth, survival, and development of human bone marrow stromal stem cells." *Blood* 105(10): 3793-3801

Kuijpers, T. W., E. Nannenberg, M. Alders, R. Bredius & R. C. Hennekam (2004). "Congenital aplastic anemia caused by mutations in the SBDS gene: a rare presentation of Shwachman-Diamond syndrome." *Pediatrics* 114(3): e387-391

Kuznetsov, S. A., P. H. Krebsbach, K. Satomura, J. Kerr, M. Riminucci, D. Benayahu & P. G. Robey (1997). "Single-colony derived strains of human marrow stromal fibroblasts form bone after transplantation in vivo." *J Bone Miner Res* 12(9): 1335-1347

Kuznetsov, S. A., M. Riminucci, N. Ziran, T. W. Tsutsui, A. Corsi, L. Calvi, H. M. Kronenberg, E. Schipani, P. G. Robey & P. Bianco (2004). "The interplay of osteogenesis and hematopoiesis: expression of a constitutively active PTH/PTHrP receptor in osteogenic cells perturbs the establishment of hematopoiesis in bone and of skeletal stem cells in the bone marrow." *J Cell Biol* 167(6): 1113-1122

Kwan, K. M., M. K. Pang, S. Zhou, S. K. Cowan, R. Y. Kong, T. Pfordte, B. R. Olsen, D. O. Sillence, P. P. Tam & K. S. Cheah (1997). "Abnormal compartmentalization of cartilage matrix components in mice lacking collagen X: implications for function." *J Cell Biol* 136(2): 459-471

Kwon, K. R., J. Y. Ahn, M. S. Kim, J. Y. Jung, J. H. Lee & I. H. Oh (2010). "Disruption of bis leads to the deterioration of the vascular niche for hematopoietic stem cells." *Stem Cells* 28(2): 268-278

Lanotte, M., D. Metcalf & T. M. Dexter (1982). "Production of monocyte/macrophage colony-stimulating factor by preadipocyte cell lines derived from murine marrow stroma." *J Cell Physiol* 112(1): 123-127

Lapid, K., Y. Vagima, O. Kollet & T. Lapidot (2008). "Egress and mobilization of hematopoietic stem and progenitor cells."

Larsson, J., M. Ohishi, B. Garrison, M. Aspling, V. Janzen, G. B. Adams, M. Curto, A. I. McClatchey, E. Schipani & D. T. Scadden (2008). "Nf2/merlin regulates hematopoietic stem cell behavior by altering microenvironmental architecture." *Cell Stem Cell* 3(2): 221-227

Lefebvre, V. & P. Smits (2005). "Transcriptional control of chondrocyte fate and differentiation." *Birth Defects Res Part C: Embryo Today: Reviews* 75(3): 200-212

Levesque, J. P. & P. J. Simmons (1999). "Cytoskeleton and integrin-mediated adhesion signaling in human CD34+ hemopoietic progenitor cells." *Exp Hematol* 27(4): 579-586

Levesque, J. P., Y. Takamatsu, S. K. Nilsson, D. N. Haylock & P. J. Simmons (2001). "Vascular cell adhesion molecule-1 (CD106) is cleaved by neutrophil proteases in the bone marrow following hematopoietic progenitor cell mobilization by granulocyte colony-stimulating factor." *Blood* 98(5): 1289-1297

Li, W., S. A. Johnson, W. C. Shelley & M. C. Yoder (2004). "Hematopoietic stem cell repopulating ability can be maintained in vitro by some primary endothelial cells." *Exp Hematol* 32(12): 1226-1237

Lo Celso, C., H. E. Fleming, J. W. Wu, C. X. Zhao, S. Miake-Lye, J. Fujisaki, D. Cote, D. W. Rowe, C. P. Lin & D. T. Scadden (2009). "Live-animal tracking of individual haematopoietic stem/progenitor cells in their niche." *Nature* 457(7225): 92-96

Lord, B. I., N. G. Testa & J. H. Hendry (1975). "The relative spatial distributions of CFUs and CFUc in the normal mouse femur." *Blood* 46(1): 65-72

Mackie, E. J., Y. A. Ahmed, L. Tatarczuch, K. S. Chen & M. Mirams (2008). "Endochondral ossification: how cartilage is converted into bone in the developing skeleton." *Int J Biochem Cell Biol* 40(1): 46-62

Maes, C., T. Kobayashi, M. K. Selig, S. Torrekens, S. I. Roth, S. Mackem, G. Carmeliet & H. M. Kronenberg (2010). "Osteoblast precursors, but not mature osteoblasts, move into developing and fractured bones along with invading blood vessels." *Dev Cell* 19(2): 329-344

Majumdar, M. K., M. A. Thiede, S. E. Haynesworth, S. P. Bruder & S. L. Gerson (2000). "Human marrow-derived mesenchymal stem cells (MSCs) express hematopoietic cytokines and support long-term hematopoiesis when differentiated toward stromal and osteogenic lineages." *J Hematother Stem Cell Res* 9(6): 841-848

Mangan, K. F., G. Chikkappa, L. Z. Bieler, W. B. Scharfman & D. R. Parkinson (1982). "Regulation of human blood erythroid burst-forming unit (BFU-E) proliferation by T-lymphocyte subpopulations defined by Fc receptors and monoclonal antibodies." *Blood* 59(5): 990-996

Mankani, M. H., S. A. Kuznetsov, G. W. Marshall & P. G. Robey (2008). "Creation of new bone by the percutaneous injection of human bone marrow stromal cell and HA/TCP suspensions." *Tissue Eng Part A* 14(12): 1949-1958

Mankani, M. H., S. A. Kuznetsov & P. G. Robey (2007). "Formation of hematopoietic territories and bone by transplanted human bone marrow stromal cells requires a critical cell density." *Exp Hematol* 35(6): 995-1004

Mansour, A., A. Anginot, S. J. Mancini, C. Schiff, G. F. Carle, A. Wakkach & C. Blin-Wakkach (2011). "Osteoclast activity modulates B-cell development in the bone marrow." *Cell Res* 21(7): 1102-1115

Matrosova, V. Y., I. A. Orlovskaya, N. Serobyan & S. K. Khaldoyanidi (2004). "Hyaluronic acid facilitates the recovery of hematopoiesis following 5-fluorouracil administration." *Stem Cells* 22(4): 544-555

Mayack, S. R., J. L. Shadrach, F. S. Kim & A. J. Wagers (2010). "Systemic signals regulate ageing and rejuvenation of blood stem cell niches." *Nature* 463(7280): 495-500

Mazzon, C., A. Anselmo, J. Cibella, C. Soldani, A. Destro, N. Kim, M. Roncalli, S. J. Burden, M. L. Dustin, A. Sarukhan & A. Viola (2011). "The critical role of agrin in the hematopoietic stem cell niche." *Blood* 118(10): 2733-2742

Meck, R. A., J. E. Haley & G. Brecher (1973). "Hematopoiesis versus osteogenesis in ectopic bone marrow transplants." *Blood* 42(5): 661-669

Medici, D., E. M. Shore, V. Y. Lounev, F. S. Kaplan, R. Kalluri & B. R. Olsen (2010). "Conversion of vascular endothelial cells into multipotent stem-like cells." *Nat Med* 16(12): 1400-1406

Medvinsky, A. & E. Dzierzak (1996). "Definitive hematopoiesis is autonomously initiated by the AGM region." *Cell* 86(6): 897-906

Mendez-Ferrer, S., D. Lucas, M. Battista & P. S. Frenette (2008). "Haematopoietic stem cell release is regulated by circadian oscillations." *Nature* 452(7186): 442-447

Mendez-Ferrer, S., T. V. Michurina, F. Ferraro, A. R. Mazloom, B. D. Macarthur, S. A. Lira, D. T. Scadden, A. Ma'ayan, G. N. Enikolopov & P. S. Frenette (2010). "Mesenchymal and haematopoietic stem cells form a unique bone marrow niche." *Nature* 466(7308): 829-834

Merzaban, J. S., M. M. Burdick, S. Z. Gadhoum, N. M. Dagia, J. T. Chu, R. C. Fuhlbrigge & R. Sackstein (2011). "Analysis of glycoprotein E-selectin ligands on human and mouse marrow cells enriched for hematopoietic stem/progenitor cells." *Blood* 118(7): 1774-1783

Mignini, F., V. Streccioni & F. Amenta (2003). "Autonomic innervation of immune organs and neuroimmune modulation." *Auton Autacoid Pharmacol* 23(1): 1-25

Miharada, K., T. Hiroyama, K. Sudo, I. Danjo, T. Nagasawa & Y. Nakamura (2008). "Lipocalin 2-mediated growth suppression is evident in human erythroid and monocyte/macrophage lineage cells." *J Cell Physiol* 215(2): 526-537

Moreaux, J., F. W. Cremer, T. Reme, M. Raab, K. Mahtouk, P. Kaukel, V. Pantesco, J. De Vos, E. Jourdan, A. Jauch, E. Legouffe, M. Moos, G. Fiol, H. Goldschmidt, J. F. Rossi, D. Hose & B. Klein (2005). "The level of TACI gene expression in myeloma cells is associated with a signature of microenvironment dependence versus a plasmablastic signature." *Blood* 106(3): 1021-1030

Moreaux, J., A. C. Sprynski, S. R. Dillon, K. Mahtouk, M. Jourdan, A. Ythier, P. Moine, N. Robert, E. Jourdan, J. F. Rossi & B. Klein (2009). "APRIL and TACI interact with syndecan-1 on the surface of multiple myeloma cells to form an essential survival loop." *Eur J Haematol* 83(2): 119-129

Morikawa, S., Y. Mabuchi, Y. Kubota, Y. Nagai, K. Niibe, E. Hiratsu, S. Suzuki, C. Miyauchi-Hara, N. Nagoshi, T. Sunabori, S. Shimmura, A. Miyawaki, T. Nakagawa, T. Suda, H. Okano & Y. Matsuzaki (2009). "Prospective identification, isolation, and systemic transplantation of multipotent mesenchymal stem cells in murine bone marrow." *J Exp Med* 206(11): 2483-2496

Morrison, S. J. & A. C. Spradling (2008). "Stem cells and niches: mechanisms that promote stem cell maintenance throughout life." *Cell* 132(4): 598-611

Muguruma, Y., T. Yahata, H. Miyatake, T. Sato, T. Uno, J. Itoh, S. Kato, M. Ito, T. Hotta & K. Ando (2006). "Reconstitution of the functional human hematopoietic microenvironment derived from human mesenchymal stem cells in the murine bone marrow compartment." *Blood* 107(5): 1878-1887

Murayama, E., K. Kissa, A. Zapata, E. Mordelet, V. Briolat, H. F. Lin, R. I. Handin & P. Herbomel (2006). "Tracing hematopoietic precursor migration to successive hematopoietic organs during zebrafish development." *Immunity* 25(6): 963-975

Nagasawa, T. (2006). "Microenvironmental niches in the bone marrow required for B-cell development." *Nat Rev Immunol* 6(2): 107-116

Nagasawa, T., S. Hirota, K. Tachibana, N. Takakura, S. Nishikawa, Y. Kitamura, N. Yoshida, H. Kikutani & T. Kishimoto (1996). "Defects of B-cell lymphopoiesis and bone-marrow myelopoiesis in mice lacking the CXC chemokine PBSF/SDF-1." *Nature* 382(6592): 635-638

Nagasawa, T., H. Kikutani & T. Kishimoto (1994). "Molecular cloning and structure of a pre-B-cell growth-stimulating factor." *Proc Natl Acad Sci U S A* 91(6): 2305-2309

Naiyer, A. J., D. Y. Jo, J. Ahn, R. Mohle, M. Peichev, G. Lam, R. L. Silverstein, M. A. Moore & S. Rafii (1999). "Stromal derived factor-1-induced chemokinesis of cord blood CD34(+) cells (long-term culture-initiating cells) through endothelial cells is mediated by E-selectin." *Blood* 94(12): 4011-4019

Nakamura, E., M. T. Nguyen & S. Mackem (2006). "Kinetics of tamoxifen-regulated Cre activity in mice using a cartilage-specific CreER(T) to assay temporal activity windows along the proximodistal limb skeleton." *Dev Dyn* 235(9): 2603-2612

Nathan, D. G., L. Chess, D. G. Hillman, B. Clarke, J. Breard, E. Merler & D. E. Housman (1978). "Human erythroid burst-forming unit: T-cell requirement for proliferation in vitro." *J Exp Med* 147(2): 324-339

Naveiras, O., V. Nardi, P. L. Wenzel, P. V. Hauschka, F. Fahey & G. Q. Daley (2009). "Bone-marrow adipocytes as negative regulators of the haematopoietic microenvironment." *Nature* 460(7252): 259-263

Nilsson, S. K., M. S. Dooner, C. Y. Tiarks, H. U. Weier & P. J. Quesenberry (1997). "Potential and distribution of transplanted hematopoietic stem cells in a nonablated mouse model." *Blood* 89(11): 4013-4020

Nilsson, S. K., H. M. Johnston & J. A. Coverdale (2001). "Spatial localization of transplanted hemopoietic stem cells: inferences for the localization of stem cell niches." *Blood* 97(8): 2293-2299

North, T. E., M. F. de Bruijn, T. Stacy, L. Talebian, E. Lind, C. Robin, M. Binder, E. Dzierzak & N. A. Speck (2002). "Runx1 expression marks long-term repopulating hematopoietic stem cells in the midgestation mouse embryo." *Immunity* 16(5): 661-672

O'Connor, B. P., V. S. Raman, L. D. Erickson, W. J. Cook, L. K. Weaver, C. Ahonen, L. L. Lin, G. T. Mantchev, R. J. Bram & R. J. Noelle (2004). "BCMA is essential for the survival of long-lived bone marrow plasma cells." *J Exp Med* 199(1): 91-98

Ohneda, O., C. Fennie, Z. Zheng, C. Donahue, H. La, R. Villacorta, B. Cairns & L. A. Lasky (1998). "Hematopoietic stem cell maintenance and differentiation are supported by embryonic aorta-gonad-mesonephros region-derived endothelium." *Blood* 92(3): 908-919

Ohta, M., T. Sakai, Y. Saga, S. Aizawa & M. Saito (1998). "Suppression of hematopoietic activity in tenascin-C-deficient mice." *Blood* 91(11): 4074-4083

Omatsu, Y., T. Sugiyama, H. Kohara, G. Kondoh, N. Fujii, K. Kohno & T. Nagasawa (2010). "The essential functions of adipo-osteogenic progenitors as the hematopoietic stem and progenitor cell niche." *Immunity* 33(3): 387-399

Osmond, D. G. (1990). "B cell development in the bone marrow." *Semin Immunol* 2(3): 173-180

Ottersbach, K. & E. Dzierzak (2005). "The murine placenta contains hematopoietic stem cells within the vascular labyrinth region." *Dev Cell* 8(3): 377-387

Patt, H. M., M. A. Maloney & M. L. Flannery (1982). "Hematopoietic microenvironment transfer by stromal fibroblasts derived from bone marrow varying in cellularity." *Exp Hematol* 10(9): 738-742

Peerani, R. & P. W. Zandstra (2010). "Enabling stem cell therapies through synthetic stem cell-niche engineering." *J Clin Invest* 120(1): 60-70

Peled, A., V. Grabovsky, L. Habler, J. Sandbank, F. Arenzana-Seisdedos, I. Petit, H. Ben-Hur, T. Lapidot & R. Alon (1999). "The chemokine SDF-1 stimulates integrin-mediated arrest of CD34(+) cells on vascular endothelium under shear flow." *J Clin Invest* 104(9): 1199-1211

Petit, I., M. Szyper-Kravitz, A. Nagler, M. Lahav, A. Peled, L. Habler, T. Ponomaryov, R. S. Taichman, F. Arenzana-Seisdedos, N. Fujii, J. Sandbank, D. Zipori & T. Lapidot (2002). "G-CSF induces stem cell mobilization by decreasing bone marrow SDF-1 and up-regulating CXCR4." *Nat Immunol* 3(7): 687-694

Petrini, M., S. Pacini, L. Trombi, R. Fazzi, M. Montali, S. Ikehara & N. G. Abraham (2009). "Identification and purification of mesodermal progenitor cells from human adult bone marrow." *Stem Cells Dev* 18(6): 857-866

Pfeiffer, C. A. (1948). "Development of bone from transplanted marrow in mice." *Anat Rec* 102(2): 225-243

Pilia, G., R. M. Hughes-Benzie, A. MacKenzie, P. Baybayan, E. Y. Chen, R. Huber, G. Neri, A. Cao, A. Forabosco & D. Schlessinger (1996). "Mutations in GPC3, a glypican gene, cause the Simpson-Golabi-Behmel overgrowth syndrome." *Nat Genet* 12(3): 241-247

Ponomaryov, T., A. Peled, I. Petit, R. S. Taichman, L. Habler, J. Sandbank, F. Arenzana-Seisdedos, A. Magerus, A. Caruz, N. Fujii, A. Nagler, M. Lahav, M. Szyper-Kravitz, D. Zipori & T. Lapidot (2000). "Induction of the chemokine stromal-derived factor-1 following DNA damage improves human stem cell function." *J Clin Invest* 106(11): 1331-1339

Purton, L. E., S. Dworkin, G. H. Olsen, C. R. Walkley, S. A. Fabb, S. J. Collins & P. Chambon (2006). "RARgamma is critical for maintaining a balance between hematopoietic stem cell self-renewal and differentiation." *J Exp Med* 203(5): 1283-1293

Raaijmakers, M. H., S. Mukherjee, S. Guo, S. Zhang, T. Kobayashi, J. A. Schoonmaker, B. L. Ebert, F. Al-Shahrour, R. P. Hasserjian, E. O. Scadden, Z. Aung, M. Matza, M. Merkenschlager, C. Lin, J. M. Rommens & D. T. Scadden (2010). "Bone progenitor dysfunction induces myelodysplasia and secondary leukaemia." *Nature* 464(7290): 852-857

Rafii, S., R. Mohle, F. Shapiro, B. M. Frey & M. A. Moore (1997). "Regulation of hematopoiesis by microvascular endothelium." *Leuk Lymphoma* 27(5-6): 375-386

Rafii, S., F. Shapiro, R. Pettengell, B. Ferris, R. L. Nachman, M. A. Moore & A. S. Asch (1995). "Human bone marrow microvascular endothelial cells support long-term proliferation and differentiation of myeloid and megakaryocytic progenitors." *Blood* 86(9): 3353-3363

Rafii, S., F. Shapiro, J. Rimarachin, R. L. Nachman, B. Ferris, B. Weksler, M. A. Moore & A. S. Asch (1994). "Isolation and characterization of human bone marrow microvascular endothelial cells: hematopoietic progenitor cell adhesion." *Blood* 84(1): 10-19

Roach, H. I. (1992). "Trans-differentiation of hypertrophic chondrocytes into cells capable of producing a mineralized bone matrix." *Bone Miner* 19(1): 1-20

Roach, H. I. & J. Erenpreisa (1996). "The phenotypic switch from chondrocytes to bone-forming cells involves asymmetric cell division and apoptosis." *Connect Tissue Res* 35(1-4): 85-91

Roach, H. I., J. Erenpreisa & T. Aigner (1995). "Osteogenic differentiation of hypertrophic chondrocytes involves asymmetric cell divisions and apoptosis." *J Cell Biol* 131(2): 483-494

Roberts, R., J. Gallagher, E. Spooncer, T. D. Allen, F. Bloomfield & T. M. Dexter (1988). "Heparan sulphate bound growth factors: a mechanism for stromal cell mediated haemopoiesis." *Nature* 332(6162): 376-378

Rodgers, K. D., J. D. San Antonio & O. Jacenko (2008). "Heparan sulfate proteoglycans: A GAGgle of skeletal-hematopoietic regulators." *Dev Dyn* 237(10): 2622-2642

Rosati, R., G. S. Horan, G. J. Pinero, S. Garofalo, D. R. Keene, W. A. Horton, E. Vuorio, B. de Crombrugghe & R. R. Behringer (1994). "Normal long bone growth and development in type X collagen-null mice." *Nat Genet* 8(2): 129-135

Rossi, D. J., D. Bryder, J. M. Zahn, H. Ahlenius, R. Sonu, A. J. Wagers & I. L. Weissman (2005). "Cell intrinsic alterations underlie hematopoietic stem cell aging." *Proc Natl Acad Sci U S A* 102(26): 9194-9199

Rossi, D. J., C. H. Jamieson & I. L. Weissman (2008). "Stems cells and the pathways to aging and cancer." *Cell* 132(4): 681-696

Rouleau, M. F., J. Mitchell & D. Goltzman (1990). "Characterization of the major parathyroid hormone target cell in the endosteal metaphysis of rat long bones." *J Bone Miner Res* 5(10): 1043-1053

Ryan, D. H., B. L. Nuccie, C. N. Abboud & J. M. Winslow (1991). "Vascular cell adhesion molecule-1 and the integrin VLA-4 mediate adhesion of human B cell precursors to cultured bone marrow adherent cells." *J Clin Invest* 88(3): 995-1004

Sacchetti, B., A. Funari, S. Michienzi, S. Di Cesare, S. Piersanti, I. Saggio, E. Tagliafico, S. Ferrari, P. G. Robey, M. Riminucci & P. Bianco (2007). "Self-renewing osteoprogenitors in bone marrow sinusoids can organize a hematopoietic microenvironment." *Cell* 131(2): 324-336

Sagar, B. M., S. Rentala, P. N. Gopal, S. Sharma & A. Mukhopadhyay (2006). "Fibronectin and laminin enhance engraftibility of cultured hematopoietic stem cells." *Biochem Biophys Res Commun* 350(4): 1000-1005

Sanchez, M. J., A. Holmes, C. Miles & E. Dzierzak (1996). "Characterization of the first definitive hematopoietic stem cells in the AGM and liver of the mouse embryo." *Immunity* 5(6): 513-525

Schmid, T. M. & T. F. Linsenmayer (1985). "Immunohistochemical localization of short chain cartilage collagen (type X) in avian tissues." *J Cell Biol* 100(2): 598-605

Schofield, R. (1978). "The relationship between the spleen colony-forming cell and the haemopoietic stem cell." *Blood Cells* 4(1-2): 7-25

Schwaller, J., P. Schneider, P. Mhawech-Fauceglia, T. McKee, S. Myit, T. Matthes, J. Tschopp, O. Donze, F. A. Le Gal & B. Huard (2007). "Neutrophil-derived APRIL concentrated in tumor lesions by proteoglycans correlates with human B-cell lymphoma aggressiveness." *Blood* 109(1): 331-338

Schweitzer, K. M., A. M. Drager, P. van der Valk, S. F. Thijsen, A. Zevenbergen, A. P. Theijsmeijer, C. E. van der Schoot & M. M. Langenhuijsen (1996). "Constitutive expression of E-selectin and vascular cell adhesion molecule-1 on endothelial cells of hematopoietic tissues." *Am J Pathol* 148(1): 165-175

Shackney, S. E., S. S. Ford & A. B. Wittig (1975). "Kinetic-microarchitectural correlations in the bone marrow of the mouse." *Cell Tissue Kinet* 8(6): 505-516

Siczkowski, M., D. Clarke & M. Y. Gordon (1992). "Binding of primitive hematopoietic progenitor cells to marrow stromal cells involves heparan sulfate." *Blood* 80(4): 912-919

Siebertz, B., G. Stocker, Z. Drzeniek, S. Handt, U. Just & H. D. Haubeck (1999). "Expression of glypican-4 in haematopoietic-progenitor and bone-marrow-stromal cells." *Biochem J* 344 (Pt 3): 937-943

Sipkins, D. A., X. Wei, J. W. Wu, J. M. Runnels, D. Cote, T. K. Means, A. D. Luster, D. T. Scadden & C. P. Lin (2005). "In vivo imaging of specialized bone marrow endothelial microdomains for tumour engraftment." *Nature* 435(7044): 969-973

Song, J., M. J. Kiel, Z. Wang, J. Wang, R. S. Taichman, S. J. Morrison & P. H. Krebsbach (2010). "An in vivo model to study and manipulate the hematopoietic stem cell niche." *Blood* 115(13): 2592-2600

Spiegel, A., A. Kalinkovich, S. Shivtiel, O. Kollet & T. Lapidot (2008). "Stem cell regulation via dynamic interactions of the nervous and immune systems with the microenvironment." *Cell Stem Cell* 3(5): 484-492

Spiegel, A., S. Shivtiel, A. Kalinkovich, A. Ludin, N. Netzer, P. Goichberg, Y. Azaria, I. Resnick, I. Hardan, H. Ben-Hur, A. Nagler, M. Rubinstein & T. Lapidot (2007). "Catecholaminergic neurotransmitters regulate migration and repopulation of immature human CD34+ cells through Wnt signaling." *Nat Immunol* 8(10): 1123-1131

Spiegel, A., E. Zcharia, Y. Vagima, T. Itkin, A. Kalinkovich, A. Dar, O. Kollet, N. Netzer, K. Golan, I. Shafat, N. Ilan, A. Nagler, I. Vlodavsky & T. Lapidot (2008). "Heparanase regulates retention and proliferation of primitive Sca-1+/c-Kit+/Lin- cells via modulation of the bone marrow microenvironment." *Blood* 111(10): 4934-4943

Spooncer, E., J. T. Gallagher, F. Krizsa & T. M. Dexter (1983). "Regulation of haemopoiesis in long-term bone marrow cultures. IV. Glycosaminoglycan synthesis and the stimulation of haemopoiesis by beta-D-xylosides." *J Cell Biol* 96(2): 510-514

Spranger, J., G. K. Hinkel, H. Stoss, W. Thoenes, D. Wargowski & F. Zepp (1991). "Schimke immuno-osseous dysplasia: a newly recognized multisystem disease." *J Pediatr* 119(1 (Pt 1)): 64-72

Srinivas, V. & I. M. Shapiro (2006). "Chondrocytes embedded in the epiphyseal growth plates of long bones undergo autophagy prior to the induction of osteogenesis." *Autophagy* 2(3): 215-216

Sugiyama, T., H. Kohara, M. Noda & T. Nagasawa (2006). "Maintenance of the hematopoietic stem cell pool by CXCL12-CXCR4 chemokine signaling in bone marrow stromal cell niches." *Immunity* 25(6): 977-988

Sweeney, E., M. Campbell, K. Watkins, C. A. Hunter & O. Jacenko (2008). "Altered endochondral ossification in collagen X mouse models leads to impaired immune responses." *Dev Dyn* 237(10): 2693-2704

Sweeney, E., D. Roberts, T. Corbo & O. Jacenko (2010). "Congenic mice confirm that collagen X is required for proper hematopoietic development." *PLoS One* 5(3): e9518

Taichman, R. S. & S. G. Emerson (1994). "Human osteoblasts support hematopoiesis through the production of granulocyte colony-stimulating factor." *J Exp Med* 179(5): 1677-1682

Taichman, R. S., M. J. Reilly & S. G. Emerson (1996). "Human osteoblasts support human hematopoietic progenitor cells in vitro bone marrow cultures." *Blood* 87(2): 518-524

Taichman, R. S., M. J. Reilly, R. S. Verma & S. G. Emerson (1997). "Augmented production of interleukin-6 by normal human osteoblasts in response to CD34+ hematopoietic bone marrow cells in vitro." *Blood* 89(4): 1165-1172

Tavassoli, M. (1984). "Hemopoiesis in ectopically implanted bone marrow." *Kroc Found Ser* 18: 31-54

Tavassoli, M. & W. H. Crosby (1968). "Transplantation of marrow to extramedullary sites." *Science* 161(836): 54-56

Tavassoli, M. & R. Khademi (1980). "The origin of hemopoietic cells in ectopic implants of spleen and marrow." *Experientia* 36(9): 1126-1127

Tavassoli, M., A. Maniatis & W. H. Crosby (1974). "Induction of sustained hemopoiesis in fatty marrow." *Blood* 43(1): 33-38

Tavassoli, M. & L. Weiss (1971). "The structure of developing bone marrow sinuses in extramedullary autotransplant of the marrow in rats." *Anat Rec* 171(4): 477-494

Teixido, J., M. E. Hemler, J. S. Greenberger & P. Anklesaria (1992). "Role of beta 1 and beta 2 integrins in the adhesion of human CD34hi stem cells to bone marrow stroma." *J Clin Invest* 90(2): 358-367

Tokoyoda, K., T. Egawa, T. Sugiyama, B. I. Choi & T. Nagasawa (2004). "Cellular niches controlling B lymphocyte behavior within bone marrow during development." *Immunity* 20(6): 707-718

Touw, I. & B. Lowenberg (1983). "No stimulative effect of adipocytes on hematopoiesis in long-term human bone marrow cultures." *Blood* 61(4): 770-774

Trentin, J. J. (1971). "Determination of bone marrow stem cell differentiation by stromal hemopoietic inductive microenvironments (HIM)." *Am J Pathol* 65(3): 621-628

Trinchieri, G., M. Murphy & B. Perussia (1987). "Regulation of hematopoiesis by T lymphocytes and natural killer cells." *Crit Rev Oncol Hematol* 7(3): 219-265

Verfaillie, C. M. (1993). "Soluble factor(s) produced by human bone marrow stroma increase cytokine-induced proliferation and maturation of primitive hematopoietic progenitors while preventing their terminal differentiation." *Blood* 82(7): 2045-2053

Verma, S., J. H. Rajaratnam, J. Denton, J. A. Hoyland & R. J. Byers (2002). "Adipocytic proportion of bone marrow is inversely related to bone formation in osteoporosis." *J Clin Pathol* 55(9): 693-698

Visnjic, D., I. Kalajzic, G. Gronowicz, H. L. Aguila, S. H. Clark, A. C. Lichtler & D. W. Rowe (2001). "Conditional ablation of the osteoblast lineage in Col2.3deltatk transgenic mice." *J Bone Miner Res* 16(12): 2222-2231

Visnjic, D., Z. Kalajzic, D. W. Rowe, V. Katavic, J. Lorenzo & H. L. Aguila (2004). "Hematopoiesis is severely altered in mice with an induced osteoblast deficiency." *Blood* 103(9): 3258-3264

Viviano, B. L., L. Silverstein, C. Pflederer, S. Paine-Saunders, K. Mills & S. Saunders (2005). "Altered hematopoiesis in glypican-3-deficient mice results in decreased osteoclast differentiation and a delay in endochondral ossification." *Dev Biol* 282(1): 152-162

Walkley, C. R., M. L. Fero, W. M. Chien, L. E. Purton & G. A. McArthur (2005). "Negative cell-cycle regulators cooperatively control self-renewal and differentiation of haematopoietic stem cells." *Nat Cell Biol* 7(2): 172-178

Walkley, C. R., G. H. Olsen, S. Dworkin, S. A. Fabb, J. Swann, G. A. McArthur, S. V. Westmoreland, P. Chambon, D. T. Scadden & L. E. Purton (2007). "A microenvironment-induced myeloproliferative syndrome caused by retinoic acid receptor gamma deficiency." *Cell* 129(6): 1097-1110

Walkley, C. R., J. M. Shea, N. A. Sims, L. E. Purton & S. H. Orkin (2007). "Rb regulates interactions between hematopoietic stem cells and their bone marrow microenvironment." *Cell* 129(6): 1081-1095

Wei, L., K. Kanbe, M. Lee, X. Wei, M. Pei, X. Sun, R. Terek & Q. Chen (2010). "Stimulation of chondrocyte hypertrophy by chemokine stromal cell-derived factor 1 in the chondro-osseous junction during endochondral bone formation." *Dev Biol* 341(1): 236-245

Weinstein, R., M. A. Riordan, K. Wenc, S. Kreczko, M. Zhou & N. Dainiak (1989). "Dual role of fibronectin in hematopoietic differentiation." *Blood* 73(1): 111-116

Weiss, L. (1976). "The hematopoietic microenvironment of the bone marrow: an ultrastructural study of the stroma in rats." *Anat Rec* 186(2): 161-184

Wittig, O., J. Paez-Cortez & J. Cardier (2009). "Liver Sinusoidal Endothelial Cells Promote B Lymphopoiesis from Primitive Hematopoietic Cells." *Stem Cells Dev*

Wolf, N. S. (1979). "The haemopoietic microenvironment." *Clin Haematol* 8(2): 469-500

Wolf, N. S. & J. J. Trentin (1968). "Hemopoietic colony studies. V. Effect of hemopoietic organ stroma on differentiation of pluripotent stem cells." *J Exp Med* 127(1): 205-214

Wu, J. Y., L. E. Purton, S. J. Rodda, M. Chen, L. S. Weinstein, A. P. McMahon, D. T. Scadden & H. M. Kronenberg (2008). "Osteoblastic regulation of B lymphopoiesis is mediated by Gs{alpha}-dependent signaling pathways." *Proc Natl Acad Sci U S A* 105(44): 16976-16981

Xie, Y., T. Yin, W. Wiegraebe, X. C. He, D. Miller, D. Stark, K. Perko, R. Alexander, J. Schwartz, J. C. Grindley, J. Park, J. S. Haug, J. P. Wunderlich, H. Li, S. Zhang, T.

Johnson, R. A. Feldman & L. Li (2009). "Detection of functional haematopoietic stem cell niche using real-time imaging." *Nature* 457(7225): 97-101

Yagi, K., K. Tsuji, A. Nifuji, K. Shinomiya, K. Nakashima, B. DeCrombrugghe & M. Noda (2003). "Bone morphogenetic protein-2 enhances osterix gene expression in chondrocytes." *J Cell Biochem* 88(6): 1077-1083

Yin, T. & L. Li (2006). "The stem cell niches in bone." *J Clin Invest* 116(5): 1195-1201

Yokota, T., K. Oritani, I. Takahashi, J. Ishikawa, A. Matsuyama, N. Ouchi, S. Kihara, T. Funahashi, A. J. Tenner, Y. Tomiyama & Y. Matsuzawa (2000). "Adiponectin, a new member of the family of soluble defense collagens, negatively regulates the growth of myelomonocytic progenitors and the functions of macrophages." *Blood* 96(5): 1723-1732

Yoshimoto, M., T. Shinohara, T. Heike, M. Shiota, M. Kanatsu-Shinohara & T. Nakahata (2003). "Direct visualization of transplanted hematopoietic cell reconstitution in intact mouse organs indicates the presence of a niche." *Exp Hematol* 31(8): 733-740

Youn, D. Y., D. H. Lee, M. H. Lim, J. S. Yoon, J. H. Lim, S. E. Jung, C. E. Yeum, C. W. Park, H. J. Youn, J. S. Lee, S. B. Lee, M. Ikawa, M. Okabe, Y. Tsujimoto & J. H. Lee (2008). "Bis deficiency results in early lethality with metabolic deterioration and involution of spleen and thymus." *Am J Physiol Endocrinol Metab* 295(6): E1349-1357

Zhang, C. C. & H. F. Lodish (2008). "Cytokines regulating hematopoietic stem cell function." *Curr Opin Hematol* 15(4): 307-311

Zhang, J., C. Niu, L. Ye, H. Huang, X. He, W. G. Tong, J. Ross, J. Haug, T. Johnson, J. Q. Feng, S. Harris, L. M. Wiedemann, Y. Mishina & L. Li (2003). "Identification of the haematopoietic stem cell niche and control of the niche size." *Nature* 425(6960): 836-841

Zhang, Y., A. Harada, H. Bluethmann, J. B. Wang, S. Nakao, N. Mukaida & K. Matsushima (1995). "Tumor necrosis factor (TNF) is a physiologic regulator of hematopoietic progenitor cells: increase of early hematopoietic progenitor cells in TNF receptor p55-deficient mice in vivo and potent inhibition of progenitor cell proliferation by TNF alpha in vitro." *Blood* 86(8): 2930-2937

Zhu, J., R. Garrett, Y. Jung, Y. Zhang, N. Kim, J. Wang, G. J. Joe, E. Hexner, Y. Choi, R. S. Taichman & S. G. Emerson (2007). "Osteoblasts support B-lymphocyte commitment and differentiation from hematopoietic stem cells." *Blood* 109(9): 3706-3712

Zuckerman, K. S., R. K. Rhodes, D. D. Goodrum, V. R. Patel, B. Sparks, J. Wells, M. S. Wicha & L. A. Mayo (1985). "Inhibition of collagen deposition in the extracellular matrix prevents the establishment of a stroma supportive of hematopoiesis in long-term murine bone marrow cultures." *J Clin Invest* 75(3): 970-975

Zuckerman, K. S. & M. S. Wicha (1983). "Extracellular matrix production by the adherent cells of long-term murine bone marrow cultures." *Blood* 61(3): 540-547

Zweegman, S., J. Van Den Born, A. M. Mus, F. L. Kessler, J. J. Janssen, T. Netelenbos, P. C. Huijgens & A. M. Drager (2004). "Bone marrow stromal proteoglycans regulate megakaryocytic differentiation of human progenitor cells." *Exp Cell Res* 299(2): 383-392

9

Molecular Mechanisms Underlying Bone Marrow Homing of Hematopoietic Stem Cells

Aysegul Ocal Sahin and Miranda Buitenhuis
Department of Hematology and Erasmus MC Stem Cell Institute for Regenerative Medicine, Erasmus MC, Rotterdam, The Netherlands

1. Introduction

The formation of blood cells, also called hematopoiesis, is a complex process that occurs in the bone marrow and depends on correct regulation of hematopoietic cell fate decisions. Aberrant regulation of hematopoiesis can result in the development of severe malignant and non-malignant hematological disorders, including leukemia. Hematopoietic stem cell transplantation is the most powerful treatment modality for a large number of those malignancies. Successful hematopoietic recovery after transplantation depends on homing of hematopoietic stem cells to the bone marrow and subsequent lodging of those cells in the bone marrow microenvironment.

Homing is a rapid, coordinated process in which circulating hematopoietic stem and progenitor cells actively enter the bone marrow within a few hours after transplantation (Figure 1). Rolling and firm adhesion of those cells to endothelial cells in small marrow sinusoids is followed by trans-endothelial migration across the endothelium/extracellular matrix barrier. Finally, in irradiated recipients, hematopoietic stem cells anchor to their specialized niches within the bone marrow compartment near osteoblasts and initiate long-term repopulation (Lo Celso et al., 2009). In absence of available niches in, for example, non-irradiated recipients, HSCs tend to be more randomly distributed throughout the bone marrow (Lo Celso et al., 2009). Since the first bone marrow transplantation decades ago, research has focused on understanding the mechanisms underlying homing of hematopoietic stem cells to the bone marrow. This chapter will focus on recent studies that have extended our understanding of the molecular mechanisms underlying adhesion, migration and bone marrow homing of hematopoietic stem cells.

2. Selectins and bone marrow homing

A first step in the process of bone marrow homing is initial tethering and rolling of hematopoietic stem and progenitor cells along the endothelial wall of blood vessels. It has been demonstrated that selectins play an important role in bone marrow homing of hematopoietic stem and progenitor cells by regulating these processes. Intravital microscopy in bone marrow sinoids and venules of mice deficient for individual selectins revealed that rolling of hematopoietic progenitor cells involves both P and E-selectin, but not L-selectin (Mazo et al.,

1998). Similarly, coating of a surface with immobilized P- or E-Selectin was sufficient to induce rolling of human CD34+ hematopoietic progenitor cells under flow conditions (Xia et al., 2004). A next step in bone marrow homing is transendothelial migration. This process requires firm adhesion of hematopoietic stem and progenitor cells to endothelial cells. Although CD34+ hematopoietic progenitor cells are capable of binding to fluid-phase P- and E-selectin (Xia et al., 2004), in vitro adhesion to bone marrow derived endothelial cells under static conditions has been shown not to depend on E-selectin (Naiyer et al., 1999). Transwell experiments performed to study the importance of E-selectin in migration of human hematopoietic progenitor cells through a confluent layer of bone marrow derived endothelial cells, precultured with IL-1B to induce E-selectin expression, yielded contradictory results. While Naiyer et al. have demonstrated with blocking antibodies that E-selectin is important for transendothelial migration (Naiyer et al., 1999), no significant inhibition in transendothelial migration could be observed by Voermans et al. who performed similar experiments (Voermans et al., 2000). Transplantation of lethally irradiated recipient mice deficient for both P-and E-selectin with wild type bone marrow cells resulted in reduced recruitment of hematopoietic progenitors to the bone marrow and enhanced levels of circulating hematopoietic progenitors, indicating that selectins indeed play an important role in bone marrow homing (Frenette et al., 1998).

Ligands for E-selectin include the PSGL-1 glycoform CLA, CD43 and the CD44 glycoform HCELL (Dimitroff et al., 2001; Merzaban et al., 2011). These ligands are all expressed on mouse Lin-Sca-1+c-Kit+ hematopoietic stem and progenitor cells and human CD34+ hematopoietic progenitor cells (Merzaban et al., 2011). Immune precipitation experiments revealed that although E-selectin can bind to CLA and CD43 in both mouse and human cells, the interaction between E-selectin and CD44 only occurs in human cells (Merzaban et al., 2011). These studies indicate that the molecular mechanism underlying bone marrow homing may be different for mouse and human hematopoietic stem cells. This hypothesis was confirmed by the observation that human CD34+ hematopoietic progenitor cells exhibit a stronger E-selectin binding capacity compared to mouse Lin-Sca-1+c-kit+ cells (Merzaban et al., 2011). In contrast to PSGL-1 which is also expressed in mature hematopoietic cells, CD44 appears to be predominantly expressed on primitive human CD34+ hematopoietic progenitor cells (Dimitroff et al., 2001). Rolling experiments performed under physiological flow conditions revealed that CD44 mediates E-selectin-dependent rolling interactions over a wider shear range in comparison to PSGL-1 and promotes rolling interactions on human bone marrow endothelial cells (Dimitroff et al., 2001). Silencing of CD44 expression in human cells with shRNAs was sufficient to decrease E-selectin binding under physiologic shear conditions, while enforced CD44 expression in Lin-Sca-1+c-kit+ cells conversely increased E-selectin adherence, resulting in improved bone marrow homing in vivo (Merzaban et al., 2011). In addition, treatment of mice with blocking antibodies against CD44 resulted in an increase in committed progenitors in the peripheral blood, suggesting that CD44 is important for lodging of hematopoietic progenitors in the bone marrow (Vermeulen et al., 1998). It has also been demonstrated that the selectin ligands must be alpha1-3 fucosylated to form glycan determinants such as sialyl Lewis x (sLe(x)). Inadequate alpha1-3 fucosylation of umbilical cord blood derived CD34+CD38-/low cells resulted in reduced interaction with both E-selectin and P-selectin, while increasing the level of cell-surface sLe(x) determinants augmented binding to fluid-phase P- and E-selectin, improved cell rolling on P- and E-selectin under flow and enhanced engraftment of human hematopoietic cells in bone marrows of irradiated NOD/SCID mice (Xia et al., 2004).

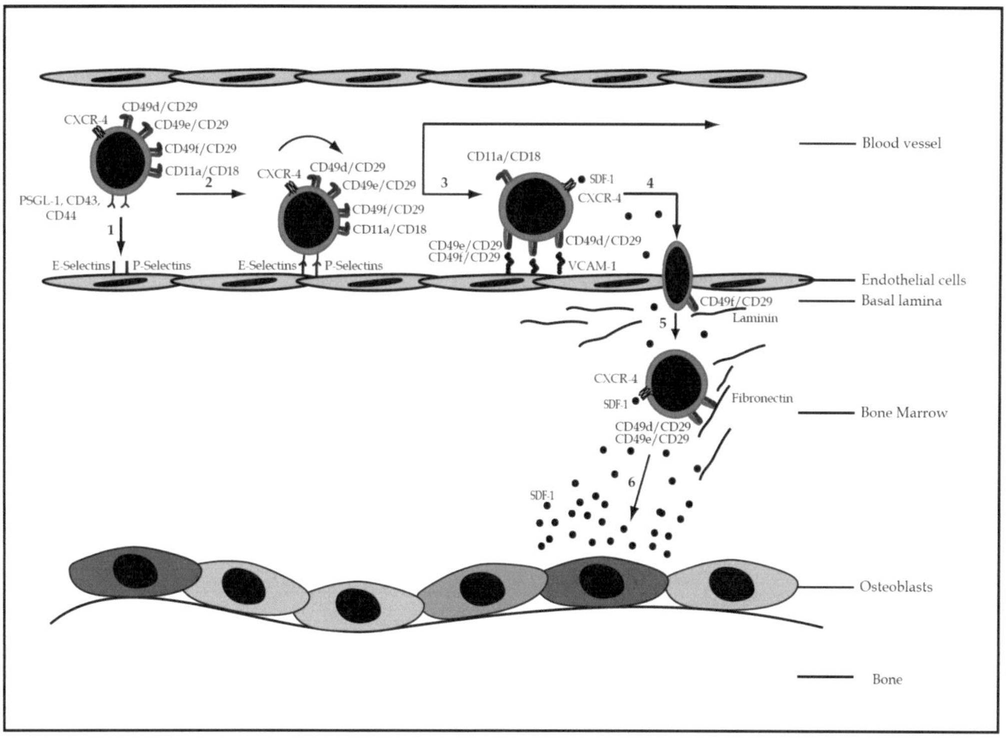

Fig. 1. Homing of hematopoietic Stem Cells to the bone marrow. 1) Initial tethering and 2) rolling are the first steps in bone marrow homing. These processes are mediated by both E- and P-selectin. 3) SDF-1 mediated integrin activation induces firm adhesion of the hematopoietic stem cells to the endothelial wall. 4) Firmly attached hematopoietic stem cells can subsequently transmigrate through the endothelial layer and 5) basal lamina, consisting of fibronectin, collagen and laminin. Integrins involved in these steps are CD49d/CD29, CD49e/CD29 and CD49f/CD29. 6) Finally, hematopoietic stem cells migrate towards the SDF-1 gradient to the osteoblasts.

3. Integrins and bone marrow homing

Integrins are, in addition to selectins, also implicated in playing an important role in regulation of bone marrow homing. Several in vitro studies with blocking antibodies have, for example, shown that both CD49d/CD29 (α4β1 or VLA-4) and CD11a/CD18 (αLβ2 or LFA-1) play an important role in adhesion of hematopoietic stem and progenitor cells to endothelial cells and subsequent transendothelial migration (Imai et al., 1999; Peled et al., 2000; Voermans et al., 2000). In addition, spontaneous migration of CD34+ hematopoietic progenitors underneath a bone marrow derived stromal cell layer, was found to be significantly inhibited by a peptide that blocks CD49d/CD29 integrin binding (Burger et al., 2003). However, adhesion of $CD34^{+}$ cells to fibronectin was found to be primarily dependent on CD49e/CD29 (α5β1 or VLA-5) and not CD49d/CD29 (Peled et al., 2000). In addition, chemotaxis of peripheral blood CD34+ progenitor cells on recombinant fibronectin appears to be mediated, at least in part, by CD49e/CD29 (Carstanjen et al., 2005). The importance of both CD49d/CD29 and CD49e/CD29 in directional migration through the basal lamina, which is composed of the extracellular matrix proteins laminin, collagen, and fibronectin, has been examined utilizing a three dimensional extra cellular matrix-like gel. In contrast to the dominant role of CD49e/CD29 in facilitating static adhesion to fibronectin, SDF-1–induced directional migration of CD34+ cells was found to be dependent on both CD49d/CD29 and CD49e/CD29 integrins (Peled et al., 2000). These studies suggest that both CD49d/CD29 and CD49e/CD29 play an important role in migration of hematopoietic stem and progenitor cells in general. However, transwell migration experiments with endothelial cells from different origin showed that CD49d/CD29 is only involved in migration of hematopoietic progenitors through a confluent layer of bone marrow derived, but not human umbilical vein derived, endothelial cells (Peled et al., 2000). In addition, inhibition of CD49e/CD29 alone was not sufficient to inhibit migration through both types of endothelial cells. However, an additive effect was observed when antibodies for CD11a/CD18, CD49d/CD29 and CD49e/CD29 were mixed together (Peled et al., 2000). These results suggest that the mechanisms underlying hematopoietic stem cell migration through endothelial walls of blood vessels depends on the origin of the endothelial cells and the VCAM-1 expression level.

As described above, deletion of both P- and E-selectin in recipient mice significantly reduced bone marrow homing after transplantation of wild type HPCs. Treatment of these mice with a blocking antibody against VCAM-1, thereby prohibiting interaction with CD49d/CD29, was sufficient to further reduce bone marrow homing after transplantation (Frenette et al., 1998), suggesting that both selectins and integrins are important for optimal bone marrow homing. In addition, the capacity of cells either deficient for CD49d (Scott et al., 2003) or pretreated with CD49d antibodies (Vermeulen et al., 1998; Papayannopoulou et al., 2001; Qian et al., 2006; Carstanjen et al., 2005) to migrate to bone marrow has been shown to be impaired resulting in delayed short-term engraftment (Scott et al., 2003). Furthermore, treatment of mice with blocking antibodies against CD49d resulted in an increase in the number of committed progenitors in the peripheral blood, suggesting that CD49d is also important for lodging of hemaopoietic progenitors in the bone marrow (Vermeulen et al., 1998). Since antibodies directed against mouse CD49d can bind to both CD49d/CD29 and CD49d/ITGB7 ($\alpha4\beta7$), and CD49d/ITGB7 is also expressed on mouse Lin-Sca-1+c-Kit+ cells, it was hypothesized that in addition to CD49d/CD29, CD49d/ITGB7 could also be involved in bone marrow homing. Indeed, inhibition of CD49d/ITGB7or its substrate MadCam-1 significantly reduced, but not completely abrogated, bone marrow homing after transplantation (Katayama et al., 2004). In contrast, other integrins, including CD11a, appear not be involved in bone marrow homing (Vermeulen et al., 1998). Transplantation studies with hematopoietic stem cells deficient for CD18 indicated that also CD18 is not essential for bone marrow homing. However, since inhibition of CD49d/CD29 in CD18 deficient hematopoietic stem cells resulted in more dramatic reduction in bone marrow homing in comparison to inhibition of CD49d/CD29 in wild type mice, it was suggested that CD18 can contribute to bone marrow homing when the function of CD49d/CD29 is compromised (Papayannopoulou et al., 2001). In addition to CD49d, CD49e/CD29 has also been implicated in playing a role in regulation of bone marrow homing. Treatment of hematopoietic progenitors with an antibody directed against CD49e/CD29 was sufficient to partially reduce homing of those cells to the bone marrow but not to the spleen (Wierenga et al., 2006; Carstanjen et al., 2005). Another integrin implicated in regulation of bone marrow homing is CD49f (α6). In contrast to CD49d that appears to primarily be involved in bone marrow homing of short-term repopulating hematopoietic stem cells, CD49f is thought to be important for homing of both short-term and long-term stem cells (Qian et al., 2006). In contrast, similar experiments with fetal liver cells revealed that, in contrast to CD49d which appeared to be important for homing of both hematopoietic stem and progenitor cells, CD49f is only important for homing of hematopoietic progenitors but not stem cells (Qian et al., 2007). These studies indicate that CD49d and Cd49f play differential roles during homing of cord blood and fetal liver derived hematopoietic stem and progenitor cells (Qian et al., 2007). In contrast, bone marrow homing was not affected in a more recent study in which also mouse bone marrow derived hematopoietic stem and progenitor cells pretreated with blocking antibodies directed against CD49f were transplanted in recipient mice (Bonig et al., 2009). In addition, blocking CD49f in human and primate bone marrow derived hematopoietic stem and progenitor cells, but not mobilized peripheral blood or cord blood derived cells that express little or no CD49f, resulted in enhanced bone marrow homing in a xenogeneic transplant model and significantly improved engraftment levels (Bonig et al., 2009). Finally, intravenous injection of anti-CD49f antibodies, in contrast to antibodies against CD49d integrin, did not mobilize progenitors or enhance cytokine-induced mobilization by G-CSF, suggesting that CD49f is not essential for lodging of hematopoietic stem and progenitor cells in the bone marrow (Qian et al., 2006). Additional research is required to investigate whether or not CD49f regulates bone marrow homing.

4. Chemoattractants involved in migration of hematopoietic stem cells

Chemoattractants play an important role in directing migration of hematopoietic stem and progenitor cells to the bone marrow. Several studies have demonstrated that Stromal cell Derived Factor 1 (SDF-1), also known as CXC chemokine ligand 12 (CXCL12) (Tashiro et al., 1993) acts as a chemoattractant for hematopoietic stem and progenitor cells and is important for their transendothelial migration (Aiuti et al., 1997; Naiyer et al., 1999; Mohle et al., 1998; Kim & Broxmeyer, 1998; Glass et al., 2011). Further investigation, utilizing a large panel of CC and CXC chemokines, suggested that the only chemokine capable of inducing migration of murine hematopoietic stem and progenitor cells appears to be SDF-1 (Liesveld et al., 2001; Wright et al., 2002). Although the chemokine receptors CCR3 and CCR9 were also expressed at mRNA level, their ligands could not induce migration (Wright et al., 2002). Similarly, examination of a panel of chemokines and cytokines in transendothelial migration assays revealed that SDF-1 is also important for migration of human hematopoietic progenitors through a confluent layer of endothelial cells (Liesveld et al., 2001). However, to a lesser extent, also other chemokines and cytokines, including CCL2 (MCP-1), CCL5 (RANTES), CXCL10 (IP-10), IL-8 and SCF could also induce transendothelial migration (Liesveld et al., 2001). In addition, LTD4, a ligand for CysLT(1), a G protein-coupled receptor recognizing inflammatory mediator of the cysteinyl leukotriene family, which is highly expressed in hematopoietic progenitors, has been demonstrated to up-regulate CD49d/CD29 and CD49e/CD29 dependent adhesion of hematopoietic progenitors (Boehmler et al., 2009) and to induce chemotaxis and in vitro transendothelial migration (Bautz et al., 2001). Recently, a role for the proteolysis-resistant bioactive lipids sphingosine-1-phosphate and ceramide-1-phosphate in regulation of bone marrow homing has been suggested. Conditioning of mice for transplantation resulted in enhanced levels of these lipids in the bone marrow. In addition, both lipids appear to be chemoattractants for hematopoietic stem and progenitor cells (Kim et al., 2011).

The role of SDF-1 in migration of hematopoietic stem and progenitor cells will be discussed below in more detail.

4.1 SDF-1 and bone marrow homing

SDF-1 is produced by several types of bone marrow cells (Maekawa & Ishii, 2000). In the adult human bone marrow, SDF-1 was found to be expressed by endothelial cells and along the endosteum region (Peled et al., 2000; Ponomaryov et al., 2000). SDF-1 plays an important role in many processes, including immune surveillance, proliferation, differentiation and survival of many cell types (Aiuti et al., 1997; Bleul et al., 1996; Bleul et al., 1998; Cashman & Eaves, 2000; Lataillade et al., 2000). In addition, SDF-1 is considered to be essential for migration of hematopoietic stem cells to the bone marrow (Imai et al., 1999; Peled et al., 1999a; Wright et al., 2002). To date, two receptors for SDF-1 have been identified, of which CXCR4 (LESTR/fusin), a seven-transmembrane domain G-protein coupled receptor, appears to be the most prominent (Heesen et al., 1997; Loetscher et al., 1994). CXCR4 is expressed by a variety of cell types, including hematopoietic stem and progenitor cells, T lymphocytes, endothelial, stromal and neuronal cells (Nagasawa et al., 1996; Ma et al., 1998; Mohle et al., 1998; Loetscher et al., 1994). Recently, CXCR7, another SDF-1 receptor, has been identified (Tarnowski et al., 2010). However, CXCR7 is expressed at low levels in normal human $CD34^{+}$ hematopoietic stem and progenitor cells and does not appear to be important for migration of those cells. In contrast, CXCR7 is highly expressed in several human myeloid leukemic cell lines and is thought to play a role in adhesion and, to a lesser extent, also in migration of those cells (Tarnowski et al., 2010).

Mouse transplantation studies have been performed to investigate the importance of SDF-1 in migration of hematopoietic stem cells to the bone marrow. Pre-treatment of human $CD34{+}CD38^{-/low}$ cells with a blocking antibody against CXCR4 has, for example, been demonstrated to be sufficient to impair their capacity to home to the bone marrow of immune deficient NOD/SCID mice or β2m deficient NOD/SCID mice (Peled et al., 1999b; Kollet et al., 2001; Kollet et al., 2002; Oberlin et al., 1996). In addition, up-regulation of CXCR4 expression by incubation with hematopoietic cytokines (SCF and IL-6) (Peled et al., 1999b) or over-expression of CXCR4 by viral transduction (Brenner et al., 2004; Kahn et al., 2004) resulted in enhanced bone marrow homing of human CD34+ and CD34+CD38- cells in NOD/SCID mice, which correlated with enhanced engraftment levels 6 weeks after transplantation (Peled et al., 1999b; Kollet et al., 2001; Kollet et al., 2002). Similarly, fetal liver hematopoietic stem and progenitor cells deficient for CXCR4 displayed a reduced bone marrow homing capacity compared to wild type cells (Ma et al., 1998). In addition to bone marrow homing, SDF-1 also appears to play a critical role in retention of hematopoietic stem cells in the hematopoietic stem cell niche. Enhancing the level of SDF-1 in plasma, but not bone marrow, utilizing adenoviral vectors (Hattori et al., 2001) or sulfated glycans (Sweeney et al., 2000; Frenette & Weiss, 2000; Sweeney et al., 2002) resulted in mobilization of CXCR4 expressing hematopoietic stem and progenitor cells (Hattori et al., 2001; Sweeney et al., 2002). Similarly, treatment of C3H/HeJ mice or healthy human volunteers with AMD3100, a selective CXCR4 antagonist, enhanced the number of HSCs and neutrophils in peripheral blood, again suggesting a role for CXCR4 and SDF-1 in HSC retention in BM (Broxmeyer et al., 2005).

4.2 Regulation of SDF-1 activity

Several proteolytic enzymes have been implicated in negatively regulating migration of hematopoietic stem cells by cleaving and inactivating SDF-1, including matrix metalloproteinases (MMP) 2/9 (Heissig et al., 2002; Sweeney et al., 2002; McQuibban et al., 2001), CD26 (Christopherson et al., 2002), carboxypeptidase M (Marquez-Curtis et al., 2008), carboxypeptidase N (Davis et al., 2005), neutrophil elastase (Petit et al., 2002; Levesque et al., 2002), cathepsin G (Petit et al., 2002; Levesque et al., 2002) and cathepsin K (Kollet et al., 2006). Cleavage of SDF-1 by several individual MMPs at Ser^4-Leu^5 bond of SDF-1 N-terminal domain has, for example, been demonstrated to result in reduced binding capacity of SDF-1 for CXCR-4 and reduced chemoattractant activity for hematopoietic stem and progenitor cells (McQuibban et al., 2001; Cho et al., 2010). Another protein involved in regulation of the activity of SDF-1 is the membrane-bound extracellular peptidase CD26 (DPPIV). It has been shown that a small number of umbilical cord blood derived CD34+CXCR4+ cells express CD26 and can therefore cleave the N-terminal part of SDF-1 at 2-proline (Christopherson et al., 2002). Functional studies showed that truncated SDF-1 lacks the ability to induce migration of CD34+ cells. In addition, inhibition of endogenous CD26 activity appears to be sufficient to enhance the migratory capacity of CD34+ cells towards SDF-1, indicating that CD26 abrogates SDF-1 induced migration of hematopoietic progenitors (Christopherson et al., 2002; Christopherson et al., 2003; Christopherson et al., 2006).

A third class of SDF-1 inhibitors includes the carboxypeptidases M and N (Marquez-Curtis et al., 2008; Davis et al., 2005). Carboxypeptidase N, which is present in human serum and plasma (Davis et al., 2005), can efficiently and specifically cleave SDF-1 at the carboxy-terminal lysine (K68) resulting in reduced SDF-1 activity and inhibition of SDF-1 mediated induction of migration of hematopoietic progenitors (Davis et al., 2005). In contrast, carboxypeptidase M is a membrane bound zinc-dependent peptidase that cleaves carboxy-terminal basic residues. This particular carboxypeptidase is expressed by stromal cells and CD34+ cells from both bone marrow and mobilized peripheral blood (Skidgel & Erdos, 1998; Marquez-Curtis et al., 2008). Carboxypeptidase M mediated cleavage of SDF-1 results in reduced chemotactic activity of hematopoietic stem and progenitor cells, which can be rescued by addition of the carboxypeptidase inhibitor DL-2-mercaptomethyl-3-guanidino-ethylthiopropanoic acid (Marquez-Curtis et al., 2008).

Whereas high SDF-1 expression in the bone marrow is essential for normal bone marrow homing of hematopoietic stem and progenitor cells and lodging of those cells in the hematopoietic stem cell niche, during mobilization SDF-1 levels should conversely be decreased. Upon administration of G-CSF, which is used to mobilize HSPCs, an accumulation of various proteolytic enzymes including MMP-9, neutrophil elastase and cathepsin G or K (Petit et al., 2002; Levesque et al., 2002) has been observed in mouse bone marrow which correlated with a gradual decrease in SDF-1 in the bone marrow, but not circulation (Petit et al., 2002). In addition, also an enhanced SDF-1 plasma level was shown to result in up-regulation of MPP-9 in bone marrow cells and mobilization of hematopoietic stem and progenitor cells (Heissig et al., 2002). The importance of MPP-9 for mobilization of hematopoietic stem cells was demonstrated utilizing MMP-9 deficient mice. A high SDF-1 level in plasma was not sufficient to induce mobilization of hematopoietic progenitors in these mice (Heissig et al., 2002). In addition, in primary myelofibrosis, which is a chronic myeloproliferative neoplasm characterized by constitutive mobilization of hematopoietic stem and progenitor cells into the peripheral blood (Migliaccio et al., 2008), both a high level of truncated SDF-1 and enhanced levels of proteases, including dipeptidyl peptidase-IV (CD26), neutrophil elastase, matrix metalloproteinase-2 (MMP-2), MMP-9, and cathepsin G have been observed (Cho et al., 2010). Taken together, these studies demonstrated that SDF-1 plays an important role in integrin-mediated firm arrest of human HSPCs, facilitate their transendothelial migration, and regulate bone marrow homing and retention of HSPCs in the hematopoietic stem cell niche.

4.3 Molecular mechanisms underlying SDF-1 mediated regulation of migration

To understand the molecular mechanism underlying migration of hematopoietic stem and progenitor cells, research has focused on identifying the downstream effectors of SDF-1 and CXCR4. SDF-1 has been demonstrated to induce the activity of the integrins CD11a/CD18 (Peled et al., 2000) and CD49/CD29 (Hidalgo et al., 2001; Peled et al., 2000) on CD34+ cells which allows interaction with their substrates ICAM-1 and VCAM-1, respectively.

Small guanosine triphosphatases (GTPases) that belong to the Ras superfamily of GTPases , including Rho, Rac and Cdc42, have been demonstrated to be involved in SDF-1 mediated homing and migration of hematopoietic stem and progenitor cells (Fuhler et al., 2008; del Pozo et al., 1999). The activity of Rho GTPases can be induced by tyrosine kinase receptors (Taylor & Metcalfe, 2000; Timokhina et al., 1998), integrin receptors (del Pozo et al., 2004) and chemokine receptors including SDF-1 (Cancelas et al., 2005; del Pozo et al., 1999; Fuhler et al., 2008; Shirvaikar et al., 2011). It has been demonstrated in *in vitro* assays that SDF-1 induced chemo-attraction is mediated, at least in part, by Rac (del Pozo et al., 1999; Shirvaikar et al., 2011; Wysoczynski et al., 2005). In addition, analysis of Rac2 deficient mice revealed that Rac2 is essential for lodging of HSPCs in the bone marrow. Deletion of Rac2 resulted, for example, in reduced adhesion and enhanced mobilization of hematopoietic stem cells to the circulation. Furthermore, Rac2 deficiency resulted in enhanced SDF-1 induced migration of hematopoietic stem and progenitor cells (Yang et al., 2001). An enhanced activation of Cdc42 and Rac1 was observed in these cells, suggesting a compensatory role of Cdc42 and Rac1 with regard to migration, but not adhesion (Yang et al., 2001). In addition, it was shown that SDF-1 mediated Rac activation is impaired in CD34+ cells from MDS patients. CD34+ cell from patients with myelodysplastic syndrome exhibit reduced F-actin polymerization and migration towards SDF-1 compared to normal CD34+ cells (Fuhler et al., 2008). While pharmacological inhibition of Rac1 activity in a human myeloblastic cell line (HL-60) with NSC23766 was sufficient to abrogate SDF-1 induced actin assembly and migration, over-expression of active Rac in HL-60 cells conversely restored both F-actin polymerization and migration, suggesting that Rac is essential for SDF-1–induced migration in these cells (Fuhler et al., 2008). Although over-expression of active Rac in CD34+ cells from patients with myelodysplastic syndrome resulted in increased F-actin polymerization and enhanced motility, directional migration toward SDF-1 was not improved (Fuhler et al., 2008). These studies suggest that SDF-1 mediated induction of Rac activity is important for migration of both normal and malignant hematopoietic progenitors (Fuhler et al., 2008). The role of the hematopoietic-specific guanine nucleotide exchange factor Vav1, which is an upstream regulator of Rac activity, in localization and engraftment of hematopoietic stem and progenitor cells has also been investigated. Deletion of Vav1 in hematopoietic stem cells has been demonstrated to result in impaired responses to SDF1α, dysregulated Rac/Cdc42 activation and a reduction of in vitro migration. In addition, intravital microscopy assays revealed that transplantation of Vav1 deficient hematopoietic stem and progenitor cells results in impaired early localization near nestin(+) perivascular mesenchymal stem cells after transplantation (Sanchez-Aguilera et al., 2011). Recently, another upstream regulator of Rac activity has been identified. In contrast to Rac, the activity of R-Ras, a member of the Ras family, is inhibited upon SDF-1 stimulation. Deletion of R-Ras resulted in enhanced levels of Rac1/2 activity, while expression of a constitutively active R-Ras mutant resulted in down-regulation of Rac1-activity. Deletion of R-Ras in hematopoietic stem and progenitor cells resulted in increased directional migration. This phenotype could be reversed by inhibition of Rac. Furthermore, R-Ras deficient mice showed enhanced responsiveness to G-CSF for progenitor cell mobilization and exhibited decreased bone marrow homing (Sanchez-Aguilera et al., 2011).

Another important mediator of hematopoietic progenitor cell migration is the GTPase Rho (Bug et al., 2002; Ghiaur et al., 2006; Gottig et al., 2006). It has been demonstrated that SDF-1 mediated release of intracellular Ca^{2+} stores requires activation of Rho GTPases, but not Rac or Cdc42 (Henschler et al., 2003). Depletion of intracellular Ca^{2+} resulted in reduced SDF-1 induced migration and bone marrow homing of hematopoietic progenitors (Henschler et al., 2003). In addition, over-expresssion of dominant negative RhoA by retroviral transduction in mouse cells (C57BL/6J mice) resulted in decreased migration of hematopoietic progenitor cells towards SDF-1 and reduced integrin-mediated adhesion (Henschler et al., 2003). Furthermore, over-expression of RhoH, a GTPase deficient type of Rho (Sahai & Marshall, 2002), in hematopoietic stem and progenitor cells resulted in impaired activation of Rac GTPases, defective actin polymerization and impaired chemotaxis. In contrast, inhibition of RhoH expression in these cells conversely stimulated SDF-1–induced migration in vitro (Gu et al., 2005). In addition, it has been demonstrated that Epac1, a nucleotide exchange protein for the GTPase Rap1, which is directly activated by cAMP, can also improve the adhesive and migratory capacity CD34+ hematopoietic progenitor cells (Carmona et al., 2008), suggesting that Rap1 may also play a role in bone marrow homing.

Endolyn (CD164), a type I integral transmembrane silomucin (Chan et al., 2001; Zannettino et al., 1998), which is recruited to CXCR4 upon SDF-1 stimulation (Forde et al., 2007) was shown to play an important role in SDF-1 mediated migration of human CD133+ hematopoietic stem and progenitor cells (Forde et al., 2007). Inhibition of CD164 in CD133+ cells with 103B2, a specific mAb, resulted in a reduction of migration towards SDF-1, but not CCL1, CCL5, CCL17, CCL19, CCL20, CCL21, CCL22 and CXCL3. A similar inhibition in SDF-1 mediated migration of CD133+ cells was observed after siRNA mediated knock-down of CD164 (Forde et al., 2007). Knock-down of CD164 resulted in a significant reduction in SDF-1 mediated activation of PI3K and PKCζ (Forde et al., 2007). Both PI3K and PKCζ have been implicated in playing an important role in SDF-1 mediated migration of $CD34^+$ cells. Inhibition of PKCζ, for example, reduced SDF-1 induced migration of CD34+ cells and reduced engraftment levels after transplantation (Petit et al., 2005). Furthermore, injection of inhibitory PKCζ pseudosubstrate peptides resulted in mobilization of murine progenitors to the circulation, suggesting an important role for PKCζin SDF-1-dependent regulation of hematopoietic stem and progenitor cell motility and localization (Petit et al., 2005) The role of PI3K in regulation of bone marrow homing will be discussed in the next section.

In addition to regulating the activity of downstream effectors, SDF-1 has also been demonstrated to regulate the expression of specific target genes. Stimulation of peripheral blood mononuclear, Jurkat or HeLa cells has, for example, been demonstrated to result in a rapid increase in expression of the ubiquitin-specific protease 17 (USP17) (de la Vega et al., 2011). A role for this protease in regulation of migration of hematopoietic progenitor cells has been examined in vitro. Inhibition of USP17 in these cells showed decreased chemotaxis towards SDF-1, whereas over-expression of USP17 conversely resulted in increased chemotaxis. Interestingly, CXCR4 levels were not affected by inhibition or over-expression of USP17, suggesting that USP17 modulates the down-stream signaling of the CXCR4 receptor. shRNA mediated inhibition of USP17 expression resulted in decreased polymerization of actin and tubulin and reduced membrane ruffling. In addition, upon SDF-1 stimulation, the GTPases, RAC1, Cdc42 and RhoA were not transported to the plasma membrane, thereby prohibiting their activiation (de la Vega et al., 2011). In addition, CD9, a member of the tetraspanin superfamily (Boucheix et al., 1991) that is widely expressed in hematopoietic and non-hematopoietic cells, has been shown to be a SDF-1 responsive gene. Microarray analysis with human umbilical cord blood derived CD34+ cells revealed that short-term exposure to SDF-1 resulted in up-regulation of CD9 mRNA expression both in $CD34^+$ $CD38^+$ and $CD34^+$ $CD38^{-/low}$ cells (Leung et al., 2011). A role for CD9 in migration and adhesion of human cord blood derived hematopoietic stem and progenitor cells was investigated utilizing a neutralizing CD9 antibody (Leung et al., 2011). Although actin polymerization was not affected, the calcium influx and transendothelial migration towards a SDF-1 gradient was reduced by this antibody (Leung et al., 2011). In contrast, adhesion of progenitor cells to fibronectin and human umbilical vein endothelial cells was enhanced (Leung et al., 2011). Transplantation experiments revealed that in NOD/SCID mice, pre-treatment of human CD34+ cells with a neutralizing CD9 antibody resulted in inhibition of homing to bone marrow and spleen. However, enhanced CD9 expression in CD34+ cells with ingenol 3,20-dibenzoate (IDB), a protein kinase C agonist which was shown to induce CD9 expression in CD34+ cells, did not result in enhanced bone marrow homing (Desmond et al., 2011).

5. The PI3K/PKB signalling module and bone marrow homing

Correct regulation of the Phosphatidylinositol-3-Kinase (PI3K) / Protein Kinase B (PKB/c-Akt) signaling module is essential for multiple processes during hematopoiesis. Phosphatidylinositol 4,5 bisphosphate ($PI(4,5)P_2$, the most important substrate for PI3K, can be phosphorylated upon extracellular stimulation, resulting in the formation of phosphatidylinositol 3,4,5 trisphosphate ($PI(3,4,5)P_3$) (Hawkins et al., 2006). $PI(3,4,5)P_3$ subsequently serves as an anchor for pleckstrin homology (PH) domain-containing proteins, such as Protein Kinase B (PKB/ c-akt) (Burgering & Coffer, 1995). Activation of PI3K and its downstream effector Protein Kinase B (PKB/c-Akt) has been observed in leukemic cell lines stimulated with SDF-1 (Ganju et al., 1998). A positive role for PI3K/PKB in regulation of SDF-1 induced migration of hematopoietic stem cells was therefore suggested. However, it has been shown that Protein Phosphatase 2A plays an important role in positively regulating SDF-1 mediated migration of human hematopoietic progenitors by inhibition of PKB activity (Basu et al., 2007). Similarly, inhibition of PKB activity in $CD34^+$ cells for over 24 hours appears to be sufficient to reduce their adhesion to bone marrow derived stromal cells and to induce their basal migratory capacity (Buitenhuis et al., 2010). Transwell migration experiments through a confluent layer of human umbilical vein endothelial cells revealed that the observed reduction in firm adhesion does not ameliorate the induced migratory capacity of $CD34^+$ cells pre-treated with a PKB inhibitor (Buitenhuis et al., 2010). In addition, ectopic expression of constitutively active PKB in CD34+ cells conversely induced firm adhesion and reduced the basal level of migration. Although it cannot be excluded that transient activation of PI3K/PKB activity by SDF-1 is important for induction of migration, these studies suggest that prolonged activation of PKB activity is detrimental for migration of $CD34^+$ cells. The role of PI3K in regulation of bone marrow homing was initially examined utilizing mice deficient for SHIP (SH2-containing inositol-5'-phosphatase), a negative regulator of PI3K (Damen et al., 1996). Transplantation of lethally irradiated recipients with HSCs from SHIP deficient mice resulted in diminished repopulation, suggesting that constitutive activation of PI3K impairs the ability of HSCs to home to and to be retained in the hematopoietic stem cell niche in the bone marrow. Assessment of bone marrow homing revealed that $SHIP^{-/-}$ hematopoietic stem and progenitor cells indeed traffic to the bone marrow and spleen with significantly reduced efficiency compared to wild type cells. Although it is evident that constitutive activation of PI3K plays a critical role in regulation of hematopoiesis per se (Buitenhuis et al., 2008), these results indicate that the inability of SHIP deficient hematopoietic stem cells to engraft and sustain long-term hematopoiesis can be, at least partially, explained by their impaired ability to home to the bone marrow (Desponts et al., 2006). Deletion of Phosphate and tensin homologue (PTEN), another critical negative regulator of PI3K signaling that dephosphorylates $PI(3,4,5)P_3$ resulting in the formation of $PI(4,5)P_2$ (Maehama & Dixon, 1998) only decreased bone marrow homing when PTEN deficient HSCs were transplanted into non-irradiated recipients. These results suggest that, although PTEN deficient hematopoietic stem cells are capable of migrating to the bone marrow, their performence is reduced compared to competeting wild-type hematopoietic stem cells when vacant niches are limited (Zhang et al., 2006). Although both PTEN and SHIP act on the main product of PI3K activity, $PI(3,4,5)P_3$, the products generated are distinct, which could explain the differences between SHIP and PTEN deficient hematopoietic stem cells in terms of bone marrow homing (Dowler et al., 2000; Golub & Caroni, 2005). Recent findings demonstrated that, similar to deletion of SHIP, constitutive activation of PKB in human hematopoietic progenitors cells is sufficient to significantly inhibit homing of these cells to the bone marrow and spleen of β2 microglobulin -/- NOD/SCID mice (Buitenhuis et al., 2010). In contrast, although transplantation of C57 BL/6 mice with bone marrow cells from *5-fluorouracil* treated mice that ectopically expressed constitutively active PKB resulted in reduced engraftment levels, bone marrow homing was only modestly impaired 18 hours after transplantation (Kharas et al., 2010). To investigate whether inhibition of PKB activity would be sufficient to conversely improve bone marrow homing, human hematopoietic progenitor cells, pre-treated with a PKB inhibitor for 24 or 48 hours, were injected into recipient mice. Flow cytometric analysis, 22 hours after transplantation, revealed that transient inhibition of PKB activity prior to transplantation is sufficient to improve bone marrow homing (Buitenhuis et al., 2010). In addition, while constitutive activation of PKB appears to be detrimental for bone marrow homing, engraftment levels and hematopoietic recovery, inhibition of PKB activity prior to transplantation, resulting in an induction of bone marrow homing, conversely enhanced engraftment levels in recipient mice. Together, these studies demonstrated that correct regulation of PI3K/PKB is essential for migration of hematopoietic stem and progenitor cells to the bone marrow after transplantation, which is essential for optimal engraftment and hematopoietic recovery (Buitenhuis et al., 2010; Desponts et al., 2006; Kharas et al., 2010).

The molecular mechanisms underlying PKB mediated regulation of migration and bone marrow homing are, thus far, incompletely understood. Although PKB mediated inhibition of migration has been demonstrated to involve RAC1 (Farooqui et al., 2006), NFAT (Yiu & Toker, 2006; Yoeli-Lerner et al., 2005) and $p27^{Kip1}$ (Baldassarre et al., 2005; Viglietto et al., 2002; Wu et al., 2006) in non-hematopoietic cell lines, their importance for migration of hematopoietic stem and progenitor cells remains to be investigated. As described above, adhesion and migration of HSCs depend on correct integrin and selectin expression and regulation of integrin activity. PKB and its downstream effector GSK-3 have initially been shown to play an important role in recycling of the CD49e/CD29 and CD51/CD61 ($\alpha v\beta 3$) integrins to the membrane in NIH 3T3 fibroblasts, resulting in enhanced cell spreading and adhesion (Roberts et al., 2004). Ectopic expression of PKB in human hematopoietic stem and progenitor cells has been demonstrated to enhance the level of CD49d, while inhibition of PKB activity conversely reduces expression of both CD49d and CD18 (Buitenhuis et al., 2010), providing a potential mechanism by which PKB induces adhesion and inhibits migration. Although it is evident that integrins play an important role in adhesion and migration of cells, the importance of these molecules in PKB mediated inhibition of migration remains to be investigated. In addition, CXCR4 expression has been demonstrated to be reduced in SHIP deficient hematopoietic stem cells, suggesting that activation of PI3K also impairs their response to SDF-1(Zhang et al., 2006).

6. Conclusion

Allogeneic HSC transplantation is the preferred treatment modality for a number of hematological malignancies. To allow normal long-term hematopoiesis to occur after transplantation, correct regulation of homing of hematopoietic stem and progenitor cells to the bone marrow and subsequent lodging of those cells into the hematopoietic stem cell niche is essential. As described above, this is a coordinated multistep process that is regulated by chemokines, integrins and selectins. Initial tethering and rolling of hematopoietic stem and progenitor cells along the endothelial wall of blood vessels are the first steps in this process. It has been demonstrated that both P and E-selectin play an important role in rolling of HSCs. In addition to selectins, integrins are also implicated in playing an important role in regulation of bone marrow homing. Both studies with blocking antibodies and knockout mice have revealed that CD49d/CD29, CD49e/CD29, CD49f, and CD49d/ITGB7 play an important role in adhesion of hematopoietic stem and progenitor cells to endothelial cells and subsequent transendothelial migration. In addition, both CD49d/CD29 and CD49e/CD29 integrins appear to be involved in mediation of SDF-1–induced directional migration of CD34+ cells through the basal lamina. In addition, although, under normal circumstances, CD18 appears not to be essential for bone marrow homing of hematopoietic stem cells, CD18 can contribute to bone marrow homing when the function of CD49d/CD29 is compromised. Although multiple chemokines are capable of inducing transendothelial migration of hematopoietic stem cells, the chemokine SDF-1 appears to be the most prominent chemokine involved in bone marrow homing. In addition, SDF-1 also appears to play a critical role in retention of hematopoietic stem cells in the hematopoietic stem cell niche. Regulation of SDF-1 activity by a variety of proteolytic enzymes has been demonstrated to play an important role in migration of hematopoietic stem cells to and from the bone marrow. The molecular mechanism underlying SDF-1 mediated regulation of HSC migration has been investigated extensively. Thus far, multiple downstream effectors have been identified, including CD164, the GTPases Rac, Rho, and Cdc42, and the signalling molecules PI3K and PKCζ. In addition, the SDF-1 responsive genes CD9, USP17, both implicated in regulation of hematopoietic stem cell migration, have been indentified. Finally, SDF-1 has been demonstrated to induce the activity of integrins which allows interaction with their substrates. Although activation of PI3K and its downstream effector Protein Kinase B (PKB/c-Akt) has been observed in leukemic cell lines stimulated with SDF-1, suggesting a positive role for PI3K/PKB in regulation of SDF-1 induced migration of hematopoietic stem cells, the above described studies clearly implicate the PI3K/PKB signalling module in playing a critical role in negatively regulating migration of HSCs and bone marrow homing.

7. References

Aiuti A, Webb IJ, et al. (1997). The chemokine SDF-1 is a chemoattractant for human CD34+ hematopoietic progenitor cells and provides a new mechanism to explain the mobilization of CD34+ progenitors to peripheral blood *Journal of Experimental Medicine*, Vol. 185, No. 1, pp. 111-20, 0022-1007

Baldassarre G, Belletti B, et al. (2005). p27(Kip1)-stathmin interaction influences sarcoma cell migration and invasion *Cancer Cell*, Vol. 7, No. 1, pp. 51-63, ISSN 1535-6108.

Basu S, Ray NT, et al. (2007). Protein phosphatase 2A plays an important role in stromal cell-derived factor-1/CXC chemokine ligand 12-mediated migration and adhesion of CD34+ cells *Journal of Immunology*, Vol. 179, No. 5, pp. 3075-85, ISSN 0022-1767.

Bautz F, Denzlinger C, et al. (2001). Chemotaxis and transendothelial migration of CD34(+) hematopoietic progenitor cells induced by the inflammatory mediator leukotriene D4 are mediated by the 7-transmembrane receptor CysLT1 *Blood*, Vol. 97, No. 11, pp. 3433-40, ISSN 0006-4971.

Bleul CC, Fuhlbrigge RC, et al. (1996). A highly efficacious lymphocyte chemoattractant, stromal cell-derived factor 1 (SDF-1) *Journal of Experimental Medicine*, Vol. 184, No. 3, pp. 1101-9, ISSN 0022-1007.

Bleul CC, Schultze JL, et al. (1998). B lymphocyte chemotaxis regulated in association with microanatomic localization, differentiation state, and B cell receptor engagement *Journal of Experimental Medicine*, Vol. 187, No. 5, pp. 753-62, ISSN 0022-1007.

Boehmler AM, Drost A, et al. (2009). The CysLT1 ligand leukotriene D4 supports alpha4beta1- and alpha5beta1-mediated adhesion and proliferation of CD34+ hematopoietic progenitor cells *Journal of Immunology*, Vol. 182,No. 11, pp. 6789-98, ISSN 1550-6606.

Bonig H, Priestley GV, et al. (2009). Blockade of alpha6-integrin reveals diversity in homing patterns among human, baboon, and murine cells *Stem Cells Develoment*, Vol. 18, No. 6, pp. 839-44, ISSN 1557-8534.

Boucheix C, Benoit P, et al. (1991). Molecular cloning of the CD9 antigen. A new family of cell surface proteins *Journal of Biological Chemistry*, Vol. 266, No. 1, pp. 117-22, ISSN 0021-9258.

Brenner S, Whiting-Theobald N, et al. (2004). CXCR4-transgene expression significantly improves marrow engraftment of cultured hematopoietic stem cells *Stem Cells*, Vol. 22, No. 7, pp. 1128-33, ISSN 1066-5099.

Broxmeyer HE, Orschell CM, et al. (2005). Rapid mobilization of murine and human hematopoietic stem and progenitor cells with AMD3100, a CXCR4 antagonist *Journal of Experimental Medicine*, Vol. 201, No. 8, pp. 1307-18, ISSN 0022-1007.

Bug G, Rossmanith T, et al. (2002). Rho family small GTPases control migration of hematopoietic progenitor cells into multicellular spheroids of bone marrow stroma cells *Journal of Leukocyte Biology*, Vol. 72, No. 4, pp. 837-45, ISSN 0741-5400.

Buitenhuis M, van der Linden E, et al. (2010). Protein kinase B (PKB/c-akt) regulates homing of hematopoietic progenitors through modulation of their adhesive and migratory properties *Blood*, Vol. 116, No. 13, pp. 2373-84, ISSN 1528-0020.

Buitenhuis M, Verhagen LP, et al. (2008). Protein kinase B (c-akt) regulates hematopoietic lineage choice decisions during myelopoiesis *Blood*, Vol. 111, No. 1, pp. 112-21, ISSN 0006-4971.

Burger JA, Spoo A, et al. (2003). CXCR4 chemokine receptors (CD184) and alpha4beta1 integrins mediate spontaneous migration of human CD34+ progenitors and acute myeloid leukaemia cells beneath marrow stromal cells (pseudoemperipolesis) *British Journal of Haematology*, Vol. 122, No. 4, pp. 579-89, ISSN 0007-1048.

Burgering BM and Coffer PJ. (1995). Protein kinase B (c-Akt) in phosphatidylinositol-3-OH kinase signal transduction *Nature*, Vol. 376, No. 6541, pp. 599-602, ISSN 0028-0836.

Cancelas JA, Lee AW, et al. (2005). Rac GTPases differentially integrate signals regulating hematopoietic stem cell localization *Nature Medicine*, Vol. 11, No. 8, pp. 886-91, ISSN 1078-8956.

Carmona G, Chavakis E, et al. (2008). Activation of Epac stimulates integrin-dependent homing of progenitor cells *Blood*, Vol. 111, No. 5, pp. 2640-6, ISSN 0006-4971.

Carstanjen D, Gross A, et al. (2005). The alpha4beta1 and alpha5beta1 integrins mediate engraftment of granulocyte-colony-stimulating factor-mobilized human hematopoietic progenitor cells *Transfusion*, Vol. 45, No. 7, pp. 1192-200, ISSN 0041-1132.

Cashman JD and Eaves CJ. (2000). High marrow seeding efficiency of human lymphomyeloid repopulating cells in irradiated NOD/SCID mice *Blood*, Vol. 96, No. 12, pp. 3979-81, ISSN 0006-4971.

Chan JY, Lee-Prudhoe JE, et al. (2001). Relationship between novel isoforms, functionally important domains, and subcellular distribution of CD164/endolyn *Journal of Biological Chemistry*, Vol. 276, No. 3, pp. 2139-52, ISSN 0021-9258.

Cho SY, Xu M, et al. (2010). The effect of CXCL12 processing on CD34+ cell migration in myeloproliferative neoplasms *Cancer Research*, Vol. 70, No. 8, pp. 3402-10, ISSN 1538-7445.

Christopherson KW, 2nd, Cooper S, et al. (2003). Cell surface peptidase CD26/DPPIV mediates G-CSF mobilization of mouse progenitor cells *Blood*, Vol. 101, No. 12, pp. 4680-6, ISSN 0006-4971.

Christopherson KW, 2nd, Hangoc G, et al. (2002). Cell surface peptidase CD26/dipeptidylpeptidase IV regulates CXCL12/stromal cell-derived factor-1 alpha-mediated chemotaxis of human cord blood CD34+ progenitor cells *Journal of Immunology*, Vol. 169, No. 12, pp. 7000-8, ISSN 0022-1767.

Christopherson KW, 2nd, Uralil SE, et al. (2006). G-CSF- and GM-CSF-induced upregulation of CD26 peptidase downregulates the functional chemotactic response of

CD34+CD38- human cord blood hematopoietic cells *Experimental Hematology*, Vol. 34, No. 8, pp. 1060-8, ISSN 0301-472X.

Damen JE, Liu L, et al. (1996). The 145-kDa protein induced to associate with Shc by multiple cytokines is an inositol tetraphosphate and phosphatidylinositol 3,4,5-triphosphate 5-phosphatase *Procedures National Academy Sciencies U S A*, Vol. 93, No. 4, pp. 1689-93, ISSN 0027-8424.

Davis DA, Singer KE, et al. (2005). Identification of carboxypeptidase N as an enzyme responsible for C-terminal cleavage of stromal cell-derived factor-1alpha in the circulation *Blood*, Vol. 105, No. 12, pp. 4561-8, ISSN 0006-4971.

de la Vega M, Kelvin AA, et al. (2011). The deubiquitinating enzyme USP17 is essential for GTPase subcellular localization and cell motility *Nature Communications*, Vol. 2, pp. 259, ISSN 2041-1723.

del Pozo MA, Alderson NB, et al. (2004). Integrins regulate Rac targeting by internalization of membrane domains *Science*, Vol. 303, No. 5659, pp. 839-42, ISSN 1095-9203.

del Pozo MA, Vicente-Manzanares M, et al. (1999). Rho GTPases control migration and polarization of adhesion molecules and cytoskeletal ERM components in T lymphocytes *European Journal of Immunology*, Vol. 29, No. 11, pp. 3609-20, ISSN 0014-2980.

Desmond R, Dunfee A, et al. (2011). CD9 up-regulation on CD34+ cells with ingenol 3,20-dibenzoate does not improve homing in NSG mice *Blood*, Vol. 117, No. 21, pp. 5774-6, ISSN 1528-0020.

Desponts C, Hazen AL, et al. (2006). SHIP deficiency enhances HSC proliferation and survival but compromises homing and repopulation *Blood*, Vol. 107, No. 11, pp. 4338-45, ISSN 0006-4971.

Dimitroff CJ, Lee JY, et al. (2001). CD44 is a major E-selectin ligand on human hematopoietic progenitor cells *Journal of Cell Biology*, Vol. 153, No. 6, pp. 1277-86, ISSN 0021-9525.

Dowler S, Currie RA, et al. (2000). Identification of pleckstrin-homology-domain-containing proteins with novel phosphoinositide-binding specificities *Biochemical Journal*, Vol. 351, No. Pt 1, pp. 19-31, ISSN 0264-6021.

Farooqui R, Zhu S, et al. (2006). Glycogen synthase kinase-3 acts upstream of ADP-ribosylation factor 6 and Rac1 to regulate epithelial cell migration *Experimental Cell Research*, Vol. 312, No. 9, pp. 1514-25, ISSN 0014-4827.

Forde S, Tye BJ, et al. (2007). Endolyn (CD164) modulates the CXCL12-mediated migration of umbilical cord blood CD133+ cells *Blood*, Vol. 109, No. 5, pp. 1825-33, ISSN 0006-4971.

Frenette PS, Subbarao S, et al. (1998). Endothelial selectins and vascular cell adhesion molecule-1 promote hematopoietic progenitor homing to bone marrow *Procedures National Academy Sciences U S A*, Vol. 95, No. 24, pp. 14423-8, ISSN 0027-8424.

Frenette PS and Weiss L. (2000). Sulfated glycans induce rapid hematopoietic progenitor cell mobilization: evidence for selectin-dependent and independent mechanisms *Blood*, Vol. 96, No. 7, pp. 2460-8, ISSN 0006-4971.

Fuhler GM, Drayer AL, et al. (2008). Reduced activation of protein kinase B, Rac, and F-actin polymerization contributes to an impairment of stromal cell derived factor-1 induced migration of CD34+ cells from patients with myelodysplasia *Blood*, Vol. 111, No. 1, pp. 359-68, ISSN 0006-4971.

Ganju RK, Brubaker SA, et al. (1998). The alpha-chemokine, stromal cell-derived factor-1alpha, binds to the transmembrane G-protein-coupled CXCR-4 receptor and activates multiple signal transduction pathways *Journal of Biological Chemistry*, Vol. 273, No. 36, pp. 23169-75, ISSN 0021-9258.

Ghiaur G, Lee A, et al. (2006). Inhibition of RhoA GTPase activity enhances hematopoietic stem and progenitor cell proliferation and engraftment *Blood*, Vol. 108, No. 6, pp. 2087-94, ISSN 0006-4971.

Glass TJ, Lund TC, et al. (2011). Stromal cell-derived factor-1 and hematopoietic cell homing in an adult zebrafish model of hematopoietic cell transplantation *Blood*, Vol. 118, No. 3, pp. 766-74, ISSN 1528-0020.

Golub T and Caroni P. (2005). PI(4,5)P2-dependent microdomain assemblies capture microtubules to promote and control leading edge motility *Journal of Cell Biology*, Vol. 169, No. 1, pp. 151-65, ISSN 0021-9525.

Gottig S, Mobest D, et al. (2006). Role of the monomeric GTPase Rho in hematopoietic progenitor cell migration and transplantation *European Journal of Immunology*, Vol. 36, No. 1, pp. 180-9, ISSN 0014-2980.

Gu Y, Jasti AC, et al. (2005). RhoH, a hematopoietic-specific Rho GTPase, regulates proliferation, survival, migration, and engraftment of hematopoietic progenitor cells *Blood*, Vol. 105, No. 4, pp. 1467-75, ISSN 0006-4971.

Hattori K, Heissig B, et al. (2001). Plasma elevation of stromal cell-derived factor-1 induces mobilization of mature and immature hematopoietic progenitor and stem cells *Blood*, Vol. 97, No. 11, pp. 3354-60, ISSN 0006-4971.

Hawkins PT, Anderson KE, et al. (2006). Signalling through Class I PI3Ks in mammalian cells *Biochemical Society Transactions*, Vol. 34, No. Pt 5, pp. 647-62, ISSN 0300-5127.

Heesen M, Berman MA, et al. (1997). Alternate splicing of mouse fusin/CXC chemokine receptor-4: stromal cell-derived factor-1alpha is a ligand for both CXC chemokine receptor-4 isoforms *Journal of Immunology*, Vol. 158, No. 8, pp. 3561-4, ISSN 0022-1767.

Heissig B, Hattori K, et al. (2002). Recruitment of stem and progenitor cells from the bone marrow niche requires MMP-9 mediated release of kit-ligand *Cell*, Vol. 109, No. 5, pp. 625-37, ISSN 0092-8674.

Henschler R, Piiper A, et al. (2003). SDF-1alpha-induced intracellular calcium transient involves Rho GTPase signalling and is required for migration of hematopoietic progenitor cells *Biochemical Biophysical Research Communnications*, Vol. 311, No. 4, pp. 1067-71, ISSN 0006-291X.

Hidalgo A, Sanz-Rodriguez F, et al. (2001). Chemokine stromal cell-derived factor-1alpha modulates VLA-4 integrin-dependent adhesion to fibronectin and VCAM-1 on bone marrow hematopoietic progenitor cells *Experimental Hematology*, Vol. 29, No. 3, pp. 345-55, ISSN 0301-472X.

Imai K, Kobayashi M, et al. (1999). Selective transendothelial migration of hematopoietic progenitor cells: a role in homing of progenitor cells *Blood*, Vol. 93, No. 1, pp. 149-56, ISSN 0006-4971.

Kahn J, Byk T, et al. (2004). Overexpression of CXCR4 on human CD34+ progenitors increases their proliferation, migration, and NOD/SCID repopulation *Blood*, Vol. 103, No. 8, pp. 2942-9, ISSN 0006-4971.

Katayama Y, Hidalgo A, et al. (2004). Integrin alpha4beta7 and its counterreceptor MAdCAM-1 contribute to hematopoietic progenitor recruitment into bone marrow following transplantation *Blood*, Vol. 104, No. 7, pp. 2020-6, ISSN 0006-4971.

Kharas MG, Okabe R, et al. (2010). Constitutively active AKT depletes hematopoietic stem cells and induces leukemia in mice *Blood*, Vol. 115, No. 7, pp. 1406-15, ISSN 1528-0020.

Kim CH and Broxmeyer HE. (1998). In vitro behavior of hematopoietic progenitor cells under the influence of chemoattractants: stromal cell-derived factor-1, steel factor, and the bone marrow environment *Blood*, Vol. 91, No. 1, pp. 100-10, ISSN 0006-4971.

Kim CH, Wu W, et al. (2011). Conditioning for hematopoietic transplantation activates the complement cascade and induces a proteolytic environment in bone marrow: a novel role for bioactive lipids and soluble C5b-C9 as homing factors *Leukemia*, ISSN 1476-5551.

Kollet O, Dar A, et al. (2006). Osteoclasts degrade endosteal components and promote mobilization of hematopoietic progenitor cells *Nature Medicine*, Vol. 12, No. 6, pp. 657-64, ISSN 1078-8956.

Kollet O, Petit I, et al. (2002). Human CD34(+)CXCR4(-) sorted cells harbor intracellular CXCR4, which can be functionally expressed and provide NOD/SCID repopulation *Blood*, Vol. 100, No. 8, pp. 2778-86, ISSN 0006-4971.

Kollet O, Spiegel A, et al. (2001). Rapid and efficient homing of human CD34(+)CD38(-/low)CXCR4(+) stem and progenitor cells to the bone marrow and spleen of NOD/SCID and NOD/SCID/B2m(null) mice *Blood*, Vol. 97, No. 10, pp. 3283-91, ISSN 0006-4971.

Lataillade JJ, Clay D, et al. (2000). Chemokine SDF-1 enhances circulating CD34(+) cell proliferation in synergy with cytokines: possible role in progenitor survival *Blood*, Vol. 95, No. 3, pp. 756-68, ISSN 0006-4971.

Leung KT, Chan KY, et al. (2011). The tetraspanin CD9 regulates migration, adhesion, and homing of human cord blood CD34+ hematopoietic stem and progenitor cells *Blood*, Vol. 117, No. 6, pp. 1840-50, ISSN 1528-0020.

Levesque JP, Hendy J, et al. (2002). Mobilization by either cyclophosphamide or granulocyte colony-stimulating factor transforms the bone marrow into a highly proteolytic environment *Experimental Hematology*, Vol. 30, No. 5, pp. 440-9, ISSN 0301-472X.

Liesveld JL, Rosell K, et al. (2001). Response of human CD34+ cells to CXC, CC, and CX3C chemokines: implications for cell migration and activation *Journal of Hematotherapy and Stem Cell Research*, Vol. 10, No. 5, pp. 643-55, ISSN 1525-8165.

Lo Celso C, Fleming HE, et al. (2009). Live-animal tracking of individual haematopoietic stem/progenitor cells in their niche *Nature*, Vol. 457, No. 7225, pp. 92-6, ISSN 1476-4687.

Loetscher M, Geiser T, et al. (1994). Cloning of a human seven-transmembrane domain receptor, LESTR, that is highly expressed in leukocytes *Journal of Biological Chemistry*, Vol. 269, No. 1, pp. 232-7, ISSN 0021-9258.

Ma Q, Jones D, et al. (1998). Impaired B-lymphopoiesis, myelopoiesis, and derailed cerebellar neuron migration in CXCR4- and SDF-1-deficient mice *Procedures National Academy Sciences U S A*, Vol. 95, No. 16, pp. 9448-53, ISSN 0027-8424.

Maehama T and Dixon JE. (1998). The tumor suppressor, PTEN/MMAC1, dephosphorylates the lipid second messenger, phosphatidylinositol 3,4,5-trisphosphate *Journal of Biological Chemistry*, Vol. 273, No. 22, pp. 13375-8, ISSN 0021-9258.

Maekawa T and Ishii T. (2000). Chemokine/receptor dynamics in the regulation of hematopoiesis *Internal Medicine*, Vol. 39, No. 2, pp. 90-100, ISSN 0918-2918.

Marquez-Curtis L, Jalili A, et al. (2008). Carboxypeptidase M expressed by human bone marrow cells cleaves the C-terminal lysine of stromal cell-derived factor-1alpha: another player in hematopoietic stem/progenitor cell mobilization? *Stem Cells*, Vol. 26, No. 5, pp. 1211-20, ISSN 1549-4918.

Mazo IB, Gutierrez-Ramos JC, et al. (1998). Hematopoietic progenitor cell rolling in bone marrow microvessels: parallel contributions by endothelial selectins and vascular cell adhesion molecule 1 *Journal of Experimental Medicine*, Vol. 188, No. 3, pp. 465-74, ISSN 0022-1007.

McQuibban GA, Butler GS, et al. (2001). Matrix metalloproteinase activity inactivates the CXC chemokine stromal cell-derived factor-1 *Journal of Biological Chemistry*, Vol. 276, No. 47, pp. 43503-8, ISSN 0021-9258.

Merzaban JS, Burdick MM, et al. (2011). Analysis of glycoprotein E-selectin ligands on human and mouse marrow cells enriched for hematopoietic stem/progenitor cells *Blood*, ISSN 1528-0020.

Migliaccio AR, Martelli F, et al. (2008). Altered SDF-1/CXCR4 axis in patients with primary myelofibrosis and in the Gata1 low mouse model of the disease *Experimental Hematology*, Vol. 36, No. 2, pp. 158-71, ISSN 0301-472X.

Mohle R, Bautz F, et al. (1998). The chemokine receptor CXCR-4 is expressed on CD34+ hematopoietic progenitors and leukemic cells and mediates transendothelial migration induced by stromal cell-derived factor-1 *Blood*, Vol. 91, No. 12, pp. 4523-30, ISSN 0006-4971.

Nagasawa T, Hirota S, et al. (1996). Defects of B-cell lymphopoiesis and bone-marrow myelopoiesis in mice lacking the CXC chemokine PBSF/SDF-1 *Nature*, Vol. 382, No. 6592, pp. 635-8, ISSN 0028-0836.

Naiyer AJ, Jo DY, et al. (1999). Stromal derived factor-1-induced chemokinesis of cord blood CD34(+) cells (long-term culture-initiating cells) through endothelial cells is mediated by E-selectin *Blood*, Vol. 94, No. 12, pp. 4011-9, ISSN 0006-4971.

Oberlin E, Amara A, et al. (1996). The CXC chemokine SDF-1 is the ligand for LESTR/fusin and prevents infection by T-cell-line-adapted HIV-1 *Nature*, Vol. 382, No. 6594, pp. 833-5, ISSN 0028-0836.

Papayannopoulou T, Priestley GV, et al. (2001). Molecular pathways in bone marrow homing: dominant role of alpha(4)beta(1) over beta(2)-integrins and selectins *Blood*, Vol. 98, No. 8, pp. 2403-11, ISSN 0006-4971.

Peled A, Grabovsky V, et al. (1999a). The chemokine SDF-1 stimulates integrin-mediated arrest of CD34(+) cells on vascular endothelium under shear flow *Journal of Clinical Investigation*, Vol. 104, No. 9, pp. 1199-211, ISSN 0021-9738.

Peled A, Kollet O, et al. (2000). The chemokine SDF-1 activates the integrins LFA-1, VLA-4, and VLA-5 on immature human CD34(+) cells: role in transendothelial/stromal migration and engraftment of NOD/SCID mice *Blood*, Vol. 95, No. 11, pp. 3289-96, ISSN 0006-4971.

Peled A, Petit I, et al. (1999b). Dependence of human stem cell engraftment and repopulation of NOD/SCID mice on CXCR4 *Science*,Vol. 283,No. 5403,pp. 845-8, ISSN 0036-8075.

Petit I, Goichberg P, et al. (2005). Atypical PKC-zeta regulates SDF-1-mediated migration and development of human CD34+ progenitor cells *Journal of Clinical Investigation*, Vol. 115, No. 1, pp. 168-76, ISSN 0021-9738.

Petit I, Szyper-Kravitz M, et al. (2002). G-CSF induces stem cell mobilization by decreasing bone marrow SDF-1 and up-regulating CXCR4 *Nature Immunology*, Vol. 3, No. 7, pp. 687-94, ISSN 1529-2908.

Ponomaryov T, Peled A, et al. (2000). Induction of the chemokine stromal-derived factor-1 following DNA damage improves human stem cell function *Journal of Clinical Investigation*, Vol. 106, No. 11, pp. 1331-9, ISSN 0021-9738.

Qian H, Georges-Labouesse E, et al. (2007). Distinct roles of integrins alpha6 and alpha4 in homing of fetal liver hematopoietic stem and progenitor cells *Blood*, Vol. 110, No. 7, pp. 2399-407, ISSN 0006-4971.

Qian H, Tryggvason K, et al. (2006). Contribution of alpha6 integrins to hematopoietic stem and progenitor cell homing to bone marrow and collaboration with alpha4 integrins *Blood*, Vol. 107, No. 9, pp. 3503-10, ISSN 0006-4971.

Roberts MS, Woods AJ, et al. (2004). Protein kinase B/Akt acts via glycogen synthase kinase 3 to regulate recycling of alpha v beta 3 and alpha 5 beta 1 integrins *Molecular Cell Biology*, Vol. 24, No. 4, pp. 1505-15, ISSN 0270-7306.

Sahai E and Marshall CJ. (2002). RHO-GTPases and cancer *Nature Reviews Cancer*, Vol. 2, No. 2, pp. 133-42, ISSN 1474-175X.

Sanchez-Aguilera A, Lee YJ, et al. (2011). Guanine nucleotide exchange factor Vav1 regulates perivascular homing and bone marrow retention of hematopoietic stem and progenitor cells *Procedures Nationall Academy Sciences U S A*, Vol. 108, No. 23, pp. 9607-12, ISSN 1091-6490.

Sauer G, Windisch J, et al. (2003). Progression of cervical carcinomas is associated with down-regulation of CD9 but strong local re-expression at sites of transendothelial invasion *Clinical Cancer Research*, Vol. 9, No. 17, pp. 6426-31, ISSN 1078-0432.

Scott LM, Priestley GV, et al. (2003). Deletion of alpha4 integrins from adult hematopoietic cells reveals roles in homeostasis, regeneration, and homing *Molecular Cell Biology*, Vol. 23, No. 24, pp. 9349-60, ISSN 0270-7306.

Skidgel RA and Erdos EG. (1998). Cellular carboxypeptidases *Immunology Reviews*, Vol. 161, pp. 129-41, ISSN 0105-2896.

Sweeney EA, Lortat-Jacob H, et al. (2002). Sulfated polysaccharides increase plasma levels of SDF-1 in monkeys and mice: involvement in mobilization of stem/progenitor cells *Blood*, Vol. 99, No. 1, pp. 44-51, ISSN 0006-4971.

Sweeney EA, Priestley GV, et al. (2000). Mobilization of stem/progenitor cells by sulfated polysaccharides does not require selectin presence *Procedures National Academy Sciences U S A*, Vol. 97, No. 12, pp. 6544-9, ISSN 0027-8424.

Tarnowski M, Liu R, et al. (2010). CXCR7: a new SDF-1-binding receptor in contrast to normal CD34(+) progenitors is functional and is expressed at higher level in human malignant hematopoietic cells *European Journal of Haematology*, Vol. 85, No. 6, pp. 472-83, ISSN 1600-0609.

Tashiro K, Tada H, et al. (1993). Signal sequence trap: a cloning strategy for secreted proteins and type I membrane proteins *Science*, Vol. 261, No. 5121, pp. 600-3, ISSN 0036-8075.

Taylor ML and Metcalfe DD. (2000). Kit signal transduction *Hematology/Oncology Clinics North America*, Vol. 14, No. 3, pp. 517-35, ISSN 0889-8588.

Timokhina I, Kissel H, et al. (1998). Kit signaling through PI 3-kinase and Src kinase pathways: an essential role for Rac1 and JNK activation in mast cell proliferation *EMBO Journal*, Vol. 17, No. 21, pp. 6250-62, ISSN 0261-4189.

Vermeulen M, Le Pesteur F, et al. (1998). Role of adhesion molecules in the homing and mobilization of murine hematopoietic stem and progenitor cells *Blood*, Vol. 92, No. 3, pp. 894-900, ISSN 0006-4971.

Viglietto G, Motti ML, et al. (2002). Cytoplasmic relocalization and inhibition of the cyclin-dependent kinase inhibitor p27(Kip1) by PKB/Akt-mediated phosphorylation in breast cancer *Nature Medicine*, Vol. 8, No. 10, pp. 1136-44, ISSN 1078-8956.

Voermans C, Rood PM, et al. (2000). Adhesion molecules involved in transendothelial migration of human hematopoietic progenitor cells *Stem Cells*, Vol. 18, No. 6, pp. 435-43, ISSN 1066-5099.

Wierenga PK, Weersing E, et al. (2006). Differential role for very late antigen-5 in mobilization and homing of hematopoietic stem cells *Bone Marrow Transplantation*, Vol. 38, No. 12, pp. 789-97, ISSN 0268-3369.

Wright DE, Bowman EP, et al. (2002). Hematopoietic stem cells are uniquely selective in their migratory response to chemokines *Journal of Experimental Medicine*, Vol. 195, No. 9, pp. 1145-54, ISSN 0022-1007.

Wu FY, Wang SE, et al. (2006). Reduction of cytosolic p27(Kip1) inhibits cancer cell motility, survival, and tumorigenicity *Cancer Research*, Vol. 66, No. 4, pp. 2162-72, ISSN 0008-5472.

Wysoczynski M, Reca R, et al. (2005). Incorporation of CXCR4 into membrane lipid rafts primes homing-related responses of hematopoietic stem/progenitor cells to an SDF-1 gradient *Blood*, Vol. 105, No. 1, pp. 40-8, ISSN 0006-4971.

Xia L, McDaniel JM, et al. (2004). Surface fucosylation of human cord blood cells augments binding to P-selectin and E-selectin and enhances engraftment in bone marrow *Blood*, Vol. 104, No. 10, pp. 3091-6, ISSN 0006-4971.

Yang FC, Atkinson SJ, et al. (2001). Rac and Cdc42 GTPases control hematopoietic stem cell shape, adhesion, migration, and mobilization *Procedures National Academy Sciences U S A*, Vol. 98, No. 10, pp. 5614-8, ISSN 0027-8424.

Yiu GK and Toker A. (2006). NFAT induces breast cancer cell invasion by promoting the induction of cyclooxygenase-2 *Journal of Biological Chemistry*, Vol. 281, No. 18, pp. 12210-7, ISSN 0021-9258.

Yoeli-Lerner M, Yiu GK, et al. (2005). Akt blocks breast cancer cell motility and invasion through the transcription factor NFAT *Molecular Cell*, Vol. 20, No. 4, pp. 539-50, ISSN 1097-2765.

Zannettino AC, Buhring HJ, et al. (1998). The sialomucin CD164 (MGC-24v) is an adhesive glycoprotein expressed by human hematopoietic progenitors and bone marrow stromal cells that serves as a potent negative regulator of hematopoiesis *Blood*, Vol. 92, No. 8, pp. 2613-28, ISSN 0006-4971.

Zhang J, Grindley JC, et al. (2006). PTEN maintains haematopoietic stem cells and acts in lineage choice and leukaemia prevention *Nature*, Vol. 441, No. 7092, pp. 518-22, ISSN 1476-4687.

10

Searching for the Key to Expand Hematopoietic Stem Cells

Jeanne Grosselin[1,2], Karine Sii-Felice[1,2],
Philippe Leboulch[1,2,3] and Diana Tronik-Le Roux[1,2]
[1]*CEA, Institute of Emerging Diseases and Innovative Therapies (iMETI), Fontenay-aux-Roses,*
[2]*Inserm U962 and University Paris 11, CEA-iMETI, Fontenay-aux-Roses,*
[3]*Harvard Medical School and Genetics Division, Brigham & Women's Hospital, Boston,*
[1,2]*France*
[3]*USA*

1. Introduction

Stem cells are characterized by their capacity to self renew and differentiate into progressively restricted cells that ultimately become limited to a specific cell fate. The two broad types of mammalian stem cells are: embryonic stem cells and adult stem cells.

Embryonic stem cells (ESC) are mostly derived from the undifferentiated inner mass cells of a blastocyst. These cells give rise during development of the embryo to all derivatives of the three primary germ layers: ectoderm, endoderm and mesoderm. They do not contribute to the extra-embryonic membranes or the placenta. Ex-vivo, they can be cultured for extended periods of time and under the appropriate conditions, they can be also directed to differentiate into many specialized types of cells. These particular features are being exploited to use ESC as starting material for treatment of degenerative diseases and replacement of damaged organs. Although their potential is great, the promise of ESC-derived therapies will be unfulfilled unless several challenges are overcome. For example, the quite small production of ESC-derived cells obtained or the active immune rejection of the ESC-derived graft.

Unlike embryonic stem cells, the adult stem cells are already partially specialized. They have been found in most self-renewing tissues, including the skin, the brain, the intestinal epithelium and the hematopoietic system and have the primary role of maintaining and repairing the tissue in which they are found. They are located deep within organs in specialized areas known as the "stem cell niche" (Scadden, 2006). This microenvironment allows for their survival, self renewal, regulated proliferation and maintenance of their quiescence for long periods of time until the moment in which they are activated. *Ex vivo,* however, the capacity of stem cells to self-renew is limited, they exhibit poor survival and consequently their numbers sharply declines during experimental manipulation.

One of the more intriguing but highly debated areas of stem cell biology was the phenomenon described as plasticity or transdifferentiation. Numerous reports expressed opposing views about this ability of stem cells to cross organ/tissue boundaries. These discrepancies have now been mostly passed over by current research showing that cell populations of one lineage might produce cells from other lineages by changing gene expression in response to micro-environmental cues (Jang and Sharkis, 2005; Theise, 2010).

Owing to their unique characteristic of plasticity, self-renewal capacity and potential to generate functional cell types, stem cells are particularly attractive for developing therapeutic settings that range from drug discovery protocols to cell transplantation and regenerative therapies. Nevertheless, several challenges including the need to identify the signals that influence the stem cell fate decisions and the application of this information towards the design of stem cell bioprocesses have to be overcome to accomplish the transition from fundamental science to functional technologies.

1.1 Hematopoietic stem cells

Hematopoietic stem cells (HSC) are probably the best characterized adult stem cell and often serve as a paradigm for other stem cells. Even though no morphological criteria to unequivocally identify such cells exist, HSC have been proven to be invaluable in the clinic. They are the only stem cells used routinely in cell based therapies, to treat numerous hematologic and non-hematologic malignancies as well as a range of both inherited and acquired diseases. This is typically due i) to the availability of a straight forward purification protocols using cell surface antigen selection and ii) to the possibility to perform reconstitution assays that rely on their clonal ability to reconstitute the entire hematopoietic system following transplantation into myeloablated recipients (Fig. 1). The same cell surface antigens, however, do not always conform to the same stem cell functional phenotype (Simonnet et al., 2009) and therefore the transplantation procedure constitutes undoubtedly the "gold standard" method for proving that a cell is indeed an HSC.

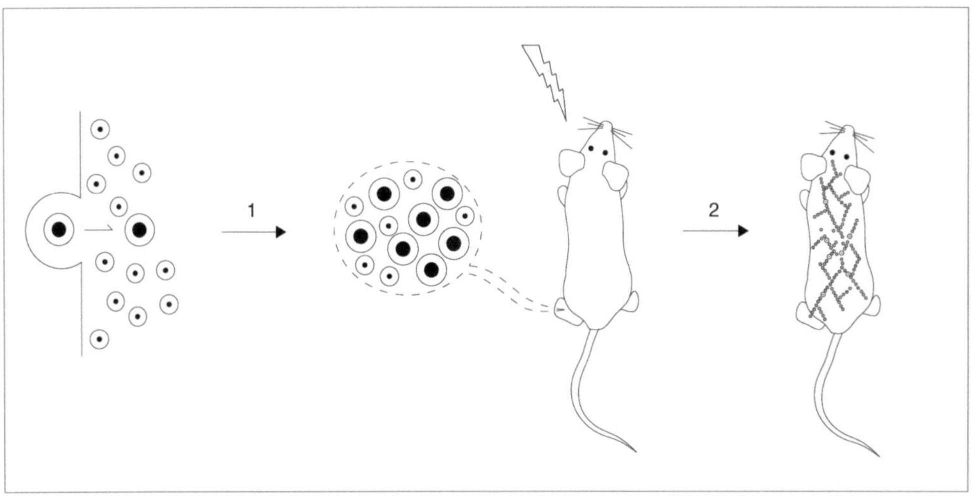

Fig. 1. A diagrammatic representation of a stem cell in its micro-environment and one stem cell induced to move out of the niche where it will undergo development. Following BM removal and cell surface antigen selection (1), cells are cultured *in vitro* and infused in a myeloablated mouse. Several weeks after (2), blood cells are regenerated in the transplanted mouse.

To maintain the steady-state of the stem cell compartment and to allow the regeneration of hematopoietic cells after transplantation or after hematopoietic injury, HSC divide asymmetrically or symmetrically. In an asymmetric self-renewing division, the two daughter cells adopt different fates, resulting in only one cell maintaining stem-cell properties. The symmetric self-renewing division refers to the process whereby both daughter cells retain stem cell properties. This type of cell division expands the stem-cell pool and is therefore critical for sustaining the HSC compartment and thus is a requirement for lifelong hematopoiesis.

The HSC fate decisions are dependent on concomitantly intrinsic HSC fate determinants and extrinsic signals delivered by the bone marrow (BM) niches were HSC resides. These niches are small cavities formed by heterogeneous types of cells, named stroma, that are positioned close to the BM longitudinal axis of the femur with more differentiated cells disposed in a graduated manner as the central longitudinal axis of the bone is approached. The attachment of HSC to the stroma via a network of adhesion molecules provide an environment that optimally balances signals that control self-renewal, proliferation and differentiation. Under normal physiological conditions, HSC are kept in a relatively low proliferative, quiescent state, protecting them from stress and preventing their depletion due to excessive proliferation (Jang and Sharkis, 2007). Recent data imply that these areas where HSC reside are hypoxic (Parmar et al., 2007).

To take advantage of the HSC plasticity capacities for therapeutic use, HSC may be withdrawn from their original niches, and placed on a novel non-hematopoietic environment. Once located in this novel medium, the reprogramming of the cell genome occurs and directs and/or contributes to their conversion into unrelated cell types (Fig. 2). The unexpected flexibility of HSC to produce non-hematopoietic cells was described for several cells/tissues (Quesenberry et al., 2010) including liver cells (Almeida-Porada et al., 2010; Jang et al., 2004), neurones (Mezey et al., 2000), lung epithelial (Abe et al., 2003) or connective tissues (Ogawa et al., 2010).

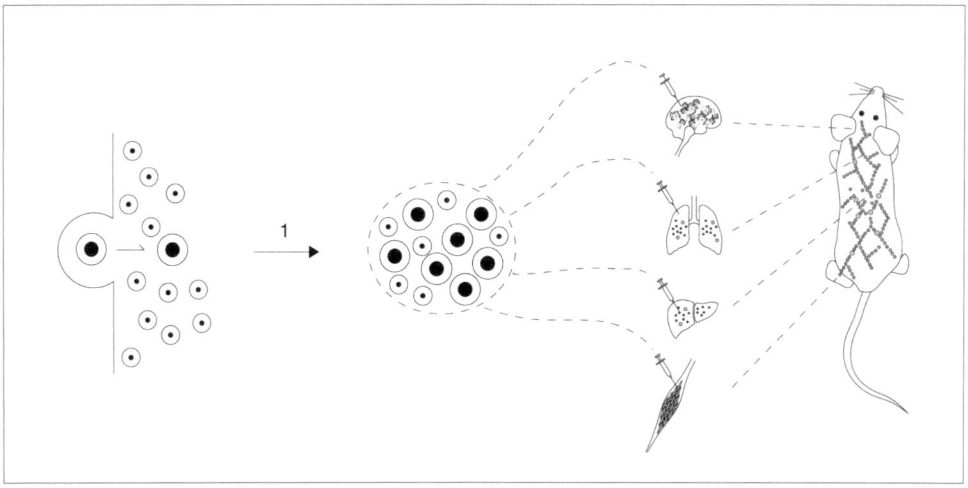

Fig. 2. A schematic representation of HSC plasticity. Hematopoietic cell are removed from the femoral bone (1) and the HSC-enriched population is cultured *in vitro*. Following an optional genetic modification, cells may be used to generate *in vivo* non-hematopoietic cell types.

The development of HSC-based therapies however, is to some extent prevented by the scarce representation of HSC in the BM and their finite lifespan *ex vivo*. Increasing their utilisation needs enhancement of hematopoietic stem cells availability or *de novo* generation of HSC. This presumes i) the development of robust methods to efficiently control HSC regulatory processes; ii) the therapeutic *in vivo* or *in vitro* expansion of HSC number and iii) the utilisation of optimized protocols to generate available HSC from ESC or IPSC.

2. Physiological pathways involved in the regulation of stem cells

HSC fate decisions are supported by the orchestration of several pathways such as Wnt, Notch and Hedgehog pathways that critically balance cell cycling and quiescence, leading to proliferation and apoptosis, self-renewal or differentiation (Fig. 3). The ultimate decision is dependent on hundreds of inputs including concentrations of different growth factors, cytokines, hormones, oxygen levels that must be integrated to subsequently activate these different signal transduction cascades. Understanding their regulation might led to the effective and more spread out utilization of HSC in clinical settings. The most relevant aspects of these pathways are briefly resumed below.

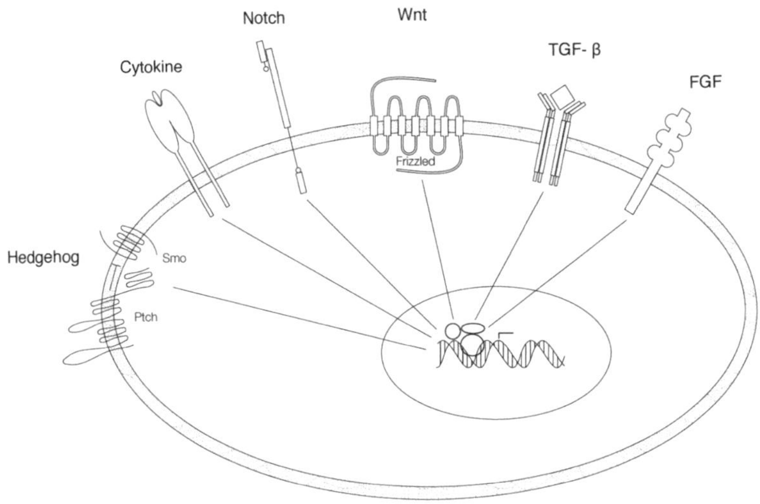

Fig. 3. A schematic representation of signaling pathways collectively influencing stem cell fate.

2.1 The hedgehog (Hh) signaling

In adult tissues, Hh signaling is involved in the maintenance of stem cells, regeneration and tissue repair where it governs processes like cell proliferation, cell renewal and differentiation. The three Hh ligand homologues: Sonic Hh, Indian Hh, and Desert Hh bind interchangeably the two related twelve-pass membrane Patched (Ptc) receptors. They relieve the inhibition of smoothened (SMO), a serpentine receptor resembling G protein–coupled receptors allowing activation of a family of zinc-finger transcription factors called GLI and the modification of the expression Hh target genes (Kasper et al., 2009).

The role of Hh signaling in HSC is controversial. Bhardwaj et al provided evidence for a role of Hh signaling in HSC (Bhardwaj et al., 2001). In this study, suppression of Hh signaling inhibited proliferation of HSC and addition of soluble SHh induced expansion of hematopoietic repopulating cells (Bhardwaj et al., 2001). More recent reports confirmed that suppression of the Hh pathway leads to a severe defect in HSC functions (Merchant et al., 2010; Trowbridge et al., 2006) whereas others reported that this pathway can be dispensable for HSC biology (Gao et al., 2009; Hofmann et al., 2009). In *Ptc1*+/−mice, which have increased Hh activity, activation of the Hh signaling pathway induces expansion of primitive blood cells under homeostatic conditions. However, when HSC are challenged to regenerate the blood system, persistent Hh activation leads to HSC exhaustion (Trowbridge et al., 2006). Furthermore, Indian Hh gene transfer can confer enhanced hematopoietic support ability to BM stromal cells, suggesting that it is involved in the interaction between HSC and the stromal cells. This leads to an increase in proliferation and repopulating capacity of primitive hematopoietic cells (Kobune et al., 2004). These results suggest a role for Hh signaling in balancing homeostasis and regeneration *in vivo*. In contrast, other reports show that Hh signaling is dispensable for adult HSC functions (Gao et al., 2009; Hofmann et al., 2009). In these studies conditional deletion of SMO, the only non redundant component of the Hh cascade, or pharmacologic inhibition of Hh signaling have no apparent effect on adult hematopoietic, including peripheral blood count, number or cell cycle status of stem or progenitor cells, hematopoietic colony-forming potential or long-term repopulating activity in *in vivo* assays. In agreement with this notion, genome-wide transcriptome analysis revealed that silencing the Hh signaling does not significantly alter the HSC-specific gene expression "signature." Taken together, these conflicting data suggest that Hh signaling may influence HSC through more complex networks such as cell-niche interactions.

2.2 Fibroblast growth factor (FGF) signaling

FGF belongs to a family of heparin-binding polypeptides that shows multiple functions, including effects on cell proliferation, differentiation and survival (Baird, 1994). Twenty-four members of the FGF family have been identified in human and mice. FGFs bind and activate their cognate FGFRs that are encoded by four genes (FGFR1- 4). This results in receptor dimerization, tyrosine kinase autophosphorylation, and recruitment of signaling complexes. The FGF signal transduction proceeds by one, or a combination, of three main pathways: Ras/mitogen-activated protein kinase (MAPK) signaling; planar cell polarity/calcium; phosphoinotitide-3-kinase (PI3K)/Akt (extensively reviewed by Bottcher and Niehrs, 2005). Both FGF-1 and FGF-2 support HSC expansion when unfractionated mouse BM cells are cultured in serum-free medium (de Haan et al., 2003; Yeoh et al., 2006). Crcareva et al. confirmed that FGF-1 stimulates ex-vivo expansion of HSC (Crcareva et al., 2005). Conditional derivatives of FGFR-1 have also been used to support short-term HSC expansion and long-term HSC survival (Weinreich et al., 2006). This factor seems to also support *ex vivo* expansion of murine and human HSC in combination with other cytokines, i.e stem cell factor [SCF], thrombopoietin [TPO], insulin-like growth factor-2 [IGF-2], and fibroblast growth factor-1 [FGF-1] (Zhang and Lodish, 2005). Moreover, a recent study showed that addition of SCF, TPO, and FGF-1 to a mesenchymal stem cells (MSC) culture stimulates proliferation, maintenance of primitive immunophenotype, and expansion of CFU-initiating cells. This supports the notion that expansion of HSC requires complex stimulation of different signal cascades activated by soluble growth factors as well as adhesion proteins (Walenda et al., 2011).

2.3 Notch signaling

The Notch pathway is also an evolutionarily conserved mechanism that plays a fundamental role in regulating cell-fate decisions (Bolos et al., 2007). Four types of Notch receptors (Notch 1-4) and five Notch ligands (Jagged 1 and 2, Delta 1, 3 and 4) have been identified in vertebrates. Notch ligands are single-pass transmembrane proteins consisting of multiple EGF-like repeats and a characteristic DSL (Delta, Serrate, and LAG-2) domain (see for review Ohishi et al., 2003; Shimizu et al., 2000). One characteristic of this signaling pathway is the dual role of Notch as both a transmembrane receptor and a transcription factor in a system where no second messengers are used (Matsuno et al., 1995). Notch can have opposite functions in different self-renewing organs indicating that the outcome of Notch activation depends to a great extent on the cell context and the specific growth factors present in the microenvironment. For example, activation of Notch1 by Delta ligands 1 and 4 is required for inducing T-cell and inhibiting B-cell differentiation whereas Notch2 activation by Jagged1, and possibly Delta1, acts on HSC (Han et al., 2002; Radtke et al., 1999; Varnum-Finney et al., 2011).

A role for Notch in hematopoietic was initially suggested by detection of the human Notch1 gene in $CD34^+$ or lineage $(Lin)^-CD34^+$ hematopoietic cells (Milner et al., 1994). Transduction of murine HSC with a retrovirus expressing a constitutively active form of Notch1 induced the emergence of an immortalized pluripotent cytokine-dependent cell line capable of both myeloid and lymphoid repopulation *in vivo*, thereby demonstrating a role for Notch in HSC self-renewal (Varnum-Finney et al., 2000). Similar results were obtained using an immobilized form of the Notch ligand Delta-1 since incubation of murine HSC with immobilized Delta-1 and cytokines led to a several-log expansion of cells capable of short-term *in vivo* reconstitution (Varnum-Finney et al., 2003).

In contrast to the murine studies, only a modest or no increase in the progenitor numbers was achieved by expressing activated Notch-1 in human $CD34^+$ cord blood cells (Carlesso et al., 1999; Chadwick et al., 2007) or by incubation with Delta-1 (Jaleco et al., 2001), Delta-4 (Lauret et al., 2004) or Jagged-1 (Karanu et al., 2000; Karanu et al., 2001; Walker et al., 1999). This contrast with other reports showing that incubation of human cord blood cells with the immobilized Delta-1 combined with fibronectin fragments and cytokines induce a 100-fold increase in the number of $CD34^+$ cells compare to controls (Ohishi et al., 2002) and a 16-fold increase in SCID Repopulating Cells (SRC) number compared to uncultured cells. *In vivo* transplanted cells persisted 9 weeks post-transplantation and in secondary recipients, suggesting the presence of both long-term and short-term repopulating cells following culture of human cord blood cells on Delta-1 ligand (Delaney et al., 2010). The SRC enhancement by relatively low density of immobilized ligand and the preference to promote differentiation toward the T-cell lineage at higher ligand density revealed important ligand dose-dependent effects of Notch signaling (Delaney et al., 2005).

The engineered Notch ligand approach for *ex vivo* expansion of human cord blood cells is now under clinical investigation (http://clinicaltrials.gov/ct2/show/record/NCT00343798). In this phase 1 clinical trial, patients undergoing a myeloablative double cord blood transplantation are receiving one non-manipulated cord blood unit along with a second cord blood unit that has undergone Notch-mediated *ex vivo* expansion. These cells were safely infused and led to a significant reduction in the time needed for neutrophil recovery (16 days in patients receiving the expanded unit, compared to 26 days in patients of the concurrent cohort). Similarly, preliminary evaluation of time needed for platelet recovery compared favourably in those patient receiving the expanded cell product compared with those receiving non-manipulated cells (Dahlberg et al., 2011). In addition, comparable overall survival and graft-versus-host disease risk of patient receiving non-manipulated cells was observed within the average follow-up of 354 days. The expanded cell population may also have retained long-term repopulating capacities as two patients display *in vivo* persistence of cultured donor cells. The lack of *in vivo* persistence in the remaining patients may either be due to loss of stem cell self-renewal capacity during *ex vivo* culture or to immune mediated rejection. Indeed, it has been well documented that in most of the patients who received two non-manipulated cord blood units for transplantation, only one contributes to persistent long-term engraftment. The mechanism responsible for this single donor dominance remains yet to be defined. Larger phase II/III studies are required to evaluate whether co-infusion of this expanded cell product decreases the occurrence of serious infection, improves survival, or affects duration of hospital stay (Delaney et al., 2010).

2.4 The transforming growth factor beta (TGFβ) superfamily

The TGFβ superfamily consist of a large collection of secreted proteins that regulate cell growth, differentiation, apoptosis, cellular homeostasis, and other functions in both the adult organism and the developing embryo. The more than 30 TGFβ family ligands are organized into three subgroups (reviewed in (Lyssiotis et al., 2011)). The TGFβ (which comprises SMAD and Activin/Nodal ligands), bone morphogenetic protein (BMP), and the growth differentiation factors (GDF). The TGFβ signaling leads to the phosphorylation of Smads by activated receptors resulting in their partnering with the common signaling transducer Smad4, and translocation to the nucleus. Once activated, Smads regulate diverse biological effects by partnering with transcription factors resulting in cell-state specific modulation of transcription (Kaivo-Oja et al., 2003).

A significant number of studies have demonstrated that TGFβ inhibits proliferation of both murine and human HSC *in vitro*. It was suggest that TGFβ induces quiescence in HSC since its neutralization was showed to release early hematopoietic progenitors cells from quiescence (Hatzfeld et al., 1991; Yamazaki et al., 2009). In agreement with studies performed *in vitro*, injection of TGFβ1 into the femoral artery of mice effectively inhibits proliferation of multipotent hematopoietic progenitors in the BM, establishing an inhibitory role of TGFβ1 also *in vivo* (Goey et al., 1989). Despite a key role *in vitro*, TGFβ did not seem to provide the necessary signals that maintain quiescence and the stem cell pool *in vivo* (Larsson et al., 2005).

To block the entire Smad signaling pathway, the Smad7 was overexpressed in murine HSC using a retroviral gene transfer approach. Forced expression of Smad7 significantly increased the self-renewal capacity of HSC *in vivo* (Blank et al., 2006). In a similar approach using human hematopoietic cells, overexpression of Smad7 resulted in a shift from lymphoid-dominant engraftment toward the myeloid lineage, and an increase of the myeloid-committed clonogenic progenitor frequency in NOD-SCID mice (Chadwick et al., 2005). Instead, Smad4-deficient HSC displayed a significantly reduced repopulative capacity of primary and secondary recipients (Karlsson et al., 2007). Because overexpression of Smad7 versus deletion of Smad4 would be anticipated to yield similar hematopoietic phenotypes, it is conceivable that Smad4 functions as a positive regulator of self-renewal independently of its role as a central mediator of the canonical Smad pathway. In the context of adult hematopoiesis, a high concentration of BMP-4 was shown to promote maintenance of human cord blood cells *in vitro*, while lower concentration of BMP4, BMP2 and BMP7 induced proliferation and differentiation of HSC (Bhatia et al., 1999).

2.5 Wingless-type (Wnt) pathway

Wnt proteins are secreted morphogens necessaries for basic developmental processes, such as cell-fate specification, progenitor-cell proliferation and the control of asymmetric cell division, in many different species and organs (Bejsovec, 2005; Moon et al., 2004). Wnt proteins bind to cell surface receptors of the Frizzled family which can translocate the signals to the nucleus and function as transcriptional activators through intracellular β-catenin. Different Wnt pathways are known but their clear separation and their independence remain controversial. There is one canonical pathway that acts on the stability of β-catenin and interacts with T cell transcription factors in the nucleus. There are many non-canonical pathways like the PCP and Wnt/Calcium pathways. The most distinctive differences between the canonical and non-canonical pathways include the specific ligands activating each pathway, ß-Catenin, LRP5/6 co-receptor, and Dsh-DEP domain independence, respectively, and the ability of the non-canonical pathways to inhibit the canonical pathway. Ligands that activate the non-canonical pathways are Wnt4, Wnt5a, and Wnt11.

Recent evidence based on genetic models suggests that canonical Wnt signaling, regulates HSC self-renewal. Active β-catenin promotes HSC proliferation and inhibits differentiation (Kirstetter et al., 2006; Scheller et al., 2006) whereas deficiency in β-catenin inhibits HSC self-renewal (Cobas et al., 2004; Luis et al., 2009; Zhao et al., 2007). Moreover, purified Wnt3a treatment of adult HSC increases self-renewal of murine HSC, as determined by *in vivo* reconstituting assays (Willert et al., 2003) and of human Lin^-CD34^+ cells as measured by immunophenotype and colony assays (Van Den Berg et al., 1998).

The role of the non-canonical pathways is not well defined, but surprisingly, their activation and consequently inhibition of the canonical pathway, appears also to be able to expand HSC. Murdoch et al. demonstrated that injecting mice with Wnt5a conditioned media prior to transplant of human umbilical cord blood cells increased engraftment more than 3-fold (Murdoch et al., 2003). Furthermore, culturing $Lin^-Sca\text{-}1^+c\text{-}Kit^+$ (LSK) cells with recombinant murine Wnt5a resulted in an enhancement of hematopoietic reconstitution in a BM transplant assay. Wnt5a seems to activate the non-canonical signaling pathways leading to a 3.5- fold more HSC in G0 phase (Nemeth et al., 2007).

Overexpression of Wnt4 led to a modest increase in HSC frequency as measured by phenotype and limiting dilution transplant assays and Wnt4-/- mice showed decreased frequencies of HSC in BM. Similar to the results obtained using Wnt5a, overexpression of Wnt4 led to an increase in the percentage of HSC in G0 (Louis et al., 2008). Whether Wnt4 and Wnt5a inhibit the canonical pathway in a similar fashion remains to be elucidated. These results show the importance of a balanced regulation of these two overlapping Wnt signaling pathways.

2.6 Cross-talk between these pathways

The individual contribution of these pathways to the hematopoietic development of HSC have been extensively addressed (Cerdan and Bhatia, 2010). However, there are many potential intersections along them and therefore the impact of their collective contribution towards influencing the fate of HSC should be carefully considered. Some of these intersection points are resumed below.

Ducan et al. provide a model for how HSC may integrate multiple signals to maintain the stem cell state. They showed that although the proliferation and survival of HSC exposed to Wnt proteins seem unaffected when Notch signaling is impaired, their ability to remain undifferentiated is substantially altered (Duncan et al., 2005). These results demonstrated that the Notch pathway is imperative in maintaining HSC in an undifferentiated state. These findings do not preclude the possibility that a stronger Wnt signal, such as activated β-catenin, may be able to overcome the consequences of loss of Notch signaling. Moreover, Wnt3a regulates the expression of established Notch target genes (Duncan et al., 2005) and the inhibition of GSK-3, a downstream target of Wnt signaling that affects HSC fate through mechanisms involving both Wnt and Notch target genes (Trowbridge et al., 2006). These findings suggest that these pathways could play a role in HSC self renewal using a common network of regulatory circuits with Wnt enhancing proliferation and survival, and Notch preventing differentiation (Blank et al., 2008).

Furthermore, there is substantial evidence for the cross-talk between the Wnt signaling pathway and FGFs and TGF-b by means of the association between Smad4 and Hox proteins. *Homeobox* (*hox*) genes encode transcription factors that function as regulators of hematopoiesis and are frequently dysregulated in human leukemia, particularly acute myeloid leukemia (Kroon et al., 1998). Recently, Wang et al described a mechanism whereby TGF-β/BMP inhibited the BM transformation capacity of HoxA9 and HoxA9-Nup98 fusion protein through a Smad4-dependent mechanism. Accordingly, Smad4 was shown to interact directly with HoxA9 and Nup98-HoxA9 fusion protein, thus precluding their DNA binding capacity and subsequent transcriptional activity (Wang et al., 2006). Smad4 also seems to participate in other signaling cascades such as Wnt or Notch (Itoh et al., 2004; Labbe et al., 2000).

These studies show the high interdependence between the different pathways, and the impact of their collective contribution on HSC self-renewal. This should be carefully considered when trying to expand HSC for clinical purposes.

2.7 Epigenetic control and HSC self-renewal

Epigenetic modifications, in addition to the intracellular pathways described in the previous section also play an essential role in regulating self-renewal, differentiation and tissue development. They induce gene expression regulation and can be grouped into three main categories: i) DNA methylation, ii) Histone modifications and iii) Nucleosome positioning. Recent studies suggest that epigenetic mechanisms contribute to establish the HSC unique characteristics. The following is a description of some of these examples.

2.7.1 Methylation of DNA

The most widely studied epigenetic modification in humans is cytosine methylation. DNA methylation occurs almost exclusively in the context of CpG dinucleotides that tend to cluster in regions called CpG islands. A group of enzymes, the DNA methyltransferases (DNMTs) tightly regulate both the initiation and maintenance of these methyl marks. DNA methylation can inhibit gene expression by various mechanisms. Methylated DNA can promote the recruitment of methyl-CpG-binding domain proteins which in turn recruit histone-modifying and chromatin-remodeling complexes to methylated sites. DNA methylation can also directly inhibit transcription by precluding the recruitment of DNA binding proteins from their target sites. In contrast, unmethylated CpG islands generate a chromatin structure favorable for gene expression (Portela and Esteller, 2010).

Methylation is controlled by at least 3 DNMTs: DNMT3a and DNMT3b for *de novo* methylation and DNMT1 for methylation maintenance. Conditionally disruption of *Dnmt3a*, *Dnmt3b*, or both *Dnmt3a* and *Dnmt3b* (*Dnmt3a/Dnmt3b*) showed that Dnmt3a and Dnmt3b function as *de novo* DNA methyltransferases during differentiation of hematopoietic cells. Unexpectedly, *in vitro* colony assays showed that both myeloid and lymphoid lineage differentiation potentials were maintained in Dnmt3a-, Dnmt3b-, and Dnmt3a/Dnmt3b-deficient HSC. However, Dnmt3a/Dnmt3b-deficient HSC, but not Dnmt3a- or Dnmt3b-deficient HSC, were incapable of long-term reconstitution in transplantation assays, suggesting a role for DNA methylation by Dnmt3a and Dnmt3b in HSC self-renewal (Tadokoro et al., 2007).

Conditional disruption of Dnmt1 in the mouse hematopoietic system revealed defects in self-renewal, niche retention, and in the ability of cells to give rise to multilineage hematopoiesis. Loss of Dnmt1 had specific impact on myeloid progenitor cells, causing enhanced cell cycling and inappropriate expression of mature lineage genes (Trowbridge et al., 2009). Consistent with these results, Broske et *al.* showed that Dnmt1 is essential for HSC self-renewal but dispensable for homing, cell cycle control and suppression of apoptosis but also implicated Dnmt1 in lymphoid differentiation (Broske et al., 2009).

2.7.2 Histone modifications and nucleosome positioning

A nucleosome is a histone octamer composed by a histone H3-H4 tetramer and two H2A-H2B dimers, around which DNA, 147 base pairs in length, is wrapped in 1.75 superhelical turns. Nucleosomes are connected by the so-called linker DNA and the histone H1. Histones post-transcriptional modifications, including acetylation, methylation, phosphorylation, ubiquitination, SUMOylation and ADP-ribosylation, occur predominantly in histone tails. They have important roles in transcriptional regulation as they can provide either an ON or OFF signature which result in the tight regulation of gene expression but display also important roles in DNA repair, DNA replication, alternative splicing and chromosome condensation. Nucleosomes act as barriers to transcription. They block access of activators and transcription factors to their sites on DNA and inhibit the elongation of the transcripts. The packaging of DNA into nucleosomes appears to affect all stages of transcription, thereby regulating gene expression. Nucleosome positioning plays also an important role in shaping the methylation landscape (Portela and Esteller, 2010).

Polycomb group (PcG) and Trithorax group (TrxG) proteins have emerged as key players in gene regulation and are thought to function coordinately to orchestrate DNA accessibility. These epigenetic regulators act antagonistically to either promote (TrxG) or repress (PcG) transcription through regulation of specific amino acid modifications in histones. It is not known how the PcG and TrxG proteins switch and balance between transcriptionally silenced heterochromatin (for example, enriched in histone H3 lysine 27 trimethylation, H3K27me3) and transcriptionally competent euchromatin (for example, enriched in histone H3 lysine 4 trimethylation, H3K4me3), respectively, during development.

In vertebrates, polycomb group proteins participate mainly in two complexes, Polycomb Repressive Complex (PRC) 1 and PRC2. Probably the best example of a chromatin-associated factor involved in self-renewal is BMI1, which is a component of PRC1. BMI1 is expressed in HSC and its expression decreases upon differentiation towards myeloid or erythroid cells, but is retained within the lymphoid compartments. Upon deletion of BMI1, no changes in the number of HSC in the fetal liver were observed, but in postnatal BMI1$^{-/-}$ mice, the number of HSC was markedly reduced. Targeted deletion of BMI1 in murine HSC impaired their competitive repopulation capacity (Park et al., 2003). *In vitro*, BMI1$^{-/-}$ HSC proliferated poorly and displayed an accelerated loss of multilineage differentiation potential and overexpression of BMI1 enhanced the self-renewal of HSC and enhanced their engraftment potential (Iwama et al., 2004).

Overexpression of BMI1 in cord blood CD34^{+} cells resulted in stem cell maintenance. After an *in vitro* culture period of 10 days, BMI1-overexpressing cells display a much better engraftment in NOD-SCID mice. Although the mechanisms involved remain to be elucidated, it was observed in single-cell assays that the percentage of CD34^{+}/CD38^{-} HSC undergoing apoptosis was reduced, whereas the percentage of quiescent HSC not undergoing cell cycle progression was increased upon BMI1 overexpression (Rizo et al., 2008). Lentiviral downmodulation of BMI1 in human cord blood CD34^{+} cells impaired long-term expansion, progenitor-forming capacity and stem cell frequencies, both in cytokine-driven liquid cultures and in BM stromal cocultures. This was associated with higher expression of p14ARF and p16INK4A and enhanced apoptosis, which coincided with increased levels of intracellular reactive oxygen species (ROS) and reduced FOXO3A expression (Rizo et al., 2008).

Another example of a chromatin-associated factor involved in self-renewal is the mixed lineage leukemia (MLL) protein, which encodes a trithorax-group chromatin regulator. Using *Mll*-deficient ESC to generate chimeras, Ernst et al. showed a cell-intrinsic requirement for MLL in the generation of lymphoid and myeloid populations in adult animals (Ernst et al., 2004). Moreover, MLL is often fused to the AF9 protein in leukemia and have been reported to impart leukaemia stem cell properties on committed hematopoietic progenitors. The leukemia stem cells generated can maintain the global identity of the progenitor from which they arose while activating a limited stem-cell- or self-renewal-associated programme (Krivtsov et al., 2006). Moreover, this MLL-AF9 fusion drives high-level expression of multiple *Hox* genes and can overcome Bmi1-deficiency to establish leukemic stem cells (Smith et al., 2011).

The studies described in this section establish that epigenetic alterations can modulate the self-renewal process. Epigenetic state in stem cells can be stably heritable or can be erased (partly or completely) by cell division. These changes might facilitate the transition of a progenitor cell to a self-renewing stem cell, or might prompt a stem cell to differentiate, divide or lose its ability to self-renew.

3. Compounds modifying HSC capacities

As described in the previous section, the strategy for stem-cell expansion involves activation of regulators that encourage HSC self-renewal and/or inhibition of pathways that mediate, differentiation or apoptosis by using primarily genetic modification approaches. An alternative strategy might imply pharmacological intervention by using a variety of small molecules. The term "small molecule" refers to a molecular entity that interacts with one or more molecular targets and effects a change in biological state while having minimal side effects. These small molecules, defined by a known structure, may be chemicals, proteins, small interfering RNAs or antibodies. Some of the most effective compounds for *ex vivo* maintaining or expanding HSC are reviewed below.

3.1 Regulation by cytokines

Cytokines are secreted proteins that regulate many aspects of hematopoiesis, such as, immune responses and inflammation. Numerous attempts have been made to use classic hematopoietic cytokines for the purpose of expanding HSC *in vitro*. Many interleukins, including interleukin (IL)-3, IL-6, and IL-11, Flt-3 ligand, TPO and SCF have extensively been investigated. In most cases, efforts to expand HSC have failed because of differentiation of HSC and subsequent loss of their reconstitution capacity. The combination of these molecules has however allowed maintaining HSC in culture for several days allowing their use in protocols for gene or cell therapies. Here we describe some examples of cytokines that were used to maintain HSC levels in culture.

3.1.1 Thrombopoietin (TPO)

TPO, acting through its receptor c-MPL, is the chief cytokine that regulates megakaryocyte production. However, several studies suggest that TPO can act to increase the *ex-vivo* expansion of HSC (Sitnicka et al., 1996). This effect was far more effective when used in combination with other cytokines including SCF, fms-like tyrosine kinase 3 ligand (FLT3-L), IL-3 or IL-6. Human cord blood cells expanded with this cytokine cocktail were shown to provide good short- and long-term platelet recovery and lymphomyeloid reconstitution in NOD-SCID mice (Ohmizono et al., 1997; Pineault et al., 2010). Further, a non peptidyl molecule agonist of c-MPL, NR-101, was found to be more efficient than TPO in expanding HSC. Indeed, 7 days culture of human cord blood $CD34^+$ or $CD34^+CD38^-$, treated with NR-101 induced a 2-fold increase in their number compare to TPO and a 2.9-fold or 2.3-fold increase in SRC numbers compared to freshly isolated $CD34^+$ cells or TPO-expanded cells respectively. As it was not more efficient than TPO in inducing megakaryocyte expansion, its effect seemed to be HSC specific. NR-101 treatment appeared to persistently activate STAT5 and to induce a long-term accumulation of HIF-1α (Nishino et al., 2009).

3.1.2 Angiopoietin-like 5 (ANGPLT5) and insulin-like growth factor binding protein 2 (IGFBP2)

Soluble growth factors, such as ANGPLT5 and IGFBP2, produced by the endothelium may enhance HSC expansion *ex vivo* when used with conventional cytokines. Although the addition of ANGPLT5 and/or IGFBP2 to a 10 days-human CD133+ cord blood cells culture has no effect upon the total nucleated cells number *in vitro*, it significantly enhances *in vivo* repopulation of NOD-SCID mice 2 months post-transplantation as well as secondary transplantation (Zhang et al., 2008a). These results were confirmed recently using human cord blood $CD34^+CD133^+$ cells cultured for 10 days in the presence of IGFBP2 and ANGPLT5. Expanded cells were shown to be capable of long-term multi-lineage and multi-site hematopoiesis in serial reconstitution in NSG mice (Drake et al., 2011).

3.1.3 Pleiotropin (Ptn)

Pleiotropin, which have mitogenic and angiogenic activities, has been found to be essential for maintenance of murine HSC. Mice transplanted with LSK $CD34^-$ cells treated with Ptn and a standard cocktail of cytokines showed 6-fold increase in HSC frequency compared to cells treated with cytokines alone. *In vivo,* systemic administration of Ptn was found to increase the number of BM LSK cells both in irradiated and nonirradiated mice, suggesting a role for this factor in the *in vivo* regeneration of HSC. Treatment of human cord blood $Lin^-CD34^+CD38^-$ cells with Ptn for 7 days induced a 4-fold increase in CFC content and a 3- or 7-fold improved engraftment at 4 or 7 weeks respectively in NOD-SCID mice compared with controls. This factor may activate the PI3-Kinase/AKT and Notch pathways by alleviating activation of its receptor, RPTP-β/ξ (Himburg et al., 2010).

3.2 Transcription factors: The HOX- family

3.2.1 HOXB4

The homeobox gene family member HoxB4 is the most investigated transcription factor for its potential to increase the self-renewal potential of HSC. HOXB4 belongs to a large family of transcription factors that share a highly conserved DNA-binding domain, the homeodomain. In mammals, there are 39 *Hox* genes grouped in four clusters referred to as A, B, C and D. In the hematopoietic system, 16 different *Hox* genes are transcribed during normal hematopoiesis. Primitive subpopulations express primarily genes of the A and B cluster (Giampaolo et al., 1995; Pineault et al., 2002; Sauvageau et al., 1994). Mice transplanted with marrow overexpressing HOXB4 resulted in a 47-fold increase of the competitive repopulating unit (CRU) numbers and did not develop leukemic transformation (Sauvageau et al., 1995). *HOXB4* overexpression in mouse HSC cultured for 14 days induced a primitive cell-specific growth advantage contrary to a progressive depletion of HSC usually observed under these conditions. Total cell growth (mostly mature cells) was enhanced by 2-fold, progenitors by 3-fold and HSC by 1000-fold in cells overexpressing HOXB4 (Antonchuk et al., 2002).

In humans, transient overexpression of HOXB4 in hematopoietic cord blood cells, did not increase proliferation of primitive progenitors, frequency of CFC, and LTC-ICs but induced an iincrease in myeloid differentiation (Brun et al., 2003). Other studies showed that enforced high level of HOXB4 expression in human hematopoietic cord blood cells cultured for 24 hours induced a 5-10-fold increase in LTC-IC and a 4-fold increase in SRC (Buske et al., 2002). However, this HOXB4 overexpression markedly impaired the lymphoid and myeloerythroid differentiation (Schiedlmeier et al., 2003). Altogether these studies demonstrated that high levels of HOXB4 perturbed the myeloid differentiation program both *in vivo* and *in vitro* and are consistent with a dose dependant activity of HOXB4 to control the differentiation or self-renewal of HSC (Klump et al., 2005).

To increase the effect of HOXB4, a *NUP98-HOXB4* fusion gene was engeeniered since the fusion of *Hox* genes with the nucleoporine gene *NUP98* is often reported in leukemia. Ohta et *al.* observed, in a murine transplantation model, a 300-fold increase in CRUs among NUP98-HOXB4-overexpressing cells compared to only 80-fold increase with HOXB4 alone. An even higher increase (2000-fold) was observed using the *NUP98-HOXA10* fusion gene that, in contrast to HOXB4, blocks terminal differentiation and leads to a sustained output of cells with a "primitive" phenotype (Pineault et al., 2005; Pineault et al., 2004). The authors did not observe any long-term hematological defect in recipients repopulated with NUP98-HOXA10 expanded HSC (Ohta et al., 2007). However, these results contrast with those obtained by Watts et al. in a nonhuman primate stem cell transplantation model. Transplantation of comparable doses of HOXB4- and NUP98-NUP98-HOXA10-overexpressing cells revealed that HOXB4 contributed more to early hematopoiesis whereas NUP98-HOXA10 contributed more to later hematopoeisis. The emergence of a deleterious effect, such as leukaemia, could not be monitored due to the short survey period of the study (Watts et al., 2011).

In 2006, Zhang et al. investigate the ability of HOXB4 to expand HSC in a clinically relevant nonhuman primate competitive repopulation model. They found an initial 56-fold advantage for the *HOXB4*-transduced cells which decline significantly over time (Zhang et al., 2006). In addition, the first appearence of myeloid leukemia linked to HOXB4 expression were observed two years later, both in the original group of monkeys (1 out of 2) and in dogs (2 out of 2) that received cells transduced with a HOXB4 expressing vector (Zhang et al., 2008b). None of the 40 dogs and monkeys that received cells transduced with control vectors developed leukemia. Besides, a profound growth inhibition and a rapid cell differentiation was induced by siRNA knocking down HOXB4 using a cell line derived from the leukemic cells of one animal. The direct implication of HOXB4 in the development of leukemia can not be certify since analysis of the vector insertion sites in the genome of all tumors revealed insertion of the transgene near or within protooncogenes, such as *c-myb* and *PRDM16* (Zhang et al., 2008b).

To avoid the use of retroviral vectors, Amsellem et al. generate an MS-5 stromal cell line secreting HOXB4 to expand human cord blood hematopoietic cells. Using a 5-week long term culture system, they show a 4-fold increase in LTC-IC and 2.5-fold increase in SRC in NOD-SCID mice. This expansion did not appear to interfere with myeloid or lymphoid differentiation. However, the coculture system might not be suitable for clinical applications (Amsellem et al., 2003). To avoid this issue, Krosl et al. used a soluble recombinant HOXB4 protein fused to a small peptide derived from the HIV TAT protein. TAT-HOXB4 treatment of murine HSC for 4 days expanded approximately 4- to 6-fold and were 8-20 times more numerous than non treated HSC. This TAT-HOXB4 expanded population retained its normal *in vivo* potential for differentiation and long-term repopulation (Krosl et al., 2003).

The capacity of soluble HOXB4 to expand human HSC was verified using several recombinant human HOXB4 proteins. The N-terminal-tat and C-terminal histidine-tagged version of HOXB4 (T-HOXB4-H) had the highest activity in expanding CFC (10-fold) and LTC-IC (15-fold), and a 1.5- to 2.7-fold increase in SRC (Tang et al., 2009).

3.2.2 Other HOX family proteins

Surveys of *Hox* gene expression in HSC enriched populations showed dominancy of the *Hox*A cluster. In d14.5 fetal liver populations enriched for HSC, the expression of HOXA4 is a log higher than that of HOXB4. The fact that during this phase of development HSC undergo their major expansion, combined with the high homology and functional redundancy found within *Hox* paralog groups, suggests a putative role of HOXA4 to expand HSC with negligible or null oncogenic potential. HOXA4 overexpressing HSC expanded 6.6-fold after a week of culture. Although HOXA4 expressing HSC produced mature myeloid and lymphoid progeny in irradiated recipient mice, B-cell progenitors were preferentially expanded compared to myeloid progenitors (Fournier et al., 2011).

HOXC4, another member of the *Hox* family, is also expressed in proliferating hematopoietic cells suggesting a role in the control of normal proliferation. Using retroviral gene transfer in human CD34^{+} cells, Daga et *al.* showed that HOXC4 induced an *in vitro* expansion of committed cells and early hematopoietic progenitors, with the most striking effect on LTC-IC (13-fold expansion) (Daga et al., 2000). These results are consistent with those of Amsellem and Fichelson who showed a more efficient expansion of human CD34^{+}/CD38^{low} cells on MS-5 cell line secreting HOXC4 compared to those secreting HOXB4. The simultaneous presence of HOXB4 and HOXC4 seems synergize to improve expansion (Amsellem and Fichelson, 2006). However, the *in vivo* effect of HOXC4 still remains to be established.

All these observations clearly implicated Hox family proteins in HSC self renewal but further studies are required to determine if the use of these compounds could be suitable for clinical applications.

3.3 Chemical compounds

The low efficiency obtained with purified proteins and the safety concerns when attempting to expand HSC with viral vector-mediated gene transfer (Baum et al., 2003) lead to searching for alternative and safer approaches. One of these promising strategies involved the use of chemical compounds.

Chemical molecules constitute a particularly useful tool for modifying biological signaling pathways since they can be arrayed in chemical libraries for high-throughput analysis, and they can be withdrawn from the biological system once the desired effect is obtained. The use of a small molecule allows the study of the kinetics of a response in a more subtle and graduated way that is not possible with gene disruption techniques. These molecules may be further transposed into drugs for therapeutic use. Their use is rapid and cost-effective.

What are the sources of molecules available?

Historically, the pharmaceutical companies gathered the collections of molecules accumulated during the year in-house companies. These molecules can come from two different sources, one from natural origin and the other from chemically-synthesized compounds. Several companies have pooled their collections through partnerships to increase the size and diversity. At present, a large collection of oriented chemical libraries is available. In the milieu of academia, access to these collections is almost impossible unless a very restrictive partnership is framed. The number of screenable drug candidates have dramatically increased in the last years, and might account for 10 000 to 1 000 000 compounds. The difficulty to use these large collections resides in the ability to order millions of natural products, many of which are available in only limited amounts and are not yet completely characterized or even purified. Further, to identify a molecule producing the desired biological effect, different concentrations covering several orders of magnitude should be initially screened. This is why their widespread use has not yet been generalized and most discoveries to date are mainly available through the pharmaceutical industry. During the past ten years, various companies have specialized in the provision of all-purpose or targeted libraries. ChemBridge, ChemDiv, Asinex, Prestwick, Maybridge, enamine, Interbioscreen, TimTec can be mentioned as examples of commercially available collections. These libraries are relatively diverse and oriented "drug-like" (Kugawa et al., 2007). Small-molecule compounds approved for use as drugs may also be "repurposed" for new indications and studied to determine the mechanisms of their beneficial and adverse effects. A comprehensive collection of all small-molecule drugs approved for human use would be invaluable for systematic repurposing across human diseases, particularly for rare and neglected diseases, for which the cost and time required for development of a new chemical entity are often prohibitive. Major efforts are now underway to produce comprehensive collections of these small molecules amenable to high-throughput screening (Huang et al., 2011).

During the last ten years, cell-based phenotypic and pathway-specific screens using synthetic small molecules have provided new insights into stem cell biology and help to identify a number of small molecules that can be used to selectively (a) control self-renewal of embryonic and adult stem cells; (b) expand therapeutically desirable mature cell types; (c) control lineage commitment; and (d) enhance the reversion of lineage-restricted cells back to the multipotent or pluripotent state. All four practices are beginning to find application in therapeutic settings.

In this section we will focus on chemical compounds that were used to expand HSC. However, the most important question to keep in mind is whether the *in vitro* expanded cells preserve their capacities to regenerate hematopoiesis *in vivo* (Fig. 4).

3.3.1 Chromatin-modifying agents

Valproic acid (VPA) and chlamydocin are histone deacetylase (HDAC) inhibitors that exert their activity by interacting with the catalytic site of HDACs.

VPA was first studied by De Felice et al. on human $CD34^+$ cells isolated from cord blood, mobilized peripheral blood and BM. They showed that VPA preserves the CD34+ population after 1 week (40-89%) or 3 weeks (21-52%) of culture with cytokines and VPA increases H4 acetylation levels at specific sites on *HOXB4* and AC133 (De Felice et al., 2005).

In 2008, using a two step culture system, Seet et al. showed that VPA induced a 2-fold expansion of human cord blood $CD34^+CD45^+$ cells. Higher numbers of treated cells resided in the S phase compare to controls. VPA-treated cells reconstituted hematopoiesis in NOD-SCID mouse with a 6-fold higher efficiency compare to control cells. The advantage of using VPA resides on the fact that this molecule is clinically well-known since it has been used for more than 25 years to treat neurologic disorders (Seet et al., 2009). Chlamydocin, was showed to enhance Thy-1 expression on human $CD34^+$ cells and to display a 4-fold increase in SRC in NOD-SCID (Young et al., 2004).

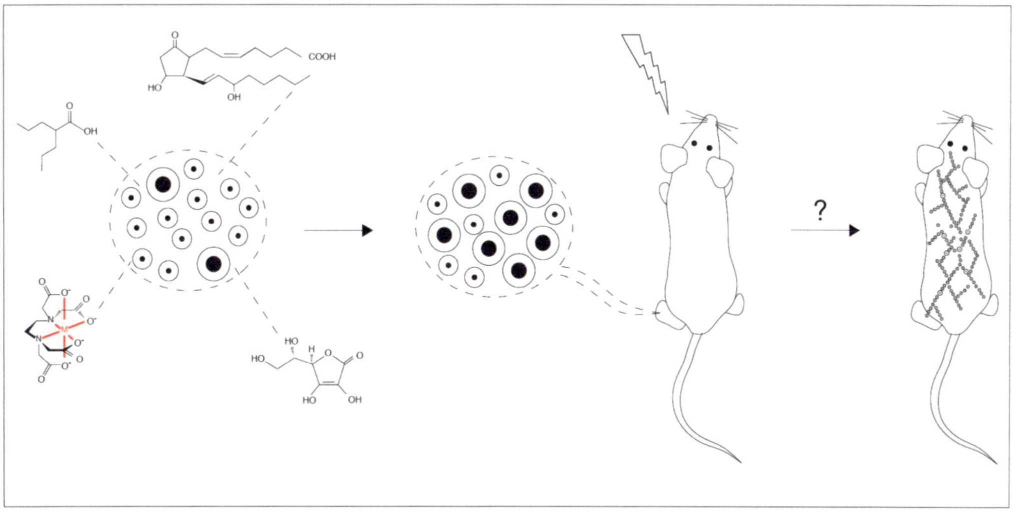

Fig. 4. A diagrammatic representation of an experimental design typology to test the effect of molecules on HSC expansion. Each molecule is added individually to the *in vitro* culture of HSC and the expansion capacities are then measured. However, infusion of the treated cells in myeloablated mice is essential to answer the question (**?**) on whether the HSC treated with the selected molecule have still the capacity to regenerate blood cells in transplanted animals.

Another HDAC inhibitor, trichostatin A (TSA), and 5-aza-2′-deoxycytidine (5azaD), a DNA methyl transferase inhibitor where shown to act in synergy to yield a 12.5-fold increase of human $CD34^+CD90^+$ cells after 9 days of culture in comparison to the input cell numbers, a 9.8-fold increase in the numbers of CFU and a 9.6-fold increase in SRC. Several genes implicated in HSC self-renewal including *HOXB4*, *BMI1*, *GATA2*, *P21*, and *P27* were up-regulated in the 5azaD/TSA-treated cells (Araki et al., 2006; Araki et al., 2007).

3.3.2 Copper chelator tetraethylenepentamine (TEPA)

Several clinical observations have suggested that copper plays a role in regulating HSC development. Peled et al. reported that modulation of cellular copper content might shift the balance between self-renewal and differentiation (Peled et al., 2005; Peled et al., 2002). This group cultured human $CD34^+$ cord blood cells with the copper chelator TEPA during extended periods of time and showed a higher percentage of early progenitors ($CD34^+CD38^-$, $CD34^+CD38^-Lin^-$) in the TEPA-treated cultures compared with controls and a 1- to 3-log-fold expansion of $CD34^+$ cells compare with that of controls. They cultured human $CD133^+$ cord blood cells during 3 weeks, in order to use a clinically suitable protocol, and found that the median output value of $CD34^+$ cells increased by 89-fold, $CD34^+CD38^-$ by 30-fold and CFU by 172-fold over the input values. Moreover, the $CD34^+$ cells expanded with TEPA appeared to show improved NOD-SCID engraftment compare to control cells (Peled et al., 2004a; Peled et al., 2004b). Based on these data, a phase 1 trial was initiated. In this study, a portion of a single cord blood unit was cultured with TEPA and cytokines for 21 days and co-infused with the remainder of the untreated cell fraction. Although this methodology showed a 219-fold expansion of total nucleated cells *in vitro*, it did not improve the time to neutrophil or platelet recovery (de Lima et al., 2008). A phase 2/3 study is under way in more than 28 centers in the United States, Europe, and Israel, to evaluate the safety and efficacy of this approach ("StemEx") in 100 patients with advanced hematologic malignancies (http://clinicaltrials.gov/ct2/show/NCT00469729).

3.3.3 Oxygen, reactive oxygen species and antioxidants

Low oxygen levels were also described to play a beneficial role on HSC expansion *in vitro*. This is consistent with the observation that protection of HSC *in vivo* is achieved by a predominantly low-oxygen environment of the stem-cell niche (Cipolleschi et al., 1993; Eliasson and Jonsson, 2010).

The positive effect of hypoxia on the survival and/or self-renewal of the HSC population *in vitro* was demonstrated quantitatively on human marrow cells with $Lin\text{-}CD34^+CD38^-$ phenotype which are enriched in SRC. A significant increase in SRC after 4 days was found in cultures under 1.5% O_2 compared to normoxic conditions. The positive effect of hypoxia on SRCs is short-lived but their engraftment into immmunocompromised mice was to some extent improved (Danet et al., 2003).

Similar studies have been performed with cord blood cells (Hermitte et al., 2006). The authors reported preferential survival of primitive HSC among cord blood $CD34^+$ cells in cultures under 0.1% O_2. After 72 hours, cells were 1.5 and 2.5 times more in quiescence (G0) at 3% and 0.1% O2. At 0.1% O2, 46.5%+/-19.1% of divided cells returned to G0 compared with 7.9%+/-0.3% at 20%. This shows a return of the cycling $CD34^+$ cells into G0, a quiescent state that characterizes steady-state HSC.

During the process of HSC purification or mobilization from the BM to the peripheral blood, the cells go across different levels of oxygenation until reach maxima in culture assays. Furthermore, cell factors added to these cultures can lead to an abnormal increase in reactive oxygen species (ROS) in the HSC and to a ROS stress that might change their properties and functions (Hao et al., 2011; Ito et al., 2006; Pervaiz et al., 2009). These ROS are unstable reactive molecular species possessing an unpaired electron that are produced continuously in cells as a byproduct of metabolism. They participate in vital signal transduction pathways but they can also oxidize DNA, proteins, and lipids leading to cell differentiation, senescence, and apoptosis. Notably, the mouse long-term repopulating HSC capacities were found in a Ros^{low} population (Jang and Sharkis, 2007). This cell population has a higher self-renewal activity than a Ros^{high} population both *in vitro* and *in vivo*. Moreover, distinct metabolic profiles of HSC reflect their location in the hypoxic niche (Simsek et al., 2010; Takubo et al., 2010).

The continual production of ROS in the *in vitro* culture (Iiyama et al., 2006) might be overcome by the addition of antioxidants. These molecules will maintain the ROS at a low level, thereby regulating the proliferation, growth, signal transduction, and gene expression of the cells (Chen et al., 2008).

Antioxidants are classified into enzyme and non-enzyme antioxidants. Enzyme antioxidants include superoxide dismutase, catalase, and glutathione peroxidase. Non-enzyme antioxidant includes vitamin C.

The application of enzyme antioxidants is limited because of the poor stability and ease of inactivation (Wojcik et al., 2010). However, when culturing mouse HSC in the presence of catalase, the number of short-term or long-term HSC with LSK immune markers was significantly increased and the stem cells begin to degenerate as the catalase is removed (Gupta et al., 2006).

Ascorbic acid (vitamin C) is a natural water-soluble antioxidant but under some conditions such as the air, heat, light, alkaline substances, enzymes and trace amount of copper oxide and iron, oxidation of vitamin C could be accelerated and the oxidative products lead to the damage of cellular DNA. The ascorbic acid 2-phosphate (AA2P), one derivative of vitamin C, is stable at 37°C in cell culture media and has no cytotoxic effect; therefore it might constitute an advantageous antioxidant (Duarte et al., 2009). Reducing oxidative stress by N-acetyl-L-cysteine (NAC) may enhance the viability and engraftment of HSC as treatment of gene corrected BM mononuclear cells or purified CD34(+) cells from FANCA patients with the reducing agent NAC showed increased CFC (Becker et al., 2010).

Although the current amplification under normal oxygen can expand a certain number of HSC, the application of glutathione for stem cell mobilization and re-infusion as well as the application of AA2P in the *in vitro* amplification culture of cells may become effective methods for protecting the hematopoietic reconstitution capacity of HSC (Hao et al., 2011). Moreover, *in vitro* culturing HSC-enriched samples under O2 concentrations that more closely resemble the BM environment (low O2 concentrations, 1–3%) might also improve their expansion and preserve proper stem cell functions for engraftment.

3.3.4 PGE2

Prostaglandin E2 (PGE2) was first identified as capable of enhancing HSC formation in zebrafish, following a high-toughput chemical screen. This effect was also tested using murine transplantation assays. When murine BM cells where briefly treated *ex vivo* by PGE2, a 3-fold increase in the CFU number and a 3.3-fold increase of SRC 6 weeks post transplantation were observed (North et al., 2007). Hoggatt et al. confirmed enhanced murine HSC engraftment following PGE2 exposure as they observed a 4-fold increase in HSC 20 weeks after transplantation. The increase in chimerism was still present in primary recipient 32 weeks post-transplant and in secondary recipients without additional PGE2 treatment. Several studies were performed to determine whether the action of PGE2 on HSC could be the result of an increase in HSC numbers, homing capability, proliferation, survival, or a combination thereof. Hoggatt et al. observed a significant increase in homing of PGE2-treated LSK cells. This was partially attributed to an increase in CXCR4 expression, a SDF1α specific receptor. This effect also occurs in human HSC, since PGE2-treated cord blood cells transplanted into NOD-SCID mice displayed an enhanced homing to marrow. In addition, PGE2 treatment increased survivin expression, reduced intracellular active caspase-3 that lead to enhanced HSC survival and increased the percentage of cycling cells (Hoggatt et al., 2009). Frish et al. treated mice *in vivo* with PGE2 by intraperitoneal injection twice a day for 16 days. They observed a significant increase of the LSK population without inhibiting their differentiation. The treatment expands preferentially the short-term-HSC/MPP subpopulation since this advantage was lost 6 weeks post-transplant in primary recipients and in secondary transplants. The disparities between these studies may be the result of the extended exposure of mice to PGE2 compared with a short term pulse used hitherto (Frisch et al., 2009).

Goessling et al. briefly treated human cord blood $CD34^+$ cells *in vitro* with dimethyl-PGE2 (dmPGE2). They showed that dmPGE2 treatment decreased apoptosis, increased 1.4-fold the CFU number and enhanced engraftment of unfractionated and $CD34^+$ cord blood cells after xenotransplantation in NOD-SCID mice. Using a non-human primate transplantation model, they found no significant enhancement of CD34+-treated cells engraftment but showed that dmPGE2 treatment had no negative impact on HSC function, including multilineage repopulation, even 1 year post-transplantation. They suggested that these results reflect suboptimal compound dosing and anticipate the use of 50μM rather than 10μM of dmPGE2 in future transplantation assays (Goessling et al., 2011). Based on these data, this brief *ex vivo* incubation with dmPGE2 is currently being tested in a phase 1 clinical trial in which adults with hematologic malignancies receive a non-myeloablative conditioning treatment followed by double-unit cord blood transplantation in which 1 of the 2 cord blood units has been incubated with dmPGE2 before infusion (http://clinicaltrials.gov/ct2/show/ NCT00890500).

3.3.5 Aryl Hydrocarbon receptor (AhR) antagonists

Using a high-throughput screen based on CD34/CD133 expression, Boitano et al identified a purine derivative (StemRegenin1 or SR1) capable of *in vitro* enhancing the levels of a $CD34^+$ cell population derived from blood of mobilized donors. SR1 added to human $CD34^+$ cells cultured for 5 weeks led to a 10-fold increase in total nucleated cells, a 47-fold increase in $CD34^+$ cells and a 65-fold increase in CFU. $CD34^+$ cord blood cells cultured in the presence of SR1 for 3 weeks revealed a 17-fold increase in SRC content in NOD-SCID Gamma (NSG) primary recipient and a 12-fold increase in the number of secondary SRC compared to input (Boitano et al., 2010). Additional screens followed by a quantitative structure-activity relationship identified three novel compounds (i.e SR2, SR3 and SR4), structurally distinct from SR1, that expand the number of human $CD34^+$ cells. Experiments that aimed to determine the ability of cord blood derived human HSC expanded with these molecules to engraft NSG mice are still undergoing (Bouchez et al., 2011). SR1, SR2, SR3 and SR4 were showed to act as antagonists of AhR signaling. Indeed, this receptor has been implicated in HSC biology and hematopoietic disease through numerous factors including c-MYC, HES-1, PU.1, C/EBP, β-catenin, CXCR4, and STAT-5 (Singh et al., 2009). However, the precise mechanism whereby an AhR inhibitor might induce HSC self-renewal remains unknown.

3.3.6 SALL4

The transcription factor SALL4 was reported to play a role in maintaining ES cell pluripotency through interaction with Oct4 and Nanog (Wu et al., 2006; Yang et al., 2010). It was recently showed that overexpression of SALL4 can expand *ex vivo* human mobilized HSC from peripheral blood (Aguila et al., 2011). SALL4-transduced cells seemed capable of *ex vivo* expansion of both, $CD34^{+}CD38^{-}$ and $CD34^{+}CD38^{+}$ cells and showed enhanced stem cell engraftment and long term repopulation capacity in NOD-SCID mice. Moreover, human $CD34^{+}$ cells cultured 3 to 4 days with a soluble SALL4 fusion protein (TAT-SALL4B) showed a 10-fold increase in total mononuclear cells, a 8-fold increase in $CD34^{+}$ cells and a 10-fold increase in the CFU number compare to controls (Aguila et al., 2011). However, *in vivo* studies with this fusion protein still have to be conducted to validate that these expanded cells are still able to reconstitute hematopoiesis in transplanted recipients.

4. *De novo* generation of HSC

Considering the interest in HSC expansion for treatment of both malignant and non-malignant diseases as well as their use in gene therapy and the difficulty to obtain *ex vivo* expansion of HSC without loss of their regeneration capacities, relevant methods to produce *de novo* HSC have emerged.

4.1 Obtaining HSC from ESC

One of these methods was initiated 20 years ago when ESC could be cultivated *in vitro* and directed to generate hematopoietic cells (Wiles and Keller, 1991). Since then, culture conditions were constantly optimized and allowed the differentiation into specific hematopoietic lineages such as erythroid and myeloid lineages, T and B lymphocytes and megakaryocytes (for review see Sakamoto et al., 2010). These protocols were then adapted to human (h) ESC. These cells like their murine counterparts, are karyotypically stable, capable of prolonged self-renewal, and might differentiate into most cell types. These properties might be exploited for therapeutic benefits to cure many human degenerative diseases and resulted in intense biomedical studies.

Different methods were established to generate hematopoietic progenitors and specific lineages from mouse ESC including embryoid bodies formation, coculture with stromal cells, and direct differentiation in coated plates using a mixture of cytokines and growth factors without stromal cells (Tian and Kaufman, 2008). These protocols were then optimized for efficient differentiation of hESC into early mesodermal cells (Bernardo et al., 2011) and for obtaining defined hematopoietic precursors from ES cells (Chiang and Wong, 2011; Salvagiotto et al., 2011).

The ultimate goal of these strategies is to produce HSC capable of robust, long-term, multilineage engraftment to alleviate blood cells diseases; however the numbers and the capacities of the *de novo* cells generated are not quite sufficient to fulfill the clinical challenge. At present, multipotent hematopoietic progenitors (short-term HSC) with limited engrafting ability in transplanted mice were obtained (Woods et al., 2011). Other groups reported efficient generation of cells that mostly produce the myeloid lineage following long term engraftment or produce CD34+ hematopoietic precursors that have phenotype similar to adult HSC but might best correspond to the embryonic stage of yolk-sac, aortogonadal-mesonephros (AGM), and/or fetal liver stage of hematopoiesis (Melichar et al., 2011; Narayan et al., 2006 and for review : Tian and Kaufman, 2008). More recently, the polycomb group protein Bmi1 was shown to promote more than 100-fold increase of hematopoietic cell development from ESC (Ding et al., 2011).

Since short-term HSC could be generated from ESC, an attractive option to increase the number of clinically competent HSC would be to find a molecule that dedifferentiate from short-term or mature hematopoietic cells to the long-term HSC population. Such a strategy might be valuable, since de-differentiation of somatic cells mediated by a chemical has been achieved in other systems. This is the case for reversine or 2-(4-morpholinoanilino)-6-cyclohexylaminopurine. This chemical compound was reported to induce reversal of mouse myoblast cell line, C2C12, to become multipotent progenitor cells, which can re-differentiate into osteoblasts and adipocytes (Chen et al., 2004). The de-differentiation activity of reversine however is not conserved across all cell lineages, since in certain cell types, it acts as a potent differentiation-inducing molecule (D'Alise et al., 2008).

To support the generation of long-term repopulating HSC from mouse ESC, other groups tested intrinsic regulators of adult HSC (Schuringa et al., 2004; Wang et al., 2005b). However, the use of many of these compounds, such as HoxB4, did not improve the expected engraftment efficiency *in vivo* (Wang et al., 2005a).

An unfavorable complication for the use of ESC in producing HSC is that lifelong use of drugs is required to prevent rejection of the transplanted cells. In order to make ESC practical for therapeutic use, it would be necessary to create a new stem cell line for each patient that needs treatment. Serious technical and ethical problems are associated with this issue.

4.2 Obtaining HSC from induced pluripotent stem cells

An alternative to the utilization of ES cells to produce *de novo* HSC arise from one of the most transformative accomplishments performed in the last years: the discovery that transient overexpression of a small number of defined transcription factors can reprogram the differentiated cells and become pluripotent populations. These cells are commonly referred to as Induced Pluripotent Stem Cells (iPSC) and have definitively broken the dogma commonly accepted that differentiated cell types generally lack the ability to revert back to a less specialized state.

4.2.1 Reprogramming somatic cells to pluripotency

The direct reprogramming of somatic cells to pluripotency was demonstrated in 2006, when Takahashi and Yamanaka converted adult mouse fibroblasts to iPSC by overexpressing four transcription factors: octamer-binding transcription factor 4 (OCT4), sex determining region Y-box 2 (SOX2), Kruppel-like factor 4 (KLF4), and cytoplasmic Myc (c-MYC) in mouse embryonic fibroblasts using retroviruses (Takahashi and Yamanaka, 2006). The transcription factors originally used for reprogramming differentiated cells are not stringently necessary to achieve this process as some of them can be replaced by other factors. Yu et al. were able to reprogram human fibroblasts with a distinct set of transcription factors comprising OCT4,

SOX2, NANOG, and LIN28 (Yu et al., 2007). Krüppel-like transcription factors (Klf2 and Klf5) and the orphan nuclear receptor, Esrrb, can replace Klf4 (Nakagawa et al., 2008 and for review see Feng et al., 2009).

The derived iPSC exhibited typical ESC morphology and were similar to ESC in their regenerative potential (Takahashi and Yamanaka, 2006) and their capacity to differentiate into cells of all three germ layers, the ectoderm, mesoderm, and endoderm. Because iPSC are generated without the need to destroy an embryo, their discovery has further energized the field of regenerative medicine and stem cell biology. Patient-specific therapeutic cells derived from induced pluripotent stem iPSC may bypass the ethical issues associated with ESC and avoid potential immunological reactions associated with allogenic transplantation. These human disease–specific iPSC provide a unique and previously unavailable resource for studying the pathophysiology of various important human diseases.

The therapeutical hope of iPSC is based on three issues: 1) The ability to generate iPSC from any tissue of the organism, and further differentiate them according to the patient needs, particularly into a wide range of primary human cell types, many of which are unavailable for routine use; 2) The ability to generate iPSC from patients with any disease; 3) The possibility of using patient-derived iPSC for drug development.

The therapeutic potential of such iPSC (schematized in Fig. 5) was demonstrated in a proof-of-principle study using a humanized sickle cell anemia mouse model (Hanna et al., 2007). In this study, mice could be rescued after transplantation with hematopoietic progenitors obtained *in vitro* from autologous iPSC. This was achieved after correction of the human sickle hemoglobin allele by gene-specific targeting.

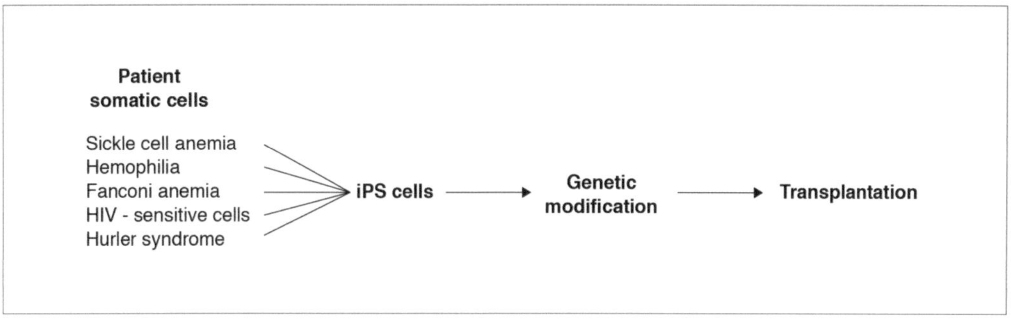

Fig. 5. Studies performed to validate the therapeutic potential of iPSC.

Also in mice, Xu et al. cured hemophilia by transplantation of cells that were generated from murine wild-type iPSC. These murine experiments suggest that human iPSC can be utilized for regenerative and therapeutic applications (Xu et al., 2009). Most recently, patient-specific iPSC have been established. Raya et al. reprogrammed dermal fibroblasts and/or epidermal keratinocytes of Fanconi anemia patients to generate iPSC, which were genetically corrected with lentiviral vectors encoding FANCA or FANCD2, to obtain hematopoietic progenitors of the myeloid and erythroid lineages that are phenotypically normal, that is, disease-free (Raya et al., 2009). Similar strategies were performed to correct the Hurler syndrome (Tolar et al., 2011) and for the production of macrophages from iPSCs which were resistant to HIV infection (Kambal et al., 2011).

The enthusiasm surrounding the clinical potential of iPSC is tempered by key issues regarding their safety, efficacy, and long-term benefits. Fully realizing the biomedical potential of iPSC in a clinical setting will require addressing certain limitations inherent to the process. First, need to find alternative strategies to remove non-viral or non-integrative vectors to overcome their potential deleterious effects. Although expression of the exogenous reprogramming factors is eventually silenced during iPSC cell generation, there is a significant risk of tumorigenesis if these exogenous genes are inadvertently reactivated. Second, it will be essential to increase the number of cells with specific phenotype. Third, it will be necessary to improve the efficiency of reprogramming (0.001–3% of cells are reprogrammed) since this is a slow and inefficient process (Jaenisch and Young, 2008; Nakagawa et al., 2008; Wernig et al., 2008).

4.2.2 Improving reprogramming somatic cells to pluripotency

To overcome the potential deleterious effects of viral vectors or oncogenes, and to improve the efficiency of the process toward a potential clinical application, a powerful alternative is offered by using small molecules. Several small molecules were reported to improve the reprogramming process by lowering the epigenetic barrier to initiate pluripotency (for reviews see Feng et al., 2009; Lyssiotis et al., 2011). Consistent with this notion, small molecules that affect reorganization of chromatin architecture, a rate limiting step during the reprogramming of a somatic genome, have been identified (Blelloch et al., 2006; Hochedlinger and Jaenisch, 2006; Huangfu et al., 2008a). In particular, the HDAC inhibitor VPA was shown to strongly increase reprogramming efficiency in the absence of c-Myc in both mouse and human cells and to allow 2-factor reprogramming (Oct4 and Sox2) of human fibroblasts in the absence of Klf4 and c-Myc (Huangfu et al., 2008b). Other epigenetic regulators such as BIX01294, a G9a histone methyltransferase inhibitor; BayK8644, an L-type calcium channel agonist and the two DNA methyltransferase inhibitors, AzaC and RG108 (summarized in Fig. 6), substantially increased reprogramming efficiency (Lukaszewicz et al., 2010; Shi et al., 2008).

The low efficiency of reprogramming (Hong et al., 2009) might also result from the accumulation of ROS (Parrinello et al., 2003). Consistent with this, Esteban et al. found that vitamin C strongly increases the reprogramming efficiency (Esteban et al., 2010). This is in line with the study reporting that hypoxic conditions improve the efficiency of iPSC production generated from mouse or human somatic cells (Yoshida et al., 2009). Co-treatment with VPA synergizes this effect. Other molecules including the MEK inhibitor PD0325901, the GSK3 inhibitor CHIR99021 combined with tranylcypromine, kenpaullone, SB-431542, and the TGF-β signaling inhibitor called RepSox (Ichida et al., 2009) were reported to enhance reprogramming or to replace viral vectors or oncogenes (Li and Ding, 2009; Pan and Thomson, 2007).

With the continued use of high-throughput screening to identify more chemicals that could assist in reprogramming, we may be closer to the goal of using a chemical-only cocktail to reprogram somatic cells to iPSC. These pluripotency gene activators may be then used in combination with specific differentiation modulators to achieve the production of the desired cell type.

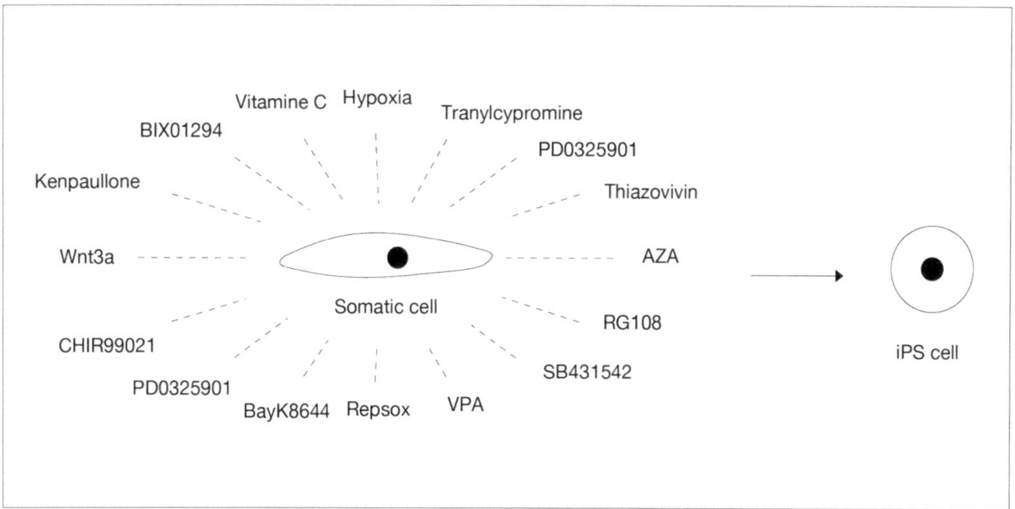

Fig. 6. Chemicals used to enhance reprogramming or to replace core reprogramming factors.

4.2.3 Obtaining HSC from induced pluripotent stem cells

Differentiation of iPSC into hematopoietic lineage have been achieved using a combination of specific cytokines and growth factors (Sakamoto et al., 2010) and have already demonstrated from both mouse and human iPSC (Lengerke et al., 2009; Niwa et al., 2009; Woods et al., 2011). However, the number of cells obtained in view of therapeutic use is still insufficient and small molecules that might expand the production of hematopoietic cells have yet to be found. Studies in this direction are beginning to emerge. For example, Wnt signaling, in particular WNT3a, mediates the stimulation of hemoangiogenic cell development and increase hematopoietic differentiation from ESC and iPSC (Wang and Nakayama, 2009; Wang et al., 2010). However, the conditions to generate human HSC capable of robust, long-term, multilineage engraftment from iPSC are still hoped for.

5. Conclusion

The ex-vivo expansion of HSC represents a promising approach to obtain large enough quantities for therapeutic intervention in cell and gene therapy protocols. Derivatives of hESCs and iPS cells are also expected to be employed as *de novo* HSC source for therapeutic settings. However, as described in the previous section, practical and ethical issues must be settled before clinical practice can begin. In both cases, the chemical biology approach using small molecules as tools or drugs holds unquestionably greater promise in the outcome of the final goal.

Even though a few molecules are being tested in clinical assays, the ideal soluble factor that enables to increase the number of rare HSC during the *ex vivo* growth culture without limiting their regeneration capacities has yet to be found. Most attempts have been unsuccessful because i) suitable expansion *in vitro* has been mostly correlated with loss of the regenerative capacities of HSC *in vivo;* ii) no straight forward method allows the association of *in vitro* observations with the *in vivo* outcome; iii) testing the *in vivo* effect of each molecule independently would be costly, time-consuming and would need an imposing number of mice which is ethically inconceivable.

In an attempt to develop new tools that might overcome some of these limitations, we have developed an innovative screening strategy to identify molecules for their potential to improve the *in vitro* HSC self-renewal and proliferation while preserving the HSC regenerative capacities *in vivo* (Sii Felice K, Grosselin J, Leboulch P, Tronik-Le Roux D, manuscript in preparation). Our approach is based on stem cells labeling with specific barcodes before exposure to the molecules (Fig. 7). Then, prior to their infusion in myeloablated mice, all the treated cells are pooled. Several weeks after transplantation, the identification of barcodes present in the blood and the BM of transplanted mice will enable the precise retrospective quantification of the initial effect of the molecule.

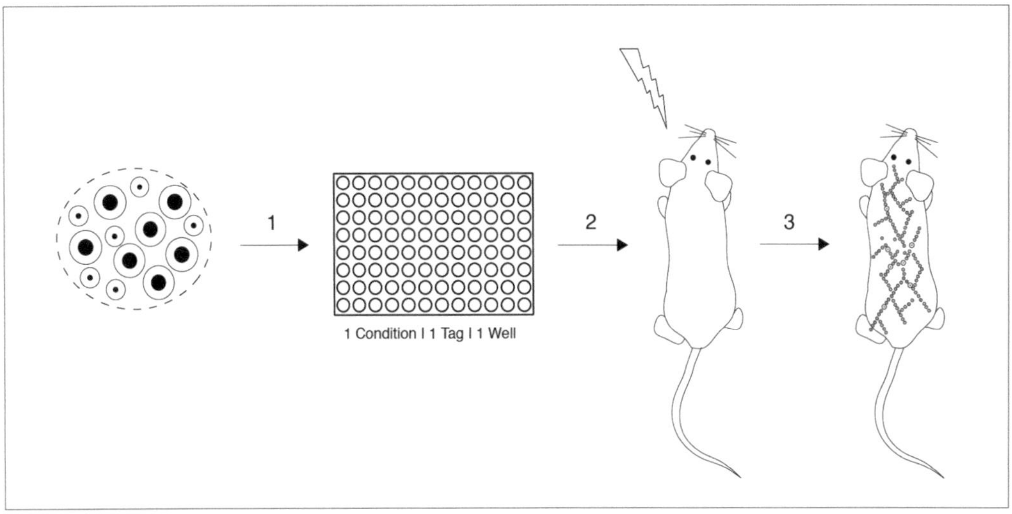

Fig. 7. Schematic representation of the strategy developed to simultaneously test dozens of molecules. Each well contains barcoded-HSC (1) treated by a particular molecule. After several days of *in vitro* culture (2), all the cells are pooled, infused in myeloablated mice. The identification of barcodes in blood and BM of transplanted mice will enable the precise retrospective quantification of the initial effect of the molecule.

This strategy might facilitate the development of high-throughput screening for fast and effective identification of small molecules that can be used to burst the production of HSC. This will undoubtedly accelerate the promise of regenerative medicine as a routine therapeutic modality for many blood diseases as well as for gene and cell therapy.

6. Acknowledgments

We are grateful to Eliane Le Roux for graphic illustrations and Matthew Uselman for critical reading of the manuscript. We sincerely apologize to our colleagues whose relevant work was omitted in this review because of space limitations.

7. References

Abe, S., Lauby, G., Boyer, C., Rennard, S. I. & Sharp, J. G. (2003). *Transplanted BM and BM side population cells contribute progeny to the lung and liver in irradiated mice.* Cytotherapy 5, 523-533.

Aguila, J. R., Liao, W., Yang, J., Avila, C., Hagag, N., Senzel, L. & Ma, Y. (2011). *SALL4 is a robust stimulator for the expansion of hematopoietic stem cells.* Blood *118*, 576-585.

Almeida-Porada, G., Zanjani, E. D. & Porada, C. D. (2010). *Bone marrow stem cells and liver regeneration.* Exp Hematol *38*, 574-580.

Amsellem, S. & Fichelson, S. (2006). *Ex vivo expansion of human hematopoietic stem cells by passive transduction of the HOXB4 homeoprotein.* J Soc Biol *200*, 235-241.

Amsellem, S., Pflumio, F., Bardinet, D., Izac, B., Charneau, P., Romeo, P. H., Dubart-Kupperschmitt, A. & Fichelson, S. (2003). *Ex vivo expansion of human hematopoietic stem cells by direct delivery of the HOXB4 homeoprotein.* Nat Med *9*, 1423-1427.

Antonchuk, J., Sauvageau, G. & Humphries, R. K. (2002). *HOXB4-induced expansion of adult hematopoietic stem cells ex vivo.* Cell *109*, 39-45.

Araki, H., Mahmud, N., Milhem, M., Nunez, R., Xu, M., Beam, C. A. & Hoffman, R. (2006). *Expansion of human umbilical cord blood SCID-repopulating cells using chromatin-modifying agents.* Exp Hematol *34*, 140-149.

Araki, H., Yoshinaga, K., Boccuni, P., Zhao, Y., Hoffman, R. & Mahmud, N. (2007). *Chromatin-modifying agents permit human hematopoietic stem cells to undergo multiple cell divisions while retaining their repopulating potential.* Blood *109*, 3570-3578.

Baird, A. (1994). *Fibroblast growth factors: activities and significance of non-neurotrophin neurotrophic growth factors.* Curr Opin Neurobiol *4*, 78-86.

Baum, C., Dullmann, J., Li, Z., Fehse, B., Meyer, J., Williams, D. A. & von Kalle, C. (2003). *Side effects of retroviral gene transfer into hematopoietic stem cells.* Blood *101*, 2099-2114.

Becker, P. S., Taylor, J. A., Trobridge, G. D., Zhao, X., Beard, B. C., Chien, S., Adair, J., Kohn, D. B., Wagner, J. E., Shimamura, A. & Kiem, H. P. (2010). *Preclinical correction of human Fanconi anemia complementation group A bone marrow cells using a safety-modified lentiviral vector.* Gene Ther *17*, 1244-1252.

Bejsovec, A. (2005). *Wnt pathway activation: new relations and locations.* Cell *120*, 11-14.

Bernardo, A. S., Faial, T., Gardner, L., Niakan, K. K., Ortmann, D., Senner, C. E., Callery, E. M., Trotter, M. W., Hemberger, M., Smith, J. C., et al. (2011). *BRACHYURY and CDX2 Mediate BMP-Induced Differentiation of Human and Mouse Pluripotent Stem Cells into Embryonic and Extraembryonic Lineages.* Cell Stem Cell *9*, 144-155.

Bhardwaj, G., Murdoch, B., Wu, D., Baker, D. P., Williams, K. P., Chadwick, K., Ling, L. E., Karanu, F. N. & Bhatia, M. (2001). *Sonic hedgehog induces the proliferation of primitive human hematopoietic cells via BMP regulation.* Nat Immunol *2*, 172-180.

Bhatia, M., Bonnet, D., Wu, D., Murdoch, B., Wrana, J., Gallacher, L. & Dick, J. E. (1999). *Bone morphogenetic proteins regulate the developmental program of human hematopoietic stem cells.* J Exp Med *189*, 1139-1148.

Blank, U., Karlsson, G. & Karlsson, S. (2008). *Signaling pathways governing stem-cell fate.* Blood *111*, 492-503.

Blank, U., Karlsson, G., Moody, J. L., Utsugisawa, T., Magnusson, M., Singbrant, S., Larsson, J. & Karlsson, S. (2006). *Smad7 promotes self-renewal of hematopoietic stem cells.* Blood *108*, 4246-4254.

Blelloch, R., Wang, Z., Meissner, A., Pollard, S., Smith, A. & Jaenisch, R. (2006). *Reprogramming efficiency following somatic cell nuclear transfer is influenced by the differentiation and methylation state of the donor nucleus.* Stem Cells *24*, 2007-2013.

Boitano, A. E., Wang, J., Romeo, R., Bouchez, L. C., Parker, A. E., Sutton, S. E., Walker, J. R., Flaveny, C. A., Perdew, G. H., Denison, M. S., et al. (2010). *Aryl hydrocarbon receptor antagonists promote the expansion of human hematopoietic stem cells.* Science *329*, 1345-1348.

Bolos, V., Grego-Bessa, J. & de la Pompa, J. L. (2007). *Notch signaling in development and cancer.* Endocr Rev *28*, 339-363.

Bottcher, R. T. & Niehrs, C. (2005). *Fibroblast growth factor signaling during early vertebrate development.* Endocr Rev *26*, 63-77.

Bouchez, L. C., Boitano, A. E., de Lichtervelde, L., Romeo, R., Cooke, M. P. & Schultz, P. G. (2011). *Small-molecule regulators of human stem cell self-renewal.* Chembiochem *12*, 854-857.

Broske, A. M., Vockentanz, L., Kharazi, S., Huska, M. R., Mancini, E., Scheller, M., Kuhl, C., Enns, A., Prinz, M., Jaenisch, R., et al. (2009). *DNA methylation protects hematopoietic stem cell multipotency from myeloerythroid restriction.* Nat Genet *41*, 1207-1215.

Brun, A. C., Fan, X., Bjornsson, J. M., Humphries, R. K. & Karlsson, S. (2003). *Enforced adenoviral vector-mediated expression of HOXB4 in human umbilical cord blood CD34+ cells promotes myeloid differentiation but not proliferation.* Mol Ther *8*, 618-628.

Buske, C., Feuring-Buske, M., Abramovich, C., Spiekermann, K., Eaves, C. J., Coulombel, L., Sauvageau, G., Hogge, D. E. & Humphries, R. K. (2002). *Deregulated expression of HOXB4 enhances the primitive growth activity of human hematopoietic cells.* Blood *100*, 862-868.

Carlesso, N., Aster, J. C., Sklar, J. & Scadden, D. T. (1999). *Notch1-induced delay of human hematopoietic progenitor cell differentiation is associated with altered cell cycle kinetics.* Blood *93*, 838-848.

Cerdan, C. & Bhatia, M. (2010). *Novel roles for Notch, Wnt and Hedgehog in hematopoesis derived from human pluripotent stem cells.* Int J Dev Biol *54*, 955-963.

Chadwick, K., Shojaei, F., Gallacher, L. & Bhatia, M. (2005). *Smad7 alters cell fate decisions of human hematopoietic repopulating cells.* Blood *105*, 1905-1915.

Chadwick, N., Nostro, M. C., Baron, M., Mottram, R., Brady, G. & Buckle, A. M. (2007). *Notch signaling induces apoptosis in primary human CD34+ hematopoietic progenitor cells.* Stem Cells *25*, 203-210.

Chen, C., Liu, Y., Liu, R., Ikenoue, T., Guan, K. L., Liu, Y. & Zheng, P. (2008). *TSC-mTOR maintains quiescence and function of hematopoietic stem cells by repressing mitochondrial biogenesis and reactive oxygen species.* J Exp Med *205*, 2397-2408.

Chen, S., Zhang, Q., Wu, X., Schultz, P. G. & Ding, S. (2004). *Dedifferentiation of lineage-committed cells by a small molecule.* J Am Chem Soc *126*, 410-411.

Chiang, P. M. & Wong, P. C. (2011). *Differentiation of an embryonic stem cell to hemogenic endothelium by defined factors: essential role of bone morphogenetic protein 4.* Development *138*, 2833-2843.

Cipolleschi, M. G., Dello Sbarba, P. & Olivotto, M. (1993). *The role of hypoxia in the maintenance of hematopoietic stem cells.* Blood *82*, 2031-2037.

Cobas, M., Wilson, A., Ernst, B., Mancini, S. J., MacDonald, H. R., Kemler, R. & Radtke, F. (2004). *Beta-catenin is dispensable for hematopoiesis and lymphopoiesis.* J Exp Med *199*, 221-229.

Crcareva, A., Saito, T., Kunisato, A., Kumano, K., Suzuki, T., Sakata-Yanagimoto, M., Kawazu, M., Stojanovic, A., Kurokawa, M., Ogawa, S., et al. (2005). *Hematopoietic stem cells expanded by fibroblast growth factor-1 are excellent targets for retrovirus-mediated gene delivery.* Exp Hematol *33*, 1459-1469.

D'Alise, A. M., Amabile, G., Iovino, M., Di Giorgio, F. P., Bartiromo, M., Sessa, F., Villa, F., Musacchio, A. & Cortese, R. (2008). *Reversine, a novel Aurora kinases inhibitor, inhibits colony formation of human acute myeloid leukemia cells.* Mol Cancer Ther *7*, 1140-1149.

Daga, A., Podesta, M., Capra, M. C., Piaggio, G., Frassoni, F. & Corte, G. (2000). *The retroviral transduction of HOXC4 into human CD34(+) cells induces an in vitro expansion of clonogenic and early progenitors.* Exp Hematol *28*, 569-574.

Dahlberg, A., Delaney, C. & Bernstein, I. D. (2011). *Ex vivo expansion of human hematopoietic stem and progenitor cells.* Blood *117*, 6083-6090.

Danet, G. H., Pan, Y., Luongo, J. L., Bonnet, D. A. & Simon, M. C. (2003). *Expansion of human SCID-repopulating cells under hypoxic conditions.* J Clin Invest *112*, 126-135.

De Felice, L., Tatarelli, C., Mascolo, M. G., Gregorj, C., Agostini, F., Fiorini, R., Gelmetti, V., Pascale, S., Padula, F., Petrucci, M. T., et al. (2005). *Histone deacetylase inhibitor valproic acid enhances the cytokine-induced expansion of human hematopoietic stem cells.* Cancer Res *65*, 1505-1513.

de Haan, G., Weersing, E., Dontje, B., van Os, R., Bystrykh, L. V., Vellenga, E. & Miller, G. (2003). *In vitro generation of long-term repopulating hematopoietic stem cells by fibroblast growth factor-1.* Dev Cell *4*, 241-251.

de Lima, M., McMannis, J., Gee, A., Komanduri, K., Couriel, D., Andersson, B. S., Hosing, C., Khouri, I., Jones, R., Champlin, R., et al. (2008). *Transplantation of ex vivo expanded cord blood cells using the copper chelator tetraethylenepentamine: a phase I/II clinical trial.* Bone Marrow Transplant *41*, 771-778.

Delaney, C., Heimfeld, S., Brashem-Stein, C., Voorhies, H., Manger, R. L. & Bernstein, I. D. (2010). *Notch-mediated expansion of human cord blood progenitor cells capable of rapid myeloid reconstitution.* Nat Med *16*, 232-236.

Delaney, C., Varnum-Finney, B., Aoyama, K., Brashem-Stein, C. & Bernstein, I. D. (2005). *Dose-dependent effects of the Notch ligand Delta1 on ex vivo differentiation and in vivo marrow repopulating ability of cord blood cells.* Blood *106*, 2693-2699.

Ding, X., Lin, Q., Ensenat-Waser, R., Rose-John, S. & Zenke, M. (2011). *Polycomb Group Protein Bmi1 Promotes Hematopoietic Cell Development from Embryonic Stem Cells.* Stem Cells Dev.

Drake, A. C., Khoury, M., Leskov, I., Iliopoulou, B. P., Fragoso, M., Lodish, H. & Chen, J. (2011). *Human CD34+ CD133+ hematopoietic stem cells cultured with growth factors including Angptl5 efficiently engraft adult NOD-SCID Il2rgamma-/- (NSG) mice.* PLoS One *6*, e18382.

Duarte, T. L., Cooke, M. S. & Jones, G. D. (2009). *Gene expression profiling reveals new protective roles for vitamin C in human skin cells.* Free Radic Biol Med *46*, 78-87.

Duncan, A. W., Rattis, F. M., DiMascio, L. N., Congdon, K. L., Pazianos, G., Zhao, C., Yoon, K., Cook, J. M., Willert, K., Gaiano, N. & Reya, T. (2005). *Integration of Notch and Wnt signaling in hematopoietic stem cell maintenance.* Nat Immunol 6, 314-322.

Eliasson, P. & Jonsson, J. I. (2010). *The hematopoietic stem cell niche: low in oxygen but a nice place to be.* J Cell Physiol 222, 17-22.

Ernst, P., Fisher, J. K., Avery, W., Wade, S., Foy, D. & Korsmeyer, S. J. (2004). *Definitive hematopoiesis requires the mixed-lineage leukemia gene.* Dev Cell 6, 437-443.

Esteban, M. A., Wang, T., Qin, B., Yang, J., Qin, D., Cai, J., Li, W., Weng, Z., Chen, J., Ni, S., et al. (2010). *Vitamin C enhances the generation of mouse and human induced pluripotent stem cells.* Cell Stem Cell 6, 71-79.

Feng, B., Ng, J. H., Heng, J. C. & Ng, H. H. (2009). *Molecules that promote or enhance reprogramming of somatic cells to induced pluripotent stem cells.* Cell Stem Cell 4, 301-312.

Fournier, M., Lebert-Ghali, C. E., Krosl, G. & Bijl, J. J. (2011). *HOXA4 Induces Expansion of Hematopoietic Stem Cells In Vitro and Confers Enhancement of Pro-B-Cells In Vivo.* Stem Cells Dev.

Frisch, B. J., Porter, R. L., Gigliotti, B. J., Olm-Shipman, A. J., Weber, J. M., O'Keefe, R. J., Jordan, C. T. & Calvi, L. M. (2009). *In vivo prostaglandin E2 treatment alters the bone marrow microenvironment and preferentially expands short-term hematopoietic stem cells.* Blood *114*, 4054-4063.

Gao, J., Graves, S., Koch, U., Liu, S., Jankovic, V., Buonamici, S., El Andaloussi, A., Nimer, S. D., Kee, B. L., Taichman, R., et al. (2009). *Hedgehog signaling is dispensable for adult hematopoietic stem cell function.* Cell Stem Cell 4, 548-558.

Giampaolo, A., Pelosi, E., Valtieri, M., Montesoro, E., Sterpetti, P., Samoggia, P., Camagna, A., Mastroberardino, G., Gabbianelli, M., Testa, U. & et al. (1995). *HOXB gene expression and function in differentiating purified hematopoietic progenitors.* Stem Cells *13 Suppl 1*, 90-105.

Goessling, W., Allen, R. S., Guan, X., Jin, P., Uchida, N., Dovey, M., Harris, J. M., Metzger, M. E., Bonifacino, A. C., Stroncek, D., et al. (2011). *Prostaglandin E2 enhances human cord blood stem cell xenotransplants and shows long-term safety in preclinical nonhuman primate transplant models.* Cell Stem Cell 8, 445-458.

Goey, H., Keller, J. R., Back, T., Longo, D. L., Ruscetti, F. W. & Wiltrout, R. H. (1989). *Inhibition of early murine hemopoietic progenitor cell proliferation after in vivo locoregional administration of transforming growth factor-beta 1.* J Immunol *143*, 877-880.

Gupta, R., Karpatkin, S. & Basch, R. S. (2006). *Hematopoiesis and stem cell renewal in long-term bone marrow cultures containing catalase.* Blood *107*, 1837-1846.

Han, H., Tanigaki, K., Yamamoto, N., Kuroda, K., Yoshimoto, M., Nakahata, T., Ikuta, K. & Honjo, T. (2002). *Inducible gene knockout of transcription factor recombination signal binding protein-J reveals its essential role in T versus B lineage decision.* Int Immunol *14*, 637-645.

Hanna, J., Wernig, M., Markoulaki, S., Sun, C. W., Meissner, A., Cassady, J. P., Beard, C., Brambrink, T., Wu, L. C., Townes, T. M. & Jaenisch, R. (2007). *Treatment of sickle cell anemia mouse model with iPS cells generated from autologous skin.* Science *318*, 1920-1923.

Hao, Y., Cheng, D., Ma, Y., Zhou, W. & Wang, Y. (2011). *Antioxidant intervention: a new method for improving hematopoietic reconstitution capacity of peripheral blood stem cells.* Med Hypotheses *76*, 421-423.

Hatzfeld, J., Li, M. L., Brown, E. L., Sookdeo, H., Levesque, J. P., O'Toole, T., Gurney, C., Clark, S. C. & Hatzfeld, A. (1991). *Release of early human hematopoietic progenitors from quiescence by antisense transforming growth factor beta 1 or Rb oligonucleotides.* J Exp Med *174*, 925-929.

Hermitte, F., Brunet de la Grange, P., Belloc, F., Praloran, V. & Ivanovic, Z. (2006). *Very low O2 concentration (0.1%) favors G0 return of dividing CD34+ cells.* Stem Cells *24*, 65-73.

Himburg, H. A., Muramoto, G. G., Daher, P., Meadows, S. K., Russell, J. L., Doan, P., Chi, J. T., Salter, A. B., Lento, W. E., Reya, T., et al. (2010). *Pleiotrophin regulates the expansion and regeneration of hematopoietic stem cells.* Nat Med *16*, 475-482.

Hochedlinger, K. & Jaenisch, R. (2006). *Nuclear reprogramming and pluripotency.* Nature *441*, 1061-1067.

Hofmann, I., Stover, E. H., Cullen, D. E., Mao, J., Morgan, K. J., Lee, B. H., Kharas, M. G., Miller, P. G., Cornejo, M. G., Okabe, R., et al. (2009). *Hedgehog signaling is dispensable for adult murine hematopoietic stem cell function and hematopoiesis.* Cell Stem Cell *4*, 559-567.

Hoggatt, J., Singh, P., Sampath, J. & Pelus, L. M. (2009). *Prostaglandin E2 enhances hematopoietic stem cell homing, survival, & proliferation.* Blood *113*, 5444-5455.

Hong, H., Takahashi, K., Ichisaka, T., Aoi, T., Kanagawa, O., Nakagawa, M., Okita, K. & Yamanaka, S. (2009). *Suppression of induced pluripotent stem cell generation by the p53-p21 pathway.* Nature *460*, 1132-1135.

Huang, R., Southall, N., Wang, Y., Yasgar, A., Shinn, P., Jadhav, A., Nguyen, D. T. & Austin, C. P. (2011). *The NCGC pharmaceutical collection: a comprehensive resource of clinically approved drugs enabling repurposing and chemical genomics.* Sci Transl Med *3*, 80ps16.

Huangfu, D., Maehr, R., Guo, W., Eijkelenboom, A., Snitow, M., Chen, A. E. & Melton, D. A. (2008a). *Induction of pluripotent stem cells by defined factors is greatly improved by small-molecule compounds.* Nat Biotechnol *26*, 795-797.

Huangfu, D., Osafune, K., Maehr, R., Guo, W., Eijkelenboom, A., Chen, S., Muhlestein, W. & Melton, D. A. (2008b). *Induction of pluripotent stem cells from primary human fibroblasts with only Oct4 and Sox2.* Nat Biotechnol *26*, 1269-1275.

Ichida, J. K., Blanchard, J., Lam, K., Son, E. Y., Chung, J. E., Egli, D., Loh, K. M., Carter, A. C., Di Giorgio, F. P., Koszka, K., et al. (2009). *A small-molecule inhibitor of tgf-Beta signaling replaces sox2 in reprogramming by inducing nanog.* Cell Stem Cell *5*, 491-503.

Iiyama, M., Kakihana, K., Kurosu, T. & Miura, O. (2006). *Reactive oxygen species generated by hematopoietic cytokines play roles in activation of receptor-mediated signaling and in cell cycle progression.* Cell Signal *18*, 174-182.

Ito, K., Hirao, A., Arai, F., Takubo, K., Matsuoka, S., Miyamoto, K., Ohmura, M., Naka, K., Hosokawa, K., Ikeda, Y. & Suda, T. (2006). *Reactive oxygen species act through p38 MAPK to limit the lifespan of hematopoietic stem cells.* Nat Med *12*, 446-451.

Itoh, F., Itoh, S., Goumans, M. J., Valdimarsdottir, G., Iso, T., Dotto, G. P., Hamamori, Y., Kedes, L., Kato, M. & ten Dijke Pt, P. (2004). *Synergy and antagonism between Notch and BMP receptor signaling pathways in endothelial cells.* Embo J *23*, 541-551.

Iwama, A., Oguro, H., Negishi, M., Kato, Y., Morita, Y., Tsukui, H., Ema, H., Kamijo, T., Katoh-Fukui, Y., Koseki, H., et al. (2004). *Enhanced self-renewal of hematopoietic stem cells mediated by the polycomb gene product Bmi-1.* Immunity *21*, 843-851.

Jaenisch, R. & Young, R. (2008). *Stem cells, the molecular circuitry of pluripotency and nuclear reprogramming.* Cell *132*, 567-582.

Jaleco, A. C., Neves, H., Hooijberg, E., Gameiro, P., Clode, N., Haury, M., Henrique, D. & Parreira, L. (2001). *Differential effects of Notch ligands Delta-1 and Jagged-1 in human lymphoid differentiation.* J Exp Med *194*, 991-1002.

Jang, Y. Y., Collector, M. I., Baylin, S. B., Diehl, A. M. & Sharkis, S. J. (2004). *Hematopoietic stem cells convert into liver cells within days without fusion.* Nat Cell Biol *6*, 532-539.

Jang, Y. Y. & Sharkis, S. J. (2005). *Stem cell plasticity: a rare cell, not a rare event.* Stem Cell Rev *1*, 45-51.

Jang, Y. Y. & Sharkis, S. J. (2007). *A low level of reactive oxygen species selects for primitive hematopoietic stem cells that may reside in the low-oxygenic niche.* Blood *110*, 3056-3063.

Kaivo-Oja, N., Bondestam, J., Kamarainen, M., Koskimies, J., Vitt, U., Cranfield, M., Vuojolainen, K., Kallio, J. P., Olkkonen, V. M., Hayashi, M., et al. (2003). *Growth differentiation factor-9 induces Smad2 activation and inhibin B production in cultured human granulosa-luteal cells.* J Clin Endocrinol Metab *88*, 755-762.

Kambal, A., Mitchell, G., Cary, W., Gruenloh, W., Jung, Y., Kalomoiris, S., Nacey, C., McGee, J., Lindsey, M., Fury, B., et al. (2011). *Generation of HIV-1 resistant and functional macrophages from hematopoietic stem cell-derived induced pluripotent stem cells.* Mol Ther *19*, 584-593.

Karanu, F. N., Murdoch, B., Gallacher, L., Wu, D. M., Koremoto, M., Sakano, S. & Bhatia, M. (2000). *The notch ligand jagged-1 represents a novel growth factor of human hematopoietic stem cells.* J Exp Med *192*, 1365-1372.

Karanu, F. N., Murdoch, B., Miyabayashi, T., Ohno, M., Koremoto, M., Gallacher, L., Wu, D., Itoh, A., Sakano, S. & Bhatia, M. (2001). *Human homologues of Delta-1 and Delta-4 function as mitogenic regulators of primitive human hematopoietic cells.* Blood *97*, 1960-1967.

Karlsson, G., Blank, U., Moody, J. L., Ehinger, M., Singbrant, S., Deng, C. X. & Karlsson, S. (2007). *Smad4 is critical for self-renewal of hematopoietic stem cells.* J Exp Med *204*, 467-474.

Kasper, M., Jaks, V., Fiaschi, M. & Toftgard, R. (2009). *Hedgehog signaling in breast cancer.* Carcinogenesis *30*, 903-911.

Kirstetter, P., Anderson, K., Porse, B. T., Jacobsen, S. E. & Nerlov, C. (2006). *Activation of the canonical Wnt pathway leads to loss of hematopoietic stem cell repopulation and multilineage differentiation block.* Nat Immunol *7*, 1048-1056.

Klump, H., Schiedlmeier, B. & Baum, C. (2005). *Control of self-renewal and differentiation of hematopoietic stem cells: HOXB4 on the threshold.* Ann N Y Acad Sci *1044*, 6-15.

Kobune, M., Ito, Y., Kawano, Y., Sasaki, K., Uchida, H., Nakamura, K., Dehari, H., Chiba, H., Takimoto, R., Matsunaga, T., et al. (2004). *Indian hedgehog gene transfer augments hematopoietic support of human stromal cells including NOD/SCID-beta2m-/- repopulating cells.* Blood *104*, 1002-1009.

Krivtsov, A. V., Twomey, D., Feng, Z., Stubbs, M. C., Wang, Y., Faber, J., Levine, J. E., Wang, J., Hahn, W. C., Gilliland, D. G., et al. (2006). *Transformation from committed progenitor to leukaemia stem cell initiated by MLL-AF9.* Nature *442*, 818-822.

Kroon, E., Krosl, J., Thorsteinsdottir, U., Baban, S., Buchberg, A. M. & Sauvageau, G. (1998). *Hoxa9 transforms primary bone marrow cells through specific collaboration with Meis1a but not Pbx1b.* Embo J *17*, 3714-3725.

Krosl, J., Austin, P., Beslu, N., Kroon, E., Humphries, R. K. & Sauvageau, G. (2003). *In vitro expansion of hematopoietic stem cells by recombinant TAT-HOXB4 protein.* Nat Med *9*, 1428-1432.

Kugawa, F., Watanabe, M. & Tamanoi, F. (2007). *Chemical Biology/Chemical genetics/chemical genomics: importance of chemical library.* Chem-Bio Informatics Journal *7*, 49-68.

Labbe, E., Letamendia, A. & Attisano, L. (2000). *Association of Smads with lymphoid enhancer binding factor 1/T cell-specific factor mediates cooperative signaling by the transforming growth factor-beta and wnt pathways.* Proc Natl Acad Sci U S A *97*, 8358-8363.

Larsson, J., Blank, U., Klintman, J., Magnusson, M. & Karlsson, S. (2005). *Quiescence of hematopoietic stem cells and maintenance of the stem cell pool is not dependent on TGF-beta signaling in vivo.* Exp Hematol *33*, 592-596.

Lauret, E., Catelain, C., Titeux, M., Poirault, S., Dando, J. S., Dorsch, M., Villeval, J. L., Groseil, A., Vainchenker, W., Sainteny, F. & Bennaceur-Griscelli, A. (2004). *Membrane-bound delta-4 notch ligand reduces the proliferative activity of primitive human hematopoietic CD34+CD38low cells while maintaining their LTC-IC potential.* Leukemia *18*, 788-797.

Lengerke, C., Grauer, M., Niebuhr, N. I., Riedt, T., Kanz, L., Park, I. H. & Daley, G. Q. (2009). *Hematopoietic development from human induced pluripotent stem cells.* Ann N Y Acad Sci *1176*, 219-227.

Li, W. & Ding, S. (2009). *Small molecules that modulate embryonic stem cell fate and somatic cell reprogramming.* Trends Pharmacol Sci.

Louis, I., Heinonen, K. M., Chagraoui, J., Vainio, S., Sauvageau, G. & Perreault, C. (2008). *The signaling protein Wnt4 enhances thymopoiesis and expands multipotent hematopoietic progenitors through beta-catenin-independent signaling.* Immunity *29*, 57-67.

Luis, T. C., Weerkamp, F., Naber, B. A., Baert, M. R., de Haas, E. F., Nikolic, T., Heuvelmans, S., De Krijger, R. R., van Dongen, J. J. & Staal, F. J. (2009). *Wnt3a deficiency irreversibly impairs hematopoietic stem cell self-renewal and leads to defects in progenitor cell differentiation.* Blood *113*, 546-554.

Lukaszewicz, A. I., McMillan, M. K. & Kahn, M. (2010). *Small molecules and stem cells. Potency and lineage commitment: the new quest for the fountain of youth.* J Med Chem *53*, 3439-3453.

Lyssiotis, C. A., Lairson, L. L., Boitano, A. E., Wurdak, H., Zhu, S. & Schultz, P. G. (2011). *Chemical control of stem cell fate and developmental potential.* Angew Chem Int Ed Engl *50*, 200-242.

Matsuno, K., Diederich, R. J., Go, M. J., Blaumueller, C. M. & Artavanis-Tsakonas, S. (1995). *Deltex acts as a positive regulator of Notch signaling through interactions with the Notch ankyrin repeats.* Development *121*, 2633-2644.

Melichar, H., Li, O., Ross, J., Haber, H., Cado, D., Nolla, H., Robey, E. A. & Winoto, A. (2011). *Comparative study of hematopoietic differentiation between human embryonic stem cell lines.* PLoS One *6*, e19854.

Merchant, A., Joseph, G., Wang, Q., Brennan, S. & Matsui, W. (2010). *Gli1 regulates the proliferation and differentiation of HSCs and myeloid progenitors.* Blood *115*, 2391-2396.

Mezey, E., Chandross, K. J., Harta, G., Maki, R. A. & McKercher, S. R. (2000). *Turning blood into brain: cells bearing neuronal antigens generated in vivo from bone marrow.* Science *290*, 1779-1782.

Milner, L. A., Kopan, R., Martin, D. I. & Bernstein, I. D. (1994). *A human homologue of the Drosophila developmental gene, Notch, is expressed in CD34+ hematopoietic precursors.* Blood *83*, 2057-2062.

Moon, R. T., Kohn, A. D., De Ferrari, G. V. & Kaykas, A. (2004). *WNT and beta-catenin signaling: diseases and therapies.* Nat Rev Genet *5*, 691-701.

Murdoch, B., Chadwick, K., Martin, M., Shojaei, F., Shah, K. V., Gallacher, L., Moon, R. T. & Bhatia, M. (2003). *Wnt-5A augments repopulating capacity and primitive hematopoietic development of human blood stem cells in vivo.* Proc Natl Acad Sci U S A *100*, 3422-3427.

Nakagawa, M., Koyanagi, M., Tanabe, K., Takahashi, K., Ichisaka, T., Aoi, T., Okita, K., Mochiduki, Y., Takizawa, N. & Yamanaka, S. (2008). *Generation of induced pluripotent stem cells without Myc from mouse and human fibroblasts.* Nat Biotechnol *26*, 101-106.

Narayan, A. D., Chase, J. L., Lewis, R. L., Tian, X., Kaufman, D. S., Thomson, J. A. & Zanjani, E. D. (2006). *Human embryonic stem cell-derived hematopoietic cells are capable of engrafting primary as well as secondary fetal sheep recipients.* Blood *107*, 2180-2183.

Nemeth, M. J., Topol, L., Anderson, S. M., Yang, Y. & Bodine, D. M. (2007). *Wnt5a inhibits canonical Wnt signaling in hematopoietic stem cells and enhances repopulation.* Proc Natl Acad Sci U S A *104*, 15436-15441.

Nishino, T., Miyaji, K., Ishiwata, N., Arai, K., Yui, M., Asai, Y., Nakauchi, H. & Iwama, A. (2009). *Ex vivo expansion of human hematopoietic stem cells by a small-molecule agonist of c-MPL.* Exp Hematol *37*, 1364-1377 e1364.

Niwa, A., Umeda, K., Chang, H., Saito, M., Okita, K., Takahashi, K., Nakagawa, M., Yamanaka, S., Nakahata, T. & Heike, T. (2009). *Orderly hematopoietic development of induced pluripotent stem cells via Flk-1(+) hemoangiogenic progenitors.* J Cell Physiol *221*, 367-377.

North, T. E., Goessling, W., Walkley, C. R., Lengerke, C., Kopani, K. R., Lord, A. M., Weber, G. J., Bowman, T. V., Jang, I. H., Grosser, T., et al. (2007). *Prostaglandin E2 regulates vertebrate haematopoietic stem cell homeostasis.* Nature *447*, 1007-1011.

Ogawa, M., Larue, A. C., Watson, P. M. & Watson, D. K. (2010). *Hematopoietic stem cell origin of connective tissues.* Exp Hematol *38*, 540-547.

Ohishi, K., Katayama, N., Shiku, H., Varnum-Finney, B. & Bernstein, I. D. (2003). *Notch signaling in hematopoiesis.* Semin Cell Dev Biol *14*, 143-150.

Ohishi, K., Varnum-Finney, B. & Bernstein, I. D. (2002). *Delta-1 enhances marrow and thymus repopulating ability of human CD34(+)CD38(-) cord blood cells.* J Clin Invest *110*, 1165-1174.

Ohmizono, Y., Sakabe, H., Kimura, T., Tanimukai, S., Matsumura, T., Miyazaki, H., Lyman, S. D. & Sonoda, Y. (1997). *Thrombopoietin augments ex vivo expansion of human cord blood-derived hematopoietic progenitors in combination with stem cell factor and flt3 ligand.* Leukemia *11*, 524-530.

Ohta, H., Sekulovic, S., Bakovic, S., Eaves, C. J., Pineault, N., Gasparetto, M., Smith, C., Sauvageau, G. & Humphries, R. K. (2007). *Near-maximal expansions of hematopoietic stem cells in culture using NUP98-HOX fusions.* Exp Hematol *35*, 817-830.

Pan, G. & Thomson, J. A. (2007). *Nanog and transcriptional networks in embryonic stem cell pluripotency.* Cell Res *17*, 42-49.

Park, I. K., Qian, D., Kiel, M., Becker, M. W., Pihalja, M., Weissman, I. L., Morrison, S. J. & Clarke, M. F. (2003). *Bmi-1 is required for maintenance of adult self-renewing haematopoietic stem cells.* Nature *423*, 302-305.

Parmar, K., Mauch, P., Vergilio, J. A., Sackstein, R. & Down, J. D. (2007). *Distribution of hematopoietic stem cells in the bone marrow according to regional hypoxia.* Proc Natl Acad Sci U S A *104*, 5431-5436.

Parrinello, S., Samper, E., Krtolica, A., Goldstein, J., Melov, S. & Campisi, J. (2003). *Oxygen sensitivity severely limits the replicative lifespan of murine fibroblasts.* Nat Cell Biol *5*, 741-747.

Peled, T., Glukhman, E., Hasson, N., Adi, S., Assor, H., Yudin, D., Landor, C., Mandel, J., Landau, E., Prus, E., et al. (2005). *Chelatable cellular copper modulates differentiation and self-renewal of cord blood-derived hematopoietic progenitor cells.* Exp Hematol *33*, 1092-1100.

Peled, T., Landau, E., Mandel, J., Glukhman, E., Goudsmid, N. R., Nagler, A. & Fibach, E. (2004a). *Linear polyamine copper chelator tetraethylenepentamine augments long-term ex vivo expansion of cord blood-derived CD34+ cells and increases their engraftment potential in NOD/SCID mice.* Exp Hematol *32*, 547-555.

Peled, T., Landau, E., Prus, E., Treves, A. J., Nagler, A. & Fibach, E. (2002). *Cellular copper content modulates differentiation and self-renewal in cultures of cord blood-derived CD34+ cells.* Br J Haematol *116*, 655-661.

Peled, T., Mandel, J., Goudsmid, R. N., Landor, C., Hasson, N., Harati, D., Austin, M., Hasson, A., Fibach, E., Shpall, E. J. & Nagler, A. (2004b). *Pre-clinical development of cord blood-derived progenitor cell graft expanded ex vivo with cytokines and the polyamine copper chelator tetraethylenepentamine.* Cytotherapy *6*, 344-355.

Pervaiz, S., Taneja, R. & Ghaffari, S. (2009). *Oxidative stress regulation of stem and progenitor cells.* Antioxid Redox Signal *11*, 2777-2789.

Pineault, N., Abramovich, C. & Humphries, R. K. (2005). *Transplantable cell lines generated with NUP98-Hox fusion genes undergo leukemic progression by Meis1 independent of its binding to DNA.* Leukemia *19*, 636-643.

Pineault, N., Abramovich, C., Ohta, H. & Humphries, R. K. (2004). *Differential and common leukemogenic potentials of multiple NUP98-Hox fusion proteins alone or with Meis1.* Mol Cell Biol *24*, 1907-1917.

Pineault, N., Cortin, V., Boyer, L., Garnier, A., Robert, A., Therien, C. & Roy, D. C. (2010). *Individual and synergistic cytokine effects controlling the expansion of cord blood CD34(+) cells and megakaryocyte progenitors in culture.* Cytotherapy *13*, 467-480.

Pineault, N., Helgason, C. D., Lawrence, H. J. & Humphries, R. K. (2002). *Differential expression of Hox, Meis1, & Pbx1 genes in primitive cells throughout murine hematopoietic ontogeny.* Exp Hematol *30*, 49-57.

Portela, A. & Esteller, M. (2010). *Epigenetic modifications and human disease.* Nat Biotechnol *28*, 1057-1068.

Quesenberry, P. J., Dooner, M. S. & Aliotta, J. M. (2010). *Stem cell plasticity revisited: the continuum marrow model and phenotypic changes mediated by microvesicles.* Exp Hematol *38*, 581-592.

Radtke, F., Wilson, A., Stark, G., Bauer, M., van Meerwijk, J., MacDonald, H. R. & Aguet, M. (1999). *Deficient T cell fate specification in mice with an induced inactivation of Notch1.* Immunity *10*, 547-558.

Raya, A., Rodriguez-Piza, I., Guenechea, G., Vassena, R., Navarro, S., Barrero, M. J., Consiglio, A., Castella, M., Rio, P., Sleep, E., et al. (2009). *Disease-corrected haematopoietic progenitors from Fanconi anaemia induced pluripotent stem cells.* Nature *460*, 53-59.

Rizo, A., Dontje, B., Vellenga, E., de Haan, G. & Schuringa, J. J. (2008). *Long-term maintenance of human hematopoietic stem/progenitor cells by expression of BMI1.* Blood *111*, 2621-2630.

Sakamoto, H., Tsuji-Tamura, K. & Ogawa, M. (2010). *Hematopoiesis from pluripotent stem cell lines.* Int J Hematol *91*, 384-391.

Salvagiotto, G., Burton, S., Daigh, C. A., Rajesh, D., Slukvin, II, & Seay, N. J. (2011). *A defined, feeder-free, serum-free system to generate in vitro hematopoietic progenitors and differentiated blood cells from hESCs and hiPSCs.* PLoS One *6*, e17829.

Sauvageau, G., Lansdorp, P. M., Eaves, C. J., Hogge, D. E., Dragowska, W. H., Reid, D. S., Largman, C., Lawrence, H. J. & Humphries, R. K. (1994). *Differential expression of homeobox genes in functionally distinct CD34+ subpopulations of human bone marrow cells.* Proc Natl Acad Sci U S A *91*, 12223-12227.

Sauvageau, G., Thorsteinsdottir, U., Eaves, C. J., Lawrence, H. J., Largman, C., Lansdorp, P. M. & Humphries, R. K. (1995). *Overexpression of HOXB4 in hematopoietic cells causes the selective expansion of more primitive populations in vitro and in vivo.* Genes Dev *9*, 1753-1765.

Scadden, D. T. (2006). *The stem-cell niche as an entity of action.* Nature *441*, 1075-1079.

Scheller, M., Huelsken, J., Rosenbauer, F., Taketo, M. M., Birchmeier, W., Tenen, D. G. & Leutz, A. (2006). *Hematopoietic stem cell and multilineage defects generated by constitutive beta-catenin activation.* Nat Immunol *7*, 1037-1047.

Schiedlmeier, B., Klump, H., Will, E., Arman-Kalcek, G., Li, Z., Wang, Z., Rimek, A., Friel, J., Baum, C. & Ostertag, W. (2003). *High-level ectopic HOXB4 expression confers a profound in vivo competitive growth advantage on human cord blood CD34+ cells, but impairs lymphomyeloid differentiation.* Blood *101*, 1759-1768.

Schuringa, J. J., Wu, K., Morrone, G. & Moore, M. A. (2004). *Enforced activation of STAT5A facilitates the generation of embryonic stem-derived hematopoietic stem cells that contribute to hematopoiesis in vivo.* Stem Cells *22*, 1191-1204.

Seet, L. F., Teng, E., Lai, Y. S., Laning, J., Kraus, M., Wnendt, S., Merchav, S. & Chan, S. L. (2009). *Valproic acid enhances the engraftability of human umbilical cord blood hematopoietic stem cells expanded under serum-free conditions.* Eur J Haematol *82*, 124-132.

Shi, Y., Desponts, C., Do, J. T., Hahm, H. S., Scholer, H. R. & Ding, S. (2008). *Induction of pluripotent stem cells from mouse embryonic fibroblasts by Oct4 and Klf4 with small-molecule compounds.* Cell Stem Cell *3*, 568-574.

Shimizu, K., Chiba, S., Saito, T., Kumano, K. & Hirai, H. (2000). *Physical interaction of Delta1, Jagged1, & Jagged2 with Notch1 and Notch3 receptors.* Biochem Biophys Res Commun *276*, 385-389.

Simonnet, A. J., Nehme, J., Vaigot, P., Barroca, V., Leboulch, P. & Tronik-Le Roux, D. (2009). *Phenotypic and functional changes induced in hematopoietic stem/progenitor cells after gamma-ray radiation exposure.* Stem Cells *27*, 1400-1409.

Simsek, T., Kocabas, F., Zheng, J., Deberardinis, R. J., Mahmoud, A. I., Olson, E. N., Schneider, J. W., Zhang, C. C. & Sadek, H. A. (2010). *The distinct metabolic profile of hematopoietic stem cells reflects their location in a hypoxic niche.* Cell Stem Cell *7*, 380-390.

Singh, K. P., Casado, F. L., Opanashuk, L. A. & Gasiewicz, T. A. (2009). *The aryl hydrocarbon receptor has a normal function in the regulation of hematopoietic and other stem/progenitor cell populations.* Biochem Pharmacol *77*, 577-587.

Sitnicka, E., Lin, N., Priestley, G. V., Fox, N., Broudy, V. C., Wolf, N. S. & Kaushansky, K. (1996). *The effect of thrombopoietin on the proliferation and differentiation of murine hematopoietic stem cells.* Blood *87*, 4998-5005.

Smith, L. L., Yeung, J., Zeisig, B. B., Popov, N., Huijbers, I., Barnes, J., Wilson, A. J., Taskesen, E., Delwel, R., Gil, J., et al. (2011). *Functional crosstalk between Bmi1 and MLL/Hoxa9 axis in establishment of normal hematopoietic and leukemic stem cells.* Cell Stem Cell *8*, 649-662.

Tadokoro, Y., Ema, H., Okano, M., Li, E. & Nakauchi, H. (2007). *De novo DNA methyltransferase is essential for self-renewal, but not for differentiation, in hematopoietic stem cells.* J Exp Med *204*, 715-722.

Takahashi, K. & Yamanaka, S. (2006). *Induction of pluripotent stem cells from mouse embryonic and adult fibroblast cultures by defined factors.* Cell *126*, 663-676.

Takubo, K., Goda, N., Yamada, W., Iriuchishima, H., Ikeda, E., Kubota, Y., Shima, H., Johnson, R. S., Hirao, A., Suematsu, M. & Suda, T. (2010). *Regulation of the HIF-1alpha level is essential for hematopoietic stem cells.* Cell Stem Cell *7*, 391-402.

Tang, Y., Chen, J. & Young, N. S. (2009). *Expansion of haematopoietic stem cells from normal donors and bone marrow failure patients by recombinant hoxb4.* Br J Haematol *144*, 603-612.

Theise, N. D. (2010). *Stem cell plasticity: recapping the decade, mapping the future.* Exp Hematol *38*, 529-539.

Tian, X. & Kaufman, D. S. (2008). *Differentiation of embryonic stem cells towards hematopoietic cells: progress and pitfalls.* Curr Opin Hematol *15*, 312-318.

Tolar, J., Park, I. H., Xia, L., Lees, C. J., Peacock, B., Webber, B., McElmurry, R. T., Eide, C. R., Orchard, P. J., Kyba, M., et al. (2011). *Hematopoietic differentiation of induced pluripotent stem cells from patients with mucopolysaccharidosis type I (Hurler syndrome).* Blood *117*, 839-847.

Trowbridge, J. J., Scott, M. P. & Bhatia, M. (2006). *Hedgehog modulates cell cycle regulators in stem cells to control hematopoietic regeneration.* Proc Natl Acad Sci U S A *103*, 14134-14139.

Trowbridge, J. J., Snow, J. W., Kim, J. & Orkin, S. H. (2009). *DNA methyltransferase 1 is essential for and uniquely regulates hematopoietic stem and progenitor cells.* Cell Stem Cell *5*, 442-449.

Van Den Berg, D. J., Sharma, A. K., Bruno, E. & Hoffman, R. (1998). *Role of members of the Wnt gene family in human hematopoiesis.* Blood *92*, 3189-3202.

Varnum-Finney, B., Brashem-Stein, C. & Bernstein, I. D. (2003). *Combined effects of Notch signaling and cytokines induce a multiple log increase in precursors with lymphoid and myeloid reconstituting ability.* Blood *101*, 1784-1789.

Varnum-Finney, B., Halasz, L. M., Sun, M., Gridley, T., Radtke, F. & Bernstein, I. D. (2011). *Notch2 governs the rate of generation of mouse long- and short-term repopulating stem cells.* J Clin Invest *121*, 1207-1216.

Varnum-Finney, B., Xu, L., Brashem-Stein, C., Nourigat, C., Flowers, D., Bakkour, S., Pear, W. S. & Bernstein, I. D. (2000). *Pluripotent, cytokine-dependent, hematopoietic stem cells are immortalized by constitutive Notch1 signaling.* Nat Med *6*, 1278-1281.

Walenda, T., Bokermann, G., Ventura Ferreira, M. S., Piroth, D. M., Hieronymus, T., Neuss, S., Zenke, M., Ho, A. D., Muller, A. M. & Wagner, W. (2011). *Synergistic effects of growth factors and mesenchymal stromal cells for expansion of hematopoietic stem and progenitor cells.* Exp Hematol *39*, 617-628.

Walker, L., Lynch, M., Silverman, S., Fraser, J., Boulter, J., Weinmaster, G. & Gasson, J. C. (1999). *The Notch/Jagged pathway inhibits proliferation of human hematopoietic progenitors in vitro.* Stem Cells *17*, 162-171.

Wang, L., Menendez, P., Shojaei, F., Li, L., Mazurier, F., Dick, J. E., Cerdan, C., Levac, K. & Bhatia, M. (2005a). *Generation of hematopoietic repopulating cells from human embryonic stem cells independent of ectopic HOXB4 expression.* J Exp Med *201*, 1603-1614.

Wang, N., Kim, H. G., Cotta, C. V., Wan, M., Tang, Y., Klug, C. A. & Cao, X. (2006). *TGFbeta/BMP inhibits the bone marrow transformation capability of Hoxa9 by repressing its DNA-binding ability.* Embo J *25*, 1469-1480.

Wang, Y. & Nakayama, N. (2009). *WNT and BMP signaling are both required for hematopoietic cell development from human ES cells.* Stem Cell Res *3*, 113-125.

Wang, Y., Umeda, K. & Nakayama, N. (2010). *Collaboration between WNT and BMP signaling promotes hemoangiogenic cell development from human fibroblast-derived iPS cells.* Stem Cell Res *4*, 223-231.

Wang, Y., Yates, F., Naveiras, O., Ernst, P. & Daley, G. Q. (2005b). *Embryonic stem cell-derived hematopoietic stem cells.* Proc Natl Acad Sci U S A *102*, 19081-19086.

Watts, K., Zhang, X., Beard, B., Chiu, S. Y., Trobridge, G. D., Humphries, R. K. & Kiem, H. P. (2011). *Differential Effects of HOXB4 and NUP98-HOXA10hd on Hematopoietic Repopulating Cells in a Nonhuman Primate Model. Hum Gene Ther.*

Weinreich, M. A., Lintmaer, I., Wang, L., Liggitt, H. D., Harkey, M. A. & Blau, C. A. (2006). *Growth factor receptors as regulators of hematopoiesis.* Blood *108*, 3713-3721.

Wernig, M., Lengner, C. J., Hanna, J., Lodato, M. A., Steine, E., Foreman, R., Staerk, J., Markoulaki, S. & Jaenisch, R. (2008). *A drug-inducible transgenic system for direct reprogramming of multiple somatic cell types.* Nat Biotechnol *26*, 916-924.

Wiles, M. V. & Keller, G. (1991). *Multiple hematopoietic lineages develop from embryonic stem (ES) cells in culture.* Development *111*, 259-267.

Willert, K., Brown, J. D., Danenberg, E., Duncan, A. W., Weissman, I. L., Reya, T., Yates, J. R., 3rd, & Nusse, R. (2003). *Wnt proteins are lipid-modified and can act as stem cell growth factors.* Nature *423*, 448-452.

Wojcik, M., Burzynska-Pedziwiatr, I. & Wozniak, L. A. (2010). *A review of natural and synthetic antioxidants important for health and longevity.* Curr Med Chem *17*, 3262-3288.

Woods, N. B., Parker, A. S., Moraghebi, R., Lutz, M. K., Firth, A. L., Brennand, K. J., Berggren, W. T., Raya, A., Belmonte, J. C., Gage, F. H. & Verma, I. M. (2011). *Brief report: efficient generation of hematopoietic precursors and progenitors from human pluripotent stem cell lines.* Stem Cells *29*, 1158-1164.

Wu, Q., Chen, X., Zhang, J., Loh, Y. H., Low, T. Y., Zhang, W., Zhang, W., Sze, S. K., Lim, B. & Ng, H. H. (2006). *Sall4 interacts with Nanog and co-occupies Nanog genomic sites in embryonic stem cells.* J Biol Chem *281*, 24090-24094.

Xu, D., Alipio, Z., Fink, L. M., Adcock, D. M., Yang, J., Ward, D. C. & Ma, Y. (2009). *Phenotypic correction of murine hemophilia A using an iPS cell-based therapy.* Proc Natl Acad Sci U S A *106*, 808-813.

Yamazaki, S., Iwama, A., Takayanagi, S., Eto, K., Ema, H. & Nakauchi, H. (2009). *TGF-beta as a candidate bone marrow niche signal to induce hematopoietic stem cell hibernation.* Blood *113*, 1250-1256.

Yang, J., Gao, C., Chai, L. & Ma, Y. (2010). *A novel SALL4/OCT4 transcriptional feedback network for pluripotency of embryonic stem cells.* PLoS One *5*, e10766.

Yeoh, J. S., van Os, R., Weersing, E., Ausema, A., Dontje, B., Vellenga, E. & de Haan, G. (2006). *Fibroblast growth factor-1 and -2 preserve long-term repopulating ability of hematopoietic stem cells in serum-free cultures.* Stem Cells *24*, 1564-1572.

Yoshida, Y., Takahashi, K., Okita, K., Ichisaka, T. & Yamanaka, S. (2009). *Hypoxia enhances the generation of induced pluripotent stem cells.* Cell Stem Cell *5*, 237-241.

Young, J. C., Wu, S., Hansteen, G., Du, C., Sambucetti, L., Remiszewski, S., O'Farrell, A. M., Hill, B., Lavau, C. & Murray, L. J. (2004). *Inhibitors of histone deacetylases promote hematopoietic stem cell self-renewal.* Cytotherapy *6*, 328-336.

Yu, J., Vodyanik, M. A., Smuga-Otto, K., Antosiewicz-Bourget, J., Frane, J. L., Tian, S., Nie, J., Jonsdottir, G. A., Ruotti, V., Stewart, R., et al. (2007). *Induced pluripotent stem cell lines derived from human somatic cells.* Science *318*, 1917-1920.

Zhang, C. C., Kaba, M., Iizuka, S., Huynh, H. & Lodish, H. F. (2008a). *Angiopoietin-like 5 and IGFBP2 stimulate ex vivo expansion of human cord blood hematopoietic stem cells as assayed by NOD/SCID transplantation.* Blood *111*, 3415-3423.

Zhang, C. C. & Lodish, H. F. (2005). *Murine hematopoietic stem cells change their surface phenotype during ex vivo expansion.* Blood *105*, 4314-4320.

Zhang, X. B., Beard, B. C., Beebe, K., Storer, B., Humphries, R. K. & Kiem, H. P. (2006). *Differential effects of HOXB4 on nonhuman primate short- and long-term repopulating cells.* PLoS Med *3*, e173.

Zhang, X. B., Beard, B. C., Trobridge, G. D., Wood, B. L., Sale, G. E., Sud, R., Humphries, R. K. & Kiem, H. P. (2008b). *High incidence of leukemia in large animals after stem cell gene therapy with a HOXB4-expressing retroviral vector.* J Clin Invest *118*, 1502-1510.

Zhao, C., Blum, J., Chen, A., Kwon, H. Y., Jung, S. H., Cook, J. M., Lagoo, A. & Reya, T. (2007). *Loss of beta-catenin impairs the renewal of normal and CML stem cells in vivo.* Cancer Cell *12*, 528-541.

Part 3

Hematopoietic Stem Cells in Aging and Disease

11

Insights Into Stem Cell Aging

A. Herrera-Merchan, I. Hidalgo, L. Arranz and S. Gonzalez
Stem Cell Aging Group, Foundation Spanish National Cardiovascular Research Centre Carlos III. (CNIC), Madrid, Spain

1. Introduction

The increase in average life expectancy in many developed countries is generating an aging society and an associated increase in age-related health problems. Mammalian aging occurs in part because of a decline in the restorative capacity of tissue stem cells. The use of stem cells in regenerative medicine promises to revolutionize the treatment of acute and chronic degenerative conditions, and stem cell research holds the key to the development of such therapies. The hallmark of adult stem cells is their ability to both self-renew and differentiate into multiple lineages. This demands a complex and still poorly understood network of molecular interactions between diverse cell-intrinsic regulators of self-renewal, such as certain Polycomb proteins and the tumor suppressor $p16^{INK4a}$, both of which are absolutely required for the maintenance of certain stem cell population. Recent studies have begun to elucidate the molecular mechanisms underlying how stem cells decide between life and death, and highlight the importance of balance in their aging pathways.

2. Aging and stem cells

Recent advances in medicine research programs, and a better health care planning, have great influences in people living in many Western countries, increasing both quality of life and average lifespan. With the extension of lifetime, there is increasing interest in slowing or reversing the negative effects of aging. The fascinating discovery of tissue-resident adult stem and progenitor cells in recent years has led to an explosion of interest in the development of novel stem cell-based therapies to improve endogenous regenerative capacity or to repair damaged and diseased tissues.

A major function of stem cells and their differentiation hierarchies may be to preserve the DNA integrity of the whole organism. When mutations occur despite certain error-prevention capacities, potent tumor-suppressor mechanisms such as senescence and apoptosis eliminate the damaged stem cell, limiting its replicative expansion. However, when unrepaired genetic lesions in stem cells are passed on to their differentiated daughters, and accumulate with aging, it is required replacement of dead and non-functional cells with newly differentiated cells derived from stem- and progenitor-cells. To date, the best-studied adult tissue stem cell type is the hematopoietic stem cell (HSC), which gives rise to all of the mature blood cells, throughout the life of the organism. Hematopoiesis in mammals occurs in distinct temporal waves shifting from the extraembryonic yolk sac and fetal liver in embryos to bone marrow in adults. Primitive HSCs are the "true" stem cells, also termed the long-term repopulating HSCs (LT-HSCs), because they replenish the pool of blood cells by both maintaining the stem cells and allowing daughter cells to differentiate into the lymphoid, myeloid, and erythroid lineages. The daily replenishment of blood cells is achieved in large part by divisions and subsequent stepwise differentiation of cells descendants of LT-HSC pool, namely short term repopulating HSC (ST-HSC), and slightly more committed hematopoietic progenitor cells (MPP-HSC). The relative quiescence of LT-HSCs pool protects their genomic integrity by reducing the rounds of DNA replication and thus the probability of acquiring DNA damage that might compromise multilineage differentiation potential and/or render them malignant over time, though they appear to age with the host (Orkin and Zon, 2008). The rapid turnover of the hematopoietic system and the availability of advanced methods to study HSCs by different markers have led to this system being widely used as a model of the effects of aging on stem cell functionality (Figure 1). It is worthy to mention that although some aspects of aging may be shared by all somatic stem cell fractions, the mechanisms of aging are likely to differ between stem cell populations located in specific tissues (for example, intestine, muscle and bone marrow).

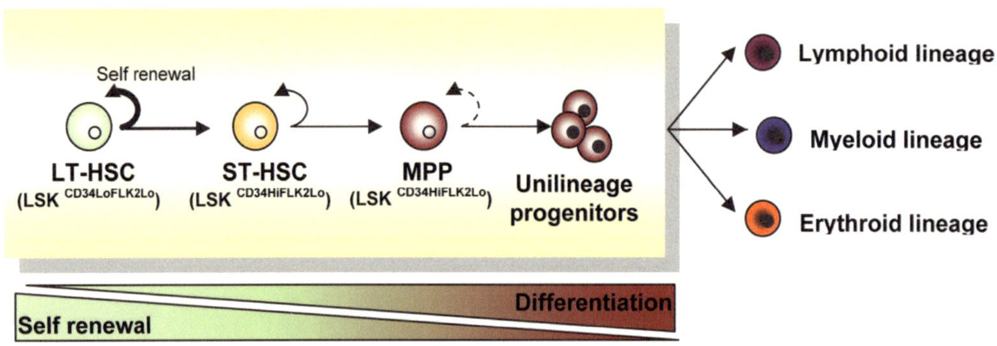

Fig. 1. The hierarchically primitive cells of the hematopoietic system. Long-Term hematopoietic stem cells (LT-HSC) maintain hematopoiesis by coordinating self-renewal, and production of short-term HSC (ST-HSC), and subsequently, the multipotent progenitors (MPP), which have an incredible capacity to divide and make other types of cells as they mature, although a limited ability to self-renew. Ultimately, this generates an array of mature blood cells with different functions: lymphoid blood cells (the B-cells; T-cells; natural killer or NK cells; plasma cells; dendritic cells and others), and erythroid and myeloid blood cells (the erythrocytes or red blood cells; megakaryocytes or platelet producing cells; granulocytes such as neutrophils, eosinophils, and basophils; and monocytes which make macrophages). The stem and progenitor cells can be purified to near-homogeneity by surface markers. For example, LT-HSCs express low levels of lineage markers, high levels of Sca1 and CD117/c-KIT receptor, and low levels of CD34 (LSK CD34 lo). With limited renewal potential, the ST-HSC pool has a similar surface immunophenotype to LT-HSC except that it has higher levels of CD34 (LSK CD34 hi). As ST-HSC in turn proliferates to form more differentiated MPP, they increase expression of another surface marker, FLK2 (LSK CD34 hi Flk2 hi).

3. The evidence for stem cell aging

A growing body of evidence shows that the capacity of stem cells to maintain tissue homeostasis declines with age, and suggests that this decline may account for many age-related phenotypes and diseases (Kirkwood and Austad, 2000). Significantly, engraftment of HSCs are capable of serial passages through a succession of mouse recipients, outliving the donor mouse (Ross et al., 1982; Siminovitch et al., 1964), though it is not possible to exceed up to five successful passages, and the recipients do not restore the hematopoietic system to the normal state (Gordon and Blackett, 1998). On the other hand, telomere length in blood cells of the transplanted recipient are 1-2 kb shorter than those in the donor, when evaluated several years following transplantation (Allsopp et al., 2003), which indicates that the level of telomerase is insufficient to prevent progressive telomere shortening in HSC. On the other hand, immunophenotypic characterization of hematopoietic stem- and progenitor-cell subsets diverges from function in old animals. The engraftment efficiency of immunophenotypically selected long-term HSCs from old mice approximately is threefold lower than that of the equivalent population from young mice (Morrison et al., 1997; Yilmaz et al., 2006). Also, age-related changes in stem-cell function include myeloid-biased differentiation and decreased homing ability (Liang et al., 2005). In conclusion, it has been extensively proved that the properties of HSCs change in several ways as they age, but still is poorly known which are the changes in the intrinsic and extrinsic factors involved that regulate the self-renewal and multilineage differentiation capacities of these regenerative cells (Huang et al., 2007).

Although the stem- and progenitor-cell proliferation guarantees tissue repair, and thereby regeneration, it can also develop hyperproliferative diseases, like cancer, risk that is moderated by tumor-suppressors mechanisms. For example, while the increased expression of tumor suppressors with age (p53, $p16^{INK4a}$) inhibits the development of cancer (inducing apoptosis or/and senescence) (Krishnamurthy et al., 2004; Ressler et al., 2006), over time it may have a negative effect on stem cell functionality, reducing capacity for self-renewal or differentiation, and ultimately leading to aging phenotypes (Beausejour and Campisi, 2006; Rodier et al., 2007). Thus, it is thought that many of the same mechanisms that contribute to cellular aging also act as suppressors of neoplastic growth (Campisi, 2005) (Figure 2). We will therefore need a better understanding of age-related changes in stem cell function by altering genetically the expression of tumor suppressors, which may improve effective longevity-promoting therapies.

4. Self-renewal regulators in adult HSCs

Stem cells are crucial for the homeostatic maintenance of mature, functional cells in many tissues throughout the lifetime of the animal, and this pool of stem cells must itself be maintained (Muller-Sieburg and Sieburg, 2008). This is achieved by self-renewal, a specialized cell division in which one or both daughter cells remain undifferentiated and retain essentially the same replication potential of the parent. The self-renewal program must involve the activity of dedicated regulatory genes (Gazit et al., 2008); but although the phenotypic and functional properties of HSCs have been characterized extensively, we have only just begun to understand how self-renewal is regulated.

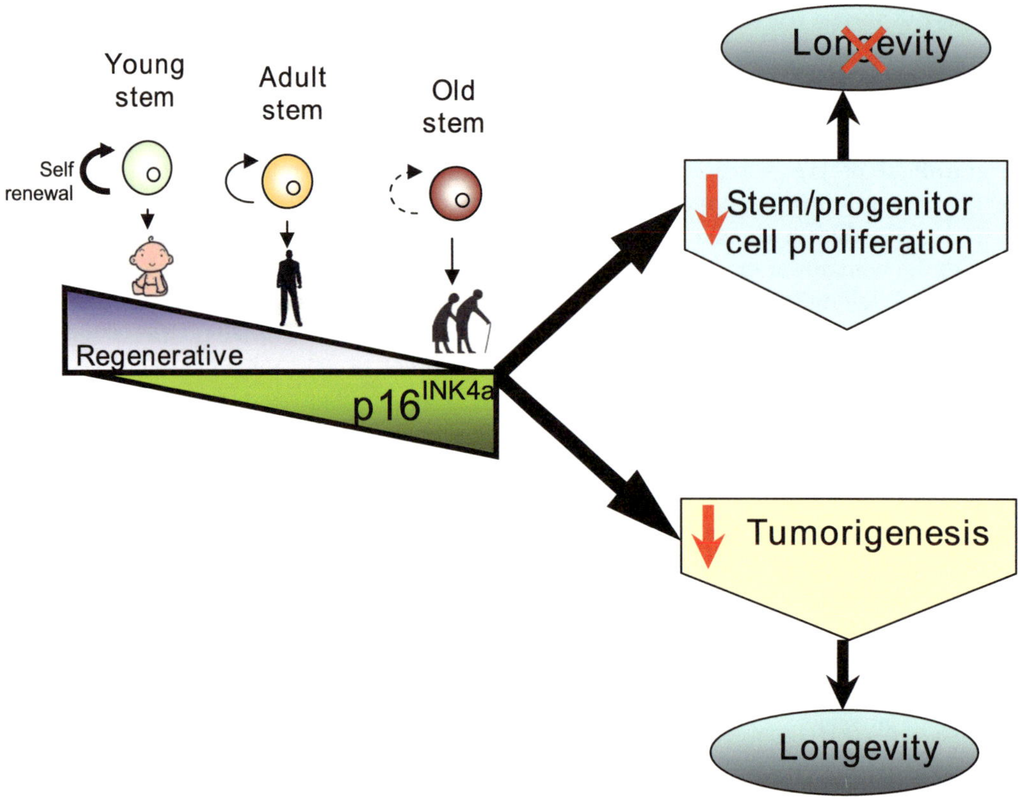

Fig. 2. Potential stem cell stage: interplay between aging and cancer. During normal aging, stem cells accumulate DNA damage as the consequence of endogenous (telomere dysfunction, oxidative stress) or exogenous (oxidative stress, g-irradiation, UV light, and others) attacks. This provokes subsequent stress-dependent changes (for example, accumulation of the products from the *INK4a/ARF* locus or telomere shortening), which activates checkpoint responses that result in apoptosis or cellular senescence. If these events occur in stem/progenitor cells, there is a decrease in the overall number and/or functionality of both stem and progenitor cells, leading an alteration of tissue homeostasis and regenerative capacity–a phenomenon that might contribute to aging and aged-related pathologies. If, instead, DNA mutations that inactivate these checkpoint pathways accumulate (for instance, loss of $p16^{INK4a}$ or reactivation of telomerase), then cancer can arise.

Polycomb complex in the maintenance of stemness. PcG proteins regulate self-renewal and lineage restriction in stem cells by inducing reversible chromatin modifications. PcG proteins have attracted increasing attention in stem cell and cancer stem cell research, given that it is now widely recognized that dynamic reprogramming of cells, for instance during differentiation, requires alterations to the epigenetic status of genes (Valk-Lingbeek et al., 2004). These features makes them interesting subjects for stem cell research, since it is conceivable that dynamic reprogramming of cells, for instance during differentiation, requires alterations in the epigenetic state of gene expression programs. The two major multiprotein PcG complexes identified to date, PRC1 and PRC2, function in a cooperative manner to maintain gene silencing (Pietersen et al., 2008) (Table 1). PRC2 initiates silencing, whereas PRC1 maintains and stabilizes gene repression. PRC2 contains histone methyltransferases (HMTs) that methylate lysines 9 and 27 on histone H3 and lysine 26 on histone H1. Deletion of PRC2 genes in mice results in early embryonic death, underscoring their importance in development. PRC1 recognizes the H3 lysine 27 methyl group added by PRC2 (Valk-Lingbeek et al., 2004), and subsequently the monoubiquityl-ligase activity of the PRC1 proteins Bmi1 or Ring1A/B toward histone H2A generates uH2AK119, which prevents access of the transcription machinery and facilitates chromatin compaction (Wang et al., 2004). Mouse mutants of most PRC1 members, in spite of displaying homeotic transformations, survive until birth as a result of partial functional redundancy provided by homologues, an exception being Ring1B-deficient mice (Voncken et al., 2003).

PRC2 is recruited to target genes by the cofactor jARID2 (jumonji/ARID domain-containing 2). Paradoxically, jARID2 also seems to inhibit PRC2 methyltransferase activity and may therefore regulate both the targeting and fine-tuning of PRC2 activity in stem cells and during differentiation (Panning, 2010). Once PRC1 recognizes and binds the H3K27me3 mark added by PRC2, it recruits additional proteins to establish the repressed chromatin configuration (Jones and Baylin, 2007). Gene promoters marked with H3K27me3 in ESCs are significantly more likely than other promoters to become methylated in cancer (Schlesinger et al., 2007). Moreover, the PcG targets in normal prostate cells are the same as those that become methylated in prostate cancer (Gal-Yam et al., 2008). Thus, altered chromatin structure does not always result in changes in gene expression associated with disease. Rather, disease results from the replacement of PcG repressive histone marks with methylation directly on DNA, which locks the chromatin in an inactive state, a process called epigenetic switching (Gal-Yam et al., 2008). Although the mechanism underlying predisposition of PcG targets to DNA methylation is not fully understood, the PRC1 component Cbx7 (chromobox homologue 7) was recently shown to interact directly with DNA (cytosine-5)-methyltransferase (DNMT)1 and DNMT3B at PcG target genes, establishing a link between histone and DNA methylation (Mohammad et al., 2009).

Among PcG proteins, the PRC1 component Bmi1 is a fundamental self-renewal regulator, being required for self-renewal of all postnatal stem cell populations studied to date (Molofsky et al., 2003; Park et al., 2003; van der Lugt et al., 1994). Bmi1 was originally described as a proto-oncogene that induces B and T cell leukemias (van Lohuizen et al., 1991), and is overexpressed in several human cancers, including mantle cell lymphoma, colorectal carcinoma, liver carcinomas, non-small-cell lung cancer, and cerebral tumors such as medulloblastomas (Martin-Perez et al., 2010). This evidence has strongly influenced cancer research, supporting the above-mentioned theory that cancer is essentially a stem cell disorder (Reya et al., 2001). The self-renewal function of Bmi1 in adult stem cells relies largely on the silencing of one of its targets, the locus encoding the p16INK4a and ARF tumor suppressors (Molofsky et al., 2006). Deletion of p16INK4a and/or ARF partially rescues the self-renewal defects observed in various stem cell populations from Bmi1-null mice. Nevertheless, as described by the authors, this rescue is incomplete and thus other major Bmi1 regulated genes must exist. Candidates for additional Bmi1 targets in the context of self renewal are the Hox (homeobox) genes. A subset of Hox genes has been implicated in mammalian brain development, and several of them are highly expressed in neurospheres formed in vitro from cultured neural stem cells of the subventricular zone of

PcG	Mouse	Human	Function	Hematopoietic defect
PRC1	Cbx2/M33 Cbx4/Mpc2 Cbx8/Pc3	CBX2/HPC1 CBX4/HPC2 CBX8/HPC3	Binds trimethylated H3K27	Hypoplasia of spleen and thymus, maturation arrest in T cell development
	Bmi1	BMI1	Co-factor of E3 ubiquitin ligase (RING1A/B) and compacts polynucleosome	Postnatal pancytopenia Impaired HSC self-renewal, hypoplasia of spleen and thymus. Maturation arrest in T and B cell development
	Ring1/Ring1a Rnf2/Ring1b	RING1/RING1A RNF2/RING1B	E3 ubiquitin ligase for H2AK119	Decreased bone marrow cells and increased myeloid progenitors
PRC2	Eed	EED	Stimulates histone methyltransferase activity of Ezh1/2	Myelo- and lymphoproliferative disease
	Suz12	SUZ12	Stimulates histone methyltransferase activity of Ezh1/2	Enhanced HSC activity
	Ezh1/Enx2 Ezh2/Enx1	EZH1 EZH2	Catalytic subunit of H3K27 histone methyltransferase	Maturation arrest of T cells at the early CD4, CD8 double negative stage in thymus and of B cells with impaired rearrangement of the IgH gene $Ezh2^{-/-}$ Lethal in HSC

Adapted from Takaaki Konuma et al. Develop. Growth Differ. (2010) 52, 505–516

Table 1. Principcal components of the Polycomb Group Complexes and their hematopoietic defects in mutant mice.

Bmi-null mice (Molofsky et al., 2006). More recently, Bmi1 and Ring1A were shown to play essential roles in H2A ubiquitylation and Hox gene silencing. Knockout of Bmi1 results in significant loss of H2A ubiquitylation and an upregulation of HoxC13 expression, whereas Ezh2-mediated H3-K27 methylation is not affected (Cao et al., 2005). Similar findings have been described for the HoxC5 gene. However, considering that PcG proteins modify the chromatin of large sets of genes (Kirmizis et al., 2004), a great number of additional targets are likely to exist. For instance, both PRC2 and Bmi1 have recently been shown to play roles in the repression of E-cadherin expression (Yang et al., 2010). Interestingly, PcG genes have been shown to have a tumor suppressive function. In *Drosophila*, PcG proteins repress JAK/STAT and Notch signaling activity, whose activation drives disc cell overproliferation (Classen et al., 2009). Specifically, the *Drosophila* complex Psc (posterior sex combs), which includes Bmi1, and Suz12 (suppressor of zeste 12) play a tumor suppressive role mediated by Wnt repression in follicle stem cells(Li et al., 2010). In mammals, Eed (embryonic ectoderm development protein) displays tumor suppressive activity in the mouse hematopoietic system (Richie et al., 2002). Thus, PcG genes have been suggested to behave either as proto-oncogenes or as tumor suppressors depending on the tissue, cell context, developmental stage and gene dosage.

Bmi1 is regulated by Sonic Hedgehog, providing a direct connection between PcG and a major stem cell-specific pathway (Leung et al., 2004). Furthermore, activation of either Hedgehog or Notch signaling has been shown to increase Bmi1 expression, whereas siRNA knockdown of Bmi1 abrogates the effects of Hedgehog or Notch signaling on sphere formation, a functional readout of stemness. Thus the effects of Hedgehog and Notch signaling on stem cell self-renewal appear to be largely dependent on Bmi1. A complex regulation of Bmi1 is suggested by the fact that distinct Bmi1 regulators have been found in different types of cancer, for example Twist1 in head and neck squamous cell carcinoma and the Zeb1 (zinc finger E-box binding homeobox 1) – miR-200 pathway in pancreatic cancers (Wellner et al., 2009). Furthermore, a single PcG function can be regulated by multiple factors, for example Snail1 regulates E-cadherin silencing by PRC2, whereas the action of PRC1 on this target is regulated by Twist1 (Yang et al., 2010).

In summary, PcG proteins, in particular Bmi1, are essential for self-renewal and proliferative potential, which are crucial for the maintenance of stemness, acting as a critical failsafe mechanism against loss of stem cells in response to senescence signals. In turn, Bmi1 must be finely-regulated to prevent uncontrolled replicative expansion and tumor induction. Despite the importance of PcG proteins, we are only beginning to unravel how these master regulators are themselves regulated to achieve an appropriate balance between ensuring stem cell longevity and preventing tumorigenesis.

The tumor suppressors $p16^{INK4a}$ and ARF. Cell-cycle regulators such as the *INK4/ARF* locus appear to play an important role in the reaction of adult stem cells to stress and aging. The *INK4/ARF* locus plays a central role in tumor suppression, reflected in its inactivation in almost 50% of human cancers (Sharpless, 2005). Indeed, this locus is regarded as one of the most important anti-oncogenic defenses of the mammalian genome, comparable in importance only to p53. The remarkable feature of the *INK4/ARF* locus is that it encodes three tumor suppressors in a genomic segment of about 50 kb: $p16^{INK4a}$, its related family member $p15^{INK4b}$, and *ARF* (called $p19^{ARF}$ in mice and $p14^{ARF}$ in humans). The actions of $p16^{INK4a}$, $p15^{INK4b}$ and ARF are well understood. Both $p16^{INK4a}$ and $p15^{INK4b}$ inhibit the kinase activity of CDK4/6-cycD complexes, thus contributing to the maintenance of the active, growth-suppressive form of the retinoblastoma (Rb) family of proteins. ARF contributes to the stability of p53 by inhibiting the p53-degrading activity of MDM2. Through the activation of Rb and p53, the *INK4/ARF* locus is able to induce cell senescence and cell death (Gil and Peters, 2006; Lowe and Sherr, 2003). These tumor suppressors have taken on additional importance given recent evidence that at least one product of the locus, $p16^{INK4a}$, also contributes to the decline in the replication potential of self-renewing cells during the aging of stem cells. The expression of $p16^{INK4a}$ is relatively low in the HSCs of young mice, but is upregulated with age or in response to cellular stresses (Janzen et al., 2006). Although the number of immunophenotypic HSCs increases with age in wild-type animals, HSC functionality is impaired. In particular, the HSC compartment of old animals is more rapidly exhausted by serial transplantation than that of young animals. In contrast, aging has the opposite effect on $p16^{INK4a}$-/- HSCs, with $p16^{INK4a}$-/- HSCs from old animals substantially outperforming young $p16^{INK4a}$-/- HSCs in serial transplantation assays (Janzen et al., 2006). In fact, old $p16^{INK4a}$-/- HSCs perform as well as young wild-type HSCs in this assay. Thus $p16^{INK4a}$ compromises HSC functionality in older mice. Similar results were obtained in studies of $p16^{INK4a}$-/- neuronal stem cells and pancreatic islets (Krishnamurthy et al., 2006; Molofsky et al., 2006), revealing a general role for $p16^{INK4a}$ in the regulation of stem cell and progenitor cell aging. Therefore, on one face of this coin, $p16^{INK4a}$ acts as a potent tumor suppressor that promotes longevity by suppressing the development of cancer, while on the flipside, the increase of $p16^{INK4a}$ levels with age impairs the proliferation of stem or progenitor cells, ultimately reducing longevity. Thus, $p16^{INK4a}$ seems to balance an equilibrium reducing cancer incidence, but also contributing to aging by decreasing stem cell self-renewal and proliferation. These observations suggest the provocative but as yet unproven notion that mammalian aging results in part from the beneficial effects of tumor suppressor proteins (Figure 2).

The transcription factor p53. Besides $p16^{INK4a}$ tumor suppressor, p53 is also a tumor suppressor that influences stem cell self-renewal, tissue regenerative capacity, age-related disease, and cancer, which activity is lost in nearly half of all human cancers(Toledo and Wahl, 2006). The p53 protein is normally inactive, due in part to its rapid degradation by the specific ubiquitin ligase Mdm2. A multitude of stresses converge on p53 through complex, and partially understood, signaling pathways that stabilize and modify p53. The analysis of the effect of p53 in aging has revealed a dual role that seems to depend on the intensity of p53 activity. Overexpression of short isoforms of p53 in mice have greater protection against tumor development than wild-type mice, while at the same time they show signs of premature aging (Maier et al., 2004; Tyner et al., 2002). However, mouse models of increased wild-type p53 activity do not present premature aging. In particular, bacterial artificial chromosome transgenic mice that bear a third copy of the p53 locus show a decreased cancer incidence but normal longevity and normal onset of aging phenotypes (Garcia-Cao et al., 2002; Matheu et al., 2007; Matheu et al., 2004). An additional mouse model, the super-INK4a/ARF mice, with an extra copy of the entire *INK4a/ARF* locus (being ARF an activator of p53), show a significantly reduced incidence of cancer, although the mice aged normally (Matheu et al., 2004). To investigate whether the concomitant expression of both tumor suppressors had a synergistic effect, mice that bear a third copy of the p53 locus and a third copy of the *INK4/ARF* locus show increased longevity and delayed aging in a manner that cannot be explained by their reduced incidence of cancer (Matheu et al., 2007). Therefore, and though the effects of p53 and *INK4/ARF* locus expression in aging are context and dosage dependent, these results suggest that under physiological aging (labeled by moderate increase of still regulated p53 activity), the damaged cells are eliminated by either triggering their self-destruction (by apoptosis) or by pulling them out of the proliferative pool (by inducing senescence). In contrast, by massive DNA damage, the presence of uncontrolled activity of p53 results in excessive elimination of cells by p53 that exhausts the capacity of tissue regeneration leading to premature aging.

The *INK4/ARF* locus and age-associated phenotypes. $p16^{INK4a}$ and ARF may also be broadly important to diseases of aging beyond their function in stem cells. Specifically, three research consortia that undertook genome-wide association studies across large, carefully annotated patient samples have reported an association between single nucleotide polymorphisms (SNPs) near to *INK4a/ARF* locus and frailty (Melzer et al., 2007), atherosclerotic heart disease (ASHD)(Helgadottir et al., 2007) (McPherson et al., 2007), and type-2 diabetes (Saxena et al., 2007; Zeggini et al., 2008) in large human cohorts. However, few of the associated SNPs near the locus, and associated with these phenotypes, are not in linkage disequilibrium with each other, which suggests that more than one polymorphism near the locus influences these aging phenotypes. Therefore, although these studies do not pinpoint specific polymorphisms that affect the risks of age-related diseases, there are only four genes in the vicinity of the mapped polymorphisms: $p16^{INK4a}$, ARF, $p15^{INK4b}$, and ANRIL (a noncoding RNA). More relative data suggest specific links: $p16^{INK4a}$ expression increases with age in pancreatic β cells, and $p16^{INK4a}$ deficiency increases β-cell regenerative capacity(Krishnamurthy et al., 2006), providing a mechanism by which polymorphisms that affect $p16^{INK4a}$ expression or activity might affect risk for type-2 diabetes. It remains unclear whether these polymorphisms influence the risk of frailty and heart disease through their effects on tissue regenerative capacity or by mechanisms that are completely independent of stem/progenitor cells. Nevertheless, in light of the murine genetic studies that link *INK4a/ARF* locus and stem cell function, proteins encoded by the locus are the strongest candidates to mediate the effects of these polymorphisms on the incidence of these common diseases that are associated with aging.

5. Conclusions

The regenerative capacity of many stem cells declines functionally with age and, this decline triggers in part many age-related symptoms, and the development of certain diseases. Recent evidences have demonstrated that certain tumor suppressors, like *$p16^{INK4a}$*, also suppresses the proliferation of stem or progenitor cells in the bone marrow, pancreas and brain. Thus, $p16^{INK4a}$ seems to balance equilibrium reducing cancer incidence, which promotes longevity, but also decreasing stem cell self-renewal and proliferation, compromising tissue regeneration and repair, which are likely to reduce longevity. These observations allow us to suggest the provocative but unproved hypothesis that mammalian aging results in part from the beneficial efforts of tumor suppressor proteins to interdict cancer. In this stage, characterization of how stem cells age, such as **the characterization of reliable biomarkers,** deregulated signaling pathways, loss of self-renewal or acquisition of defects in differentiation of stem cells, will contribute to understand the age-associated pathophysiological decline. Likewise, it is also essential to figure out the cellular and molecular components of stem cell niches, how the niche changes during aging, and whether senescent stem or support cells alter the niche. In summary, the rescue, treatment, or replacement of aged and dysfunctional adult stem and progenitor cells may provide novel avenues to treat diverse devastating premature aging and age-related disorders including hematopoietic and immune disorders, heart failure and cardiovascular diseases, neurodegenerative, muscular and gastrointestinal diseases, atherosclerosis and aggressive and lethal cancers.

6. Acknowledgement

This work was supported by the Human Frontiers Science Program Organization, the Spanish Ministries of Science and Innovation (SAF2010-15386) and Health (FIS PI06/0627). We thank Simon Bartlett for editing assistance. The CNIC is supported by the Ministry of Science and Innovation and the Pro-CNIC Foundation.

7. Abbreviatons

PcG, Polycomb Group, PRC1, Polycomb repressive complex 1, PRC2, polycomb repressive complex 2.

8. References

Allsopp, R.C., Morin, G.B., DePinho, R., Harley, C.B., and Weissman, I.L. (2003). Telomerase is required to slow telomere shortening and extend replicative lifespan of HSCs during serial transplantation. Blood *102*, 517-520.

Beausejour, C.M., and Campisi, J. (2006). Ageing: balancing regeneration and cancer. Nature *443*, 404-405.

Campisi, J. (2005). Senescent cells, tumor suppression, and organismal aging: good citizens, bad neighbors. Cell *120*, 513-522.

Cao, R., Tsukada, Y., and Zhang, Y. (2005). Role of Bmi-1 and Ring1A in H2A ubiquitylation and Hox gene silencing. Mol Cell *20*, 845-854.

Classen, A.K., Bunker, B.D., Harvey, K.F., Vaccari, T., and Bilder, D. (2009). A tumor suppressor activity of Drosophila Polycomb genes mediated by JAK-STAT signaling. Nat Genet *41*, 1150-1155.

Gal-Yam, E.N., Egger, G., Iniguez, L., Holster, H., Einarsson, S., Zhang, X., Lin, J.C., Liang, G., Jones, P.A., and Tanay, A. (2008). Frequent switching of Polycomb repressive marks and DNA hypermethylation in the PC3 prostate cancer cell line. Proc Natl Acad Sci U S A *105*, 12979-12984.

Garcia-Cao, I., Garcia-Cao, M., Martin-Caballero, J., Criado, L.M., Klatt, P., Flores, J.M., Weill, J.C., Blasco, M.A., and Serrano, M. (2002). "Super p53" mice exhibit enhanced DNA damage response, are tumor resistant and age normally. Embo J *21*, 6225-6235.

Gazit, R., Weissman, I.L., and Rossi, D.J. (2008). Hematopoietic stem cells and the aging hematopoietic system. Semin Hematol *45*, 218-224.

Gil, J., and Peters, G. (2006). Regulation of the INK4b-ARF-INK4a tumour suppressor locus: all for one or one for all. Nat Rev Mol Cell Biol *7*, 667-677.

Gordon, M.Y., and Blackett, N.M. (1998). Reconstruction of the hematopoietic system after stem cell transplantation. Cell Transplant *7*, 339-344.

Helgadottir, A., Thorleifsson, G., Manolescu, A., Gretarsdottir, S., Blondal, T., Jonasdottir, A., Sigurdsson, A., Baker, A., Palsson, A., Masson, G., *et al.* (2007). A common variant on chromosome 9p21 affects the risk of myocardial infarction. Science *316*, 1491-1493.

Huang, X., Cho, S., and Spangrude, G.J. (2007). Hematopoietic stem cells: generation and self-renewal. Cell Death Differ *14*, 1851-1859.

Janzen, V., Forkert, R., Fleming, H.E., Saito, Y., Waring, M.T., Dombkowski, D.M., Cheng, T., DePinho, R.A., Sharpless, N.E., and Scadden, D.T. (2006). Stem-cell ageing modified by the cyclin-dependent kinase inhibitor p16INK4a. Nature *443*, 421-426.

Jones, P.A., and Baylin, S.B. (2007). The epigenomics of cancer. Cell *128*, 683-692.

Kirkwood, T.B., and Austad, S.N. (2000). Why do we age? Nature *408*, 233-238.

Kirmizis, A., Bartley, S.M., Kuzmichev, A., Margueron, R., Reinberg, D., Green, R., and Farnham, P.J. (2004). Silencing of human polycomb target genes is associated with methylation of histone H3 Lys 27. Genes Dev *18*, 1592-1605.

Krishnamurthy, J., Ramsey, M.R., Ligon, K.L., Torrice, C., Koh, A., Bonner-Weir, S., and Sharpless, N.E. (2006). p16INK4a induces an age-dependent decline in islet regenerative potential. Nature *443*, 453-457.

Krishnamurthy, J., Torrice, C., Ramsey, M.R., Kovalev, G.I., Al-Regaiey, K., Su, L., and Sharpless, N.E. (2004). Ink4a/Arf expression is a biomarker of aging. J Clin Invest *114*, 1299-1307.

Leung, C., Lingbeek, M., Shakhova, O., Liu, J., Tanger, E., Saremaslani, P., Van Lohuizen, M., and Marino, S. (2004). Bmi1 is essential for cerebellar development and is overexpressed in human medulloblastomas. Nature *428*, 337-341.

Li, X., Han, Y., and Xi, R. (2010). Polycomb group genes Psc and Su(z)2 restrict follicle stem cell self-renewal and extrusion by controlling canonical and noncanonical Wnt signaling. Genes Dev *24*, 933-946.

Liang, Y., Van Zant, G., and Szilvassy, S.J. (2005). Effects of aging on the homing and engraftment of murine hematopoietic stem and progenitor cells. Blood *106*, 1479-1487.

Lowe, S.W., and Sherr, C.J. (2003). Tumor suppression by Ink4a-Arf: progress and puzzles. Curr Opin Genet Dev *13*, 77-83.

Maier, B., Gluba, W., Bernier, B., Turner, T., Mohammad, K., Guise, T., Sutherland, A., Thorner, M., and Scrable, H. (2004). Modulation of mammalian life span by the short isoform of p53. Genes Dev *18*, 306-319.

Martin-Perez, D., Piris, M.A., and Sanchez-Beato, M. (2010). Polycomb proteins in hematologic malignancies. Blood *116*, 5465-5475.

Matheu, A., Maraver, A., Klatt, P., Flores, I., Garcia-Cao, I., Borras, C., Flores, J.M., Viña, J., Blasco, M.A., and Serrano, M. (2007). Delayed aging through damage protection by the Arf/p53 pathway. Nature *in press.*

Matheu, A., Pantoja, C., Efeyan, A., Criado, L.M., Martin-Caballero, J., Flores, J.M., Klatt, P., and Serrano, M. (2004). Increased gene dosage of Ink4a/Arf results in cancer resistance and normal aging. Genes Dev *18*, 2736-2746.

McPherson, R., Pertsemlidis, A., Kavaslar, N., Stewart, A., Roberts, R., Cox, D.R., Hinds, D.A., Pennacchio, L.A., Tybjaerg-Hansen, A., Folsom, A.R., *et al.* (2007). A common allele on chromosome 9 associated with coronary heart disease. Science *316*, 1488-1491.

Melzer, D., Frayling, T.M., Murray, A., Hurst, A.J., Harries, L.W., Song, H., Khaw, K., Luben, R., Surtees, P.G., Bandinelli, S.S., *et al.* (2007). A common variant of the p16(INK4a) genetic region is associated with physical function in older people. Mech Ageing Dev *128*, 370-377.

Mohammad, H.P., Cai, Y., McGarvey, K.M., Easwaran, H., Van Neste, L., Ohm, J.E., O'Hagan, H.M., and Baylin, S.B. (2009). Polycomb CBX7 promotes initiation of heritable repression of genes frequently silenced with cancer-specific DNA hypermethylation. Cancer Res *69*, 6322-6330.

Molofsky, A.V., Pardal, R., Iwashita, T., Park, I.K., Clarke, M.F., and Morrison, S.J. (2003). Bmi-1 dependence distinguishes neural stem cell self-renewal from progenitor proliferation. Nature *425*, 962-967.

Molofsky, A.V., Slutsky, S.G., Joseph, N.M., He, S., Pardal, R., Krishnamurthy, J., Sharpless, N.E., and Morrison, S.J. (2006). Increasing p16INK4a expression decreases forebrain progenitors and neurogenesis during ageing. Nature *443*, 448-452.

Morrison, S.J., Wright, D.E., Cheshier, S.H., and Weissman, I.L. (1997). Hematopoietic stem cells: challenges to expectations. Curr Opin Immunol *9*, 216-221.

Muller-Sieburg, C., and Sieburg, H.B. (2008). Stem cell aging: survival of the laziest? Cell Cycle *7*, 3798-3804.

Orkin, S.H., and Zon, L.I. (2008). Hematopoiesis: an evolving paradigm for stem cell biology. Cell *132*, 631-644.

Panning, B. (2010). Fine-tuning silencing. Cell Stem Cell *6*, 3-4.

Park, I.K., Qian, D., Kiel, M., Becker, M.W., Pihalja, M., Weissman, I.L., Morrison, S.J., and Clarke, M.F. (2003). Bmi-1 is required for maintenance of adult self-renewing haematopoietic stem cells. Nature *423*, 302-305.

Pietersen, A.M., Evers, B., Prasad, A.A., Tanger, E., Cornelissen-Steijger, P., Jonkers, J., and van Lohuizen, M. (2008). Bmi1 regulates stem cells and proliferation and differentiation of committed cells in mammary epithelium. Curr Biol *18*, 1094-1099.

Ressler, S., Bartkova, J., Niederegger, H., Bartek, J., Scharffetter-Kochanek, K., Jansen-Durr, P., and Wlaschek, M. (2006). p16INK4A is a robust in vivo biomarker of cellular aging in human skin. Aging Cell *5*, 379-389.

Reya, T., Morrison, S.J., Clarke, M.F., and Weissman, I.L. (2001). Stem cells, cancer, and cancer stem cells. Nature *414*, 105-111.

Richie, E.R., Schumacher, A., Angel, J.M., Holloway, M., Rinchik, E.M., and Magnuson, T. (2002). The Polycomb-group gene eed regulates thymocyte differentiation and suppresses the development of carcinogen-induced T-cell lymphomas. Oncogene *21*, 299-306.

Rodier, F., Campisi, J., and Bhaumik, D. (2007). Two faces of p53: aging and tumor suppression. Nucleic Acids Res *35*, 7475-7484.

Ross, E.A., Anderson, N., and Micklem, H.S. (1982). Serial depletion and regeneration of the murine hematopoietic system. Implications for hematopoietic organization and the study of cellular aging. J Exp Med *155*, 432-444.

Saxena, V., Ondr, J.K., Magnusen, A.F., Munn, D.H., and Katz, J.D. (2007). The countervailing actions of myeloid and plasmacytoid dendritic cells control autoimmune diabetes in the nonobese diabetic mouse. J Immunol *179*, 5041-5053.

Schlesinger, Y., Straussman, R., Keshet, I., Farkash, S., Hecht, M., Zimmerman, J., Eden, E., Yakhini, Z., Ben-Shushan, E., Reubinoff, B.E., *et al.* (2007). Polycomb-mediated methylation on Lys27 of histone H3 pre-marks genes for de novo methylation in cancer. Nat Genet *39*, 232-236.

Sharpless, N.E. (2005). INK4a/ARF: a multifunctional tumor suppressor locus. Mutat Res *576*, 22-38.

Siminovitch, L., Till, J.E., and McCulloch, E.A. (1964). Decline in Colony-Forming Ability of Marrow Cells Subjected to Serial Transplantation into Irradiated Mice. J Cell Physiol *64*, 23-31.

Toledo, F., and Wahl, G.M. (2006). Regulating the p53 pathway: in vitro hypotheses, in vivo veritas. Nat Rev Cancer *6*, 909-923.

Tyner, S.D., Venkatachalam, S., Choi, J., Jones, S., Ghebranious, N., Igelmann, H., Lu, X., Soron, G., Cooper, B., Brayton, C., *et al.* (2002). p53 mutant mice that display early ageing-associated phenotypes. Nature *415*, 45-53.

Valk-Lingbeek, M.E., Bruggeman, S.W., and van Lohuizen, M. (2004). Stem cells and cancer; the polycomb connection. Cell *118*, 409-418.

van der Lugt, N.M., Domen, J., Linders, K., van Roon, M., Robanus-Maandag, E., te Riele, H., van der Valk, M., Deschamps, J., Sofroniew, M., van Lohuizen, M., *et al.* (1994). Posterior transformation, neurological abnormalities, and severe hematopoietic defects in mice with a targeted deletion of the bmi-1 proto-oncogene. Genes Dev *8*, 757-769.

van Lohuizen, M., Verbeek, S., Scheijen, B., Wientjens, E., van der Gulden, H., and Berns, A. (1991). Identification of cooperating oncogenes in E mu-myc transgenic mice by provirus tagging. Cell *65*, 737-752.

Voncken, J.W., Roelen, B.A., Roefs, M., de Vries, S., Verhoeven, E., Marino, S., Deschamps, J., and van Lohuizen, M. (2003). Rnf2 (Ring1b) deficiency causes gastrulation arrest and cell cycle inhibition. Proc Natl Acad Sci U S A *100*, 2468-2473.

Wang, L., Brown, J.L., Cao, R., Zhang, Y., Kassis, J.A., and Jones, R.S. (2004). Hierarchical recruitment of polycomb group silencing complexes. Mol Cell *14*, 637-646.

Wellner, U., Schubert, J., Burk, U.C., Schmalhofer, O., Zhu, F., Sonntag, A., Waldvogel, B., Vannier, C., Darling, D., zur Hausen, A., *et al.* (2009). The EMT-activator ZEB1 promotes tumorigenicity by repressing stemness-inhibiting microRNAs. Nat Cell Biol *11*, 1487-1495.

Yang, M.H., Hsu, D.S., Wang, H.W., Wang, H.J., Lan, H.Y., Yang, W.H., Huang, C.H., Kao, S.Y., Tzeng, C.H., Tai, S.K., *et al.* (2010). Bmi1 is essential in Twist1-induced epithelial-mesenchymal transition. Nat Cell Biol *12*, 982-992.

Yilmaz, O.H., Kiel, M.J., and Morrison, S.J. (2006). SLAM family markers are conserved among hematopoietic stem cells from old and reconstituted mice and markedly increase their purity. Blood *107*, 924-930.

Zeggini, E., Scott, L.J., Saxena, R., Voight, B.F., Marchini, J.L., Hu, T., de Bakker, P.I., Abecasis, G.R., Almgren, P., Andersen, G., *et al.* (2008). Meta-analysis of genome-wide association data and large-scale replication identifies additional susceptibility loci for type 2 diabetes. Nat Genet *40*, 638-645.

12

Hematopoietic Stem Cell in Acute Myeloid Leukemia Development

Sérgio Paulo Bydlowski and Felipe de Lara Janz
University of Sao Paulo School of Medicine
Laboratory of Genetics and Molecular Hematology, São Paulo, SP, Brazil

1. Introduction

Hematopoietic stem cells (HSCs) are multipotent stem cells defined by their ability to self-renewal, differentiation and maintenance of all blood cell types in the hematological system during the entire lifetime of the organism. This physiological process, called hematopoiesis, is controlled by several complex interactions between genetic processes in blood progenitor cells and bone marrow microenvironment. Hemostasis is maintained by a delicate balance between processes such as self-renewal, proliferation and differentiation versus apoptosis and cell-cycle arrest in HSCs. Over the last two decades, several studies have been made to understand possible mechanisms of cell malignancy and tumor growth, both in solid and hematological cancers. Leukemias are hematological malignancies that arise from cancer stem cells (CSCs). Neoplastic transformation of hematopoietic stem or progenitor cells occurs by unbalanced critical mechanisms. Blood cancers, as acute myeloid leukemia (AML), are sustained by leukemic stem cells (LSCs) which, like normal HSCs, present a range of biological characteristics that enable their long-term survival. AML is a well studied hematological cancer type characterized by an accumulation of clonal myeloid progenitor cells that do not differentiate normally. However, there is still no consensus about the mechanisms by which the HSCs transformation occurs. In this chapter, the hematopoietic stem cells and leukemic stem cells will be focused on leukemia development, mainly in AML.

2. Hematopoietic stem cells

Hematopoietic stem cells are a well characterized stem cell type which has been used in bone marrow transplantation for treatment of hematological malignancies as well as non-malignant disorders (Warner et al, 2004). In fact, bone marrow (BM) has been, for many years, transplanted as an unfractionated cell pool, until researchers discovered which cellular components were responsible for the engraftment of the donor hematopoietic and immune systems in marrow-ablated patients.

HSCs present self-renewal potential and differentiation capacity into blood lineages. The self-renewal concept means that when stem cells divide, 50% of the daughter cells, on average, is committed with a cell lineage; the remaining 50% do not differentiate; therefore the process maintains the same number of stem cells. This is accomplished by the so-called asymmetric cell division, so that each dividing stem cell originates one new stem cell and one differentiated cell (Gordon, 2005) (Figure 1). In the symmetric division, the stem cells originate 100% of identical stem cells.

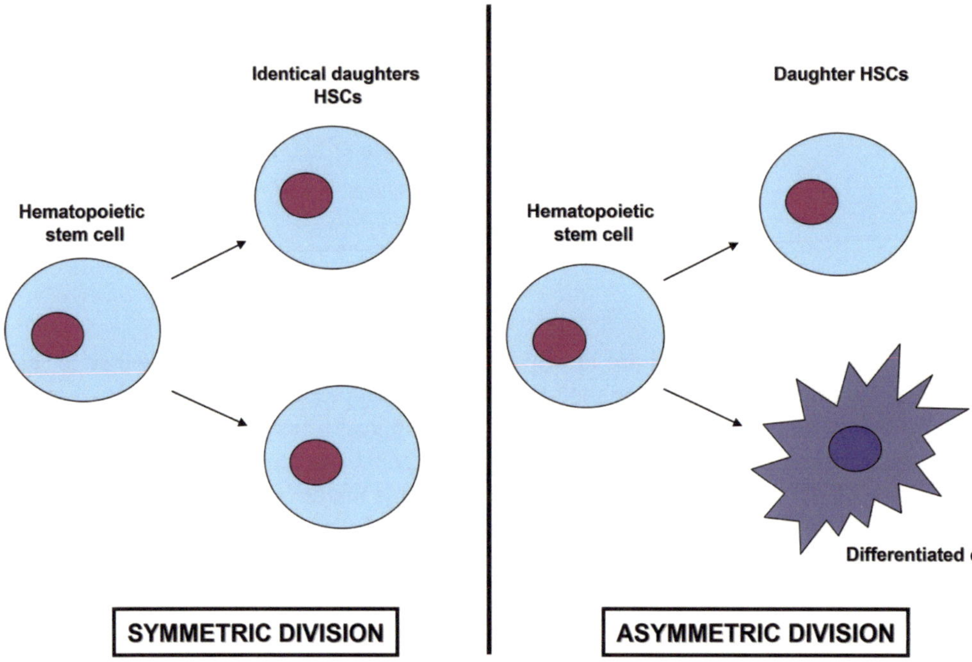

Fig. 1. Schematic illustration of two different types of HSCs division. On the left, the symmetric division in which mitosis originates two identical stem cell daughters. On the right, the characteristic stem cells asymmetric division where each dividing stem cell forms one new stem cell and one differentiated cell.

HSCs are classified as multipotent stem cells due to their ability to differentiate in lymphoid as well as myeloid cells types; however, some studies showed that transplanted bone marrow cells can contribute to the repair and regeneration of a spectrum of other tissue cell types including those from brain, muscle, lung and liver.

Lymphoid cell lineage includes T and B cells, while megakaryocytes, erythrocytes, granulocytes and macrophages belong to the myeloid lineage. These two lineages derive from different progenitor cells. Common lymphoid progenitors (CLPs) can differentiate into all types of lymphocytes without noticeable myeloid potential under physiological conditions. Similarly, common myeloid progenitors (CMPs) can give rise to all classes of myeloid cells with no or extensively low levels of B-cell potential (Kondo, 2010). It is likely that differences in the expression levels of transcription factors determine the lineage affiliation of a differentiating cell (Figure 2). The transcription factors PU.1 and GATA-1 have been implicated in myeloid and erythroid/megakaryocyte lineage differentiation, respectively (Gordon, 2005).

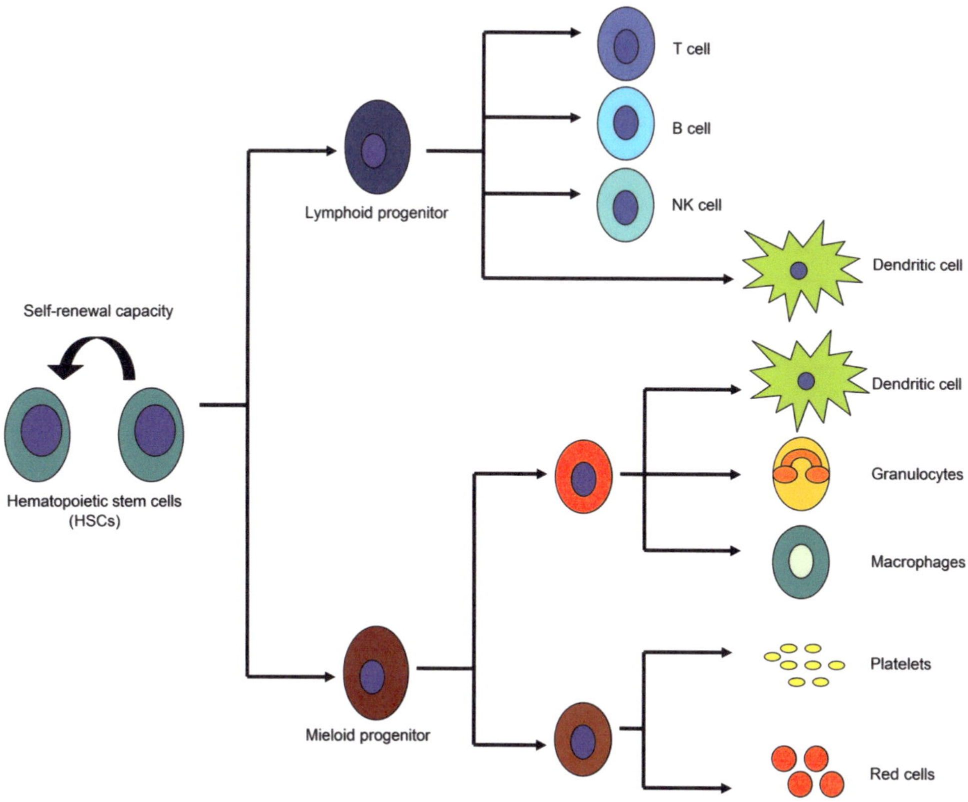

Fig. 2. HSCs differentiation pathways. HSCs could differentiate into specific lymphoid and myeloid cell types. Common lymphoid progenitors (CLPs) can differentiate into all types of lymphocytes and common myeloid progenitors (CMPs) can give rise to all classes of myeloid cells (megakaryocytes, erythrocytes, granulocytes and macrophages) (Adapted from Du et al., 2008).

2.1 Characterization

Morphologically, hematopoietic stem cells are undifferentiated and resemble small lymphocytes. Normally, a large fraction is quiescent, in the G_0 phase of the cell cycle, which protects them from the action of cell cycle-dependent drugs. The quiescent state of stem cells is maintained by transforming growth factor-β (TGF-β). The activity of TGF-β is mediated by p53, a tumor suppressor gene that regulates cell proliferation and targets the cyclin-dependent kinase inhibitor p21 (Gordon, 2005). Quiescence of HSCs is critical not only for protecting the stem cell compartment and sustaining stem cell pools during long periods of time, but also by minimizing the accumulation of replication-associated mutations.

Quiescence regulation in HSCs is also of great importance for understanding the pathophysiological origins of many related disorders. Interestingly, many of the intrinsic transcriptional factors that maintain HSCs quiescence are found to be associated with leukemias. For example, chromosomal translocations resulting in the fusion of FoxOs and myeloid/lymphoid or mixed lineage leukemia have been reported in acute myeloid leukemias.

The majority of normal HSCs are present among the CD34+/CD38− bone marrow cell fractions; some HSCs are also observed among CD34−/Lin− cells; CD34+/CD38+ cell fractions contain some HSCs but endowed with short-term repopulating activity. Other recognized marker is the tyrosine kinase receptor c-kit (CD117), concomitantly with the lack of terminal differentiation markers (as CD4 and CD8; Figure 3) (Rossi et al., 2011).

Primitive HSCs populations show low fluorescence ratios after Hoechst 33342 and Rhodamine 123 staining; these cells are described as side population (SP). SP cells demonstrate high expression of ATP binding cassette (ABC) transporters as P-glycoprotein (P-gp/ABCB1), breast cancer resistance protein (BCRP/ABCG2) and lung resistance protein (LRP) (Huls et al., 2009). MDR1 has been implicated in the protection of cells against apoptotic cell death induced by a variety of methods including growth factor deprivation, UV irradiation, ionizing radiation, or tumor necrosis factor-α treatment. BCRP is a half-transporter and characterized as a novel stem cell transporter. Like MDR1, enforced overexpression of BCRP in human MCF-7 breast cancer cells confers a broad spectrum of drug resistance, and elevated levels of expression of BCRP have been reported to be associated with acute myeloid leukemia.

Since ABC transporter function is associated with both normal and aberrant hematopoiesis, it is important to fully characterize the function of this class of transporter proteins in hematopoietic cell differentiation and to define the underlying mechanisms.

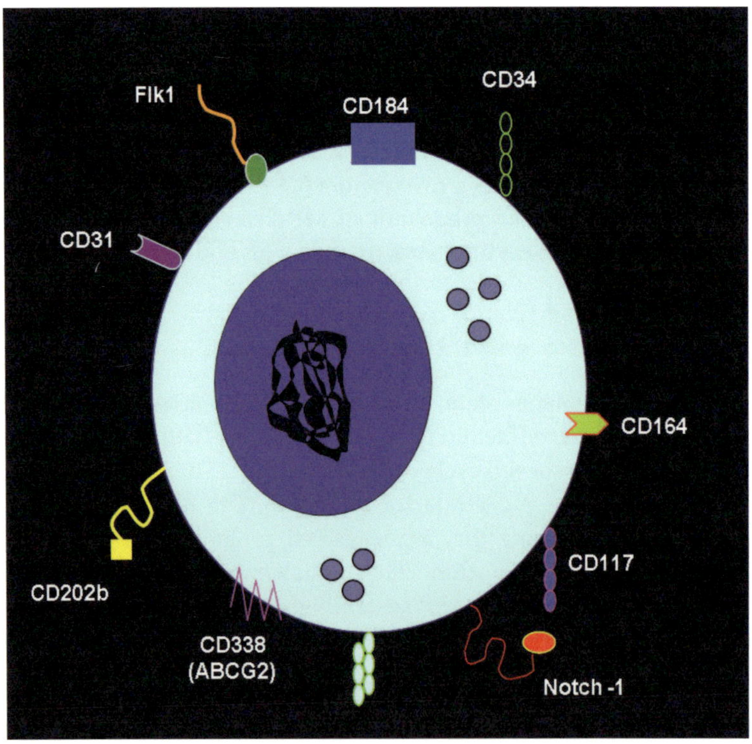

Fig. 3. HSCs main surface markers. HSCs express typical antigens as: CD34, CD117, CD164, CD202b, CD31, Flk-1, CD184, CD338 or ABCG2, Notch-1 concomitantly with the lack of terminal differentiation markers (CD4 and CD8).

2.2 Classification

According to its hematopoietic repopulation capacity, the hematopoietic stem cell pool can be subdivided into three main groups:

a. short-term HSCs, capable of generating clones of differentiating cells for only 4-6 weeks;
b. intermediate-term HSCs, capable of sustaining a differentiating cell progeny for 6-8 months before becoming extinct;
c. long-term HSCs, capable of maintaining hematopoiesis indefinitely (Testa, 2011).

2.3 HSC sources

HSCs can be harvested from healthy donors either by bone marrow aspiration, peripheral stem cell mobilization or from umbilical cord blood (Dick, 2003). HSCs located in the bone marrow present an estimated frequency of 0.01% of total nucleated cells and can be collected by iliac crest puncture and then separated from the other blood cells by magnetic beads or cell sorting.

Umbilical cord blood (UCB) is a source of the rare but precious primitive HSCs and progenitor cells that can reconstitute the hematopoietic system in patients with malignant and nonmalignant disorders treated with myeloablative therapy. UCB cells possess an enhanced progenitor cell proliferation capacity and self-renewal *in vitro*. UCB is usually discarded and it exists in almost limitless supply. The blood remaining in the delivered placenta is safely and easily collected and stored. The predominant collection procedure currently involves a relatively simple venipuncture, followed by gravity drainage into a standard sterile anti-coagulant-filled blood bag, using a closed system, similar to that utilized for whole blood collection (Bojanic & Golubic Cepulic, 2006).

Peripheral blood hematopoietic stem cells (PBSCs) have numerous advantages in comparison with traditionally used bone marrow. PBSCs collection by leukapheresis procedure is simple and better tolerated than bone marrow harvest. PBCSs are mobilized by myelosupressive chemotherapy or/and hematopoietic growth factors. Leukapheresis product contains PBSCs along with committed lineage of progenitors and precursors which contribute to faster hematopoietic recovery.

Unfortunately, the expansion of HSCs *in vitro* is difficult to achieve because the proliferation is accompanied by differentiation. This is presumably caused by a lack of appropriate cues that are provided *in vivo* by the microenvironment. The most excellent defined culture medium for HSCs expansion is supplemented with cytokines such as fetal liver tyrosine kinase-3 ligand (FLT3-L), stem cell factor (SCF), interleukin-3 (IL-3) and thrombopoietin (TPO). Interestingly, mesenchymal stem cells (MSCs), which are characterized by multi-differentiation potential, are important players of the bone marrow HSCs niche. In recent years, MSCs have been shown to support HSCs maintenance and engraftment (Jing et al., 2010).

3. Factors involved in hematopoiesis

Hematopoiesis is a highly coordinated process wherein HSCs differentiate into mature blood cells supported by a physical environment called **niche** (Figure 3). The bone marrow niche is the most important post-natal microenvironment in which HSCs proliferate, mature and give rise to myeloid and lymphoid progenitors. BM is present in the medullary cavities of all animal bones. Unlike secondary lymphoid organs such as spleen with distinct gross structures including red and white pulp, BM has no clear structural features, except for the endosteum that contains osteoblasts. The endosteum region comes in contact with calcified hard bones and provides a special microenvironment to HSCs, which is necessary for the maintenance of HSC activity (Kondo, 2010).

Within the niche, HSCs are believed to receive support and growth signals originating from several sources, including: fibroblasts, endothelial and reticular cells, adipocytes, osteoblasts and mesenchymal stem cells. The main function of the niche is to integrate local changes in nutrients, oxygen, paracrine and autocrine signals and to change HSCs quiescence, trafficking, and/or expansion in response to signals from the systemic circulation (Broner & Carson, 2009). Although the nature of true MSCs remains misunderstood, CXC chemokine ligand 12 (CXCL12) - expressing CD146 MSCs were recently reported to be self-renewing progenitors that reside on the sinusoidal surfaces and contribute to organization of the sinusoidal wall structure, produce angiopoietin-1 (Ang-1), and are capable of generating osteoblasts that form the endosteal niche (Konopleva & Jordan, 2011).

These CXCL12 reticular cells may serve as a transit pathway for shuttling HSCs between the osteoblastic and vascular niches where essential but different maintenance signals are provided. Cytokines and chemokines produced by bone marrow MSCs concentrate in particular niches secondary to varying local production and through the effects of cytokine-binding glycosaminoglycans. Of these, CXCL12/stromal cell–derived factor-1 alpha positively regulates HSCs homing, while transforming growth factors FMS-like tyrosine kinase 3 (Flt3) ligand and Ang-1 function act as quiescence factors.

CXCL12-CXCR4 signaling is involved in homing of HSCs into BM during ontogeny as well as survival and proliferation of colony-forming progenitor cells. The CXCR4-selective antagonist–induced mobilization of HSCs into the peripheral blood further indicates a role for CXCL12 in retaining HSCs in hematopoietic organs. BM engraftment involves subsequent cell-to-cell interactions through the BMSC-produced complex extracellular matrix. Thus, vascular cell adhesion molecule-1 (VCAM-1) or fibronectin is critical for adhesion to the BM derived MSCs.

In this way, the control of hematopoietic stem cell proliferation kinetics is critically important for the regulation of correct hematopoietic cells production. These control mechanisms could be classified in intrinsic or extrinsic to the stem cells, or a combination of both.

Extrinsic control means that self-renewal and differentiation can be controlled by external factors, such as cell–cell interactions in the hematopoietic microenvironment or cytokines as SCF (stem cell factor) and its receptor c-kit, Flt-3 ligand, TGF-β, TNF-α and others. Cytokines regulate a variety of hematopoietic cell functions through the activation of multiple signal transduction pathways. The major pathways relevant to cell proliferation and differentiation are the Janus kinase (Jak)/signal transducers and activators of transcription (STATs), the mitogen-activated protein (MAP) kinase and the phosphatidylinositol (PI) 3-kinase pathways.

Yet, in intrinsic control, the expression of other transcription factors has been shown to be essential for hematopoietic cell development from the earliest stages, as: SCL (stem cell leukaemia hematopoietic transcription factor); GATA-2; gene products involved in cell cycle control, such as the cyclin dependent kinase inhibitors (CKIs) p16, p21 and p27.

Notch-1-Jagged pathway may serve to integrate extracellular signals with intracellular signalling and cell cycle control. Notch-1 is a surface receptor on hematopoietic stem cell membranes that binds to its ligand, Jagged, on stromal cells. This results in cleavage of the cytoplasmic portion of Notch-1, which can then act as a transcription factor (Gordon, 2005).

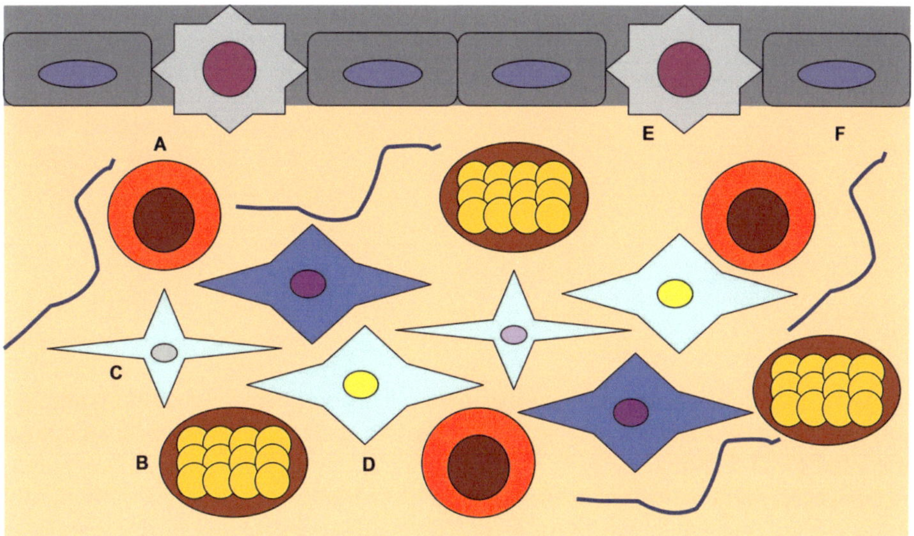

Fig. 4. Schematic representation of main bone marrow niche cells: A- hematopoietic stem cells, B- adipocytes, C- mesenchymal stem cells, D- reticular cells, E- osteoclasts, and F- osteoblasts.

4. Leukemia and leukemic development

Leukemia is the consequence of stepwise genetic alterations that confer both proliferative and survival advantage, as well as self-renewal capacity to the malignant cells (Lane et al., 2011). When the HSCs processes of self-renewal and differentiation become deregulated or uncoupled, leukemias can result, characterized by an accumulation of immature blast cells that fail to differentiate into functional cells. Two types of abnormal events can lead to leukemia. First, a normal stem cell acquires several mutations (from different types of genetic events) and, due to epigenetic changes that alter its growth control, the resistance to apoptosis is increased, interfering with the ability of its progeny to differentiate. Second, partially differentiated cells restore gene expression patterns that allow them to reacquire the unique self-renewal properties of stem cells while also interferes with their subsequent ability to differentiate (Testa, 2011).

Hanahan & Weinberg (2000) described the rules that govern the transformation of normal cells into a malignant cell. The six main properties that define malignant cells are: self-sufficiency in growth signals; insensitivity to growth inhibitory signals; evasion of programmed cell death (apoptosis); limitless replicative potential; sustained angiogenesis; and tissue invasion and metastasis.

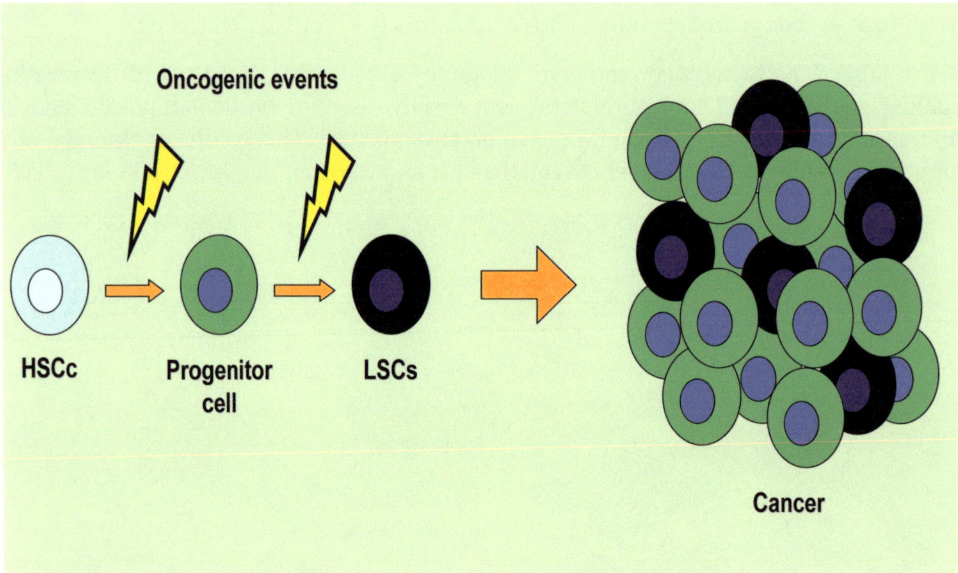

Fig. 5. Cancer stem cell hypothesis. A normal stem cell acquires several mutations and in consequence, by epigenetic changes that alter its growth control, its resistance to apoptosis increased and the ability of its progeny to differentiate is changed. Partially differentiated cells restore gene expression patterns that allow these cells to reacquire the unique self-renewal properties of stem cells while also interfere with their subsequent ability to differentiate.

The "cancer stem cell hypothesis" has gained considerable interest in recent years. This theory states that cells in a tumor are organized as a hierarchy similar to that of normal tissues, and are maintained by a small subset of tumor cells that are ultimately responsible for tumor formation and growth. These cells, defined as "cancer stem cells" (CSCs) or "tumor initiating cells" (TICs), possess several key properties of normal tissue stem cells including self-renewal, unlimited proliferative potential (i.e., the ability of a cell to renew itself indefinitely in an undifferentiated state), infrequent or slow replication, resistance to toxic xenobiotics, high DNA repair capacity, and the ability to give rise to daughter cells that differentiate. However, the major difference between cancer growth and normal tissue renewal is that whereas normal transit amplifying cells usually differentiate and die, at various levels of differentiation, the cancer transit-amplifying cells fail to differentiate normally and instead, accumulate (i.e. they undergo maturation arrest), resulting in cancer growth (Soltanian & Matin, 2011).

In the last years, studies have also clearly demonstrated that leukemia populations are highly heterogeneous and that the disease is propagated by a subpopulation of leukemia stem cells (LSC). LSCs, like normal hematopoietic stem cells, possess a range of biological characteristics that enable their long-term survival. Therefore, LSCs reside in a mostly quiescent state, and as a consequence, the overall activity of many chemotherapeutic agents that function by targeting cycling cells is likely diminished (Konopleva & Jordan, 2011).

LSCs infiltrate the bone marrow and interfere with the normal HSC-microenvironment homeostasis. Available data indicate that LSCs also interact with the hematopoietic microenvironment to maintain self-renewal and to mitigate the effects of cytotoxic chemotherapy (Lane et al., 2011)

4.1 Leukemia stem cells characterization

The immunophenotype and isolation of LSCs were first described by Lapidot et al.(1994) from primary human AML samples and, later, studies have shown that LSCs can be defined as expressing CD34, CD382, HLA-DR2, CD902, CD117 and CD123. Some of these markers are also detected in HSCs, but the expression of CD123 seems to be leukemic-specific (Blair et al., 1998).

Another LSC specific antigen is C-type lectin-like molecule-1, CLL-1. This antigen was demonstrated to be capable of identifying residual leukemic CD34+CD38− cells in clinical remission bone marrow samples. However, more recent data indicate that the phenotype of LSCs may be somewhat variable from patient to patient and that, in some cases, more than one phenotypically distinct subpopulation may possess LSC activity (Konopleva & Jordan, 2011).

Expression of Oct-4 is another similarity between normal and cancer stem cells. Oct-4, a member of the family of POU-domain transcription factors, is expressed in pluripotent embryonic stem and germ cells. Oct-4 mRNA is normally found in totipotent and pluripotent stem cells of pregastrulation embryos (Soltanian & Matin, 2011). Expression of this factor plays a crucial role in maintaining the self-renewing, cancer stem-cell-like, and chemoradioresistant properties in lung cancer-derived CD133+ cells (Chen et al., 2008).

Oct-4 gene product is expressed in several types of adult pluripotent stem cells including kidney, breast, epithelial, pancreatic, mesenchymal, gastric and liver, as well as in tumor cell lines derived from pancreas and liver (Tai et al., 2005). According to Marques et al. (2010) it is also possible that the resistance phenotype developed by leukemic cells is determined by ABC transporter expression which is probably activated by the induction of the Oct-4 transcription factor. The ABCB1, ABCG2 and ABCC1 transporters exhibit binding sites (octamer-ATGCAAAT) for the Oct-4 transcription factor. The presence of these binding sites in the gene promoter of these transport proteins suggests that the transporter regulation pathways may be initiated at the Oct-4 recognized binding sites. However, the presence of Oct-4 alone is not always sufficient for induction of transporter genes. Transporter expression levels are often dependent upon Oct-4 interactions with other transcription factors.

4.2 Genetic pathways of LSCs

4.2.1 Wnt/Catenin

The Wnt/Catenin signaling has been implicated in the self-renewal of LSCs. Wnt proteins are a large family of glycoproteins that bind to Frizzled receptors and LRP5/6 coreceptors. By stabilizing the mediator β-catenin, they start a complex signaling cascade that plays a significant role in regulating cell proliferation and differentiation. Wnt cascade has appeared as a critical regulator of stem cells self-renewal. Comparing the expression of normal hematopoietic stem cells to that of AML leukemic stem cells, evidences show that the Wnt signaling pathway is aberrantly regulated in leukemic stem cells.

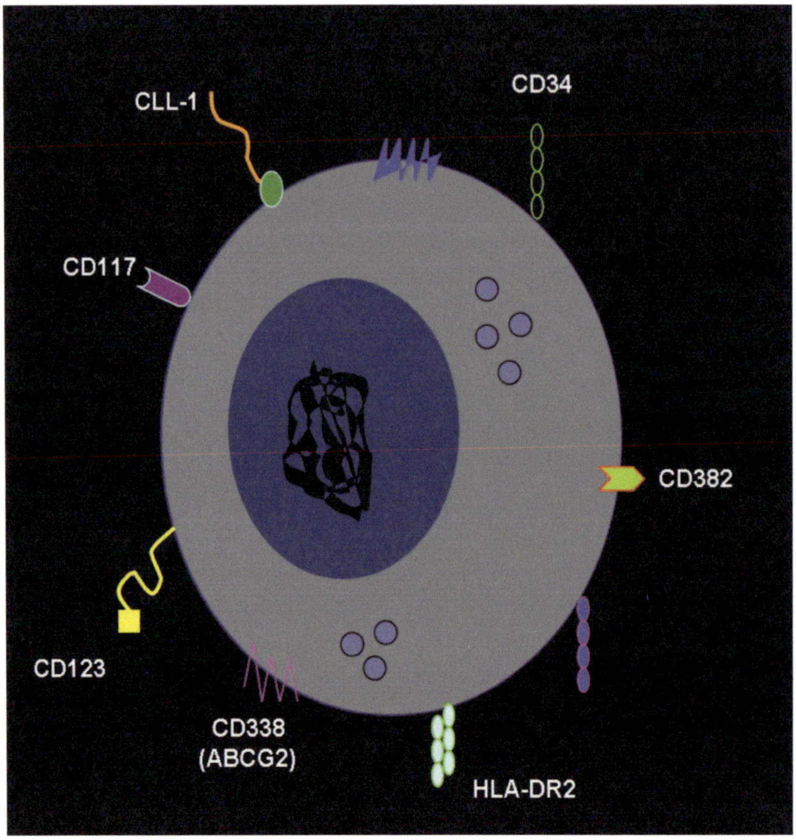

Fig. 6. LSCs main surface markers. LSCs can be defined as CD34, CD382, HLA-DR2, CD902, CD117 and CD123. The expression of CD123 seems to be leukemic-specific. Another LSC specific antigen is C-type lectin-like molecule-1, CLL-1.

4.2.2 PTEN (Akt/mTOR)

PTEN is a phosphatase that negatively regulates signaling through the PI3K pathway, attenuating proliferation and survival signals. PTEN deficiency causes an initial expansion of normal hematopoietic stem cells due to their cycling, followed by their exhaustion. In contrast to this requirement for PTEN in the maintenance of hematopoietic stem cells, leukemic stem cells arise and expand in numbers following PTEN deletion. The observation that PTEN deletion had opposite effects on normal hematopoietic stem cells compared to leukemic stem cells raised possibility for therapeutic targeting of this pathway to eliminate only the leukemic stem cells, without affecting normal hematopoietic stem cells. Since PTEN deletion causes increased AKT and mTOR activation, it seems logical that mTOR targeting by pharmacological agents, such as rapamycin, could represent an interesting option for AML treatment (Testa, 2011).

4.2.3 NF-kB

In addition to genes involved in the control of stem cell self-renewal, leukemic stem cells are expected to express, at high levels, genes involved in anti-apoptotic mechanisms. In this context, particular attention has been focused on the study of NF-kB. NF-kB plays a critical role in inflammation, anti-apoptotic responses, and carcinogenesis. High NF-kB expression was found in primitive AML blasts. In particular, the constitutive activation of NF-kB was observed in AML cell populations enriched in leukemic stem cells, but not in normal hematopoietic stem cells. According to these observations, it seemed clear that NF-kB could be a potential therapeutic target for attempting leukemia stem cell eradication (Guzman et al., 2001).

4.2.4 BMI1

BMI1 is a polycomb group protein which, together with Ring1 proteins, is part of PRC1 complex that has histone H2A-K119 ubiquitin E3 ligase activity. BMI1 has a role in HOX gene (HOXC13) silencing by H2A ubiquitylation (Cao et al, 2005). BMI1 is also known to be important in the regulation and maintenance of proliferative/self-renewal potential in both normal hematopoietic and leukemic stem cells (Park et al, 2003). Upon knockdown of BMI-1, cells lose their ability to engraft and reconstitute leukemia in mice (Bomken et al., 2010).

4.3 Xenotransplantation model of leukemia

A key component for understanding the biological mechanisms for tumor heterogeneity is the ability to functionally assess the capacity for limitless proliferative capacity for segregated populations of tumor cells. Unfortunately, for hematologic malignancies, *in vitro* culture assays are not entirely effective as a means of functionally assessing self-renewal capacity. Thus, transplantation assays in which candidate populations are assessed for their ability to establish long-term serial engraftment of recipient animals is the gold standard for assigning limitless proliferative capability, i.e. self-renewal. For murine studies, the availability of syngeneic transplantation models has been responsible in large part for our in depth understanding of the normal murine hematopoietic hierarchy.

Since the 70 s, there is ample evidence supporting the existence of a discrete compartment of slowly cycling leukemic cells that are resistant to standard chemotherapeutic agents. These cell populations were felt to represent the leukemic stem cells and though many observations were consistent with this hypothesis, there was no direct evidence that this was indeed the case. As the first direct evidence for the existence of cancer stem cells, the work of Lapidot et al. (1994) represented a milestone in the history of the leukemic stem cell model. This study identified an infrequent population of leukemic cells capable of recapitulating the human tumor in xenotransplants. A key finding was that the SCID mouse leukemia repopulating cell, SRC, possessed a phenotype that was similar to that of the normal hematopoietic stem cell (CD34+ and CD38−).

4.4 Acute myeloid leukemia

Acute myeloid leukemia (AML) is a clonal disorder characterized by arrest of differentiation in the myeloid lineage coupled with an accumulation of immature progenitors in the bone marrow, resulting in hematopoietic failure (Pollyea et al., 2011). AML is the most common acute leukemia in adults, affecting roughly three out of 100.000 people. AML patients are predominantly elderly, with a median age at diagnosis of 67 (National Cancer Institute 1975–2007).

In AML, there is wide patient-to-patient heterogeneity in the appearance of the leukemic blasts. Conventionally, AML is classified into seven French–American–British (FAB) subtypes corresponding to the maturation stage of the leukemia (Warner et al., 2004).

The discovery of leukemia-initiating cells in acute myeloid leukemias (AMLs) started with the discovery that the large majority of AML blasts do not proliferate and only a small minority is capable of forming new colonies (Testa, 2011).

A common feature to all AML cases is the arrested aberrant differentiation leading to an accumulation of more than 20% blast cells in the bone marrow (Gilliland & Tallman, 2002). More than 80% of myeloid leukemias are associated with at least one chromosomal rearrangement (Pandolfi, 2001), and over 100 different chromosomal translocations have been cloned (Gilliland & Tallman, 2002). Frequently, these translocations involve genes encoding transcription factors that have been shown to play an important role in hematopoietic lineage development. Thus, alteration of the transcriptional machinery appears to be a common mechanism leading to arrested differentiation (Pandolfi, 2001; Tenen, 2003).

Clinical investigation and experimental animal models suggest that at least two genetic alterations are required for the clinical manifestation of acute leukemia. According to the model proposed by Gilliland & Tallman (2002), cooperation between class I activating mutations and class II mutations that induce termination of differentiation give rise to AML. The class I mutations, such as mutations in the receptor tyrosine kinase genes FLT3 and KIT, RAS family members, and loss of function of neurofibromin 1, confer proliferative and/or survival advantage to hematopoietic progenitors, typically as a consequence of aberrant activation of signal transduction pathways. The class II mutations lead to a halt in differentiation via interference with transcription factors or co-activators (Frankfurt et al., 2009).

While the LSC appears to share many of the cell surface markers previously identified for HSC such as CD34, CD38, HLA-DR, and CD71, there have been several groups who have reported surface markers that are differentially expressed in the two populations. CD90 or Thy-1 is one marker that has been described to be potentially specific of the LSC compartment. Thy-1 is downregulated in normal hematopoiesis as the most primitive stem cells progress toward the progenitor stage. This finding of the lack of expression on LSC might suggest that the primitive stem cell does not contribute to the primary pathological event, or that Thy-1 expression is downregulated as a result of the leukemogenic events (Hope et al., 2003).

The interaction between CXCL12 (stromal cell–derived factor-1 alpha) and its receptor CXCR4 on leukemic progenitor cells contributes to their homing to the bone marrow microenvironment. CXCR4 levels are significantly elevated in leukemic cells from patients with AML and CXCR4 expression is associated with poor outcome (Konopleva & Jordan, 2011). Constitutive activation of the nuclear factor kappa B (NF-kB) pathway in primary human AML stem cells provided evidence that NF-kB plays a significant role in the overall survival of LSCs as well as AML cell types in general. This pathway is strongly implicated as a central target in developing LSC-specific therapies (Konopleva & Jordan, 2011).

FLT3, a member of the class III tyrosine kinase receptor family, is expressed in normal hematopoietic progenitors as well as in leukemic blasts, and it plays an important role in cell proliferation, differentiation, and survival. Activation of the FLT3 receptor by the FLT3 ligand leads to receptor dimerization and phosphorylation, and activation of downstream signaling pathways, including the Janus kinase (JAK) 2 signal transducer (JAK2), signal transducer and activator of transcription (STAT) 5, and mitogen-activated protein kinase (MAPK) pathways. Mutations in the FLT3 gene, found in approximately 40% of patients with AML, are believed to promote its autophosphorylation and constitutive activation, leading to ligand-independent proliferation (Frankfurt et al., 2009).

The adhesion receptor CD96 (TACTILE) is a transmembrane glycoprotein possessing three extracellular immunoglobulin-like domains. It is a member of the Ig gene superfamily and was first identified as a gene expressed in activated T cells. CD96 was described as a tumor marker AML stem cell (Konopleva & Jordan, 2011). Hosen et al. (2011) showed that AML-LSC can be distinguished from normal HSC by the presence of CD96 expression. This finding suggests that CD96 also may prove to be an excellent target for antibody therapy against LSC because hematopoietic progenitors are regenerated rapidly from HSC.

Another adhesion molecule, CD44, has been demonstrated to be a key regulator of AML LSCs homing to microenvironmental niches, maintaining a primitive state. CD44 mediates adhesive cell-cell and cell–complex extracellular matrix interactions through binding to its main ligand, hyaluronan, a glycosaminoglycan highly concentrated in the endosteal region. Other ligands include osteopontin, fibronectin, and selectin, all of which are involved in cell trafficking and lodgment. Beyond its adhesion function, CD44 can also transduce multiple intracellular signal transduction pathways when bound to hyaluronan or to specific function-activating monoclonal antibodies (Konopleva & Jordan, 2011).

5. Conclusions

Adult hematopoietic stem cells are undifferentiated cells capable of self-renewal and differentiation potential in several cell types that comprise the hematopoietic tissue. These cells have been used in bone marrow transplantation for treatment of hematological malignancies as well as non-hematological diseases. The HSC is the main component in the process of hematopoiesis which, together with the cells that make up the bone marrow stromal environment and other intrinsic and extrinsic factors, orchestrates the entire production of progenitors and terminally differentiated blood cells.

However, when this process of cell production is unbalanced, leading to an exacerbated and uncontrolled proliferation of blood progenitor cells, leukemia may develop.The ultimate challenge in coming years will be to understand the stem cell 'programme', particularly the control of self-renewal, in an attempt to develop novel, stem cell-directed therapies. An improved understanding of clonal evolution will be critical if we are to ensure that cancers are not able to evolve mechanisms to evade the new directed therapies. However, reducing the risk of relapse and minimizing long-term side effects should always remain the ultimate goal of understanding the CSCs.

With little doubt, the leukemia stem cell model has had the greatest clinical impact on our understanding and treatment of Philadelphia chromosome positive leukemias. Effective targeted agents and the ability to follow the impact of therapy on critical rare sub-populations of the malignant clone has greatly advanced our understanding of chronic myeloid leukemia.

For AML, the clinical impact of the leukemic stem cell model is less clear. The ability to isolate and characterize rare LSC populations has had a significant importance on our understanding of the biology of AML. In the past decade, we have gained considerable insight into the properties that distinguish leukemic stem cells from their normal counterparts and some of the rules that govern the leukemic hierarchy.

Despite the wide variance in techniques and to some degree expression profiles, common signaling pathways have been shown to play a role not only in AML stem cells, but also in cancer stem cells in general. These include BMI1, Wnt, Sonic Hedgehog, Notch, and NF-kB. These pathways are being evaluated for their role in LSC biology and agents targeting these pathways are making their way through the pre-clinical focus.

6. References

Becker, M. W. & Jordan, C. T. (2011). Leukemia stem cells in 2010: Current understanding and future directions. *Blood Reviews* Vol.25, No. 2, (March 2011), pp. 75–81.

Blair, A.;Hogge, D.E. & Sutherland, H.J. (1998). Most acute myeloid leukemia progenitor cells with long-term proliferative ability in vitro and in vivo have the phenotype CD34þ/CD71-/HLA-DR-. *Blood* Vol.92, No. 11, (December 1998), pp. 4325–4335.

Bojanić, I. & Golubić Cepulić, B. (2006).Umbilical cord blood as a source of stem cells. *Acta Medica Croatica* Vol.60, No. 3, (June 2006),pp. 215-225.

Bomken, S., Fiser, K., Heidenreich, O. (2010). Vormoor, J. Understanding the cancer stem cell. *British Journal of Cancer* Vol.103, No. 4, (August 2010), pp. 439 - 445.

Broner, F. & Carson, M C. *Topics in bone biology*. Springer. 2009; 4: pp. 2-4. New York, USA.

Buting, K. D. (2002). ABC transporters as phenotypic markers and functional regulators of stem cells. *Stem Cells* Vol.20, No. 12, (December 2002), pp. 11-20.

Cao, R., Tsukada, Y., Zhang, Y. (2005). Role of Bmi-1 and Ring1A in H2A ubiquitylation and Hox gene silencing. *Molecular Cell* Vol.22, No. 6, (December 2005),pp. 845-854.

Chen, Y.C., Hsu, H.S., Chen, Y.W., Tsai, T.H., How, C.K., Wang, C.Y., Hung,S.C.,Chang,Y.L.,Tsai,M.L.,Lee,Y.Y.,Ku,H.H.,Chiou,S.H.(2008). Oct-4 expression maintained cancer stem-like properties in lung cancer-derived CD-133 positive cells. *PLoS ONE* Vol.3, No. 7, (July 2008), pp. 2637.

Dick, J.E. (2003). Stem cells: Self-renewal writ in blood. *Nature* Vol.423, No. 6937, (May 2003), pp. 231–233.

Du, W., Adam, Z., Rani, R., Zhang, X., Pang, Q. (2008). Oxidative stress in Fanconi anemia hematopoiesis and disease progression. *Antioxidants & Redox Signaling* Vol.10, No. 11 (November 2008), pp. 1909-1921.

Frankfurt, O., Licht, J.D., Tallman, M.S. (2007). Molecular characterization of acute myeloid leukemia and its impact on treatment. *Current Opinion in Oncology* Vol.19, No. 6, (November 2007), pp. 635–649.

Gilliland, D.G.& Tallman, M.S. (2002). Focus on acute leukemias. *Cancer Cell.* Vol.1, No. 5, (June 2002), pp. 417-420.

Gordon, M. *Stem cells and haemopoiesis.* In: Hoffbrand, V., Catovsky, D., Tuddenham, E.G., 5th ed. Blackwell Publishing, (2005): Differential niche and Wnt requirements during acute myeloid leukemia. pp. 1-12. New York.

Guzman ML, Neering SJ, Upchurch D, Grimes B, Howard DS, Rizzieri DA, Luger SM, Jordan CT (2001). Nuclear factor-KB is constitutively activated in primitive human acute myelogenous leukemia cells. *Blood*: 2301–2307.

Hanahan, D. & Weinberg, R.A. (2000). The hallmarks of cancer. *Cell* Vol.100, No. 1, (January 2000), 57-70.

Hope, K. J., Jin, L., Dick, JE. (2003). Human Acute Myeloid Leukemia Stem Cells. *Archives of Medical Research* Vol.34, No. 6, (November-December 2003), pp. 507–514.

Hosen, N., Park, C.Y., Tatsumi, N., Oji, Y., Sugiyama, H., Gramatzki, M., Krensky, A.M., Weissman, I.L. (2007). CD96 is a leukemic stem cell-specific marker in human acute myeloid leukemia. *Proceedings of National Academy of Science of U S A* Vol.104, No. 26, (June 2007), pp. 11008-11013.

Huls, M., Russel, F.G., Masereeuw, R. (2009). The role of ATP binding cassette transporters in tissue defense and organ regeneration. *Journal of Pharmacogycal Experimental Therapy* Vol.328, No. 1, (January 2009), pp. 3-9.

Jing, D., Fonseca, A.V., Alakel, N., Fierro, F.A., Muller, K., Bornhauser, M., Ehninger, G., Corbeil, D., Ordemann, R. (2010). Hematopoietic stem cells in co-culture with mesenchymal stromal cells-modeling the niche compartments in vitro. *Haematologica* Vol.95, No. 4, (April 2010), pp. 542550.

Kohler, B.A., Ward, E., McCarthy, B.J., Schymura, M.J., Ries, L.A., Eheman, C., Jemal, A., Anderson, R.N., Ajani, U.A., Edwards, B.K. (2011). Annual report to the nation on the status of cancer, 1975-2007, featuring tumors of the brain and other nervous system. *Journal of National Cancer Institute* Vol.103, No. 9, (May 2011), pp. 714-736.

Kondo, M. (2010). Lymphoid and myeloid lineage commitment in multipotent hematopoietic progenitors. *Immunology Reviews* Vol. 238, No. 1, (January 2010), pp. 37-46.

Konopleva, M.Y. & Jordan, CT. (2011). Leukemia Stem Cells and Microenvironment. *Biology and Therapeutic Targeting* Vol.29, No. 5, (May 2011), pp. 591-599.

Lane, S.W., Wang, Y.J., Lo Celso, C., Ragu, C., Bullinger, L., Sykes, S.M., Ferraro, F., Shterental, S., Lin, C.P., Gilliland, D.G., Scadden, D.T., Armstrong, S.A., Williams, D.A. (2011). Differential niche and Wnt requirements during acute myeloid leukemia progression. *Blood* (July 2011), *in press.*

Lapidot, T, Sirard, C., Vormoor, J., Murdoch, B., Hoang, T., Caceres-Cortes, J., Minden, M., Paterson, B., Caliguri, M.A., Dick, J.E. (1994). A cell initiating human acute myeloid leukaemia after transplantation into SCID mice. *Nature* Vol.367, No. 6464, (February 1994), pp. 645–648.

Marques, D.S., Sandrini, J.Z., Boyle, R.T., Marins, L.F., Trindade, G.S. (2010). Relationships between multidrug resistance (MDR) and stem cell markers in human chronic myeloid leukemia cell lines. *Leukemia Research* Vol.34, No. 6, (June 2010), pp. 757–762.

Pandolfi, P.P. (2001). In vivo analysis of the molecular genetics of acute promyelocytic leukemia. *Oncogene* Vol.20, No. 40, (September 2001), pp. 5726-5735.

Pollyea, D.A., Kohrt, H.E., Medeiros, B.C. (2011). Acute myeloid leukaemia in the elderly: a review. *British Journal of Haematology* Vol.152, No. 5, (March 2011), pp. 524-542.

Rossi, L., Challen, G.A., Sirin, O., Lin, K.K., Goodell, M.A. (2011). Hematopoietic Stem Cell Characterization and Isolation. *Methods in Molecular Biology*. Vol.750, No. 2, (2011), pp. 47-59.

Soltanian, S. & Matin, M. (2011). Cancer stem cells and cancer therapy. *Tumor Biology* Vol.32, No. 3, (June 2011), pp. 425–440.

Tai, M.H., Chang, C.C., Olson, L.K., Tosko, J.E. (2005). Oct-4 expression in adult human stem cells: evidence in support of the stem cell theory of carcinogenesis. *Carcinogenesis* Vol.26, No. 2, (February 2005),pp. 495–502.

Tenen, D.G. (2003). Disruption of differentiation in human cancer: AML shows the way. *Nature Reviews of Cancer* Vol.3, No. 2, (February 2003), pp. 89-101.

Testa, U. (2011). Leukemia stem cells. *Annals of Hematology* Vol.90, No. 3, (March 2011), pp. 245–271.

Warner, J., Wang, J.C., Hope, K.J., Jin, L., Dick, J.E. (2004). Concepts of human leukemic development. *Oncogene* Vol.23, No. 43, (September 2004), pp. 7164–7177.

13

From HSC to B-Lymphoid Cells in Normal and Malignant Hematopoiesis

Rosana Pelayo[1], Elisa Dorantes-Acosta[1,2],
Eduardo Vadillo[1] and Ezequiel Fuentes-Panana[3]
[1]*Oncology Research Unit, Oncology Hospital, Mexican Institute for Social Security, Mexico City*
[2]*Leukemia Clinic, Mexican Children's Hospital 'Federico Gómez', Mexico City*
[3]*Research Unit on Parasitic and Infectious Diseases, Pediatric Hospital, Mexican Institute for Social Security, Mexico City*
Mexico

1. Introduction

Development of B-lymphoid cells is a highly ordered multi-step process that, in adult mammals, starts in bone marrow in a pool of self-renewing multipotential hematopoietic stem cells, which gradually commit to the lymphoid lineage and advance through high regulated differentiation pathways until formation of mature functional cells. Over the last few years, exceptional advances have been recorded in identifying primitive progenitors that lay the foundations of the lymphoid program while losing myeloid potential, along with patterns of transcriptional activity controlling lineage fate decisions and environmental cues that influence the differentiation pathway during normal hematopoiesis. Multicolor flow cytometry, controlled cell cultures, genetic marking systems, microarray technologies and xenotransplantation approaches are being extensively used to address fundamental questions on this regard. Of special interest is the stem cell research with relevance to hierarchy and early events in malignant lymphopoiesis, and to new insights into perspectives that may allow progress in means to protect and sustain the immune system during chemotherapy, inflammation, infection, and following hematopoietic transplantation. In this book chapter, we focus on the hierarchical structure of the early lymphoid system, the current knowledge about intrinsic and microenvironmental factors regulating the differentiation of lymphoid progenitors, and the emerging research to understand malignant lymphoid development.

2. The early steps in the lymphoid development

Mature blood cells are constantly replaced from a unique cell population of hematopoietic stem cells (HSC) residing in specialized niches within the bone marrow (BM), where the hematopoietic system is organized as a hierarchy of cell types that gradually lose multiple alternate potentials while commit to lineage fates and gain specialized functions (Baba et al., 2004; Seita & Weissman, 2010). HSC possess two major characteristics: they are capable of maintaining their constant number by self-renewal and they are in charge of producing all mature blood cells through differentiation processes (Figure 1). Furthermore, HSC are mitotically inactive (quiescent) and divide very slow and intermittently under normal conditions, but are capable of proliferation and differentiation during recovery from chemotherapy or stress circumstances (Takizawa et al, 2011; Mayani, 2010; Passegue et al., 2005; Pelayo et al., 2006b). Movement into and out of a resting state might be crucial for ensuring that the correct number of new hematopoietic cells is produced.

The lymphoid pathway proceeds through critical stages of differentiation of HSC to multipotential early progenitors (MPP), which upon progressive loss of self-renewal capacity, give rise to oligopotent progenitors. Downstream, the production of lineage-committed precursors is crucial for cell maturation. Current knowledge about development of the lymphoid system is based, in great part, on the work done in animal models, demonstrating that lymphoid specification begins in the fraction of lymphoid-primed multipotent progenitors (LMPP). A series of studies using the transgenic RAG-GFP mouse (Igarashi et al., 2002) permitted us to determine that RAG^{+} early lymphoid progenitors (ELP) are capable of differentiating into T, B, NK and conventional dendritic cells (cDC) (Pelayo et al., 2005a; Pelayo et al., 2006a; Welner et al., 2008a). Studies using defined co-cultures and short-term reconstitution assays have shown that ELP are also good producers of plasmacytoid dendritic cells (pDC) and of interferon-producing killer dendritic cells (IKDC), both being key components of the innate immune response to infections (Pelayo et al., 2005b; Welner et al., 2007). At the same time, ELP give rise to committed oligopotent common lymphoid progenitors (CLP), which are responsible for B- and NK- precursor cells production. CLP and lineage precursors have substantially lost the possibility of differentiating into the rest of the lineages.

Due to ethical reasons and technical limitations, human hematopoietic stem cell research has been slower than it has been in mouse models. In humans, the early hematopoietic progenitors are confined in bone marrow to a cellular compartment that expresses CD34 (Blom & Spits, 2006). The fraction of multipotent stem cells is characterized by the phenotype $Lin^{-}CD34^{+}CD38^{-/lo}CD10^{-}CD45RA^{-}$, whereas that of probably the earliest lymphoid progenitors is $Lin^{-}CD34^{+}CD38^{-/lo}CD45RA^{+}CD10^{+}$ and has been recently designated as multi-lymphoid progenitor (MLP) (Doulatov et al., 2010). According to Doulatov's studies, MLP may be directly derived from HSC. However, a precise precursor-product relationship needs to be determined (Figure 1). A description that fully matches the definition of mouse ELP is still missing, but cells with $Lin^{-}CD34^{+}CD38^{+}CD45RA^{+}CD7^{+}CD10^{+}$ phenotype seem to represent good candidates (Blom & Spits, 2006; and our unpublished observations). $Lin^{-}CD34^{+}CD38^{+}CD45RA^{+}CD10^{+}$ B/NK cells, which differentiate principally into B & NK cells, are considered the counterparts of CLP in mice (Figures 1 & 2) (Doulatov et al., 2010). Of special importance is the fact that increasing levels of CD10 correspond to B-lineage specification (Ichii et al., 2010). Downstream, the differentiation of fully committed precursors gives rise to B cells that eventually are exported to peripheral lymphoid tissues (see B cell development sections below).

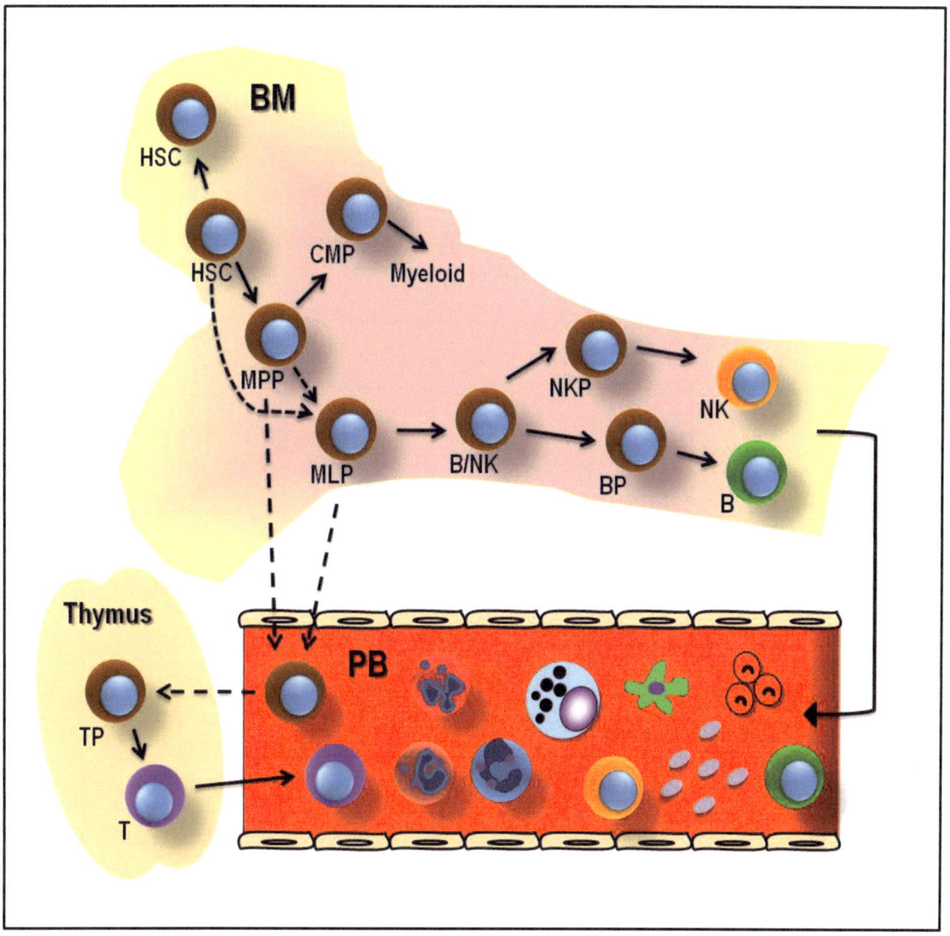

Fig. 1. Early lymphoid development in humans. Within bone marrow (BM), self-renewing hematopoietic stem cells (HSC) give rise to multipotent progenitors (MPP), which have the ability to differentiate into common myeloid progenitors (CMP) and into multi-lymphoid progenitors (MLP). MLP might alternately derive from HSC. NK and B-lymphoid cells are produced from B/NK-derived lineage committed precursors. Mature hematopoietic cells are exported to peripheral blood (PB). Early progenitor cells may colonize the thymus via circulation, and initiate the T-lymphoid development pathway. NKP, natural killer cell precursor; BP, B cell precursor; TP, T cell precursor.

The rigorous purification of human HSC and progenitor cell populations based on their surface phenotype has promoted the study of their biology in adult bone marrow, cord blood and G-CSF-mobilized peripheral blood (Figure 2). Importantly, some of their properties, including cell frequencies, developmental capacities, cell cycle status, transcription factors networks and growth factors production, show substantial differences between newborns and adults (Mayani, 2010). According to literature, we have found that most hematopoietic progenitors are more abundant in cord blood than in the adult tissues bone marrow and mobilized peripheral blood (Mayani, 2010). The implications of these discrepancies during haematological neoplastic diseases are not as yet clear.

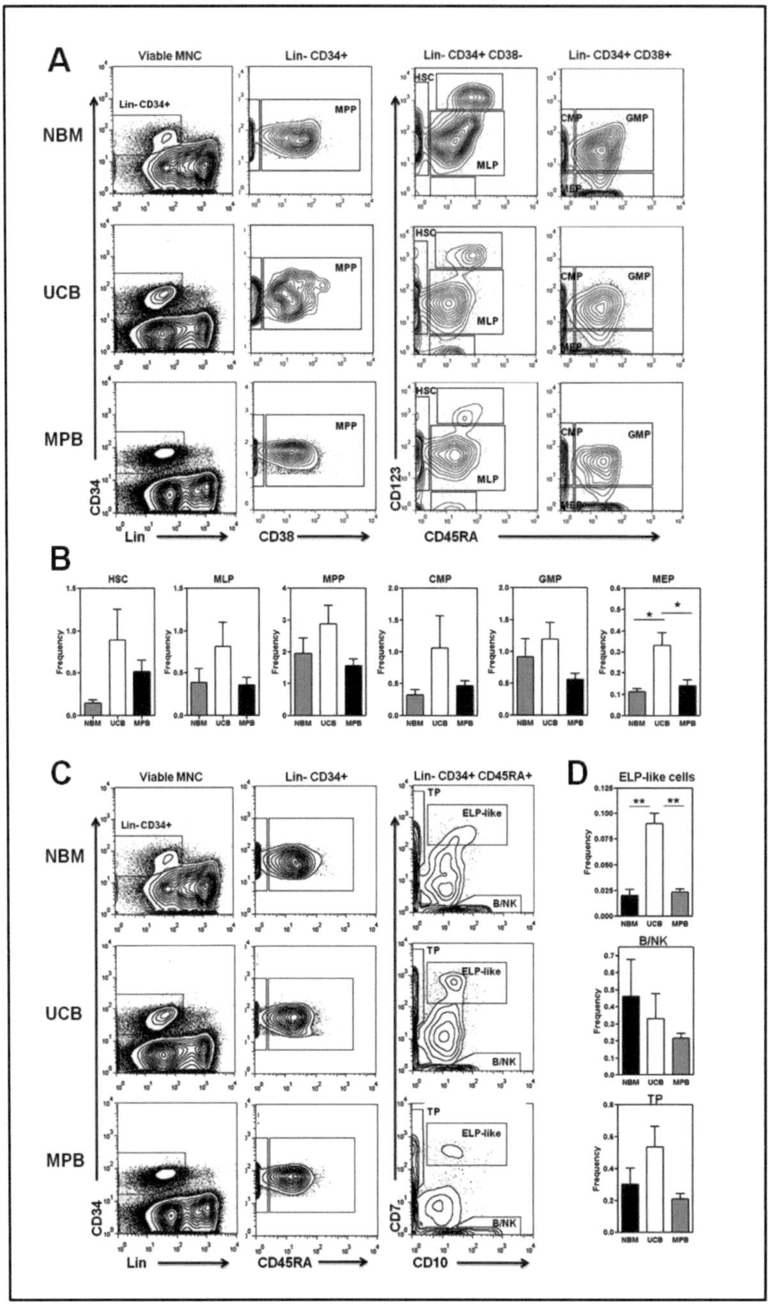

Fig. 2. Prospective identification of human myeloid and lymphoid progenitor cells by flow cytometry. HSC and early progenitor cells reside in the Lin- CD34+ fraction of adult normal bone marrow (NBM), as well as in umbilical cord blood (UCB) and mobilized peripheral blood (MPB). Based on the surface expression of CD38, CD123 and CD45RA, multilymphoid progenitor cells (MLP) and most of the myeloid progenitors can be recognized (A). Further fractionation of Lin-CD34+CD45RA+ cells into CD7 & CD10-expressing cells allows the

identification of T-cell progenitors (TP), B/NK progenitors and ELP-like cells (C). Cell frequencies for each population from the different sources are shown (B and D panels). CMP, common myeloid progenitor; GMP, granulocyte & monocyte progenitor; MEP, megakaryocyte & erythrocyte progenitor. The identity and functions of Lin$^-$CD34$^+$CD38$^-$ CD45RA$^+$CD123^{hi} cells still need more investigation.

During biological contingencies -chemotherapy, infections and transplantation procedures-, the replenishment of the innate immune system from hematopoietic stem/progenitor cells appears to be critical. Interestingly, these seminal cells can proliferate in response to stress conditions and systemic infection by using mechanisms that apparently involve interferons and tumor necrosis factors, among others (Baldridge et al., 2011). Moreover, they are capable of self/non-self discrimination through Toll-like receptors (TLR), which recognize microbial components. Mouse stem cells and early B-cell progenitors express and use TLR, a mechanism that facilitates their differentiation to the innate immune system (Nagai et al, 2006; Welner et al., 2008b; Welner et al., 2009). Recent work suggests that, as in mice, human primitive cells, including MLP, also express functional TLR (Kim et al., 2005; Sioud & Fløisand, 2007; De Luca et al, 2009; Doulatov et al, 2010). In shape with those findings, we have found that BM lymphoid progenitor-enriched fractions display TLR9 (Figure 3) and their differentiation potentials bias toward NK and DC production upon TLR9 ligation (RP & EV, unpublished observations). Thus, plasticity in primitive cells is vulnerable to extrinsic agents that can modify early cell fate decisions during infections or stress, suggesting that the stages of lineage restrictions are less abrupt than previously assumed (Welner et al., 2008a).

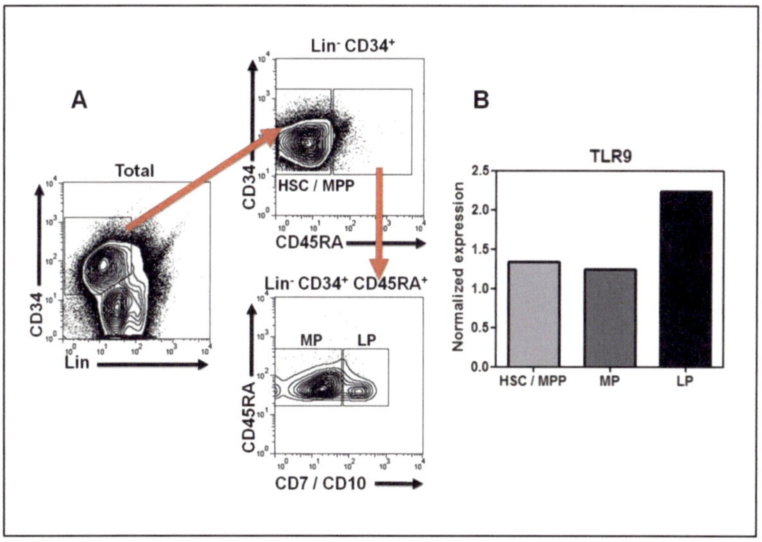

Fig. 3. Lymphoid progenitors from human bone marrow express TLR9. Adult bone marrow is fractionated according to cell surface expression of lineage markers, CD34, CD45RA and CD7/CD10 (A). Lin$^-$CD34$^+$CD45RA$^-$ HSC/MPP, Lin$^-$CD34$^+$CD45RA$^+$CD7/CD10$^-$ myeloid progenitors (MP) and Lin$^-$CD34$^+$CD45RA$^+$CD7/CD10$^+$ lymphoid progenitors (LP) were tested for their intracellular expression of TLR9 by flow cytometry using a specific anti-TLR9 antibody (B).

3. The B cell antigen receptor (BCR) and bone marrow B cell development

The main function of mature immunocompetent B cells is to make antibodies upon recognition of particular new or recurrent antigens by the B cell receptor (BCR). The BCR is a membrane-bound complex of proteins, consisting of a heterodimer of identical pairs of immunoglobulin (Ig) heavy and light chains, which are responsible for the clonal diversity of the B cell repertoire and the antigen identification, but are unable to generate signals and trigger biological responses after antigen binding. This function is mediated by the disulfide-coupled heterodimer of Igα (CD79a) and Igβ (CD79b), which is non-covalently associated with the Ig antigen recognition unit (Figure 4A). Igα/Igβ signaling is dependent on distinct tyrosine-based activation motifs localized in the cytoplasmic tails of these proteins. It is the sequential expression and assembly of the BCR components that defines each developmental stage of the B cell pathway, and, therefore, each stage is characterized by a particular form of BCR, reflecting the progression of receptor assembly (Fuentes-Pananá et al., 2004a).

To achieve BCR clonal diversity, the Ig heavy and light chain genes are composed of constant and variable regions. The variable region is formed by a series of segments V (variable), D (diversity) and J (joining) (Figure 4B), which are brought together by a highly ordered process of VDJ recombination accomplished by the products of the recombinase-associated genes 1 and 2 (RAG1 and RAG2) occurring first in the heavy and then in the light chain loci (Thomas et al., 2009). ProB and PreB stages are characterized by rearrangements of the Ig heavy and light chains, respectively (Figure 5) (Fuentes-Pananá et al., 2004b), and further divided according to the status of the recombination. In mice, ProB-A is the sub-stage during which the heavy chain is in germ line state, whereas during ProB-B the heavy chain D and J fragments are recombined, and in ProB-C, V-DJ is recombined. In large PreB cells, the preBCR is already expressed in surface and the light chain V and J fragments are in germ line state, while in small PreB cells light chain V-J is recombined (Hardy et al., 1991). These stages are better known in humans as Early ProB or Pre-proB (A), ProB (B), PreB I (C), large and small PreB II (Figure 5). In the ProB stage Igα and Igβ are expressed at cell surface in association with chaperon proteins such as calnexin (the proBCR). As soon as the heavy chain is successfully recombined, it is assembled with Igα and Igβ and the surrogate light chains λ5 and VpreB to form the preBCR. Surface expression of this receptor marks the transition to the preB stage (Figure 5) (Fuentes-Pananá et al., 2004a; 2004b).

In addition to their VDJ recombination status and pattern of surface marker expression, ProB and PreB stages can be recognized by their proliferative state (Hardy et al., 1991). RAG-1 and RAG-2 enzymes are tightly regulated during the cell cycle, being highly active in G_0 and degraded before the cell enters S phase (Li et al., 1996). By assuring that proliferation and recombination are mutually exclusive mechanisms, the developing B cell guarantees that no events of non-homologous recombination will occur during DNA replication, thus avoiding an increase in the mutation rate.

3.1 Self-recognition and peripheral B cell development

Once the mature BCR is present in the surface of immature B cells, it is finally able to interact with conventional polymorphic ligands, and selection at this stage is designed to test the receptor-ligand interaction. Intimate contact between the immature B cell and the stromal cells of the bone marrow allows those receptors capable of recognizing self-antigens to be identified and eliminated through a variety of mechanisms collectively termed "tolerance". Non-self-reactive B cells exit to the periphery and reach the spleen where they are again tested for reactivity against self-antigens before they transition to the mature stage (Figure 5) (von Boehmer & Melchers, 2010). Three main mechanisms of B-cell tolerance are known: receptor editing, deletion of auto-reactive clones (negative selection) and anergy. Only those B cells that carry receptors without self-specificity are allowed to exit the bone marrow and become mature B cells in peripheral lymphoid organs.

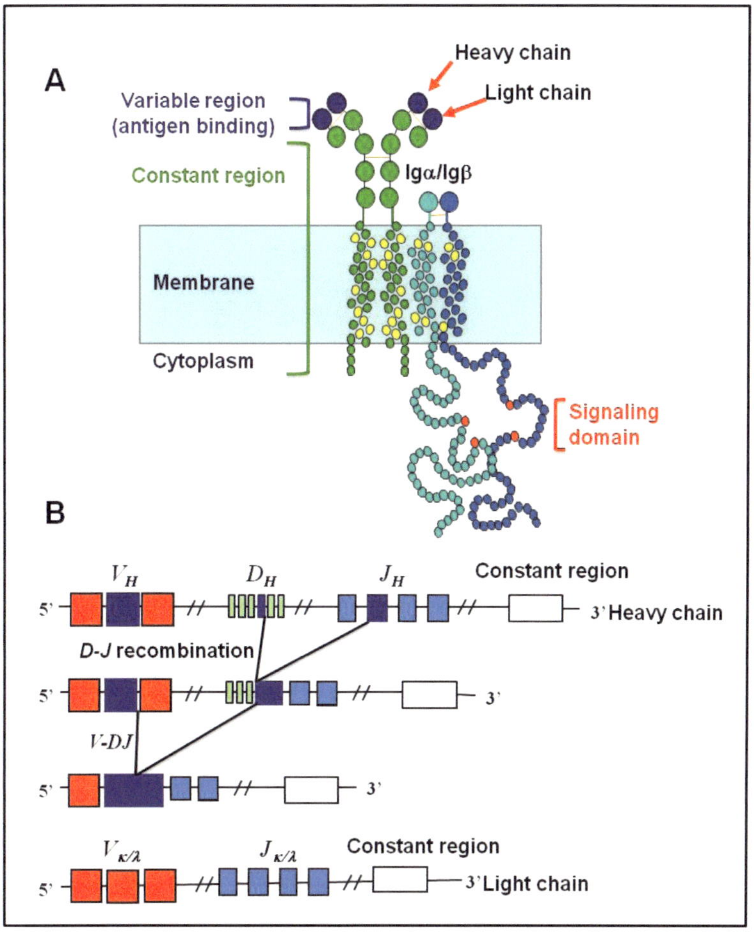

Fig. 4. The B cell antigen receptor (BCR). A) Heavy and light chains are comprised of variable regions where VDJ recombination occurs (shown in dark blue) and constant regions (green). The signaling domains are present in the cytoplasmic leaflet of Igα and Igβ. B) Variable regions are formed by a number of segments termed V (variable), D (diversity) and J (joining) within the heavy chain, and by segments V and J within the light chain, which are brought together by a VDJ recombination process. Randomly, D and J segments recombine at first, followed by V segments joining the DJ fragment (shown in dark blue squares is an example of segment choice). This mechanism is responsible for the extensive repertoire of BCR specificities.

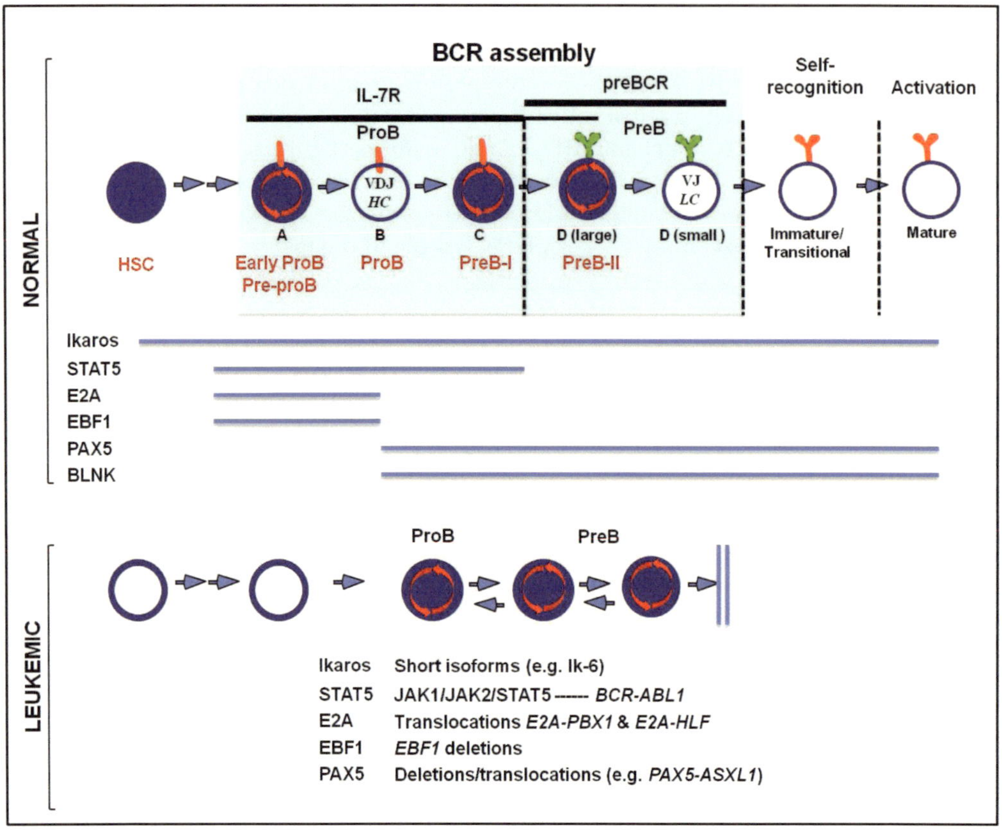

Fig. 5. Normal and leukemic B cell development. B cell stages can be divided according to the main processes guiding development: receptor assembly, self-recognition and activation (top panel). Receptor assembly occurs in bone marrow (light blue box) by VDJ recombination in the Pro-B and Pre-B stages, whereas self-recognition starts in bone marrow and ends in periphery, and activation takes place at peripheral level. Nomenclature for each sub-stage in mice is shown in black letters while the most common nomenclature for their counterparts in humans is shown in red letters. The dashed lines separating all stages indicate checkpoints at which signaling from the preBCR and BCR is required for positive selection and progression along the B-cell maturation pathway. The proBCR, preBCR, and mature receptor are also illustrated in their respective stages. Replication and recombination processes are mutually exclusive as denoted by the circular arrows and VDJ signs inside the cell. The replication stages are also frequently compromised in pediatric B cell acute leukemia. Black lines under IL-7R and preBCR indicate the stages where these receptors are most required. The differential thickness in the IL-7R line shows the sub-stages where a higher (nanograms) or lower (picograms) concentration of the IL-7 is required. Homeostatic and leukemic expression of transcription factors along the B cell pathway are shown in the middle and bottom panels. Blue bars mark normal gene expression, and the most common modified forms of the transcription factors associated with B cell acute lymphoblastic leukemia are revealed. HSC, hematopoietic stem cell.

On the basis of their cell-surface phenotype, peripheral immature B cells are further divided into transitional 1 (T1, $AA4^{+}IgM^{high}CD23^{-}$) and transitional 2 (T2, $AA4^{+}IgM^{high}CD23^{+}$). T1 cells inhabit the spleen's red pulp and give rise to T2 cells (Allman et al., 2001). There is an additional population designated T3, but it is controversial whether this is a population in line in the progression to the mature stage or whether it represents a population of anergic cells (Merrell et al., 2006).

3.2 Innate and adaptive mature B cell populations

Following antigen binding, mature B cells activate pathways that lead to proliferation and further differentiation into antibody-producing B cells (plasma cells) or memory B cells. In the spleen, mature B cells are sub-divided into follicular (FO, $AA4.1^{-}CD21^{int}CD23^{high}$) and marginal zone (MZ, $AA4.1^{-}CD21^{high}CD23^{-}$) B cells according to both their location and their cell-surface phenotype. A distinct subset of mature B cells is preferentially present in the peritoneal cavity; these are known as B1 cells [$B220^{+}CD11b^{+}CD5^{+}$ (B1a) or $CD5^{-}$ (B1b)]. Among them, FO B cells are responsible for adaptive antibody responses, whereas MZ and B1 mature populations respond rapidly to antigenic stimulus but do not go through germinal-center reactions and thus their response can be independent of T cell help (Martin et al., 2001). Therefore, MZ and B1 B cells are thought to be part of an innate-like response. The origin of both of these populations is not well understood. While MZ B cells share part of FO pathway, the fetal liver was thought to originate a large fraction of the adult B1 B cells (Tung et al., 2006). Recently, a novel developmental model suggests that some B1 cell progenitors can be produced in bone marrow (Esplin et al., 2009).

3.3 Regulation of B lineage commitment: The critical role of preBCR tonic signaling, IL-7R and transcription factors in context

Limitation of lineage choice during development is regulated by a combination of signaling pathways and transcription factors (TF). In mice, the main receptor controlling the ProB stage is the IL-7R, which is composed of a α chain (IL-7Rα) and the common cytokine receptor γ chain (γc). Deletion of IL-7Rα or γc leads to developmental arrest at the early ProB stage (von Freeden-Jeffrey et al., 1995; Cao et al., 1995).

IL-7 activates the major signaling pathway JAK–STAT, with STAT5 being the essential mediator of IL-7 signals in early B cell development (Yao et al., 2006).

By the other hand, an important characteristic of the developmental process that distinguishes B and T lymphocytes from other cell lineages is the continuous selection of these lymphoid cells for their ability to express a competent, non-self receptor. B cells that fail to express a receptor are eliminated. Thus, BCR and BCR-like receptors must generate active permissive signals that allow differentiation through the different developmental stages. Because the preBCR lacks of the light chain and therefore of the capacity to bind polymorphic ligands, it has been proposed that this receptor is able to signal constitutively and independently on ligand, an activity also known as tonic signaling. Although there is little understanding of how tonic signals are generated, the view is supported by receptor-less B cells able to differentiate into mature B cells by expression of a chimeric construct of Igα and Igβ positioned in the cell surface membrane (Bannish et al., 2001).

Once the preBCR is expressed at the end of the proB stage, it can take over many of the functions performed by the IL-7 receptor signaling. Both receptors act individually and together to allow B cell development (Figure 5). Like IL-7R, the preBCR promotes mechanisms of positive selection, survival and proliferation (Ramadani et al., 2010; Yasuda et al., 2008). The CCND3 gene, which encodes for cyclin D3, is essential for PreB cell expansion and integrates IL-7R and preBCR signals (Cooper et al., 2006).

Downstream the IL-7 and preBCR receptors, a handful of transcription factors (TF) are critical for commitment to the B cell lineage and early development; these include E2A/TCF3 (immunoglobulin enhancer binding factors E12/E47/transcription factor 3), EBF1 (Early B cell Factor 1) and PAX5 (Paired box 5) (Figure 5). Loss of E2A and EBF1 blocks entry into the B cell lineage, while loss of PAX5 redirects B cells into other lineages (Nutt et al., 1999; O'Riordan & Grosschedl, 1999). Acting together with E2A, EBF1 and STAT5, one of the main molecular functions of PAX5 is to allow VDJ recombination (Hsu et al., 2004). Also, E2A, PAX5, IKZF1 and RUNX1, among other TF, are responsible for RAG expression (Kuo & Schlissel, 2009). Moreover, IL-7R signaling fulfills an essential role in early B cell development, with STAT5 participating in the activation of the B cell regulatory genes E2A, EBF1 and PAX5. E2A encodes two TF via alternative splicing, E12 and E47. In mice lacking the E2A gene, the B cell lineage is lost, there is no heavy chain recombination, and the expression of the B cell-restricted genes EBF1, PAX5, CD79A/B and VPREB1 (CD179A) is also affected.

Enforced expression of EBF1 and PAX5 is sufficient to overcome the developmental block in mice deficient in E2A, IL-7 or IL-7Rα, further illustrating the transcriptional hierarchy of the B cell-specific program triggered by IL-7 receptor signaling (Nutt & Kee, 2007). EBF1 acting together with PAX5 drives the expression of many genes critical for early B cell development and B cell function, including FOXO1, MYCN, LEF1, BLNK, CD79A (MB-1), RAG2, CD19 and CR2 (CD21) (Nutt & Kee, 2007; Smith & Sigvardsson, 2004).

Although PAX5 is a positive regulator of B-cell specific genes, also functions as a repressor of non B-lineage genes such as M-CSFR, NOTCH1 and FLT3 (Cobaleda et al., 2007) so B cell development is unidirectional and mostly irreversible in homeostatic conditions.

Also important for lymphoid development are members of the Ikaros family of TFs, mainly IKZF1 (which encodes Ikaros) and IKZF3 (which encodes Aiolos). Ikaros activates B cell genes and represses genes that are unrelated to the B lineage. Expression of IKZF1 and IKZF3 is regulated by alternative splicing, which produces long isoforms (Ik-1, Ik-2, Ik-3, Aio-1, Aio-3, Aio-4 and Aio-6) that efficiently bind to DNA, and short isoforms (Ik-4, Ik-5/7, Ik-6, Ik8, Aio-2, Aio-5) that are unable to bind DNA with high affinity and do not activate transcription (Liippo et al., 2001). Ikaros is activated in early stages of lymphopoiesis and is required for both early and late events in lymphocyte differentiation. Aiolos is not required during the early specification of the B and T lineages but is essential during further B cell maturation. They also act in concert to promote preB cell cycle exit and transition to small PreB stage (Ma et al., 2010).

3.4 Human B cell development

Selection processes operating on developing B cells are similar in all mammals. Thus, early B cell development in humans is also mainly guided by VDJ recombination and by the proliferative expansion of clones that have successfully completed the rearrangement of their receptors, whereas late development is led by mechanisms of tolerance to self-antigens. All these processes in humans are less well understood than are their counterparts in mice. Importantly, human B cells can still be generated in severe combined immunodeficiency (SCID) patients with mutations in the IL-7R gene, suggesting that IL-7 signaling is not essential for human B cell development (Puel et al., 1998) although a recent study has demonstrated that in vitro human B cell production is dependent on IL-7 (Parrish et al., 2009). The fine regulatory mechanism separating proliferation and differentiation might explain why the proliferating ProB and PreB sub-stages are the ones generally found to be compromised in human pediatric B-cell acute lymphoblastic leukemia (B-cell ALL) and why this disease is characterized by leukemic blast cells that are often unable to progress through the differentiation pathway. This tendency to be arrested in proliferative states might result in an increased rate of mutations, leading to formation of neoplastic cells. Supporting the later, mice expressing B cell mutants in the adaptor protein BLNK are arrested in the large PreB stage and often develop B cell malignancies (Flemming et al., 2003). Proliferative stages occur in the early ProB, PreB-I and large PreB-II fractions (Figure 5).

4. Acute lymphoblastic leukemia

Acute lymphoblastic leukemia (ALL) is a disorder characterized by the monoclonal and/or oligoclonal proliferation of hematopoietic precursor cells of the lymphoid series within the bone marrow. At present, ALL is the most frequent malignancy in children worldwide and a serious problem of public health, constituting 25% of all childhood cancers and 75%-85% of the cases of childhood leukemias (Perez-Saldivar et al., 2011). Near to 80% of ALL cases have precursor B-cell immunophenotype, while approximately 15% show T-cell immunophenotype. Even when a relatively high efficiency of therapeutic agents has been demonstrated (Pieters & Carroll, 2010), there has been a slight but gradual increase in the incidence of ALL in the past 25 years, and appears to be highest in Hispanic population, which also show superior rates of high risk patients (Fajardo-Gutiérrez et al., 2007; Abdullaev et al., 2000; Perez-Saldivar et al., 2011; Mejía-Aranguré et al., 2011). Factors such as drug resistance, minimal residual disease, cell lineage switch, and the rise of mixed lineages often put the success of treatment at risk and change the prognosis of the illness. The molecular mechanism involved in these phenomena and the identities of the target hematopoietic populations have not been completely defined, due in part, to the fact that neither the precise origin of the disease, nor the susceptibility of primitive leukemic cells to extrinsic factors, is known.

4.1 The origin of ALL

Over the last two decades, cancer stem cells (CSC) have been defined as cells within a tumor that possess the capacity to self-renew and to cause heterogeneous lineages of cancer cells that comprise the tumor (Clarke et al., 2006). According to MF Greaves, who proposed the original hypothesis for leukemogenesis, multiple consecutive carcinogenic hits in hematopoietic cells may drive the malignant transformation (Greaves, 1993; Greaves & Wiernels, 2003), where the second oncogenic event on pre-leukemic clones could be indirectly promoted by delayed infections (Greaves, 2006; Mejía-Aranguré et al., 2011). Our general current view suggests the occurring of oncogenic lesions in early development or in a primitive cell that result in the abnormal differentiation of leukemic stem cells. Among the various factors that hit the HSC fraction, anomalous microenvironmental cues may contribute to trigger and support the leukemic behaviour of precursor cells (Figure 6).

Although CSC in myeloid leukemias have been strictly depicted as the responsible cells for tumour maintenance, which clearly keep the biological hierarchy within the hematopoietic structure (Dick, 2008), identification of a rare primitive and malignant cell with intrinsic stem cell properties and the ability to recapitulate the acute lymphoblastic leukemia has been more complicated (Bomken et al.,2010), particularly due to the genetic diversity of the disease and the lack of appropriate *in vitro* and *in vivo* models.

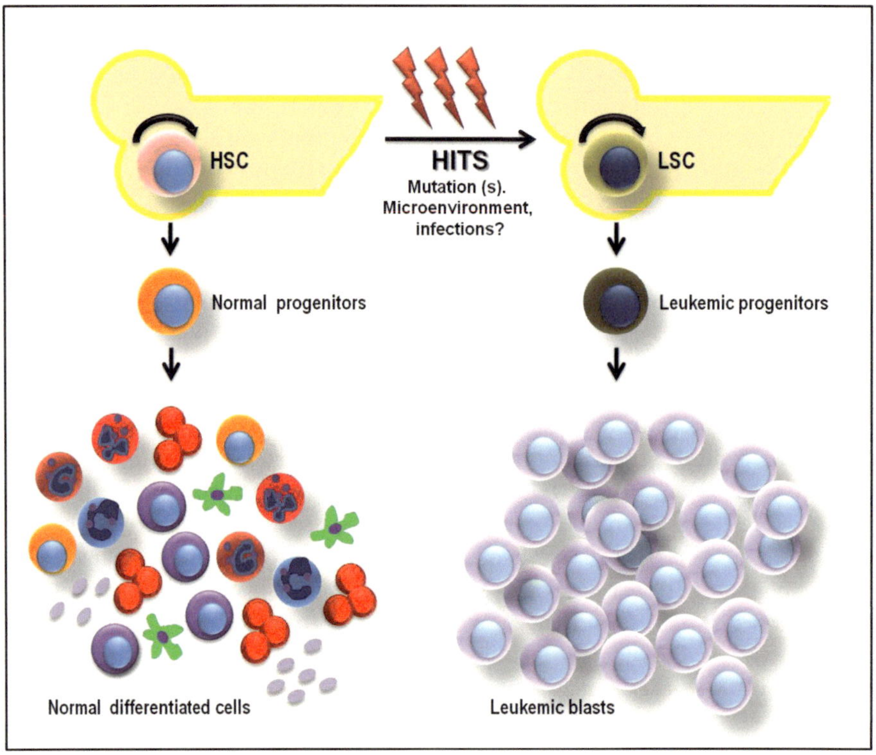

Fig. 6. Leukemic stem cell model. Normal hematopoietic stem cells (HSC) give rise to progenitors and mature blood cells within a hierarchical structure in the bone marrow. As a result of multiple and consecutive oncogenic hits on HSC including genetic and microenvironmental alterations, a malignant counterpart (the leukemic stem cell, LSC) emerge, which maintains some degree of developmental potential, generating the leukemic progenitor and blast cells.

Cell culture systems revealing alterations in early hematopoiesis, the existence of leukemic clones with unrelated DJ rearrangements and cytogenetic abnormalities on cells lacking lineage markers, have strongly suggested the participation of primitive cells in ALL. Moreover, data showing cells with immature phenotypes capable of engrafting and reconstituting leukemia in immunodeficient mice, lead to believe that, as in AML & CML, the hierarchy structure of the hematopoietic system is kept in ALL, and infant B cell-leukemia initiating cells have undifferentiated characteristics (Espinoza-Hernandez et al., 2001; Cobaleda et al., 2000; Cox et al., 2004; Cox et al., 2009). To characterize ALL progenitor cells, Blair and colleagues have purified by flow cytometry a number of cell fractions based on the expression of CD34 and the B-lymphoid marker CD19. Regardless the risk stratum of the patient, $CD34^{+}CD19^{-}$ cells, but not committed B precursors, were able to reconstitute the disease in NOD/SCID models (Cox et al., 2004). Moreover, $CD133^{+}CD38^{-}CD19^{-}$ primitive cells residing in ALL BM are suggested to be the leukemia-initiating cells and responsible of drug-resistant residual disease (Cox et al., 2009). However, recent studies have remarkably shown that precursor blasts can also reestablish leukemic phenotypes *in vivo*, conferring them stem cell properties (Heidenreich & Vormoor, 2009; Bomken et al., 2010). Using novel intrafemoral xenotransplantation strategies, Vormoor's Lab has found that all differentiation stages of B precursor cells within $CD34^{+}CD19^{+}$ and $CD34^{-}CD19^{+}$ fractions are able to successfully engraft and recapitulate the original patient's disease in long-term systems, suggesting that committed cells in ALL do not lose the self-renewal stem cell property while they mature (le Viseur et al., 2008) (Figure 5), though their multi-lineage potential is uncertain.

These discordant results unveil that key questions regarding leukemic stem cells and the earliest steps of the lymphoid program in ALL still to be solved. Recently, the combination of clonal studies and alterations on genetic copies along with xenotransplant models, showed unsuspected genetic diversity, supporting multiclonal evolution of leukemogenesis rather than lineal succession (Dick, 2008). Thus, a less rigid structure of CSC models should further take account of functional plasticity and clonal evolution to understand CSC biology and to develop novel, stem/progenitor cell-directed therapies (Bomken et al., 2010).

4.2 Genes, cytogenetic alterations and transcription factors in B-cell leukemogenesis

The leukemogenic program is characterized by arrest of differentiation pathways, increased cell proliferation, enhanced self-renewal, decreased apoptosis rates and telomere maintenance. It is thought that together these alterations result in production of highly proliferative clones of immature leukemic blast cells with intrinsic survival advantage and limitless replicative potential (Warner et al., 2004).

Gain or loss of function of transcription factors such as E2A, EBF1, PAX5 and Ikaros affect homeostatic B cell lymphopoiesis in murine models, and are often associated with malignant transformation in humans, supporting conserved roles for these TFs and their activating signaling pathways (Figure 5) (Pérez-Vera et al., 2011).

A high frequency of ALL patients has genetic lesions -mostly chromosomal translocations- associated with leukemic cells. E2A is often translocated with several partners, including PBX1 [t(1;19)(q23;p13)] and HLF [t(17;19)(q22;p13)], which are detected in 5-6% and 1% of ALL children, respectively. E2A-PBX1 is a potent transcriptional activator of the WNT16 oncogene (McWhirter et al., 1999), while E2A-HLF functions as a survival factor of early B cells by activating expression of the anti-apoptotic genes SNAI2 (SLUG) and LMO2. Accordingly, gene silencing of LMO2 in an E2A-HLF^{pos} cell line induced apoptotic cell death (Hirose et al., 2010). RUNX1 is also a frequent target for chromosomal rearrangements and mutations in ALL. 25% of children and 2% of adults of ALL patients carry the ETV6/RUNX1 fusion as a result of the translocation t(12;21)(p12;q21), which may play a role regulating the B lineage-specific transcriptional program at an early stage (Durst & Hiebert, 2004). SNP array analysis of ETV6-RUNX1 samples has recently identified multiple additional genetic alterations, but the role of these lesions in leukemogenesis remains undetermined (van der Weyden et al., 2011).

Genome-wide analysis has recorded abnormalities in PAX5 and EBF1 in up to 32% of children and 30% of adults with B ALL, and in 35% of relapsed cases (Mullighan et al., 2007). Currently, five PAX5 fusions have been identified with the gene partners LOC392027 (7p12.1), SLCO1B3 (12p12), ASXL1 (20q11.1), KIF3B (20q11.21) and C20orf112 (20q11.1), with the resulting chimeric proteins expressing lower levels of PAX5 and its target genes (An et al., 2008). EBF1 alterations are common in patients with poor outcomes and are particularly frequent (25%) in relapsed children (Harvey et al., 2010).

The MLL (mixed lineage leukemia) gene is often rearranged in leukemias with myeloid and lymphoid phenotype, probably indicating a very early multipotent progenitor origin. More than 50 fusions involving MLL have been documented. Among them, the MLL-AF4 [t(4;11)(q21;q23)] translocation is present in 80% of infant, 2% of children, and 5-10% of adult ALL (McCarthy, 2010).

The BCR-ABL1 translocation [t(9;22)(q34;q11), also known as Philadelphia chromosome] is found in 5% of pediatric and 25% of adult B cell ALL. An important consequence for this translocation is the over-expression of STAT5. STAT5 inactivation results in cell cycle arrest and apoptosis of BCR-ABL^{pos} malignant B cells and BCR–ABL1^{pos} STAT5 knockout mice do not develop leukemia (Malin et al., 2010). Interestingly, genome-wide analysis of B cell ALL has identified mutations in the STAT5 upstream regulators JAK1 and JAK2 in up to 10% of patients, and patients BCR-ABL^{pos} or with JAK1&2 mutations have a similar gene expression profile and prognosis (Malin et al., 2010). JAK2 mutations lead over-expression of CRLF2 (also known as thymic stromal lymphopoietin receptor) which forms a heterodimeric complex with the IL-7R (Harvey et al., 2010). In a subset of cases, CRLF2 promotes constitutive dimerization and cytokine-independent proliferation. Finally, high expression levels of the short Ikaros isoforms, particularly the dominant negative Ik-6, are also associated with high risk leukemia (Sun et al., 1999). Most of the BCR-ABL^{pos} B ALL patients have deletions in IKZF1 and increased levels of the short isoforms; however, Ik-6 has also been found to be elevated in BCR-ABL^{neg} patients (Mullighan et al., 2008). It has been proposed that the high level of Ikaros short isoform expression is due to genetic lesions. Supporting this idea, IKZF1 somatic deletions have been found in a number of recurrences and are strongly associated with minimal residual disease (Mullighan et al., 2009). A summary of homeostatic and leukemic expression of transcription factors along the B cell pathway is shown in Figure 5.

Despite these important advances in the definition of genetic abnormalities that are prevalent in ALL, the disease is heterogeneous at the molecular level, and possibly it is the result of combination of genetic and epigenetic alterations. Furthermore, high frequencies of ALL cases seem not to be associated to intrinsic genetic abnormalities, opening the possibility of microenvironmental cues leading to disease.

4.3 Leukemic microenvironmental cues?

The complexity of leukemogenesis increases when we consider the indubitable influence of the bone marrow microenvironment in the hematopoietic development, which is a network of cells (mesenchymal cells, osteoblasts, fibroblasts, adipocytes, endothelial cells, etc) and their products (extracellular matrix molecules, cytokines and chemokines) that support hematopoiesis. Under physiological conditions, the appropriate production of mature blood cells throughout life is sustained by special niches that provide stem and progenitor cells with regulatory signals essential for their maintenance, proliferation and differentiation (Nagasawa et al., 2011). Among secreted factors, CXCL12, FLT3-L, interleukin 7 and stem cell factor are critical for commitment to the lymphoid program and normal B cell development is supported by two stage specific cellular niches within central bone marrow: a CXCL12/SDF1 expressing niche, and a IL-7 expressing niche. B cell precursors are thought to move from one to another as differentiation progresses (Tokoyoda et al., 2004; Nagasawa, 2006). The role of the bone marrow microenvironment in carcinogenesis has been conceived through three possible mechanisms: competition of tumor cells for normal HSC niches, which may allow their maintenance and survival; manipulation of the environment to promote tumor progression and disruption of hematopoietic-niche communication that drives oncogenesis (Raaijmakers, 2011). Although these potential mechanisms are tempting, their contribution to ALL remains formally unexplored. It has been proposed by Sipkins and colleagues that the leukemic cells derived-tumor microenvironment impairs the behavior of normal hematopoietic cells (Colmone et al., 2008). Furthermore, a number of alterations have been recorded in the marrow microenvironment of ALL, including chromosomal aberrations in mesenchymal stem cells, anomalous expression of adhesion molecules, abnormal levels of CXCR4 and growth factors, as well as prevalence of pro-inflammatory cytokines (Menendez et al., 2009; Geijtenbeek et al., 1999; Juarez et al., 2009; and our unpublished results). Whether an abnormal microenvironment anticipates the leukemic stage or is a consequent fact, is still an open issue.

5. Conclusion

Much has been learned about identity, function and intercommunication of seminal cells within the hematopoietic system from animal models. However, our understanding of the hierarchy and regulation of human stem/progenitor cells is still incomplete and the hematopoietic charts have been in constant re-construction over the last few years. Furthermore, while it has long been recognized that intrinsic abnormalities in primitive hematopoietic cells may cause hematological disorders, it has also become clear that changes in both cell composition and function of the bone marrow microenvironment might govern stem cell activity and lead to disease. Future progress in these areas will be decisive to suggest novel classification, prognosis and treatment venues.

6. Acknowledgment

R.P. is recipient of funding from the National Council of Science and Technology, CONACYT (grant CB-2010-CO1-152695) and from the Mexican Institute for Social Security, IMSS (grant FIS/IMSS/852). EDA and EV are scholarship holders from CONACYT.

7. References

Abdullaev, F., Rivera-Luna, R., Roitenburd-Belacortu, V., & Espinosa-Aguirre, J. (2000). Pattern of childhood cancer mortality in Mexico. *Arch Med Res.* Vol. 31, No.5, (September-October 2000), pp. (526-31), ISSN 0188-4409

Allman, D., Lindsay, R. C., DeMuth, W., Rudd, K., Shinton, S. A., & Hardy R. R. (2001). Resolution of three nonproliferative immature splenic B cell subsets reveals multiple selection points during peripheral B cell maturation. *Journal of immunology*, Vol.167, No.12, (December 2001), pp. (6834-6840), ISSN 0022-1767

An, Q., Wright S. L., Konn Z. J., Matheson E., Minto L., Moorman A. V., Parker H., Griffiths M., Ross F. M., Davies T., Hall A. G., Harrison C. J., Irving J. A., & Stretfford J. C. (2008). Variable breakpoints target PAX5 in patients with dicentric chromosomes: a model for the basis of unbalanced translocations in cancer. *Proceedings of the National Academy of Science USA*, Vol. 105, No. 44, (October 2008), pp. (17050-17054), ISSN 0027-8424

Baba, Y. Pelayo, R.,& Kincade, PW. (2004). Relationships between hematopoietic stem cells and lymphocyte progenitors. *TRENDS in Immunology*, Vol.25, No.12, (December 2004), pp. 645-649, ISSN 1471-4906

Baldridge, MT., King, KY.,& Goodell, MA. (2011). Inflammatory signals regulate hematopoietic stem cells. *Trends in Immunology*, Vol.32, No.2, (February 2011), pp.57-65, ISSN 1471-4906

Bannish, G., Fuentes-Pananá, E. M., Cambier, J. C., Pear, W. S., & Monroe, J. G. (2001). Ligand-independent signaling functions for the B lymphocyte antigen receptor and their role in positive selection during B lymphopoiesis. *Journal of Experimental Medicine*, Vol.194, No.11 (December 2001), pp. (1583-1596), ISSN: 0022-100

Blom, B. & Spits, H. (2006). Development of human lymphoid cells. *Annual Reviews of Immunology*. Vol. 24, pp. (287-320), ISSN 0732-0582

Bomken, S., Fiser, K., Heidenreich, O., & Vormoor, J.,(2010). Understanding the cancer stem cell. *Br J Cancer*. Vol. 103 No.4, (August 2010), pp (439-45), ISSN 1532-1827

Cao X., Shores E. W., Hu-Li J., Anver M. R., Kelsall B. L, Russell S. M., Drago J., Noguchi M., Grinberg A., & Bloom E. T. (1995). Defective lymphoid development in mice lacking expression of the common cytokine receptor □ chain. *Immunity*, Vol. 2, No. 3, (March 1995), pp. (223-238), ISSN 1074-7613

Clarke, M., Dick, J, Dirks, P., Eaves, C., Jamieson, C., Jones, D., Visvader, J., Weissman, I., & Wahl, M., (2006) Cancer stem cells perspectives on current status and future directions: AACR Workshop on cancer stem cells. *Cancer Res.* Vol.66, No.19, (October 2006) pp (9339-44), ISSN 1538-7445

Cobaleda, C., Gutiérrez-Cianca, N., Pérez-Losada, J., Flores, T., García-Sanz, R., González, M., & Sánchez-García, I. (2000) A primitive hematopoietic cell is the target for the leukemic transformation in human philadelphia-positive acute lymphoblastic leukemia. *Blood* Vol.95, No.3, (February 2000) pp. (1007-1013), ISSN 0006-4971

Cobaleda, C., Schebesta, A., Delogu, A., & Busslinger M. (2007). Pax5: the guardian of B cell identity and function. *Nature Immunology*, Vol. 8, No. 4, (April 2007), pp. (463-470), ISSN 1529-2908

Colmone, A., Amorim, M., Pontier, A., , Wang, S., Jablonski, E., & Sipkins, D., (2008) Leukemic cells create bone marrow niches that disrupt the behavior of normal hematopoietic progenitor cells. *Science*. Vol.322, No.5909, (December 2008), pp. (1861-5), ISSN 1095-9203

Cooper A. B., Sawai C. M., Sicinska E., Powers S. E., Sicinski P., Clark M. R., & Aifantis I. (2006). A unique function for cyclin D3 in early B cell development. *Nature Immunology*, Vol. 7, No. 5, (May 2006), pp. (489-497), ISSN 1529-2908

Cox, CV., Diamanti, P., Evely, RS., Kearns, PR., & Blair A. (2009) Expression of CD133 on leukemia-initiating cells in childhood ALL. *Blood* Vol.113, No.14, (April 2009), pp. (3287-3295), ISSN 1528-0020

Cox, CV., Evely, R., Oakhill, A., Pamphilon, D., Goulden, N.,& Blair, A., (2004) Characterization of acute lymphoblastic leukemia progenitor cells. *Blood.* Vol.104, No.9, (November 2004), pp. (2919-25), ISSN 0006-4971

De Luca, K. Frances-Duvert, V. Asensio, M. Ihsani, R., Debien, E., Taillardet, M., Verhoeyen, E., Bella, C., Lantheaume, S., Genestier, L., & Defrance, T. (2009). The TLR1/2 agonist PAM(3)CSK(4) instructs commitment of human hematopoietic stem cells to a myeloid cell fate. *Leukemia*. Vol. 23, No. 11, pp. 2063-74, ISSN 0887-6924

Dick, J., (2008) Stem cell concepts renew cancer research. *Blood*. Vol.112, No.13, (December 2008), pp. (4793-807), ISSN 1528-0020

Doulatov, S., Notta, F., Eppert, K., Nguyen, LT., Ohashi, PS., & Dick, JE. (2010). Revised map of the human progenitor hierarchy shows the origin of macrophages and dendritic cells in early lymphoid development. *Nat Immunol*, Vol.11, No.7, (July 2010), pp.585-593, ISSN 1529-2908

Durst, KL., & Hiebert, SW. (2004). Role of RUNX family members in transcriptional repression and gene silencing. *Oncogene*, Vol.23, No.24, (May 2004), pp. (4220-4224), ISSN 0950-9232

Espinoza-Hernández, L., Cruz-Rico, J., Benítez-Aranda, H., Martínez-Jaramillo, G., Rodríguez-Zepeda, MC., Vélez-Ruelas, MA., Mayani, H. (2001) In vitro characterization of the hematopoietic system in pediatric patients with acute lymphoblastic leukemia. *Leukemia Research* Vol.25, No.4, (April 2001) pp (295-303), ISSN 0145-2126

Esplin B.L., Welner R.S., Zhang Q., Borghesi L.A., & Kincade P.W. (2009). A differentiation pathway for B1 cells in adult bone marrow. *Proceedings of the National Academy of Science USA,* Vol. 106, No.14, (April 2009), pp. (5773-5778), ISSN 0027-8424

Fajardo-Gutiérrez, A., Juárez-Ocaña, S., González-Miranda, G., Palma-Padilla, V., Carreón-Cruz, R., Ortega-Alvárez, M., & Mejía-Arangure, J., (2007) Incidence of cancer in children residing in ten jurisdictions of the Mexican Republic: importance of the Cancer registry (a population-based study). *BMC Cancer.* Vol.7, No.68, (April 2007), pp (68-82), ISSN 1471-2407

Flemming A., Brummer T., Reth M., & Jumaa H. (2003). The adaptor protein SLP-65 acts as a tumor suppressor that limits pre-B cell expansion. *Nature Immunology,* Vol. 4, No. 1, (November 2003), pp. (38-43), ISSN 1529-2908

Fuentes-Pananá, E. M. Bannish, G., & Monroe, J. G. (2004a). Basal B-cell receptor signaling in B lymphocytes: mechanisms of regulation and role in positive selection, differentiation, and peripheral survival. *Immunological Reviews,* Vol.197, No.1, (January 2004), pp. (26-40), ISSN 0105-2896

Fuentes-Pananá E. M., Bannish G., Monroe J. G. (2004b). Basal Iga/Igb signals trigger the coordinated initiation of preBCR-dependent processes. *Journal of Immunology,* Vol.173, No.2, (July 2004), pp. 1000-1011, ISSN 0022-1767

Geijtenbeek, TB., van Kooyk, Y., van Vliet, SJ., Renes, MH., Raymakers, RA., & Figdor, CG. (1999) High frequency of adhesion defects in B-lineage acute lymphoblastic leukemia. *Blood.* Vol.94. No.15, (July 1999) pp. (754-64), ISSN 0006-4971

Greaves, M., (1993) Stem cell origins of leukaemia and curability. *Br J Cancer.* Vol.67, No.3, (March 1993), pp. (413-23), ISSN 0007-0920

Greaves, M.(2006) Infection, immune responses and the aetiology of childhood leukaemia. *Nat Rev Cancer*. Vol.6, No.3, (March 2006), pp. (193-203), ISSN 1474-175X

Greaves, MF., & Wiemels, J.(2003) Origins of chromosome translocations in childhood leukaemia. *Nat Rev Cancer*. Vol.3, No.9, (September 2003) pp. (639-49), ISSN 1474-175X

Hardy, R. R., Carmack, C. E., Shinton, S. A., Kemp, J. D.,& Hayakawa, K. (1991). Resolution and characterization of pro-B and pre-pro-B cell stages in normal mouse bone marrow. *Journal of Experimental Medicine,* Vol.173, No.5, (May 1991), pp. (1213-1225), ISSN: 1932-6203

Harvey, RC., Mullighan, CG., Wang, X., Dobbin, KK., Davidson, GS., Bedrick, EJ., Chen, I M., Atlas, SR., Kang, H., Ar, K., Wharton, W., Murphy, M., Devidas, M., Carroll, AJ., Borowitz, MJ., Bowman, WP., Downing, JR., Relling, M., Yang, J., Bhojwani, D., Carroll, WL., Camitta, B., Reaman, GH., Smith, M., Hunger, SP., & Willman, CL. (2010). Identification of novel cluster groups in pediatric high-risk B-precursor acute lymphoblastic leukemia with gene expression profiling: correlation with genome-wide DNA copy number alterations, clinical characteristics, and outcome. *Blood,* Vol. 116, No. 23, (December 2010), pp. (4874-4884), ISSN 0006-4971

Heidenreich, O., & Vormoor, J., (2009) Malignant stem cells in childhood ALL: the debate continues! *Blood.* Vol.113, No.18, (April 2009), pp. (4476-7), ISSN 1528-0020

Hirose, K., Inukai T., Kikuchi, J., Furukawa, Y., Ikawa, T., Kawamoto, H., Oram, SH., Göttgens, B., Kiyokawa, N., Miyagawa, Y., Okita, H., Akahane, K., Zhang, X., Kuroda, I., Honna, H., Kagami, K., Goi, K., Kurosawa, H., Look, AT., Matsui, H., Inaba, T., & Sugita, K. (2010). Aberrant induction of LMO2 by the E2A-HLF chimeric transcription factor and its implication in leukemogenesis of B-precursor ALL with t(17;19). *Blood,* Vol.116, No.6, (August 2010), pp. (962-970), ISSN 0006-4971

Hsu L. Y., Liang H. E., Johnson K., Kang C., & Schlissel M. S. (2004). Pax5 activates immunoglobulin heavy chain V to DJ rearrangement in transgenic thymocytes. *Journal of Experimental Medicine,* Vol. 199, No. 6, (March 2004), pp. (825-830), ISSN 0022-1007

Ichii, M., Oritani, K., Yokota, T., Zhang, Q., Garrett, KP., Kanakura, Y., & Kincade, PW. (2010) The density of CD10 corresponds to commitment and progression in the human B lymphoid lineage *PloS One.* Vol.5 No9. (September 2010) pp e12954, ISSN 1932-6203

Igarashi, H., Gregory, S.C., Yokota, T., Sakaguchi, N., & Kincade, P.W. (2002). Transcription from the RAG1 locus marks the earliest lymphocyte progenitors in bone marrow. *Immunity,* Vol. 17, No. 2, pp. (117-130), ISSN 1074-7613

Juarez, JG., Thien, M., De la Pena, A., Baraz, R, Bradstock, KF., & Bendall, LJ. (2009) CXCR4 mediates the homing of B cell progenitor acute lymphoblastic leukaemia cells to the bone marrow via activation of p38MAPK. *Br J Haematol.* Vol.145, No.4, (May 2009) pp. (491-9), ISSN 1365-2141

Kim, JM., Kim, NI., Oh, YK., Kim, YJ., Youn, J., & Ahn, MJ. (2005). CpG oligodeoxynucleotides induce IL-8 expression in CD34+ cells via mitogen-activated protein kinase-dependent and NF-kB-independent pathways. *International Immunol,* Vol.17, pp.1525-1531, ISSN 0953-8178

Kuo T. C., & Schlissel M. S. (2009). Mechanisms controlling expression of the RAG locus during lymphocyte development. *Current Opinion in Immunology*, Vol. 21, No. 2, (April 2009), pp. (173–178), ISSN 0952-7915

le Viseur, C., Hotfilder, M., Bomken, S., Wilson, K., Röttgers, S., Schrauder, A., Rosemann, A., Irving, J., Stam, R., Shultz, L., Harbott, J., Jürgens, H., Schrappe, M., Pieters, R., & Vormoor, J., (2008) In childhood acute lymphoblastic leukemia, blasts at different stages of immunophenotypic maturation have stem cell properties. *Cancer* Martin F., Oliver A. M., Kearney J. F. (2001). Marginal Zone and B1 B Cells Unite in the Early Response against T-Independent Blood-Borne Particulate Antigens. *Immunity*, Vol.14, No.5, (May 2001), pp. (617–629), ISSN 1074-7613

Li, Z., Dordai D. I., Lee, J., & Desiderio, S. (1996). A conserved degradation signal regulates RAG-2 accumulation during cell division and links V(D)J recombination to the cell cycle. *Immunity*, Vol.5, No.6, (December 1996), pp. (575-589), ISSN 1074-7613

Liippo J., Nera K. P., Veistinen E., Ländesmäki A., Postila V., Kimby E., Riikonen P., Hammarström L., Pelkonen J., & Lassila O. (2001). Both normal and leukemic B lymphocytes express multiple isoforms of the human aiolos gene. *European Journal of Immunology*, Vol. 31, No. 12, (December 2001), pp. (3469-3474), ISSN 0014-2980

Ma S., Pathak S., Mandal M., Trinh L., Clark M. R., & Lu R. (2010). Ikaros and Aiolos inhibit pre-B-cell proliferation by directly suppressing c-Myc expression. *Molecular and Cellular Biology*, Vol. 30, No. 17, (September 2010), pp. (4149-4158), ISSN 0270-7306

Malin, S., McManus, S., & Busslinger, M. (2010). STAT5 in B cell development and leukemia. *Current Opinion in Immunology*, Vol.22, No.2, (April 2010), pp. (168-176), ISSN 0952-7915

Martin F., Oliver A. M., Kearney J. F. (2001). Marginal Zone and B1 B Cells Unite in the Early Response against T-Independent Blood-Borne Particulate Antigens. *Immunity*, Vol. 14, No. 5, (May 2001), pp. (617–629), ISSN 1074-7613

Mayani, H. (2010). Biological differences between neonatal and adult human hematopoietic stem/progenitor cells. *Stem Cells and Dev*, Vol.19, No.3, pp.285-298, ISSN 1547-3287

McCarthy, N. (2010). Leukaemia: MLL makes friends and influences. *Nature Reviews in Cancer*, Vol. 10, No. 8, (August 2010), pp. (529), ISNN 1474-175X

McWhirter J. R., Neuteboom S. T., Wancewicz E. V., Monia B. P., Downing J. R., & Murre C. (1999). Oncogenic homeodomain transcription factor E2A-Pbx1 activates a novel WNT gene in pre-B acute lymphoblastoid leukemia. *Proceedings of the National Academy of Science USA*, Vol.96, No.20, (September 1999), pp. (11464-11469), ISSN 0027-8424

Mejia-Arangure, JM.; (2011). *Childhood Acute Leukemias in Hispanic Population: Differences by Age Peak and Immunophenotype*, InTech, ISBN México City, México. In Press

Menendez, P., Catalina, P., Rodríguez, R., Melen, GJ., Bueno, C., Arriero, M., García-Sánchez, F., Lassaletta, A., García-Sanz, R., & García-Castro, J. (2009) Bone marrow mesenchymal stem cells from infants with MLL-AF4+ acute leukemia harbor and express the MLL-AF4 fusion gene. *J Exp Med*. Vol.206, No.13, (December 2009) pp 3131-41. ISSN 1540-9538

Merrell, K. T., Benschop, R. J., Gauld, S. B., Aviszus, K., Decote-Ricardo, D., Wysocki, L. J. & Cambier, J.C. (2006). Identification of Anergic B Cells within a Wild-Type Repertoire. *Immunity*, Vol.25, No.6 (December 2006), pp (953–962), ISSN 1074-7613

Mullighan, C G., Goorha, S., Radtke, I., Miller, CB., Coustan-Smith, E., Dalton, JD., Girtman, K., Mathew, S., Ma, J., Pound, SB., Su, X., Pui, CH., Relling, MV., Evans, WE., Shurtleff, SA., & Downing, JR. (2007). Genome-wide analysis of genetic alterations in acute lymphoblastic leukemia. *Nature,* Vol. 446, No. 7137, (April 2007), pp. (758-764), ISSN 0028-0836

Mullighan, CG., Miller, CB., Radtke, I., Phillips, LA., Dalton, J., Ma, J., White, D., Hughes, TP., Le Beau, MM., Pui, CH., Relling, MV., Shurtleff, SA., & Downing, JR. (2008). BCR-ABL1 lymphoblastic leukaemia is characterized by the deletion of ikaros. *Nature,* Vol. 453, No. 7191, (May 2008), pp. (110-114), ISSN 0028-0836

Mullighan, CG., Su, X., Zhang, J., Radtke, I., Phillips, LA., Miller, CB., Ma, J., Liu, W., Cheng, C., Schulman, BA., Harvey, RC., Chen, IM., Clifford, RJ., Carroll, WL., Reaman, G., Bowman, WP., Devidas, M., Gerhard, DS., Yang, W., Relling, MV., Shurtleff, SA., Campana, D., Borowitz, MJ., Pui, CH., Smith, M., Hunger, SP., Willman, CL., & Downing JR; Children's Oncology Group. (2009). Deletion of IKZF1 and prognosis in acute lymphoblastic leukemia. *New England Journal of Medicine,* Vol. 360, No.5, (January 2009), pp. (470-480), ISSN 0028-4793

Nagai, Y. Garrett, K. Ohta, S. Bahrun, U., Kouro, T., Akira, S., Takatsu, K., & Kincade, PW.. (2006). Toll-like receptors on hematopoietic progenitor cells stimulate innate immune system replenishment. *Immunity.* Vol. 24, No. 6, pp. 801-12, ISSN 1074-7613

Nagasawa ,T. (2006) Microenvironmental niches in the bone marrow required for B-celldevelopment. *Nat Rev Immunol.* Vol.6, No.2, (February 2006) pp. (107-16), ISSN 1474-1733

Nagasawa, T., Omatsu, Y., & Sugiyama, T. (2011) Control of hematopoietic stem cells by the bone marrow stromal niche: the role of reticular cells. *Trends Immunol.* Vol.32, No.7, (July 2011), pp. (315-20), ISSN 1471-4981

Nutt S. L., Heavey B., Rolink A. G., & Busslinger M. (1999). Commitment to the B-lymphoid lineage depends on the transcription factor Pax5. *Nature,* Vol. 401, No. 6753, (October 1999), pp. (556-562), ISSN 0028-0836

Nutt S. L., & Kee B. L. 2007. The transcriptional regulation of B cell lineage commitment. *Immunity,* Vol. 26, No. 6, (June 2007), pp. (715-725), ISSN 1074-7613

O'Riordan M., & Grosschedl R. (1999). Coordinate regulation of B cell differentiation by the transcription factors EBF and E2A. *Immunity,* Vol. 11, No. 1, (July 1999), pp. (21-31), ISSN 1074-7613

Parrish Y. K., Baez I., Milford T. A., Benitez A., Galloway N., Rogerio J. W., Sahakian E., Kagoda M., Huang G., Hao Q. L., Sevilla Y., Barsky L. W., Zielinska E., Price M. A., Wall N. R., Dovat S., & Payne K. J. (2009). IL-7 dependence in human B lymphopoiesis increases during progression of ontogeny from cord blood to bone marrow. *Journal of Immunology,* Vol. 182, No. 7, (April 2009), pp. (4255-4266), ISSN 0022-1767

Passegué, E., Wagers, AJ., Giuriato, S., Anderson, WC.,& Weissman IL (2005). Global analysis of proliferation and cell cycle gene expression in the regulation of hematopoietic stem and progenitor cell fates. *J Exp Med* Vol. 202 No. 11 (December 2005) pp (1599-1611), ISSN 0022-1007

Pelayo, R., Hirose, J., Huang, J., Garrett, KP., Delogu, A., Busslinger, M., & Kincade PW. (2005b). Derivation of 2 categories of plasmacytoid dendritic cells in murine bone marrow. *Blood,* Vol.105, No.11, (Jun 2005), pp. 4407-4415, ISSN 0006-4971

Pelayo, R., Miyazaki, K., Huang, J., Garrett, KP., Osmond, DG., & Kincade. PW.. (2006b). Cell cycle quiescence of early lymphoid progenitors in adult bone marrow. *Stem Cells,* Vol.24, No.12, (December 2006), pp.2703-2713, ISSN 1549-4918

Pelayo, R., Welner, RS., Nagai, Y., & Kincade PW. (2006a). Life before the pre-B cell receptor checkpoint: specification and commitment of primitive lymphoid progenitors in adult bone marrow. *Semin Immunol*, Vol.18, No.1, pp. 2-11, ISSN 1044-5323

Pelayo, R., Welner, R., Perry, SS., Huang, J., Baba, Y., Yokota, T,, & Kincade, PW.. (2005a). Lymphoid progenitors and primary routes to becoming cells of the immune system. *Curr Opin Immunol*, Vol.17, No.2, pp. 100-107, ISSN 0952-7915

Perez-Saldivar, M., Fajardo-Gutierrez, A., Bernaldez-Rios, R., Martinez-Avalos, A., Medina-Sanson, A., Espinosa-Hernandez, L., Flores-Chapa, J., Amador-Sanchez, R., Penaloza-Gonzalez, J., Alvarez-Rodriguez,F., Bolea-Murga, V., Flores-Lujano, J., Rodriguez-Zepeda, M., Rivera-Luna, R., Dorantes-Acosta, E., Jimenez-Hernandez, E., Alvarado-Ibarra, M., Velazquez-Avina, M., Torres-Nava, J., Duarte-Rodriguez, D., Paredes-Aguilera, R., Del Campo-Martinez, M., Cardenas-Cardos, R., Alamilla-Galicia, P., Bekker-Mendez, V., Ortega-Alvarez, M., & Mejia-Arangure, J. (2011) Childhood acute leukemias are frequent in Mexico City: descriptive epidemiology. *BMC Cancer*. Vol.11, No.1 (August 2011) In Press, ISSN 1471-2407

Pérez-Vera P., Reyes-León A., & Fuentes-Panañá E. M. (2011). Signaling proteins and trasncription factors in normal and malignant early B cell development. *Bone Marrow Research*, Vol. 2011, (no date), Article ID 502751, ISSN 2090-2999

Pieters, R., & Carroll, W. (2010) Biology and treatment of acute lymphoblastic leukemia. *Hematol Oncol Clin North Am.* Vol.14, No.1, (February 2010), pp. (1-18), ISSN 1558-1977

Puel A, Ziegler S. F., Buckley R. H., Leonard W. J. (1998). Defective IL7R, expression in T-B+NK+ severe combined immunodeficiency. *Nature Genetics*, Vol. 20, No. 4, (December 1998), pp. (394-397), ISSN 1061-4036

Raaijmakers, MH. (2011) Niche contributions to oncogenesis: emerging concepts and implications for the hematopoietic system. *Haematologica*. Vol.96, No.7, (July 2011) pp. (1041-8), ISSN 1592-8721

Ramadani F., Bolland D. J., Garcon F., Emery J. L., Vanhaesebroeck B., Corcoran A. E., & Okkenhaug K. (2010). The PI3K isoforms p110 alpha and p110 delta are essential for pre-B cell receptor signaling and B cell development. *Science Signaling*, Vol. 10, No. 134, (August 2010), pp. (ra60), ISSN 1945-0877

Seita,J. & Weissman, IL. (2010). Hematopoietic stem cell: self-renewal versus differentiation. *WIREs Systems Biology and Medicine*, Vol.2, (November/December 2010), pp. 640-653, ISSN 1939-005X

Sioud, M. & Fløisand, Y. (2007). TLR agonists induce the differentiation of human bone marrow CD34+ progenitors into CD11c+ CD80/86+ DC capable of inducing a Th1-type response. *European Journal of Immunology*. Vol. 30, No. 10, pp. 2834-46, ISSN 0014-2980

Smith E., & Sigvardsson M. J. (2004). The roles of transcription factors in B lymphocyte commitment, development, and transformation. *Journal of Leukocyte Biology*, Vol. 75, No. 6, pp. (973-981), ISSN 0741-5400

Sun, L., Heerema, N., Crotty, L., Wu, X., Navara, C., Vassilev, A., Sensel, M., Reaman, GH., & Uckun, FM. (1999). Expression of dominant-negative and mutant isoforms of the antileukemic transcription factor ikaros in infant acute lymphoblastic leukemia. *Proceedings of the National Academy of Science USA*, Vol.96, No.2, (January 1999), pp. (680-685), ISSN 0027-8424

Takizawa, H. Regoes, R. Boddupalli, C. Bonhoeffer, S.,& Manz, MG.. (2011). Dynamic variation in cycling of hematopoietic stem cells in steady state and inflammation. *Journal of Experimental Medicine*. Vol. 208, No. 2, pp. 273-84, ISSN 0022-1007

Thomas, L. R., Cobb, R. M., & Oltz, E. M. (2009). Dynamic regulation of antigen receptor gene assembly. *Advances in Experimental Medicine and Biology*. Vol.650, pp. (103-115), ISSN 0065-2598

Tokoyoda, K., Egawa, T., Sugiyama, T., Choi, BI., & Nagasawa, T. (2004) Cellular niches controlling B lymphocyte behavior within bone marrow during development. *Immunity*. Vol.20, No.6, (June 2004), pp. (707-18), ISSN 1074-7613

Tung J. W., Mrazek M. D., Yang Y., Herzenberg L. A., & Herzenberg L. A. (2006). Phenotypically distinct B cell development pathways map to the three B cell lineages in the mouse. *Proceedings of the National Academy of Science USA*, Vol. 103, No. 16, (April 2006), pp. (6293–6298), ISSN 0027-8424

van der Weyden, L., Giotopoulos, G., Rust, A., Matheson, L., van Delft, F., Kong, J., Corcoran, A., Greaves, M., Mullighan, C., Huntly, B., & Adams, D. (2011) Modeling the evolution of ETV6-RUNX1-induced B-cell precursor acute lymphoblastic leukemia in mice. *Blood*. Vol.118, No.4, (May 2011), pp. (1041-51), ISSN 1528-0020

von Boehmer, H., & Melcher, F. (2010). Checkpoints in lymphocyte development and autoimmune disease. *Nature Immunology*, Vol.11, No.1, (January 2010), pp. (14-20), ISSN: 1529-2908

von Freeden-Jeffrey U., Vieira P., Lucian L. A., McNeil T., Burdach S. E., & Murray R. (1995). Lymphopenia in interleukin (IL)-7 gene-deleted mice identifies IL-7 as a non-redundant cytokine. *Journal of Experimental Medicine*, Vol. 181, No. 4, (April 1995), pp. (1519-1526), ISSN 0022-100

Warner, J., Wang, J., Hope, K., Jin, L., & Dick, J.(2004) Concepts of human leukemic development. *Oncogene*. Vol.23, No.43, (September 2004) pp. (7164-77), ISSN 0950-9232

Welner, RS., Pelayo, R., Garrett, KP., Chen, X., Perry, SS., Sun, XH., Kee, BL., & Kincade PW. (2007). Interferon-producing dendritic cells (IKDC) arise via a unique differentiation pathway from primitive c-kit^{hi}CD62L^{+} lymphoid progenitors. *Blood*, Vol.109, No. 11, (Jun 2007), pp. 4825-4831, ISSN 0006-4971

Welner, RS., Pelayo, R., & Kincade, PW. (2008a). Evolving views on the genealogy of B cells. *Nat Rev Immunol*, Vol.8, No.2, (February 2008), pp. 95-106, ISSN 1474-1733

Welner, R.S., Pelayo,R., Nagai,Y., Garrett,K.P., Whest,T.R., Carr,D.J., Borghesi,L.A., Farrar,M.A., & Kincade, P.W. (2008b). Lymphoid precursors are directed to produce dendritic cells as a result of TLR9 ligation during Herpes infection. *Blood*, Vol.112, No.9, (November 2008), pp.3753-3761, ISSN 0006-4971

Welner, RS., Esplin, BL., Garrett, KP., Pelayo, R., Luche, H., Fehling, HJ., & Kincade PW.(2009) Asynchronous RAG-1 expression during B lymphopoiesis. *J Immunol.*

Yao Z., Cui Y., Watford W. T., Bream J. H., Yamaoka K., Hissong B. D., Li D., Durum S. K., Jiang Q., Bhandoola A., Hennighausen L., & O´Shea J. J. (2006). Stat5a/b are essential for normal lymphoid development and differentiation. *Proceedings of the National Academy of Science USA*, Vol. 103, No. 4, (January 2006), pp. (1000-1005), ISSN 0022-1767

Yasuda, T., Sanjo, H., Pagès, G., Kawano, Y., Karasuyama, H., Pouysségur, J., Ogata M., & Kurosaki T. (2008). Erk kinases link pre-B cell receptor signaling to transcriptional events required for early B cell expansion. *Immunity*, Vol.28, No.4, (April 2008), pp. (499-508), ISSN 1074-7613

14

Distribution of SDF1-3'A, GNB3 C825T and MMP-9 C-1562T Polymorphisms in HSC CD34+ from Peripheral Blood of Patients with Hematological Malignancies

Ben Nasr Moufida[2] and Jenhani Faouzi[1,2]
[1]*Cellular Immunology and Cytometry and Cellular Therapy Laboratory, National Blood Transfusion Center,*
[2]*Immunology Unit research, Faculty of Pharmacy, Monastir*
Tunisia

1. Introduction

Mobilized peripheral blood stem cells (MPBSC) have nearly replaced bone marrow (BM). So, they become the primary source of hematopoietic grafts especially for patients with hematological malignancies undergoing aggressive myelosuppressive or myeloablative chemotherapy. It allows faster engraftment and equivalent disease-free survival compared with bone marrow cells [Siena S et al, 2000; To LB et al, 1997; Roberto M. Lemoli and Alessandra D'Addio, 2008].

Some reports suggested that hematopoietic stem cell mobilization involves a complex interplay between adhesion molecules, cytokines, proteolytic enzymes such as MMP-9 and MMP-2, stromal cells and chemokines among them (e.g,; SDF-1/CXCR4) play a central role [Roberto M. Lemoli and Alessandra D'Addio, 2008; Tsevee Lapidot and Isabelle Petit, 2002]. It has been reported that increased secretion of SDF-1 downmodulates CXCR4 on CD34+ cells, thus preventing the homing of hematopoietic progenitors to the bone marrow [Signoret N et al, 1997]. Moreover, Dlubek D et al, have observed a negative correlation between mobilization capacity and a reduced expression of CXCR4 on mobilized HPC CD34+ in the leukapheresis product [Dlubek D et al, 2006].

These data suggested a central role for CXCR4 and SDF-1 on mobilization of hematopoietic stem cell as well as their homing to the bone marrow [Dlubek D et al, 2006].

The reason for poor mobilization of hematopoietic stem cells that occur in many donors or patients is fully recognized and patients' characteristics (age, BMI, mobilization regimen, diagnosis and clinical status or ulterior therapy) did not explain the whole thing.

Benboubker and his colleagues identified an association of a polymorphism in the SDF-1 gene, designated as SDF1-3'A, with the rate of mobilization of HPCs CD34+ into peripheral blood [Benboubker L et al, 2001]. Hence, we hypothesized that individual genetic factors might explain, at least in part, this variability and that polymorphism analysis can be used to anticipate CD34+ cells mobilization.

So, identifying SNPs predictive of poor or good response to G-CSF or any mobilization regimen, in terms of number of CD34+ cells mobilized, might be useful in discussing the possibility of using a different mobilizing agent or a different source of CD34+ cells for auto-HSCT and allo-HSCT.

In this issue, we proposed to study the distribution of three genetic polymorphisms: SDF1-3'A, MMP-9 C-1562T and GNB3 C825T in Tunisian patients with malignant hematological diseases who underwent stem cell mobilization for autologous transplantation compared to a group of healthy allogenic PBPC donors.

2. Materials and methods

2.1 Study population

250 subjects (144 men, 106 women) admitted to the Cellular Immunology and Cytometry and Cellular Therapy Laboratory of National Blood Transfusion Center of Tunis -Tunisia, for autologous PBPC mobilization were enrolled.

Our patients can be divided in 4 subgroups distributed as follows: Group 1: 85 Non-Hodgkin's Lymphoma (57 men, 28 women) which comprises 80 Diffuse B Cell Lymphoma, 4 Mantle Cell Lymphoma and a patient with Follicular Lymphoma.

Group 2: 87 Multiple Myeloma (48 men, 39 women).

Group 3: 63 Hodgkin's disease (31 men, 32 women).

Group 4: composed of 15 patients with Acute Myeloid Leukemia (9 men, 6 women).

Besides, a group composed of 41 subjects (24 men, 17 women) with mean age of 32 years (range 12-63 years) designated for peripheral blood stem cells (PBSC) mobilization. They were visiting the Cellular Immunology and Cytometry and Cellular therapy Laboratory of National Blood Transfusion Center of Tunis–Tunisia as allogenic donors for stem cell transplantation.

Then, a group of 165 healthy blood donors visiting the Blood Transfusion Service of National Blood Transfusion Center of Tunis -Tunisia served as a control group was enrolled in the study. Whole details concerning the subjects will be resumed in Table 1.

Written informed consent was obtained from all subjects according to a protocol approved by the ethical committee for scientific and medical research of the National Blood Transfusion Center and National Bone marrow transplantation center of Tunis (Tunisia) in accordance with the Declaration of Helsinki.

Circulating hematopoietic progenitors CD34+ were evaluated daily by flow cytometry and PBSC collections or apheresis were begun when peripheral CD34+ cells were ~20 cells/µl. Apheresis was usually performed daily using continuous flow blood cell separators COBE SPECTRA and MCS+.

2.2 DNA extraction and genotyping

Genomic DNA was prepared from EDTA anticoagulated peripheral blood by using a common salting-out procedure [Miller SA et al, 1988].

	PATIENTS				PBSC DONORS		
	Total	<2x10e6 CD34+/kg	>2x10e6 CD34+/kg	p	Total	<3x10^6 CD34/kg	≥3x10^6 CD34/kg
Age (years) Median	40.58				33.25 (12-63)	32.25 (15-57)	33.5 (12-63)
Range	12-64						
Male	144	27	117	NS	24	6	11
Female	106	26	80		17	6 6	18
Diagnosis							
NHL (non Hodgkin's lymphoma)		25	60				
Diffuse large Cell Lymphoma	80						
FL(follicular lymphoma)	1						
ML (mantle Cell lymphoma)	4						
Hodgkin's Disease	63	14	49				
Multiple Myeloma	87	12	77				
AML (acute myeloid leukemia)	15	7	8				
Prior radiotherapy	62	19	23				
Prior chemotherapy	250						
time from last chemotherapy to mobilization							
< 1 month	121	-					
1 to 2 months	20						
2 to 3 months	4						
> 3 months	5						
Chemomobilization							
Rituximab ESHAP/ rituximab DSHAP	59	-					
rituximab CHOP	2						
ICE/ RICE	21						
Others	168						
Mobilization regimen							
growth factor only							
Lenograstim (Granocyte®)	80	-					
filgrastim (Neupogen®)	75	-					
G/C [endoxan+ G-CSF]	95	-					

Table 1. patients and healthy allogenic PBPC donors charachteristics Abbreviations: G-CSF, granulocyte colony-stimulating factor; G/C, G-CSF- chemotherapy; ICE, ifosfamide, carboplatin, etoposide; ESHAP/DHAP, etoposide, cytarabine, methylprednisolone,

2.3 Genotyping

The reaction mixture consisted of 1µl PCR buffer 10x, 2 mM of MgSo4, 0.2 mM of each dNTP, 400mM of each primer, and 0,5units/reaction Taq DNA polymerase (Bio Basic Inc).

The reaction conditions were: For SDF1-3'A an initial denaturation at 95°C for five minutes, then 35 cycles at 94°C for 30 seconds, at 58°C for 30 seconds, at 72°C for 1min, and finally extension at 72°C for 7 minutes.

All specimens were examined for the presence of amplifiable DNA. PCR products were digested with 10units HpaII/reaction (Fermentas) at 37°C for overnight [Benboubker L et al, 2001] (figure 1).

For, MMP-9 C-1562T, PCR conditions as above, with annealing temperature at 67°C. PCR products were digested with 10units Hin1II/reaction (Fermentas) at 37°C for overnight [Zhang B et al, 1999; Toru Ogata et al, 2005] (figure 2).

For GNB3 C825T, the PCR-reaction began with denaturation at 95°C for 5 min, followed by 35 cycles of denaturation at 94°C (for 30 seconds), annealing at 55°C (30 s), extension at 72°C (1min), and a final extension at 72°C (7 min). PCR products were digested with BseDI at 60°C (4 h), separated on 2% agarose gels, and visualized under UV illumination [Cheng-Ho Tsai MD et al, 2000] (figure 3)

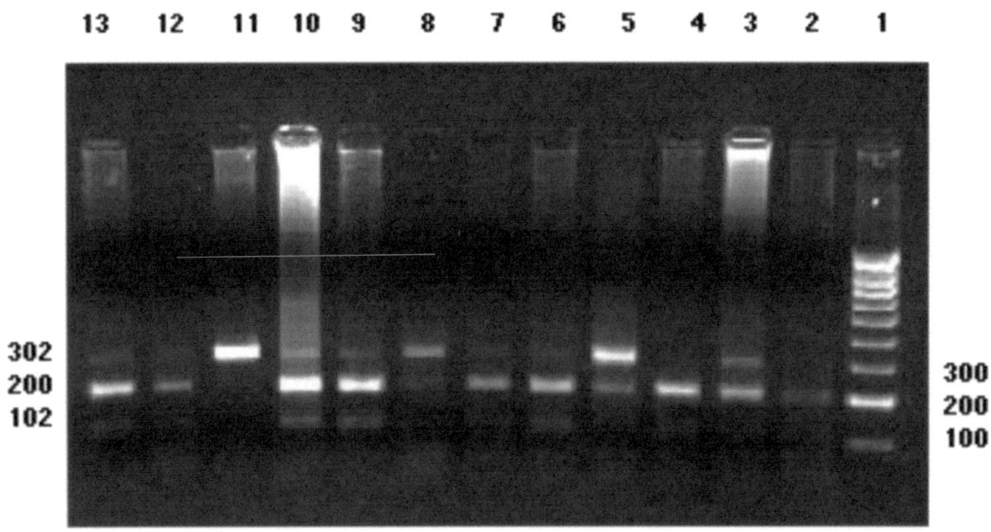

Fig. 1. SDF-1 genotyping by PCR-RFLP analysis followed by separation on 2% agarose gel as described in text. Lane 1, 100pb ladder; lanes 2 and 4, G/G; lanes 3 and 5, G/A; lane 11, A/A

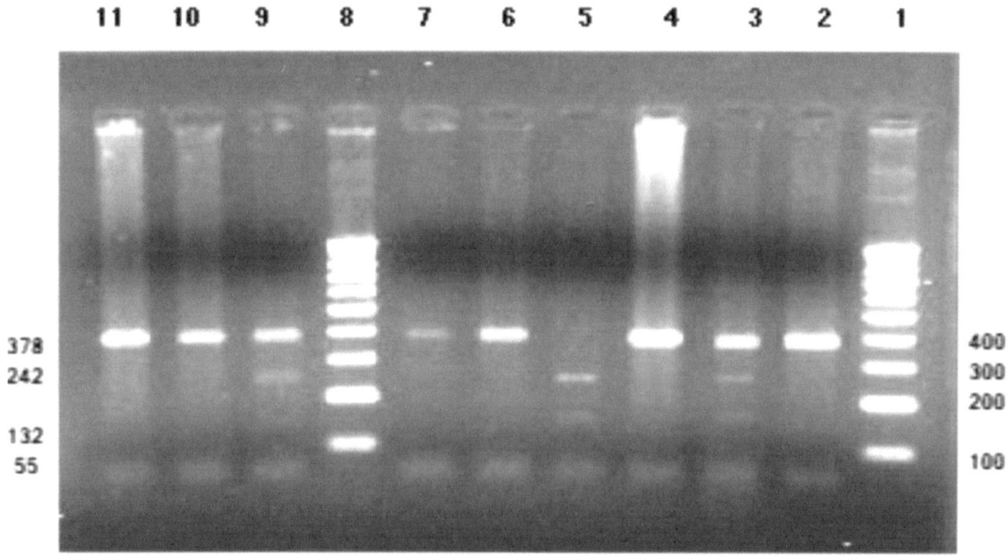

Fig. 2. MMP-9 genotyping by PCR-RFLP analysis followed by separation on 2% agarose gel as described in text. Lanes 1 and 8, 100 pb ladder; lanes 2 and 6, C/C; lanes 3 and 9, C/T; lane 5, T/T.

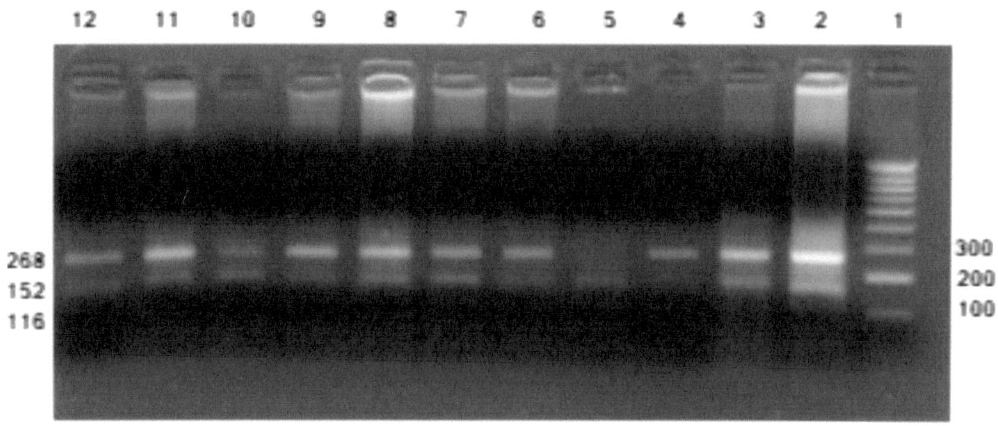

Fig. 3. GNB3 C825T genotyping by PCR-RFLP analysis followed by separation on 2% agarose gel as described in text. Lane 1, 100 pb ladder; lanes 2 and 3, C/T; lane 4, T/T; lane 5, C/C.

Gene	Variant	rs #	Forward and Reverse PCR Primer Sequences 5'–3'	PCR Product lenght, pb	Restriction enzyme	Restriction Fragments Generated, bp
SDF-1	SDF1-3'A	rs1801157	5'-CAGTCAACCTGGGCAAAGCC-3'	302	HpaII	G allele: 102 and 200 bp
			5'-AGCTTTGGTCCTGAGAGTCC-3'			A allele: 302 bp
MMP-9	C-1562T	rs3918242	5'-ATGCTCATGCCCGTAATCCT-3'	435	Hin1II	C allele: 378- and 55-bp
			5'-TGGGAAAAACCTGCTAACAACT-3'			T allele: 242, 132 and 55pb
GNB3	C825T	rs5443	5'-TGACCCACTTGCCACCCGTGC-3'	268	BseDI	C allele: 153- and 116-bp
			5'-GCAGCAGCCAGGGCTGGC-3'			T allele:268pb

Table 2. All genotyping Details, corresponding to each polymorphism studied are provided.

2.4 Statistical analysis

Allele and genotype frequencies of the studied polymorphisms in patients and healthy controls were formulated by direct counting. Statistical analysis was performed using SPSS software (SPSS 16.0 for windows; SPSS Inc., Chicago, IL.).

The allele frequencies of SDF1-3'A, GNB3 C825T and MMP-9C-1562T polymorphisms were tested for the Hardy-Weinberg equilibrium of the whole group or subgroups of patients and were compared to the respective frequencies of the control group using the Pearson chi-square test or Fisher's exact test when appropriate. The same test was applied to compare the genotype frequency between patients and controls. Association of the allelic frequencies with the clinico-pathologic parameters was evaluated by $\chi 2$ test. The odds ratios (OR) and 95% confidence intervals (CI) were calculated too. P<0.05 was required for statistical significance.

3. Results

3.1 Patient's distributions according to their CD34+ cell yield and failure rates

Overall 83% of patients included in this study collected $\geq 2x10^6$ CD34+ cells/kg after a maximum of 4 aphereses, among them 20% collected 2-5x10^6 CD34cells/kg, and 63% collected $\geq 5x10^6$ CD34 cells/kg. Beside, 10% are remobilizers as they did not achieve the threshold of CD34+ cell yield of 2x10^6 CD34/kg within 4 apheresis days and are subjects to another mobilization protocol. Among them, the group of NHL represented the highest rate (40%), the lower ones, the group of MM and AML, which represented respectively 19% and 13%. By contrast, others are designed as first mobilizers (90%) since they have already collected $\geq 2x10^6$CD34+ cells/kg after a maximum of 4 aphereses days. Amongst them the group of multiple myeloma was the most frequent (40%), thereafter the group of Non-Hodgkin's lymphoma (34%) and Hodgkin's disease with 26%. For the patients included in this study, mobilization failure was defined as <2x10^6 CD34+ cells/kg obtained within 4 apheresis days. So, especially MM patients collected $\geq 5x10^6$CD34+ cells/kg and contained the highest CD34+ cell yield (8,89x10^6 CD34/kg for MM, and 5,51x10^6 CD34/kg for the others patients). Furthermore, the fact that MM patients had higher yield of CD34+ cells compared to NHL and HD is likely since that NHL and HL patients are frequently more heavily pretreated with cytotoxic chemotherapy than patients with MM [Iskra Pusic et al, 2008] (figure 4).

3.2 Analysis of the studied polymorphisms in the 4 subgroups of patients according to disease: A comparison between healthy donors of PBSC and patients

According to this study, SDF1-3'A and MMP-9 C-1562 T polymorphisms were significantly different between the patients and healthy controls (table 3). Particularly, we found significant differences in all the allelic and genotypic frequencies of the SDF1-3'A polymorphism in the MM group (p<0.05; OR=3.245 CI (95%) [1.830-5.753] for A allele; p= 0.017; OR= 3.324 CI (95%) [1.182-9.348]; p= 0.009; OR= 2.072 CI (95%) [1.200-3.580] for AA and GA genotypes, respectively).

Concerning the MMP-9 C-1562 T polymorphism its distribution was significantly different in the same MM group of patients compared to the control group, significant differences were observed exclusively for the T allele (p=0.041; OR=2.295 CI (95%) [1.020-5.168]) and also for the CC and CT genotypes (p= 0.039; p= 0.004; Table 3).

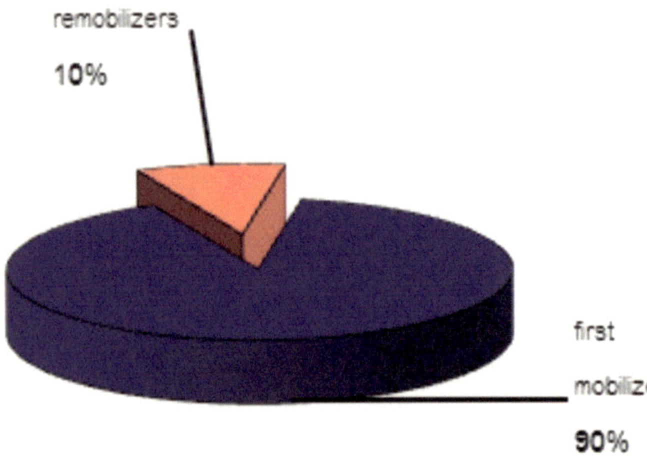

A number of first mobilization and remobilization in database

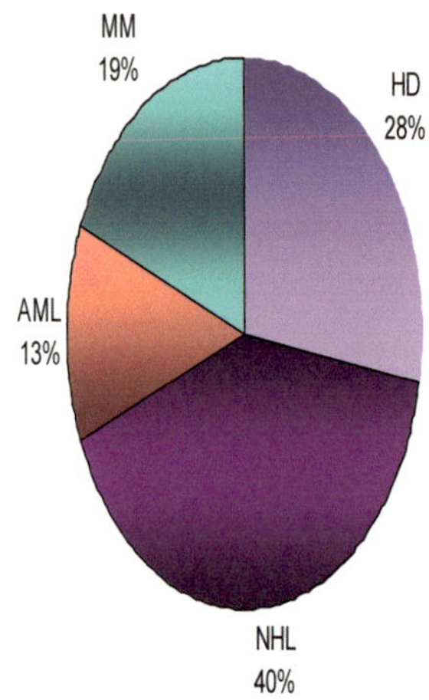

Distribution of remobilizers in the 4 subgroup of patients

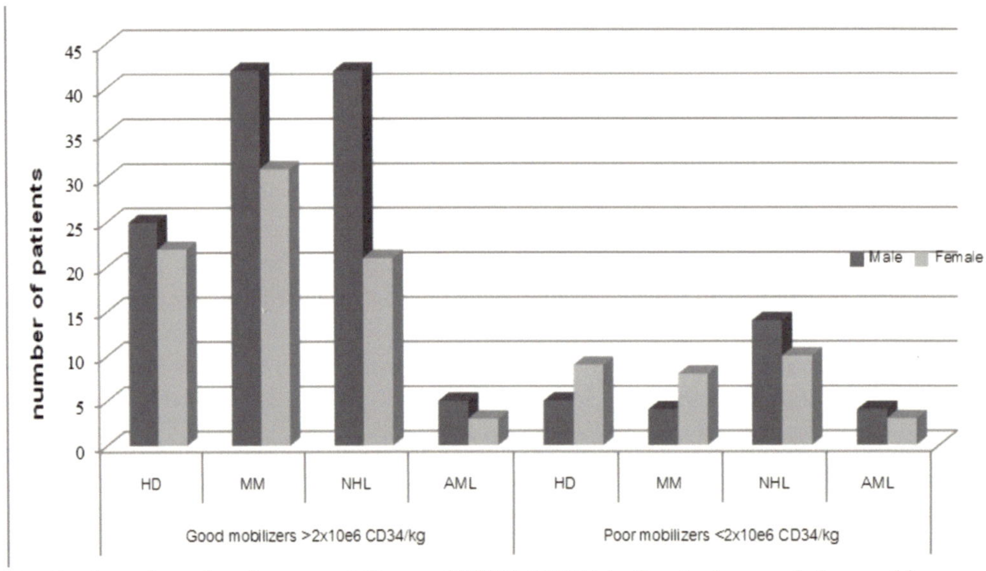

Distribution of good and poor mobilizers of PBPC CD34+ in the study population and by sex

Fig. 4. Overview of autologous stem cell transplantation database by disease as well as the distribution of good/poor mobilizers of PBSC CD34+ within the study population and by sex is already represented

	MM		P	OR	CI	NHL		P	OR	CI	HD		P	OR	CI	AML		P	OR	CI
SDF-1																				
Alleles	N	af				N	af				N	af				N	af			
A	75	0.436	<0.05	3.24	[1.830-5.753]	60	0.34	0.019	1.88	[1.104-3.209]	40	0.345	NS			9	0.3	NS		
G	97	0.464	0.013	0.30	[0.113-0.810]	114	0.65	NS			76	0.655	0.027§	0.27	[0.096-0.802]	21	0.7	NS		
Genotypes	N	gf				N	gf				N	gf				N	gf			
GG	22	0.256	<0.05	0.31	[0.176-0.567]	32	0.36	0.029	0.54	[0.315-0.941]	26	0.449	NS			7	0.467	NS		
GA	53	0.616	0.009	2.07	[1.200-3.580]	50	0.57	NS			24	0.413	NS			7	0.467	NS		
AA	11	0.128	0.017	3.32	[1.182-9.348]	5	0.05	NS			8	0.138	0.028§	3.62	[1.199-10.972]	1	0.066	NS		
MMP-9																				
Alleles	N'	af				N'	af				N'	af				N'	af			
T	14	0.16	0.041	2.29	[1.020-5.168]	21	0.22	<0.05	4.05	[1.901-8.646]	14	0.16	0.041	2.29	[1.020-5.168]	21	0.223	<0.05	4,05	[1,901-8,646]
C	74	0.84	NS			73	0.77	NS			74	0.84	NS			73	0.777	NS		
Genotypes	N'	gf				N'	gf				N'	gf				N'	gf			
CC	31	0.705	0.039	0.43	[0.192-0.971]	27	0.57	<0.05	0.24	[0.115-0.521]	31	0.705	0.039	0.43	[0.192-0.971]	27	0.574	<0.05	0.24	[0.115-0.521]
CT	12	0.272	0.004	3.5	[1.435-8.535]	19	0.40	<0.05	6.33	[2.754-14.567]	12	0.272	0.004	3.5	[1.435-8.535]	19	0.404	<0.05	6.33	[2.754-14.567]
TT	1	0.023	NS			1	0.02	NS			1	0.023	NS			1	0.021	NS		
GNB3																				
Alleles	N''	af				N''	af				N''	af				N''	af			
A	42	0.72	NS			30	0.45	NS			20	0.52	NS			6	0.6	NS		
G	25	0.084	NS			36	0.55	NS			18	0.48	NS			4	0.4	NS		
Genotypes	N''	gf				N''	gf				N''	gf				N''	gf			
GG	2	0.069	NS			8	0.24	NS			1	0.052	NS			0	0	NS		
GA	21	0.72	NS			20	0.6	NS			16	0.84	NS			4	0.8	NS		
AA	6	0.206	NS			5	0.15	NS			2	0.1057	NS			1	0.2	NS		

Table 3. Allele and genotype frequencies of *SDF-1, GNB3* and *MMP-9 polymorphisms* in the groups of patients with MM, NHL, Hodgkin's' Disease and AML

In table 3 are provided: all genotypic and allelic frequencies according to each polymorphism studied and corresponding to all patients. Distribution of genotypic and allelic frequencies by each disease included in this study. Then, all frequencies are calculated by statistical software SPSS 16.0 as well as p value and odd ratios (OR) are provided.

For the group of NHL, the distribution of the SDF1-3'A polymorphism was significantly different between patients and healthy controls especially for the A allele which seemed to be associated to this disease (p=0,019). Moreover, a decrease in GG genotype frequency compared to the control group was observed too reaching a statistically significance (p=0.029).

Concerning the MMP-9 C-1562T polymorphism, like the MM group, high significant differences were seen especially for the T allele (P<0.05; OR=4.055; CI (95%) [1.901-8.646]) and CT genotypes (P<0.05; OR=6.333; CI (95%) [2.754-14.567]). Similar results were obtained concerning the distribution of the MMP-9 C-1562T polymorphism in the group of Hodgkin's disease where significant differences were found in the T allele and CT genotype frequencies (p<0.05; Table 3).

While, the distribution of the SDF1-3'A polymorphism was not significantly different between the group of patients with AML and the control group, MMP-9 C-1562T distribution was significantly different essentially for the T allele (p=0.019, OR= 7.298, CI (95%) [1.511-35.249]) and the CT genotypes (p=0.004, OR= 12.444, CI (95%) [2.485-62.319]) Table 3.

So the presence of the MMP-9 C-1562T might be associated with this disease.

When considering the GNB3 C825T polymorphism, we observed that the TT genotype was more frequent in patient with MM and NHL with respectively 20.69% and 15.15% compared to the Hodgkin's disease group (only 10.52%). Whereas, the CC genotype was more frequent in the NHL group (24.24%) (Table 3).

3.3 Association of the SDF1-3'A allele with a good mobilizing capacity

As the clinicians have defined mobilization failure as <2x106 CD34+ cells/kg obtained within 4 apheresis days, two mainly group of patients emerged: the subjects with a good capacity of mobilization who collected ≥2x106 CD34+ cells/kg obtained within 4 apheresis days. Others with a poor mobilizing capacity and didn't collect 2x106 CD34+ cells/kg within 4 apheresis days. For the healthy allogenic PBSC donors, the mobilization failure was defined as <3x106 CD34+ cells/kg obtained within 4 apheresis days.

When considering the SDF1-3'A polymorphism, significant difference was observed in the SDF1-3'A allele carriers and GG carriers (p=0.023). A higher concentration of CD34+ cells in the leukapheresis products was detected in SDF1-3'A positive patients compared to GG homozygous subjects

Besides, a lower increase in the GG genotypes was observed in the "poor" mobilizer group compared to the "good" ones reaching a statistical significance (p=0.023; OR =0.494; CI (95%) [0.268-0.912]) (Table 4).

Thus, the SDF1-3'A allele carriers, especially the SDF1-3'AA homozygous individuals in the group of healthy allogenic PBSC donors had a better mobilization potential (table 4).

	Patients (285*, 161**)		Healthy controls (165*, 124**)		P	OR	CI	Good mobilizers (219*, 126**)		Poor mobilizers (66*, 35**)		P	OR	CI
SDF-1														
Alleles	N	af	N	af				N	af	N	af			
A	218	0.383	80	0.24	<0.05	2.185	[1.477-3.232]	177	0.404	41	0.31	0.009	0.451	[0.252-0.807]
G	352	0.617	250	0.76	0.029	0.38	[0.163-0.885]	261	0.596	91	0.69	NS		
Genotypes	N	gf	N	gf				N	gf	N	gf			
GG	97	0.34	90	0.545	<0.05	0.425	[0.287-0.630]	67	0.306	30	0.4	0.007	2.161	[1.208-3.868]
GA	158	0.555	70	0.424	0.008	1.686	[1.145-2.482]	127	0.58	31	0.47	NS		
AA	30	0.105	5	0.031	0.003	3.875	[1.476-10.171]	25	0.114	5	0.076	NS		
MMP-9														
Alleles	N'	af	N'	af				N'	af	N'	af			
T	66	0.205	26	0.105	<0.05	3.428	[1.914-6.141]	49	0.195	17	0.243	NS		
C	256	0.795	222	0.895	NS			203	0.805	53	0.757	NS		
Genotypes	N'	gf	N'	gf				N'	gf	N'	gf			
CC	99	0.615	105	0.847	<0.05	0.286	[0.160-0.512]	81	0.643	18	0.514	NS		
CT	58	0.36	12	0.097	<0.05	5.307	[2.697-10.444]	41	0.325	17	0.486	0.038	0.438	[0.198-0.968]
TT	4	0.025	7	0.056	NS			4	0.032	0	0	NS		

Abbreviations: OR, odds ratio; af, allele frequency; gf, genotype frequency; CI, confidence interval (CI=95%); Corrected *p* value; NS, not significant; *, for SDF-1 polymorphism; ** for MMP-9 polymorphism, Good mobilizers (>$2X10^6$ CD34/kg), Poor mobilizers (<$2x10^6$CD34/kg)

Table 4. Allele and genotype frequencies of *SDF-1,* and *MMP-9* polymorphisms in mobilized peripheral blood patients and healthy controls

In this table are provided:

All genotypic Allelic frequencies designed as "gf" and allelic frequencies designed as "af" of SDF1-3'A and MMP-9 C-1562T polymorphisms in all the study populations (all patients), then in a group of healthy blood donors (as control group)

Then when, dividing the whole patients according to their mobilization capacity into: good mobilizers (>$2X10^6$ CD34/kg), and poor mobilizers (<$2x10^6$CD34/kg).

OR designed as odd radio and p value of all genotypic and allelic frequencies are provided in the table by using statistical software (SPSS 16.0) as it was mentioned above in section materials and methods-statistical analysis.

However, when considering the group of remobilizers in our study population we have observed that 48% of subjects were GG, 12% were AA and 40% were GA. This led us to consider a probable association of the GG genotypes to mobilization failure.

For the MMP-9 C-1562T polymorphism, significant difference was obtained with CT genotypes between the two groups (p=0.004; OR= 0.297; CI (95%) [0.125-0.703]).

For the GNB3 C825T polymorphism, we didn't observe any difference between the 2 groups of poor and good mobilizers.

This let us consider that there's no association between GNB3 C825T polymorphism and the capacity of mobilization of hematopoietic stem cells.

For the group of healthy PBSC donors, and with respect to our classification according to mobilization failure (<3x106 CD34/kg within 4 apheresis days), we have found an important association of SDF1-3'A distribution with higher mobilization yield of hematopoietic stem cells CD34+ reaching a higher statistical significance (p=0.001; OR=12.6; table 5).

Besides, we have observed a similar increase in the SDF1-3'G allele in the intermediate to poor mobilizers' subgroup reaching a statistical significance (p=0.035; OR=1.25; table 4). Similarly, the association was already observed when comparing the genotypic frequencies between the two subgroups.

The AA genotype was absent in the poor mobilizer subgroup, then was highly increased in the other subgroup reaching a statistical significance (p=0.035; OR=1.25).

While, the GG genotype was more represented in the poor mobilizers and the differences were significant too (p=0.001; OR=0.079; table 4).

Healthy allogenic PBSC Donors						
Good mobilizers		Poor mobilizers		*P*	OR	CI
N	af	N	af			
32	0,55	5	0,208	*0,001§*	12,6	[2,407-65,953]
26	0,45	19	0,792	*0,035§*	1,25	[1,045-1,495]
N	gf	N	gf			
3	0,10	7	0,583	*0,001§*	0,079	[0,015-0,415]
20	0,69	5	0,417	*NS*		
6	0,20	0	0	*0,035§*	1,25	[1,045-1,495]
N'	af	N'	af			
7	0,17	6	0,23	*0,633 (NS)*		
33	0,82	20	0,77	*NS*		
N'	gf	N'	gf			
13	0,65	7	0,54	*NS*		
7	0,35	6	0,46	*NS*		
0	0	0	0	*NS*		

Abbreviations: OR, odds ratio; af, allele frequency; gf, genotype frequency; CI, confidence interval (CI=95%); Corrected *p* value; NS, not significant; *, for SDF-1 polymorphism; ** for MMP-9 polymorphism, for healthy allogenic PBPC donors: Good mobilizers (>3X10^6 CD34/kg), Poor mobilizers (<3x10^{6}CD34/kg)

Table 5. Allele and genotype frequencies of *SDF-1,* and *MMP-9* polymorphisms in mobilized peripheral blood of healthy allogenic PBSC donors

4. Discussion

In the present study, we investigated the effect of polymorphisms in the genes SDF-1, GNB3 and MMP-9 on the outcome of mobilization of peripheral blood stem cells for autologous transplantation by using a PCR-RFLP analysis.

We observed a significant association for SDF-1 and MMP-9 polymorphisms exclusively in patients with MM, NHL and Hodgkin's disease suggesting that these polymorphisms are fair candidate gene variants to these 3 hematological diseases.

In fact, Association of these polymorphisms to cancer has been previously reported by many investigators [De Oliveira KB et al, 2009; Rabkin CS et al, 1999].

Our results were in agreement with other studies suggesting that SDF1-3'A polymorphism is a genetic determinant of NHL [Gabriela Gonçavales de Olivera Cavassin et al, 2004]. Furthermore; as the SDF1-3'A polymorphism is situated in the mRNAs of 3'UTR region (untranslated region) which has been identified as an important regulator of the mRNA transcript, as well as the translated product [Catia Andreassi and Antonella Riccio, 2004; Marilyn Kozak, 2004; Gavin S. Wilkie et al, 2003].

The second polymorphism studied encoded for MMP-9, J. Arai et al, have reported that SDF-1 mRNAs abundantly expressed in stromal cells from the lymph nodes of patients with malignant lymphoma, so that 3'A carriers NHL are good candidates for presenting proliferation of neoplasic cells in the lymph nodes since that SDF-1 variant is associated with an increase of SDF-1 levels [J. Arai et al, 2000; Gabriela Gonçavales de Olivera Cavassin et al, 2004].

De Oliveira KB et al, when studying distribution of SDF1-3'A polymorphism have reported also a significant difference in genotype distribution between NHL patients (GG: 51.4%; GA: 47.1%; AA: 1.5%) compared to healthy controls (GG: 65.6%; GA: 28.9%; AA: 5.5%). Whereas, they didn't find any significant differences in genotypes distributions with breast cancer and Hodgkin's lymphoma [De Oliveira KB et al, 2009].

Moreover, previous reports on AIDS related non-Hodgkin's lymphoma (NHL) demonstrated that the CXCL12–3'A chemokine variant was associated with approximate doubling of the NHL risk in heterozygotes and an approximately fourfold increase in homozygotes [Rabkin CS et al, 1999; A Zafiropoulos et al, 2004]. Hence, this might let us suggest the possible role of such variant in the pathogenesis of NHL.

In this present work, we did not find a significant association between SDF1-3'A polymorphism and our group of patients with AML, this could be due to the lower number of patients (15 patients).

However, Dommange et al, have reported the implication of SDF1-3'A polymorphism in the clinical representation of acute myeloid leukemia in 86 patients with AML, as an association between this polymorphism and the risk of tissue infiltration by malignant cell was established by an increased release of the blast from the bone marrow in the blood in the SDF1-3'A carriers suggesting that this SDF-1 variant is associated with clinical representation of AML [A Zafiropoulos et al, 2004].

MMP-9 is a zinc-dependent proteinase, which is involved in numerous physiological and pathological processes. In the present study, we reported the distribution of the functional

MMP-9 polymorphism -1567 C/T in the promoter region of the MMP-9 gene in group of patients with some haematological malignancies as well as in patients undergoing stem cell mobilization.

Then, we observed that the T allele was highly associated to the susceptibility to the four diseases studied (table 3). We have to investigate either this variant have major influence on the circulating levels of MMP-9.

Concerning the group of MM, we observed a significant association in all allelic and genotypic frequencies of SDF1-3'A polymorphism with statistical differences when compared to control. Hence, as increased angiogenesis was related to the pathogenesis of MM, and because SDF-1 chemokine induces increased VEGF production, which is responsible for an angiogenic activity [Florence Dommange et al, 2006], we hypothesize that the SDF1-3'A polymorphism might increase SDF-1 protein which would have a role in developing angiogenesis and in the pathogenesis of the disease.

On the other hand, frequent distribution of the SDF-1 3'A allele in multiple myeloma patients confirms the implication of SDF-1 in hematopoietic stem cells. This logical consequence of the widely distribution of SDF-1 3'A allele proving that multiple myeloma patient's could be considered as good mobilizers.

For the GNB3 polymorphism we've observed that the TT genotype and the T allele frequencies are more frequent especially in patients with MM (0.72 for Tallele frequency) and NHL (0.45 for Tallele frequency) compared to healthy donors of PBSC (peripheral blood stem cells) (Table 3) which is far from the others populations [Maggie C.Y et al, 2004] . Then, suggesting the possible relation with these diseases.

Maggie et al when studying the ethnic differences in the linkage disequilibrium and distribution of single-nucleotide polymorphisms in 35 candidate genes for cardiovascular diseases have reported that the frequency of the T allele of GNB3 polymorphism in Chinese population is about 0.545. Then, such frequency is far from those of the French and of the Spanish population (0.329 and 0.359) and more closer to our result in Tunisian population [Yair Gazitt & Cagla Akai, 2004].

When interesting to the capacity of mobilization which was largely demonstrated to vary from a subject to another, several studies have focused on such phenomena and have reported that 10–30% of patients with hematological malignancies fail to mobilize PBSC [Ingrid G. Winkler & Jean-Pierre Levesque, 2006] and either a small proportion of normal donors (1–5%) fail to mobilize sufficient CD34+ cells.

Besides, many reports suggest that numerous factors are related to poorer mobilization including age, gender, type of growth factor, dose of the growth factor and in the autologous setting patient's diagnosis, chemotherapy regimen and number of previous chemotherapy cycles or radiation [Sugrue MW et al, 2001].

In our study we were interested in the possible implication of some genetic factors in mobilization and as we've found an association with the SDF-1 3'A variant only, then we supposed that this polymorphism is the only predictor of mobilization capacity of PBSC CD34+.

In fact, when analyzing the distribution of the two functional polymorphisms SDF-1 G801A and MMP-9 C-1562T considering the two groups of "good" and "poor" mobilizers, we've found an association only with SDF1-3'A polymorphism. While no association with capacity of mobilization was observed with GNB3 C825T and MMP-9 C-1562T polymorphisms.

When observing the distribution of the two polymorphisms not only when considering the mobilization capacity but also in relation to each studied disease enrolled in this work we've found that the good mobilizer group was mainly composed of MM patients. Whereas the poor mobilizer group contains Hodgkin's disease who are considered in previous studies as hard-to-mobilize patients [Benboubker L et al, 2001; Patrick J Stiff, 1999].

The fact that multiple myeloma patients mobilized better PBSC CD34+ (peripheral blood stem cells) than the others groups seem to be related to their ulterior chemotherapy (dexamethasone + thalidomide) and didn't receive any radiation therapy unlike the HD and NHL groups.

In the good mobilizer group composed of patients needing fewer apheresis than the other group, genotypes frequencies for the GG,GA, AA represented respectively 30.6%, 58% and 11.4%, and corresponded respectively to 45.5%, 47% and 7.6% in the poor mobilizer's group, and significant differences were found for GG genotype (p=0.007) and for A allele (p=0.009).

This confirms on the one hand that the SDF1-3'A allele was associated with good mobilizing capacity not only in the group of patients but for instance in the group of healthy allogenic PBSC donors (see table 5). Thus, our results regarding patients undergoing autologous transplantation of haematopoietic stem cells concur with those reported by Benboubker et al [Bogunia-Kubik K et al, 2009].

Moreover this deduction is already found in the group of healthy allogenic transplantation donors as it was reported in the present study and by Bogunia-Kubik K et al who have suggested that the SDF1-3'A allele was associated with a higher yield of CD34+ cells from healthy donors of PBPC for allogeneic haematopoietic SCT (stem cell transplantation) compared to GG homozygotes [Patrick J Stiff, 1999].

Recent studies by the same group underlined an association of the SDF1-3'A allele with faster granulocyte and platelet recovery after transplantation. Therefore they suggested that the SDF-1 gene polymorphism could be a useful tool of prognostic value for recipients of autologous haematopoietic stem cells [A. Gieryng et al, 2010]. The allelic variant SDF1-3'A is a result of the SNP rs1801157, which is located in a highly demethylated area of the 3'UTR region. This SNP confers a G to A transition in the nucleotide position 801, resulting in a loss of a methylation site, which could affect the methylating effect of G-CSF [Nagler A et al, 2004], and leading to a more decreased SDF-1 expression in healthy individuals carrying the polymorphism.

So, it's of interest to investigate either this variant have major influence on the circulating levels of SDF-1 and its mRNA expression, one of our future's interests.

Further studies examining how these three polymorphisms interact with disease risk factors are needed.

Interestingly, the possible implication of others genes involved of homing and migration process of CD34+ cells and for instance VCAM-1 to higher or lower mobilization yield of PBPC might emphasize new strategies for poor mobilizers subjects and lead to the identification of new biomarkers and/or therapeutic targets.

5. Conclusion

In the present study, we observed a significant association for CXCL12 and MMP-9 polymorphisms exclusively in patients with MM, NHL and Hodgkin's disease suggesting that these polymorphisms are fair candidate gene variants to these 3 hematological diseases.

Furthermore we've confirmed that the SDF1-3'A allele was highly associated to a good mobilizing capacity especially in the group of healthy allogenic PBSC donors where the analysis not biased by background disease or chemotherapy.

Besides, we suggested a possible association of GG genotypes to poorer mobilization is already deduced.

6. Acknowledgment

We thank all participant and all patients in this work

7. References

Catia Andreassi and Antonella Riccio. (2004). To localize or not to localize: mRNA fate is in 3'UTR ends. Trends in Cell Biology; Vol.19 No.9

J. Arai, M Yasukawa, Y. Yakushijin, T. Miyazaki, S. Fujita. (2000). Stromal cells in lymph nodes attract B-lymphoma cells via production of stromal cell-derived factor-1. Eur. J. Haematol; 64:323-32.

Benboubker L, Watier H, Carion A, Georget MT, Desbois I, Colombat P, et al. (2001). Association between the SDF1-3_A allele and high levels of CD34+ progenitor cells mobilized into peripheral blood in humans. Br J Haematol; 113:247–50.

Belvisi MG and Bottomley KM. (2003). The role of matrix metalloproteinases (MMPs) in the pathophysiology of chronic obstructive pulmonary disease (COPD): a therapeutic role for inhibitors of MMPs? *Inflamm. Res.* 52: 95-100.

Bogunia-Kubik K, Gieryng A, Dlubek D, Lange A. (2009). The CXCL12-3'A allele is associated with a higher mobilization yield of CD34 progenitors to the peripheral blood of healthy donors for allogeneic transplantation. Bone Marrow Transplant:273-8.

Dlubek D, Drabczak-Skrzypek D, Lange A. (2006). Low CXCR4 membrane expression on CD34+ cells characterizes cells mobilized to blood. Bone Marrow Transplant 37:19.

Florence Dommange,* Guillaume Cartron, Claire Espanel, Nathalie Gallay, Jorge Domenech, Lotfi Benboubker, et al for the GOELAMS Study Group. (2006). CXCL12polymorphism and malignant cell dissemination/tissue infiltration in acute myeloid leukemia. FASEB J; 20: 1296–1300

De Oliveira KB, Oda JM, Voltarelli JC, Nasser TF, Ono MA, Fujita TC, et al. (2009). CXCL12 rs1801157 polymorphism in patients with breast cancer, Hodgkin's lymphoma, and non-Hodgkin's lymphoma. J Clin Lab Anal, 23(6):387-93.

Gabriela Gonçavales de Olivera Cavassin, Fernando Luiz De Luca, Nayara Delgado André, Dimas Tadeu Covas, Maria Helena Pelegrinelli Fungaro, Júlio César Voltarelli, and Maria Angelica Ehara Watanabe. (2004). Molecular investigation of the stromal cell-derived factor-1 chemokine in lymphoid leukemia and lymphoma patients from Brazil. Blood Cells, Molecules, and Disease; 33: 90-93.

Yair Gazitt, Cagla Akai. (2004). Mobilization of myeloma cells involves SDF-1/CXCR4 signaling and downregulation of VLA-4. Stem Cells; 22:65-73.

A. Gieryng, K. Bogunia-Kubik, and A. Lange. (2010). *CXCL12* Gene Polymorphism and Hematologic Recovery After Transplantation of Peripheral Blood Progenitor Cells. *Transplantation Proceedings*, 42, 3280–3283

Marilyn Kozak. (2004). How strong is the case for regulation of the initiation step of translation by elements at the 3′ end of eukaryotic mRNAs? Gene; 343: 41–54

Tsevee Lapidot and Isabelle Petit. (2002). Current understanding of stem cell mobilization: the roles of chemokines, proteolytic enzymes, adhesion molecules, cytokines and stromal cells. Experimental Hematology; 30: 973-81.

Roberto M. Lemoli and Alessandra D'Addio. (2008). Hematopoietic stem cell mobilization. Haematologica; 93(3): 321.

Maggie C.Y., Ng Ying Wang, Wing-Yee So, Suzanne Cheng, Sophie Visvikis, Robert Y.L. Zee et al. (2004). Ethnic differences in the linkage disequilibrium and distribution of single-nucleotide polymorphisms in 35 candidate genes for cardiovascular diseases. Genomics, 83; 559–65.

Miller SA, Dykes DD, Polesky HF. (1988). A simple salting out procedure for extracting DNA from human nucleated cells. Nucleic Acids Res; 16:1215.

Nagler A, kollenstein-Ilan A, Amiel A, Avivi L. (2004). Granulocyte-colony stimulating factor generates epigenetic and genetic alterations in lymphocytes of normal volunteers donors of stem cells. Exp Hematol, 32 (1): 122-30.

Iskra Pusic, Shi Yuan Jiang, Scott Landua, Geoffrey L. Uy, Michael P. Rettig, Amanda F. Cashen, et al. (2008). Impact of Mobilization and Remobilization Strategies on Achieving Sufficient Stem Cell Yields for Autologous Transplantation. Biology of Blood and Marrow Transplantation; 14:1045-56.

Rabkin CS, Yang Q, Goedert JJ, et al. (1999). Chemokine and chemokine receptor gene variants and risk of non-Hodgkin's lymphoma in human immunodeficiency virus-1-infected individuals. Blood; 93(6):1838–42.

Siena S, Schiavo R, Pedrazzoli P, Carlo-Stella C. (2000). Therapeutic relevance of CD34+ cell dose in blood cell transplantation for cancer therapy. J Clin Oncol. 18:1360–77.

Signoret N, Oldridge J, Pelchen-Matthews A, et al. (1997). Phorbol esters and SDF-1 induce rapid endocytosis and down modulation of the chemokine receptor CXCR4. J Cell Biol 139:651

Stetler-Stevenson WG. (2001). The role of matrix metalloproteinases in tumour invasion, metastasis, and angiogenesis. Surg Oncol Clin N Am; 10:383–92.

Patrick J Stiff. (1999). Management strategies for the hard-to-mobilize patient. Bone Marrow Transplantation; (23), Suppl. 2: 29–33

Sugrue MW, Willians K, Pollok BH, et al. (2001). Characterization and outcome of "hard to mobilize" lymphoma patient sundergoing autologous stem cell transplantation. Leukemia Lymphoma; 39:509–19

To LB, Haylock DN, Simmons PJ, Juttner CA. (1997). The biology and clinical uses of blood stem cells. Blood; 89:2233–58.

Toru Ogata, Hidenori Shibamura, Gerard Tromp, Moumita Sinha, MStat, Katrina A. B. Goddard et al. (2005). Genetic analysis of polymorphisms in biologically relevant candidate genes in patients with abdominal aortic aneurysms. J Vasc Surg; 41: 1036–42.

Cheng-Ho Tsai, Hung-I Yeh, Yusan Chou, Hsin-Fu Liu, Tzu-Yao Yang , Jyh-Chwan Wang et al. (2000). G protein b3 subunit variant and essential hypertension in Taiwan – a case–control study. International Journal of Cardiology; 73 :191–95.

Gavin S. Wilkie, Kirsten S. Dickson and Nicola K. Gray. (2003). Regulation of mRNA translation by 5′- and 3′-UTR-binding factors .TRENDS in Biochemical Sciences; 28:182-88.

Ingrid G. Winkler and Jean-Pierre Levesque. (2006). Mechanisms of hematopoietic stem cell mobilization:When innate immunity assails the cells that make blood and bone. Experimental Hematology; 34:996–1009.

A Zafiropoulos, N Crikas, A M Passam and D A Spandidos. (2004). CXCL12–3′A in the development of sporadic breast cancer. J. Med. Genet; 41; e59

Zhang B, Ye S., Herrmann SM, Eriksson P, de Maat M, Evans A, et al. (1999). Functional polymorphism in the regulatory region of gelatinase B gene in relation to severity of coronary atherosclerosis. Circulation; 99:1788– 94.

15

Hematopoietic Derived Fibrocytes: Emerging Effector Cells in Fibrotic Disorders

Carolina García-de-Alba, Moisés Selman and Annie Pardo
Instituto Nacional de Enfermedades Respiratorias,
Universidad Nacional Autónoma de México
México

1. Introduction

Fibrocytes constitute a unique population of mesenchymal progenitor cells from hematopoietic origin. They display a unique spectrum of immune and molecular characteristics such as the simultaneous expression of mesenchymal (collagen types I and III, fibronectin), leukocyte (CD45), monocyte (CD14), and hematopoietic stem cell (CD34) markers. Fibrocytes were initially described in the context of wound repair and since their original description in 1994, our understanding and knowledge of this novel cell population has grown considerably. They have the potential to differentiate into fibroblasts and myofibroblasts among other mesenchymal cells such adipocytes, osteoblasts, and chondrocytes. Fibrocytes are a rich source of inflammatory cytokines, growth factors, and chemokines that provide important intercellular signals within the local tissue microenvironment. Moreover, fibrocytes possess the immunological features typical of an antigen-presenting cell (APC), and they have the capacity for the presentation of antigens to naïve T-cells.

The aim of this chapter is to present a comprehensive overview over the history and recent findings on the biology of fibrocytes as well as their putative participation in fibrotic disorders.

2. History

After an injury occurs, a number of extracellular and intercellular responses are initiated and coordinated in order to restore the tissue integrity and homeostasis. Wound healing is a dynamic, interactive process in which cellular components of the immune system, the blood coagulation cascade and the inflammatory pathways are activated. The cells involved including neutrophils, monocytes, lymphocytes, dendritic cells, endothelial cells, keratinocytes and fibroblasts undergo marked changes in gene expression and phenotype, leading to cell proliferation, differentiation and migration (Singer & Clark 1999; Arabi et al., 2007; Gurtner et al., 2008)

Tissue fibroblasts play a key role not only in normal reparative processes, but also in pathological fibrotic processes. In the past decade it has been established that fibroblasts/myofibroblasts, which participate in repair and fibrosis have their origin not only in the fibroblasts already present in the injured tissues, but also may derive from other sources such as mesenchymal and hematopoietic stem cells (Hinz et al., 2007). The notion of a monocytic fibroblast precursor was first proposed more than a hundred years ago by James Paget, and probably represents the first observations of cells with the molecular features of circulating fibrocytes (Herzog & Bucala 2010). Afterward in the early 1960´s the hypothesis of the blood borne origin of fibroblast appeared again in the literature; of particular significance are the observations of Petrakis and co-workers who reported the in vivo differentiation of human leukocytes into fibroblasts, histiocytes and adipocytes in subcutaneous diffusion chambers (Petrakis et al., 1961). More recently, it was demonstrated that bone-marrow (BM) contributes to the expansion of the fibroblast population in multiple organs and tissues, including skin, stomach and esophagus using mouse transplantation models, and in human liver fibrosis (Direkze et al 2003, 2004 and Forbes et al 2004). Regarding the lung, a pioneer work published in 2004 described that the collagen-producing fibroblasts in experimental pulmonary fibrosis are derived from BM progenitor cells (Hashimoto et al., 2004). While these studies documented the BM origin of at least part of the tissue fibroblasts during injury, they did not resolve whether these BM derived fibroblasts were from hematopoietic stem cells (HSCs) or mesenchymal stem cells. Later, through a model of transplantation of clones of cells derived from a single HSC from transgenic enhanced green fluorescent protein (EGFP) mice, it was clearly demonstrated that fibrocytes are derived from HSCs (Ebihara et al., 2006).

The circulating fibrocyte was first described in 1994 by Bucala, in a model of wound healing response, with the surgical implantation of wound chambers into the subcutaneous tissues of mice. The implantation resulted in a rapid influx of peripheral blood cells such as neutrophils, monocytes, and lymphocyte subpopulations within 24 hr. They noticed that 10% of the cells present in the wound chamber, were spindle shaped cells and expressed collagen I, and CD34, (Bucala et al., 1994). The idea that these cells were of circulating origin arose from the observation that their arrival in to the wound chamber was much faster than would be expected by entry of fibroblasts from the surrounding tissue, since the fibroblasts would have to migrate across the permeable plastic layer, enter the wound chamber, and begin matrix deposition, (Bucala, 2008). Hence, the entrance of large numbers of fibroblast-like cells simultaneously with circulating inflammatory cells suggested that this cell population was from peripheral blood origin and not exclusively by slow migration from adjacent connective tissue (Bucala et al., 1994). This new leukocyte sub-population was termed "fibrocytes", which combines the greek "kytos" referring to cell, and "fibro", which is from the latin denoting fiber. This nomenclature may lead to some overlap as the term "fibrocyte" is also used in histopathologic literature as a synonym for "mature" fibroblasts, and to name a cell constituent of the inner ear spiral ligament, (Quan et al., 2004).

Now it is known that fibrocytes are a hematopoietic stem cell source of fibroblasts/myofibroblasts that participate in the mechanisms of wound healing and fibrosis in many organs (Schmidt et al., 2003; Mori et al., 2005; Ebihara et al., 2006; Andersson-Sjöland et al., 2008; El-Asrar et al., 2008; Strieter et al., 2009).

3. Purification and culture

The current methods and techniques employed for the isolation and characterization of peripheral blood fibrocytes are based mainly in the derivation of these cells from the buffy coat of peripheral blood obtained from human or animal sources. Circulating fibrocytes comprise the ~0.1-0.5% of the non-erythrocytic cells in the peripheral blood and they can be quantified and analyzed by flow cytometry (Bucala et al., 1994, Moeller et al., 2009).

Fibrocytes can be obtained and/or differentiated in vitro from the complete peripheral blood mononuclear cell (PBMC) population as well as from an enriched CD14+ population (Abe et al., 2001, Pilling et al., 2009, García-de-Alba et al., 2010). Accordingly, fibrocytes represent one of the variety of cell types that can differentiate from monocytes, including macrophages, osteoclasts and dendritic cells (Wu & Madri 2010; Seta et al., 2010; Castiello et al., 2011).

The fibrocytes obtained from human or mouse blood, either from PBMCs or CD14+ enriched cells, are grown commonly in Dubelcco´s Modified Eagle Medium (DMEM) supplemented with 20% human AB serum (HAB) or fetal calf or bovine serum without the addition of any other growth factors. Some authors have reported the use of RPMI instead of DMEM with good results (Curnow et al., 2010). The resulting fibrocyte population (≥95% pure) is then characterized based on the combined expression of extracellular surface markers including cluster of differentiation (CD) antigens, major histocompatibility complex (MHC)-like molecules, extracellular matrix protein (ECM) markers, and chemokines receptors expression patterns (Metz, 2003) (table 1).

Marker type	Function
Extracellular matrix proteins	
Collagen I and III	Extracellular matrix
Type I pro-collagen	Collagen I precursor
Prolyl -4-hydroxilase	Collagen hydroxiproline
α-smooth muscle actin	Contractile element
Vimentin	Intermediate Filament
CD markers	
CD11a (LFA-1)	L subunit of integrin LFA-1, adhesion molecule
CD11b (Mac 1)	M subunit of integrin CR3, adhesion molecule
CD13	Pan-myeloyd antigen
CD34	Hematopoietic stem cell antigen, endothelial cell
CD45	Leukocyte common antigen
CD54 (ICAM)	Intracellular adhesion molecule binds LFA-1 and Mac-1
CD58 (LFA-3)	Adhesion molecule, binds CD2
CD80 (B7-1)	Co-stimulatory molecule binds CD28
CD86 (B7-2)	Co-stimulatory molecule binds CD28 and CTLA-4
MHC-related markers	
MHC class II	
HLA-DP	Major histocompatability molecule for antigen presentation
HLA-DQ	
HLA-DR	
Chemokine receptors	
CCR3	Receptor for secondary lymphoid chemokine (SLC)
CCR5 (CD195)	Receptor for RANTES (CCL5), MIP-1α and MIP-1β
CCR7 (CD197)	Receptor for CCL19 and CCL21
CXCR4	Receptor for CXCL12 (SDF-1α)

Table 1. Human fibrocytes surface and intracellular phenotype. Reviewed in Bucala et al., 1994, Chesney et al., 1998, Abe et al., 2001, Hartlapp et al., 2001

It was previously reported that the differentiation of fibrocytes is inhibited by serum amyloid P (SAP), a major constituent of serum, (Pilling et al., 2003) and more recently it was described that in the absence of serum the process of differentiation of fibrocytes can be accelerated with cells with the spindle-shaped morphology appearing in culture after only ~2-5 days, compared to ~8-14 days when fibrocytes are cultured with serum supplemented medium (Curnow et al., 2010). They also reported a difference in the ability of serum free and serum complemented fibrocytes to differentiate from PBMC and CD14+ peripheral blood cells, with more efficient generation of fibrocytes from PBMC cultured without serum, and from CD14+ cells when these were cultured in the presence of serum complemented medium (Curnow et al., 2010).

Cell population obtained regardless of the initial method for enrichment (PBMC or CD14+ enriched cell culture) are cells expressing a combination of CD45 or other haematopoietic markers (CD34, CD11b), as well as collagen I and III, with an elongated spindle-shaped morphology, making clear that the cells differentiated under both conditions can be classified as fibrocytes, based on the current definition: spindle-shaped cell that expresses both haematopoietic and mesenchymal cell markers, (Bucalla et al., 1994, C. Metz 2003),

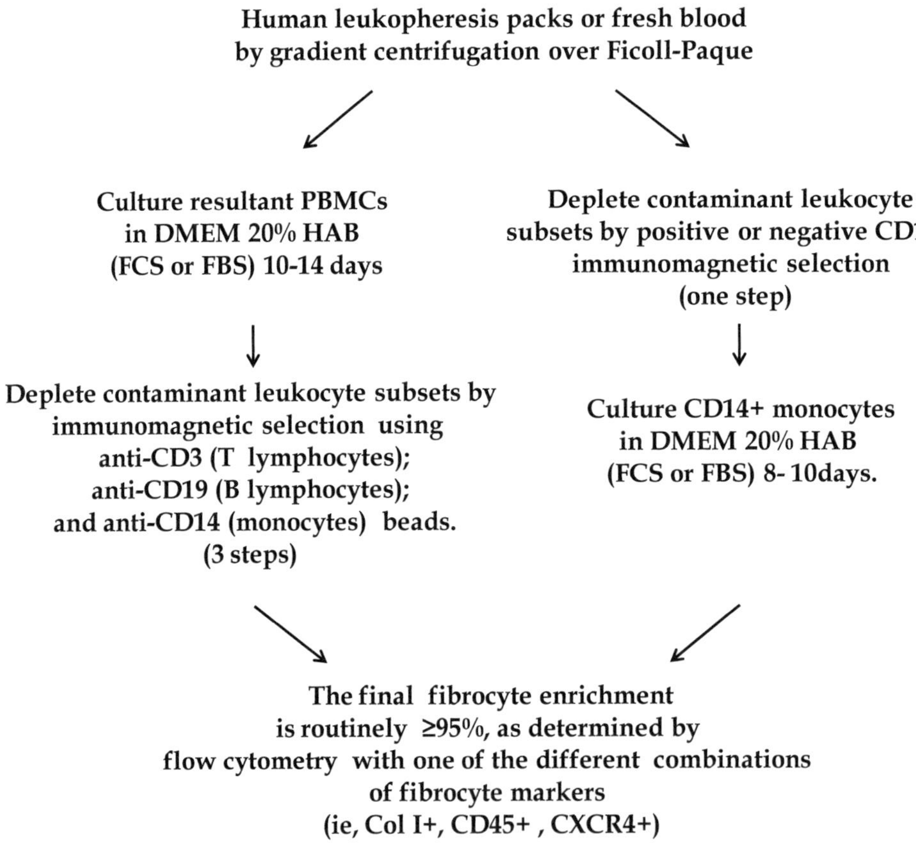

Fig. 1. Schematic description of the two most common methods for fibrocytes culture and enrichment.

3.1 Flow cytometry analysis

Flow cytometry is a critical technique for the characterization and quantification of circulating fibrocytes after their enrichment and in vitro differentiation as well as for fibrocytes obtained directly from fresh blood samples.

Cell preparation. Flow analysis requires a single cells suspension. Ice cold 0.05% EDTA in PBS or trypsin-EDTA 0.05% are recommended to detach the cells from the plastic surface, just covering it for 1-2 min at 37° C. Since trypsin is toxic for the cells, they must be observed closely to adjust and change the timing of the trypsin digestion. Immediately, media complemented with 10% serum is added to neutralize the enzymatic activity of the trypsin present in the buffer (normal human AB serum, FCS or FBS). Horizontal shear force can be applied, or cells can be gently scraped if needed for harvesting and they are immediately washed in cold PBS. The number of dead cells should be estimated by trypan blue exclusion.

Number of cells required for staining. Approximately 2.5-5 x 10^5 cells with a minimum volume of 300ul of staining buffer (1% BSA-PBS) in polystrene tubes 12X75 are needed for the analysis of the in vitro cultured fibrocytes; fewer cells mean longer collection time and potentially more background noise. For the analysis of circulating fibrocytes from fresh blood samples, ~ 0.5- 1 × 10^6 cells in 300ul of staining buffer (1% BSA-PBS) in polystrene tubes 12X75 are needed, since the normal percentage of this cells in the circulation is 0.1-0.5% of the total leukocytes it is better to analyze at least 50,000 events, the use of high performance flow cytometers is recommended.

Protocol for staining cell surface and intracellular antigens for fibrocytes analysis. The following steps are the same for both cell types (fresh PBMC´s or cultured cells).

Cells are centrifuged and resuspended in staining buffer (1% BSA-PBS). The optimum amount of buffer to incubate depends on the protocol suggested by the antibody´s manufacturer technical data sheet, commonly 100µl is an adequate volume. Cells are incubated with the corresponding fluorochrome-labeled antibodies for surface markers (i. e., CD45, CD34, CD11b, and CXCR4) and then fixed and permeabilized with a commercial kit recommended for this purpose (i.e., BD Cytofix/Cytoperm™ Fixation/Permeabilization Solution Kit, BD Biosciences) prior to staining with the anti-collagen antibody or its corresponding isotype control. It is important to consider that isotype controls are critical in the analysis of these cells, since they have to be permeabilized and fixed and a high percentage of nonspecific binding can occur. Also non-stained cells treated with the same process are required as control to discriminate collagen fibers autofluorescence. The number of markers that can be analyzed depend on the capacity (i.e., lasers, filters) of the cytometer to be used, at least 2- 3 fibrocyte markers (i.e., CD45+/Collagen I or CD45+/CXCR4+/Collagen I) are needed to meet the minimum criteria of the fibrocytes definition.

3.2 Fibrocyte to myofibroblast differentiation

Fibrocytes increase the expression of α-smooth muscle actin (α-SMA) spontaneously in culture, and gradually loose the expression of CD34 and CD45 over time, which likely reflects terminal differentiation or other phenomena related specifically to a particular tissue microenviroment (Schmidt et al., 2003, Mori et al., 2005, Bucala, 2008). The differentiation of fibrocytes into myofibroblasts can be enhanced by transforming growth factor (TGF)-β or endothelin-1, which results in an increment in the synthesis of collagen and the myofibroblast marker α-SMA (Schmidt et al., 2003; Bucala, 2008).

For myofibroblast differentiation as described in the literature (Hong et al., 2007): a population of enriched fibrocytes has to be previously obtained with one of the techniques described above. The percentage of enrichment needs to be verified by flow cytometry in each culture to ensure reproducibility of the results.

Fibrocytes are treated with serum-free DMEM with 10 ng/ml TGF-β1 for 3 weeks, refreshing TGF-β1 supplemented medium every 48-72hs. If the objective is to analyze changes in the pattern of gene and protein expression, time curves should be performed previously since these effects might be different depending on the gene or protein of interest.

The signaling pathways that are activated by TGF-β1 to induce α-SMA transcription and thus fibrocyte differentiation to myofibroblast-like cells include Smad2/3 and stress-activated protein kinase (SAPK)/c-Jun N-terminal kinase (JNK) - mitogen-activated protein kinase (MAPK). Interestingly, it was reported that treatment with troglitazone (TGZ, a synthetic agonist of peroxisome proliferator-activated receptor gamma: PPARγ), inhibits TGF-β1 induced α-SMA expression and this effect is modulated through attenuation of the SAPK/JNK activity leading to decreased Smad2/3 levels and transactivation activity, (Hong et al., 2007).

3.3 Adipocyte differentiation

Hong et al., demonstrated in a model of differentiation of human circulating adipogenic progenitors to adipocytes in SCID mice, that fibrocytes, in the presence of specific environmental characteristics can give rise to adipocytes. By gene microarray analysis they found a significant up-regulation of specific mature adipocyte genes and proteins after fibrocyte differentiation to adipocyte, including fatty acid binding protein 4 (FABP4), leptin, and PPARγ; remarkably certain genes, such as those involved in cell motility, chemotaxis, or metalloproteinase activity where also upregulated in the process of differentiation to adipocytes. These findings indicate that fibrocytes may retain unique functions for motility and chemoattractive activity that might allow them to participate in migration and trafficking despite their differentiation into adipocytes (Hong et al., 2005). Differentiation of fibrocytes into adipocytes appears to be mediated by PPARγ that leads to lipid accumulation and induction of aP2 gene expression (Rival et al., 2004). By contrast, this process is inhibited by TGF-β through SAPK/JNK pathway activation (Hong et al., 2007).

For adipocyte differentiation: fibrocytes are treated with PBM culture media (Cambrex Bio Science) supplemented with 10 M troglitazone. Culture media has to be changed every 48 h for 21 days. Following 21 days in culture, the cells accumulate lipids in intracellular vacuoles. Oil Red O Staining can be used to confirm fibrocytes differentiation to adipocytes (Hong et al., 2007).

3.4 Osteoblast and chondrocyte differentiation

Osteoblasts and chondrocytes, which are derived from a common mesenchymal precursor cell, are critical in bone and cartilage formation respectively (Knothe et al., 2010). It has recently been reported that fibrocytes possess the ability to differentiate into chondrocytes and osteoblasts in vitro when the appropriate combination of cytokines and growth factors are used (Choi et al., 2010). These findings, taken together with their capacity to differentiate into myofibroblasts and adipocytes, indicate that fibrocyte may differentiate toward several types of mesenchymal cell types and that this process is influenced by a complex profile of cytokines within the local microenvironment of the host tissue or tissue injury.

Induction of the differentiation of fibrocytes to osteoblasts: Purified fibrocytes are seeded at a concentration of 1×10^5 cells/well in a fibronectin-coated 12 well plate, they are treated with osteogenic basal media (this media is commercially available) supplemented with dexamethasone, ascorbate, mesenchymal cell growth supplement (MCGS), l-glutamine, 1× Penicillin/Streptomycin, and β-glycerophosphate. β-glycerophosphate is critical to stimulate calcified matrix formation in combination with the effects of dexamethasone and ascorbate. Cells are cultured during 21 days with media replacement every 3 days (Choi et al., 2010).

Induction of the differentiation of fibrocytes to chondrocytes: Purified fibrocytes are seeded at the concentration of 5×10^4 cells/tube in 15ml sterile polypropylene tubes, followed by centrifugation at ~300×g for 10min to form pellets. Supernatant has to be carefully removed in order not to disrupt the fibrocyte micromass pellet. Fibrocytes are additioned with chondrogenic differentiation cocktail: basal chondrogenic media (also commercially available) supplemented with 1×10^{-7}M dexamethasone, 0.1 M ascorbate, l-glutamine, Penicillin/Streptomycin, 1 M sodium pyruvate, proline and 10 ng/ml of TGF-β3. Cells are cultured for 21 days with media replacement every 2 - 3 days (Barry et al., 2001; Choi et al., 2010). Fig. 2.

4. Fibrocytes participation in repair processes

Wound repair is a complex process that results from the coordinated release of cytokines, chemokines, and growth factors, leading successively to the recruitment and activation of different cells into the injured site from the very initial phases of repair (Gurtner GC et al., 2008). Fibrocytes have been postulated as important players of the tissue repair process since they have the ability to rapidly home to sites of tissue together with the infiltrating inflammatory cells that act to prevent infection and degrade damaged connective tissue components (Bucala et al., 1994).

Fibrocytes secrete proinflammatory cytokines such as tumor necrosis factor alpha (TNFα), interleukin (IL)-6, IL-8, IL-10, macrophage inflammatory protein-1α/β (MIP-1α/β) CC-chemokine ligands (CCL) -3 and -4 in response to IL-1β which is an important mediator of wound healing response (Chesney et al., 1998). The fibrocyte products MIP-1α , MIP-1β, and monocyte chemotactic protein-1 (MCP-1) are potent T cell chemoattractants and may act to specifically recruit CD4+ T cells into the tissue repair microenvironment; moreover, the fibrocytes increase the cell surface expression of leukocyte adhesion molecules, such as intercellular adhesion molecule 1 (ICAM1), which would enhance leukocyte trafficking (Chesney et al., 1998). Interestingly, in addition to these functions, fibrocytes may play an early and important role in the initiation of antigen-specific immunity. Thus, it has been demonstrated that peripheral blood fibrocytes: express the surface proteins required for antigen presentation, including class II major histocompatability complex molecules: HLA-

DP, -DQ, and -DR; the costimulatory molecules CD80 and CD86, and the adhesion molecules CD11a, CD54, and CD58. Fibrocytes are potent stimulators of antigen-specific T cells in vitro, and migrate to lymph nodes and sensitize naïve T cells in situ (Chesney et al., 1997). Likewise, fibrocytes may also participate in the development of the innate immune response; in porcine models, specific in vitro stimulation of fibrocytes for TLR 2, 4, 7 or TLR3 leads rapidly to the translocation of the NF-kB transcription factor and the production of high levels of IL-6 (Balmelli et al., 2007); on the other hand, exposure to innate immune stimulation in the form of TLR agonists induces an increased expression of MHC class I and II molecules and of the co-stimulatory proteins CD80 and CD86 on fibrocytes, which enables these cells to function as antigen-presenting cells for the activation of cytotoxic CD8+ T cells. All these findings indicate that fibrocytes may recognize a large variety of pathogens such as viruses or bacteria and could be part of the initiation of innate immune responses (Balmelli et al., 2005 and 2007).

Blood vessel formation during normal physiological processes, such as wound healing, is highly regulated by a delicate balance between pro- and antiangiogenic factors. As mentioned, circulating fibrocytes have been shown to migrate to early wound sites where angiogenesis occurs, fibrocytes produce and secrete active matrix metalloproteinase 9 and 2 (MMP-9: gelatinase B; MMP-2: gelatinase A) (Hartlapp I et al., 2001, García-de-Alba et al., 2010), which are implicated in the proteolysis of the basement membrane early during the invasion stage of angiogenesis. In addition, cultured fibrocytes constitutively secrete vascular endothelial growth factor (VEGF), basic fibroblast growth factor (bFGF), platelet derived growth factor (PDGF), insulin growth factor (IGF-I) and hematopoietic factors as granulocyte monocyte-colony stimulating factor (GM-CSF) that induce endothelial cell migration, proliferation, and alignment of endothelial cells into tubular-like structures in vitro. In like manner cultured fibrocytes (and fibrocyte-conditioned media) showed the ability to promote angiogenesis in vivo using a Matrigel implant model, (Hartlapp I et al., 2001).

Interestingly, it has been reported that Th2 cytokines (IL-4 and IL-13) induce, whereas Th1 cytokines (IFN-γ and IL-12) inhibit the CD14+ monocyte to fibrocyte differentiation. When added together the profibrocyte activities of IL-4 and IL-13 and the fibrocyte-inhibitory activities of IFN-γ and IL-12 counteract each other in a concentration-dependent form. By contrast, the fibrocyte-inhibitory activity of the plasma protein serum amyloid P (SAP) dominates over the profibrocyte activities of IL-4 and IL-13. These results might indicate that the complex mix of cytokines and plasma proteins present in inflammatory lesions, wounds, and fibrosis will influence fibrocyte differentiation (Shao et al., 2008). Consistent with this data, it was recently reported that CD14+ monocytes can differentiate in vitro into two different subtypes of fibrocytes depending on the presence or absence of serum in the culture media, which could resemble the changes in serum protein concentrations that occur during tissue repair, inflammation and its resolution (Curnow et al., 2010).

Fibrocytes also contribute to normal wound healing by serving as the contractile force of wound closure via α-smooth muscle actin expression (Abe et al., 2001; Metz, 2003), and secreting components of the extracellular matrix (collagen I, collagen III, fibronectin) (Abe et al., 2001; Bucala et al., 1994). Interestingly, it has been reported that the capacity to produce collagen of fibrocytes from normal subjects or from burn patients is less than that of fibroblasts (dermal and lung fibroblasts) (Wang et al., 2007; García-de-Alba et al., 2010), which raises the question if fibrocytes main contribution to the process of tissue repair is only a direct participation in the production of ECM components. In this context, it is important to emphasize that fibrocytes secrete paracrine growth factors such as connective tissue growth factor (CTGF), PDGF, FGF and TGF-β1 that induce proliferation, migration and differentiation of fibroblasts to myofibroblasts in culture (Chesney et al., 1998, Wang et al., 2007). These findings suggest that the predominant role of fibrocytes in scarring could be the regulation of the functions of local fibroblasts.

Proteins secreted by fibrocytes	Pattern of Expression
Growth factors	
Platelet-Derived Growth Factor A (PDGF-α)	Constitutive
Fibroblast Growth Factor basic (bFGF)	
Granulocyte-Monocyte Colony Stimulating Factor (GM-CSF)	
Insulin Growth Factor 1 (IGF1)	
Vascular Endothelial Cell Growth Factor (VEGF)	
Transforming Growth Factor-beta1 (TGF-β1)	
Connective Tissue Growth Factor (CTGF)	
Hepatocyte Growth Factor (HGF)	
Cytokines	
Tumoral Necrosis Factor-alpha (TNF-α)	Induced by IL-1β stimulation
IL-6	Induced by IL-1β or TNF-α stimulation
IL-10	Induced by IL-1β or TNF-α stimulation
IL-1α	Constitutive
Chemokines	
IL-8	Constitutive; ↑ with TGF-β1 or IL-1β
GROα	
MIP-1α	
MIP-1β	
MCP-1	
MMPs	
MMP-2 MMP-9	Constitutive; ↑ with TGF-β1
MMP-7 MMP-8	Constitutive; ↓ with TGF-β1

Table 2. Fibrocytes pattern of expression for diverse proteins.

4.1 Migration and homing

Fibrocytes trafficking from the bone marrow and circulation to the organs or site of lesion is given through several chemokines. Human fibrocytes express diverse chemokine receptors, including CCR3, CCR5, CCR7, and CXCR4; whereas mouse fibrocytes express CXCR4, CCR7, and CCR2 (Abe et al., 2001; Phillips et al., 2004; Moore et al., 2005; Mehrad et al., 2009). Secondary lymphoid tissue chemokine (SLC/CCL21) and its receptor CCR7 was the first chemokine-chemokine receptor system described to induce the recruitment of fibrocytes as a mechanism of migration to wound sites (Abe et al., 2001). In humans as in mice CCL21 is constitutively abundant in lymphoid tissues, particularly in the lymph nodes and spleen but it is also expressed at lower levels in some non-lymphoid tissues, including the kidneys and lungs (Gunn et al., 1998; Abe et al., 2001; Sakai N et al., 2006).

CXCR4 is an important chemokine receptor for stem and immune cell migration, high levels of CXCL12, which is the only known ligand for CXCR4, were found in the lungs and plasma of patients with IPF and these levels correlated with circulating fibrocyte concentrations (Mehrad B et al., 2007; Andersson-Sjöland A et al., 2008).

Recently (Mehrad et al., 2009) reported that most (but not all) freshly isolated human fibrocytes expressed CXCR4, whereas 46% expressed CCR2 and 9% expressed CCR7. Approximately 30% were CCR2/CXCR4+ and most CCR7+ cells also expressed CCR2, but there was no overlap between CXCR4+ and CCR7+ receptors.

It has been reported an association between serum concentration of MCP-1 and high levels of CD45/pro-Col-I+ fibrocytes in the circulation of scleroderma patients with interstitial lung disease (ILD) or in healthy aging subjects, suggesting that MCP-1 may be also involved in mobilization of fibrocytes into the peripheral blood. (S. Mathai et al., as cited in Herzog & Bucala , 2010),

Thus, fibrocytes can use different chemokine–chemokine receptor axis for tissue homing and this might be related to the type of process (acute or chronic) or to the organs involved; however, the mechanisms implicated in the migration through basement membranes and extracellular matrix and subsequent tissue homing remain unclear. In this context, it was recently reported that fibrocytes express several MMP's (MMP- 2, 7, 8 and 9) (Fig 3) that may facilitate the process of migration of fibrocytes from the circulation to the tissues in response to chemokine gradients (Garcia-de-Alba et al., 2010). In this work it was showed that fibrocytes transmigration towards CXCL12 or PDGF through collagen I coated migration chambers, was highly associated with the collagenase MMP8, while migration through a combination of proteins of basal membrane was facilitated by gelatinases MMP2 and MMP9. Thus, these MMPs may ease cell migration by breaking down matrix barriers or affecting the state of cell-matrix interactions and also may play an important role in the remodelling of ECM. Interestingly, PDGF showed to be a more potent chemotactic agent when migration was given through collagen I coated chambers, possibly indicating that when fibrocytes have arrived to lung interstitium, PDGF plays an important role as a chemoattractant through lung parenchyma.

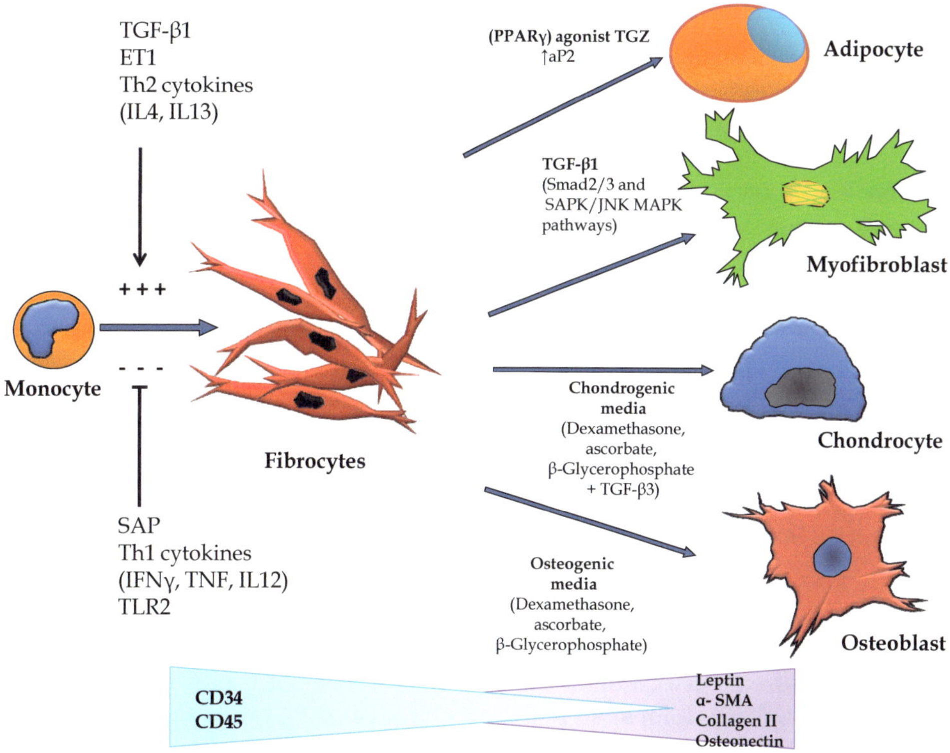

Fig. 2. Schematic summary of the mediators and inhibitors of CD14+monocyte to fibrocyte differentiation, and fibrocytes differentiation to other mesenchymal cells. TGZ: trogliotazone. A crosstalk between PPARγ and TGF-β1 exists, where they can strongly inhibit each other signaling, making clear that a complex and critical balance exists between both of them. It is noteworthy that the expression of hematopoietic markers decrease as fibrocytes differentiate into other mesenchymal cells, while specific markers for that given cell increase their expression during differentiation.

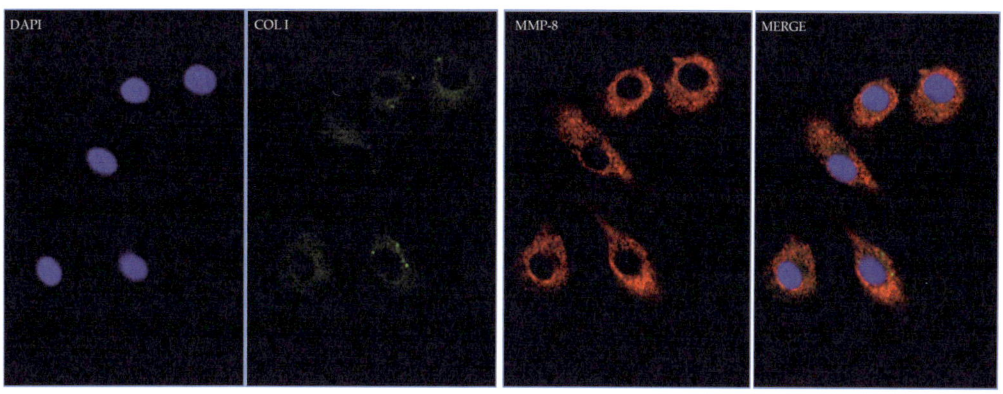

Fig. 3. Fluorescent immunocytochemistry showing a group of fibrocytes positive for collagen I and MMP-8 staining.

5. Role of fibrocytes in the pathogenesis of fibrotic disorders

In contrast to acute inflammatory reactions, which are characterized by rapidly resolving events; fibrosis typically results from chronic unsolved inflammation or aberrant epithelial activation (King, Pardo & Selman, 2011). Despite having distinct etiological and clinical manifestations, fibrotic remodelling is characterized by fibroblast/myofibroblast activation, and excessive extracellular matrix accumulation leading to scarring formation and progressive dysfunction of a given organ.

Fibrocytes have become the focus of research of a wide variety of focal and diffuse fibrosing disorders in diverse organs including lung, heart, liver, and kidney (Barth et al., 2005; Sakai et al., 2006, 2008, 2010; Andersson-Sjöland et al., 2008; Scholten et al., 2011); primarily because of their ability to home into tissues and secret extracellular matrix components. More recently however, a large and varied amount of new knowledge about fibrocytes biology has emerged, rising new hypothesis that have enriched the understanding of these cells and their participation in fibrotic diseases.

5.1 Pulmonary fibrosis

Pulmonary fibrosis is the final result of a numerous and heterogeneous group of disorders known as interstitial lung diseases (ILD). Lung fibrotic remodeling is characterized by fibroblast/myofibroblast activation, and excessive extracellular matrix accumulation leading to progressive destruction of the lung architecture and usually terminal outcome (Pardo & Selman, 2002). Idiopathic pulmonary fibrosis (IPF), the most common form of the idiopathic interstitial pneumonias, is a chronic, progressive, irreversible, and usually lethal lung disease of unknown cause (King, Pardo & Selman 2011). IPF is characterized by the presence of clusters of fibroblasts and myofibroblasts circumscribed from surrounding cells (fibroblastic foci), which represent sites of active fibrogenesis (Selman, King & Pardo, 2001).

During a long time, proliferation of local (resident) fibroblasts and differentiation to myofibroblasts were considered the main source of extracellular matrix deposition in pulmonary fibrosis. The first report of the possible participation of mesenchymal stem cells in the pathogenesis of pulmonary fibrosis, described that collagen-producing cells with fibroblast characteristics were derived from BM progenitor cells, in a model of bleomycin induced pulmonary fibrosis (Hashimoto et al., 2004). The mice in this model were engrafted with BM from GFP transgenic mice that allow to easily follow the fate of these BM-derived cells. Though this group did not prove that these cells were actually fibrocytes, they recognize the possibility of this premise. Not much later, a work that showed that human CD45+Col I+CXCR4+ circulating fibrocytes were able to migrate to the lung of mice treated with bleomycin was published (Phillips et al., 2004). These authors also described that maximal intrapulmonary recruitment of CD45+Col I+CXCR4+ fibrocytes directly correlated with increased collagen deposition in the lungs. Likewise, they identified a second fibrocyte population that is CD45+Col I+CCR7+ and also traffics to the lungs of bleomycin-treated mice; interestingly the absolute number of CCR7+ fibrocytes found in the fibrotic lung was two to three fold lower than the number of CXCR4+ fibrocytes present under similar conditions, indicating that CXCR4 predominates for the recruitment of fibrocytes to injured lungs (Phillips et al., 2004).

Fibrocyte recruitment to damaged lungs has been proved to be mediated by several chemokine/chemokine receptor interactions. Thus, in a model of fluorescein isothiocyanate (FITC)-induced lung fibrosis, it was demonstrated that significantly higher numbers of fibrocytes are present in the airspaces of fluorescein isothiocyanate-injured $CCR2^{+}/^{+}$ mice compared to $CCR2^{-}/^{-}$ mice (Moore et al., 2005; 2006). Fibrocytes isolated from the lung expressed CCR2 and migrated toward CCL2 and CCL12 ligands. Interestingly, CCL2 stimulated collagen secretion by lung fibrocytes, which differentiated towards a myofibroblast phenotype, transition that was associated with loss of CCR2 expression (Moore et al 2005).

Importantly, interruption of the chemokine axis attenuated both fibrocyte accumulation and pulmonary fibrosis (Phillips et al., 2004; Moore et al., 2006), strengthening the notion that these chemokine/chemokine receptor axis are the main responsible of fibrocytes trafficking to the lungs; however, under which biological/pathological conditions one or other chemokine/chemokine receptor system is activated, or if they represent redundant mechanisms, yet remains to be elucidated.

Recently several independent research groups have identified fibrocytes in different forms of fibrotic human lung disease. In an initial study, it was reported that circulating fibrocytes expressing CXCR4 and both lung and plasma levels of CXCL12 were elevated in IPF patients (Mehrad 2007). CXCL12 levels showed a positive correlation with higher number of circulating fibrocytes in the peripheral blood of these patients. Later, Andersson-Sjöland et al., evaluated the presence of fibrocytes in the lung of patients with idiopathic pulmonary fibrosis by immunofluorescence and confocal microscopy. Fibrocytes were identified with different combinations of markers in most fibrotic lungs; interestingly, no fibrocytes were identified in normal lungs. They also found a positive correlation between the abundance of fibroblastic foci and the amount of lung fibrocytes and a negative correlation between plasma levels of CXCL12 with lung function tests (lung diffusing capacity for carbon monoxide and oxygen saturation on exercise) (Andersson-Sjöland et al., 2008). These findings indicate that circulating fibrocytes may contribute to the expansion of the fibroblast/myofibroblast population in idiopathic pulmonary fibrosis.

On the other hand, as mentioned earlier in this chapter, fibrocytes constitutively synthesize and release to the medium important amounts of MMP-2, MMP-7, MMP-8, and MMP-9 (García-de-Alba et al., 2010).

MMPs consist of a large family of zinc endoproteases, collectively capable of degrading all ECM components (Pardo et al., 2006). However, ECM represents only a fraction of their proteolytic targets, and moreover, a given MMP can act on various proteins and, in turn, affect a variety of processes. Gelatinases (MMP-2 and MMP-9) have been found up-regulated in human pulmonary fibrosis and animal models of lung fibrosis (Swiderski et al., 1998; Selman et al., 2000; Oikonomidi et al., 2009). The overexpression of MMP-2 and MMP-9 has been mainly associated with their capacity to provoke disruption of alveolar epithelial basement membrane and enhanced fibroblast invasion into the alveolar spaces (Ruiz et al., 2003; Pardo et al., 2006). In the case of fibrocytes, these MMPs may facilitate the process of migration from the circulation to the interstitial and alveolar spaces in response to SDF-1/CXCL12 synthesized by alveolar epithelial cells (Andersson-Sjöland et al., 2008; García-de-Alba et al., 2010). TGF-β1–stimulated fibrocytes significantly increase gene and protein expression of both MMP-2 and MMP-9 in vitro. Another putative pathogenic role of these two enzymes is that cell surface localized MMP-2 and MMP-9 can activate latent TGF-β, and this constitutes a mechanism that may operate in normal tissue remodeling as well as in fibrosis, tumor growth, and invasion (Yu et al., 2000). Fibrocytes also synthesize MMP-7 and MMP-8; the presence of MMP-7 is interesting because this metalloproteinase has been associated with pulmonary fibrosis since is one of the most up-regulated genes in IPF and display several profibrotic activities (Zuo et al., 2002, Pardo et al., 2005). Moreover, MMP-7 and MMP-1 have been related to alveolar and bronchiolar cell migration over different matrices during IPF lung remodeling (Oikonomidi, 2009). In addition, MMP-7 cleave E-cadherin, which may influence several aspects of cell behavior, such as epithelial-to-mesenchymal transition, which is a well-recognized event that recently has gained importance as a mechanism in the pathogenesis of fibrosis (Lochter et al.,1997; Noe et al., 2001; Hinz et al., 2007). MMP-8 or collagenase-2 specifically degrades fibrillar collagen types I, II and III, and is known to play an important regulatory role in both acute and chronic inflammation (Prikk et al., 2001). In the context of fibrocytes it seems to have an important role in the transmigration of these cells through collagen I (García-de-Alba et al., 2010).

Recently it has been suggested that circulating fibrocytes could have a role as biomarkers for disease severity in IPF; Moeller and coworkers quantified circulating fibrocytes from patients with IPF and found that high percentages of these cells in blood were predictive of poor clinical outcomes; they compared fibrocyte levels in peripheral blood from patients with idiopathic pulmonary fibrosis (stable and during an exacerbation), patients with acute respiratory distress syndrome, and normal controls. Fibrocytes were significantly elevated in patients with stable idiopathic pulmonary fibrosis compared with normal controls, but showed a prominent increase during acute exacerbations of the disease. The number of fibrocytes in patients with acute respiratory distress syndrome was not significantly different from patients with stable idiopathic pulmonary fibrosis or normal controls (Moeller et al., 2009). These data suggest that serial measurements of fibrocyte percentages may predict acute exacerbations (Moore, 2009). This work was the first to bring up the notion that fibrocytes measurements may be a useful biomarker in this disease but larger studies are needed to confirm this hypothesis.

Finally, a recently published work exploring senescence-accelerated prone mice found increased levels of CXCR4 expressing fibrocytes in the blood of these mice when compared to wild type controls. The senescence-prone mice also displayed increased lung fibrosis when exposed to intratracheal bleomycin, suggesting the possibility that the increased number of fibrocytes contributed to disease. This is an interesting observation since IPF is considered an age related disease (Selman M et al., 2010). Actually, unpublished data from Mathai et al (Mathai et al., as cited in Herzog & Bucala 2010) indicates that the blood of healthy aged individuals contain increased concentrations of CD45+/Col-1+ fibrocytes and high circulating levels of MCP-1 and IL-13, suggesting that fibrocytes may be associated with certain aging processes.

5.2 Asthma

Asthma is an inflammatory disorder of the conducting airways which undergo distinct structural and functional changes, leading to non-specific bronchial hyper-responsiveness

(BHR) and airflow obstruction. It is among the commonest chronic conditions in Western countries affecting 1 from 7 children and 1 from 12 adults (Holgate et al., 2009).

It has long been known that architectural and structural remodeling occur in the airways of asthmatic patients. These changes include increased collagen (type III and IV) and fibronectin deposition, increased thickness of subepithelial basement membrane, angiogenesis, and fibrosis. All these processes collectively contribute to severe alterations of the normal bronchial architecture in response to the inflammatory tissue injury, leading to progressive airway obstruction and a permanent impairment in respiratory function (Holgate et al., 2009, Hamid & Tulic 2009). Pathologic examination of these tissues demonstrates subepithelial fibrosis and myofibroblast accumulation. Fibrocytes have been identified in the airways of patients with asthma, and it has been reported that allergen exposure induced an increment of fibrocyte-like cells in the bronchial mucosa of patients with allergic asthma (Shmidt et al., 2003). In a mouse model of allergic asthma, fibrocytes were recruited into the bronchial tissue following allergen exposure and differentiated into myofibroblasts providing evidence for the first time that these cells might be a source of myofibroblasts in allergic asthma (Shmidt et al., 2003). Nihlberg and his group showed that fibrocytes in patients with mild asthma were primarily localized, either individually or in clusters, close to the epithelium and to blood vessels. Fibrocyte numbers correlated to the thickness of the basement membrane, supporting that these cells may participate in airway wall remodeling. The increase number of fibrocytes expressing α-SMA seen in patients with increment in the basement membrane thickness may indicate a more differentiated phenotype (Nihilberg et al., 2006). More recently, in two different works, fibrocytes percentages in peripheral blood were shown to be increased in patients with asthma with chronic airway obstruction and severe refractory asthma (Saunders et al., 2008; Chun-Hua et al., 2009). Additionally, a yearly decline in lung function has been significantly associated with the percentage of circulating fibrocytes in patients with chronic obstructive asthma (Saunders et al., 2008).

5.3 Renal fibrosis

Renal tubulo-interstitial fibrosis is a non-specific process, representing the common end-stage for kidney diseases, regardless of their etiology. The histological characteristics include the presence of tubular atrophy and dilation, interstitial leukocyte infiltration, accumulation of fibroblasts, and increased interstitial matrix deposition (Strutz et al., 2006). Fibrocytes have also been implicated in the pathogenesis of renal fibrosis in diverse models. For example, in an experimental model of unilateral ureteral obstruction, fibrocytes appeared in injured parenchyma in a time dependent fashion. Thus, a remarkable number of fibrocytes dual-positive for CD45 or CD34 and type I collagen infiltrated the interstitium, reaching a peak at day 7. Morphological interstitial fibrosis and collagen content were reduced by almost 50% in mice treated with anti-CCL21 antibodies 7 days after ureteral ligation. A similar reduction was observed in CCR7-null mice (Sakai et al., 2006). Interestingly, most fibrocytes were positive for CCR7 and CCL21, and the blockade of CCR7 reduced the number of infiltrating fibrocytes indicating that for this organ, CCR7/CCL21 might be the main recruitment axis. The same investigators showed later that fibrocytes might contribute to fibrosis by an angiotensin II dependent pathway (Sakai et al., 2008). Using two models of renal fibrosis (unilateral ureteral obstruction and chronic angiotensin II infusion), angiotensin II type 2 receptor (AT2R)-deficient mice developed increased renal fibrosis and fibrocyte infiltration and a concomitant upregulation of procollagen type I compared with wild-type mice. Pharmacologic inhibition of angiotensin II type 1 receptor (AT1R) with valsartan reduced the degree of renal fibrosis and the number of fibrocytes in both the kidney and the bone marrow. In isolated human fibrocytes, inhibition of AT2R signalling increased the angiotensin II-stimulated expression of type I collagen, whereas inhibition of AT1R decreased collagen synthesis. These results suggest that AT1R/AT2R signalling may contribute to the pathogenesis of renal fibrosis by at least two fibrocytes-related mechanisms: by regulating the number of fibrocytes in the bone marrow, and by activation of these cells in the tissues (Sakai et al. 2008).

More recently, the presence of fibrocytes was investigated by immunohistochemistry in kidney biopsy specimens from 100 patients with chronic disease; in addition 6 patients with thin basement membrane disease were studied as a disease control. In patients with chronic kidney disease, the infiltration of fibrocytes was observed mainly in the interstitium and their numbers were higher than that in patients with thin basement membrane disease. Moreover, there was an inverse correlation between the number of interstitial fibrocytes and kidney function at the time of biopsy (Sakai et al., 2010). These results suggest that fibrocytes may be involved in the pathogenesis of human chronic kidney disease though the mechanisms involved in their participation are yet to be studied.

CD34+spindle-shaped cells have also been detected in tubulointerstitial lesions in patients with renal interstitial fibrosis. Although in this work the complete phenotype corresponding to fibrocytes was not documented, it is possible that the described CD34+ cells were actually fibrocytes (Okona et al., 2003).

5.4 Liver fibrosis

Hepatic fibrogenesis represents a wound-healing response of liver to a variety of insults. The net accumulation of extracellular matrix (ECM) in liver injury arises from increased synthesis by activated hepatic stellate cells and other hepatic fibrogenic cell types, as well as from bone marrow and circulating fibrocytes (Guo & Friedman, 2007).

Fibrocytes participation in liver fibrosis is a growing field of research and has been assessed in different models. In a murine model of bile duct ligation-induced liver fibrosis, investigators found bone marrow derived collagen-expressing GFP+ cells in the liver of chimeric mice (Kisseleva et al., 2006). The majority of these bone marrow derived cells co-expressed collagen-GFP+ and CD45+, suggesting that collagen-producing fibrocytes were recruited from the bone marrow to the damaged liver (Kisseleva et al., 2006). Later, fibrocyte migration in response to liver injury was investigated using bone marrow (BM) from chimeric mice expressing luciferase (Col-Luc-wt) or green fluorescent protein (Col-GFP-wt) under control of the α1(I) collagen promoter and enhancer, respectively. Migration of CD45+Col I+ fibrocytes was regulated by chemokine receptors CCR2 and CCR1. In addition to CCR2 and CCR1, egress of BM CD45+Col I+ cells was regulated by TGF-β and liposaccharide in vitro and in vivo. Interestingly, development of liver fibrosis was also increased in aged mice and correlated with high numbers of liver fibrocytes (Kisseleva et al., 2011). However, it is unknown what proportion of tissue myofibroblasts/fibrocytes are derived from bone marrow or circulating fibrocytes, whether myofibroblasts of these origins transition through a stellate cell phenotype, and what happens to activated myofibroblasts from various sources when liver injury resolves (Guo & Friedman, 2007).

5.5 Cardiovascular disease

Deposition and remodeling of connective tissue in the heart plays a critical role in cardiac repair and response to injury. Fibrosis also occurs on a reactive basis around coronary vessels (perivascular fibrosis) and in the interstitial space (Haudek et al., 2006). It is generally considered that both reactive and reparative fibrosis may contribute to adverse remodeling. A number of studies have supported the contribution of bone marrow progenitor cells or fibrocytes to remodeling in diverse areas of the cardiovascular system where fibrotic response seems to be a common feature.

In a mouse model of fibrotic ischemia/reperfusion cardiomyopathy (I/RC) it was observed a prolonged elevation of MCP-1, and concomitantly a population of small spindle-shaped fibroblasts with a distinct phenotype appeared in the sites of lesion. These cells were highly proliferative and expressed collagen I and α-smooth muscle actin as well as CD34, and CD45; these cells represented 3% of all non myocyte live cells. Haudek and coworkers confirmed the bone marrow origin of these cells creating a chimeric mice expressing lacZ; I/RC injury resulted in a large population of spindle-shaped fibroblasts containing lacZ. Interestingly, the administration of SAP in vivo markedly reduced the number of proliferative spindle-shaped fibroblasts and completely prevented I/RC-induced fibrosis and global ventricular dysfunction (Haudek et al., 2006). Similar results were reported later, in a model induced by Ang-II. Ang-II infusion resulted in the appearance of bone marrow-derived CD34+/CD45+ fibroblasts that expressed collagen type I and the cardiac fibroblast marker DDR2 while local fibroblasts were CD34−/CD45−. Genetic deletion of MCP-1 (MCP-1-deficient mice) prevented the Ang-II-induced cardiac fibrosis and the appearance of CD34+/CD45+ fibroblasts. Interestingly, Ang-II-treated hearts showed induction of types I and III collagens, TGF-β1, and TNF mRNA expression. Apparently the differentiation of a CD34+/CD45+ fibroblast precursor population in the heart is induced by Ang-II and mediated by MCP-1 (Haudek et al., 2010).

Neointimal hyperplasia is a common feature of various cardiovascular diseases such as atherosclerosis, postangioplasty restenosis and transplant arteriopathy. Neointima usually consists of smooth muscle cells and deposited extracellular matrix. In an in vivo ovine model of carotid artery synthetic patch graft, circulating leukocytes were shown to express collagen and α-SMA. Importantly, these cells also expressed markers unique to fibrocytes (CD34, CD45, vimentin; Varcoe et al., 2006), suggesting an association between intimal hyperplasia and fibrocyte migration. In other work performed in a rat model of transplant vasculopathy, accelerated transplant vasculopathy was associated with increased levels of host-endothelial chimerism and increased neointimal smooth muscle cell proliferation; moreover, accelerated transplant vasculopathy was associated with increased frequency of circulating CD45+vimentin+ fibrocytes (Onauta et al., 2009).

CD34+ fibrocyte-like cells are detectable in normal mitral valves. In cases of myxomatous degeneration CD34+ fibrocytes make up the majority of mitral valve stromal cells (Barth et al., 2005). Since major factors in the development of myxomatous valve degeneration are the MMP-9 and collagen I and III, which are secreted by CD34+ fibrocytes, they propose that these cells might be involved in the pathogenesis of myxomatous mitral valve (Barth et al., 2005).

5.6 Skin disease

Fibrocytes are thought to play a role in skin repair by several mechanisms such as the secretion of ECM, antigen presentation, cytokine production, angiogenesis, and wound closure (Metz, 2003). After the original work by Bucala, several groups examined the participation of fibrocytes in the wound healing process. Mori and coworkers examined the phenotype of fibrocytes and myofibroblasts present in the wounded skin of BALB/c mice and observed that during wound healing, between 4 and 7 days post-wounding, more than 50% of the cells present at the site of injury were CD13+/collagen I+ fibrocytes that could be isolated at an early stage of the healing process from digested fragments of wounded tissue by fluorescence-activated cell sorting (Mori et al., 2005). Fibrocytes have been identified in postburn hypertrophic scar tissue but were absent from normal skin, moreover, the number of fibrocytes was higher in hypertrophic than in mature scar tissue (Yang et al., 2005). It is noteworthy that over time the expression of CD34 on fibrocytes present in these wounds decreases, whereas the expression of proline-4-hydroxylase (an enzyme involved in collagen synthesis) increases in both hypertrophic or keloid scars (Aiba and Tagami, 1997). This finding has been corroborated by other authors (Abe et al., 2001; Phillips et al., 2004) and it's an important feature to be considered for the analysis of these cells in organ fibrosis. In other words, it seems that fibrocytes, once in the tissues, progressively lose their typical markers and can be difficult to identify.

Also, the participation of fibrocytes in wound healing of human skin has been postulated as a useful marker for wound age determination in the legal pathology area. In an interesting study (Ishida et al., 2009) a double-color immunofluorescence analysis was carried out using anti-CD45 and anti-collagen type I antibodies to examine the time-dependent appearance of fibrocytes in 53 human skin wounds with different wound ages. Fibrocytes were initially observed in wounds aged 4 days, and their number increased in lesions proportionally with advances in wound age. These findings imply that human skin wounds containing fibrocytes are at least 4 days old. Moreover, a fibrocyte number of over 10 indicates a wound age between 9 and 14 days. Fibrocytes numbers, evaluated with these markers (CD45+/Col I+) showed a marked decrease from day 17 to 21 which was the longest time of evaluation, exposing the need to use other parameters to confirm the wound ages since fibrocytes numbers in day 4 were similar to numbers in day 17-21.

Yang and his group reported high percentages of fibrocytes present in the cultures of peripheral blood mononuclear cells obtained from burn patients compared with controls (89.7 +/- 7.9% versus 69.9 +/- 14.7%, $p < 0.001$) and this percentages were consistently higher in patients with more than 30% extent of burn; moreover, they found a positive correlation between the levels of serum TGF-β1 and the percentage of fibrocytes developed in the cultures of PBMC derived from these patients (Yang et al., 2002). Interestingly, it has been postulated that the principal role of fibrocytes in burn injury as well as in hypertrophic scars is the regulation of the function of local fibroblasts. Thus, dermal fibroblasts treated with conditioned medium obtained from burn patient fibrocytes, but not by those derived from normal subjects, showed an increase in cell proliferation and migration (Wang et al., 2007). Furthemore, it has been suggested that fibrocytes can be reprogrammed by changes in the culture media, and that this reprogrammed fibrocytes have the ability to increase cell proliferation and MMP-1 expression in dermal fibroblasts (Medina, A & Ghahar, A. 2010). These findings have opened a new research line worthy of follow up.

5.7 Nephrogenic systemic fibrosis

Fibrocytes have been also identified in the skin of patients with cutaneous fibrosing diseases, such as nephrogenic systemic fibrosis. Nephrogenic systemic fibrosis (NSF) is a recently described cutaneous fibrosing disorder that exhibits pathologic similarities with scleroderma but occurs exclusively in patients with renal insufficiency who have received gadolinium containing magnetic resonance contrast agents. The onset of the disease varies from days to several months following exposure to gadolinium-based contrast. It is a debilitating disease characterized by the development of discolored plaques on the skin of the extremities and trunk. Over time, contractures develop and complete loss of range of motion can occur (Cowper & Bucala, 2003; Cowper et al., 2008). Skin biopsies from patients with this disease have revealed an important accumulation of CD34, pro-Col-I+ fibrocytes in the dermis with abundant connective tissue matrix production; it is noteworthy that in vitro studies revealed that gadolinium may decrease the ability of endogenous mediators, such as SAP and IL-12, to inhibit fibrocyte outgrowth (Vakil et al., 2009). The reason for why fibrocytes are present in high numbers and are such a prominent feature of the dermatopathology of NSF remains unclear, but may be due to the acute and abrupt development of skin fibrosis (Bucala, 2008).

6. Opportunities for research and therapeutic targets

The study of fibrocytes and their participation in the pathogenesis of chronic inflammation and fibroproliferative diseases presents both important challenges and opportunities for researchers. To advance this field, detailed molecular characterization of these cells and establishment of defined experimental strategies in animals and humans will be necessary to catalyze progress in this area of investigation. Recent studies and emerging concepts have significantly improved our understanding of the participation of fibrocytes in health and disease and so have opened the door to new hypotheses and approaches aimed at therapeutic targets and strategies.

One of the main therapeutic targets, suggested since the initial works on fibrocyte biology research, was the serum amyloid P (SAP), a member of the pentraxin family of proteins. In this context, it was first demonstrated that SAP could inhibit the differentiation of monocytes into fibrocytes (Pilling et al., 2003). SAP binds to Fcγ receptors through which apparently mediates its anti-fibrotic activities affecting peripheral blood monocyte differentiation and activation states (Lu et al., 2008). In a rat model of bleomycin-induced lung injury it was shown that purified rat SAP could suppress development of lung fibrosis which correlated with reduced fibrocyte numbers within the lung tissue (Pilling et al., 2007). More recently, SAP ability to reduce fibrosis was tested in models of renal and lung fibrosis where this therapeutic potential was confirmed. In both models, the mechanisms through which SAP exerts its antifibrotic effect seemed to be independent of monocyte to fibrocyte differentiation (Casraño et al., 2009; Murray 2010). Further analysis of this molecule and its potential as antifibrogenic therapy is needed to identify all the mechanisms involved in its effect as well as the feasibility of its use in human disease.

Several chemokines are abundantly expressed in experimental models of fibrosis and in the human disease (Agostini & Gurrieri 2006). Regarding fibrocytes, several studies have focused on the role of chemokines in recruiting these cells to the injured lung. In human IPF, the

CXCL12/CXCR4 axis may be of particular significance (Andersson-Sjoland et al., 2008). As mentioned human circulating fibrocytes express CXCR4 and α-SMA, and can traffic toward the unique CXCR4 ligand, CXCL12 (Mehrad et al., 2007; Andersson-Sjoland et al., 2008). Supporting a major role of this axis in the lung disease, it was demonstrated that the administration of neutralizing anti-CXCL12 antibodies to bleomycin-treated mice resulted in a significant reduction of fibrocyte lung homing and collagen deposition, but interestingly without affecting the numbers of other leukocyte populations in the lungs (Phillips et al., 2004). These data suggest that blocking or interfering with chemokine/chemokine receptor networks may help to diminish or stop fibrocyte recruitment in fibrotic lung disorders. Recently, two groups have explored this hypothesis. Xu et al., 2007 used an antagonist of the receptor CXCR4 (TN14003) in a model of bleomycin-induced pulmonary fibrosis. Intraperitoneal treatment of mice with TN14003 attenuated the development of lung fibrosis and blocked in vitro migration of bone marrow derived stem cells towards CXCL12 or lung homogenates of bleomycin treated mice. Likewise, Song and coworkers showed that intraperitoneal treatment of mice with AMD3100 (Plerixafor, which is a small synthetic specific inhibitor of CXCR4), resulted in decreased levels of CXCL12 in the bronchoalveolar fluid and decreased numbers of fibrocytes in the lungs of mice treated with bleomycin (Song et al 2010). Collagen deposition and pulmonary fibrosis were also attenuated by treatment with AMD3100 (Song et al., 2010). Though the initial results seem to be optimistic, this is still an area of active research, and further studies are needed to elucidate whether pharmacologic inhibition of the CXCR4/CXCL12 axis could modify the lung fibrotic process in human disease.

The potential use of circulating fibrocytes as biomarkers in fibrosing diseases is a window of opportunity that has to be explored; diverse groups have reported differences in the percentages of circulating fibrocytes between healthy controls and patients (Mehard et al., 2007; Moeller et al., 2008; Chun-Hua et al., 2008). An increase in the percentages of circulating fibrocytes was demonstrated in a cohort of 51 patients with stable IPF, compared to healthy controls, but more important, a huge increase was observed during an acute exacerbation, a highly lethal process in IPF. Moreover, the number of fibrocytes returned to the values of stable IPF in the few patients that recovered. In general, fibrocyte numbers were an independent predictor of early mortality (Moeller et al., 2008).

However, higher number of patients should be evaluated and larger longitudinal studies should be done in order to establish if differences in percentages of circulating fibrocytes as well as changes in the percentages of circulating fibrocytes in a given patient with a given disease may predict outcome. The possibility of using differences in the percentages of circulating fibrocytes as biomarkers for disease diagnostic, outcome, or therapeutic response is an important biomedical area of research that needs attention.

Fibrocytes are progenitor cells capable to differentiate not only into myofibroblasts but also in other mesenchymal cells (Hong et al., 2005 and 2007; Choi et al., 2010). The ability of fibrocytes to undergo differentiation to osteoblasts and chondrocytes like cells when treated with specific cytokines and defined media raises the opportunity of their use for regenerative therapy related to bone or articular cartilage repair. Hypothetically, circulating fibrocytes could be isolated from the patient's own blood, processed for differentiation into osteoblasts or chondrocytes, followed by transplantation into the damaged tissue. Tissue engineering is a growing field in the biomedical sciences, and the role of fibrocytes in regenerative therapy has to be assessed with future studies in the area.

7. References

Aarabi, S; Longaker, MT & Gurtner, GC. (2007). Hypertrophic scar formation following burns and trauma: new approaches to treatment. *PLoS Medicine*, Vol 4, pp: e234 (1464-1470).

Abe, R; Donnelly, S. C.; Peng, T; Bucala, R. and Metz, C. N. (2001). Peripheral blood fibrocytes: differentiation pathway and migration to wound sites. *Journal of Immunology*. 166: 7556–7562

Agostini, C & Gurrieri, C. (2006). Chemokine/cytokine cocktail in idiopathic pulmonary fibrosis. *Proccedings of the American Thoracic Society*, Vol. 3, pp:357–363

Andersson-Sjöland, A; Garcia-de Alba, C; Nihlberg, K; Becerril, C; Ramirez, R; Pardo, A; Westergren-Thorsson, G and Selman M. (2008). Fibrocytes are a potential source of lung fibroblasts in idiopathic pulmonary fibrosis. *The International Journal of Biochemistry & Cell Biology*, Vol. 40, pp: 2129–2140

Balmelli, C, Ruggli N, McCullough, K. & Summerfield, A. (2005). Fibrocytes are potent stimulators of anti-virus cytotoxic T cells. *Journal of Leukocyte Biology*, Vol. 77, pp: 923–933

Balmelli, C; Alves, MP; Steiner, E; Zingg, D; Peduto, N; Ruggli, N; Gerber, H; McCullogh, K and Summerfield, A. (2007). Responsiveness of fibrocytes to Toll-like receptor danger signals. *Immunobiology, Vol.* 212, pp: 693–699

Barry, F; Boynton, RE; Liu, B and Murphy, JM. (2001). Chondrogenic differentiation of mesenchymal stem cells from bone marrow: differentiation-dependent gene expression of matrix components. *Experimental Cell Research*, Vol. 268, No. 2, pp: 189-200.

Barth, PJ; Köster, H & Moosdorf R. (2005). CD34+ fibrocytes in normal mitral valves and myxomatous mitral valve degeneration. *Pathology, Research & Practice*. Vol. 201, No. 4, pp: 301-304.

Broekema, M; Harmsen, MC; van Luyn, MJ; Koerts, JA; Petersen, AH; van Kooten, TG; van Goor, H; Navis, G & Popa ER.(2007). Bone marrow-derived myofibroblasts contribute to the renal interstitial myofibroblast population and produce procollagen I after ischemia/reperfusion in rats. *Journal of the American Society of Nephrology*, Vol. 18, pp: 165–175.

Bucala, R; Spiegel, L. A; Chesney, J; Hogan, M. & Cerami, A. (1994) Circulating fibrocytes define a new leukocyte subpopulation that mediates tissue repair. *Molecular Medicine, Vol.* 1, pp: 71–81

Bucala, R. (2008). Circulating fibrocytes: cellular basis for NSF. *Journal American College of Radiology*; Vol. 5, pp: 36–39

Bujak, M; Dobaczewski, M; Gonzalez-Quesada, C; Xia, Y; Leucker, T; Zymek, P; Veeranna, V; Tager, AM; Luster, AD; & Frangogiannis. NG. (2009). Induction of the CXC chemokine interferon-gamma-inducible protein 10 regulates the reparative response following myocardial infarction. *Circulatory Research*, Vol. 105, pp: 973-983.

Castaño, AP; Lin, SL; Surowy, T; Nowlin, BT; Turlapati, SA; Patel, T; Singh, A; Li, S; Lupher, ML Jr & Duffield, JS. (2009). Serum amyloid P inhibits fibrosis through Fc gamma R-dependent monocyte-macrophage regulation in vivo. *Science Translational Medicine*, Vol.1, Num. 5, pp: 5ra13.

Castielo, L; Sabatino, M; Jin, P; Clayberger, C; Marincola, FM; Krensky, AM & Stroncek, DF. (2011). Monocyte-derived DC maturation strategies and related pathways: a transcriptional view. *Cancer Immunology and Immunotherapy*, Vol. 60, No. 4, pp: 457-466.

Chesney, J; Bacher, M; Bender, A & Bucala, R. (1997) The peripheral blood fibrocyte is a potent antigen-presenting cell capable of priming naive T cells in situ. *Proceedings of the National Academy Sciences USA* Vol. 94, pp. 6307–6312.

Chesney, J; Metz, C; Stavitsky, A. B; Bacher, M & Bucala, R. (1998) Regulated production of type I collagen and inflammatory cytokines by peripheral blood fibrocytes. *Journal of Immunology*, Vol. 160, pp: 419–425

Choi, ES; Pierce, EM; Jakubzick, C; Carpenter, KJ; Kunkel, SL; Evanoff, ;, Martinez, FJ; Flaherty, KR; Moore, BB; Toews, GB; Colby, TV; Kazerooni, EA; Gross, BH; Travis, WD & Hogaboam, CM. (2007) Focal interstitial CC chemokine receptor 7 (CCR7) expression in idiopathic interstitial pneumonia. *Journal of Clinical Pathology*. Vol.59, No. 1, pp: 28-39

Choi, YH; Burdick, MD & Strieter, RM . (2010) Human circulating fibrocytes have the capacity to differentiate osteoblasts and chondrocytes. *The International Journal of Biochemistry & Cell Biology* Vol 42, pp: 662–671.

Chu, PY; Mariani, J; Finch, S; McMullen, JR; Sadoshima, J; Marshall, T & Kaye, DM. (2010). Bone marrow-derived cells contribute to fibrosis in the chronically failing heart. *American Journal of Pathology*, Vo1. 176, pp: 1735-1742.

Cowper, SE & Bucala, R. (2003). Nephrogenic fibrosing dermopathy: suspect identified, motive unclear. *American Journal of Dermatopathology*. Vol.25, pp: 358

Cowper, SE; Rabach, M & Girardi, M. (2008) Clinical and histological findings in nephrogenic systemic fibrosis. *European Journal of Radiology*, Vol. 66, pp: 191–199.

Curnow, SJ; Fairclough, M; Schmutz, C; Kissane, S; Denniston, A;, Nash, K; Buckley, CD; Lord, JM & Salmon M.. (2010) Distinct Types of Fibrocyte Can Differentiate from Mononuclear Cells in the Presence and Absence of Serum. *PLoS ONE*, Vol. 5, No. 3, pp: e9730. doi:10.1371/journal.pone.0009730

Direkze, NC; Forbes, SJ; Brittan, M; Hunt, T; Jeffery, R; Preston, SL; Poulsom, R; Hodivala-Dilke, K; Alison, M & Wright, NA. (2003) Multiple organ engraftment by bone-marrow-derived myofibroblasts and fibroblasts in bone-marrow-transplanted mice. *Stem Cells, Vol* 21, No. 5, pp: 514-20.

Direkze, NC; Hodivala-Dilke, K; Jeffery, R; Hunt, T; Poulsom, R; Oukrif, D; Alison, MR & Wright, NA. (2004). Bone marrow contribution to tumor-associated myofibroblasts and fibroblasts. *Cancer Research*. Vol. 64, No.23, pp: 8492-8495.

Ebihara, Y; Masuya, M; Larue, AC; Fleming, PA; Visconti, RP; Minamiguchi, H; Drake, CJ & Ogawa, M: (2006) Hematopoietic origins of fibroblasts: II. In vitro studies of fibroblasts, CFU-F, and fibrocytes. *Experimental Hematology*, Vol 34, No. 2, pp: 219-229

El-Asrar, AM; Struyf, S; Van Damme, J & Geboes K. (2008). Circulating fibrocytes contribute to the myofibroblast population in proliferative vitreoretinopathy epiretinal membranes. *Brithish Journal Ophthalmology*, Vol. 92, pp:699-704

Forbes, SJ; Russo, FP; Rey, V; Burra, P; Rugge, M; Wright, NA & Alison MR. (2004) A significant proportion of myofibroblasts are of bone marrow origin in human liver fibrosis. *Gastroenterology, Vol.* 126, No. 4, pp: 955-963

García-de-Alba, C,; Becerril, C; Ruiz, V; González, Y; Reyes, S; García-Alvarez, J; Selman, M & Pardo A. (2010) Expression of matrix metalloproteases by fibrocytes: possible role in migration and homing. *American Journal of Respiratory and Critical Care Medicine.* Vol. 182, No. 9, pp: 1144-1152.

Gunn, MD; Tangemann, DK; Tam, C; Cyster, JG; Rosen, SD & Williams LT. (1998). A chemokine expressed in lymphoid high endothelial venules promotes the adhesion and chemotaxis of naïve T lymphocytes. *Proceedings of the National Academy Sciences USA, Vol.* 95, pp: 258–263.

Gurtner, GC; Werner, S; Barrandon, Y & Longaker, MT. (2008) Wound repair and regeneration. *Nature,* Vol 453, pp: 314-321.

Hamid, Q & Tulic, M. (2009). Immunobiology of Asthma. *The Annual Review of Physiology, Vol. 71, pp: 489–507*

Hartlapp, I; Abe, R; Saeed, R. W; Peng, T; Voelter, W; Bucala R. & Metz, CN. (2001). Fibrocytes induce an angiogenic phenotype in cultured endothelial cells and promote angiogenesis in vivo. *FASEB J.* 15: 2215–2224

Hashimoto, N; Jin, H; Liu, T; Chensue, SW & Phan SH.(2004) Bone marrow-derived progenitor cells in pulmonary fibrosis. *Journal of Clinical Investigation, Vol.* 113, pp: 243–252.

Haudek, SB; Xia, Y; Huebener, P; Lee, JM; Carlson, S; Crawford, JR; Pilling, D; Gomer, RH; Trial, J; Frangogiannis, NG & Entman, ML. (2006). Bone marrow-derived fibroblast precursors mediate ischemic cardiomyopathy in mice. *Proccedings of the National Academy of Sciences USA*, Vol. 103, pp:18284-18289.

Haudek, SB; Cheng, J; Du, J; Wang, Y; Hermosillo-Rodriguez, J; Trial, J; Taffet, GE & Entman ML. (2010) Monocytic fibroblast precursors mediate fibrosis in angiotensin-II-induced cardiac hypertrophy. *Journal of Molecular Cell Cardiology,* Vol. 49, No. 3, pp:499-507.

Herzog, EL; Chai, L & Krause, DS.(2003). Plasticity of marrow-derived stem cells. *Blood,* Vol. 102, pp: 3483–3493.

Herzog, EL & Bucala, R. (2010) Fibrocytes in health and disease. *Experimental Hematology.* Vol.38, No. 7, pp: 548-556.

Hinz, B; Phan, SH; Thannickal, VJ; Galli, A; Bochaton-Piallat, ML & Gabbiani G. (2007) The myofibroblast: one function, multiple origins. *American Journal of Pathology* , Vol. 170, No. 6, pp: 1807-1816.

Holgate, ST; Arshad, HS; Roberts, GC; Howarth, PH; Thurner, P & Davies DE. (2009) A new look at the pathogenesis of asthma. *Clinical Science.* Vol.118, No.7, pp:439-50.

Hong, KM; Belperio, JA; Keane, MP; Burdick, MD & Strieter RM. (2007). Differentiation of Human Circulating Fibrocytes as Mediated by Transforming Growth Factor-β1 and Peroxisome Proliferator-activated Receptor γ. *The Journal of Biological Chemistry* Vol. 282, No. 31, pp. 22910–22920.

Hong, KM; Burdick, MD; Phillips, RJ; Heber, D & Strieter RM. (2005). Characterization of human fibrocytes as circulating adipocyte progenitors and the formation of human adipose tissue in SCID mice. *FASEB Journal. Vol.* 14, pp:2029-2031

Ishida, Y; Kimura, A; Takayasu, T; Eisenmenger, W & Kondo T. (2009). Detection of fibrocytes in human skin wounds and its application for wound age determination. *International Journal of Legal Medicine,* Vol. 123, pp: 299–304

Kisseleva, T; Uchinami, H; Feirt, N; Quintana-Bustamante, O; Segovia, JC; Schwabe, RF & Brenner, DA. (2006). Bone marrow-derived fibrocytes participate in pathogenesis of liver fibrosis. *Journal of Hepatology;* Vol. 45, pp: 429–438

King, TE Jr; Pardo, A & Selman M. (2011). Idiopathic Pulmonry Fibrosis. *The Lancet. PMID: 21719092*

Knothe Tate, ML; Falls, TD; McBride, SH; Atit, R & Knothe UR. (2008) Mechanical modulation of osteochondroprogenitor cell fate. *International Journal of Biochemestry and Cell Biology.* Vol 40, No.12, pp: 2720-2738.

Lochter, A; Galosy, S; Muschler, J; Freedman, N; Werb, Z & Bissell, MJ. (1997). Matrix metalloproteinase stromelysin-1 triggers a cascade of molecular alterations that leads to stable epithelial-to-mesenchymal conversion and a premalignant phenotype in mammary epithelial cells. *Journal of Cell Biology, Vol.* 139, pp:1861–1872.

Lu, J; Marnell, LL; Marjon, KD; Mold, C; Du Clos, TW & Sun P. (2008). Structural recognition and functional activation of FcgammaR by innate pentraxins. *Nature,* Vol. 456, No. 7224, pp: 989-92

Medina, A & Ghahar, A. (2010) Fibrocytes can be reprogrammed to promote tissue remodeling capacity of dermal fibroblasts. *Mollecular Cell Biochemestry,* Vol. 344, pp:11–21.

Mehrad, B; Burdick, MD; Zisman, DA; Keane, MP; Belperio, JA & Strieter RM. (2007). Circulating peripheral blood fibrocytes in human fibrotic interstitial lung disease. *Biochemestry Biophysics Research Community;* Vol. 353, pp:104–108

Mehrad, B; Burdick, MD & Strieter RM. (2009). Fibrocyte CXCR4 regulation as a therapeutic target in pulmonary fibrosis. *International Journal of Biochemestry and Cell Biology,* Vol. 41, pp: 1708-1718.

Metz, C.. (2003). Fibrocytes: a unique cell population implicated in wound healing. *Cellular and Molecular Life Sciences,* Vol. 60, pp: 1342–1350.

Moeller, A; Gilpin, S; Ask, K; Cox, G; Cook, D; Gauldie, J; Margetts, P; Farkas, L; Dobranowski, J; Boylan, C; O'Byrne, P; Strieter, R & Kolb M. (2009) Circulating Fibrocytes Are an Indicator of Poor Prognosis in Idiopathic Pulmonary Fibrosis. *American Journal of Respiratory and Critical Care Medicine,* Vol. 179, pp: 588–594,

Moore, BB; Kolodsick, JE; Thannickal, VJ; Cooke, K; Moore, TA; Hogaboam, C; Wilke, CA & Toews GB. (2005). CCR2-mediated recruitment of fibrocytes to the alveolar space after fibrotic injury. *American Journal of Pathology,* Vol. 166, pp: 675–684.

Moore, BB; Murray, L; Das, A; Wilke, CA; Herrygers, AB & Toews, GB.(2006). The role of CCL12 in the recruitment of fibrocytes and lung fibrosis. *American Journal of Respiratory Cell and Molecular Biology.* Vol. 35 No. 2, pp: 175-181

Moore, B. (2009). Fibrocytes as Potential Biomarkers in Idiopathic Pulmonary Fibrosis. *International Journal of Respiratory and Critical Care Medicine.* Vol 179, pp: 524-525

Mori, L; Bellini, A; Stacey, MA; Schmidt, M & Mattoli S. (2005) Fibrocytes contribute to the myofibroblast population in wounded skin and originate from the bone marrow. *Experimental Cell Research,* Vol 304, pp: 81–90

Murray, LA; Rosada, R; Moreira, AP; Joshi, A; Kramer, MS, Hesson, DP; Argentieri, RL; Mathai, S; Gulati, M; Herzog, EL & Hogaboam CM. (2010) Serum Amyloid P Therapeutically Attenuates Murine Bleomycin-Induced Pulmonary ia Its Effects on

Macrophages. *PLoS ONE,* Vol. 5, No. 3, pp: e9683. doi:10.1371/journal.pone. 0009683

Nihlberg, K; Larsen, K; Hultgårdh-Nilsson, A; Malmström, A; Bjermer, L & Westergren-Thorsson G. (2007) Tissue fibrocytes in patients with mild asthma: a possible link to thickness of reticular basement membrane? *Respiratory Research.* Vol.7, pp:50

Noe, V; Fingleton, B; Jacobs, K; Crawford, HC; Vermeulen, S; Steelant, W; Bruyneel, E; Matrisian, LM & Mareel M. (2001). Release of an invasion promoter E-cadherin fragment by matrilysin and stromelysin-1. *Journal of Cell Sciences* Vol. 114, pp:111-118.

Oikonomidi, S; Kostikas, K; Tsilioni, I; Tanou, ;, Gourgoulianis, KI & Kiropoulos, TS.(2009). Matrix metalloproteinases in respiratory diseases: from pathogenesis to potential clinical implications. *Current Medical Chemestry, Vol.* 16, pp: 1214–1228.

Okona, K; Szumera, A; & Kuzniewski M. (2003). Are CD34+ cells found in renal interstitial fibrosis? *American Journal of Nephrology,* Vol. 23, pp: 409–414.

Onuta, G; van Ark, J; Rienstra, H; Boer, MW; Klatter, FA; Bruggeman, CA; Zeebregts, CJ; Rozing, J & Hillebrands JL. (2010). Development of transplant vasculopathy in aortic allografts correlates with neointimal smoothmuscle cell proliferative capacity and fibrocyte frequency. *Atherosclerosis,* Vol. 209, pp: 393-402

Pardo, A & Selman, M. (2002). Molecular mechanisms of pulmonary fibrosis. *Frontiers in Biosciences.* Vol. 7, pp: d1743-1761

Pardo, A& Selman, M. (2006) Matrix metalloproteases in aberrant fibrotic tissue remodeling. *Proceedings of the American Thoracic Society;* Vol. 3, No. 4, pp: 383-388

Petrakis, NL; Davis, M & Lucia, SP. (1961) The in vivo differentiation of human leukocytes into histiocytes, fibroblasts and fat cells in subcutaneous diffusion chambers. *Blood,* Vol 17, pp: 109-18.

Pilling, D; Buckley, CD; Salmon, M & Gomer, RH. (2003). Inhibition of fibrocyte differentiation by serum amyloid P. *Journal of Immunology,* Vol. 171, No. 10, pp: 5537-5546

Pilling, D; Roife, D; Wang, M; Ronkainen, SD; Crawford, JR; Travis, EL & Gomer RH. (2007). Reduction of bleomycin-induced pulmonary fibrosis by serum amyloid P. *Journal of Immunology, Vol.* 179, pp: 4035–4044

Pilling, D; Vakil, V & Gomer RH. (2009). Improved serum-free culture conditions for the differentiation of human and murine fibrocytes. *Journal Immunology Methods, Vol.* 351, pp: 62-70

Phillips, RJ; Burdick, MD; Hong, K; Lutz, MA; Murray, LA; Xue, YY, Belperio, JA; Keane, MP & Strieter RM.(2004). Circulating fibrocytes traffic to the lungs in response to CXCL12 and mediate fibrosis. *Journal of Clinical Investigation,* Vol. 114, pp: 438–46.

Prikk, K; Maisi, P; Pirila, E; Sepper, R; Salo, T; Wahlgren, J & Sorsa T. (2001). In vivo collagenase-2 (MMP-8) expression by human bronchial epithelial cells and monocytes/macrophages in bronchiectasis. *Journal of Pathology,* Vol. 194, pp: 232–238

Rabkin, E; Aikawa, M; Stone, JR; Fukumoto, Y; Libby, P & Schoen, FJ. (2001). Activated interstitial myofibroblasts express catabolic enzymes and mediate matrix remodeling in myxomatous heart valves. *Circulation,* Vol. 104, pp: 2525-2532.

Rival Y, Stennevin A, Puech L, Rouquette A, Cathala C, Lestienne F, Dupont-Passelaigue E, Patoiseau JF, Wurch T, Junquéro D. (2004) Human adipocyte fatty acid-binding protein (aP2) gene promoter-driven reporter assay discriminates nonlipogenic peroxisome proliferator-activated receptor gamma ligands. *Journal of Pharmacology and Experimental Therapy*. Vol 311, No. 2, pp:467-475

Roufosse, C; Bou-Gharios, G; Prodromidi, E; Alexakis, C; Jeffery, R; Khan, S; Otto, WR; Alter, J; Poulsom, R & Cook HT. (2006). Bone marrow-derived cells do not contribute significantly to collagen I synthesis in a murine model of renal fibrosis. *Journal of the American Society of Nephrology*. Vol.17, pp: 775–782.

Ruiz, V; Ordoñez, RM; Berumen, J; Ramírez, R; Uhal, B; Becerril, C; Pardo, A & Selman, M. (2003). Unbalanced collagenases/TIMP-1 expression and epithelial apoptosis in experimental lung fibrosis. *American Journal of Physiology Lung Cell Molecular Physiology*, Vol. 285, pp: L1026–L1036.

Sakai, N; Wada, T; Yokoyama, ; Lipp, M; Ueha, S; Matsushima, K & Kaneko S. (2006) Secondary lymphoid tissue chemokine (SLC/CCL21)/CCR7 signaling regulates fibrocytes in renal fibrosis. *Proceedings of the National Academy Sciences USA;* Vol 103, No. 38, pp:14098-14103.

Sakai, N; Wada, T; Matsushima, K; Bucala, R; Iwai, M; Horiuchi, M & Kaneko S. (2008) The renin-angiotensin system contributes to renal fibrosis through regulation of fibrocytes. *Journal of Hypertens*ion. Vol.26, No.4, pp:780-90.

Sakai, N; Furuichi, K; Shinozaki, Y; Yamauchi, H; Toyama, T; Kitajima, S; Okumura, T; Kokubo, S; Kobayashi, M; Takasawa, K; Takeda, S; Yoshimura, M; Kaneko, S & Wada, T. (2010). Fibrocytes are involved in the pathogenesis of human chronic kidney disease. *Human Pathology*. Vol; 41, No. 5, pp:672-67

Selman, M; Ruiz, V; Cabrera, S; Segura, L; Ramirez, R; Barrios, R & Pardo A. (2000). TIMP-1, -2, -3 and -4 in idiopathic pulmonary fibrosis. A prevailing non degradative lung microenvironment? *American Journal of Physiology* Vol. 279, pp: L562–L574.

Selman, M; King, TE & Pardo A. (2001) Idiopathic pulmonary fibrosis: prevailing and evolving hypotheses about its pathogenesis and implications for therapy. *Annals of Internal Medicine*, Vol. 134, No. 2, pp: 136-51

Selman, M; Rojas, M; Mora, AL & Pardo A. (2010). Aging and interstitial lung diseases: unraveling an old forgotten player in the pathogenesis of lung fibrosis. *Seminars in Respiratory and Critical Care Medicine*, Vol. 31, pp: 607-17

Seta, N & Kuwana, M (2010). Derivation of multipotent progenitors from human circulating CD14+ monocytes. *Experimental Hematology*, Vol. 38, No. 7, pp: 557-563.

Singer, AJ & Clark RA. (1999) Cutaneous wound healing. *New England Journal of Medicine*, Vol 341, pp 738–746

Shao, DD; Suresh, R; Vakil, V; Gomer, RH & Pilling D. (2008) Pivotal Advance: Th-1 cytokines inhibit, and Th-2 cytokines promote fibrocyte differentiation influence fibrocyte differentiation. *Journal of Leukocyte Biology, Vol.* 83, pp: 1323–1333

Schmidt, M; Sun, G; Stacey, MA; Mori, L & Mattoli S. (2003). Identification of Circulating Fibrocytes as Precursors of Bronchial Myofibroblasts in Asthma. *Journal of Immunology*, Vol. 170, pp: 380–389.

Scholten, D; Reichart, D; Paik, YH; Lindert, J; Bhattacharya, J; Glass, CK; Brenner,DA & Kisseleva T. (2011) Migration of Fibrocytes in Fibrogenic Liver Injury. *American Journal of Pathology*, Vol. 179, pp: 189–198

Song, JS; Kang, CM; Kang, HH; Yoon, HK; Kim, YK; Kim, KH; Moon, HS & Park SH. (2010). Inhibitory effect of CXC chemokine receptor 4 antagonist AMD3100 on bleomycin induced murine pulmonary fibrosis. *Experimental Mollecular Medicine*, Vol.42, pp: 465-472

Strieter, RM; Keeley, EC; Hughes, MA; Burdick, MD & Mehrad B. (2009) The role of circulating mesenchymal progenitor cells (fibrocytes) in the pathogenesis of pulmonary fibrosis. *Journal of Leukocyte Biology, Vol* 86, No. 5, pp: 1111-1118.

Strutz, F & Zeisberg, M. (2006). Renal fibroblasts and myofibroblasts in chronic kidney disease. Journal of the American Society of Nephrology, Vol. 17, pp:2992–2998.

Swiderski, RE; Dencoff, JE; Floerchinger, CS; Shapiro, SD & Hunninghake, GW. (1998). Differential expression of extracellular matrix remodeling genes in a murine model of bleomycin-induced pulmonary fibrosis. *American Journal of Pathology*, Vol. 152, pp: 821–828

Varcoe, RL; Mikhail, M; Guiffre, A;, Pennings, G; Vicaretti, M; Hawthorne, WJ; Fletcher, JP & Medbury, HJ. (2006). The role of the fibrocyte in intimal hyperplasia. *Journal of Thrombosis and Haemostasis*,Vol. 4, pp:1125-1133.

Vakil, V; Sung, JJ; Piecychna, M; Crawford, JR; Kuo, P; Abu-Alfa, AK; Cowper, SE; Bucala, R & Gomer RH. (2009). Gadolinium-containing magnetic resonance image contrast agent promotes fibrocyte differentiation. *Journal of Magnetic Resonance and Imaging*. Vol. 30, No. 6, pp: 1284-8.

Wang, JF; Jiao, H; Stewart, TL; Shankowsky, HA; Scott, PG & Tredget EE.(2007) Fibrocytes from burn patients regulate the activities of fibroblasts. *Wound Repair and Regeneration, Vol.* 15, pp: 113–121

Wang, J; Jiao, H; Stewart, TL; Shankowsky, HA; Scott, PG, & Tredget EE. (2007). Improvement in postburn hypertrophic scar after treatment with IFN-alpha2b is associated with decreased fibrocytes. *Journal of Interferon Cytokine Research*. Vol. 11, pp: 921-930.

Wu, Y & Madri J. (2010). Insights into monocyte-driven osteoclastogenesis and its link with hematopoiesis: regulatory roles of PECAM-1 (CD31) and SHP-1. *Critical Review in Immunology*, Vol. 30, No. 5, pp: 423-33.

Xu, J; Mora, A; Shim, H; Stecenko, A; Brigham, KL & Rojas M. (2007) Role of the SDF-1/CXCR4 axis in the pathogenesis of lung injury and fibrosis. *Am J Respir Cell Mol Biol.*, Vol. 37, No. 3, pp:291-299.

Xu, J; Gonzalez, ET; Iyer, SS; Mac, V; Mora, AL; Sutliff, RL; Reed, A,; Brigham, KL; Kelly, P & Rojas M. (2009) Use of senescence-accelerated mouse model in bleomycin-induced lung injury suggests that bone marrow-derived cells can alter the outcome of lung injury in aged mice. *Journal of Gerontology A Biological Science Medical Science*. Vol. 64, No. 7, pp: 731–739

Yang, L,; Scott, PG; Giuffre, J; Shankowsky, HA; Ghahary, A & Tredget, EE. (2002). Peripheral blood fibrocytes from burn patients: identification and quantification of fibrocytes in adherent cells cultured from peripheral blood mononuclear cells. *Laboratory Investigation*. Vol. 82, No. 9, pp: 1183-1192.

Yang, L; Scott, PG; Dodd, C; Medina, A; Jiao, H; Shankowsky, HA; Ghahary, A & Tredget, EE. (2005). Identification of fibrocytes in postburn hypertrophic scar. *Wound Repair And Regeneration*. Vol.4, pp:398-404.

Yu, Q & Stamenkovic I. (2000). Cell surface-localized matrix metalloproteinase-9 proteolytically activates TGF-b and promotes tumor invasion and angiogenesis. *Genes Development, Vol.* 14, pp: 163–176.

Zulli, A; Buxton, BF; Black, MJ & Hare, DL.. (2005). CD34 Class III positive cells are present in atherosclerotic plaques of the rabbit model of atherosclerosis. *Histochemestry and Cell Biology*, Vol. 124, pp: 517-522.

Zuo, F; Kaminski , N; Eugui, E; Allard, J; Yakhini, Z; Ben-Dor, A; Lollini, L; Morris, D; Kim, Y; DeLustro, B ; Sheppard D; Pardo, A; Selman, M & Heller, RA. (2002). Gene expression analysis reveals matrilysin as a key regulator of pulmonary fibrosis in mice and humans. *Proccedings of the National Academy of Sciences USA*, Vol.99, pp: 6292–6297.

Part 4

Hematopoietic Stem Cell Therapy

16

Hematopoietic Stem Cells Therapeutic Applications

Carla McCrave
Children's Mercy Hospital in Kansas City, MO, USA

1. Introduction

Hematopoietic stem cell transplantation (HSCT) has become an established treatment for malignant hematological diseases, solid malignancies and non-malignant diseases (figure 1). Newer indications for HSCT have emerged because of better understanding of human immunology, tumor biology and immunotherapy (table 1). Novel approaches have resulted in increase number of transplants as well as significant reductions in the morbidity and mortality associated with HSCT. These include more suitable donors with the addition of unrelated cord blood units (single & double) and partially matched family members; and novel conditioning regimens (reduced & non-myeloablative) that allow patients with significant co-morbidities to undergo transplantation. On the other hand, the introduction of alternative therapies, such as imatinib (tyrosine kinase inhibitor) for chronic myelogenous leukemia (CML), has challenged well established indications. This chapter summarizes the current indications for HSCT in pediatrics and address recent clinical developments in the field of HSCT.

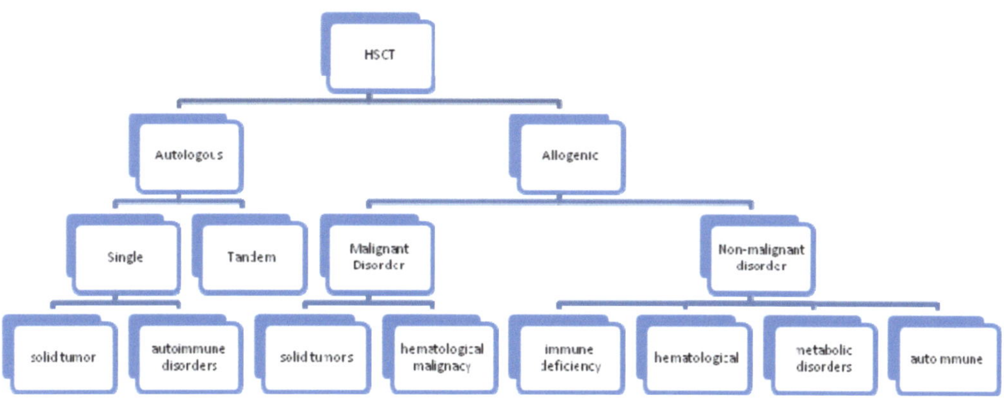

Fig. 1. Indications for HSCT

ALL • In CR 1[a] • In CR 2 • In CR 3 or further AML in CR I or further CML Myelodysplastic syndromes Hodgkin and non-Hodgkin lymphoma Selected types of solid tumors[b] Bone marrow failure syndromes (acquired & congenital) Thalassemia major Sickle cell disease Infantile malignant osteopetrosis SCID	Immunodeficiency with hyper IgM Leukocyte adhesion deficiency Omenn syndrome Chediak-Higashi syndrome X-linked lymphoproliferative disease Kostmann syndrome Chronic granulomatosis disease Glanzmann thromboasthenia Bernard-Soulier syndrome Familial hemophagocytic lymphohistiocytosis Selected types of mucopolysaccharidoses, Selected types of peroxisomal and lysosomal disorders Selected types of life-threatening autoimmune disorders resistant to conventional treatments

Abbreviations: ALL= acute lymphoblastic leukemia; AML= acute myeloblastic leukemia; CML= chronic myeloid leukemia; CR1, 2, 3= first, second and third complete remission; SCID= severe combined immunodeficiency.

[a]Patients at high risk of recurrence (that is, t (9; 22) or t (4; 11); T-ALL with poor prednisone response, high levels of minimal residual disease).

[b]Stage IV neuroblastoma, renal cell carcinoma, very high risk Ewing sarcoma.

Table 1. Main Indications to allogeneic hematopoietic SCT in childhood

2. Indications for hematopoietic stem cell transplantation (HSCT) in pediatrics

There are two types of HSCT: autologous and allogeneic. Autologous HSCT consists of removal, storage and reinfusion of patients own hematopoietic stem cells as a way to restore the patient's depleted bone marrow after high dose myeloablative therapy (figure 2). Allogeneic HSCT consists of transferring both immature and mature blood cells to a patient from the bone marrow, peripheral blood or umbilical cord blood of a sibling, relative or an unrelated donor (figure 2) as a way to restore the patients bone marrow with a new immune system after a conditioning regimen (non-myeloablative or myeloablative chemotherapy). The success of an allo-HSCT is limited by the toxicity associated with the conditioning regimens, graft versus host disease (GVHD) and the development of opportunistic infections. New concepts and interventions over the last two decades have resulted in reduction of the morbidity and mortality associated with allo-HSCT. These include the utilization of reduced intensity regimens, more effective GVHD prophylaxis, new sources of progenitor hematopoietic stem cells, donor lymphocyte infusions and better prophylaxis and treatment for infectious diseases.

The decision to transplant or not to transplant should be determined on individual basis and several factors should be considered including the disease status, age, prior treatments and responses, donor availability and evolving alternative therapies.

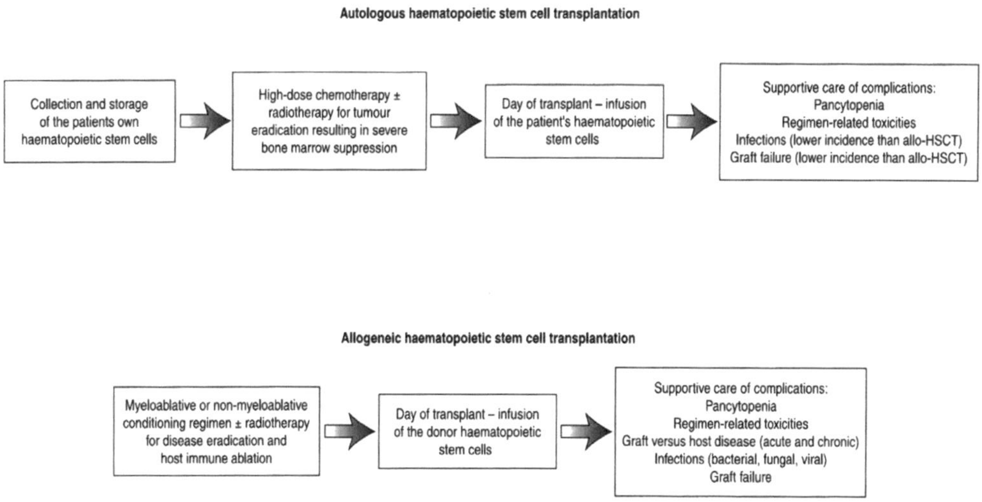

Fig. 2. Major Key Steps of HSCT.

2.1 Leukemias

2.1.1 Acute myeloid leukemia (AML)

Despite intensive chemotherapy, less than half of all patients with AML will survive in the long term (Creutzin, 2005; Gibson, 2005). Treatment outcome of pediatric AML is not as favorable as in ALL. AML treatment failure is due primarily to disease recurrence, although treatment-related mortality remains an important cause of treatment failure. Improvement in AML outcomes have been due primarily to intensification of therapy and improved supportive care guidelines. In AML, treatment intensity is an important determinator of outcome, and many studies have focused on the role of HSCT as post-remission intensification, utilizing both autologous as well as allogeneic HSCT. Allogeneic HSCT may provide a graft versus leukemia effect in pediatric AML. This is supported by a study from Bader et al that showed that preemptive immunotherapy following HSCT in patients with increasing (mixed chimerism) may lead to improved outcome. In another study Neudorf et al reported that children treated with allogeneic-HSCT in the children's cancer group 2891 study who developed acute graft versus host disease (GVHD) had fewer relapses (Bader, 2004; Neudorf et al., 2004).

The American society of bone marrow transplant position statement for the treatment of AML in children indicates that allogeneic HSCT should be recommended in the first complete remission because transplant has better overall survival and leukemia-free survival compared with chemotherapy alone (ASBMT, 2007; Oliansky, 2007). However, the role of allogeneic-HSCT in complete remission one (CR1) is declining because of the better outcome with modern multiagent chemotherapy and better methods of identifying patients that have low risk features at diagnosis and therefore are more likely to be cured with conventional chemotherapy. Recent AML trials (MRC-AML-12 & AML 0531) have shown that prognostic factors like cytogenetic and response to induction therapy are highly predictive of determining patients that are high risk at diagnosis and therefore would benefit from allogeneic-HSCT in CR1, while sparing lower risk patients the potential toxicities associated with an allogeneic-

HSCT (Ljungman, 2009). Recent analysis by several cooperative groups has now identified relapse risk group parameters based on cytogenetics abnormalities and early response to treatment : Low risk is defined as inversion (16)/t(16;16) or t(8,21). Down syndrome patients are also included in this low risk group; High risk is defined as monosomy 7, monosomy 5,5q deletions, or greater than 15% blasts at the end of induction I but who achieve complete remission after induction II, or high FLT3-ITD alleic ratio; Intermediate risk includes all other patients with no cytogenetic information available. This risk group is used to determine which patients should receive a HSCT in CR1.

Currently, HSCT is not recommended as frontline therapy for low-risk patients with AML in CR1, as they have an overall survival of 60% with conventional chemotherapy and HSCT has not been demonstrated to improve outcome for patients in CR1 (Gibson, 2005). HSCT is also not indicated for Myeloid Leukemia of Down Syndrome because HSCT is associated with excess toxicity with or without therapeutic gain (Lange et al., 1998). In addition, HSCT is also not indicated for acute promyelocytic leukemia (APL) due to excellent cure rates with conventional chemotherapy. However, for the few patients with APL who relapse or have persistent minimal residual disease, the prognosis is less favorable and HSCT might be a recommended choice (Oliansky et al., 2007). Allogeneic-HSCT from an HLA-identical sibling is an option for patients defined as intermediate risk. Allogeneic-HSCT from an HLA-identical sibling or an unrelated donor in CR1 is indicated for children with high risk AML including infant AML, therapy-related AML and children with M0 or M7 as it was proven to be more efficient than chemotherapy in some comparative studies with an event free survival ranging from 55 to 72% (Gibson, 2005). Regarding the use of haploidentical HSCT for AML, results in children with AML undergoing haploidentical HSCT have shown some effect of natural killer alloreactivity, suggesting that haploidentical HSCT may have a role in early phase very high AML patients (Marks et al., 2006).

HSCT also has an important role in the treatment of relapsed AML because outcome is poor with chemotherapy alone. Marrow transplantation in early first untreated relapse or CR2 results in a two-year EFS rate of 30-40%(Besinger,1995; Schimitz, 1998). Analyzes that attempt to compare outcome based on treatment have shown a survival advantage for patients who receive marrow transplants compared with chemotherapy alone, particularly for patients with longer first remission (Besinger, 1996).Therefore, allogeneic-HSCT from an unrelated or related donor is indicated in children with relapse AML in CR2, as it may provide long-term survival, particularly those in first relapse that are in remission.

Autologous HSCT has been used as consolidation in children with AML in CR1 after induction therapy and represents a valid alternative for high-risk children lacking a matched sibling donor. Nevertheless, results of pediatric studies comparing autologous HSCT with chemotherapy are conflicting. The use of peripheral blood stem cells in children with AML given autologous HSCT is infrequent. Further prospective clinical trials are needed to address the pivotal clinical question of whether autologous HSCT is better than chemotherapy or allograft as consolidation treatment for childhood AML in first CR (Miano et al., 2007).

2.1.2 Acute lymphoblastic leukemia (ALL)

ALL is not a uniform disease, but consists of different subtypes with different clinical prognostic and cytogenetic features. The prognosis of childhood ALL has improved dramatically over the past quarter of a century. Currently, over 2500 children in the United States are diagnosed each year with ALL and almost 95% attain a clinical remission after three or four drug induction chemotherapy (Clavell, 1986; Pui, 1998; Reiter, et al., 1994; Rivera, 1993). Over 83% of children with newly diagnosed ALL treated with multi-agent chemotherapy with or without clinical radiotherapy are alive and disease free at 5 years (Gaynon, 2000; Silverman, 2001; Vilmer, 2000).

Despite recent advances in the diagnosis and treatment of childhood ALL, there are several subpopulations of patients that have molecular biological markers or chromosomal abnormalities and biological factors that include poor prednisone response and resistance to initial chemotherapy including persistence minimal residual disease, that makes them very high risk of failing current multi-agent chemotherapy regimens. These very high risk patients require alternative treatment strategies to prevent progression and/or relapse of their disease (Kersey, 1997; Pui,1995). Table 2 defines the very high risk ALL patients.

The indication for HSCT from a match sibling or an unrelated donor for children with ALL in CR1 is limited to the subpopulation of patients that have clinical and biological features that identifies them as very high risk of relapse, as most studies quote an event-free survival (EFS) of less than 50% and a relapse rate of up to 50% (Reiter et al., 1994; Rivera, 1993). Children's oncology group conducted a clinical research study from 1993 to 1996 to investigate the toxicity and efficacy of HSCT in newly diagnosed children with very high risk features of ALL at diagnosis and/or during initial induction chemotherapy and their findings support the current indication of HSCT for very high risk ALL in CR1, especially patients with primary induction failure and Philadelphia chromosome positive ALL (Satwani, 2007).

HSCT should also be considered as an option for relapse ALL. The decision to perform an allogeneic matched related or unrelated donor HSCT for patients with relapse ALL depends on many factors which can be considered strong predictors of outcome as suggested from a number of literature reports. Different sites of relapse and the duration of first remission may be the most important factors predicting outcome after a first relapse. Patients with late relapse (over 6 months from therapy withdrawal) may have relatively good outcome with conventional chemotherapy alone (Borgmann et al., 1995; Ritchey, 1999; Uderzo et al., 1990). In contrast, children who relapse (isolated/combined medullary) during therapy or within 30 months of diagnosis seem to benefit more from HSCT than chemotherapy with an event-free survival rates of 40-50% reported for patients in CR2 who underwent a HSCT (Kawakami et al.,1990).

It has been difficult to compare outcomes of patients treated with chemotherapy or HSCT, since patient populations are not necessarily equivalent. Patients with aggressive disease die earlier and may not be included in studies of marrow transplantation, resulting in selection bias (Tichelli et al., 1999). To address this question, matched-pair analyses have been performed for ALL CR2 patients treated with chemotherapy or HSCT (Dreger et al., 1997; Novotny et al., 1998). For patients with early first relapse, HSCT resulted in significantly better EFS rates at 5 years compared with chemotherapy alone (40% vs 17%; $p<0.001$) (Novotny et al., 1998). Marrow transplantation was associated with a reduced risk of relapse that was not negated by increased treatment related deaths. The difference between chemotherapy and HSCT for patients who experienced a late marrow relapse (45% DFS vs

65%) (Chessells et al., 1986; Hoogerbrugge et al., 1995) was evident but not statistically significant.

Another factor to consider when deciding whether HSCT is an option for relapse ALL is the phase of leukemia at the time of transplant because it is also highly predictive for the risk of leukemia relapse and death from non-relapse causes. In particular, patients transplanted in relapse with over 30% circulating blast, have very poor survival following HSCT (Kessinger, 1989). Patients transplanted in remission compared to those in relapse have a two to five fold reduction in risk of relapse (p=0.0001) (36).

In summary the current opinion is that the earlier the relapse the more difficult is to obtain and maintain a second complete remission, so HSCT should be consider as an elective therapeutic option in order to eradicate a resistant disease. Relapse patients who fail to achieve remission prior to transplant have very poor outcome, so HSCT should not be undertaken.

Any one or more of the following:
- Cytogenetics t(9;22) (q34, q11) or BCR-ABL molecular rearrangement t(4;11) (q21, q23) or 11q23 molecular rearrangement Hypodiploidy (≤44 chromosomes)
- Age ≥10 years and WBC ≥200 x10^9 /L
- Induction failure (day 28 M2 or M3 BM)
- Infant ALL (2-12 months) with any one or more of the following: CD10 negative (CALLA) ALL phenotype WBC ≥100 x10^9 /L at diagnosis Day 14 M2 or M3 BM

Table 2. Ultra High-Risk Criteria of Childhood ALL in CR1.

2.1.3 Chronic myelogenous leukemia (CML)

CML is rare in childhood and accounts for less than 10% of all childhood leukemia. The treatment of CML has undergone dramatic changes in recent years. Before introduction of HSCT, the standard treatment approach for chronic phase CML was single-agent chemotherapy such as busulfan, hydroxyurea and interferon-alpha, however, treatment rarely produced a true complete remission. After 1980's, allogenic-HSCT was introduced as the only curative therapy for patients with CML. Five large multi-institutional retrospective studies have shown a high rate of long-term disease free survival (55-75% after myeloablative allogeneic HSCT), but survival was accompanied by significant treatment-related mortality, especially when unrelated donor allografts were used (Creutzig, 1996; Cwynarski et al., 2003; Millot et al., 2003; Weisdorf et al., 2002). From the 1980's to 2000, allogeneic HSCT was the treatment of choice for younger patients in first chronic phase if an HLA-matched donor was available. Before 1999, CML was the most frequent indication for allogeneic HSCT worldwide. With the approval of imatinib by the FDA in 2001, this tyrosine kinase inhibitor soon became the frontline therapy for newly diagnosed CML patients and transplant rates in CML dropped quickly worldwide (Muramatsu et al., 2010).

Dramatic responses to oral imatinib administration were observed in adult patients with CML (Druker et al., 2001; Hughes et al., 2003). However, clinical experience with imatinib in the pediatric population is limited. Several studies have shown that treatment with imatinib has resulted in prolonged molecular response with limited drug toxicity with comparable results with those in adult patients (Millot et al., 2006). Imatinib is now implemented in the primary treatment regimen for children, but the paucity of evidence on its ability to result in permanent cure and the potential complications that may arise from long-term treatment with imatinib have prevented imatinib from superseding HSCT as the primary means of curative treatment in children. The results of allogeneic HSCT in children with CML are similar to those observed in adults; HSCT-related complications such as transplant-related mortality and graft versus host disease remain significant challenges.

There is a general consensus for the need for HSCT in patients with imatinib resistance or those with advance-phase (accelerated and blast phase). (Table 3). However, issues such as when to undertake HSCT in chronic-phase CML pediatric patients or how best to treat patients who have relapsed after HSCT are still controversial. When considering HSCT vs imatinib in pediatric CML patients in early chronic phase, one must consider that the objective for treatment of childhood CML is not palliation, but cure. Hence, the possible adverse effects that stem from long-term tyrosine kinase weigh more heavily in the childhood CML population. HSCT still remains an important treatment option especially for younger patients with CML depending on physician and patient preferences. As a result of multiple clinical trials in adults that have documented great results with the use of imatinib in CML in chronic phase (87% of patients treated with imatinib showed complete cytogenetic response at 18 months with 3.3% disease progression) (O'Brian et al., 2003), this results have been applied to children, and imatinib is now also the front-line treatment for childhood CML.

World Health Organization (WHO) Criteria	International Bone Marrow Transplant Registry Criteria
Accelerated phase	Accelerated phase
1) Persistent or increasing WBC (>10×10^9/L) and/or persistent or	1) Leucocyte count difficult to control with hydroxyurea or busulfan
increasing splenomegaly	2) Rapid leucocyte doubling time (<5 days)
2) Persistent thrombocytosis (>1,000×10^9/L) uncontrolled by therapy	3) PB or marrow blasts ≥10%
3) Persistent thrombocytopenia (<100×10^9/L) unrelated to therapy	4) PB or marrow blasts and promyelocytes ≥20%
4) Clonal cytogenetic evolution occuring after the initial diagnostic karyotype	5) PB basophils and eosinophils ≥20%
5) Peripheral blood (PB) basophils ≥20%	6) Anemia or thrombocytopenia unresponsive to hydroxyurea or busulfan
6) 10-19% myeloblasts in the PB or bone marrow (BM)	7) Persistent thrombocytopenia
Blast phase	8) Clonal evolution
1) Blasts equal or are greater than 20% or the PB WBC or the nucleated cells of	9) Progressive splenomegaly
the BM,	10) Development of myelofibrosis
or	Blast phase
2) Extramedullary blast proliferation	1) ≥30% blasts in the PB, marrow, or both
3) Accumulation of blats occupy focal but significant areas of the BM	2) Extramedullary infiltrates of leukemic cells

Adapted from Swerdlow, 2008; Speck, 1984.

Table 3. Definition of Accelerated Phase and Blast Phase Chronic Myeloid Leukemia (by WHO2008 and IBMTR Criteria)

The evaluation of the response to tyrosine kinase treatment is made through hematologic, cytogenetic and molecular testing (table 4). The overall evaluation should lead to a classification of treatment response as optimal, suboptimal or failure (table 5). For patients in early chronic phase who achieve an optimal response, the drug should be continued until allogeneic HSCT is undertaken. In those patients who fail to respond, second-generation tyrosine kinase inhibitors and HSCT need to be considered. In suboptimal responders, imatinib may be continued, possibly at a higher dosage, or second-generation tyrosine kinase inhibitors may be introduced (Lee & Chung, 2011) Prospective cooperative studies are needed to address this complex issue in young patients with CML.

Complete hematologic response

1. Complete normalization of peripheral blood counts with leukocyte count <10x10^9/L
2. Platelet count <450x10^9/L
3. No immature cells, such as blasts, promyelocytes, metamyelocyte
4. No signs or symptoms of disease with disappearance of palpable splenomegaly

Partial hematologic response

- Same as those for complete hematologic response, except for

1. persistence of immature cells or
2. platelet count <50% of the pretreatment count but >450xx10^9/L
3. persistent splenomegaly but <50% of the pretreatment extent

Cytogenetic response (in patients with complete hematologic response)

1. Complete response; No Ph-positive metaphase cells
2. Major response; 0-35% Ph-positive metaphase cells (complete+partial)
3. Partial response; 1-34% Ph-positive metaphase cells
4. Minor response; 35-90% Ph-positive metaphase cells

Molecular response

1. Complete molecular response; bcr-abl mRNA undetectable by RT-PCR
2. Major molecular response; $\geq$ 3-log reduction of bcr-abl mRNA

Adapted from Faderls, et al, 1999.

Table 4. Criteria for Cytogenetic and Hematologic Remission in CML.

Time	Response			
	Optimal	Suboptimal	Failure	Warning
At diagnosis	NA	NA	NA	Advanced phase
3 months	CHR	Less than CHR	No hematologic response ; Stable disease or disease progression	
6 months	At least PCyR (Ph+≤ 35%)	Less than PCyR (Ph+> 35%)	No CyR (Ph+> 95%)	
9 months	At least PCyR (Ph+≤ 35%)	Less than PCyR (Ph+> 35%)	No CyR (Ph+> 95%)	
12 months	CCyR	PCyR (Ph+, 1-35%)	Less than PCyR (Ph+> 35%)	Less than MMolR
18 months	MMolR	Less than MMolR	Less than CCyR	
Any time during treatment	Stable or improving MMolR[†]	Loss of MMolR, mutations in bcr-abl kinase domain	Disease progression, mutations in bcr-abl kinase domain	Disease progression, new clonal cytogen-etic abnormalities

Abbreviations: NA, not applicable; CHR, complete hematologic response, CyR; Cytogenetic response, PCyR; partial cytogenetic response, CCyR; complete cytogenetic response, MMolR; major molecular response, Ph+; philadelphia chromosome positive

[†]MMolR indicated a ratio of BCR-ABL1 to ABL1 or other housekeeping genes of ≤0.1% on the international scale.

Table 5. Recommendation for Definitions of Treatment Response to Imatinib Used in Early Chronic Phase. Modified from Suttorp M, et al, 2011 and Baccarani M et al, 2009.

2.2 Lymphomas

2.2.1 Non hodgkin's lymphoma (NHL)

Children suffering from NHL(Burkitt, lymphoblastic, diffuse large B cell and anaplastic large cell lymphoma) even with stages III/IV have excellent results when treated with first-line chemotherapy and radiation therapy. Long term EFS is between 60-90%(Cairo et al., 2007; Gerrad et al., 2008; Link, 1997; Patte et al., 2007). However, for refractory or recurrent Burkitt's, diffuse large cell and lymphoblastic lymphoma, the long term survival is only 10-20% (Atr, 2001; Cairo, 2003) In contrast, for refractory or recurrent anaplastic large lymphoma, up to 60% of patients may achieve long-term survival (53).

Several studies have shown that patients with chemosensitive recurrent diseases can achieve long-term disease free survival after HSCT. In a recent study by Thomas Gross published in 2010, he examined the role of HSCT for patients less than 18 years with the four different histologic subtypes receiving autologous or allogeneic HSCT (sibling & unrelated) from 1990-2005. To date this is the largest study done for refractory/relapse NHL. He concluded that EFS rates were lower for patients not in complete remission at HSCT, regardless of donor type. After adjusting for disease status, 5-year EFS were similar after allogeneic and autologous HSCT for diffuse large B cell (50% vs 52%), Burkitt's (31% vs 27% and anaplastic large cell lymphoma (46% vs 35%). However, EFS was higher for lymphoblastic lymphoma after allogeneic HSCT (40% vs 4% p<0.01). Predictors of EFS for progressive or recurrent disease after HSCT included disease status at HSCT and use of allogeneic donor for lymphoblastic lymphoma.

HSCT (auto & allo) can be effective in salvaging children and adolescents with refractory or recurrent NHL and results are superior if complete remission can be achieved prior to HSCT. Allogeneic donor is preferred for patients with lymphoblastic lymphoma.

2.2.2 Hodgkin's disease (HD)

Autologous HSCT is the standard therapy for patients with HD in first chemosensitive relapse or second complete remission (CR) as shown by two prospective randomized clinical trials (Linch et al., 1993; Schmitz et al., 2002)

Currently, there is no indication for autologous HSCT in first CR, even in patients with bad prognostic features at diagnosis (Federico et al., 2003; Proctor et al., 2002).

For primary refractory patients or for patients in chemorefractory relapse, autologous HSCT has only a small chance of inducing long-term remission (Lazarus et al., 1999; Sweetenham et al., 1999). As part of a clinical protocol for patients with resistant HD, autologous HSCT might be considered as an initial debulking therapy to be followed by an allogeneic HSCT as consolidation therapy (Carella et al., 2000).

Allogeneic HSCT has mainly been used as salvage therapy for multiply relapsed or refractory HD. A retrospective analysis indicates that reduced intensity conditioning allogeneic HSCT can improve the outcome of HD patients that relapse after an autologous HSCT (Thomson et al., 2008). Its impact in the long term outcome of these patients has still to be prospectively evaluated. HSCTs from HLA-identical sibling donors and well-matched unrelated donors give a similar outcome (Anderlini et al., 2008).

2.3 Myelodysplastic syndrome (MDS)

MDS is rare in children an allogeneic HSCT from a sibling donor or a well-matched unrelated donor is currently the only curative therapy that is available for children with de novo MDS, JMML or secondary MDS. MDS is a heterogeneous disorder, characterized by a clonal stem cell disease with ineffective hematopoiesis which is morphologically abnormal. MDS in children differs from MDS in adults, as children more frequently suffer from hypocellular MDS. De novo MDS can be further classified as refractory cytopenia (RC; previously known as refractory anemia or RA), RA with excess of blast (RAEB) and RAEB in transformation (RAEBt).

The European working party on myelodysplastic syndrome (EWOG-MDS) reported their retrospective results on 63 children with RC (Kardos et al., 2003). Over 40% of patients had hypocellular marrows. Almost 50% of children with monosomy 7 progressed to advanced MDS within 2 years from diagnosis. By contrast, patients with hypocellular RC with a normal karyotype, may experience a long stable course before progression to generalized marrow failure occurs. Therefore, in patients with monosomy 7, HSCT should be performed soon after the diagnosis has been established. This is also advised for patients with advanced MDS (RAEB or RAEBt), and for patients with hypercellular RC, or with other clonal aberrations. In some patients the differentiation between hypocellular RC with a normal karyotype and aplastic anemia may be difficult, and in such patients a "watch and wait" strategy may be considered with repeated bone marrow evaluation before a final decision on diagnosis and therapy is made.

After the introduction of the new WHO definition of acute myeloid leukemia, which lowered the threshold to diagnose AML from 30 to 20% blasts, there has been a debate whether RAEBt should be classified and treated as MDS or AML (VArdiman, 2002). One approach is to build in some observation time to assess progression, and to look for signs indicative of AML, such as organomegaly or non-random chromosomal aberrations such as t(8;21) or inversion(16).

Another relevant question in this respect is whether patients with advanced MDS benefit from pre-HSCT chemotherapy or not. Current results indicate this is not the case, as outcome did not differ according to blast percentage <5%, 5-19% or >20% in directly transplanted patients (Stary, 2005).

In summary, patients diagnosed with advanced MDS should be treated with allogeneic-HSCT, which may even include less suitable donors such as mismatched or haploidentical donors if this is the only available choice for a particular patient.

2.4 Solid tumors

Neuroblastoma (stage IV beyond the age of 1 year, or high risk factors in lower stage) is still the only indication where the benefit of high-dose therapy with autologous HSCT has been shown by randomized trials (Ladenstein et al., 2008; Matthay et al., 2009).

Although to date the published results do not show an unequivocal benefit for consolidation with high-dose therapy, children and adolescents with solid tumors might undergo autologous HSCT after high-dose chemotherapy within clinical research trial, preferably as part of first -line treatment strategies in the following situations:

- Neuroblastoma (high risk, >CR1)
- Ewing's sarcoma (high risk or >CR1).
- Brain tumors: children with medulloblastoma and high-grade gliomas responsive to chemotherapy in an attempt to avoid or postpone radiotherapy.
- Soft tissue sarcoma: stage IV or in responding relapse.
- Germ cell tumors: after a relapse or with progressive disease.
- Wilm's tumor: relapse.
- Osteogenic sarcoma: the value of HSCT is not yet clear.

In general, allogeneic HSCT cannot be recommended in children with solid tumors. Allogeneic HSCT may be undertaken in the context of a clinical protocol in specialized centers.

2.5 Bone marrow failure (BMF)

BMF syndromes include a broad group of diseases of varying etiologies in which hematopoiesis is abnormal or completely arrested in one or more cell lines. BMF can be acquired aplastic anemia (AA) or can be congenital, as part of such syndromes as Fanconi anemia (FA), Diamond Blackfan anemia (DBA), and Shwachman Diamond syndrome (SDS). The estimated incidence of BMF is 2 per million in Europe, with higher rates in Asia, perhaps resulting from environmental factors.

2.5.1 Acquired severe aplastic anemia (AA)

HSCT using an HLA-matched related donor is the treatment of choice for severe acquired aplastic anemia, resulting in long-term survival rates of over 90% If an HLA-compatible family donor is not available, most patients are treated with high-dose immunosuppression, using antithymocyte globulin (ATG) plus cyclosporine, with or without granulocyte colony-stimulating factor (G-CSF). Approximately 70-80% of patients respond to immunosuppression, although the actuarial 10-year survival rate is about 40%. Marrow transplantation from unrelated donors is reserved for those patients who do not respond to or who relapse after immunosuppressive therapy.

2.5.2 Inherited bone marrow failure syndromes (IBMFS)

IBMFS should be considered for all patients presenting with AA, regardless of the presence or absence of characteristic physical findings. IBMFS require specific approaches to management. Sensitive and specific diagnostic tests, including identification of mutations in specific genes, are available for many disorders.

2.5.2.1 Fanconi anemia (FA)

FA is the most common IBMFS and consists of a complex disorder of increased sensitivity to DNA damage characterized by congenital anomalies, progressive BMF, and high risk of MDS, malignant transformation to acute leukemia and solid tumors. Significantly, a large percentage of affected persons (25% to 40%) have no visible anomalies, and FA cannot be excluded without specific testing for mutagen sensitivity. BMF in FA typically presents between the ages of 5 and 10 years, with an actuarial risk of developing bone marrow failure of 50% to 90% by age 40 years (Kutler et al., 2003; Rosenberg, 2008). The median age of patients who develop AML is 14 years (Alter, 2003), and cumulative incidence of hematologic malignancy by age 40 years is 22% to 33% (Kutler et al., 2003; Rosenberg, 2008). Symptomatic transfusion, G-CSF, and androgens can be used to treat cytopenias; however, HSCT is the only current definitive therapy to restore normal hematopoiesis.

Commonly agreed-upon indications for HSCT in these patients include evidence of severe marrow failure as manifested by an ANC less then 1000 x 10^9/L with or without G-CSF support, or hemoglobin of less than 8 g/dl or platelet count less than 50,000 x 10^9/L or requirement of blood transfusion on regular basis. HSCT is also indicated for FA patients with evidence of progression to MDS or AML. Patients with FA who have an HLA-identical related donor, early HSCT is now the first-line treatment of choice for BMF, and preferably before transfusion dependence develops, to limit the risk of graft failure.

Preparative regimens for HSCT in FA patients are modified from standard approaches because of the chromosomal instability present in all FA cells, including nonhematopoietic tissues. In vitro studies have shown that FA cells are hypersensitive to DNA cross-linking agents, such as cytoxan (Berger, 1980). In addition, patients with FA are at increased risk of severe GVHD compared with patients with severe AA because of defective DNA repair mechanisms, leading to prolonged tissue damage after targeting by an alloreactive response (Guardiola et al., 2004).

Elaine Gluckman's group at St Louis, Paris investigated the use of reduced-dose cytoxan (20 to 40mg/kg) and reduced-dose thoracoabdominal irradiation or total body irradiation (TBI) (400-450 cGy) and reported a long-term survival of 58.5% after sibling donor transplantation, although with high incidences of aGVHD (55%) and cGVHD (70%). Later series modified the Gluckman regimen with the addition of ATG, resulting in less aGVHD and cGVHD and improved survival (Ayas et al., 2001). A recent series of 35 FA patients undergoing matched-related HSCT using this regimen along with peri transplantation ATG reported an excellent 10-year actuarial survival of 89%, with aGVHD in 23% of cases and cGVHD in 12% of cases (Farzin et al., 2007).

These studies have used low dose radiation because patients with FA have an increased risk of posttransplantation malignancy, but what about avoiding radiation altogether? A recent retrospective review of experience with matched related HSCT in FA patients in Saudi Arabia by Ayas et al (Ayas et al., 2008) found significantly greater OS in patients receiving non radiation, low dose cytoxan and ATG regimens compared with those undergoing preparative regimens with cytoxan and additional thoracoabdominal radiation (72.5% vs 96.9%; p=0.013). The availability of fludarabine, a highly immunosuppressive nucleoside analog that is well tolerated by patients with FA, has allowed the elimination of radiation with good results. Tan et al in 2006, recently reported an actual OS of 82%, transplant related mortality of 9% and minimal GVHD in a cohort of 11 patients who underwent transplantation with low dose cytoxan, fludarabine and ATG with T cell-depleted bone marrow or umbilical cord cells.

HSCT from an unrelated donor for patients with FA remains a key treatment strategy. Historically, outcomes of alternative donor transplantation in FA have been discouraging, with high incidences of graft failure, aGVHD and cGVHD and organ toxicity related to preparative regimens. Many regimens have been looked at over the years for unrelated transplants including increasing the dose of radiation, adding ATG without significant improvement in overall survival. The advent of fludarabine based preparative regimens has resulted in considerable progress, improving engraftment without significant toxicity attributable to the drug. However, although fludarabine regimens have had some success in treating FA, concerns regarding reduced intensity conditioning (RIC) regimens persist; residual FA cells that survive the preparative regimen may present as AML as much as 10 years later (Ayas et al., 2001).Despite these data, (Chaudhury et al., 2008), in a study of 18 high-risk patients with transfusion dependent AA, MDS and AML receiving either related mismatched or unrelated matched or mismatched HSCT using fludarabine, TBI and cytoxan for preparative regimens with T-cell depleted stem cell sources, found 100% engraftment, OS 72.2% and DFS of 66.6% with a median follow up of 4.2 years, suggesting that a RIC preparative regimen might be sufficient to control malignancy in FA. Cord blood is an alternative stem cells source for patients with FA who lack an HLA-matched unrelated bone marrow donor, as umbilical cord blood transplant has decreased incidence of GVHD.

Despite the improved survival, identifying the ideal time for HSCT in FA patients requiring alternative donor transplantations remains challenging, given the still-significant peri transplantation mortality and the possibility of long lasting androgen response or survival with AA for a significant period without progression to MDS/AML. Referral and transplantation before exposure to large amounts of blood products or prolonged periods of severe neutropenia are likely to lead to the best outcomes.

2.5.2.2 Shwachman-diamond syndrome (SDS)

SDS is a rare autosomal recessive disorder characterized by exocrine pancreatic insufficiency, skeletal abnormalities and BMF with a predisposition to MDS and leukemia, especially AML. Although most patients with SDS have some hematologic abnormalities, most of them do not require HSCT. In the largest reported series, 20% of cases developed pancytopenia and 6% progressed to MDS (Ginzberg et al., 1999).

HSCT is the only curative treatment for bone marrow dysfunction associated with SDS. However, the timing of HSCT remains a subject of controversy, and the apparent lack of genotype-phenotype correlation makes selection of patients for early preemptive HSCT difficult at present. In addition, SDS patients, like FA patients, have increased toxicity with intensive conditioning regimens. Overall, the available literature on HSCT in SDS patients is limited and consists mainly of case reports (Cesaro et al., 2001; Fleitz et al., 2002). Preliminary data indicates that HSCT with reduced intensity conditioning is feasible in patients with SDS and is associated with excellent donor cell engraftment and modest morbidity.

2.5.2.3 Dyskeratosis congenita (DS)

DC is a disorder of diverse inheritance with chromosomal instability related to a defect in telomere maintenance, characterized by a triad of reticulate skin pigmentation, mucosal leukoplakia and nail dystrophy, along with BMF. Between 80% and 90% of persons with DC will develop hematopoietic abnormalities by age 30 years, and BMF is the leading cause of early mortality in this population (Dokal, 2000). In addition, DC patients are at increased risk for MDS/AML and solid tumors, especially squamous cell carcinomas, as well as progressive pulmonary fibrosis (Dokal, 2000).

Allogeneic HSCT remains the only curative approach for marrow failure in patients with DC; however outcomes have been poor due to early and late complications. Initial attempts at HSCT in DC patients with myeloablative regimens had poor results, with significant morbidity and mortality, including increased incidences of chronic pulmonary and vascular complications, likely related to these patients underlying tendency to develop restrictive pulmonary disease. Non-myeloablative transplants using low-dose Cytoxan and fludarabine and ATG have produced successful engraftment and good short term outcomes, largely in case reports (de laFuente, 2007). Regardless of the potential reduction in toxicity associated with non-myeloablative regimens, preexisting conditions characteristic of DC (e.g. pulmonary disease) may ultimately limit the effectiveness of HSCT in DC patients.

2.5.2.4 Diamond-blackfan anemia (DBA)

DBA is a rare inherited form of pure red blood cell aplasia that presents early in infancy. Mutations in one of a number of ribosomal proteins have been identified in approximately 50% of DBA patients, implicating ribosomal biogenesis or function in the disorder. Clinically, DBA is associated with macrocytosis, reticulocytopenia, and normal marrow cellularity with erythroblastopenia. Characteristically, these patients have elevated fetal hemoglobin and erythrocyte adenosine deaminase activity, and up to 35% have an associated congenital anomaly, with craniofacial and thumb abnormalities the most common.

Corticosteroids remain the mainstay of initial therapy in DBA, with 80% response rate. Only 20% of patients achieve remission; 40% require continued therapy with steroids, which can have significant side effects, and another 40% remain transfusion and chelation dependent (Vlachos et al., 2008). Steroid-intolerant or transfusion-dependent patients may be considered for HSCT, which although curative for DBA, remains controversial, because most of these patients can achieve long-term survival with supportive therapy alone.

A series of 36 patients from the DBA registry who underwent HSCT (main indication transfusion dependence) yielded 5-year survival rates of 72.7% in matched sibling donor recipients and 19% in alternative donor recipients (p=0.01) (Lipton et al., 2006). Similar results were reported in an international bone marrow transplant registry series of 61 patients with DBA undergoing HSCT with conventional cytoxan containing preparative regimens; 3-year survival was 76% after sibling donor transplantation compared with 39% after alternative donor transplantation (Roy et al., 2005). In both studies, the alternative donor recipients were more likely to have received a TBI-containing regimen or to have a longer time from diagnosis to transplantation, suggesting that TBI should be avoided. In addition, patients with DBA have an increased risk of malignancy compared to the general population, another reason why TBI-containing regimens should be avoided in this population. There also are encouraging case reports of successful HSCT in DBA with RIC fludarabine containing preparative regimens; however, the data are scanty and reflect short follow-up times; further study is needed in this area (Berndt, 2004; Ostronoff, 2004).

2.5.2.5 Congenital Amegakaryocytic Thrombocytopenia (CAMT)

CAMT is a rare autosomal recessive disorder caused by mutations in the thrombopoietin receptor. It is usually diagnosed early in childhood, presenting with isolated nonimmune thrombocytopenia with decreased marrow megakaryocytes. Approximately 50% of CAMT patients develop marrow aplasia, and some develop MDS or leukemia.

Although transient responses to steroids, cyclosporine and growth factors in CAMT have been documented, HSCT remains the only curative treatment. Good short-term survival has been reported after matched related donor HSCT in small case series. Reports of unrelated donor HSCT are largely case reports and describe significant engraftment challenges.

2.6 Immunodeficiencies

Primary cellular immunodeficiencies are a group of inherited disorders characterized by severe impairment of the innate or adaptive immune systems, which generally leads to early death from infectious complications. These disorders can be further categorized by the cell lineage primarily affected (table 6). Supportive care can extend the life span of patients affected by these diseases, definitive cure is generally only achieved by allogeneic hematopoietic stem cell transplantation, though recent advances in gene therapy hold significant promise that this may soon be a viable alternative. Allogeneic HSCT is indicated for severe primary immunodeficiencies from both HLA-identical and alternative donors.

Absent T- and B-lymphocyte function	*Defective T and B lymphocytes*	*Dysfunctional T lymphocytes with predisposition to HLH*	*Absent or dysfunctional granulocytes*
SCID	Wiskott–Aldrich syndrome	Familial HLH (defects in *perforin*, *MUNC*, etc.)	Severe congenital neutropenia
	HIGM1	Chediak–Higashi syndrome	Leukocyte adhesion disorder
		Griscelli syndrome	Chronic granulomatous disease
		XLP	

Abbreviations: HIGM1 = hyper IgM syndrome (CD40 ligand deficiency); HLH = hemophagocytic lymphohistiocytosis; XLP = X-linked lymphoproliferative disease.

Table 6. Primary Immunodeficiencies Potentially Treated with HSCT.

2.6.1 Severe combined immunodeficiency (SCID)

SCID is a rare disorder caused by a group of genetic disorders with a shared phenotype of deficient T and B lymphocyte infunction (with or without abnormal natural killer (NK) cell development) that leads to early death from recurrent infections in affected children (table 7). Except for those patients with SCID due to deficiency of adenosine deaminase (ADA), for which replacement enzyme exists, the only curative therapy for SCIS is allogeneic HSCT. However, early results with gene insertion into autologous hematopoietic stem cells for children with x-linked SCID and ADA deficiency (Cavazzana-Calco, 2007) suggest that eventually this will become a more common form of curative treatment for many primary immunodeficiency diseases.

Name	*Defect*	*Phenotype*	*Special*
X-linked	Common γ chain	T-B + NK –	
JAK3 deficiency	Janus kinase 3	T-B + NK –	
Rag 1 or 2	Recombinase-activating proteins 1 or 2	T-B – NK +	Frequently associated with Omenn's syndrome: autoreactive GVHD
Artemis deficiency	Artemis (also known as DCLRE1C)	T-B – NK +	Athabascan-speaking Native Americans, radiosensitive
Ligase 4 deficiency	Ligase 4	T-B – NK +	Radiosensitive
IL-7Rα deficiency	IL-7 receptorα	T-B + NK +	
CD45 deficiency	CD45	T-B + NK +	
CD3δ deficiency	CD3δ subunit	T-B + NK +	
CD3ε deficiency	CD3ε subunit	T-B + NK +	
CD3ς deficiency	CD3ς subunit	T-B + NK +	
Cartilage hair hypoplasia	Endoribonuclease	T-B + NK +	Dwarfism, hypoplastic hair Finnish, Amish
p56lck deficiency	p56lck Protein tyrosine kinase	T-B + NK +	
ADA deficiency	Adenosine deaminase	T-B – NK –	
PNP deficiency	Purine nucleoside phosphorylase	T-B – NK –	Neurologic dysfunction, ataxia
Reticular dysgenesis	Unknown	T-B – NK –	Impaired myeloid and erythroid development, sensorineural deafness
ZAP70 deficiency	ς-chain-associated protein kinase	CD4 +, CD8 – B +, NK +	
Bare lymphocyte Syndrome type II	HLA class II	CD4 –(mild), CD8 + B +, NK +	North African
SCID with bowel atresia	Unknown	CD4 +, CD8 +, B + NK +	

Abbreviations: ADA = adenosine deaminase; DCLREIC = DNA cross-link repair enzyme 1C; HLA = human leukocyte antigen.

Table 7. Genetic Sub-Types of Severe Combined Immunodeficiency.

HSCT should be done as soon as the diagnosis is confirmed because these patients are at risk of developing a life-threatening infection, particularly pulmonary infections. For all stem cell sources, successful outcomes are more likely to be achieved when the patient is still very young, preferably less than 6 months of age.(Buckley et al., 1999), demonstrated that infants transplanted less than 3.5 months of age had a 95% overall survival compared to only 76% overall survival in older children. The preferred choice of stem cell donor for a patient with SCID is an HLA-identical sibling, in which the overall survival now exceeds 90%, if the transplant is performed promptly. In patients without a matched sibling, the choice is whether to use an immediately available T cell depleted haplocompatible family member or to perform a search for an HLA matched unrelated donor or cord blood unit. Table 8 lists the reports on transplantation with different stem cell sources.

2.6.2 Wiskott-Aldrich syndrome (WAS)

WAS is characterized by a trial of thrombocytopenia with small platelets, eczema and recurrent infections. The T cell immunodeficiency predisposes to the development of autoimmune phenomena and lymphoma. Affected males rarely survive past the second decade of life. The only curative strategy is allogeneic HSCT. The international bone marrow registry and national marrow donor program demonstrated in 170 patients that while the 5-year OS of patients transplanted from HLA-identical siblings was 87%, the results for unrelated HSCT were significantly related to the age at transplant (Filipovich et al., 2001). Unrelated donors less than 5 years of age had an 85% 5-year OS, while all 15 patients greater than 5 years of age died (Filipovich et al., 2001). Haploidentical related transplants have been less successful with an OS of 45-52%.

	Year	*Reference*	*MRD*	*Haplo*	*Haplo*	*MUD*	*MUD*	*Cord*
Conditioning			None	None	MA	MA	RI	MA
Dror *et al.*	1993	3	—	67% (12)	50% (12)	—	—	—
Buckley *et al.*	1999	5	100% (12)	78% (77)	—	—	—	66% (3)
Bertrand *et al.*	1999	4	—	46% (50)	54% (129)	—	—	—
Dalal *et al.*	2000	15	—	—	—	67% (9)	—	—
Knutsen and Wall	2000	16	—	—	—	—	—	88% (8)
Antoine *et al.*	2003	6	81% (104)	—	—	63% (28)	—	—
Rao *et al.*	2005	17	—	—	—	71% (7)	83% (6)	—
Bhattacharya *et al.*	2005	18	—	—	—	—	—	80% (10)[a]
Grunebaum *et al.*	2006	14	92% (13)	—	53% (40)	81% (41)	—	—

Stem cell source: MRD (matched related donor) vs haplo (haplocompatible family donor) vs MUD (matched unrelated donor) vs Cord (unrelated cord blood stem cells). Conditioning: none vs MA (myeloablative) vs RI (reduced intensity). Percentage indicates overall survival (absolute number of patients).
[a]Some patients received no conditioning.

Table 8. Survival Following HSCT For SCID Based on Stem Cell Source and Conditioning Regimen.

2.6.3 Familial hemophagocytic lymphohistiocytosis (HLH)

Familial HLH is characterized by episodes of fever, hepatosplenomegaly and cytopenias. An autosomal recessive defect in one of the several genes including those encoding perforin or Munc 13, causes reduced NK and T cell cytotoxicity. This leads to a widespread accumulation of lymphocytes and mature macrophages with hypercytokinemia. Familial HLH is invariably fatal. The only curative strategy for treatment of familial HLH is allogeneic HSCT. A report from a multicenter prospective trial, HLH-94, demonstrated a 62% 3-year EFS in 65 children undergoing allogeneic HSCT with a variety of stem cell sources (Henter et al., 2002).

2.6.4 Chronic granulomatous disease (CGD)

CGD is characterized by recurrent pyogenic infections in patients with normal neutrophil numbers. A defect in one of the four genes encoding subunits of the nicotinamide adenine dinucleotide phosphate-oxidase complex leads to insufficient production of free protons from which to make hydrogen peroxide. With good supportive care, including therapy with interferon gamma, affected individuals can live up to the fourth decade of life, but suffer early mortality from recurrent pulmonary infections.

Allogeneic HSCT is the only curative strategy. A report from the European group for Blood and Marrow Transplantation demonstrated in 23 patients that myeloablative conditioning prior to matched sibling HSCT can be safely performed (85%OS), especially if the patient were free of infection at the time of HSCT (100% OS) (Seger et al., 2002). Given the current success rates, some favor transplantation in all patients with CGD who have an appropriate donor at the earliest opportunity.

Recent data, (Kuhn's et al., 2010) showed that patients with very low superoxide production had worse long-term survival than those with higher levels of NADPH oxidase activity suggesting that these patients should be considered appropriate candidates for early HSCT, particularly if a sibling matched donor is available. An increased alkaline phosphatase level, a history of liver abscesses, and a decrease in platelet count reflecting portal hypertension are adverse prognostic indicators (Feld et al., 2008). These patients might also be considered for early transplantation. Even with improved survival and longevity caused by better infection and inflammation management, complications and their consequences can accumulate over time. However, HSCT is probably better before infections and inflammatory damage accumulates. Transplantation has aloes reversed some of the inflammatory and autoimmune complications associated with CGD and might prevent their development (Seger et al., 2002). Allogeneic HSCT has improved dramatically over the last decade because of improved conditioning regimens and GVHD prophylaxis, high-resolution sequence-based matching and improved pre transplantation, peri transplantation and post transplantation management and as a result it has become a successful and sensible option for many patients with CGD.

2.7 Inherited metabolic diseases (IMD)

IMD is a diverse group of diseases arising from genetic defects in lysosomal enzymes or peroxisomal function. The lysosome is an intracellular sorting, recycling and digestion of organic molecules. Loss of functional activity of lysosomal enzymes results in accumulation of substrates, such as glycoprotein or mucopolysaccharides (MPS). The clinical manifestations vary depending on the specific enzymatic deficiency, level of residual activity, and site of substrate accumulation.

Allogeneic HSCT can prolong life and improve its quality in patients with IMD. HSCT offers a permanent source of enzyme replacement therapy and also might mediate nonhematopoietic cell regeneration or repair. The likely processes responsible for the effectiveness of HSCT for IMD includes cytoreduction to ablate myeloid and immune elements, engraftment of donor-derived hematopoietic and immune system, donor leukocytes production of enzyme, distribution of enzyme through blood circulation, migration of cells to brain, cross blood-brain barrier, many develop microglia, replacement of enzyme in the brain by cross-correction and nonhematopoietic cell engraftment (Prasad and Kurtzberg, 2008).HSCT has been performed in almost 20 of the 40 known lysosomal storage disorders and peroxisomal storage disorders. However, the majority of transplant experience to date is in patients with MPS I (Hurler Syndrome), other MPS syndromes (MPSII, MPSIII, A & B, MPSVI), adrenal leukodystrophy (ALD), metachromic leukodystrophy (MLD), and globoid leukodystrophy (Krabbe disease), accounting for more than 80% of the cases. Table 9 identifies the IMD for which allogeneic HSCT is currently indicated or under investigation. The response to HSCT varies from disease to disease, within patients with same disease, and within different organ systems in the same patient.

2.7.1 Hurler syndrome (MPS IH)

MPS IH, the most sever phenotype of alpha-l-iduronidase deficiency, is an autosomal recessive disorder characterized by progressive accumulation of stored glycosaminoglycans (GAGs). Hurler and other phenotypes of MPS I are a broad continuous clinical spectrum. Accumulation of GAGs results in progressive, multisystem dysfunction that includes psychomotor retardation, severe skeletal malformations, life-threatening cardiopulmonary complications, and early death.

Disorder	Enzyme/Protein	HSCT Indication	Comments
Mucopolysaccharidoses			
Hurler (MPS IH)	α-L-Iduronidase	Standard therapy	
Hurler/Scheie (MPS IH/S)	α- L-Iduronidase	Optional	ERT first-line therapy
Scheie (MPS IS)	α- L-Iduronidase	Optional	ERT first-line therapy
Hunter: severe (MPS IIA)	Iduronate-2-sulfatase	Investigational	Only early or asymptomatic
Hunter: attenuated (MPS IIB)	Iduronate-2-sulfatase	Investigational	Only early or asymptomatic
Sanfilippo (MPS IIIA)	Heparan-*N*-sulfatase	Investigational	Only early or asymptomatic
Sanfilippo (MPS IIIB)	*N*-Acetylglucosaminidase	Investigational	Only early or asymptomatic
Sanfilippo (MPS IIIC)	AcetylCoA:*N*-acetyltransferase	Investigational	Only early or asymptomatic
Sanfilippo (MPS IIID)	*N*-Acetylglucosamine 6-sulfatase	Investigational	Only early or asymptomatic
Maroteaux-Lamy (MPS VI)	Arylsulfatase B	Optional	ERT first-line therapy
Sly (MPS VII)	β-Glucuronidase	Optional	
Leukodystrophies			
X-ALD, cerebral	ALD protein	Standard therapy	Not for advanced disease
MLD: early onset	ARSA	Unknown	Only early or asymptomatic
MLD: late onset	ARSA	Standard therapy	
GLD: early onset	GALC	Standard therapy	Neonate, screening diagnosis, or second case in known family; not for advanced disease
GLD: late onset	GALC	Optional	
Glycoprotein metabolic and miscellaneous disorders			
Fucosidosis	Fucosidase	Optional	
α-Mannosidosis	α-Mannosidase	Optional	
Aspartylglucosaminuria	Aspartylglucosaminidase	Optional	
Farber	Ceraminidase	Optional	
Tay-Sachs: early onset	Hexosaminidase A	Unknown	Neonate, screening diagnosis, or second case in known family
Tay-Sachs: juvenile	Hexosaminidase A	Unknown	
Sandhoff: early onset	Hexosaminidase A & B	Unknown	Neonate, screening diagnosis, or second case in known family
Sandhoff: juvenile	Hexosaminidase A & B	Unknown	
Gaucher 1 (nonneuronopathic)	Glucocerebrosidase	Optional	ERT first-line therapy
Gaucher 2 (acute neuronopathic)	Glucocerebrosidase	Unknown	
Gaucher 3 (subacute neuronopathic)	Glucocerebrosidase	Unknown	Limited benefit of ERT
Gaucher 3 (Norrbottnian)	Glucocerebrosidase	Optional	
Pompe	Glucosidase	Investigational	ERT available
Niemann-Pick: type A	Acid sphingomyelinase	Unknown	
Niemann-Pick: type B	Acid sphingomyelinase	Unknown	ERT in clinical trial
Niemann-Pick: type C	Cholesterol trafficking	Optional for C-2	
Mucolipidosis: type II (I-cell)	*N*-Acetylglucosamine-1-phosphotransferase	Investigational	Only early or asymptomatic
Wolman syndrome	Acid lipase	Optional	May be viewed as standard
MSD	Sulfatases	Investigational	

Table does not include diseases where HSCT is not indicated. Standard therapy: HSCT applied routinely. Considerable published research evidence from registries and institutions shows efficacy. Delayed diagnosis or advanced disease may preclude transplant for individual patients. Optional: HSCT is effective but other therapy is increasingly considered first choice. Or, insufficient published evidence for HSCT to be considered standard. Investigational: possible a priori reason for HSCT. Further published evidence needed to support the use of HSCT in clinical practice. Unknown: no published evidence that HSCT is beneficial.

Table 9. IMD for which HSCT may be indicated

Data from the CIBMTR and EBMT indicate that more than 500 allogeneic HSCTs have been performed worldwide for children with MPS IH since 1980, making it the most commonly transplanted IMD. HSCT is effective, resulting in increased life expectancy and improvement of clinical parameters if performed early in the disease course before the onset of irreversible damage. Donor engraftment after HSCT has resulted in improvement of the following clinical symptoms: rapid reduction of obstructive airway symptoms, and hepatomegaly; improvement in cardiovascular function as well as hearing, vision and linear growth; finally hydrocephalus is either prevented or stabilized. In addition, cerebral damage already present before HSCT seems to be irreversible, but HSCT is able to prevent progressive psychomotor deterioration and improve cognitive function (Peters, 1998; Vellodi et al., 1997).

A matched normal sibling is the preferred HSCT donor. In the past decade an unrelated cord blood (CB) has been used with increasing frequency in patients without a sibling donor. CB offers several potential advantages compared with bone marrow or peripheral blood for HSCT, including better availability, greater tolerance for HLA mismatches, lower incidence and severity of GVHD and reduced likelihood of transmitting viral infections (Staba et al., 2004; Prasad et al., 2008). The use of CB for children with MPS IH has been associated with high rates of chimerism, engraftment and overall survival (Staba et al., 2004; Prasad et al., 2008). Similar results are noted for CB in other selected IMD (Escolar et al., 2005). As a result of this data, the EBMT developed transplantation guidelines for patients with MPS IH in 2005. These guidelines are widely used today and include a standardized busulfan/cytoxan (BU/CY) conditioning regimen, an enzymatically normal matched sibling bone marrow donor if available, and if not, cord blood as the preferred graft source. A recent EUROCORD- Duke university MPS IH collaborative study showed that early transplant (i.e., within 4.6 months from diagnosis) with CB and BU/Cy conditioning was associated with improved engraftment and overall survival. Furthermore, 94% of engrafted survivors achieved full donor chimerism. (Boelens et al., 2007).

Despite the overall success from HSCT, some disease manifestations persists or can even progress after HSCT, and this includes the musculoskeletal disorders secondary to the IMD that does not resolve and often requires orthopedic surgical intervention. In addition, neurocognitive dysfunction and corneal clouding that developed before HSCT may be irreversible. The outcome of HSCT for children with MPS IH is promising, yet variable from child to child. The variability is presumably caused by factors such as genotype, age and clinical status before HSCT, donor enzyme activity level, donor chimerism (mixed or full) stem cell source (CB, BM,PB) and resultant enzyme activity level in the recipient (Aldenhoven, 2008). An international long-term follow up study involving Europe and North America is underway to evaluate the influence of these various factors. Overall progress has been made. HSCT for children with MPS IH has become a safer procedure, with recent survival rates exceeding 90%.

2.7.2 Other mucopolysaccharidosis syndromes

Compared with MPS IH, experience with HSCT for treatment of other MPS disorders is limited. Small numbers and lack of detailed functional outcome data hamper the development of specific therapy guidelines. Conceptually, the basis for the effectiveness of HSCT in these children is the same as those with MPS IH. However, the kinetics of cellular migration, differentiation, distribution, and effective enzyme delivery may differ. Also, there is wide clinical variability within and across specific MPS diseases. As with HSCT for other IMD, important factors in the outcome may be timing of transplant, graft source, and the underlying severity of the phenotype in a given child. To date, most of the published experience is in recipients of BMT (Guffon et al., 2009). Recently, survival has been reported in small cohorts undergoing CBT, but their functional outcomes are not yet published.

The role of HSCT in MPS II remains controversial because of lack of convincing evidence of neurocognitive benefit. The status of HSCT for Sanfilippo Syndrome (MPS III) is similar to that of MPS II with inadequate data and inability to make specific recommendations about timing of transplant, graft source, and potential neurological benefit. Eleven long-term survivors of BMT have been reported, but all showed declined in neurocognitive function

(Gungor,1995; Vellodi et al., 1992). On the other hand, the results of HSCT for Maroteaux-Lamy Syndrome (MPS VI) have been promising. MPS VI has multiple clinical phenotypes, but generally patients live into the second to fourth decade. HSCT in 4 patients with MPS VI lead to improvement in cardiopulmonary function, facial features, and quality of life (Herskhovitz et al., 1999). HSCT can be considered a therapeutic option for patients with MPS VI that are intolerant or fail ERT.

2.7.3 Adrenal leukodystrophy (ALD)

X-ALD is a peroxisomal disorder involving defective beta-oxidation of very long chain fatty acids (VLCFA). The affected gene in X-ALD is ABCD1 and the peroxisomal membrane protein for which it codes is ALDP. More than 500 mutations in the gene are described, but there is no relationship between the nature of the mutation and the clinical presentation of illness. X-ALD has a variable clinical presentation. Patients can be asymptomatic or present with adrenal insufficiency and/or non inflammatory axonopathy (AMN) and/or cerebral disease. The clinical course is so variable with some individuals never developing symptoms so therefore, HSCT can not be recommended based on the presence or absence of the genetic mutation. HSCT is indicated only in those patients with clear evidence of early cerebral inflammatory disease as determined by a gadolinium enhanced MRI. (Peters, 2003). Cerebral disease may manifest itself during childhood or adolescence. Approximately 40% of genetically affected boys develop childhood cerebral X-ALD. Many of the remainder develops AMN. Cerebral disease is usually progressive, although clinical stabilization without HSCT can occur. HSCT is not currently indicated for asymptomatic individuals as prophylaxis. In view of the natural history of the disease such a practice would mean that some boys would undergo HSCT (with its short-term mortality and long-term morbidity risks) who might otherwise have been healthy. Nor is indicated for those individuals with advanced cerebral disease because HSCT does not reverse and may even worsen, established disease.

In this disease, judicious timing of the transplant is paramount. Asymptomatic boys should be regularly screened for signs of inflammatory brain disease, a potential donor identified, and HSCT rapidly performed if and when such symptoms appear. The presence of brain MRI abnormalities and the presence or absence of enhancement with gadolinium has been shown to be of prognostic value. A 34-point MRI scoring system specific for X-ALD that was designed by Loes and colleagues (Loes et al., 1994; Loes et al., 2003) is now used worldwide for patient evaluation and treatment decisions. An MRI severity score as low as 1 with gadolinium enhancement in a young boy is highly predictive of subsequent progressive demyelination and is an indication for transplant. However, the identification of an HSCT donor for asymptomatic boys should not await MRI anomalies, but should done immediately after diagnosis to prevent delays if a follow up MRI indicates disease progression.

Review of the literature supports that most boys that have been transplanted from the best available donor have received full intensity chemotherapy-only preparative regimen (Peters, 2004); most unrelated donors have been adult bone marrow donors, but some CB donors have been used (Beam, 2007); donor-derived engraftment rates seem higher than seen in patients transplanted for MPS IH syndrome (86% of 93 evaluable patients at a median follow-up of 11 months; 93% of related donor transplants; 80% of unrelated donor transplant) (Peters,1998,2004); outcome is affected by disease status, donor source and HLA matching (Peters, 2004). The most common causes of death are progressive cerebral X-ALD disease and GVHD. TRM is 10% in related donors and 18% in unrelated donors. Five-year survival rates for recipients of related donor and unrelated donor transplants have been reported at 64% and 53%, respectively (Peters, 2004); and finally, survival is clearly affected by disease status at time of transplant as assessed by the number of neurologic deficits and MRI severity score. In those with 0 or 1 neurologic deficit and MRI score of less than 9, the 5-year survival was 92% compare to 45% in all other patients (Peters, 2004).

2.7.4 Globoid leukodystrophy (GLD)

GLD or Krabbe disease is an autosomal recessive lysosomal storage disorder caused by deficiency of galactocerebrosidase (GALC), an enzyme responsible for degrading beta-galactocerebroside, a major component of myelin sheath. GALC deficiency causes defective and decreased myelination and inflammation in the CNS and peripheral nervous systems from catabolic derivatives of beta-galactocerebroside such as psychosine. These changes lead to progressive deterioration in neurologic and cognitive function, resulting in spasticity, mental deterioration, blindness, deafness, seizures and early death. In the most severe "early onset or infantile" form, children develop symptoms before 6 months of age and usually die by age 2. In the "late onset" form, symptoms appear in early to late childhood, but only a few children survive into teenage years.

HSCT is the only available therapy with potential to improve neurocognitive function, increase survival and alter the natural history of the disease. Krivit and colleagues (Krivit et al., 1998) described the use of allogeneic HSCT to treat 5 patients with GLD (4 received HLA-sibling HSCT & 1 unrelated cord). Two children with late onset GLD had substantial neurologic disability and they had resolution of their symptoms after transplant. Cognition, language and memory continued to develop normally in 3 children with late-onset disease. Most children had improvement in MRI, CSF protein levels, and all had normalization of enzyme activity. These findings support the use of allogeneic HSCT for children with GLD. If a matched related donor is not available, unrelated cord blood has also been shown to be beneficial (Escolar et al., 2005).

2.7.5 Metachromatic leukodystrophy (MLD)

MLD is an autosomal recessive lysosomal disorder arising from deficiency of arylsulfatase A (ARSA) enzyme activity and characterized by increased urinary sulfatides. The clinical phenotype is a broad continuous spectrum ranging from early-infantile MLD to adult-onset forms. Clinical symptoms vary depending on timing of presentation (infantile, juvenile or adult form), but all include abnormal cognitive skills, behavioral abnormalities with adults having mental regression and psychiatric symptoms, progressive spastic disease and increased CSF protein.

The first BMT for MLD was performed more than 20 years ago. According to the EBMT and CIBMTR registries, more than 100 transplants have since been performed for this disorder. Despite this number, the lack of graft-outcome and long-term follow up studies makes it difficult to draw firm conclusions regarding the efficacy of HSCT in MLD. In addition, data suggest that outcomes are less promising than those for MPS IH. It is not clear if MLD patients, or which phenotypes, might benefit from HSCT. For presymptomatic juvenile and adult onset patients there is positive evidence. Improved transplantation techniques and the prompt availability of CB grafts may positively influence long-term outcomes. An international registry would facilitate comparative evaluation of therapeutic options, leading to improved guidelines.

3. Expanding Indications for transplant

HSCT has been explored in a number of malignant and nonmalignant diseases. Currently, research is rapidly expanding in areas not historically considered for HSCT. Also, as morbidity and mortality decrease, HSCT is being reconsidered for many diseases in which HSCT was previously considered and rejected. Several potential indications are reported in this section.

3.1 Beta-thalassemia

Thalassemias result from mutations of the globin genes that cause reduced or absent hemoglobin production, reducing oxygen delivery. To treat the anemia and restore oxygen delivery to tissues, chronic lifelong transfusions are required in those who have thalassemia major. However, this promotes progressive iron overload and organ damage. The only definitive cure for thalassemia is to correct the genetic defect by HSCT. Transplantation is recommended early, if an allogeneic healthy related sibling donor or a related CB is available. Several studies have suggested that umbilical cord blood transplant (UCBT) recipients benefit from a lower risk of GVHD (Gluckman, 1997; Wagner, 1995) and a recent analysis comparing 113 children who received a UCBT from a compatible sibling with 2052 HLA-identical sibling marrow transplant recipients showed that children receiving UCB experienced a significantly reduced risk of developing aGVHD and cGVHD (Rocha, 2000).

Prior to transplant, the patient should be assigned to 1 of 3 Pesaro risk class to assess risk factors for BMT. This classification is based upon clinical features of thalassemia that include: (1) adherence to a program of regular iron chelation therapy, (2) the presence or absence of hepatomegaly and (3) the presence or absence of portal fibrosis observed by liver biopsy. The conditioning regimen is uniform for classes 1 and 2 patients, but is modified for those who have class 3 features due to an increased risk of transplant-related mortality (Lucarelli, 1990). As a result of this risk classification and the development of new conditioning regimens, the outcome of thalassemia patients have improved with thalassemia-free survival and EFS over 70% reported worldwide. When stratifying patients, initially those with Pesaro Class 1 characteristics < 17 years had a superior thalassemia-free survival; however, recent updates show that outcomes are very similar across all three risk categories after employing risk-based conditioning regimens (Bhatia, 2008). Unrelated donor transplants are also used in selected patients (Bhatia, 2008). Following transplant, iron overload may still be a problem; consequently, chelation or phlebotomy may still be necessary.

3.2 Sickle cell disease (SCD)

SCD contrasts with thalassemia major by its variable course of clinical severity. Its typical clinical manifestation include anemia, severe painful crisis, acute chest syndrome, splenic sequestration, stroke (clinically overt and silent), chronic pulmonary and renal dysfunction, growth retardation, neuropsychological deficits and premature death. Historically, the mainstays of treatment are both preventive and supportive. The three major therapeutic options available for children affected with SCD are: chronic blood transfusion, hydroxyurea and HSCT. Of these options, only HSCT affords patients the possibility of cure. The use of transplantation for the treatment of patients with SCD has been considered for many years. However, because of the morbidity and mortality of HSCT, it was considered too risky. Recently, due to advances in supportive care and immunosuppressive therapy, transplant is again being considered for SCD. The preliminary experience of HSCT for beta-thalassemia major has in part provided the rationale for extending this treatment to sickle cell anemia. Walter et al (Walter et al., 1996) used selection criteria similar to that applied to patients with beta thalassemia major and chose patients with debilitating clinical events, including stroke, recurrent acute chest syndrome and recurrent painful vaso-occlusive crises, but selected children rather than adults and before the development of permanent end organ damage. These recommendations are associated with significant morbidity and early mortality among patients with SCD and are the criteria upon which most early studies using HSCT are based.

Three major clinical series account for most of the experience of HSCT for SCD (Bernaudin et al, 2007; Walters et al, 2000; Vermylen et al, 1998). In all three series, the majority of patients received HLA-identical sibling donor allograft and all patients received the same conditioning regimen (busulfan 14-16mg/kg with cytoxan 200) and GVHD prophylaxis (ATG, cyclosporine and methotrexate). The results of these three studies were very similar. OS was 92-94% and EFS was 82-86% with a median follow-up range of 0.9-17.9 years. TRM from all three series was also similar and was approximately 7% with infections as the chief cause. Similarly, the incidence of aGVHD > grade II was approximately 15-20%. The rate of cGVHD was 20% in Vermylen et al study compared to 12 and 13.5% in the Walters et al and Bernaudin et al reports, respectively. While HSCT is curative in patients with SCD, only 14-18% of patients have a matched family donor. The use of unrelated donors in HSCT for SCD is under development. There are several limitations which restrict the uniform utilization of allogeneic adult donors that include donor availability, and the high risk of severe aGVHD. The use of unrelated cord blood transplantation is also being considered and recent studies have shown promising results, although g raft rejection and aGVHD still remain issues. In addition, efforts to expand the application of HSCT for SCD have been restricted not only by lacking suitable donors, but also by the risk of significant toxicity from the myeloablative conditioning regimen. With the advent of lower intensity conditioning regimens which rely on less myeloablation and more immunosuppression, many of the long-term effects, such as growth and endocrine dysfunction observed after myeloablative conditioning regimens, may be ameliorated.

3.3 Autoimmune disease

Autoimmune diseases are often controlled with treatments that act on the immune system. However, these therapies are usually not curative. Recently many autoimmune diseases have been treated with HSCT. The goal of autologous HSCT is to reset the immune system. Studies on thymic lymphocytes after auto HSCT have shown that, after a burst sustained by pre transplant memory cells, the organ is repopulated by likely harvest-derived naïve T cells, and also the T-lymphocyte repertoire may significantly differ before and after autografting, thus suggesting the possibility of achieving an immune resetting through autologous HSCT (Isaacs, 2004; Sun, 2004). Allogeneic probably results in the highest potential for cure. However, there is higher morbidity and mortality caused by GVHD. Marmont summarize several allogeneic transplant cases in which the patient achieved full post transplant donor chimerism but their autoimmune disease still relapsed. A European database, the International Autoimmune Disease Stem Cell Project Database, was established in 1996. The database contains 600 patients, most treated with autologous HSCT; 15% of the patients registered are children. Some of the autoimmune diseases in children that were treated with HSCT are juvenile idiopathic arthritis, immune cytopenias, systemic sclerosis, systemic lupus and Crohn's disease (Rabusin, 2008).

3.4 Other non-malignant disease

3.4.1 Autosomal recessive osteopetrosis (ARO)

ARO is a rare genetic bone disease in which a deficit in bone resorption by osteoclasts leads to increased bone density and secondary defects. The disease is often lethal early in life unless treated with HSCT. However, recently the dissection of the molecular bases of the disease has shown that ARO is genetically heterogeneous and has revealed the presence of subsets of patients which do not benefit from HSCT, highlighting the importance of molecular diagnosing ARO to identify and establish the proper therapies for better prognosis (Villa, 2008). EBMT conducted a retrospective analysis of 122 children who had received an allogeneic HSCT for ARO between 1980 and 2001. The actuarial probabilities of 5 years disease free survival were 73% for recipients of a genotype HLA-identical HSCT (n=40), 43% for recipients of a phenotype HLA-identical or one HLA antigen mismatch graft from a related donor (n=21), 40% for recipients of a graft from a matched unrelated donor (n=20) and 24% for patients who received a graft from an HLA-haplotype-mismatch related donor (n=41). Causes of death after HSCT were graft failure and early-TRM complications. Conservation of vision was better in children transplanted before the age of 3 months (Driessen, 2003). HSCT is the only curative treatment for ARO and should be offered as early as possible.

3.4.2 Congenital erythropoietic porphyria (CEP)

CEP is a rare autosomal recessive disorder of porphyrin metabolism in which the genetic defect is the deficiency of uroporphyrinogen III cosynthase (UIIIC). Deficiency of this enzyme results in an accumulation of high amounts of uroporphyrin I in all tissues leading to hemolytic anemia, splenomegaly, erythrodontia, bone fragility, exquisite photosensitivity and mutilating skin lesions. The vital prognosis is very bad and until now, no treatment seems to be efficient. Bone marrow transplantation seems to be able to correct the enzymatic deficit that causes the disease because it is located in the bone marrow. A few cases of patients have been reported to be cured of the disease with stem cell transplantation (Shaw, 2001). HSCT should be strongly considered because this is currently the only known curative therapy.

3.4.3 Immune dysregulation, polyendocrinopathy, enteropathy, X-linked (IPEX) syndrome

IPEX syndrome is a rare, fatal autoimmune disorder caused by mutations in the forkhead box protein 3 (FOXP3) genes leading to the disruption of signaling pathways involved in regulatory T-Lymphocyte function. Patients with IPEX syndrome often present in early infancy and without therapeutic intervention, affected male patients usually die within the first or second year of life. These patients require supportive therapy including parental nutrition, insulin, antibiotics and blood transfusions. Immunosuppressive therapy has been used with variable improvement in symptoms. Correction of the dysregulated immune system can be achieved by allogeneic HSCT using a suitable donor. Although, HSCT is the only viable option for long-term survival, patients are usually very ill to tolerate traditional myeloablative conditioning regimens. Recent studies reported the successful outcome of HSCT using a low-intensity, nonmyeloablative conditioning regimen in 2 patients with IPEX syndrome and significant pre transplant risk factors (Burroughs, 2010; Rao, 2007).

3.4.4 Epidermolysis bullosa (EM)

EB is a group of blistering skin disorders resulting from mutations in genes encoding protein components of the cutaneous basement membrane zone. HSCT has been shown to ameliorate the deficiency of the skin-specific structural protein in children with EB (Fujita, 2010; Tolar, 2011).

4. Conclusion

The indications for HSCT are continually changing and expanding rapidly beyond the traditional use as a treatment for malignant and nonmalignant diseases. The inclusion of cord blood as a source of stem cells and the availability of reduce intensity regimens has allowed us to expand the indications for HSCT to patients who otherwise would not meet accepted criteria for conventional HSCT. The field of HSCT is continually growing and a great deal of additional research is needed to continue to improve our outcomes. This is an exciting time in HSCT with many new avenues becoming available for patients.

5. References

[1] The role of cytotoxic therapy with hematopoietic stem cell transplantation in the therapy of acute myeloid leukemia in children. *Biol Blood Marrow Transplant.* Apr 2007;13(4):500-501.

[2] Aldenhoven M, Boelens JJ, de Koning TJ. The clinical outcome of Hurler syndrome after stem cell transplantation. *Biol Blood Marrow Transplant.* May 2008; 14(5):485-498.

[3] Anderlini P, Saliba R, Acholonu S, et al. Fludarabine-melphalan as a preparative regimen for reduced-intensity conditioning allogeneic stem cell transplantation in relapsed and refractory Hodgkin's lymphoma: the updated M.D. Anderson Cancer Center experience. *Haematologica.* Feb 2008; 93(2):257-264.

[4] Apperley J. CML in pregnancy and childhood. *Best Pract Res Clin Haematol.* Sep 2009; 22(3):455-474.

[5] Arndt C, Tefft M, Gehan E, et al. A feasibility, toxicity, and early response study of etoposide, ifosfamide, and vincristine for the treatment of children with rhabdomyosarcoma: a report from the Intergroup Rhabdomyosarcoma Study (IRS) IV pilot study. *J Pediatr Hematol Oncol.* Mar-Apr 1997; 19(2):124-129.

[6] Atra A, Gerrard M, Hobson R, Imeson JD, Hann IM, Pinkerton CR. Outcome of relapsed or refractory childhood B-cell acute lymphoblastic leukaemia and B-cell non-Hodgkin's lymphoma treated with the UKCCSG 9003/9002 protocols. *Br J Haematol.* Mar 2001; 112(4):965-968.

[7] Ayas M, Al-Jefri A, Al-Seraihi A, Elkum N, Al-Mahr M, El-Solh H. Matched-related allogeneic stem cell transplantation in Saudi patients with Fanconi anemia: 10 year's experience. *Bone Marrow Transplant.* Aug 2008; 42 Suppl 1:S45-S48.

[8] Ayas M, Solh H, Mustafa MM, et al. Bone marrow transplantation from matched siblings in patients with fanconi anemia utilizing low-dose cyclophosphamide, thoracoabdominal radiation and antithymocyte globulin. *Bone Marrow Transplant.* Jan 2001; 27(2):139-143.

[9] Baccarani M, Cortes J, Pane F, Niederwieser D, Saglio G, Apperley J, et al. Chronic myeloid leukemia: an update of concepts and management recommendations of European Leukemia Net. *J Clin Oncol 2009; 27:6041-51.*

[10] Bader P, Kreyenberg H, Hoelle W, et al. Increasing mixed chimerism defines a high-risk group of childhood acute myelogenous leukemia patients after allogeneic stem cell transplantation where pre-emptive immunotherapy may be effective. *Bone Marrow Transplant.* Apr 2004; 33(8):815-821.

[11] Beam D, Poe MD, Provenzale JM, et al. Outcomes of unrelated umbilical cord blood transplantation for X-linked adrenoleukodystrophy. *Biol Blood Marrow Transplant.* Jun 2007; 13(6):665-674.

[12] Bensinger WI, Buckner CD, Demirer T, Storb R, Appelbaum FA. Transplantation of allogeneic peripheral blood stem cells. *Bone Marrow Transplant.* Mar 1996; 17 Suppl 2:S56-57.

[13] Bensinger WI, Weaver CH, Appelbaum FR, et al. Transplantation of allogeneic peripheral blood stem cells mobilized by recombinant human granulocyte colony-stimulating factor. *Blood.* Mar 15 1995; 85(6):1655-1658.

[14] Berger R, Bernheim A, Gluckman E, Gisselbrecht C. In vitro effect of cyclophosphamide metabolites on chromosomes of Fanconi anaemia patients. *Br J Haematol.* Aug 1980;45(4):565-568.

[15] Bernaudin F, Socie G, Kuentz M, et al. Long-term results of related myeloablative stem-cell transplantation to cure sickle cell disease. *Blood.* Oct 1 2007; 110(7):2749-2756.

[16] Berndt A, Helwig A, Ehninger G, Bornhauser M. Successful transplantation of CD34+ selected peripheral blood stem cells from an unrelated donor in an adult patient with Diamond-Blackfan anemia and secondary hemochromatosis. *Bone Marrow Transplant.* Jan 2005; 35(1):99-100.

[17] Bhatia M, Walters MC. Hematopoietic cell transplantation for thalassemia and sickle cell disease: past, present and future. *Bone Marrow Transplant.* Jan 2008;41(2):109-117.

[18] Boelens JJ, Rocha V, Aldenhoven M, et al. Risk factor analysis of outcomes after unrelated cord blood transplantation in patients with hurler syndrome. *Biol Blood Marrow Transplant.* May 2009; 15(5):618-625.

[19] Borgmann A, Hartmann R, Schmid H, et al. Isolated extramedullary relapse in children with acute lymphoblastic leukemia: a comparison between treatment results of chemotherapy and bone marrow transplantation. BFM Relapse Study Group. *Bone Marrow Transplant.* Apr 1995; 15(4):515-521.

[20] Buckley RH, Schiff SE, Schiff RI, et al. Hematopoietic stem-cell transplantation for the treatment of severe combined immunodeficiency. *N Engl J Med.* Feb 18 1999;340(7):508-516.

[21] Burroughs LM, Torgerson TR, Storb R, et al. Stable hematopoietic cell engraftment after low-intensity nonmyeloablative conditioning in patients with immune dysregulation, polyendocrinopathy, enteropathy, X-linked syndrome. *J Allergy Clin Immunol.* Nov 2010; 126(5):1000-1005.

[22] Cairo MS, Gerrard M, Sposto R, et al. Results of a randomized international study of high-risk central nervous system B non-Hodgkin lymphoma and B acute lymphoblastic leukemia in children and adolescents. *Blood.* Apr 1 2007;109(7):2736-2743.

[23] Cairo MS, Sposto R, Hoover-Regan M, et al. Childhood and adolescent large-cell lymphoma (LCL): a review of the Children's Cancer Group experience. *Am J Hematol.* Jan 2003; 72(1):53-63.

[24] Cairo MS, Sposto R, Perkins SL, et al. Burkitt's and Burkitt-like lymphoma in children and adolescents: a review of the Children's Cancer Group experience. *Br J Haematol.* Feb 2003;120(4):660-670.

[25] Carella AM, Cavaliere M, Lerma E, et al. Autografting followed by nonmyeloablative immunosuppressive chemotherapy and allogeneic peripheral-blood hematopoietic stem-cell transplantation as treatment of resistant Hodgkin's disease and non-Hodgkin's lymphoma. *J Clin Oncol.* Dec 1 2000; 18(23):3918-3924.

[26] Cavazzana-Calvo M, Fischer A. Gene therapy for severe combined immunodeficiency: are we there yet? *J Clin Invest.* Jun 2007; 117(6):1456-1465.

[27] Cesaro S, Guariso G, Calore E, et al. Successful unrelated bone marrow transplantation for Shwachman-Diamond syndrome. *Bone Marrow Transplant.* Jan 2001; 27(1):97-99.

[28] Chaudhury S, Auerbach AD, Kernan NA, et al. Fludarabine-based cytoreductive regimen and T-cell-depleted grafts from alternative donors for the treatment of high-risk patients with Fanconi anaemia. *Br J Haematol.* Mar 2008;140(6):644-655.

[29] Chessells JM, Rogers DW, Leiper AD, et al. Bone-marrow transplantation has a limited role in prolonging second marrow remission in childhood lymphoblastic leukaemia. *Lancet.* May 31 1986; 1(8492):1239-1241.

[30] Clavell LA, Gelber RD, Cohen HJ, et al. Four-agent induction and intensive asparaginase therapy for treatment of childhood acute lymphoblastic leukemia. *N Engl J Med.* Sep 11 1986; 315(11):657-663.

[31] Creutzig U, Ritter J, Zimmermann M, Klingebiel T. [Prognosis of children with chronic myeloid leukemia: a retrospective analysis of 75 patients]. *Klin Padiatr.* Jul-Aug 1996;208(4):236-241.

[32] Creutzig U, Zimmermann M, Ritter J, et al. Treatment strategies and long-term results in paediatric patients treated in four consecutive AML-BFM trials. *Leukemia.* Dec 2005; 19(12):2030-2042.

[33] Cwynarski K, Roberts IA, Iacobelli S, et al. Stem cell transplantation for chronic myeloid leukemia in children. *Blood.* Aug 15 2003; 102(4):1224-1231.

[34] de la Fuente J, Dokal I. Dyskeratosis congenita: advances in the understanding of the telomerase defect and the role of stem cell transplantation. *Pediatr Transplant.* Sep 2007; 11(6):584-594.

[35] Dokal I. Dyskeratosis congenita in all its forms. *Br J Haematol.* Sep 2000; 110(4):768-779.

[36] Dreger P, Glass B, Uharek L, Zeis M, Schmitz N. Allogenic transplantation of mobilized peripheral blood progenitor cells: towards tailored cell therapy. *Int J Hematol.* Jul 1997;66(1):1-11.

[37] Druker BJ, Talpaz M, Resta DJ, et al. Efficacy and safety of a specific inhibitor of the BCR-ABL tyrosine kinase in chronic myeloid leukemia. *N Engl J Med.* Apr 5 2001;344(14):1031-1037.

[38] Escolar ML, Poe MD, Provenzale JM, et al. Transplantation of umbilical-cord blood in babies with infantile Krabbe's disease. *N Engl J Med.* May 19 2005; 352(20):2069-2081.

[39] Farzin A, Davies SM, Smith FO, et al. Matched sibling donor haematopoietic stem cell transplantation in Fanconi anaemia: an update of the Cincinnati Children's experience. *Br J Haematol.* Feb 2007; 136(4):633-640.

[40] Federico M, Bellei M, Brice P, et al. High-dose therapy and autologous stem-cell transplantation versus conventional therapy for patients with advanced Hodgkin's lymphoma responding to front-line therapy. *J Clin Oncol.* Jun 15 2003; 21(12):2320-2325.

[41] Feld JJ, Hussain N, Wright EC, et al. Hepatic involvement and portal hypertension predict mortality in chronic granulomatous disease. *Gastroenterology.* Jun 2008; 134(7):1917-1926.

[42] Filipovich AH, Stone JV, Tomany SC, et al. Impact of donor type on outcome of bone marrow transplantation for Wiskott-Aldrich syndrome: collaborative study of the International Bone Marrow Transplant Registry and the National Marrow Donor Program. *Blood.* Mar 15 2001;97(6):1598-1603.

[43] Fleitz J, Rumelhart S, Goldman F, et al. Successful allogeneic hematopoietic stem cell transplantation (HSCT) for Shwachman-Diamond syndrome. *Bone Marrow Transplant.* Jan 2002; 29(1):75-79.

[44] Fujita Y, Abe R, Inokuma D, et al. Bone marrow transplantation restores epidermal basement membrane protein expression and rescues epidermolysis bullosa model mice. *Proc Natl Acad Sci U S A.* Aug 10 2010; 107(32):14345-14350.

[45] Gaynon PS, Trigg ME, Heerema NA, et al. Children's Cancer Group trials in childhood acute lymphoblastic leukemia: 1983-1995. *Leukemia.* Dec 2000; 14(12):2223-2233.

[46] Gerrard M, Cairo MS, Weston C, et al. Excellent survival following two courses of COPAD chemotherapy in children and adolescents with resected localized B-cell non-Hodgkin's lymphoma: results of the FAB/LMB 96 international study. *Br J Haematol.* Jun 2008; 141(6):840-847.

[47] Gibson BE, Wheatley K, Hann IM, et al. Treatment strategy and long-term results in paediatric patients treated in consecutive UK AML trials. *Leukemia.* Dec 2005; 19(12):2130-2138.

[48] Gibson BE, Wheatley K, Hann IM, et al. Treatment strategy and long-term results in paediatric patients treated in consecutive UK AML trials. *Leukemia.* Dec 2005;19(12):2130-2138.

[49] Ginzberg H, Shin J, Ellis L, et al. Shwachman syndrome: phenotypic manifestations of sibling sets and isolated cases in a large patient cohort are similar. *J Pediatr.* Jul 1999; 135(1):81-88.

[50] Gluckman E, Auerbach AD, Horowitz MM, et al. Bone marrow transplantation for Fanconi anemia. *Blood.* Oct 1 1995; 86(7):2856-2862.

[51] Gluckman EG, Roch VV, Chastang C. Use of Cord Blood Cells for Banking and Transplant. *Oncologist.* 1997; 2(5):340-343.

[52] Gross TG, Hale GA, He W, et al. Hematopoietic stem cell transplantation for refractory or recurrent non-Hodgkin lymphoma in children and adolescents. *Biol Blood Marrow Transplant.* Feb 2010; 16(2):223-230.

[53] Guardiola P, Socie G, Li X, et al. Acute graft-versus-host disease in patients with Fanconi anemia or acquired aplastic anemia undergoing bone marrow transplantation from HLA-identical sibling donors: risk factors and influence on outcome. *Blood.* Jan 1 2004; 103(1):73-77.

[54] Guffon N, Bertrand Y, Forest I, Fouilhoux A, Froissart R. Bone marrow transplantation in children with Hunter syndrome: outcome after 7 to 17 years. *J Pediatr.* May 2009; 154(5):733-737.

[55] Gungor N, Tuncbilek E. Sanfilippo disease type B. A case report and review of the literature on recent advances in bone marrow transplantation. *Turk J Pediatr.* Apr-Jun 1995;37(2):157-163.

[56] Henter JI, Samuelsson-Horne A, Arico M, et al. Treatment of hemophagocytic lymphohistiocytosis with HLH-94 immunochemotherapy and bone marrow transplantation. *Blood.* Oct 1 2002; 100(7):2367-2373.

[57] Herskhovitz E, Young E, Rainer J, et al. Bone marrow transplantation for Maroteaux-Lamy syndrome (MPS VI): long-term follow-up. *J Inherit Metab Dis.* Feb 1999;22(1):50-62.

[58] Hoogerbrugge PM, Gerritsen EJ, vd Does-van den Berg A, et al. Case-control analysis of allogeneic bone marrow transplantation versus maintenance chemotherapy for relapsed ALL in children. *Bone Marrow Transplant.* Feb 1995; 15(2):255-259.

[59] Hughes TP, Kaeda J, Branford S, et al. Frequency of major molecular responses to imatinib or interferon alfa plus cytarabine in newly diagnosed chronic myeloid leukemia. *N Engl J Med.* Oct 9 2003; 349(15):1423-1432.

[60] Isaacs JD, Thiel A. Stem cell transplantation for autoimmune disorders. Immune reconstitution. *Best Pract Res Clin Haematol.* Jun 2004; 17(2):345-358.

[61] Kardos G, Baumann I, Passmore SJ, et al. Refractory anemia in childhood: a retrospective analysis of 67 patients with particular reference to monosomy 7. *Blood.* Sep 15 2003;102(6):1997-2003.

[62] Kawakami M, Tsutsumi H, Kumakawa T, et al. Levels of serum granulocyte colony-stimulating factor in patients with infections. *Blood.* Nov 15 1990; 76(10):1962-1964.

[63] Kersey JH. Fifty years of studies of the biology and therapy of childhood leukemia. *Blood.* Dec 1 1997;90(11):4243-4251.

[64] Kessinger A, Smith DM, Strandjord SE, et al. Allogeneic transplantation of blood-derived, T cell-depleted hemopoietic stem cells after myeloablative treatment in a patient with acute lymphoblastic leukemia. *Bone Marrow Transplant.* Nov 1989; 4(6):643-646.

[65] Krivit W, Shapiro EG, Peters C, et al. Hematopoietic stem-cell transplantation in globoid-cell leukodystrophy. *N Engl J Med.* Apr 16 1998;338(16):1119-1126.

[66] Kuhns DB, Alvord WG, Heller T, et al. Residual NADPH oxidase and survival in chronic granulomatous disease. *N Engl J Med.* Dec 30 2010;363(27):2600-2610.

[67] Kutler DI, Singh B, Satagopan J, et al. A 20-year perspective on the International Fanconi Anemia Registry (IFAR). *Blood.* Feb 15 2003; 101(4):1249-1256.

[68] Ladenstein R, Potschger U, Hartman O, et al. 28 years of high-dose therapy and SCT for neuroblastoma in Europe: lessons from more than 4000 procedures. *Bone Marrow Transplant.* Jun 2008; 41 Suppl 2:S118-127.

[69] Lange BJ et al. Distinctive demography, biology, and outcome of acute myeloid leukemia and myelodysplastic syndrome in children with Down syndrome: Children's Cancer Group Studies 2861 and 2891. *Blood.* 1998 Jan 15; 91 (2): 608-15.

[70] Lazarus HM, Rowlings PA, Zhang MJ, et al. Autotransplants for Hodgkin's disease in patients never achieving remission: a report from the Autologous Blood and Marrow Transplant Registry. *J Clin Oncol.* Feb 1999; 17(2):534-545.

[71] Lee JW, Chung NG. The treatment of pediatric chronic myelogenous leukemia in the imatinib era. *Korean J Pediatr.* Mar 2011; 54(3):111-116.

[72] Linch DC, Winfield D, Goldstone AH, et al. Dose intensification with autologous bone-marrow transplantation in relapsed and resistant Hodgkin's disease: results of a BNLI randomised trial. *Lancet.* Apr 24 1993; 341(8852):1051-1054.

[73] Link MP, Shuster JJ, Donaldson SS, Berard CW, Murphy SB. Treatment of children and young adults with early-stage non-Hodgkin's lymphoma. *N Engl J Med.* Oct 30 1997; 337(18):1259-1266.

[74] Lipton JM, Atsidaftos E, Zyskind I, Vlachos A. Improving clinical care and elucidating the pathophysiology of Diamond Blackfan anemia: an update from the Diamond Blackfan Anemia Registry. *Pediatr Blood Cancer.* May 1 2006;46(5):558-564.

[75] Ljungman P, Bregni M, Brune M, et al. Allogeneic and autologous transplantation for haematological diseases, solid tumours and immune disorders: current practice in Europe 2009. *Bone Marrow Transplant.* Feb 2010; 45(2):219-234.

[76] Ljungman P, Urbano-Ispizua A, Cavazzana-Calvo M, et al. Allogeneic and autologous transplantation for haematological diseases, solid tumours and immune disorders: definitions and current practice in Europe. *Bone Marrow Transplant.* Mar 2006; 37(5):439-449.

[77] Locatelli F, Zecca M, Pession A, et al. The outcome of children with Fanconi anemia given hematopoietic stem cell transplantation and the influence of fludarabine in the conditioning regimen: a report from the Italian pediatric group. *Haematologica.* Oct 2007;92(10):1381-1388.

[78] Loes DJ, Fatemi A, Melhem ER, et al. Analysis of MRI patterns aids prediction of progression in X-linked adrenoleukodystrophy. *Neurology.* Aug 12 2003; 61(3):369-374.

[79] Loes DJ, Hite S, Moser H, et al. Adrenoleukodystrophy: a scoring method for brain MR observations. *AJNR Am J Neuroradiol.* Oct 1994; 15(9):1761-1766.

[80] Mack DR, Forstner GG, Wilschanski M, Freedman MH, Durie PR. Shwachman syndrome: exocrine pancreatic dysfunction and variable phenotypic expression. *Gastroenterology.* Dec 1996; 111(6):1593-1602.

[81] Marks DI, Khattry N, Cummins M, et al. Haploidentical stem cell transplantation for children with acute leukaemia. *Br J Haematol.* Jul 2006; 134(2):196-201.

[82] Marmont AM. Allogeneic haematopoietic stem cell transplantation for severe autoimmune diseases: great expectations but controversial evidence. *Bone Marrow Transplant.* Jul 2006; 38(1):1-4.

[83] Matthay KK, Reynolds CP, Seeger RC, et al. Long-term results for children with high-risk neuroblastoma treated on a randomized trial of myeloablative therapy followed by 13-cis-retinoic acid: a children's oncology group study. *J Clin Oncol.* Mar 1 2009; 27(7):1007-1013.

[84] Miano M, Labopin M, Hartmann O, et al. Haematopoietic stem cell transplantation trends in children over the last three decades: a survey by the paediatric diseases working party of the European Group for Blood and Marrow Transplantation. *Bone Marrow Transplant.* Jan 2007; 39(2):89-99.

[85] Millot F, Esperou H, Bordigoni P, et al. Allogeneic bone marrow transplantation for chronic myeloid leukemia in childhood: a report from the Societe Francaise de Greffe de Moelle et de Therapie Cellulaire (SFGM-TC). *Bone Marrow Transplant.* Nov 2003; 32(10):993-999.

[86] Millot F, Guilhot J, Nelken B, et al. Imatinib mesylate is effective in children with chronic myelogenous leukemia in late chronic and advanced phase and in relapse after stem cell transplantation. *Leukemia.* Feb 2006; 20(2):187-192.

[87] Muramatsu H, Kojima S, Yoshimi A, et al. Outcome of 125 children with chronic myelogenous leukemia who received transplants from unrelated donors: the Japan Marrow Donor Program. *Biol Blood Marrow Transplant.* Feb 2010; 16(2):231-238.

[88] Neudorf S, Sanders J, Kobrinsky N, et al. Allogeneic bone marrow transplantation for children with acute myelocytic leukemia in first remission demonstrates a role for graft versus leukemia in the maintenance of disease-free survival. *Blood.* May 15 2004;103(10):3655-3661.

[89] Novotny J, Kadar J, Hertenstein B, et al. Sustained decrease of peripheral lymphocytes after allogeneic blood stem cell aphereses. *Br J Haematol.* Mar 1998;100(4):695-697.

[90] O'Brien SG, Guilhot F, Larson RA, et al. Imatinib compared with interferon and low-dose cytarabine for newly diagnosed chronic-phase chronic myeloid leukemia. *N Engl J Med.* Mar 13 2003;348(11):994-1004.

[91] Oliansky DM, Rizzo JD, Aplan PD, et al. The role of cytotoxic therapy with hematopoietic stem cell transplantation in the therapy of acute myeloid leukemia in children: an evidence-based review. *Biol Blood Marrow Transplant.* Jan 2007; 13(1):1-25.

[92] Ostronoff M, Florencio R, Campos G, et al. Successful nonmyeloablative bone marrow transplantation in a corticosteroid-resistant infant with Diamond-Blackfan anemia. *Bone Marrow Transplant.* Aug 2004; 34(4):371-372.

[93] Patte C, Auperin A, Gerrard M, et al. Results of the randomized international FAB/LMB96 trial for intermediate risk B-cell non-Hodgkin lymphoma in children and adolescents: it is possible to reduce treatment for the early responding patients. *Blood.* Apr 1 2007; 109(7):2773-2780.

[94] Peters C, Charnas LR, Tan Y, et al. Cerebral X-linked adrenoleukodystrophy: the international hematopoietic cell transplantation experience from 1982 to 1999. *Blood.* Aug 1 2004;104(3):881-888.

[95] Peters C, Shapiro EG, Anderson J, et al. Hurler syndrome: II. Outcome of HLA-genotypically identical sibling and HLA-haploidentical related donor bone marrow transplantation in fifty-four children. The Storage Disease Collaborative Study Group. *Blood.* Apr 1 1998; 91(7):2601-2608.

[96] Peters C, Steward CG. Hematopoietic cell transplantation for inherited metabolic diseases: an overview of outcomes and practice guidelines. *Bone Marrow Transplant.* Feb 2003; 31(4):229-239.

[97] Prasad VK, Kurtzberg J. Emerging trends in transplantation of inherited metabolic diseases. *Bone Marrow Transplant.* Jan 2008; 41(2):99-108.

[98] Prasad VK, Mendizabal A, Parikh SH, et al. Unrelated donor umbilical cord blood transplantation for inherited metabolic disorders in 159 pediatric patients from a single center: influence of cellular composition of the graft on transplantation outcomes. *Blood.* Oct 1 2008; 112(7):2979-2989.

[99] Proctor SJ, Mackie M, Dawson A, et al. A population-based study of intensive multi-agent chemotherapy with or without autotransplant for the highest risk Hodgkin's disease patients identified by the Scotland and Newcastle Lymphoma Group (SNLG) prognostic index. A Scotland and Newcastle Lymphoma Group study (SNLG HD III). *Eur J Cancer.* Apr 2002; 38(6):795-806.

[100] Pui CH. Childhood leukemias. *N Engl J Med.* Jun 15 1995; 332(24):1618-1630.

[101] Pui CH, Evans WE. Acute lymphoblastic leukemia. *N Engl J Med.* Aug 27 1998; 339(9):605-615.

[102] Rabusin M, Andolina M, Maximova N. Haematopoietic SCT in autoimmune diseases in children: rationale and new perspectives. *Bone Marrow Transplant.* Jun 2008; 41 Suppl 2:S96-99.

[103] Rao A, Kamani N, Filipovich A, et al. Successful bone marrow transplantation for IPEX syndrome after reduced-intensity conditioning. *Blood.* Jan 1 2007; 109(1):383-385.

[104] Reiter A, Schrappe M, Ludwig WD, et al. Chemotherapy in 998 unselected childhood acute lymphoblastic leukemia patients. Results and conclusions of the multicenter trial ALL-BFM 86. *Blood.* Nov 1 1994; 84(9):3122-3133.

[105] Ritchey AK, Pollock BH, Lauer SJ, Andejeski Y, Barredo J, Buchanan GR. Improved survival of children with isolated CNS relapse of acute lymphoblastic leukemia: a pediatric oncology group study. *J Clin Oncol.* Dec 1999; 17(12):3745-3752.

[106] Rivera GK, Pinkel D, Simone JV, Hancock ML, Crist WM. Treatment of acute lymphoblastic leukemia. 30 years' experience at St. Jude Children's Research Hospital. *N Engl J Med.* Oct 28 1993; 329(18):1289-1295.
[107] Rocha V, Wagner JE, Jr., Sobocinski KA, et al. Graft-versus-host disease in children who have received a cord-blood or bone marrow transplant from an HLA-identical sibling. Eurocord and International Bone Marrow Transplant Registry Working Committee on Alternative Donor and Stem Cell Sources. *N Engl J Med.* Jun 22 2000; 342(25):1846-1854.
[108] Rosenberg PS, Alter BP, Ebell W. Cancer risks in Fanconi anemia: findings from the German Fanconi Anemia Registry. *Haematologica.* Apr 2008;93(4):511-517.
[109] Roy V, Perez WS, Eapen M, et al. Bone marrow transplantation for diamond-blackfan anemia. *Biol Blood Marrow Transplant.* Aug 2005; 11(8):600-608.
[110] Satwani P, Sather H, Ozkaynak F, et al. Allogeneic bone marrow transplantation in first remission for children with ultra-high-risk features of acute lymphoblastic leukemia: A children's oncology group study report. *Biol Blood Marrow Transplant.* Feb 2007; 13(2):218-227.
[111] Schmit-Pokorny K. Expanding indications for stem cell transplantation. *Semin Oncol Nurs.* May 2009; 25(2):105-114.
[112] Schmitz N, Pfistner B, Sextro M, et al. Aggressive conventional chemotherapy compared with high-dose chemotherapy with autologous haemopoietic stem-cell transplantation for relapsed chemosensitive Hodgkin's disease: a randomised trial. *Lancet.* Jun 15 2002;359(9323):2065-2071.
[113] Seger RA, Gungor T, Belohradsky BH, et al. Treatment of chronic granulomatous disease with myeloablative conditioning and an unmodified hemopoietic allograft: a survey of the European experience, 1985-2000. *Blood.* Dec 15 2002;100(13):4344-4350.
[114] Silverman LB, Gelber RD, Dalton VK, et al. Improved outcome for children with acute lymphoblastic leukemia: results of Dana-Farber Consortium Protocol 91-01. *Blood.* Mar 1 2001;97(5):1211-1218.
[115] Staba SL, Escolar ML, Poe M, et al. Cord-blood transplants from unrelated donors in patients with Hurler's syndrome. *N Engl J Med.* May 6 2004; 350(19):1960-1969.
[116] Stary J, Locatelli F, Niemeyer CM. Stem cell transplantation for aplastic anemia and myelodysplastic syndrome. *Bone Marrow Transplant.* Mar 2005; 35 Suppl 1:S13-16.
[117] Sun W, Popat U, Hutton G, et al. Characteristics of T-cell receptor repertoire and myelin-reactive T cells reconstituted from autologous haematopoietic stem-cell grafts in multiple sclerosis. *Brain.* May 2004; 127(Pt 5):996-1008.
[118] Suttorp M, Yaniv I, Schultz KR. Controversies in the treatment of CML in children and adolescents: TKIs versus BMT? *Biol Blood Marrow Transplant.* Jan 2011; 17(1 Suppl):S115-122.
[119] Sweetenham JW, Carella AM, Taghipour G, et al. High-dose therapy and autologous stem-cell transplantation for adult patients with Hodgkin's disease who do not enter remission after induction chemotherapy: results in 175 patients reported to the European Group for Blood and Marrow Transplantation. Lymphoma Working Party. *J Clin Oncol.* Oct 1999; 17(10):3101-3109.

[120] Tan PL, Wagner JE, Auerbach AD, Defor TE, Slungaard A, Macmillan ML. Successful engraftment without radiation after fludarabine-based regimen in Fanconi anemia patients undergoing genotypically identical donor hematopoietic cell transplantation. *Pediatr Blood Cancer.* May 1 2006; 46(5):630-636.
[121] Thomson KJ, Peggs KS, Smith P, et al. Superiority of reduced-intensity allogeneic transplantation over conventional treatment for relapse of Hodgkin's lymphoma following autologous stem cell transplantation. *Bone Marrow Transplant.* May 2008; 41(9):765-770.
[122] Tichelli A, Passweg J, Hoffmann T, et al. Repeated peripheral stem cell mobilization in healthy donors: time-dependent changes in mobilization efficiency. *Br J Haematol.* Jul 1999;106(1):152-158.
[123] Tolar J, Blazar BR, Wagner JE. Concise review: Transplantation of human hematopoietic cells for extracellular matrix protein deficiency in epidermolysis bullosa. *Stem Cells.* Jun 2011;29(6):900-906.
[124] Uderzo C, Grazia Zurlo M, Adamoli L, et al. Treatment of isolated testicular relapse in childhood acute lymphoblastic leukemia: an Italian multicenter study. Associazione Italiana Ematologia ed Oncologia Pediatrica. *J Clin Oncol.* Apr 1990; 8(4):672-677.
[125] Vardiman JW, Harris NL, Brunning RD. The World Health Organization (WHO) classification of the myeloid neoplasms. *Blood.* Oct 1 2002; 100(7):2292-2302.
[126] Vellodi A, Young E, New M, Pot-Mees C, Hugh-Jones K. Bone marrow transplantation for Sanfilippo disease type B. *J Inherit Metab Dis.* 1992;15(6):911-918.
[127] Vellodi A, Young EP, Cooper A, et al. Bone marrow transplantation for mucopolysaccharidosis type I: experience of two British centres. *Arch Dis Child.* Feb 1997; 76(2):92-99.
[128] Vermylen C, Cornu G, Ferster A, et al. Haematopoietic stem cell transplantation for sickle cell anaemia: the first 50 patients transplanted in Belgium. *Bone Marrow Transplant.* Jul 1998; 22(1):1-6.
[129] Vilmer E, Suciu S, Ferster A, et al. Long-term results of three randomized trials (58831, 58832, 58881) in childhood acute lymphoblastic leukemia: a CLCG-EORTC report. Children Leukemia Cooperative Group. *Leukemia.* Dec 2000; 14(12):2257-2266.
[130] Vlachos A, Ball S, Dahl N, et al. Diagnosing and treating Diamond Blackfan anaemia: results of an international clinical consensus conference. *Br J Haematol.* Sep 2008;142(6):859-876.
[131] Wagner JE, Eapen M, MacMillan ML, et al. Unrelated donor bone marrow transplantation for the treatment of Fanconi anemia. *Blood.* Mar 1 2007; 109(5):2256-2262.
[132] Wagner JE, Kernan NA, Steinbuch M, Broxmeyer HE, Gluckman E. Allogeneic sibling umbilical-cord-blood transplantation in children with malignant and non-malignant disease. *Lancet.* Jul 22 1995;346(8969):214-219.
[133] Walters MC, Patience M, Leisenring W, et al. Barriers to bone marrow transplantation for sickle cell anemia. *Biol Blood Marrow Transplant.* May 1996; 2(2):100-104.

[134] Walters MC, Storb R, Patience M, et al. Impact of bone marrow transplantation for symptomatic sickle cell disease: an interim report. Multicenter investigation of bone marrow transplantation for sickle cell disease. *Blood.* Mar 15 2000; 95(6):1918-1924.

[135] Weisdorf DJ, Anasetti C, Antin JH, et al. Allogeneic bone marrow transplantation for chronic myelogenous leukemia: comparative analysis of unrelated versus matched sibling donor transplantation. *Blood.* Mar 15 2002; 99(6):1971-1977.

17

Hematopoietic Stem Cell Potency for Cellular Therapeutic Transplantation

Karen M. Hall, Holli Harper and Ivan N. Rich
HemoGenix, Inc
U.S.A.

1. Introduction

Potency is the quantitative measurement of biological activity of a product (European Medicines Agency (EMA), 2008). Potency provides assurance that production and manufacture demonstrate consistency and provides information on stability and performance of the product. It also allows correlation with the clinical response and can help avoid product failure or toxicity due to the improper dose of the product being administered. For biopharmaceutical products such as drugs, growth factors and cytokines, vaccines etc., measurement of potency to predict dose has been a routine procedure for many years. Cells, on the other hand, are complex living entities that are continuously in flux. The potency of cells can change depending on numerous physiological and external environmental factors. Yet, with the increased number of cellular therapeutic applications and clinical regimen involving numerous cell types, the need to reliably and reproducibly measure biological and functional activity to meet the requirements of potency and ensure patient safety is of increasing importance (EMA, 2008; U.S. Food and Drug Administration, (FDA), 2011).

Determining the potency of a stem cell therapeutic can be a daunting task, especially if knowledge of the system biology, physiology and regulation is limited. In contrast, the hematopoietic system has proven to be not only an excellent model for stem cell biology, but also a model system for proliferation and differentiation in different applications. One of these applications is stem cell transplantation, a procedure that had its origins during the 1950s, became a quantitative assay in mice in 1961 (Till & McCulloch), and a routine clinical procedure in the 1970s (Santos et al. 1972; Thomas et al. 1977; Santos, 1983) Since that time, the number of human bone marrow transplantations reached a peak in the late 1990s (National Marrow Donor Program (NMDP); Pasquini & Wang, 2010) and has been declining to be replaced by alternative stem cell sources derived from mobilized peripheral blood (Haas et al. 1990; Koerbling et al. 1990; Sohn et.al. 2002) and umbilical cord blood (Broxmeyer et al. 1989; Gluckman et al. 1989).

Regardless of the tissue source, a successful transplant of stem cells is dependent upon the ability of the transplanted stem cells to lodge or "seed" in the bone marrow and begin the process of proliferation to produce lineage-specific progenitor cells. These differentiate into functionally mature circulating neutrophils, platelets and erythroid cells, the number of which provides information on the time at which engraftment took place. Proper lympho-hematopoietic reconstitution occurs much later. The ability of the stem cells to engraft is dependent upon two primary factors. The first is the status and condition of the patient. The second is the proliferation ability and potential of the stem cells prior to being transplanted.

Proliferation ability is equivalent to the proliferation status of the stem cells at the time of testing. This parameter defines stem cell "quality". Proliferation potential, on the other hand, is the capacity or potential of the stem cells to proliferate. For a continuously proliferating system such as lympho-hematopoiesis, stem cell potential decreases from the most primitive to the most mature stem cells. Thus, the more primitive a stem cell, the greater its proliferation potential and therefore its potency. It follows that the primary goal of stem cell transplantation is to provide the patient with stem cells that exhibit varying degrees of proliferation potential or potency. In this way, the patient can be endowed with stem cells that provide both short- (Charbord, 1994; Civin et al. 1996; Leung et al. 1999; Zubair et al. 2006) and long-term (Civin et al. 1996; Leung et al. 1999; Zubair et al. 2006; Duggan et al. 2000) engraftment and reconstitution.

A product that is "balanced" to provide the correct amount of short- and long-term stem cell engraftment and reconstitution would be the ideal situation. Present technology is not, however, capable of measuring or delivering a "balanced" stem cell product. In many cases, the donor stem cell product is skewed towards a greater proportion of mature rather than primitive stem cells or visa versa. However, it is possible to quantitatively measure both stem cell quality and potency of representative stem cell populations to provide a reasonably good approximation of the overall quality and potency of the stem cell product. These parameters would then predict the potential of the stem cells to engraft and reconstitute the system.

In 2009, the U.S. Food and Drug Administration (FDA, 2009) designated umbilical cord blood as a drug because, when transplanted into a patient, it results in systemic effects. The consequence of this designation has meant that virtually every aspect from cord blood collection to transplantation must be validated and documented according to regulatory requirements. Included in this process are the tests and assays to monitor the procedures and characterize the product prior to use. Besides histocompatability testing, the most important parameter that should be measured just prior to the stem cell product being used is potency. The FDA guidance on potency for cellular therapeutic products specifically describes the regulations that define a potency assay as compliant (FDA, 2011). A potency assay must provide quantitative data demonstrating the biological activity of all "active ingredients" specific to the product. In the case of a stem cell product, the "active ingredients" are the stem cell themselves. The results must meet pre-defined acceptance and/or rejection criteria so that the test results provide information as to whether the product can be released for use. In addition, the assay(s) must include reference materials, standards and controls, since without these, the necessary validation parameters (accuracy, sensitivity, specificity, precision and robustness) cannot be measured and documented.

The present communication describes an *in vitro* assay that measures stem cell potency and quality and helps define release criteria for hematopoietic products derived from mobilized peripheral blood, umbilical cord blood or bone marrow. The assay was designed to comply with regulatory requirements. In the 3-step process, all of the data required is accumulated in the initial stem cell culture and measurement step. The data obtained provides a degree of stem cell quality and potency assurance that has hitherto not been possible using the traditional methods of total nucleated cell count (TNC), viability and viable $CD34^+$ counts, which provide no indication of stem cell functionality or growth. To illustrate the steps of the assay, a small number of mobilized peripheral blood samples are used to demonstrate the procedure for determining potency, quality and release criteria. A larger cohort of umbilical cord blood samples is then used to show the applicability of the assay.

The assay relies on two basic characteristics of stem cells, namely proliferation ability (quality) and potential (potency). It had been previously demonstrated that when hematopoietic stem cells were stimulated to proliferate in the presence of growth factors and cytokines, the intracellular ATP (iATP) concentration increased proportionately to the cell concentration plated (Rich & Hall, 2005; Rich, 2007). The steepness or slope of the cell dose response was dependent upon the primitiveness and proliferation potential of the cells being examined. Stem cells have a greater proliferation potential than lineage-specific progenitor cells (Botnick e al. 1979). It would therefore be expected that the slope of the cell dose response would be steeper for stem cells than progenitor cells. In other words, the steeper the slope of the cell dose response, the greater the proliferation potential and the greater the potency. This biological phenomenon was incorporated into an assay that first estimates the potency ratio for two stem cell populations of a sample compared to a reference standard of the same material. The information obtained from the initial culture step was then used in the second step to substantiate the correlation between stem cell potency and quality. Finally, stem cell potency and quality were combined to determine release criteria of a sample. This information is provided when the iATP is released after culture by lysis of the cells. The iATP acts as a limiting substrate for the most sensitive, non-radioactive signal detection system available. This is a luciferin/luciferase reaction that produces bioluminescence, which is measured as light in a plate luminometer (Rich, 2003). The procedure and results described in this communication lay the foundation for future studies of stem cell potency and clinical outcome that might improve the risk of graft failure (Picardi & Arcese, 2010; Querol et al. 2010) as well as safety and efficacy for the patient.

2. Materials and methods

2.1 Cells

Several cryopreserved, mobilized peripheral blood (mPB) samples from different donors were obtained from AllCells, Inc (Berkley, CA) in accordance with the company's Internal Review Board (IRB) approval. Vials of cryopreserved umbilical cord blood (UCB) samples were provided and released for research purposes by the University of Colorado Cord Blood Bank (ClinImmune, Inc) in Aurora, CO with approval by the respective Internal Review Board. Additional mPB and UCB cells were obtained from each source to use as internal reference standards.

2.2 Reference standards (RS)

The establishment of RSs is an absolute requirement for performing a potency assay. For hematopoietic cell-based therapeutics the number of cells obtained from a single donor UCB unit, mPB procedure or bone marrow aspirate are limited. This poses severe restrictions on establishing cellular reference standards. From a practical viewpoint, there are two alternatives. The first would be to establish multiple aliquots from several different donors that could be used as reference standards. Although each batch of RS would be expected to exhibit different biological activity and therefore different potency and quality characteristics, one batch would be designated as the primary RS. A second (donor specific) batch of cells would be tested against the primary RS and designated the secondary RS. Similarly, a third batch of cells would be tested against the secondary RS and designated the tertiary RS. The most recent batch of cells established as the RS would be used for every day testing until a new RS is established and tested. In this way it would always be possible to prepare a new RS and compare it against and established RS. The second alternative would be to assay a statistically significant number of samples of the same material to establish a range and mean/median potency that could be used as a "combined" RS for individual samples. This type of RS would take considerable time to establish. It would also require multiple laboratories to use the same standardized and validated assay so that results could be compared. The advantage would be that a "global" reference standard might be established for different cellular products that would allow comparison and calculation of potency ratios and quality of samples processed by individual laboratories. Release criteria for use in transplantation could also be established. The regulatory requirement for reference standards needed to measure cell potency is probably one of the most important aspects that has to be addressed by the different cellular therapeutic communities and standards organizations.

For the present study, the first alternative to establish reference standards was used. Cells designated as reference standards were prepared by separating the mononuclear cells (MNC) by density gradient centrifugation (see below), adjusting the cell concentration so that 1 million MNC were prepared in 7.5% DMSO with 10% fetal bovine serum (FBS) and medium in 1ml. The cells were frozen in ampoules using an automated rate freezer and stored in liquid nitrogen (LN2).

2.3 Preparation of cells for culture

Cryopreserved cells were thawed in a 37°C water bath and the contents transferred to a tube containing 20mL of warmed Iscove's Modified Dulbecco's Medium (IMDM) supplemented with 10% FBS. After thawing and washing the cells once, followed by resuspension in 1mL IMDM and 2% DNase, a cell count was performed on 20μL using a cell counter (Z2, Beckman Coulter, Brea, CA). Another aliquot of 20μL was stained with 7-aminoactinomycin D (7-AAD, Beckman Coulter, Brea, CA) and the viability measured by flow cytometry using an EPICS XL/MCL flow cytometer (Beckman Coulter, Brea, CA). Samples exhibiting viability below 85% were not used since cells either demonstrated poor proliferation or did not proliferate. The MNCs from each sample were fractionated on density gradient medium (Nycoprep 1.077, Axis-Shield, Accurate Chemicals and Scientific, Westbury, NY) by centrifugation for 10 min at 1,000 x g at room temperature (RT). The cells were washed in IMDM, centrifuged at 300 x g for 10 min at RT and resuspended in IMDM. This additional step removed the contaminating and dead cells and increased the viability to above 90%. Since several internal studies indicated that 7-AAD could produce false positive results with respect to cell growth potential (data not shown), all samples were assessed for the production of iATP at 2,500, 5,000 and 7,500 cells/well to substantiate metabolic cell viability and functionality as described below.

2.4 *In vitro* cell culture of 2 stem cell populations to determine potency, quality and release criteria

The instrument-based, ATP bioluminescence assay used to determine potency, quality and release (HALO-96 PQR, HemoGenix, Inc, Colorado Springs, CO) has been previously described in detail (Hall & Rich, 2009). It is summarized here for completeness. In contrast to a previous study using cord blood cells and a methylcellulose assay format (Reems et al. 2008), the assay described below is a methylcellulose-free, 96-well culture system that incorporates Suspension Expansion Culture (SEC) technology (Rich, 2007; Hall & Rich, 2009; Olaharski et al. 2009) for detecting both primitive (high proliferative potential stem and progenitor cells, HPP-SP) and more mature multipotential hematopoietic stem cells (colony-forming cell granulocyte, erythroid, macrophage megakaryocyte, CFC-GEMM). The assay was performed as follows. For each sample, the cell concentration was adjusted to 7.5 x 10^5 cells/mL and a serial dilution performed in IMDM to produce 5 x 10^5 and 2.5 x 10^5 cells/mL. From each cell dilution, 0.1mL was added to two separate tubes containing 0.9mL of master mix, one for each stem cell population being determined. After mixing, 0.1mL of the culture master mix was dispensed into 8 replicate wells of a 96-well plate to achieve the final concentrations of 2,500, 5,000 and 7,500 cells/well. The cocktail to stimulate CFC-GEMM consisted of erythropoietin, granulocyte-macrophage and granulocyte colony-stimulating factor, stem cell factor, thrombopoietin, Flt3-ligand and interleukins 3 and 6. The cocktail to stimulate the HPP-SP stem cell population contained the same growth factors/cytokines as that for CFC-GEMM, but with the addition of interleukins 2 and 7. The plates were incubated for 5 days at 37°C in a fully humidified incubator containing 5% CO_2 and 5% O_2 (Rich & Kubanek, 1982).

2.5 Assay calibration, standardization and sample processing

Prior to measuring bioluminescence of the samples after culture incubation, an ATP standard curve was performed (Rich & Hall, 2005; Reems et al. 2008; Hall & Rich, 2009). Serial dilutions from a 10μM stock concentration were prepared so that the final dilutions were 5, 1, 0.5, 0.1, 0.05, 0.01 and 0.005μM. In addition, an IMDM background and high and low ATP controls were included. Each dilution was dispensed into 4 wells (0.1mL/well) of a 96-well plate. To each well, 0.1 mL of an ATP enumeration reagent containing a lysis buffer, luciferin and luciferase was added. The contents were mixed and the plate left to incubate for 2 min in a plate luminometer (SpectraMax L, Molecular Devices, Sunnyvale, CA) after which the bioluminescence was measured as light (photons). The resulting ATP standard curve was then used to automatically interpolate the output of the luminometer in relative luminescence units (RLU) into standardized ATP concentrations (μM) using the instrument software (SoftMax Pro v5.4, Molecular Devices, Sunnyvale, CA). Inclusion of high and low controls in addition to the ATP standard curve allowed the assay to be calibrated and standardized. After performing the ATP standard curve, the sample plate(s) were removed from the incubator and allowed to attain room temperature. Thereafter, 0.1mL of the ATP enumeration reagent was dispensed into each well and the contents mixed. After 10 min incubation in the instrument or in the dark, the bioluminescence was measured and the ATP concentrations automatically interpolated from the ATP standard curve.

2.6 Assay validation and statistics

The ATP bioluminescence assay has been previously validated in accordance with bioanalytical method validation (FDA, 2001). For this specific application, the assay exhibited an accuracy (proportion of correct outcomes) of greater than 90%. Sensitivity (proportion of correctly identified positive samples) and specificity (proportion of correctly identified negative samples) were determined using receiver operator characteristic (ROC) statistics (DeLong et al. 1985) in which the area under the curve (AUC) was determined for background (no stimulatory cocktail) versus CFC-GEMM and background versus HPP-SP. For the former, the AUC was 0.752 (95% confidence intervals; 0.71-0.8; $p < 0.0001$), while for the latter the AUC was 0.73 (95% confidence intervals; 0.68-0.78, $p<0.001$). Since the AUC must be between 0.5 and 1, the results demonstrated that the assay could differentiate between sensitivity and specificity. Assay precision (reliability and reproducibility) was performed on background, CFC-GEMM and HPP-SP over a cell dose range from 2,500 to 10,000 cells/well and demonstrated coefficients of variation (CV) of 15% or less. This was in compliance with regulatory requirements (FDA, 2011). Robustness, in this case transferability of the assay from one laboratory to another, had been previously reported (Reems et al. 2008). The results demonstrated a correlation coefficient (R) between laboratories of 0.94 ($p<0.001$).

Concentrations of ATP (μM) are provided as the mean ± 1 standard deviation of 8 replicate wells. The slope of the 3-point cell dose response was obtained from the linear regression using least squares analysis (Prism version 5, GraphPad Software, LaJolla, CA). For correlations, the slope of the linear regression, goodness of fit (r^2) and correlation coefficient (R) are reported. Tests of significance for correlation were performed using the Pearson two-tailed test with an alpha of 0.05.

3. Results

The procedure for determining stem cell potency, quality and release criteria is a 3-step process. However, only the first step requires cell culture and provides all the information for the remaining steps of the procedure. The culture step involves a 3-point cell dose response for two stem cell populations (CFC-GEMM and HPP-SP) for both the RS and samples.

3.1 Step 1 – Measuring stem cell potency of mobilized peripheral blood

The first step in the procedure is illustrated in Figure 1. This shows the cell dose responses for the mature multipotential stem cell, CFC-GEMM (Figs. 1A and 1B), and the more primitive stem cell HPP-SP (Figs. 1C and 1D) from 4 different mPB samples cultured for 5 (Figs. 1A and 1C) and 7 days (Figs. 1B and 1D). The graphs demonstrate that an approximate 3-fold increase in ATP concentration occurs within 2 days when the incubation time is increased from 5 to 7 days. A 7 day incubation period allows for increased assay sensitivity as well as the ability to perform the assay to accommodate a work schedule. Since the increase is cell dose dependent, it demonstrates that the assay is directly measuring an increase in the number of cells as a result of cell proliferation. It should be noted however, that although measurements at both 5 and 7 days are on the exponential part of the growth curve for both HPP-SP and CFC-GEMM, measurement of proliferation on day 7 will exhibit slightly greater coefficients of variation (CVs) and will also include cells that have initiated differentiation. On day 5, little or no differentiation occurs (data not shown).

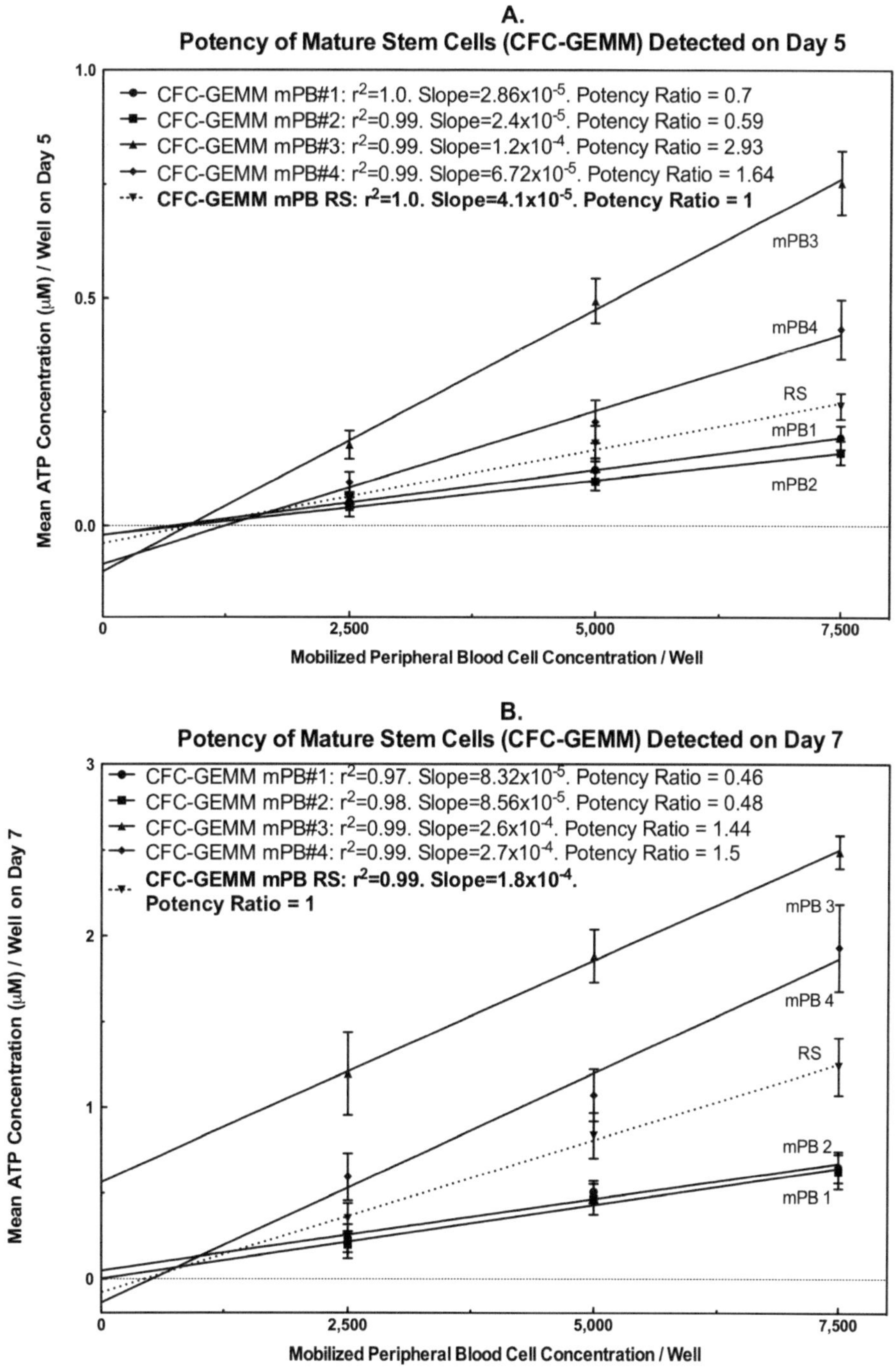

Fig. 1A and 1B. Measurement of Mobilized Peripheral Blood CFC-GEMM Stem Cell Potency on 5 and 7 Days of Culture.

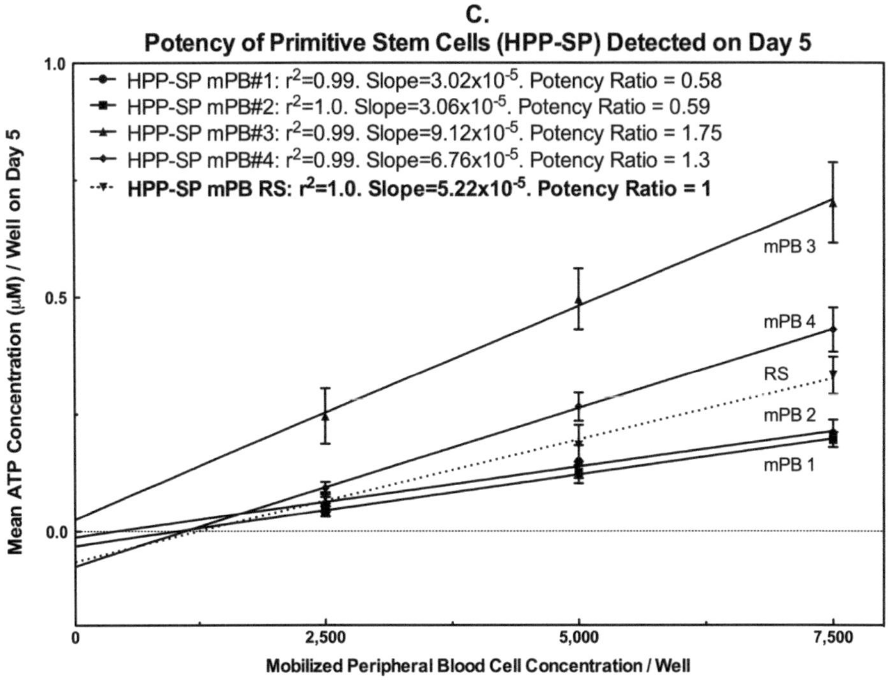

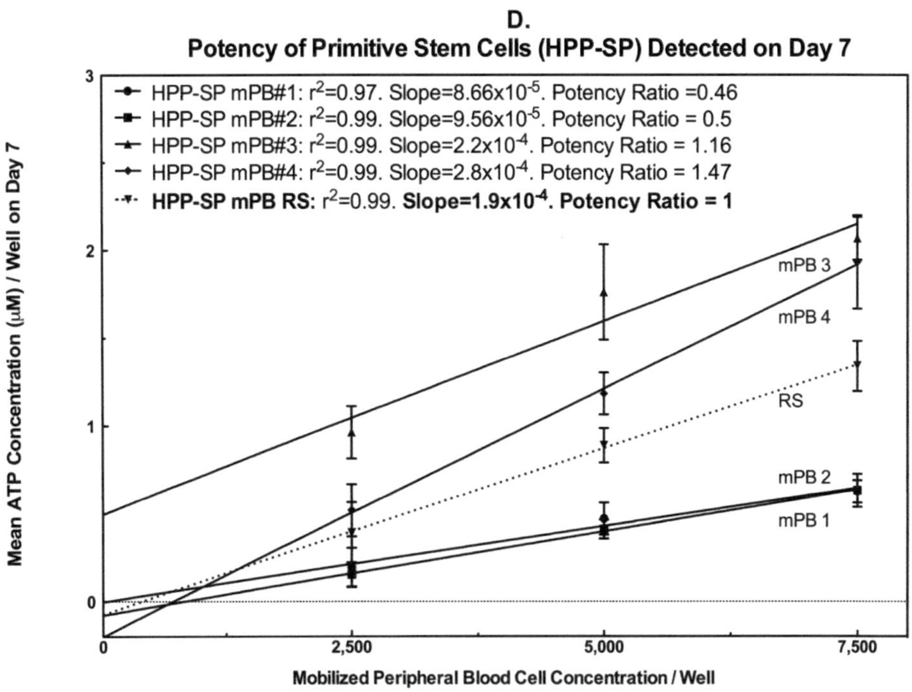

Fig. 1C and 1D. Measurement of Mobilized Peripheral Blood HPP-SP Stem Cell Potency on 5 and 7 Days of Culture.

In addition to the samples, a mPB RS was also included and allows the potency ratio to be calculated as follows:

Potency Ratio = Slope of the sample linear regression / Slope of the RS linear regression.

The potency ratio therefore provides information on the dose required to obtain the same response as the RS. Since the potency of the RS is always 1, samples with a potency ratio less than 1 will require larger cell doses to produce the same response, while potency ratios greater than 1 will require fewer cells to produce the same response as the RS.

3.2 Step 2 – The relationship between stem cell potency and quality

The results in Fig. 1 illustrate two fundamental concepts that are necessary for measuring stem cell potency and quality. The first concept is that the slope of the cell dose response should be greater for the more primitive stem cells (HPP-SP) than for the mature hematopoietic stem cells (CFC-GEMM), since the former have greater proliferation potential than the latter. The slope of the cell dose response therefore provides a direct measurement of stem cell proliferation potential. The greater the proliferation potential, the greater the potency. Thus, the slope of the cell dose response is also a direct measurement of potency. The second concept, also illustrated in Fig. 1, shows that as the slope increases at a specific cell dose, there is a concomitant increase in ATP concentration. This is a measure of stem cell quality.

The result of combining these two concepts is shown in Figures 2A and 2B for CFC-GEMM and HPP-SP stem cell populations, respectively. The figures show that when the ATP concentration at a specific cell dose (in this case 5,000 cells/well) is plotted against the slope of the cell dose response linear regression for both stem cell populations cultured for either 5 or 7 days, there is a direct correlation between stem cell potency and quality. As a result, both stem cell potency and quality have to be taken into account to determine if the stem cell product conforms to specific, but arbitrary, acceptance values and can therefore be released for use.

3.3 Step 3 – Using stem cell potency and quality to determine release criteria

Figure 3A and 3B shows the ability of CFC-GEMM and HPP-SP to proliferate at 5,000 cells/well after 5 and 7 days of culture, respectively. It had previously been found that, after 5 days in culture, an ATP concentration below 0.04μM indicated that cells could not sustain proliferation. At 7 days, this threshold was increased to 0.12μM. After 5 days of culture, the ATP concentration of samples 1 and 2 demonstrated minimal proliferation, but greater than the 0.04μM threshold. After 7 days of culture, proliferation of both samples had increased, together with samples 3 and 4. If release criteria were based solely on stem cell quality or proliferation ability, it would be assumed that all 4 samples might be acceptable for release. However, Fig. 2 demonstrates that both stem cell quality and potency have to be considered as part of the release criteria.

Figure 3C shows the cumulative potency ratios of both CFC-GEMM and HPP-SP after 5 days and 7 days (Fig. 3D) in culture. Since the potency of the CFC-GEMM and HPP-SP reference standards is always 1, samples 1 and 2 exhibited potency ratios significantly less than the reference standard. In contrast, samples 3 and 4 exhibited both high stem cell quality and potency after 5 and 7 days of culture. Based on these results, mPB samples 1 and 2 would be sub-optimal or rejected, while samples 3 and 4 would be acceptable for use.

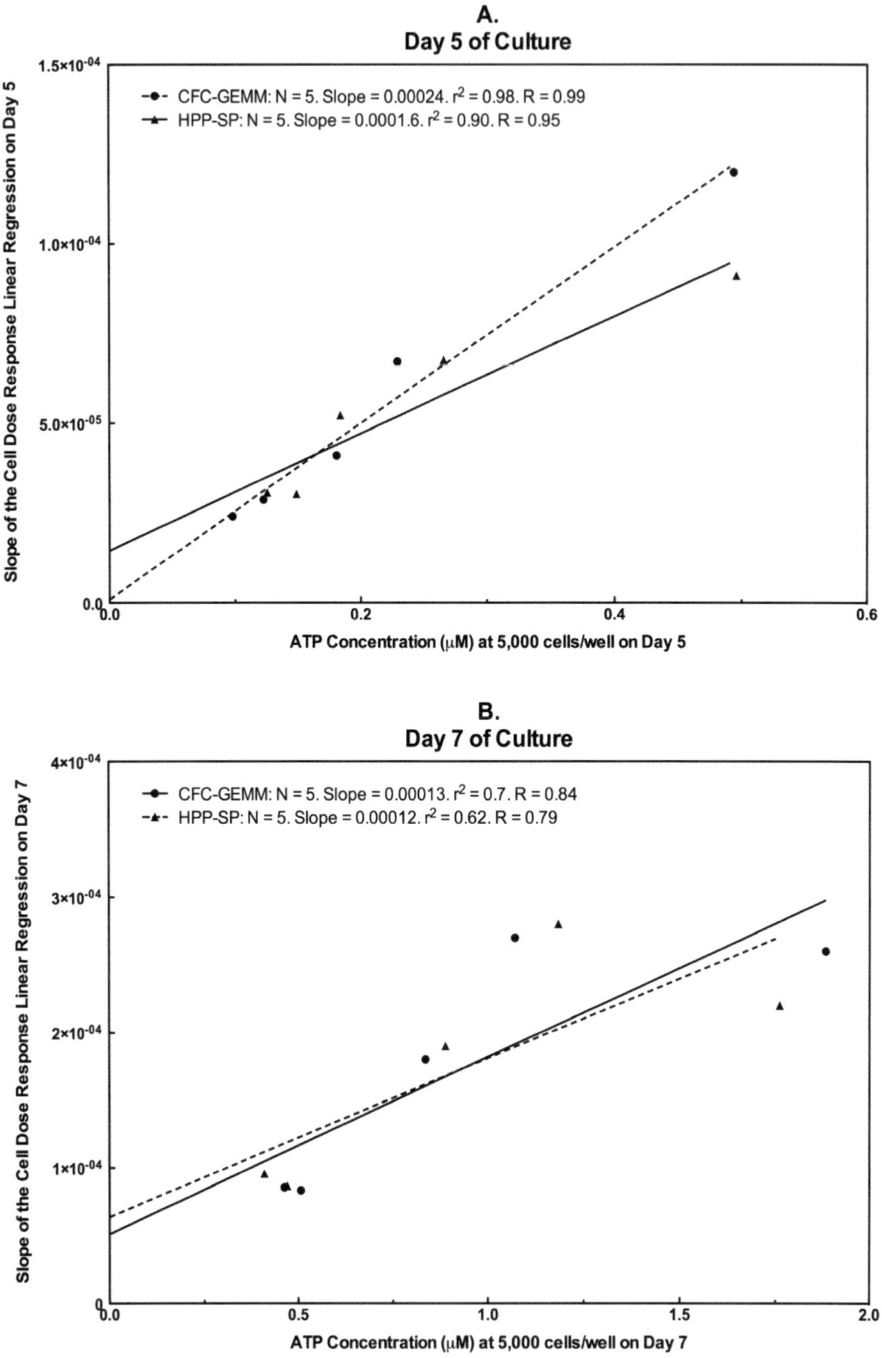

Fig. 2. Relationship Between Stem Cell Potency and Quality for Mobilized Peripheral Blood CFC-GEMM and HPP-SP Detected on Day 5 or 7 of Culture.

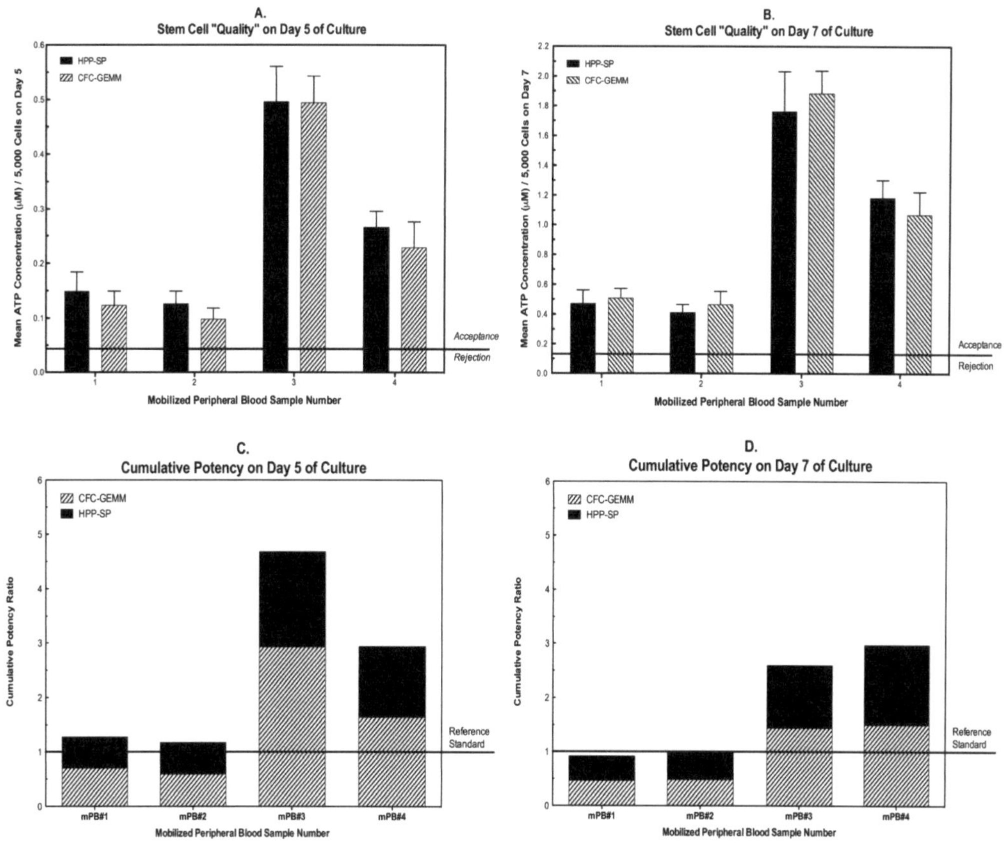

Fig. 3. Combining Mobilized Peripheral Blood Stem Cell Potency and Quality to Provide Release Criteria.

3.4 Umbilical cord blood stem cell potency, quality and release and the relationship to engraftment potential

A total of 28 UCB samples were analyzed for potency and quality using the same procedure described for mPB above, except that all assays were terminated after 5 days in culture, rather than performing both 5 and 7 day cultures. A 3-point cell dose response was performed for both the CFC-GEMM and HPP-SP stem cell populations and the slope of the linear regression was calculated for each cell dose response. The respective potency ratio for each CFC-GEMM and HPP-SP sample was then calculated using a UCB reference standard that was prepared from cord blood unit cells that did not meet the necessary criteria for storage. The slopes and potency ratios for each stem cell population are shown in Table 1. Also shown are the reported times to neutrophil and platelet engraftment. For one sample (sample 10), insufficient cells were obtained to perform a cell dose response for both stem cell populations. In two other samples (samples 18 and 25), insufficient cells were obtained after thawing to perform a HPP-SP stem cell dose response.

Sample Number	Slope for CFC-GEMM	Potency Ratio for CFC-GEMM	Slope for HPP-SP	Potency Ratio for HPP-SP	Days to Neutrophil Engraftment (>500/ul)	Days to Platelet Engraftment (>50k/ul)
1	2.94E-05	2.91	2.23E-05	0.70	28	237
2	3.14E-05	3.12	2.19E-05	0.69	14	2
3	1.41E-05	1.40	1.01E-05	0.31	6	45
4	2.09E-05	2.07	3.04E-05	0.95	30	49
5	2.19E-05	2.18	2.59E-05	0.81	17	39
6	9.28E-06	0.92	1.79E-05	0.56	12	9
7	2.25E-05	2.23	1.67E-05	0.52	17	45
8	2.50E-05	2.48	1.48E-05	0.46	22	39
9	1.52E-05	1.50	9.70E-06	0.30	56	13
10	IE	-	IE	-	43	103
11	1.77E-05	1.75	9.39E-06	0.29	34	7
12	2.83E-05	2.81	1.99E-05	0.62	20	26
13	1.09E-05	1.08	7.43E-06	0.23	NE	NE
14	8.12E-06	0.81	3.55E-06	0.11	19	183
15	6.69E-06	0.66	4.77E-06	0.15	31	122
16	1.26E-05	1.25	1.02E-05	0.32	13	39
17	1.81E-05	1.80	1.90E-05	0.60	5	40
18	3.07E-05	3.05	IE	-	NE	NE
19	2.01E-05	1.99	4.45E-05	1.39	27	38
20	2.52E-05	2.50	3.18E-05	1.00	22	62
21	1.54E-05	1.53	3.30E-05	1.03	29	55
22	2.31E-05	2.29	3.30E-05	1.03	18	23
23	1.20E-05	1.19	1.70E-05	0.53	28	70
24	1.63E-05	1.62	2.22E-05	0.70	15	46
25	1.46E-05	1.45	IE	-	26	39
26	2.12E-05	2.10	2.73E-05	0.85	28	61
27	1.37E-05	1.35	1.74E-05	0.54	37	126
28	1.50E-05	1.49	2.16E-05	0.68	114	113

IE = Insufficient cells to perform cell dose response. NE = No engraftment.

Table 1. Stem Cell Proliferation Potential / Potency Characteristics and Time to Engraftment of 28 Umbilical Cord Blood Samples.

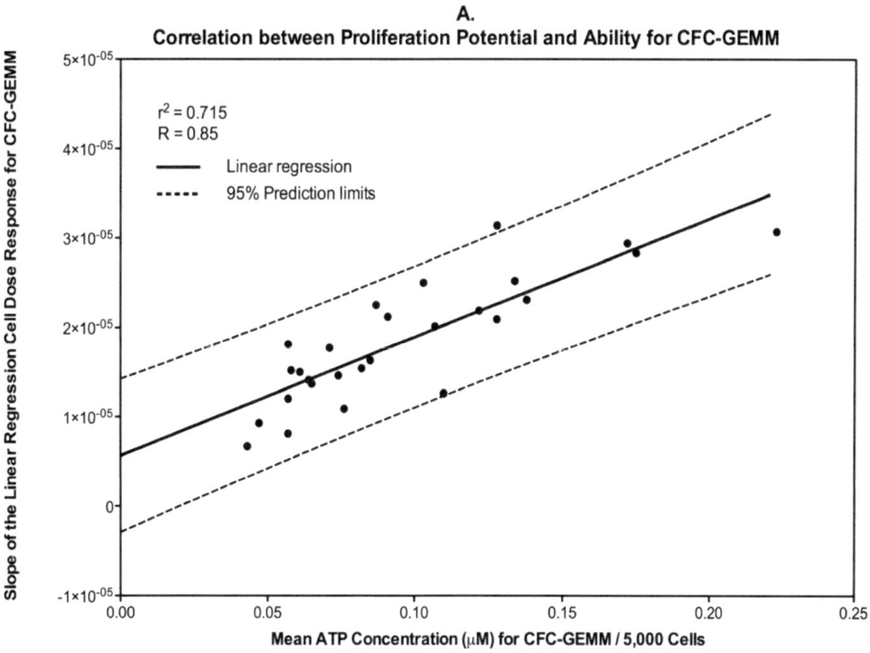

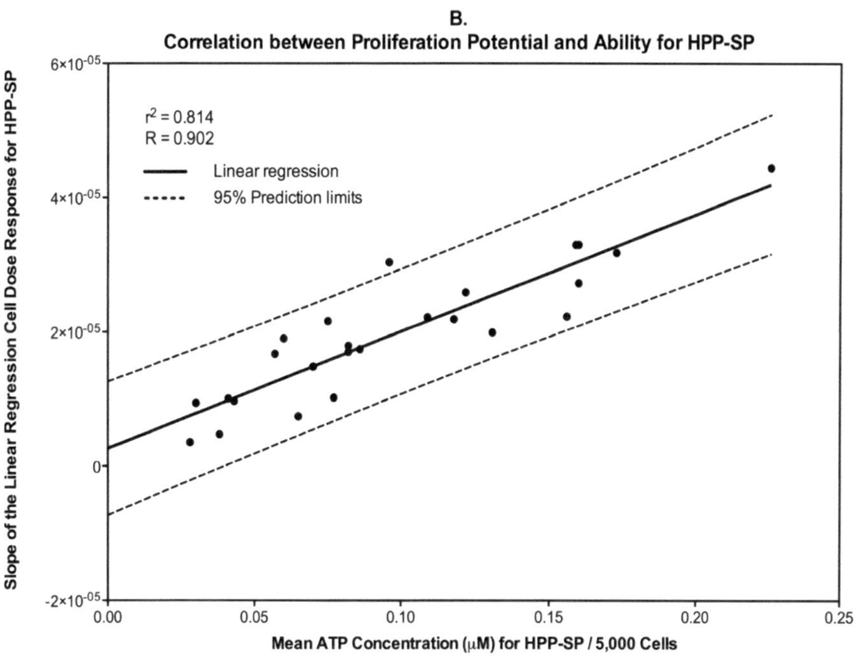

Fig. 4. Correlation Between Proliferation Potential (Potency) and Proliferation Ability (Quality) for CFC-GEMM and HPP-SP Stem Cells from Umbilical Cord Blood.

Figures 4A and 4B show the correlation of ATP concentrations at 5,000 cells/well with the slope of the UCB dose response for both CFC-GEMM and HPP-SP stem cell populations. The correlation coefficient (R) for HPP-SP was greater than that for CFC-GEMM, but the correlation for both stem cell populations was statistically significant ($p < 0.001$). The relationship between stem cell potency and quality is an indication that both parameters have to be taken into consideration when defining release criteria. Although stem cell quality could be ascertained for sample 10, insufficient cells were available to measure stem cell potency. Insufficient cells for samples 18 and 26 were also the reason why potency could not be determined for the HPP-SP stem cell population.

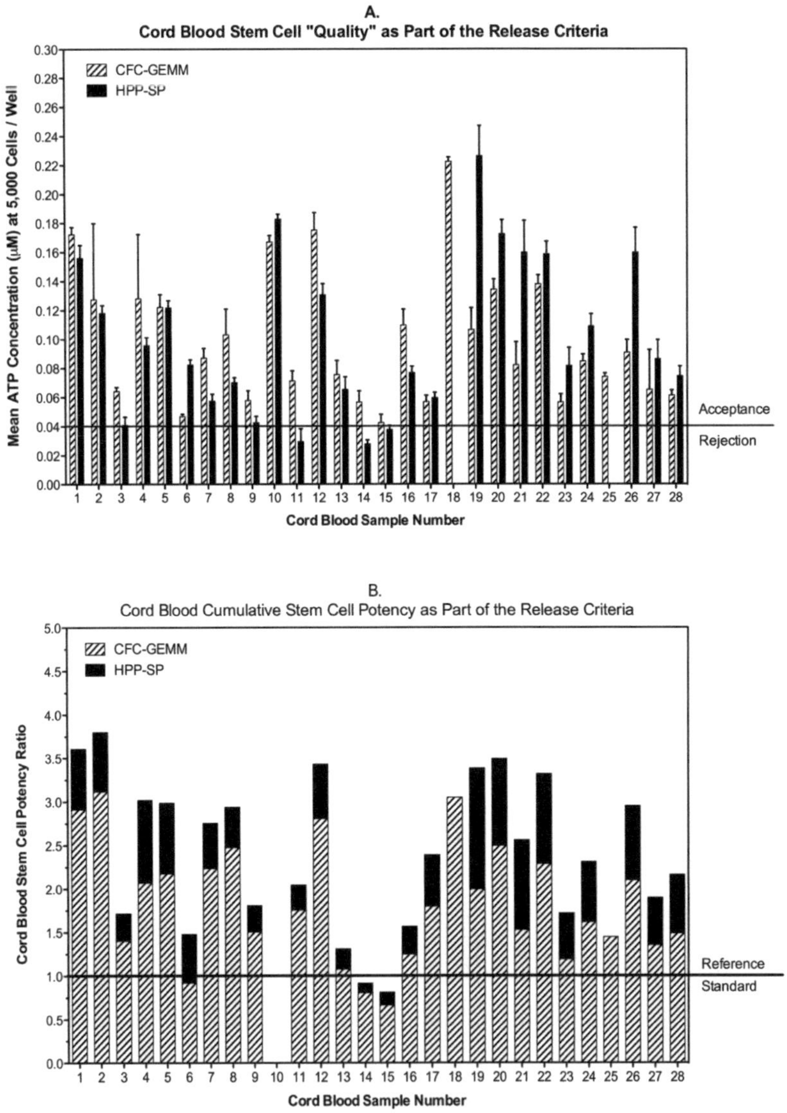

Fig. 5. Combining Umbilical Cord Blood Stem Cell Potency and Quality to Provide Release Criteria.

Figure 5A shows the stem cell quality (proliferation ability at a specific cell dose) and Fig. 5B, the cumulative stem cell potency ratio (proliferation potential measured as the slope of the linear regression of the cell dose response and compared to that of the reference standard) for each of the stem cell populations. Samples 18 and 25 only show the potency for the CFC-GEMM populations since insufficient cells were available to measure the potency of the more primitive HPP-SP population. For 21 of the samples, stem cell quality of both populations was greater than the arbitrary ATP concentration cutoff level of 0.04μM, below which cells cannot sustain proliferation. The same 21 samples also exhibited a cumulative potency above the RS potency of 1. Sample 6 exhibited a CFC-GEMM potency below the RS, while sample 13 demonstrated a CFC-GEMM potency slightly greater than the RS. However, the additional potency provided by the HPP-SP stem cell populations increased the cumulative potency above that of the RS.

It is now possible to consider the interpretation of the results. Samples 14 and 15 pose an interesting anomaly. The CFC-GEMM quality is below the ATP concentration cutoff point for both samples and slightly greater than the cutoff point for HPP-SP. However, both samples exhibit a cumulative potency below the RS. These results would indicate that both sample 14 and 15 would exhibit limited or no engraftment potential. From Table 1, the time to neutrophil engraftment for sample 14 was only 19 days while that for sample 15 was 31 days. Platelet engraftment was 183 and 122 days for sample 14 and 15, respectively. Therefore, these two samples did not agree with the reported clinical outcome. Table 1 also shows that samples 13 and 18 did not engraft. As described above, sample 13 exhibited a CFC-GEMM potency that was in a questionable range and may not have provided the necessary short-term engraftment and reconstitution. In contrast, sample 18, appeared to exhibit sufficient CFC-GEMM quality and potency, although insufficient cells did not allow information to be obtained for the primitive HPP-SP stem cell population. Despite the four sample outliers, the assay exhibits an accuracy of greater than 85%. Nevertheless, further studies that correlate *in vitro* data with more detailed clinical outcome for both engraftment and reconstitution would be prudent to ascertain a range for both stem cell quality and potency that would improve the accuracy of the assay.

3.5 Correlation between the ATP concentration and TNC, MNC, viability, CD34

There was no correlation between the ATP concentration for both cord blood stem cell populations with either dye exclusion viability or $CD34^+$ counts. This was to be expected since neither viability nor CD34 membrane expression are cell functionality or proliferation markers. However, ATP concentration did correlate with both the TNC and MNC, but only when calculated on a per kilogram patient body weight basis. These results are shown in Fig. 6A for TNC and 6B for MNC. In both cases, the ATP concentration was calculated based on the patient body weight of the number of cells transplanted. The results in Fig. 6A demonstrate that when TNC is used, a strong correlation is obtained for the CFC-GEMM, but although still statistically significant, the primitive HPP-SP stem cell population exhibited a lower correlation coefficient. In contrast, Fig. 6B shows that the correlation between the ATP concentration and the MNC, both based on kilogram body weight, for CFC-GEMM and HPP-SP is highly significant with lower variation compared to the TNC values. The results clearly demonstrate that the greater the number of cells transplanted, the greater the number of stem cells transplanted that can exhibit proliferation ability. However, the results also demonstrates that using the mononuclear cell count rather than total nucleated cell count produces a better estimate for the stem cell response. However, cell counts alone cannot be used as a potency assay and cannot replace the information and value provided by a standardized cell functionality assay.

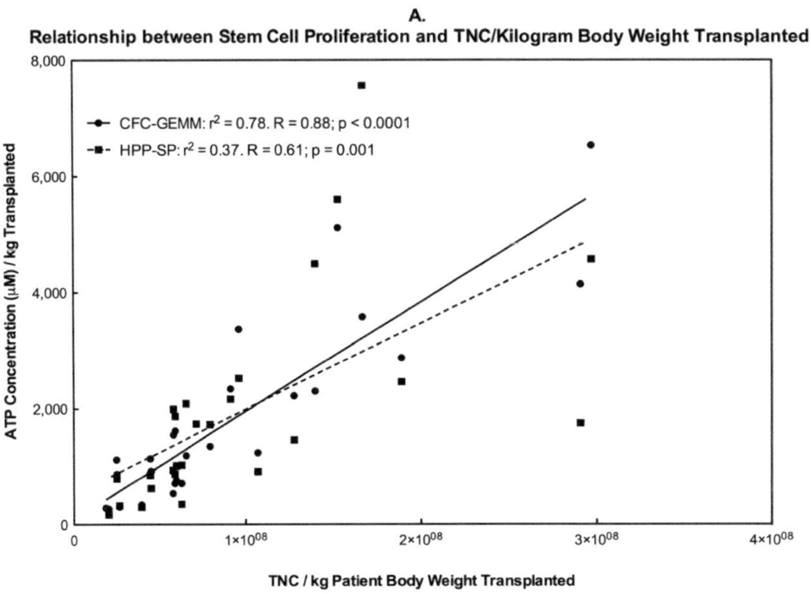

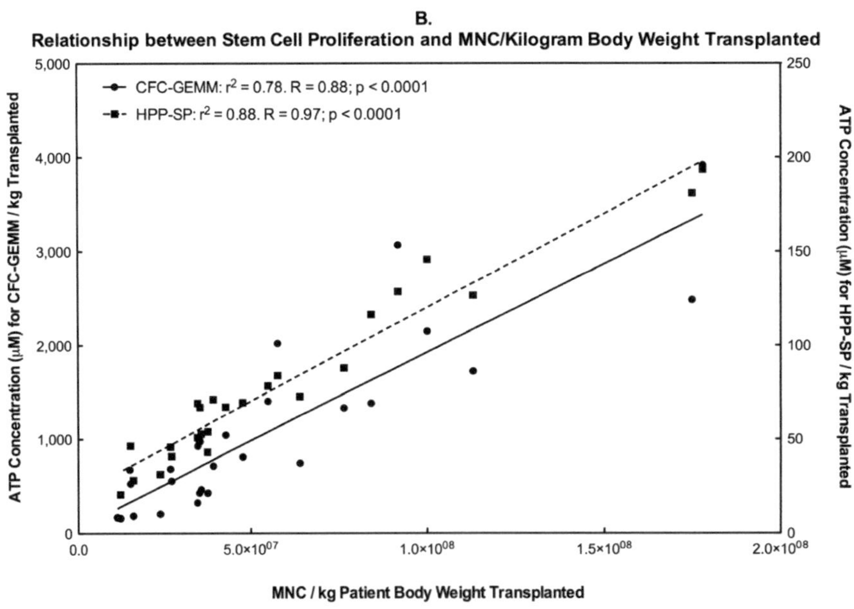

Fig. 6. Correlation between ATP Concentration as a Measure of Stem Cell Proliferation and the Number of Cord Blood Cells Transplanted Expressed as either Total Nucleated Cell Counts (TNC) or Mononuclear Cell Counts (MNC)/Kilogram Body Weight.

4. Discussion

For biopharmaceutical products, potency is measured by comparing the dose response to that of an established RS (Thorpe et al. 1999; Lansky, 1999; FDA, 2011). For these materials, a parallel dose response to that of the RS should be obtained (Thorpe et al, 1999; Gottscalk & Dunn, 2005; Jonkman & Sidik, 2009). Lack of parallelism indicates either contamination or a different material to that of the RS. When the dose response curves are parallel to the RS, the horizontal displacement to the left or right indicates a greater or lower potency, respectively. The dose of the compound can be compared and, if necessary, adjusted to that of the reference standard. In this way, the same dose can always be used with compound batches of different potency.

Cells, in particular, stem cells, pose significant differences and challenges to this paradigm. First, unlike biopharmaceutical products, where relatively large quantities of the material would be available to establish several batches of reference standards, it might be extremely difficult to establish cell reference standards (Strong et al. 2009; Rayment & Williams, 2010). There are several alternatives to establishing reference standards for cell therapeutics that have already been discussed above. However, a prerequisite for establishing reference standards and to compare results within and between laboratories is the use of a standardized and validated assay that is sufficiently robust so that it can be transferred and established in different laboratories. Lack of such an assay has been the reason why laboratories have not been able to directly compare processing, cryopreservation and thawing procedures for hematopoietic and other stem cell therapeutic products prior to use.

It might be argued that the present study should have been performed in parallel with the CFU assay. This has been the functional assay used previously in hematopoietic stem cell processing laboratories and is still used by the cord blood community today. More recently, the CFU assay has been suggested as a potency assay (Page et al. 2011a, 2011b), in addition to other parameters normally measured. These include total nucleated cell count (TNC), viability and CD34 membrane expression (FDA, 2009). There were three reasons for not performing parallel studies using the CFU assay. First, the ATP bioluminescence assay was originally derived from the methylcellulose CFU assay. Unlike the CFU assay, however, the ATP assay has undergone several major technical advances culminating in the assay used for the present study. It was also previously shown that even the methylcellulose-free format used in this study is not only equivalent to the original CFU assay, but is clearly a more reliable, reproducible and robust assay (Rich, 2007, Reems et al, 2008). Furthermore, lower sensitivity and precision (high variations) of the CFU assay, coupled with the lack of standardization (see below), would have resulted in inconclusive results. Second, unlike the CFU assay, the ATP assay can and has been validated in compliance with bioanalytical regulatory requirements (FDA, 2001). Furthermore, for an assay to be a potency assay, the regulatory agencies require demonstration of specific assay characteristics (FDA, 2011). Assay validation is just one of these characteristics, but to validate an assay, standards and controls are required. The CFU assay lacks standards and controls and cannot be validated according to regulatory requirements. Finally, it is often assumed that the CFU assay measures proliferation, whereas the CFU assay is actually a clonogenic differentiation assay. Proliferation is certainly involved in the formation of hematopoietic colonies. However, the colonies produced in methylcellulose are identified and counted by the ability of the cells producing the colonies to differentiate and mature. Therefore, the CFU assay detects differentiation ability and/or potential, but does not measure a parameter that directly correlates with the stem cell proliferation process. Based on these and other characteristics (discussed below), the CFU assay was not considered as a comparison assay for this study.

Many factors affect the quality and potency of a cellular product. First, proportions of different stem cell populations originally present in the umbilical cord blood, their quality and potency, are an inherent property of the tissue. Second, the procedures used to collect and store the cells prior to processing can affect quality and potency. Third, different stem cell processing, cryopreservation and thawing procedures not only affect quality and potency, but the proportion of stem cells remaining in the product. Finally, the decision process to use a particular product should be based on trusted results that can only be obtained from an assay(s) that is quantitative, standardized and validated to measure quality and potency of the active stem cell ingredients.

The stem cell potency assay described above is based on performing a minimum 3-point cell dose response and comparing the slope of the resulting linear regression to that of a RS of the same material. Comparison of cell dose response slopes to calculate the potency ratio was used for two reasons. First, measurement of potency of a biopharmaceutical compound usually relies on establishing parallelism between the sample and RS dose response curves. When the linear portions of the dose response curve are parallel, not only is this an indication that the sample and RS are of the same material, but also allows the potency ratio to be calculated from the horizontal displacement between the two dose response curves. If cells, and stem cells in particular, exhibit parallel cell dose response curves, this is an indication that both the sample and RS stem cells demonstrate a similar "stemness" or primitiveness. The resulting parallel displacement indicates a difference, not in potency, but in stem cell number between the sample and RS. Since hematopoiesis is a continuously proliferating system and the cells are continuously in flux, it follows that very few hematopoietic stem cell samples will exhibit exactly the same degree of primitiveness to the cells in the RS. Therefore potency measurement by parallelism will not provide a general procedure to calculate the potency ratio. Since an assay should show linearity within a specific cell concentration range, measuring the slope of the cell dose response not only demonstrates assay linearity characteristics, but also provides a direct measurement of stem cell primitiveness and proliferation potential, which in turn, is equivalent to stem cell potency. By comparing the slope of the sample cell dose response with that of the RS, the potency ratio can be calculated. Depending on the stem cell population detected, more primitive stem cells will show a steeper slope to that of mature stem cells. This procedure can then be used for any proliferating cell population. A 3-point cell dose response is the minimum number of data points that can be used to perform linear regression analysis. Although a larger number of points could also be used and would be more accurate, it was necessary to take into account that for hematopoietic tissues (and other cellular therapeutic products), cell availability is limited. Potency measurement should be performed just prior to use, since it is related to the dose that is to be administered. From a practical viewpoint, a cord blood potency determination would be performed on the limited number of cells thawed in a segment used for confirmatory testing prior to the cord blood unit being transplanted. Several publications have shown, using TNC, viability, CD34 and CFU, that segments of cryopreserved cord blood used for confirmatory testing are a representative sample of the cells in the cord blood unit (Goodwin et al. 2003; Solves et al. 2004, Rodriguez et al. 2005; Page et al. 2011b). However, in all of these cases, the potency of the active ingredients, i.e. the stem cells, were not taken into account.

A potency assay requires that all active ingredients be measured (FDA, 2011). This is relatively easy for a biopharmaceutical compound, but is impossible for a continuously proliferating system such as hematopoiesis. Although a hematopoietic stem cell tissue may contain many different cell types, the stem cells are the only cells responsible for engraftment and reconstitution. The stem cells are therefore the active ingredients for which the potency must be measured. Since the hematopoietic stem cell compartment consists of a continuum of stem cells and assays are not yet available to test each and every stem cell subpopulation, the regulatory requirement to measure all active ingredients, cannot be met at the present time. As an alternative, the quality and potency of a minimum of two different stem cell populations have been determined; the primitive HPP-SP and the more mature CFC-GEMM stem cell populations. As shown in Fig. 5, the potency of a single stem cell population would be insufficient and could lead to a false interpretation of the results. As also demonstrated in Fig. 5, even a minimum of two stem cell populations may result in a false interpretation, but the accuracy is significantly greater than if the potency of only one stem cell population was measured. It is certainly possible to reliably and reproducibly measure the quality and potency of more than two stem cell populations with present technology. However, this has to be weighed against the use of larger numbers of cells and the costs associated with testing. It should also be emphasized that potency and quality testing need to be performed on the cryopreserved sample intended for use. They provide predictive information for release. It is therefore reasonable to pose the question, is it preferable to use more cells for a predictive assay that might ensure stem cell functionality, engraftment potential and growth than to use fewer cells and not perform any assay?

In this respect, it is worth returning to the tests and assays presently used to characterize the cells in the processing laboratory. These are TNC, viability and viable $CD34^+$ counts. These three parameters have been designated as measurements of potency for umbilical cord blood (FDA, 2009), despite the fact that they do not comply with the necessary regulations for a potency assay, especially since none of the parameters are functional assays and measure the active ingredients. Of the three parameters listed above, probably the most important is the TNC dose. However, the TNC count includes a large proportion of cells that play no role in engraftment. Inclusion of these cells actually results in a dilution of the active stem cell ingredients. In contrast to using TNC, removing most of the unnecessary cells to produce an MNC fraction that contained the pool of stem cells, it was possible to demonstrate that the MNC dose used for transplantation exhibited a greater correlation with the ATP dose for both the CFC-GEMM and HPP-SP stem cell populations (Fig. 6). This result illustrates that the ATP concentration can be used as a measure of stem cell dose, which in turn is related to the potency ratio.

The potency predicts the dose of the product for the intended use. The potency of a stem cell product should predict the dose of stem cells required to achieve engraftment. In other words, stem cell potency predicts engraftment potential. This should not be confused with, and is not the same as time to engraftment. If the slope of the stem cell linear regression dose response curves or the stem cell potency ratios provided in Table 1 is plotted against the time to engraftment, no correlation will be obtained. This is because potency is entirely dependent upon stem cell proliferation potential, while time to engraftment is dependent upon the differentiation and maturation of hematopoietic progenitor cells into neutrophils, platelets and erythrocytes. This is the reason why the presence and number of progenitor cell colonies counted in the CFU assay, especially GM-CFC and Mk-CFC, relates to the appearance and number of neutrophils and platelets in the circulation of the patient (Page et al. 2011a).

Although the number of UCB transplants has increased almost exponentially since the first published UCB transplant in 1989 (Gluckman et al. 1989), approximately 20% of patients receiving an unrelated UCB transplant exhibit graft failure (Page et al. 2011a, 2011b). This has, in part, been attributed to inadequate UCB potency (Page et al. 2011a, 2011b). Previous publications have focused on the need for standardized laboratory procedures (Rich, 1997; Wagner E et al. 2006; Brand A et al. 2008). A recent publication by Spellman et al. (2011) discusses problems facing the cord blood community and the guidelines and requirements for "standardized testing methodologies" to be established. The cell-based, ATP bioluminescence assay platform described in this communication to measure both stem cell potency and quality and, in addition, help define release criteria, constitutes the next generation of assays that addresses all of the necessary requirements including, but not limited to, standardized methodology, reproducibility with limited variability between testing sites, automated testing outputs, high throughput capability and rapid turnaround time.

5. Conclusions

It is often the case that an assay will be generic and adapted to fit the intended application. The present communication describes an assay that has been specifically designed and validated for the purpose of measuring stem cell quality and potency for hematopoietic cellular therapeutic products derived from mobilized peripheral blood, umbilical cord blood and even bone marrow (data not shown). A similar potency and quality assay has also been developed for mesenchymal stem cells. The assays incorporate an instrument-based, biochemical marker in the form of ATP, the concentration of which is directly proportional to the proliferation ability and potential of the stem cell populations being measured. The bioluminescence signal detection system is the most sensitive, non-radioactive readout available allowing the assay to incorporate external standards and controls. The implementation of a fully compliant potency and quality assay specific for hematopoietic stem cell products should not only help standardize cell processing procedures, but also reduce the risk of graft failure and improve safety and efficacy for the patient.

6. References

Botnick LE, Hannon EC, Hellman S. (1979). Nature of the hematopoietic stem cell compartment and its proliferative potential. Blood Cells. 5:195-210.

Brand A, Eichler H, Szczepiorkowski ZM, Hess JR, Kekomaki E, McKenna DH, PamphilionD, Reems J, Sacher RA, Takahashi TA, van der Watering LM (2008). Viability does not necessarily reflect the hematopoietic cell potency of a cord blood unit: results ofan interlaboratory exercise. Transfusion. 48:546-549.

Broxmeyer HE, Douglas GW, Hangoc C, Cooper S, Bard J, English D, Arny M, Thomas L, Boyse EA. (1989) Human umbilical cord blood as a potential source of transplantable hematopoietic stem/progenitor cells. Proc Natl Acad Sci USA 86:3828-3832.

Charbord P. (1994). Hemopoietic stem cells: analysis of some parameters critical for engraftment. Stem Cells. 12:545-62.

Civin CI, Almeida-Porada G, Lee MJ, Olweus J, Testappen LW, Zanjani ED. (1996). Sustained, retransplantable, multilineage engraftment of highly purified adult human bone marrow stem cells in vivo. Blood. 88:4102-9.

DeLong ER, Vernon WB, Bollinger RR. (1985). Sensitivity and specificity of a monitoring test. Biometrics. 41:947-58.

Duggan PR, Guo D, Luider J, Auer I, Klassen J, Chaudhry A, Morris D, Glueck S, Brown CB, Russell JA, Stewart DA. (2000). Predictive factors for long-term engraftment of autologous blood stem cells. Bone Marrow Transplant. 26:1299-304.

European Medicines Agency (EMA) (2008). Guideline on potency testing of cell based immunotherapy medicinal products for the treatment of cancer. http://www.tga.gov.au/pdf/euguide/bwp27147506en.pdf.

FDA Guidance for Industry. (2009). Minimally manipulated, unrelated allogeneic placental/umbilical cord blood intended for hematopoietic reconstitution for specified indication.
http://www.fda.gov/downloads/BiologicsBloodVaccines/GuidanceCompliance RegulatoryInformation/Guidances/Blood/UCM187144.pdf.

Gluckman E, Broxmeyer HA, Auerbach AD, Friedman HS, Douglas GW, Devergie A, Esperou H, Thierry D, Socie G, Lehn P, Scott Cooper BS, English D, Kurtzberg J, Bard J, Boyse EA. (1989). Hematopoietic reconstitution in a patient with Fanconi's anemia by means of umbilical-cord blood from an HLA-identical sibling. N Engl J Med. 321:1174-1178.

Goodwin HS, Grunzinger LM, Regan DM, McCormick KA, Johnson CE, Oliver DA, Muecki KA, Alonso JM, Wall DA. (2003). Long term cryostorage of UC blood units: ability of the integral segment to confirm both identity and hematopoietic potential. Cytotherapy. 5:80-86.

Gottschalk PG, Dunn JR. (2005). Measuring parallelism, linearity, and relative potency in bioassay and immunoassay data. J Biopharma Stat. 15:437-463.

Haas R, Ho AD, Bredthauer U, Cayeux S, Egerer G, Knauf W & Hunstein W. (1990). Successful autologous transplantation of blood stem cells mobilized with recombinant human granulocyte-macrophage colony-stimulating factor. Exp Hematol. 18:94-98.

Hall KM & Rich IN. (2009). Bioluminescence assays for assessing potency of cellular therapeutic products, In: Cellular Therapy: Principles, Methods and Regulations, Areman EM & Loper K, 581-591. AABB. ISBN 978-1-56395-296-8. Bethesda, MD.

Jonkman JN, Sidik, K. (2009). Equivalence testing for parallelism in the four-parameter logistic model. J Biopharma Stat. 19:818-37.

Koerbling M, Holle R, Haas R, Knauf W, Doerken B, Ho AS, Kuse R, Pralle H, Fliedner TM, Hunstein W. (1990). Autologous blood stem cell transplantation in patients with advanced Hodgkin's disease and prior radiation to the pelvic site. J Clin Oncol. 8:978-985.

Lansky D. (1999). Validation of bioassay for quality control. Dev Biol Stand. 97:157-68.

Leung W, Ramirez M, Civin CI. (1999). Quantity and quality of engrafting cells in cord blood and autologous mobilized peripheral blood. Biol Blood Marrow Transplant. 5:69-76.

National Marrow Donor Program (NMDP). http://www.marrow.org/PHYSICIAN/URD_Search_and_Tx/Number_of_Allogeneic_Tx_Perfor/index.html#grafts.

Olaharski AJ, Uppal H, Cooper M, Platz S, Zabka TS, Kolaja KL. (2009). In vitro to in vivo concordance of a high throughput assay for bone marrow toxicity across a diverse set of drug candidates. Toxicol Let 188:98-103.

Page KM, Zhang L, Mendizabai A, Weasse S, Carter S, Gentry T, Balber E, Kurtzberg J. (2011a). Total colony-forming units are a strong, independent predictor of neutrophil and platelet engraftment after unrelated umbilical cord blood transplantation: A single-center analysis of 435 cord blood transplants. Biol Blood Marrow Transplant. (Jan 28. Epub ahead of print).

Page KM, Zhang L, Mendizabai A, Weasse S, Carter S, Shoulars K, Gentry T, Balber E, Kurtzberg J. (2011b). The cord blood Apgar: a novel scoring system to optimize selection of banked cord blood grafts for transplantation. (2011b). Transfusion. (Aug. 2. Epub ahead of print).

Pasquini MC & Wang Z. (2010). Current use and outcome of hematopoietic stem cell transplantation. CIBMTR Summary Slides, 2010. http://www.cibmtr.org/ReferenceCenter/SlidesReports/SummarySlides/pages/index.aspx#CiteSummarySlides.

Picardi A, Arcese W. (2010). Quality assessment of cord blood units selected for unrelated transplantation: a transplant center perspective. Transfus Apher Sci. 42:289-97.

Querol S, Gomez SG, Pagliuca A, Torrabadella M, Madrigal JA. (2010). Quality rather than quantity: the cord blood bank dilemma. Bone Marrow Transplant. 45:970-8.

Rayment EA & Williams DJ. (2010). Concise rewiew: Mind the gap: challenges in characterizing and quantifying cell- and tissue-based therapies for clinical translation. Stem Cells. 28:996-1004.

Reems J-A, Hall KM, Gebru LH, Taber G, Rich IN. (2008). Development of a novel assay to evaluate the functional potential of umbilical cord blood progenitors. Transfusion. 48:620-628.

Rich IN & Hall KM. (2005). Validation and development of a predictive paradigm for hemotoxicity using a multifunctional bioluminescence colony-forming proliferation assay. Tox Sci. 87:427-41.

Rich IN, Kubanek B. (1982). The effect of reduced oxygen tension on colony formation of erythropoietic cells in vitro. Brit J Haematol. 52:579-88.

Rich IN. (1997). Standardization of the CFU-GM assay using hematopoietic growth factors. J. Hematother. 6:191-193.

Rich IN. (2003). In vitro hematotoxicity testing in drug development: A review of past, present, and future applications. Curr Opinion Drug Disc Devel. 6:100-109.

Rich IN. (2007). High-throughput in vitro hemotoxicity testing and in vitro cross-platform comparative toxicity. Expert Opin. Drug Metab Toxicol. 3:295-307.

Rodriguez L, Garcia J, Querol S. (2005). Predictive utility of the attached segment in the quality control of a cord blood graft. Biol Blood Marrow Transplant. 11:247-251.

Santos GW, Sensenbrenner LL, Burke PJ, Mullins GM, Vias WB, Tutschka PJ & Slavin RE. (1972). The use of cyclophosphamide for clinical transplantation. Transplant Proc. 4:559-564.

Santos GW. (1983). History of bone marrow transplantation. Clin Haematol. 12:611-639.

Sohn SK, Kim JG, Seo KW, Chae YS, Jung JT, Suh JS, Lee KB. (2002). GM-CSF-based mobilized effect in normal healthy donors for allogeneic peripheral blood stem cell transplantation. Bone Marrow Transplant. 30:81-86.

Solves P, Planelles D, Mirabet V, Blasco I, Carbonell-Uberos F, Soler MA, Roig RJ. (2004) Utility of bag segment and cryovial samples for quality and confirmatory HLA typing in umbilical cord blood banking. Clin Lab Haematol. 26:413-418.

Spellman S, Hurley CK, Brady C, Phillips-Johnson L, Chow R, Laughlin M, McMannis J, Reems J-A, Regan D, Rubinstein P, Kurtzberg J. (2011). Guidelines for the development and validation of new potency assays for the evaluation of umbilical cord blood. Cyotherapy (March, Epub ahead of print).

Strong M, Farrugia A & Rebulla P. (2009). Stem cell and cellular thereapy developments. Biologicals. 37:103-107.

Thomas ES, Buckner CD, Banaji M, Clift RA, Fefer A, Flournoy N, Goodell BW, Hickman RO, Lerner KG, Neiman PE, Sale GE, Sanders JE, Singer J, Stevens M, Storb R & Weiden PI. (1977).One hundred patients with acute leukemia treated by chemotherapy, total body irradiation, and allogeneic marrow transplantation. Blood. 49:511-533.

Thorpe R. Wadhwa M, Page C, Mire-Sluis A. (1999). Bioassays for the characterization and control of therapeutic cytokines; determination of potency. Dev Biol Stand. 97:61-71.

Till JE & McCulloch EA. (1971). A direct measurement of the radiation sensitivity of normal mouse bone marrow cells. Radiat Res. 175:145-149.

U.S. Food and Drug Administration (FDA) (2001). Guidance for Industry. Bioanalytical method validation. http://www.fda.gov/downloads/Drugs/GuidanceComplianceRegulatoryInformation/Guidances/ucm070107.pdf.

U.S. Food and Drug Administration (FDA) (2011). Guidance for Industry. Potency tests for cellular and gene therapy products. http://www.fda.gov/downloads/BiologicsBloodVaccines/GuidanceComplianceRegulatoryInformation/Guidances/CellularandGeneTherapy/UCM243392.pdf.

Wagner E. Duval M, Dalle JH, Morin H, Bizier S, Champagne J, Champagne MA. (2006) Assessment of cord blood unit characteristics on the day of transplant: comparison with data issued by cord blood banks. Transfusion. 46: 1190-1198.

Zubair AC, Kao G, Daley H, Schott D, Freedman A, Ritz J. (2006). CD34(+) CD38(-) and CD34(+) HLA-DR(-) cells in BM stem cell grafts correlate with short-term engraftment but have no influence on long-term hematopoietic reconstitution after autologous transplantation. Cytotherapy. 8:399-407.

18

Detection of CMV Infection in Allogeneic SCT Recipients: The Multiple Assays

Pilar Blanco-Lobo, Omar J. BenMarzouk-Hidalgo and Pilar Pérez-Romero
Unit of Infectious Disease, Microbiology and Preventive Medicine,
Instituto de Biomedicina de Sevilla (IBiS)/CSIC/Universidad de Sevilla,
University Hospital Virgen del Rocio, Sevilla,
Spain

1. Introduction

Cytomegalovirus (CMV) end-organ disease is a serious complication after stem cell transplantation (SCT) (Boeckh M, 2003). Within the first one hundred days after SCT, 50% of recipients develop CMV infection determined by positive antigenemia and 65 to 86.5% when viral replication is determined by real-time PCR (RT-PCR) (Ljungman et al., 2006; Solano et al., 2001). Described risk factors for CMV infection concern donor type, graft source, positive CMV serostatus of donor and recipient, CD34^{+} graft selection, preconditioning regimen, GvHD prophylaxis regimen, incidence of acute and chronic GvHD and prophylaxis and treatment for GvHD (Ljungman et al., 2002; Ozdemir et al., 2007). Pre-emptive therapy is currently based on viral replication determined by either antigenemia or RT-PCR (Drew, 2007). Although antigenemia has been extensively used (Drew, 2007), RT-PCR has been shown to be more sensitive (Hakki et al., 2003; Solano et al., 2001).

The use of techniques based on nucleic acid amplification for the detection of CMV in clinical samples are in expansion and in many hospitals have replaced the use of other assays such as viral cultures or pp65 antigenemia. Several studies have assessed the performance between the different CMV viral load assays available. However, no many studies have compared the differences between the DNA extraction methods used (Fahle & Fischer, 2000; Caliendo et al., 2007; Kalpoe et al., 2004; Leruez-Ville et al., 2003; Avetisyan et al., 2006; Boeckh et al., 2009; Gerna et al., 2008; Gimeno et al., 2008). Although during the DNA extraction the majority of the methods use internal controls as a measurement of the DNA loss during the extraction procedure, in the downstream amplification not many of the assays use DNA standards that will facilitate the comparison among the different kits and standardization of the results between hospitals. In fact the availability of an optimal and efficient DNA extraction procedure that can use a broad type of samples with minimum modifications in the procedure may be practical and affordable for use in the clinical practice (Fahle & Fischer, 2000).

One of the differences between results using the methods available is type of clinical specimen used to perform the CMV DNA extraction. Samples collected vary from plasma, whole blood or leukocytes, with the optimal sample for monitoring CMV viral being controversial. Since CMV infect cells, the viral load results obtained from leukocytes isolated from peripheral blood or whole blood samples tend to be higher than the results obtained from plasma. However, it has been reported a high correlation between CMV viral load results from plasma samples and whole blood samples (Caliendo et al., 2007; Kalpoe et al., 2004; Leruez-Ville et al., 2003). Moreover, some authors believe that the presence of CMV particles in plasma is related with the level of viral replication (Kalpoe et al., 2004), representing the infectious viral particles able to spread to other host cells. Thus whole blood and plasma samples are equally suitable for testing CMV infection in SCT recipients.

Molecular techniques for CMV quantification such as RT-PCR have been shown to be useful for the rapid diagnosis of CMV infection and for monitoring clinical responses to antiviral therapy. This technique offers some advantages over others PCR methods, including increased precision, accuracy, reproducibility and a shorter turnaround time. To date, the clinical utility of using the RT-PCR test to guide preemptive therapy in transplant recipients has been mainly studied in SCT recipients (Avetisyan et al., 2006; Boeckh M, 2009; Gerna et al., 2008; Gimeno et al., 2008; Harrington et al., 2007; Kalpoe et al., 2004; Lilleri et al., 2004; Limaye et al., 2001; Machida et al., 2000; Ruell et al., 2007; Verkruyse et al., 2006). However, it has not been established a cutoff threshold for initiating antiviral therapy against CMV probably due to the significant differences between the different techniques used to determine the CMV viral load. In the absence of standardization the current clinical guidelines recommend to each individual laboratory to establish their own viral thresholds for CMV management (Kotton et al., 2010; Razonable & Emery, 2004).

2. DNA extraction methods

CMV extraction assays can be performed manually and automated. While manual extraction assays use non-corrosive reagents, are generally inexpensive and are easy to use, they require more labour intensive manipulation increasing the risk for contamination of the samples. In addition, this type of extraction procedures requires highly trained laboratory personnel to ensure reproducible results. Another limitation of the manual assays is the use of ethanol to precipitate the DNA, which may inhibit subsequent RT-PCR assays if not properly removed (Valentine-Thon, 2002). The manual assays are mostly used in research laboratories where the number of samples used at once is not high and the personnel are highly qualified for the procedures.

Automated extraction methods are not widely extended although there are commonly used in clinical services where the number of clinical samples to process every day is high. The main feature of the automated extraction systems is the increase in reproducibility of the extraction among different samples, in addition to a reduction of the risk of contamination and the high number of samples that could be performed at the same time. However, the main handicap of this technology is the elevated cost of the instruments, as well as the high-costs of instruments` reagents and maintenance and the necessary laboratory space required (Espy et al., 2006).

While recently reports have shown improvement in the sensibility obtained by the automated extraction instruments in comparison with the manual extraction kits (Gartner et al., 2004; Mengelle et al., 2011), and several studies performed on different herpesviruses have shown increased sensibility when automated extraction was performed compared to manual extraction kits (Nicholson et al.; 1997; Griffiths et al.; 1984), our laboratory recently demonstrated that the DNA extraction method from Affigene was more efficient than the automated system from Abbott providing a more accurate estimation of CMV DNA load (Gracia-Ahufinger et al., 2010). Our data proving that the manual DNA extraction method from Affigene resulted in a more efficient DNA extraction in comparison with that of an automated procedure from Abbott were somewhat surprising and are in contrast to previously published studies showing just the opposite (Kalpoe et al., 2004; Limaye et al., 2001). In this context, our data underscore the fact that the DNA extraction efficiency of distinct automated systems may not be comparable and should be thoroughly evaluated. This finding translates into critical therapeutic consequences, as patients would be treated depending on a threshold viral load, which will be different depending on the method used. In this context, these data reinforce the idea that local guidelines for the initiation of pre-emptive therapy based on commercial assays must be established as long as universally accepted standards for quantitative analysis of CMV DNAemia are not available.

3. Detection of CMV infection

CMV viral load determination is used to diagnose active CMV infection, to adopt treatment strategies to prevent CMV infection after transplantation and to monitor CMV after therapy. For this reason it is necessary to establish robust and reproducible assays to make possible to detect CMV levels within a wide range from low to very high number of copies (Abbate et al., 2008).

3.1 CMV detection using antigemia assays

The pp65 antigenemia developed in the late 1980s was the first non-cell culture based quantitative assay used in clinic to detect CMV infection (Atkinson & Emery), making obsolete the previous techniques such as shell vial assays (Gleaves et al., 1984; Nicholson et al., 1997), or the detection of early antigen fluorescent foci (DEAFF) test (Griffiths et al., 1984). The pp65 antigenemia assay is based on the detection of the pp65 phosphoprotein of CMV in peripheral blood leukocytes (Van der Bij et al., 1988), and it has been widely used for years to quantify and guiding the administration of therapy and monitoring active CMV infection of STC recipients (Bonon et al., 2005; Tormo et al., 2010). However, the antigenemia assay has many disadvantages such as, it requires quite a lot blood volume as well as intensive labour and need to process samples within 6h from the time of collection to achieve optimal sensitivity (Kim et al., 2007; Mhiri et al., 2007), it restricts the numbers of samples that can be analyzed simultaneously and it requires a high number of leukocytes (at least more than 200 leukocytes) for acceptable performance of the assay (Preiser et al., 2001), being unfeasible during periods of severe neutropenia. In addition, due to the fact that antigenemia results can be elevated after following ganciclovir treatment despite of a decrease of DNAemia levels, results using antigenemia for monitoring efficacy of the pre-emptive therapy of CMV infection in SCT recipients may be mislead (Sia et al., 2000). Other molecular techniques have reduced the turnaround time for monitoring CMV infection.

3.2 Qualitative PCR assays

In the past few years new sensitive PCR based techniques have been developed for earlier detection of CMV infection. The new assays developed were initially qualitative and they were able to detect CMV viremia in plasma of SCT recipients, and were compared with antigenemia assay (Boeckh et al., 1997; Boivin et al., 2000; Ksouri et al., 2007; Mori et al., 2000; Preiser et al., 2001). Results from Boivin et al. found a higher sensitivity in antigenemia test, while Boeckh et al. suggested a similar sensibility in both techniques. Most of these studies used in-house PCR assays (Boeckh et al., 1997; Boivin et al., 2000; Preiser et al., 2001), which made difficult to compare results and to conclude the clinical value of the methods (Solano et al., 2001).

The AMPLICOR CMV DNA PCR assay (Roche Diagnostics, Branchburg, N.J.) was the first qualitative technique commercialized. However, despite of being a more sensitive technique, antigenemia was found to be a more suitable technique both for guiding the initiation of preemptive therapy and for monitoring the efficacy of ganciclovir treatment (Solano et al., 2001).

3.3 Quantitative PCR assays

The quantitative PCR assays have demonstrated to be more suitable and clinically relevant than qualitative PCR for the detection of CMV DNA (Sia et al., 2000), providing useful information for the management of patients at high risk for developing CMV infection. Quantitative results may facilitate the establishment of a threshold for CMV viral load and the discrimination between patients who had symptomatic CMV infection and those who do not. Thus, allowing to establish the degree of viral replication and to distinguish between low and high level of CMV infection that may lead to disease after SCT (Preiser et al., 2001). Although there are many different commercially available quantitative PCR assays for CMV detection, the COBAS AMPLICOR CMV MONITOR test is one of the more commonly used in the clinical practice. This quantitative PCR developed by Roche included an internal quantification standard. The performance of the assay was found to be more sensitive compared with other qualitative tests (Boivin et al., 2000; Caliendo et al., 2001), with a lower limit of detection of 400 copies/ml of plasma and a dynamic range up to 50,000CMV DNA copies/ml. This assay has been widely used for early detection of CMV infection in a variety of clinical specimens and clinical studies (Ghisetti et al., 2004; Lehto et al., 2005; Martin-Davila et al., 2005; Piiparinen et al., 2005; Sia et al., 2000; Westall et al., 2004). However, it shows some disadvantages, due to the fact that it requires manual extraction it has a low number of sample processing (24 per run) and a long performance (approximately 8 h). In addition, the limit of detection has been established in 2.78 log10 cop/ml, value that is high especially for the early detection of CMV replication (Kerschner et al., 2011).

The use of quantitative PCR to detect CMV infection has been highly controversial regarding the specimen used (plasma, whole blood or leukocytes) for the quantification of the CMV viral load (Boeckh et al., 1997; Boivin et al., 2000; Caliendo et al., 2000; Flexman et al., 2001; Kaiser et al., 2002; Machida et al., 2000; Razonable et al., 2002; Tanaka et al., 2000; Weinberg, Schissel, & Giller, 2002). Some studies have suggested that quantitative PCR measurements for monitoring CMV viral load in whole-blood have a higher sensitivity compared to using cells or plasma during CMV disease in immunocompromised patients (Razonable et al., 2002) (von Muller et al., 2007). The authors consider that whole blood includes all the compartments in which the virus can replicate (Deback et al., 2007). In addition, Cortez et al found that quantitative PCR performed in whole blood provided a higher number of positive results (58.2% vs. 39.5%) compared to plasma (Cortez et al., 2003). However, Leruez-Ville et al. compared the performance of a RT-PCR specifically to amplify high conserved region of CMV UL93 gene in plasma and whole blood, demonstrating that both plasma and whole blood were equally suitable for monitoring active CMV infection (Leruez-Ville et al., 2003).

3.4 RT-PCR assays

In the mid 1990s become available the first two commercialized RT-PCR platforms. In the last years, different companies have tried to improve the technique including faster cycling, higher throughput and flexibility, new optical systems and more accessible software (Table 1). For example, it has been developed several versions of the LightCycler instruments such as Roche LightCycler™ PCR or SmartCycler (Cepheid) for performing sensitive, specific and rapid assays for the detection of CMV, time- and cost-effectiveness and with low contamination risk (Schaade et al., 2000). RT-PCR based on TaqMan probes and related technologies have proven higher dynamic range, precision, accuracy, reproducibility, a shorter turnaround time and a low risk of contamination, offering many advantages over quantitative-competitive PCR assays. With the use of these techniques, the quantification of CMV in clinically relevant samples could be reproducibly achieved in 2h allowing to understand CMV replication kinetics in humans (Atkinson & Emery, 2011). In addition, other advantages have been described about the use of RT-PCR to evaluate the CMV load in HSCT including the ability to test blood during episodes of neutropenia and subsequent disease that had been missed by antigenemia (Kaiser et al., 2002).

On the contrary, RT-PCR also has some disadvantages compared with conventional PCR such as the start-up expense of the assay and the incompatibility of some platforms with certain reagents (Mackay, Arden, & Nitsche, 2002).

Although most of RT-PCR assays for monitoring CMV infection in SCT recipients have been laboratory developed (Boeckh et al., 2004; Griscelli et al., 2001; Herrmann et al., 2004; Hong et al., 2004; Kalpoe et al., 2004; Leruez-Ville et al., 2003; Lilleri et al., 2004; Limaye et al., 2001; Nitsche et al., 2000; Pumannova et al., 2006; Ruell et al., 2007; Schaade et al., 2000; Tanaka et al., 2000; Tanaka et al., 2002; Yakushiji et al., 2002; Yun et al., 2003), several commercial tests are available and have been used in different clinical diagnostic laboratories. However, there are not many studies based on the application of these commercial assays in SCT recipients (Bravo et al.; Gimeno et al., 2008; Gouarin et al., 2007; Gracia-Ahufinger et al., 2010; Hanson et al., 2007). As it will be described below, these studies evaluated the suitability of the commercial assays for the surveillance of active CMV infection in these patients and compared the performance of the different tests.

LABORATORY DEVELOPED REAL TIME PCR				
ADVANTAGES			REF	
Less expensive and with the possibility of being personally established by developers.			[10, 11, 17, 18, 20, 56, 61, 63-71]	
COMMERCIAL REAL TIME PCR				
NAME	MANUFACTURER	VIRAL TARGET	ADVANTAGES	REF
COBAS Amplicor CMV Monitor	Roche	UL54	Sensitive, low limit of detection and broad dynamic range.	[40-45, 51]
LightCycler® CMV Quantitative Kit	Roche	UL54	Reasonably accurate, sensitive, specific and linear. Suitable for the detection of CMV DNA early after transplantation.	[74,75]
Artus CMV PCR test	QIAGEN	UL122	Reliable CMV diagnostic early after transplantation. High sensitivity and performance.	74
CMV R-gene™	Argene	UL83	Accurate quantification in SCT patients, good correlation with other RT-PCR assays and pp65 antigenemia. Validated with several types of specimen and DNA purification systems (automatic and manual).	73
Abbott CMV real-time PCR Kit	Abbott Diagnosis	UL122	High sensitivity and very low limit of detection (25 cps/mL). Good correlation with antigenemia and suitable to monitoring active CMV infection in SCT patients.	[15, 72]
Affigene CMV Trender	Cepheid	Not specified	Robust, reproducible and sensitive. Better analytical performance than the Abbot test and accurate estimation of the viral load.	[29,28]

Table 1. Technical advantages of the laboratory developed and commercially available RT-PCR methods.

3.4.1 Commercial RT-PCR assays

There are several commercially available RT-PCR assays developed for the detection of CMV infection in clinical samples. Most of these assays use specific targets, such as UL83, UL123 genes or the HXFL4 region (Alain et al.; Caliendo et al., 2007; Gault et al., 2001; Gouarin et al., 2007; Mengelle et al., 2003). The most common targets used for the detection of CMV by RT-PCR are the immediate early (IE) gene (Nitsche et al., 1999), the polymerase (UL54) gene (Schaade et al., 2000) , the glycoprotein B gene (UL55) (Espy et al., 2006) and the pp65 gene (UL83) (Stocher et al., 2003). Among the commercially available standardized methods to detect CMV infection, the LightCycler® CMV Quantitative Kit (Roche), the RealArt CMV LightCycler PCR reagent test (QIAGEN, Germantown, MD), CMV R-gene™ (Argene, France), Affigene CMV Trender (Cepheid, Sweden) and the Abbott CMV real-time PCR Kit (Abbott Diagnosis, USA) have been evaluated in SCT.

The LightCycler® CMV Quantitative assay (Roche) is a standardized RT-PCR test based on analyte-specific reagents (ASR) designed to detect a fragment of 240 pb within the polymerase gene (UL54) (Alain et al.). This test has been compared with the RealArt CMV LightCycler PCR reagent test (QIAGEN) that detects a fragment of 105 pb within the IE gene (Hanson et al., 2007). They made the comparison using OptiQuant CMV DNA panels (AcroMetrix Corp.) that contained four concentrations of CMV strain AD169 and with plasma specimens collected from SCT recipients. Although both tests were suitable to detect CMV DNA early after transplantation, the results using the Qiagen test showed higher sensitivity as well as a better performance at the lower standard concentration (Hanson et al., 2007).

Other remarkable CMV RT-PCR assay is the CMV R-gene™ (Argene, France) that targets the pp65 gene (UL83). This test has been evaluated in SCT recipients from four centers showing an accurate quantification, as well as a good correlation with other laboratory-developed RT-PCR assays and pp65 antigenemia, thus the authors suggest that the R-gene test is a good alternative method to diagnose and monitor CMV infection (Gouarin et al., 2007). The Affigene CMV Trender kit was developed by Cepheid and it has been shown to be robust, reproducible and sensitive enough for routine measurement of patient samples (Abbate et al., 2008). The analytical performance of this assay was also evaluated in our laboratory compared with the Abbott CMV RT-PCR Kit, using samples obtained from SCT recipients. The Affigene CMV Trender assay yielded higher viral load than the Abbot test, suggesting a better analytical performance. The comparison was also performed using the OptiQuant CMV DNA quantification panel showing that the Affigene test provides a more accurate estimation of the CMV DNA load (Gracia-Ahufinger et al., 2010). The test manufactured by Abbott, was previously evaluated in plasma samples from SCT recipients (Gimeno et al., 2008).

However the assay was compared with the antigenemia test to monitoring active CMV infection in SCT patients. Results showed a good correlation of the results but higher sensitivity for the RT-PCR assay (Gimeno et al., 2008). More recently the Abbott CMV RT-PCR assay was also evaluated in SCT recipients compared with other two commercial tests (Roche and Nanogen) (Bravo et al., 2011). The results found variations in the performance of the tests which limited to establish a common cutoff between different assays. This issue will be discussed below.

4. CMV viral load threshold for treatment initiation

The development of antiviral strategies has resulted in a large decrease in the incidence of CMV disease. Two main therapeutic strategies have been developed for the control of CMV infection and prevention of CMV disease, prophylaxis and pre-emptive therapy. Both strategies have been shown to be equally effective to protect against CMV infection. In the prophylaxis strategy antiviral treatment is administered to all patients after transplantation for a period of time between 100 and 200 days. (Boeckh M, 2003; Hebart & Einsele, 2004; Meijer et al., 2003; Zaia, 2002). In the pre-emptive strategy treatment is administered only when the CMV viral load reaches an established threshold (Gimeno et al., 2008; Machida et al., 2000). The preemptive administration of treatment consequently requires the use of highly sensitive assay for monitoring CMV viral load.

The International Consensus Guidelines on the Management of CMV after transplantation considered necessary the establishment of a universal cut-off value for initiating therapy (Razonable & Emery, 2004). Several studies have tried to establish the clinical utility of using the RT-PCR test to guide preemptive therapy in SCT recipients (Avetisyan et al., 2006; Boeckh M, 2009; Gerna et al., 2008; Gimeno et al., 2008; Harrington et al., 2007; Kalpoe et al., 2004; Lilleri et al., 2004; Limaye et al., 2001; Machida et al., 2000; Ruell et al., 2007; Verkruyse et al., 2006). However as stated earlier, there are significant differences between the different techniques used to determine the CMV viral load, thus the standardization of specific cutoff values is limited by the variations in the performance of the test, the assay design, and the diversity in the patient population studied thus results can not be extrapolated. So, it would be necessary an international reference standard between viral loads obtained with different tests. Currently, each laboratory must establish its own cutoff value and monitor clinical outcomes to verify the trigger points used.

Some authors have suggested that pre-emptive therapy should be initiated after two consecutive increased viral load values (Boeckh M, 2009; Gimeno et al., 2008). However, the inter- and intra-assay variability in some cases with variations over 30% that may represent a risk for using this treatment initiation strategy (Boeckh & Boivin, 1998; Boeckh & Nichols, 2004), with variations of the viral load less than 0.5 log may not be significant. Other authors propose considering CMV replication kinetics for the initiation of treatment. In these cases, it needs to be considered that while CMV duplicate within 1-2 days on average, the time for replication is shorter in the presence of immunosuppressant drugs, which may result in faster increase of the viral loads. In addition in these cases may be necessary taking into account the initial viral load since it is predictive of risk for developing CMV disease (Emery & Griffiths, 2000).

Some studies have established optimal cut-offs for treatment initiation, classifying patients according to the risk for developing CMV disease. A recent publication by Boeckh and Ljungman recommends initiating treatment with viral loads over 100 copies/ml in SCT recipients at high risk for CMV infection that received the transplant within the last 100 days, and 500 copies/ml for patients at low risk. For long-term SCT recipients initiation of treatment is recommended with viral loads over 1,000 copies/ml (Boeckh M, 2009).

Most of the studies to establish thresholds for the control of CMV infection have been performed using plasma for testing CMV viral load with a wide range of cut-off viral loads varying from 550 copies/ml to 10,000 copies/ml (Avetisyan et al., 2006; Gimeno et al., 2008;

Hong et al., 2004; Yakushiji et al., 2002). Other studies have established thresholds for the control of CMV infection of 1,000 copies/ml (Boeckh M, 2009; Harrington et al., 2007), and 10,000 (Gerna et al., 2008; Lilleri et al., 2004; Verkruyse et al., 2006) using whole blood. These cut-off values were defined to be protective independently of their CMV sero-status. Few studies have established a protective cut-off for CMV infection in leukocytes; one of these established a cut off of 1,000 CMV genomes copies per 200,000 leukocytes (Avetisyan et al., 2006).

Another important issue after transplantation is the optimal frequency of CMV monitoring which has not been defined for SCT. Most authors recommend a weekly periodicity increased twice a week once CMV replication is detected and during treatment administration, while treatment administration should be interrupted after two consecutive negative determinations (Boeckh M, 2009).

In summary, it has not been established a cutoff threshold for initiating antiviral therapy against CMV maybe due to differences in CMV serological status, immunosuppressive drug regimens and period of treatment. Further studies are necessary in large series of SCT recipients assessing safety of viral load thresholds.

5. Standardization of CMV viral load quantification

Since CMV viral loads (in copies per milliliter of body fluid) correlate with the development of CMV disease (Emery & Griffiths, 2000; Humar et al., 1999), the use of molecular diagnostics based on the measurement of the viral load has contributed to patients' management after transplantation for more than a decade. During preemptive administration of treatment, antiviral therapy is initiated when CMV replication reaches an established threshold in the peripheral blood (Lilleri et al., 2009), prior to develop clinical symptoms. Therefore, the use methods such as RT-PCR can be useful to determine when to initiate the preemptive therapy and its duration (Emery & Griffiths, 2000; Humar et al., 1999; Humar et al., 2002; Sia & Patel, 2000) as well as to monitor the response to the administered therapy. However, as previously stated the most important handicap of the available RT-PCR techniques is the high variability of CMV viral load results among different laboratories. Pang et al designed a comparative study among thirty laboratories to evaluate the reproducibility of in-house and commercial assays to detect CMV infection. They prepared a panel of samples with different CMV DNA concentrations that were evaluated in different laboratories, with several commercial available assays. While the intra-laboratory coefficient of variation was considered acceptable (around 17%), they inter-laboratory variability resulted higher than 140%. These authors considered that differences in viral load lower at <0.5 log10 are not considered clinically relevant. This difference limits the comparisons inter-laboratories and prevents the establishment of a determinate cut-off broadly applicable for making clinical decisions and monitoring the initiation of pre-emptive therapy (Caliendo et al., 2009). These discrepancies result into clinical therapeutic consequences, as a number of patients may receive treatment in one hospital while not in other hospital using a different assay (Gracia-Ahufinger et al., 2010).

The differences among assays are based on the method for nucleic acid extraction, the specimen type, the target genomic region, primers and probes used for amplification and detection (Caliendo et al., 2009; Ikewaki et al., 2005). Although, the variability in the results obtained with commercial assays (kits or ASRs) are proven to be lower than when using in-house developed methods (Pang et al., 2009).

The most critical factor for standardization is the lack of using an universally accepted standard for CMV quantification will make possible the comparison among results thus establishing a common management of patients in different centers (Atkinson & Emery, 2011). The reference material used as a calibrator have to be traceable and commutable to achieve accurate clinical results ensuring consistency with clinical samples (Caliendo et al., 2009). While many laboratories produce their own calibrator as an attempt of standardizing some CMV reference material is under development and in fact there are standards commercially available for quantification. For instance, the National Institute of Standards and Technology (NIST) started the development of a reference standard for CMV based on pure CMV DNA from a Towne strain from which after some modification, the final construct will be used to produce viral DNA (Wang et al., 2004). Moreover, the OptiQuant® CMV DNA Quantification Panel from AcroMetrix has been carefully formulated to mimic naturally occurring human specimens containing CMV viral DNA. It consists of cultured CMV that has been diluted in defibrinated, delipidized normal human plasma at different concentrations. The panel can be used with any test procedure designed for measuring CMV DNA in human serum or plasma and it has been widely used as a standard in several studies (Bravo et al.; Forman et al., 2011; Hanson et al., 2007; Raggam et al., 2010) to compare techniques with the same laboratory and inter-laboratory.

In the absence of standardization the current clinical guidelines recommend to each individual laboratory to establish their own viral thresholds for CMV management, (Kotton et al., 2010; Razonable & Emery, 2004), threshold that cover a wide range of viral loads varying from 200-500 copies/mL in some laboratories (Ikewaki et al., 2005; Mori et al., 2002) to 2000-5000 copies/mL in others (Humar et al., 1999).

6. Conclusions

In conclusion, the quantitative PCR assays have demonstrated to be more suitable and clinically relevant for the monitoring of the CMV viral load, and the management of patients that develop CMV infection. Although there are several commercially available RT-PCR assays developed for the detection of CMV infection in clinical samples, there are variations in the performance of these tests which limit to establish a common cutoff between different assays. Each laboratory must establish its own cutoff value and monitor clinical outcomes to verify the trigger points used. Using universally accepted standards for CMV quantification will make possible the comparison among results establishing a common management of patients in different centers. However, in the absence of standardization the current clinical guidelines recommend to each individual laboratory to establish their own viral thresholds for CMV management. Further studies are still necessary to establish standardized cut-off values in large series of transplant recipients.

7. Acknowledgement

The authors thank other members of the laboratory. P.P-R. was funded by Instituto de Salud Carlos III, Programa Miguel Servet CP05/00226.

8. References

Abbate, I.; Finnstrom, N.; Zaniratti, S.; Solmone, M. C.; Selvaggini, S.; Bennici, E.; Neri, S.; Brega, C.; Paterno, M.; & Capobianchi, M. R. (2008). Evaluation of an automated extraction system in combination with Affigene CMV Trender for CMV DNA quantitative determination: comparison with nested PCR and pp65 antigen test. *J Virol Methods* Vol.151, No.1, pp. 61-5, ISSN 0166-0934

Alain, S.; Lachaise, V.; Hantz, S.; & Denis, F. Comparison between the LightCycler CMV Quant Kit (Roche Diagnostics) with a standardized in-house Taqman assay for cytomegalovirus blood viral load quantification. *Pathol Biol (Paris)* Vol.58, No.2, pp. 156-61, ISSN 1768-3114

Atkinson, C.; & Emery, V. C. (2011). Cytomegalovirus quantification: Where to next in optimising patient management? *J Clin Virol* Vol.51, No.4, pp. 223-8, ISSN 1873-5967

Avetisyan, G.; Larsson, K.; Aschan, J.; Nilsson, C.; Hassan, M.; & Ljungman, P. (2006). Impact on the cytomegalovirus (CMV) viral load by CMV-specific T-cell immunity in recipients of allogeneic stem cell transplantation. *Bone Marrow Transplant* Vol.38, No.10, pp. 687-92, ISSN 0268-3369

Boeckh, M.; & Boivin, G. (1998). Quantitation of cytomegalovirus: methodologic aspects and clinical applications. *Clin Microbiol Rev* Vol.11, No.3, pp. 533-54, ISSN 0893-8512

Boeckh, M.; Gallez-Hawkins, G. M.; Myerson, D.; Zaia, J. A.; & Bowden, R. A. (1997). Plasma polymerase chain reaction for cytomegalovirus DNA after allogeneic marrow transplantation: comparison with polymerase chain reaction using peripheral blood leukocytes, pp65 antigenemia, and viral culture. *Transplantation* Vol.64, No.1, pp. 108-13, ISSN 0041-1337

Boeckh, M.; Huang, M.; Ferrenberg, J.; Stevens-Ayers, T.; Stensl&, L.; Nichols, W. G.; & Corey, L. (2004). Optimization of quantitative detection of cytomegalovirus DNA in plasma by real-time PCR. *J Clin Microbiol* Vol.42, No.3, pp. 1142-8, ISSN 0095-1137

Boeckh M. (2009). How we treat cytomegalovirus in hematopoietic cell transplant recipients. *Blood* Vo.113, No.23, pp. 5711-9, ISSN 1528-0020

Boeckh, M.; & Nichols, W. G. (2004). The impact of cytomegalovirus serostatus of donor & recipient before hematopoietic stem cell transplantation in the era of antiviral prophylaxis and preemptive therapy. *Blood* Vol.103, No.6, pp. 2003-8, ISSN 0006-4971

Boeckh, M.; Papanicolaou, G.; Rubin, R.; Wingard, J.R.; Zaia, J. (2003). CMV in hematopoietic stem cell transplant recipients: Current status, known challenges, and future strategies. *Biol. Blood Marrow Transplant* Vol.5, pp. 543-558, ISSN 1083-8791

Boivin, G.; Belanger, R.; Delage, R.; Beliveau, C.; Demers, C.; Goyette, N.; & Roy, J. (2000). Quantitative analysis of cytomegalovirus (CMV) viremia using the pp65 antigenemia assay and the COBAS AMPLICOR CMV MONITOR PCR test after blood & marrow allogeneic transplantation. *J Clin Microbiol* Vol.38, No.12, pp. 4356-60, ISSN 0095-1137

Bonon, S. H.; Menoni, S. M.; Rossi, C. L.; De Souza, C. A.; Vigorito, A. C.; Costa, D. B.; & Costa, S. C. (2005). Surveillance of cytomegalovirus infection in haematopoietic stem cell transplantation patients. *J Infect* Vol.50, No.2, pp. 130-7, ISSN 0163-4453

Bravo, D.; Clari, M. A.; Costa, E.; Munoz-Cobo, B.; Solano, C.; Jose Remigia, M.; & Navarro, D. (2011). Comparative Evaluation of Three Automated Systems for DNA Extraction in Conjunction with Three Commercially Available Real-Time PCR

Assays for Quantitation of Plasma Cytomegalovirus DNAemia in Allogeneic Stem Cell Transplant Recipients. *J Clin Microbiol* Vol.49, No.8, pp. 2899-904, ISSN 1098-660X

Caliendo, A. M.; Ingersoll, J.; Fox-Canale, A. M.; Pargman, S.; Bythwood, T.; Hayden, M. K.; Bremer, J. W.; & Lurain, N. S. (2007). Evaluation of real-time PCR laboratory-developed tests using analyte-specific reagents for cytomegalovirus quantification. *J Clin Microbiol* Vol.45, No.6, pp. 1723-7, ISSN 0095-1137

Caliendo, A. M.; Schuurman, R.; Yen-Lieberman, B.; Spector, S. A.; &ersen, J.; Manjiry, R.; Crumpacker, C.; Lurain, N. S.; & Erice, A. (2001). Comparison of quantitative & qualitative PCR assays for cytomegalovirus DNA in plasma. *J Clin Microbiol* Vol.39, No.4, pp. 1334-8, ISSN 0095-1137

Caliendo, A. M.; Shahbazian, M. D.; Schaper, C.; Ingersoll, J.; Abdul-Ali, D.; Boonyaratanakornkit, J.; Pang, X. L.; Fox, J.; Preiksaitis, J.; & Schonbrunner, E. R. (2009). A commutable cytomegalovirus calibrator is required to improve the agreement of viral load values between laboratories. *Clin Chem* Vol.55, No.9, pp. 1701-10, ISSN 1530-8561

Caliendo, A. M.; St George, K.; Kao, S. Y.; Allega, J.; Tan, B. H.; LaFontaine, R.; Bui, L.; & Rinaldo, C. R. (2000). Comparison of quantitative cytomegalovirus (CMV) PCR in plasma & CMV antigenemia assay: clinical utility of the prototype AMPLICOR CMV MONITOR test in transplant recipients. *J Clin Microbiol* Vol.38, No.6, pp. 2122-7, ISSN 0095-1137

Cortez, K. J.; Fischer, S. H.; Fahle, G. A.; Calhoun, L. B.; Childs, R. W.; Barrett, A. J.; & Bennett, J. E. (2003). Clinical trial of quantitative real-time polymerase chain reaction for detection of cytomegalovirus in peripheral blood of allogeneic hematopoietic stem-cell transplant recipients. *J Infect Dis* Vol.188, No.7, pp. 967-72, ISSN 0022-1899

Deback, C.; Fillet, A. M.; Dhedin, N.; Barrou, B.; Varnous, S.; Najioullah, F.; Bricaire, F.; & Agut, H. (2007). Monitoring of human cytomegalovirus infection in immunosuppressed patients using real-time PCR on whole blood. *J Clin Virol* Vol.40, No.3, pp. 173-9, ISSN 1386-6532

Drew, W. L. (2007). Laboratory diagnosis of cytomegalovirus infection and disease in immunocompromised patients. *Curr Opin Infect Dis* Vol.20, No.4, pp. 408-11, ISSN 0951-7375

Emery, V. C.; & Griffiths, P. D. (2000). Prediction of cytomegalovirus load and resistance patterns after antiviral chemotherapy. *Proc Natl Acad Sci U S A* Vol.97, No.14, pp. 8039-44, ISSN 0027-8424

Espy, M. J.; Uhl, J. R.; Sloan, L. M.; Buckwalter, S. P.; Jones, M. F.; Vetter, E. A.; Yao, J. D.; Wengenack, N. L.; Rosenblatt, J. E.; Cockerill, F. R. 3rd; & Smith, T. F. (2006). Real-time PCR in clinical microbiology: applications for routine laboratory testing. *Clin Microbiol Rev* Vol.19, No.1, pp. 165-256, ISSN 0893-8512

Fahle, G. A.; & Fischer, S. H. (2000). Comparison of six commercial DNA extraction kits for recovery of cytomegalovirus DNA from spiked human specimens. *J Clin Microbiol* Vol.38, No.10, pp. 3860-3, ISSN 0095-1137

Flexman, J.; Kay, I.; Fonte, R.; Herrmann, R.; Gabbay, E.; & Palladino, S. (2001). Differences between the quantitative antigenemia assay and the cobas amplicor monitor quantitative PCR assay for detecting CMV viraemia in bone marrow and solid organ transplant patients. *J Med Virol* Vol.64, No.3, pp. 275-82, ISSN 0146-6615

Forman, M.; Wilson, A.; & Valsamakis, A. (2011). Cytomegalovirus DNA quantification using an automated platform for nucleic Acid extraction & real-time PCR assay setup. *J Clin Microbiol* Vol.49, No.7, pp. 2703-5, ISSN 1098-660X

Gartner, B. C.; Fischinger, J. M.; Litwicki, A.; Roemer, K.; & Mueller-Lantzsch, N. (2004). Evaluation of a new automated, st&ardized generic nucleic acid extraction method (total nucleic acid isolation kit) used in combination with cytomegalovirus DNA quantification by COBAS AMPLICOR CMV MONITOR. *J Clin Microbiol* Vol.42, No.8, pp. 3881-2, ISSN 0095-1137

Gault, E.; Michel, Y.; Dehee, A.; Belabani, C.; Nicolas, J. C.; & Garbarg-Chenon, A. (2001). Quantification of human cytomegalovirus DNA by real-time PCR. *J Clin Microbiol* Vol.39, No.2, pp. 772-5, ISSN 0095-1137

Gerna, G.; Lilleri, D.; Caldera, D.; Furione, M.; Zenone Bragotti, L.; & Aless&rino, E. P. (2008). Validation of a DNAemia cutoff for preemptive therapy of cytomegalovirus infection in adult hematopoietic stem cell transplant recipients. *Bone Marrow Transplant* Vol.41, No.10, pp. 873-9, ISSN 0268-3369

Ghisetti, V.; Barbui, A.; Franchello, A.; Varetto, S.; Pittaluga, F.; Bobbio, M.; Salizzoni, M.; & Marchiaro, G. (2004). Quantitation of cytomegalovirus DNA by the polymerase chain reaction as a predictor of disease in solid organ transplantation. *J Med Virol* Vol.73, No.2, pp. 223-9, ISSN 0146-6615

Gimeno, C.; Solano, C.; Latorre, J. C.; Hern&ez-Boluda, J. C.; Clari, M. A.; Remigia, M. J.; Furio, S.; Calabuig, M.; Tormo, N.; & Navarro, D. (2008). Quantification of DNA in plasma by an automated real-time PCR assay (cytomegalovirus PCR kit) for surveillance of active cytomegalovirus infection and guidance of preemptive therapy for allogeneic hematopoietic stem cell transplant recipients. *J Clin Microbiol* Vol.46, No.10, pp. 3311-8, ISSN 1098-660X

Gleaves, C. A.; Smith, T. F.; Shuster, E. A.; & Pearson, G. R. (1984). Rapid detection of cytomegalovirus in MRC-5 cells inoculated with urine specimens by using low-speed centrifugation and monoclonal antibody to an early antigen. *J Clin Microbiol* Vol.19, No.6, pp. 917-9, ISSN 0095-1137

Gouarin, S.; Vabret, A.; Scieux, C.; Agbalika, F.; Cherot, J.; Mengelle, C.; Deback, C.; Petitjean, J.; Dina, J.; & Freymuth, F. (2007). Multicentric evaluation of a new commercial cytomegalovirus real-time PCR quantitation assay. *J Virol Methods* Vol.146, No.1-2, pp. 147-54, ISSN 0166-0934

Gracia-Ahufinger, I.; Tormo, N.; Espigado, I.; Solano, C.; Urbano-Ispizua, A.; Clari, M. A.; de la Cruz-Vicente, F.; Navarro, D.; & Perez-Romero, P. (2010). Differences in cytomegalovirus plasma viral loads measured in allogeneic hematopoietic stem cell transplant recipients using two commercial real-time PCR assays. *J Clin Virol* Vol.48, No.2, pp. 142-6, ISSN 1873-5967

Griffiths, P. D.; Panjwani, D. D.; Stirk, P. R.; Ball, M. G.; Ganczakowski, M.; Blacklock, H. A.; & Prentice, H. G. (1984). Rapid diagnosis of cytomegalovirus infection in immunocompromised patients by detection of early antigen fluorescent foci. *Lancet* Vol.2, No.8414, pp. 1242-5, ISSN 0140-6736

Griscelli, F.; Barrois, M.; Chauvin, S.; Lastere, S.; Bellet, D.; & Bourhis, J. H. (2001). Quantification of human cytomegalovirus DNA in bone marrow transplant recipients by real-time PCR. *J Clin Microbiol* Vol.39, No.12, pp. 4362-9, ISSN 0095-1137

Hakki, M.; Riddell, S. R.; Storek, J.; Carter, R. A.; Stevens-Ayers, T.; Sudour, P.; White, K.; Corey, L.; & Boeckh, M. (2003). Immune reconstitution to cytomegalovirus after allogeneic hematopoietic stem cell transplantation: impact of host factors, drug therapy, and subclinical reactivation. *Blood* Vol.102, No.8, pp. 3060-7, ISSN 0006-4971

Hanson, K. E.; Reller, L. B.; Kurtzberg, J.; Horwitz, M.; Long, G.; & Alex&er, B. D. (2007). Comparison of the Digene Hybrid Capture System Cytomegalovirus (CMV) DNA (version 2.0), Roche CMV UL54 analyte-specific reagent, and QIAGEN RealArt CMV LightCycler PCR reagent tests using AcroMetrix OptiQuant CMV DNA quantification panels and specimens from allogeneic-stem-cell transplant recipients. *J Clin Microbiol* Vol.45, No.6, pp. 1972-3, ISSN 0095-1137

Harrington, S. M.; Buller, R. S.; Storch, G. A.; Li, L.; Fischer, S. H.; Murray, P. R.; & Gea-Banacloche, J. C. (2007). The effect of quantification standards used in real-time CMV PCR assays on guidelines for initiation of therapy in allogeneic stem cell transplant patients. *Bone Marrow Transplant* Vol.39, No.4, pp. 237-8, ISSN 0268-3369

Hebart, H.; & Einsele, H. (2004). Clinical aspects of CMV infection after stem cell transplantation. *Hum Immunol* 65(5), pp. 432-6.

Herrmann, B.; Larsson, V. C.; Rubin, C. J.; Sund, F.; Eriksson, B. M.; Arvidson, J.; Yun, Z.; Bondeson, K.; & Blomberg, J. (2004). Comparison of a duplex quantitative real-time PCR assay and the COBAS Amplicor CMV Monitor test for detection of cytomegalovirus. *J Clin Microbiol* Vol.42, No.5, pp. 1909-14, ISSN 0095-1137

Hong, K. M.; Najjar, H.; Hawley, M.; & Press, R. D. (2004). Quantitative real-time PCR with automated sample preparation for diagnosis & monitoring of cytomegalovirus infection in bone marrow transplant patients. *Clin Chem* Vol.50, No.5, pp. 846-56, ISSN 0009-9147

Humar, A.; Gregson, D.; Caliendo, A. M.; McGeer, A.; Malkan, G.; Krajden, M.; Corey, P.; Greig, P.; Walmsley, S.; Levy, G.; & Mazzulli, T. (1999). Clinical utility of quantitative cytomegalovirus viral load determination for predicting cytomegalovirus disease in liver transplant recipients. *Transplantation* Vol.68, No.9, pp. 1305-11, ISSN 0041-1337

Humar, A.; Kumar, D.; Boivin, G.; & Caliendo, A. M. (2002). Cytomegalovirus (CMV) virus load kinetics to predict recurrent disease in solid-organ transplant patients with CMV disease. *J Infect Dis* Vol.186, No.6, pp. 829-33, ISSN 0022-1899

Ikewaki, J.; Ohtsuka, E.; Satou, T.; Kawano, R.; Ogata, M.; Kikuchi, H.; & Nasu, M. (2005). Real-time PCR assays based on distinct genomic regions for cytomegalovirus reactivation following hematopoietic stem cell transplantation. *Bone Marrow Transplant* Vol.35, No.4, pp. 403-10, ISSN 0268-3369

Kaiser, L.; Perrin, L.; Chapuis, B.; Hadaya, K.; Kolarova, L.; Deffernez, C.; Huguet, S.; Helg, C.; & Wunderli, W. (2002). Improved monitoring of cytomegalovirus infection after allogeneic hematopoietic stem cell transplantation by an ultrasensitive plasma DNA PCR assay. *J Clin Microbiol* Vol.40, No.11, pp. 4251-5, ISSN 0095-1137

Kalpoe, J. S.; Kroes, A. C.; de Jong, M. D.; Schinkel, J.; de Brouwer, C. S.; Beersma, M. F.; & Claas, E. C. (2004). Validation of clinical application of cytomegalovirus plasma DNA load measurement and definition of treatment criteria by analysis of correlation to antigen detection. *J Clin Microbiol* Vol.42, No.4, pp. 1498-504, ISSN 0095-1137

Kerschner, H.; Bauer, C.; Schlag, P.; Lee, S.; Goedel, S.; & Popow-Kraupp, T. (2011). Clinical evaluation of a fully automated CMV PCR assay. *J Clin Virol* Vol.50, No.4, pp. 281-6, ISSN 1873-5967

Kim, D. J.; Kim, S. J.; Park, J.; Choi, G. S.; Lee, S.; Kwon, C. D.; Ki, C.; & Joh, J. (2007). Real-time PCR assay compared with antigenemia assay for detecting cytomegalovirus infection in kidney transplant recipients. *Transplant Proc* Vol.39, No.5, pp. 1458-60, ISSN 0041-1345

Kotton, C. N.; Kumar, D.; Caliendo, A. M.; Asberg, A.; Chou, S.; Snydman, D. R.; Allen, U.; & Humar, A. (2010). International consensus guidelines on the management of cytomegalovirus in solid organ transplantation. *Transplantation* Vol.89, No.7, pp. 779-95, ISSN 1534-6080

Ksouri, H.; Eljed, H.; Greco, A.; Lakhal, A.; Torjman, L.; Abdelkefi, A.; Ben Othmen, T.; Ladeb, S.; Slim, A.; Zouari, B.; Abdeladhim, A.; & Ben Hassen, A. (2007). Analysis of cytomegalovirus (CMV) viremia using the pp65 antigenemia assay, the amplicor CMV test, and a semi-quantitative polymerase chain reaction test after allogeneic marrow transplantation. *Transpl Infect Dis* Vol.9, No.1, pp. 16-21, ISSN 1398-2273

Lehto, J. T.; Lemstrom, K.; Halme, M.; Lappalainen, M.; Lommi, J.; Sipponen, J.; Harjula, A.; Tukiainen, P.; & Koskinen, P. K. (2005). A prospective study comparing cytomegalovirus antigenemia, DNAemia and RNAemia tests in guiding pre-emptive therapy in thoracic organ transplant recipients. *Transpl Int* Vol.18, No.12, pp. 1318-27, ISSN 0934-0874

Leruez-Ville, M.; Ouachee, M.; Delarue, R.; Sauget, A. S.; Blanche, S.; Buzyn, A.; & Rouzioux, C. (2003). Monitoring cytomegalovirus infection in adult and pediatric bone marrow transplant recipients by a real-time PCR assay performed with blood plasma. *J Clin Microbiol* Vol.41, No.5, pp. 2040-6, ISSN 0095-1137

Lilleri, D.; Baldanti, F.; Gatti, M.; Rovida, F.; Dossena, L.; De Grazia, S.; Torsellini, M.; & Gerna, G. (2004). Clinically-based determination of safe DNAemia cutoff levels for preemptive therapy or human cytomegalovirus infections in solid organ and hematopoietic stem cell transplant recipients. *J Med Virol* Vol.73, No.3, pp. 412-8, ISSN 0146-6615

Lilleri, D.; Lazzarotto, T.; Ghisetti, V.; Ravanini, P.; Capobianchi, M. R.; Baldanti, F.; & Gerna, G. (2009). Multicenter quality control study for human cytomegalovirus DNAemia quantification. *New Microbiol* Vol.32, No.3, 245-53, ISSN 1121-7138

Limaye, A. P.; Huang, M. L.; Leisenring, W.; Stensl&, L.; Corey, L.; & Boeckh, M. (2001). Cytomegalovirus (CMV) DNA load in plasma for the diagnosis of CMV disease before engraftment in hematopoietic stem-cell transplant recipients. *J Infect Dis* Vol.183, No.3, pp. 377-82, ISSN 0022-1899

Ljungman, P.; Griffiths, P.; & Paya, C. (2002). Definitions of cytomegalovirus infection and disease in transplant recipients. *Clin Infect Dis* Vol.34, No.8, pp. 1094-7, ISSN 1537-6591

Ljungman, P.; Perez-Bercoff, L.; Jonsson, J.; Avetisyan, G.; Sparrelid, E.; Aschan, J.; Barkholt, L.; Larsson, K.; Winiarski, J.; Yun, Z.; & Ringden, O. (2006). Risk factors for the development of cytomegalovirus disease after allogeneic stem cell transplantation. *Haematologica* Vol.91, No.1, pp. 78-83, ISSN 1592-8721

Mackay, I. M.; Arden, K. E.; & Nitsche, A. (2002). Real-time PCR in virology. *Nucleic Acids Res* Vol.30, No.6, pp. 1292-305, ISSN 1362-4962

Machida, U.; Kami, M.; Fukui, T.; Kazuyama, Y.; Kinoshita, M.; Tanaka, Y.; K&a, Y.; Ogawa, S.; Honda, H.; Chiba, S.; Mitani, K.; Muto, Y.; Osumi, K.; Kimura, S.; & Hirai, H. (2000). Real-time automated PCR for early diagnosis and monitoring of cytomegalovirus infection after bone marrow transplantation. *J Clin Microbiol* Vol.38, No.7, pp. 2536-42, ISSN 0095-1137

Martin-Davila, P.; Fortun, J.; Gutierrez, C.; Marti-Belda, P.; C&elas, A.; Honrubia, A.; Barcena, R.; Martinez, A.; Puente, A.; de Vicente, E.; & Moreno, S. (2005). Analysis of quantitative PCR assay for CMV infection in liver transplant recipients: an intent to find the cut-off value. *J Clin Virol* Vol.33, pp. 138-144, ISSN 1386-6532

Meijer, E.; Bol&, G. J.; & Verdonck, L. F. (2003). Prevention of cytomegalovirus disease in recipients of allogeneic stem cell transplants. *Clin Microbiol Rev* Vol.16, No.4, pp. 647-57, ISSN 0893-8512

Mengelle, C.; Mansuy, J. M.; Da Silva, I.; Davrinche, C.; & Izopet, J. (2011). Comparison of 2 highly automated nucleic acid extraction systems for quantitation of human cytomegalovirus in whole blood. *Diagn Microbiol Infect Dis* Vol.69, No.2, pp. 161-6, ISSN 1879-0070

Mengelle, C.; Pasquier, C.; Rostaing, L.; S&res-Saune, K.; Puel, J.; Berges, L.; Righi, L.; Bouquies, C.; & Izopet, J. (2003). Quantitation of human cytomegalovirus in recipients of solid organ transplants by real-time quantitative PCR and pp65 antigenemia. *J Med Virol* Vol.69, No.2, pp. 225-31, ISSN 0146-6615

Mhiri, L.; Kaabi, B.; Houimel, M.; Arrouji, Z.; & Slim, A. (2007). Comparison of pp65 antigenemia, quantitative PCR and DNA hybrid capture for detection of cytomegalovirus in transplant recipients & AIDS patients. *J Virol Methods* Vol.143, No.1, pp. 23-8, ISSN 0166-0934

Mori, T.; Okamoto, S.; Watanabe, R.; Yajima, T.; Iwao, Y.; Yamazaki, R.; Nakazato, T.; Sato, N.; Iguchi, T.; Nagayama, H.; Takayama, N.; Hibi, T.; & Ikeda, Y. (2002). Dose-adjusted preemptive therapy for cytomegalovirus disease based on real-time polymerase chain reaction after allogeneic hematopoietic stem cell transplantation. *Bone Marrow Transplant* Vol.29, No.9, pp. 777-82, ISSN 0268-3369

Mori, T.; Sato, N.; Watanabe, R.; Okamoto, S.; & Ikeda, Y. (2000). Erythema exsudativum multiforme induced by granulocyte colony-stimulating factor in an allogeneic peripheral blood stem cell donor. *Bone Marrow Transplant* Vol.26, No.2, pp. 239-40, ISSN 0268-3369

Nicholson, V. A.; Whimbey, E.; Champlin, R.; Abi-Said, D.; Przepiorka, D.; Tarr&, J.; Chan, K.; Bodey, G. P.; & Goodrich, J. M. (1997). Comparison of cytomegalovirus antigenemia & shell vial culture in allogeneic marrow transplantation recipients receiving ganciclovir prophylaxis. *Bone Marrow Transplant* Vol.19, No.1, pp. 37-41, ISSN 0268-3369

Nitsche, A.; Steuer, N.; Schmidt, C. A.; L&t, O.; Ellerbrok, H.; Pauli, G.; & Siegert, W. (2000). Detection of human cytomegalovirus DNA by real-time quantitative PCR. *J Clin Microbiol* Vol.38, No.7, pp. 2734-7, ISSN 0095-1137

Nitsche, A.; Steuer, N.; Schmidt, C. A.; O.; & Siegert, W. (1999). Different real-time PCR formats compared for the quantitative detection of human cytomegalovirus DNA. *Clin Chem* Vol.45, No.11, pp. 1932-7, ISSN 0009-9147

Ozdemir, E.; Saliba, R. M.; Champlin, R. E.; Couriel, D. R.; Giralt, S. A.; de Lima, M.; Khouri, I. F.; Hosing, C.; Kornblau, S. M.; Anderlini, P.; Shpall, E. J.; Qazilbash, M. H.; Molldrem, J. J.; Chemaly, R. F.; & Komanduri, K. V. (2007). Risk factors associated with late cytomegalovirus reactivation after allogeneic stem cell transplantation for hematological malignancies. *Bone Marrow Transplant* Vol.40, No.2, pp. 125-36, ISSN 0268-3369

Pang, X. L.; Fox, J. D.; Fenton, J. M.; Miller, G. G.; Caliendo, A. M.; & Preiksaitis, J. K. (2009). Interlaboratory comparison of cytomegalovirus viral load assays. *Am J Transplant* Vol.9, No.2, pp. 258-68, ISSN 1600-6143

Piiparinen, H.; Helantera, I.; Lappalainen, M.; Suni, J.; Koskinen, P.; Gronhagen-Riska, C.; & Lautenschlager, I. (2005). Quantitative PCR in the diagnosis of CMV infection and in the monitoring of viral load during the antiviral treatment in renal transplant patients. *J Med Virol* Vol.76, No.3, pp. 367-72, ISSN 0146-6615

Preiser, W.; Brauninger, S.; Schwerdtfeger, R.; Ayliffe, U.; Garson, J. A.; Brink, N. S.; Franck, S.; Doerr, H. W.; & Rabenau, H. F. (2001). Evaluation of diagnostic methods for the detection of cytomegalovirus in recipients of allogeneic stem cell transplants. *J Clin Virol* Vol.20, No.1-2, pp. 59-70, ISSN 1386-6532

Pumannova, M.; Roubalova, K.; Vitek, A.; & Sajdova, J. (2006). Comparison of quantitative competitive polymerase chain reaction-enzyme-linked immunosorbent assay with LightCycler-based polymerase chain reaction for measuring cytomegalovirus DNA in patients after hematopoietic stem cell transplantation. *Diagn Microbiol Infect Dis* Vol.54, No.2, pp. 115-20, ISSN 0732-8893

Raggam, R. B.; Bozic, M.; Salzer, H. J.; Hammerschmidt, S.; Homberg, C.; Ruzicka, K.; & Kessler, H. H. (2010). Rapid quantitation of cytomegalovirus DNA in whole blood by a new molecular assay based on automated sample preparation & real-time PCR. *Med Microbiol Immunol* Vol.199, No.4, pp. 311-6, ISSN 1432-1831

Razonable, R. R.; Brown, R. A.; Wilson, J.; Groettum, C.; Kremers, W.; Espy, M.; Smith, T. F.; & Paya, C. V. (2002). The clinical use of various blood compartments for cytomegalovirus (CMV) DNA quantitation in transplant recipients with CMV disease. *Transplantation* Vol.73, No.6, pp. 968-73, ISSN 0041-1337

Razonable, R. R.; & Emery, V. C. (2004). Management of CMV infection and disease in transplant patients. 27-29 February 2004. *Herpes* Vol.11, No.3, pp. 77-86, ISSN 0969-7667

Ruell, J.; Barnes, C.; Mutton, K.; Foulkes, B.; Chang, J.; Cavet, J.; Guiver, M.; Menasce, L.; Dougal, M.; & Chopra, R. (2007). Active CMV disease does not always correlate with viral load detection. *Bone Marrow Transplant* Vol.40, No.1, pp. 55-61, ISSN 0268-3369

Schaade, L.; Kockelkorn, P.; Ritter, K.; & Kleines, M. (2000). Detection of cytomegalovirus DNA in human specimens by LightCycler PCR. *J Clin Microbiol* Vol.38, No.11, pp. 4006-9, ISSN 0095-1137

Sia, I. G.; & Patel, R. (2000). New strategies for prevention & therapy of cytomegalovirus infection and disease in solid-organ transplant recipients. *Clin Microbiol Rev* Vol.13, No.1, 83-121, ISSN 0893-8512

Sia, I. G.; Wilson, J. A.; Espy, M. J.; Paya, C. V.; & Smith, T. F. (2000). Evaluation of the COBAS AMPLICOR CMV MONITOR test for detection of viral DNA in specimens taken from patients after liver transplantation. *J Clin Microbiol* Vol.38, No.2, pp. 600-6, ISSN 0095-1137

Solano, C.; Munoz, I.; Gutierrez, A.; Farga, A.; Prosper, F.; Garcia-Conde, J.; Navarro, D.; & Gimeno, C. (2001). Qualitative plasma PCR assay (AMPLICOR CMV test) versus pp65 antigenemia assay for monitoring cytomegalovirus viremia & guiding preemptive ganciclovir therapy in allogeneic stem cell transplantation. *J Clin Microbiol* Vol.39, No.11, pp. 3938-41, ISSN 0095-1137

Stocher, M.; Leb, V.; Bozic, M.; Kessler, H. H.; Halwachs-Baumann, G.; O.; Stekel, H.; & Berg, J. (2003). Parallel detection of five human herpes virus DNAs by a set of real-time polymerase chain reactions in a single run. *J Clin Virol* Vol.26, No.1, pp. 85-93, ISSN 1386-6532

Tanaka, N.; Kimura, H.; Iida, K.; Saito, Y.; Tsuge, I.; Yoshimi, A.; Matsuyama, T.; & Morishima, T. (2000). Quantitative analysis of cytomegalovirus load using a real-time PCR assay. *J Med Virol* Vol.60, No.4, pp. 455-62, ISSN 0146-6615

Tanaka, Y.; K&a, Y.; Kami, M.; Mori, S.; Hamaki, T.; Kusumi, E.; Miyakoshi, S.; Nannya, Y.; Chiba, S.; Arai, Y.; Mitani, K.; Hirai, H.; & Mutou, Y. (2002). Monitoring cytomegalovirus infection by antigenemia assay & two distinct plasma real-time PCR methods after hematopoietic stem cell transplantation. *Bone Marrow Transplant* Vol.30, No.5, pp. 315-9, ISSN 0268-3369

Tormo, N.; Solano, C.; Benet, I.; Clari, M. A.; Nieto, J.; de la Camara, R.; Lopez, J.; Lopez-Aldeguer, N.; Hern&ez-Boluda, J. C.; Remigia, M. J.; Garcia-Noblejas, A.; Gimeno, C.; & Navarro, D. (2010). Lack of prompt expansion of cytomegalovirus pp65 and IE-1-specific IFNgamma CD8+ and CD4+ T cells is associated with rising levels of pp65 antigenemia and DNAemia during pre-emptive therapy in allogeneic hematopoietic stem cell transplant recipients. *Bone Marrow Transplant* Vol.45, No.3, pp. 543-9, ISSN 1476-5365

Valentine-Thon, E. (2002). Quality control in nucleic acid testing--where do we stand? *J Clin Virol* Vol.25, No.Suppl 3, pp. S13-21, ISSN 1386-6532

Van der Bij, W.; Schirm, J.; Torensma, R.; Van Son, W. J.; Tegzess, A. M.; & The, T. H. (1988). Comparison between Viremia and Antigenemia for Detection of Cytomegalovirus in Blood. *J Clin Microbiol* Vol.26, No.12, pp. 5, ISSN 0095-1137

Verkruyse, L. A.; Storch, G. A.; Devine, S. M.; Dipersio, J. F.; & Vij, R. (2006). Once daily ganciclovir as initial pre-emptive therapy delayed until threshold CMV load > or =10000 copies/ml: a safe and effective strategy for allogeneic stem cell transplant patients. *Bone Marrow Transplant* Vol.37, No.1, pp. 51-6, ISSN 0268-3369

von Muller, L.; Hinz, J.; Bommer, M.; Hampl, W.; Kluwick, S.; Wiedmann, M.; Bunjes, D.; & Mertens, T. (2007). CMV monitoring using blood cells and plasma: a comparison of apples with oranges? *Bone Marrow Transplant* Vol.39, No.6, pp. 353-7, ISSN 0268-3369

Wang, W.; Patterson, C. E.; Yang, S.; & Zhu, H. (2004). Coupling generation of cytomegalovirus deletion mutants and amplification of viral BAC clones. *J Virol Methods* Vol.121, No.2, pp. 137-43, ISSN 0166-0934

Weinberg, A.; Schissel, D.; & Giller, R. (2002). Molecular methods for cytomegalovirus surveillance in bone marrow transplant recipients. *J Clin Microbiol* Vol.40, No.11, 4203-6, ISSN 0095-1137

Westall, G. P.; Michaelides, A.; Williams, T. J.; Snell, G. I.; & Kotsimbos, T. C. (2004). Human cytomegalovirus load in plasma and bronchoalveolar lavage fluid: a longitudinal study of lung transplant recipients. *J Infect Dis* Vol.190, No.6, pp. 1076-83, ISSN 0022-1899

Yakushiji, K.; Gondo, H.; Kamezaki, K.; Shigematsu, K.; Hayashi, S.; Kuroiwa, M.; Taniguchi, S.; Ohno, Y.; Takase, K.; Numata, A.; Aoki, K.; Kato, K.; Nagafuji, K.; Shimoda, K.; Okamura, T.; Kinukawa, N.; Kasuga, N.; Sata, M.; & Harada, M. (2002). Monitoring of cytomegalovirus reactivation after allogeneic stem cell transplantation: comparison of an antigenemia assay and quantitative real-time polymerase chain reaction. *Bone Marrow Transplant* Vol.29, No.7, pp. 599-606, ISSN 0268-3369

Yun, Z.; Lewensohn-Fuchs, I.; Ljungman, P.; Ringholm, L.; Jonsson, J.; & Albert, J. (2003). A real-time TaqMan PCR for routine quantitation of cytomegalovirus DNA in crude leukocyte lysates from stem cell transplant patients. *J Virol Methods* Vol.110, No.1, pp. 73-9, ISSN 0166-0934

Zaia, J. A. (2002). Prevention of cytomegalovirus disease in hematopoietic stem cell transplantation. *Clin Infect Dis* Vol.35, No.8, pp. 999-1004, ISSN 1537-6591

19

Bone Marrow Derived Pluripotent Stem Cells in Ischemic Heart Disease: Bridging the Gap Between Basic Research and Clinical Applications

Ahmed Abdel-Latif[1], Ewa Zuba-Surma[2] and Mariusz Z. Ratajczak[3]
[1]*Gill Heart Institute and Division of Cardiovascular Medicine, University of Kentucky, and Lexington VA Medical Center, Lexington, KY*
[2]*Department of Medical Biotechnology, Faculty of Biochemistry, Biophysics and Biotechnology, Jagiellonian University, Cracow,*
[3]*Stem Cell Biology Institute, James Graham Brown Cancer Center, University of Louisville, Louisville, KY,*
[1,3]*USA*
[2]*Poland*

1. Introduction

The prevalence of ischemic heart disease and acute myocardial infarction (AMI) has increased to alarming rates in the United States and the western world (Roger *et al.*, 2011). Patients who survive the initial AMI suffer ischemic cardiomyopathy (ICM) which is often complicated by high mortality and poor overall prognosis (Braunwald *et al.*, 2000; McMurray *et al.*, 2005). Despite significant advances in medical therapy and revascularization strategies, the prognosis of patients with AMI and ischemic cardiomyopathy remains dismal (Levy D *et al.*, 2002; Roger VL *et al.*, 2004). The last decade has demonstrated significant progress and rapid translation of myocardial regenerative therapies particularly those utilizing stem cells isolated from adult tissues (Abdel-Latif *et al.*, 2007).

Studies examining the potential therapeutic use of bone marrow (BM)-derived cells in myocardial regeneration have overshadowed the growing evidence of innate cardiac reparatory mechanisms. Several studies have demonstrated the capability of cardiomyocytes to replenish through poorly understood innate mechanisms. Follow up of cardiac transplantation patients have demonstrated continuous replenishment of cardiomyocytes by recipient's derived cells through poorly understood mechanisms (Quaini *et al.*, 2002). There is growing evidence that BM-derived cells are responsible, at least in part, for organ chimerism including cardiomyocyte chimerism (de Weger *et al.*, 2008; Deb *et al.*, 2003). Animal studies have confirmed this to be a dynamic process responding to significant injury such as myocardial infarction and peaks in the peri-infarct zone (Hsieh *et al.*, 2007). Although this process appears to be robust enough to achieve the renewal of approximately 50% of all cardiomyocytes in the normal life span, very little is known about its underpinnings (Bergmann *et al.*, 2009).

Complex innate reparatory mechanisms are initiated by myocardial ischemia interacting with different elements of the immune system, the infarcted myocardium and bone marrow stem cells, culminating in BM-stem and progenitor cells' (SPCs) mobilization as we and others have demonstrated (Abdel-Latif *et al.*, 2010; Walter *et al.*, 2007; Wojakowski *et al.*, 2009). However, very little is known about the mechanisms and clinical significance of this mobilization. Animal studies show that mobilized BM-derived cells (BMCs) repopulate the infarct border, however the significance of this mobilization is unclear given the low rate of their differentiation into cardiomyocytes (Fukuhara *et al.*, 2005).

2. Isolation and functional characteristics of BM-derived pluripotent stem cells

The bone marrow acts as a reservoir for a heterogeneous pool of tissue-committed and non-committed stem cells. These populations contain progenitors that aid in the chimerism and cellular turnover of different organs as well as very rare populations of pluripotent and non-committed stem cells. The old dogma that adult tissues lack pluripotent stem cells (PSCs) has been continuously challenged during the last decade through multiple studies that isolated PSCs from adult humans' and animals' tissues. These populations were distinguished based on their morphology with small cell size, large nucleus demonstrating euchromatin and large nucleus to cytoplasm ratio. Furthermore, cell surface markers as well as nuclear transcription factors, such as SSEA1/4, Oct4 and Nanog, have been deployed.

Very small embryonic like stem cells (VSELs) represent a rare yet pluripotent population of adult stem cells. They have been initially described by Dr. Ratajczak's group in the murine BM based on their expression of Sca1 (murine stem cell marker) and lack of expression of CD45 (pan-leukocytic marker) and differentiated lineage (Lin) markers (Kucia *et al.*, 2006; Zuba-Surma *et al.*, 2008). Following their isolation from murine tissues, VSELs were subsequently isolated from human BM, umbilical cord blood (CB) and peripheral blood based on the lack of expression of Lin/CD45 and the expression of the stem cell markers CD133, CXCR4 and CD34. **Figure 1** illustrates the flow cytometry protocol for identifying and isolating VSELs from murine and human samples. VSELs were further characterized using a multi-dimensional approach comprising molecular, protein and cell imaging techniques to confirm their pluripotent features (Zuba-Surma *et al.* 2008). VSELs are morphologically similar to embryonic stem cells demonstrating small diameter compared to more committed progenitors/stem cells with large nucleus containing open-type chromatin surrounded with thin rim of cytoplasm and multiple mitochondria (Zuba-Surma *et al.* 2008).

VSELs exhibit multiple embryonic and pluripotent surface and nuclear embryonic markers such as Oct4, SSEA1/4, Nanog, and Rex1. In vivo and in vitro studies have demonstrated the capability of VSELs to differentiate into multiple cell lines across germ lines including cardiomyocytes (Kucia *et al.* 2006).

The bone marrow harbors other multi- and pluri-potent stem cell populations such as the mesenchymal stem cells (MSC) (Hattan *et al.*, 2005; Kawada *et al.*, 2004), multipotent adult progenitor cells (MAPC) (Jiang *et al.*, 2002), and marrow-isolated multilineage inducible cells (MIAMI) (D'Ippolito *et al.*, 2004). Similar populations with cardiac differentiation potential have been also isolated from skeletal muscle and other tissues (Abdel-Latif *et al.,* 2008a). However, it is conceivable that different investigators have isolated, using different methods, the same or very similar populations and named them differently. It is also possible that these populations at least in part contain VSELs which might explain their pluripotent potential.

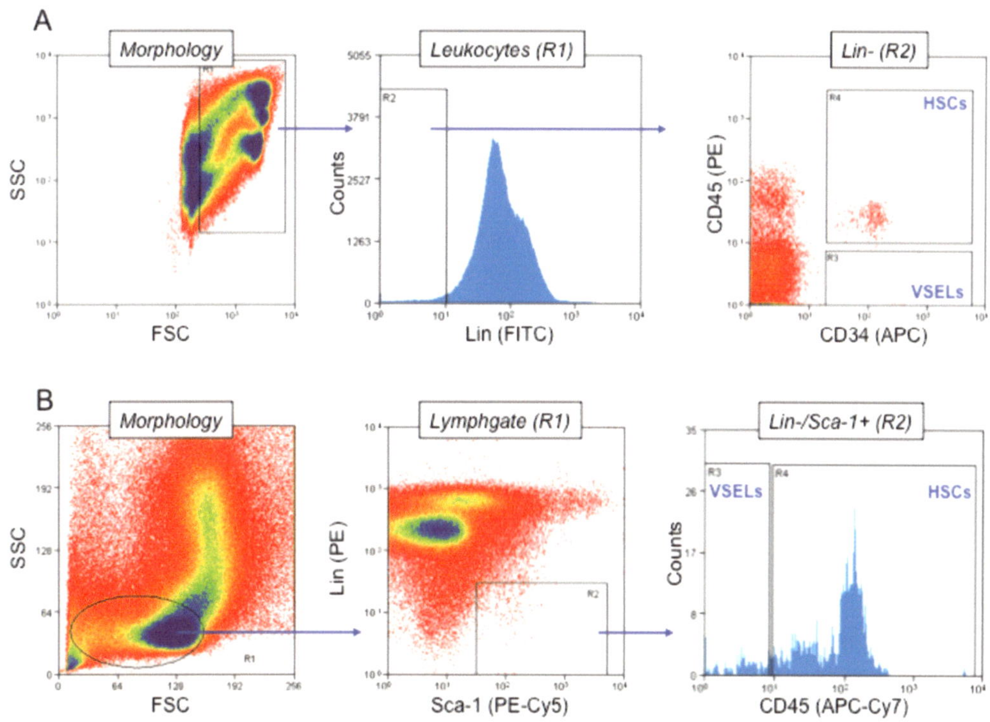

Panel A: Gating strategy for isolating human cord blood (CB)-derived VSELs. Morphology of total CB-derived nucleated cells is shown on dot-plot representing FSC and SSC parameters related to cell size and granularity/ complexity, respectively. All objects larger than 2μm are enclosed in region R1 and further visualized on histogram showing the expression of markers of mature hematopoietic cells (lineage markers; Lin). Cells not expressing differentiated hematopoietic markers (Lin- in region R2) are then analyzed for CD34 and CD45 expression. VSELs are identified as CD45-/Lin-/CD34+ cells (region R3), while hematopoietic stem cells (HSCs) as CD45+/Lin-/CD34+ cells (region R4).
Panel B: Sorting of murine bone marrow (BM)-derived VSELs. Morphology of total murine BM-derived nucleated cells is shown on dot-plot presenting FSC and SSC parameters and all objects in range of 2-10μm in diameter are included in region R1. Lymphocytic cells including stem cell fraction is further analyzed for Sca-1 and differentiated hematopoietic lineages markers (Lin) expression and only Sca-1+/Lin- cells are included in region R2. Cells from this region are further seperated based on CD45 expression. Murine VSELs are distinguished as CD45-/Lin-/Sca-1+ cells (region R3), while HSCs as CD45+/Lin-/Sca-1+ cells (region R4).

Fig. 1. Strategy for flow cytometric analysis of human and murine Very Small Embryonic-Like and hematopoietic stem cells. Briefly, BM is flushed from the femurs and tibias. Nucleated cells are isolated by lysis of red blood cells and cells are then gated on based on the cell size (>2 μm). Of note, lysis is preferred for isolating VSELs rather than Ficoll gradient that we have shown to lose some of the VSELs due to their small size.

3. BM-derived pluripotent stem cells are mobilized in the peripheral circulation following myocardial ischemia in animal models and humans

Myocardial ischemia, particularly large myocardial infarction, produce multiple stimuli include various chemokines, cytokines, kinins, bioactive lipids and members of the complement cascade, that lead to the mobilization and subsequent homing of BMSPCs. Indeed, several reports have confirmed that mobilization of stem cells originating from the BM occurs in response to myocardial ischemic injury (Grundmann *et al.*, 2007; Kucia *et al.*, 2004b; Leone *et al.*, 2005; Massa *et al.*, 2005; Shintani *et al.*, 2001; Wojakowski *et al.*, 2006) and heart failure (Valgimigli *et al.*, 2004). Similar observations were noted in patients with acute neurological ischemia (Paczkowska *et al.*, 2005) and patients with extensive skin burn (Drukala *et al.*, 2011).

Stimuli responsible for the mobilization and homing of BMSCs in the setting of myocardial ischemia show similarities and differences with those involved in hematopoietic stem cells (HSCs) homing to the BM. The role of stromal cell derived factor (SDF-1) and its receptor (CXCR4) axis in the retention of hematopoietic stem/progenitor cells (HSPCs) in bone marrow is undisputed (Kucia *et al.*, 2005; Lapidot *et al.*, 2002), however, its role in the mobilization and homing of BM-SPCs to a highly proteolytic microenvironment, such as the ischemic/infarcted myocardium, is somewhat less certain. Studies have demonstrated that multiple members of the metalloproteinases (MMP) family, such as MMP2, MMP9 and MMP13, are upreagulated in the myocardium following infarction (Peterson *et al.*, 2000). The elevated levels of the MMPs contribute to the degradation of chemokines such as SDF-1 and the byproduct of this degradation acts as an inhibitor the sole SDF-1 receptor, CXCR4 (McQuibban *et al.*, 2001; McQuibban *et al.*, 2002). In support of this hypothesis, Agrawal *et al* demonstrated that the conditional deletion of CXCR4 in cardiomyocytes did not influence the recovery of left ventricular (LV) function, reduce the scar size or alter the homing of MSCs to the myocardium following myocardial infarction (Agarwal *et al.*, 2010). Thus, there is growing evidence that other mechanisms beside the SDF-1/CXCR4 axis are contributing to the mobilization and homing of BM-SPCs in AMI and other conditions (Jalili *et al.*, 2010; Ratajczak *et al.*, 2010). These data suggest an important interplay between the complement cascade, the immune system, cathelicidins, low levels of SDF-1, and sphingosine-1 phosphate (S1P) and other bioactive lipids in the mobilization and homing of HSPCs. Our preliminary data suggest that these complex interactions might be involved in the mobilization of BM-SPCs in acute myocardial ischemia as well (unpublished data). Clinically, pharmacological modulators of S1P receptors are already approved by the FDA and can be utilized to enhance BM-SPC mobilization in the setting of ischemic heart disease. Similarly, modulation of the complement cascade can be also utilized in this process similar to their role in the mobilization of HSPCs.

The first evidence for the mobilization of CD34+ mononuclear cells in AMI was demonstrated by Shintani *et al* (Shintani *et al.* 2001). The authors demonstrated successful in vitro differentiation of circulating BM-SPCs into endothelial cells that expressed CD31, VE-cadherin and the kinase insert domain receptor (KDR) (Shintani, *et al.* 2001). Leone *et al* demonstrated that the levels of circulating CD34+ cells in the setting of AMI were higher when compared to patients with mild chronic stable angina and healthy controls. The magnitude of CD34+ cell mobilization correlated with the recovery of regional and global LV function recovery as well as other functional LV parameters (Leone *et al.* 2005). Similarly,

Wojakowski *et al* demonstrated the mobilization of multiple BM-SPCs populations in patients with AMI and found significant correlation between the number of circulating CD34+ cells and plasma SDF-1 levels (Wojakowski *et al.,* 2004). In their following publication, the authors demonstrated the correlation between circulating BM-SPCs and ejection fraction at baseline and lower brain natriuretic peptide (BNP) levels (Wojakowski *et al.* 2006). Interestingly, the mobilization of BM-SPCs is reduced by the successful revascularization of the culprit vessel in acute STEMI (Müller-Ehmsen *et al.,* 2005). However, the majority of the above mentioned studies have focused on the mobilization of partially committed stem cells such as HSPCs and endothelial progenitor cells (EPCs).

We and others have demonstrated the mobilization of pluripotent stem cells (PSCs) including VSELs in the setting of myocardial ischemia (Abdel-Latif *et al.* 2010; Wojakowski *et al.* 2009). The number of circulating VSELs was highest in patients with ST-elevation myocardial infarction (STEMI), particularly in the early phases following the injury, when compared to patients with lesser degrees of ischemia such as non STEMI (NSTEMI) and those with chronic ischemic heart disease (Abdel-Latif *et al.* 2010). The mobilization of PSCs appears to be related to the extent of myocardial ischemia and the degree of myocardial damage. Moreover, the ability of patients to mobilize PSCs in the peripheral circulation in response to AMI decreases with age, reduced global LV ejection fraction (LVEF) and diabetes supporting the notion of an age/comorbidity related decline in the regenerative capacity (Abdel-Latif, *et al.* 2010; Wojakowski, *et al.* 2009). Indeed, animal models confirm the reduction of number as well as pluripotent features of BM-derived VSELs with age (Zuba-Surma *et al.* 2008). Similarly, studies have demonstrated the reduction of number as well as functional capacity of EPCs in diabetic patients (Fadini *et al.,* 2006).

The pluripotent features of mobilized VSELs, including the presence of octamer-binding transcription factor-4 (Oct4) and stage specific embryonic antigen-4 (SSEA4), were confirmed both on the RNA and protein levels. Utilizing the capabilities of the ImageStream system, we demonstrated that circulating VSELs during AMI have very similar embryonic features similar to their BM and CB counterparts including the small size (7-8 μm), large nucleus and high nuclear-to-cytoplasm ratio (**Figure 2**). Furthermore, circulating VSELs during AMI express markers of early cardiac and endothelial progenitors that suggest that the mobilization is rather specific and that circulating VSELs are destined to aid in the myocardial regeneration following injury (Abdel-Latif *et al.* 2010; Kucia *et al.* 2004b; Wojakowski *et al.* 2009).

The above evidence suggest an innate, yet poorly understood, reparatory mechanism that culminates in the mobilization of BMSCs including pluripotent and embryonic like stem cells in acute myocardial injury. This mobilization correlates with the recovery of LV function and other LV functional parameters. Therefore, mobilization of PSCs in myocardial ischemia is a relevant and clinically significant process. Future studies aiming at selective mobilization of PSCs rather than the non-selective actions of agents such as granulocyte colony stimulating factor (G-CSF) may prove beneficial in the field of myocardial regeneration.

Indeed, there is evidence that the mobilization of CXCR4+ cells in the setting of AMI is correlated with LVEF recovery as well as myocardial reperfusion when assessed with cardiac MRI in humans (Wojakowski, *et al.* 2006).

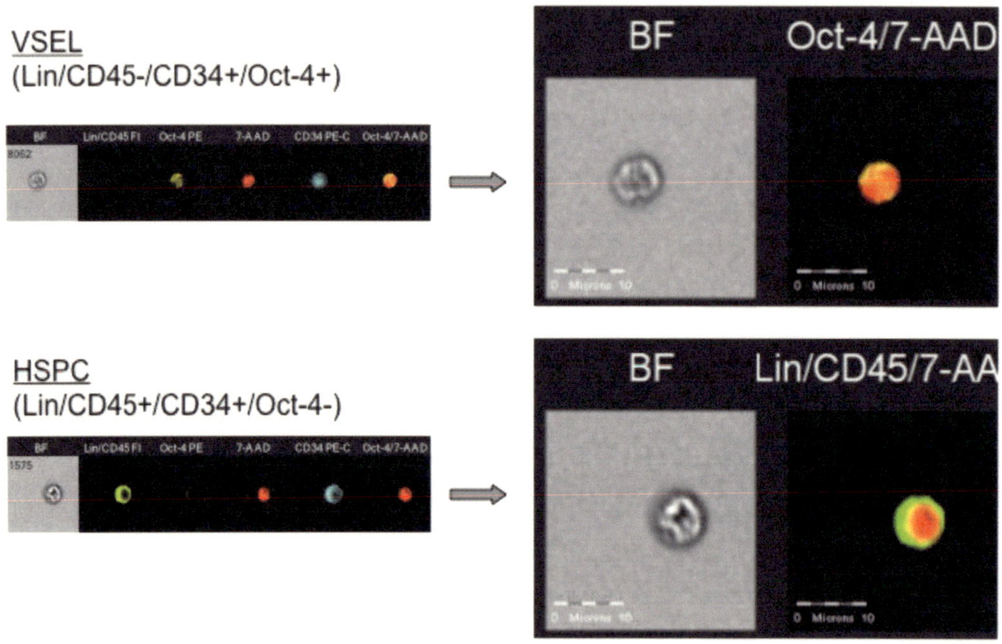

Fig. 2. Representative ImageStream images of VSEL and hematopoietic stem/ progenitor cell (HSPC) circulating in peripheral blood following acute ST-elevation myocardial infarction. Cells were stained against: 1) hematopoietic lineages markers (Lin) and CD45 to be detected in one channel (FITC, green), 2) marker of pluripotency Oct4 (PE, yellow) and 3) stem cell antigen CD34 (PE-Cy5, cyan). Nuclei are stained with 7-aminoactinomycin D (7-AAD, red). Scales represents 10 μm. VSELs are identified based on the lack of expression of both Lin and CD45 markers and positive staining for CD34 antigen and nuclear appearence of Oct-4 transcription factor (**Upper Panel**). HSCs are identified as cell expressing Lin and/ or CD45 markers as well as CD34 antigen; however, negative for Oct-4 (**Lower Panel**). BF: Bright field.

4. Therapeutic mobilization of BM-derived stem cells in myocardial regeneration

Hematologists have used the concept of BM-derived stem cell mobilization using pharmacological agents such as G-CSF for a long time. Based on the available clinical experience and safety profile of these therapies, pharmacological stem cell mobilization in the setting of AMI has gained increasing enthusiasm. Mobilized BM-SPCs are either harvested for further transplantation or allowed for spontaneous homing to the infarcted myocardium and has demonstrated various degrees of success (Engelmann *et al.*, 2006; Ince *et al.*, 2005; Zohlnhöfer *et al.*, 2006). Similar to BMCs transplantation studies, the heterogeneous methodologies of the included studies diluted the effect. The overall lack of efficacy with G-CSF BMCs mobilization in the setting of acute myocardial infarction is somewhat incongruent with the salutary effects of BMCs transplantation in humans and G-CSF therapy in animal models for myocardial regeneration.

The largest study utilizing G-CSF in the setting of acute myocardial infarction was the REVIVAL-2 trial that included 114 patients (Zohlnhöfer, *et al.* 2006). The study randomized AMI patients to 10 μg/kg of G-CSF vs. placebo and left ventricular functional parameters were assessed using cardiac MRI (CMR). The study demonstrated no significant difference in the tested parameters between patients treated with G-CSF or placebo. However, baseline characteristics in the study population showed normal or near normal LV function and therefore the expected benefit is minimal. Patient selection was a methodological flaw that plagued some of the studies that utilized G-CSF. Indeed, with careful examination of the available literature, patients with reduced LV function at baseline as well as those treated within the first 36 hours following AMI benefited the most (Abdel-Latif *et al.*, 2008b; Achilli *et al.*, 2010). On the other hand, safety concerns regarding a potentially increasing evidence of instent restenosis (Kang *et al.*, 2004) and recurrent ischemia (Hill *et al.*, 2005) have halted subsequent clinical trials. However, it is important to note that these safety concerns were not confirmed in large studies (Zohlnhöfer, *et al.* 2006) or in the cumulative meta-analyses (Abdel-Latif *et al.* 2008b).

Beyond the methodological flaws encountered in human trials, this lack of efficacy can be explained by multiple factors. While G-CSF and similar therapies mobilize a wide array of BMSPCs in the peripheral blood, homing factors may not be sufficient to guide them to the myocardial infarct zone. Indeed, the homing of c-Kit+ cells to the infarcted myocardium improved when G-CSF therapy was combined with local administration of SDF-1 (Askari *et al.*, 2003). The myocardial levels of chemoattractants peaks within 24-72 hours following injury (Kucia *et al.*, 2004a; Ma *et al.*, 2005; Wang *et al.*, 2006) and therefore delayed therapy in some human trials may have missed the homing window to the infarct zone. Similarly, the addition of Flt-3 to G-CSF therapy improved outcomes in animal models (Dawn *et al.*, 2008). Moreover, different cytokines are known to preferentially mobilize somewhat different subsets of BMCs (Hess *et al.*, 2002; Neipp *et al.*, 1998). Future studies investigating the characteristics of G-CSF-mobilized cells will be necessary to glean additional mechanistic insights in this regard.

Recently, a combined approach with stem cell mobilization and enhanced homing using therapies known to increase local SDF-1 or CXCR4 antagonists have been proposed and is currently being tested (Jujo *et al.*, 2010; Zaruba *et al.*, 2009). Going forward, the beneficial effects of BM-derived stem cell mobilization may be augmented by selective mobilization of undifferentiated BMSCs rather than differentiated inflammatory cells. It is also important to remember that some of the G-CSF arbitrated effects can be mediated by its direct effect on cardiomyocytes which are known to express G-CSF receptor (Shimoji *et al.*, 2010). G-CSF therapy may be inducing the proliferation of cardiomyocytes or the differentiation of resident cardiac stem cells. On a similar note, G-CSF therapy upregulates Akt (Ohtsuka *et al.*, 2004) and may result in reducing apoptosis of ischemic cardiomyocytes if utilized early following the acute event.

5. BM-derived stem cell transplantation for myocardial repair

The use of BM-derived cells in myocardial regeneration has moved rapidly from the basic research lab to the clinical arena. The results from these studies varied widely probably secondary to the heterogeneous methodologies used with an overall marginal benefit with

BM-derived cell transplantation compared to placebo. The majority of studies, however, utilized unselected populations of BMCs and these studies provide the longest follow-up of up to 5 years (Assmus *et al.,* 2010; Schachinger *et al.,* 2009). The underling mechanisms leading to the beneficial effect of transplanted BMCs are unclear. The observed benefits of BMCs transplantation is out of proportion of the observed rates of newly formed cardiomyocytes from BMCs' origin (Zuba-Surma *et al.,* 2011). Indeed, recent evidence suggest a primarily paracrine effect of BM-derived stem cells following their transplantation by recruiting and stimulating resident cardiac stem cells (CSCs) (Loffredo *et al.,* 2011). Furthermore, human purified CD34+ cells are a source of several growth factors including VEGF, cytokines and chemokines that may prevent apoptosis of dying cardiomyocytes and promote angiogenesis in damaged myocardium (Majka *et al.,* 2001). Cell membrane derived microvesicles or exosomes that are enriched in S1P may contribute to regeneration of myocardium and its re-vascularization (Baj-Krzyworzeka *et al.,* 2002). Hence, transplanted CD34+ cells may contribute to regeneration of damaged heart by paracrine signals and released microvesicles (Ratajczak *et al.,* 2008) and was recently confirmed by others (Sahoo *et al.,* 2011).

Long-term follow-up studies demonstrated 'catch-up phenomenon' of the placebo treated patients, thus leading to mixed results regarding the sustainability of the BMCs treatment benefit (Assmus, *et al.* 2010; Meyer *et al.,* 2009; Yousef *et al.,* 2009). The benefit of BMCs therapy is less robust among patients with chronic ischemic heart disease (IHD) (Assmus *et al.,* 2006; Strauer *et al.,* 2010). Similarly, smaller studies have demonstrated the antianginal effects of BMCs in patients with non-revascularizable severe coronary artery disease (Losordo *et al.,* 2007; Tse *et al.,* 2007).

Selected BM-derived stem cell subpopulations represent an attractive substrate for cellular therapies since they lack the inflammatory cells, which contribute to the ongoing inflammatory response at the site of myocardial infarction, contained in the unselected BMCs populations. Furthermore, highly purified stem cell populations are more likely to induce myocardial regeneration through paracrine effects or by directly differentiating into cardiomyocytes. The largest study utilizing selected BM-derived stem cell population is the REGENT study which compared selected to non-selected populations of BMCs in patients with acute ischemic heart disease and reduced LV function at baseline (Tendera *et al.,* 2009). While there were no significant differences between the groups, patients treated with selected CD34+/CXCR4+ cells showed trends of improvement in LV function when compared to controls. Other studies utilizing primitive populations of BM-SPCs such as CD133+ cells have reported improvement of LV function and perfusion (Bartunek *et al.,* 2005; Stamm *et al.,* 2004).

Nevertheless and despite the disparity in the methodologies of the conducted studies, the overall collective effect of BMCs' transplantation suggests small yet statistically significant benefit in myocardial regeneration (Abdel-Latif, *et al.* 2007; Martin-Rendon *et al.,* 2008). However, these trials have been hampered by their reliance on surrogate endpoints rather than patient important endpoints such as mortality, need for repeat revascularization, recurrent MI or re-hospitalization for congestive heart failure. While surrogate endpoints are important for mechanistic studies, patient-important endpoints are quintessential for a new therapy to achieve mainstream status.

6. Future directions

Growing evidence suggest that a multitude of BM-derived stem and pluripotent stem cells are mobilized in the peripheral blood following AMI. However, the clinical significance and the potential therapeutic use of this mobilization are still not fully understood. Circulating PSCs can be used as markers of ischemic injury in humans or as predictors of myocardial recovery following large ischemic damage. On the other hand, the therapeutic application of VSELs in myocardial regeneration has proven beneficial although the beneficial mechanisms remain elusive and are probably mainly paracrine in nature. Given the pluripotent potential of VSELs, their transplantation at smaller numbers (10,000 cells per mouse) have proven to be more beneficial than larger numbers of the more committed HSPCs (100,000 cells per mouse) indicating their greater therapeutic potential (Dawn *et al.* 2008). Current efforts directed at the ex-vivo expansion and priming of VSELs have proven to be a successful strategy in animal models and their clinical applications are pending (Dawn *et al.* 2008; Zuba-Surma, *et al.* 2011). Nuclear reprogramming has opened the door for creating patient-specific autologous pluripotent stem cells with multiple therapeutic opportunities (Takahashi *et al.*, 2006). Further studies are needed to examine the feasibility as well as the safety of inducible pluripotent stem cells (iPS) particularly their tumorigenicity and immunogenicity before they can be explored in human studies.

On the biotechnology frontier, multiple modifications of the transplanted cells (priming) and the host environment are being tested in humans to improve the efficiency of BM-SPCs' regenerative capacity. Transplanting three dimensional constructs that provide an enriched environment for the transplanted and resident stem cells are attractive modifications to the currently tested protocols [reviewed in (Mooney *et al.*, 2008)]. Similarly, the concept of multiple doses of stem cell to repair the complex process of myocardial remodeling following acute myocardial infarction is gaining traction and is very appealing. While the field of stem cell regenerative therapy for ischemic heart disease is still in its infancy, the accelerated advances in a wide array of biological and biotechnological areas have rapidly propelled the field from the bench to clinical applications.

7. Acknowledgments

Dr. Abdel-Latif is supported by the University of Kentucky Clinical and Translational Science Pilot Award and the University of Kentucky Clinical Scholar program.

Dr. Zuba-Surma is supported by the "Polish Foundation of Science" homing program grant number 2008/15.

Dr. Ratajczak is supported by NIH grant R01 CA106281, NIH R01 DK074720, and Stella and Henry Endowment.

We thank Dr. Karapetyan for her technical support for this review.

Conflict of interest: None.

8. References

Abdel-Latif A, Bolli R, Tleyjeh I, Montori V, Perin E, Hornung C, Zuba-Surma E, Al-Mallah M, and Dawn B (2007). Adult bone marrow-derived cells for cardiac repair: a systematic review and meta-analysis. *Arch Intern Med, 167:* 989-997.

Abdel-Latif A, Bolli R, Zuba-Surma EK, Tleyjeh IM, Hornung CA, and Dawn B (2008b). Granulocyte colony-stimulating factor therapy for cardiac repair after acute myocardial infarction: a systematic review and meta-analysis of randomized controlled trials. *Am Heart J, 156*: 216-226.

Abdel-Latif A, Zuba-Surma EK, Case J, Tiwari S, Hunt G, Ranjan S, Vincent RJ, Srour EF, Bolli R, and Dawn B (2008a). TGF-beta1 enhances cardiomyogenic differentiation of skeletal muscle-derived adult primitive cells. *Basic Res Cardiol, 103*: 514-524.

Abdel-Latif A, Zuba-Surma EK, Ziada KM, Kucia M, Cohen DA, Kaplan AM, Van Zant G, Selim S, Smyth SS, and Ratajczak MZ (2010). Evidence of mobilization of pluripotent stem cells into peripheral blood of patients with myocardial ischemia. *Exp Hematol, 38: 1131-1142.*

Achilli F, Malafronte C, Lenatti L, Gentile F, Dadone V, Gibelli G, Maggiolini S, Squadroni L, Di Leo C, Burba I, Pesce M, Mircoli L, Capogrossi MC, Di Lelio A, Camisasca P, Morabito A, Colombo G, and Pompilio G (2010). Granulocyte colony-stimulating factor attenuates left ventricular remodelling after acute anterior STEMI: results of the single-blind, randomized, placebo-controlled multicentre STem cEll Mobilization in Acute Myocardial Infarction (STEM-AMI) Trial. *Eur J Heart Fail, 12*: 1111-1121.

Agarwal U, Ghalayini W, Dong F, Weber K, Zou YR, Rabbany SY, Rafii S, and Penn MS (2010). Role of cardiac myocyte CXCR4 expression in development and left ventricular remodeling after acute myocardial infarction. *Circ Res, 107*: 667-676.

Askari AT, Unzek S, Popovic ZB, Goldman CK, Forudi F, Kiedrowski M, Rovner A, Ellis SG, Thomas JD, DiCorleto PE, Topol EJ, and Penn MS (2003). Effect of stromal-cell-derived factor 1 on stem-cell homing and tissue regeneration in ischaemic cardiomyopathy. *Lancet, 362*: 697-703.

Assmus B, Honold J, Schachinger V, Britten M, Fischer-Rasokat U, Lehmann R, Teupe C, Pistorius K, Martin H, Abolmaali N, Tonn T, Dimmeler S, and Zeiher A (2006). Transcoronary transplantation of progenitor cells after myocardial infarction. *N Engl J Med, 355*: 1222-1232.

Assmus B, Rolf A, Erbs S, Elsasser A, Haberbosch W, Hambrecht R, Tillmanns H, Yu J, Corti R, Mathey DG, Hamm CW, Suselbeck T, Tonn T, Dimmeler S, Dill T, Zeiher AM, and Schachinger V (2010). Clinical outcome 2 years after intracoronary administration of bone marrow-derived progenitor cells in acute myocardial infarction. *Circ Heart Fail, 3*: 89-96.

Baj-Krzyworzeka M, Majka M, Pratico D, Ratajczak J, Vilaire G, Kijowski J, Reca R, Janowska-Wieczorek A, and Ratajczak MZ (2002). Platelet-derived microparticles stimulate proliferation, survival, adhesion, and chemotaxis of hematopoietic cells. *Exp Hematol, 30*: 450-459.

Bartunek J, Vanderheyden M, Vandekerckhove B, Mansour S, De Bruyne B, De Bondt P, Van Haute I, Lootens N, Heyndrickx G, and Wijns W (2005). Intracoronary injection of CD133-positive enriched bone marrow progenitor cells promotes cardiac recovery after recent myocardial infarction: feasibility and safety. *Circulation, 112*: I178-183.

Bergmann O, Bhardwaj RD, Bernard S, Zdunek S, Barnabe-Heider F, Walsh S, Zupicich J, Alkass K, Buchholz BA, Druid H, Jovinge S, and Frisen J (2009). Evidence for cardiomyocyte renewal in humans. *Science, 324*: 98-102.

Braunwald E, and Bristow M (2000). Congestive heart failure: fifty years of progress. *Circulation, 102*: IV14-IV23.

D'Ippolito G, Diabira S, Howard G, Menei P, Roos B, and Schiller P (2004). Marrow-isolated adult multilineage inducible (MIAMI) cells, a unique population of postnatal young and old human cells with extensive expansion and differentiation potential. *J Cell Sci, 117*: 2971-2981.

Dawn B, Tiwari S, Kucia MJ, Zuba-Surma EK, Guo Y, Sanganalmath SK, Abdel-Latif A, Hunt G, Vincent RJ, Taher H, Reed NJ, Ratajczak MZ, and Bolli R (2008). Transplantation of bone marrow-derived very small embryonic-like stem cells attenuates left ventricular dysfunction and remodeling after myocardial infarction. *Stem Cells, 26*: 1646-1655.

de Weger RA, Verbrugge I, Bruggink AH, van Oosterhout MM, de Souza Y, van Wichen DF, Gmelig-Meyling FH, de Jonge N, and Verdonck LF (2008). Stem cell-derived cardiomyocytes after bone marrow and heart transplantation. *Bone Marrow Transplant, 41*: 563-569.

Deb A, Wang S, Skelding KA, Miller D, Simper D, and Caplice NM (2003). Bone marrow-derived cardiomyocytes are present in adult human heart: A study of gender-mismatched bone marrow transplantation patients. *Circulation, 107*: 1247-1249.

Drukala J, Paczkowska E, Kucia M, Mlynska E, Krajewski A, Machalinski B, Madeja Z, and Ratajczak MZ (2011). Stem Cells, Including a Population of Very Small Embryonic-Like Stem Cells, are Mobilized Into Peripheral Blood in Patients After Skin Burn Injury. *Stem Cell Rev.*

Engelmann M, Theiss H, Hennig-Theiss C, Huber A, Wintersperger B, Werle-Ruedinger A, Schoenberg S, Steinbeck G, and Franz W (2006). Autologous bone marrow stem cell mobilization induced by granulocyte colony-stimulating factor after subacute ST-segment elevation myocardial infarction undergoing late revascularization: final results from the G-CSF-STEMI (Granulocyte Colony-Stimulating Factor ST-Segment Elevation Myocardial Infarction) trial. *J Am Coll Cardiol, 48*: 1712-1721.

Fadini GP, Sartore S, Schiavon M, Albiero M, Baesso I, Cabrelle A, Agostini C, and Avogaro A (2006). Diabetes impairs progenitor cell mobilisation after hindlimb ischaemia-reperfusion injury in rats. *Diabetologia, 49*: 3075-3084.

Fukuhara S, Tomita S, Nakatani T, Yutani C, and Kitamura S (2005). Endogenous bone-marrow-derived stem cells contribute only a small proportion of regenerated myocardium in the acute infarction model. *J Heart Lung Transplant, 24*: 67-72.

Grundmann F, Scheid C, Braun D, Zobel C, Reuter H, Schwinger R, and Müller-Ehmsen J (2007). Differential increase of CD34, KDR/CD34, CD133/CD34 and CD117/CD34 positive cells in peripheral blood of patients with acute myocardial infarction. *Clin Res Cardiol, 96*: 621-627.

Hattan N, Kawaguchi H, Ando K, Kuwabara E, Fujita J, Murata M, Suematsu M, Mori H, and Fukuda K (2005). Purified cardiomyocytes from bone marrow mesenchymal stem cells produce stable intracardiac grafts in mice. *Cardiovasc Res, 65*: 293-295.

Hess DA, Levac KD, Karanu FN, Rosu-Myles M, White MJ, Gallacher L, Murdoch B, Keeney M, Ottowski P, Foley R, Chin-Yee I, and Bhatia M (2002). Functional analysis of human hematopoietic repopulating cells mobilized with granulocyte colony-stimulating factor alone versus granulocyte colony-stimulating factor in combination with stem cell factor. *Blood, 100*: 869-878.

Hill JM, Syed MA, Arai AE, Powell TM, Paul JD, Zalos G, Read EJ, Khuu HM, Leitman SF, Horne M, Csako G, Dunbar CE, Waclawiw MA, and Cannon RO, 3rd (2005). Outcomes and risks of granulocyte colony-stimulating factor in patients with coronary artery disease. *J Am Coll Cardiol, 46*: 1643-1648.

Hsieh PC, Segers VF, Davis ME, MacGillivray C, Gannon J, Molkentin JD, Robbins J, and Lee RT (2007). Evidence from a genetic fate-mapping study that stem cells refresh adult mammalian cardiomyocytes after injury. *Nat Med, 13*: 970-974.

Ince H, Petzsch M, Kleine H, Eckard H, Rehders T, Burska D, Kische S, Freund M, and Nienaber C (2005). Prevention of left ventricular remodeling with granulocyte colony-stimulating factor after acute myocardial infarction: final 1-year results of the Front-Integrated Revascularization and Stem Cell Liberation in Evolving Acute Myocardial Infarction by Granulocyte Colony-Stimulating Factor (FIRSTLINE-AMI) Trial. *Circulation, 112*: I73-80.

Jalili A, Shirvaikar N, Marquez-Curtis L, Qiu Y, Korol C, Lee H, Turner AR, Ratajczak MZ, and Janowska-Wieczorek A (2010). Fifth complement cascade protein (C5) cleavage fragments disrupt the SDF-1/CXCR4 axis: further evidence that innate immunity orchestrates the mobilization of hematopoietic stem/progenitor cells. *Exp Hematol, 38*: 321-332.

Jiang Y, Jahagirdar B, Reinhardt R, Schwartz R, Keene C, Ortiz-Gonzalez X, Reyes M, Lenvik T, Lund T, Blackstad M, Du J, Aldrich S, Lisberg A, Low W, Largaespada D, and Verfaillie C (2002). Pluripotency of mesenchymal stem cells derived from adult marrow. *Nature, 418*: 41-49.

Jujo K, Hamada H, Iwakura A, Thorne T, Sekiguchi H, Clarke T, Ito A, Misener S, Tanaka T, Klyachko E, Kobayashi K, Tongers J, Roncalli J, Tsurumi Y, Hagiwara N, and Losordo DW (2010). CXCR4 blockade augments bone marrow progenitor cell recruitment to the neovasculature and reduces mortality after myocardial infarction. *Proc Natl Acad Sci U S A, 107*: 11008-11013.

Kang HJ, Kim HS, Zhang SY, Park KW, Cho HJ, Koo BK, Kim YJ, Soo Lee D, Sohn DW, Han KS, Oh BH, Lee MM, and Park YB (2004). Effects of intracoronary infusion of peripheral blood stem-cells mobilised with granulocyte-colony stimulating factor on left ventricular systolic function and restenosis after coronary stenting in myocardial infarction: the MAGIC cell randomised clinical trial. *Lancet, 363*: 751-756.

Kawada H, Fujita J, Kinjo K, Matsuzaki Y, Tsuma M, Miyatake H, Muguruma Y, Tsuboi K, Itabashi Y, Ikeda Y, Ogawa S, Okano H, Hotta T, Ando K, and Fukuda K (2004). Nonhematopoietic mesenchymal stem cells can be mobilized and differentiate into cardiomyocytes after myocardial infarction. *Blood, 104*: 3581-3587.

Kucia M, Dawn B, Hunt G, Guo Y, Wysoczynski M, Majka M, Ratajczak J, Rezzoug F, Ildstad ST, Bolli R, and Ratajczak MZ (2004a). Cells expressing early cardiac markers reside in the bone marrow and are mobilized into the peripheral blood after myocardial infarction. *Circ Res, 95*: 1191-1199.

Kucia M, Dawn B, Hunt G, Guo Y, Wysoczynski M, Majka M, Ratajczak J, Rezzoug F, Ildstad ST, Bolli R, and Ratajczak MZ (2004b). Cells expressing early cardiac markers reside in the bone marrow and are mobilized into the peripheral blood following myocardial infarction. *Circ Res, 95*: 1191-1199.

Kucia M, Reca R, Campbell FR, Zuba-Surma E, Majka M, Ratajczak J, and Ratajczak MZ (2006). A population of very small embryonic-like (VSEL) CXCR4(+)SSEA-1(+)Oct-4+ stem cells identified in adult bone marrow. *Leukemia, 20*: 857-869.

Kucia M, Reca R, Miekus K, Wanzeck J, Wojakowski W, Janowska-Wieczorek A, Ratajczak J, and Ratajczak MZ (2005). Trafficking of normal stem cells and metastasis of cancer stem cells involve similar mechanisms: pivotal role of the SDF-1-CXCR4 axis. *Stem Cells, 23*: 879-894.

Lapidot T, and Kollet O (2002). The essential roles of the chemokine SDF-1 and its receptor CXCR4 in human stem cell homing and repopulation of transplanted immune-deficient NOD/SCID and NOD/SCID/B2m(null) mice. *Leukemia : official journal of the Leukemia Society of America, Leukemia Research Fund, U.K, 16*: 1992-2003.

Leone A, Rutella S, Bonanno G, Abbate A, Rebuzzi A, Giovannini S, Lombardi M, Galiuto L, Liuzzo G, Andreotti F, Lanza G, Contemi A, Leone G, and Crea F (2005). Mobilization of bone marrow-derived stem cells after myocardial infarction and left ventricular function. *Eur Heart J, 26*: 1196-1204.

Levy D, Kenchaiah S, Larson MG, Benjamin EJ, Kupka MJ, Ho KK, Murabito JM, and RS. V (2002). Long-term trends in the incidence of and survival with heart failure. *N Engl J Med, 347*: 1442-1444.

Loffredo FS, Steinhauser ML, Gannon J, and Lee RT (2011). Bone marrow-derived cell therapy stimulates endogenous cardiomyocyte progenitors and promotes cardiac repair. *Cell Stem Cell, 8*: 389-398.

Losordo DW, Schatz RA, White CJ, Udelson JE, Veereshwarayya V, Durgin M, Poh KK, Weinstein R, Kearney M, Chaudhry M, Burg A, Eaton L, Heyd L, Thorne T, Shturman L, Hoffmeister P, Story K, Zak V, Dowling D, Traverse JH, Olson RE, Flanagan J, Sodano D, Murayama T, Kawamoto A, Kusano KF, Wollins J, Welt F, Shah P, Soukas P, Asahara T, and Henry TD (2007). Intramyocardial transplantation of autologous CD34+ stem cells for intractable angina: a phase I/IIa double-blind, randomized controlled trial. *Circulation, 115*: 3165-3172.

Ma J, Ge J, Zhang S, Sun A, Shen J, Chen L, Wang K, and Zou Y (2005). Time course of myocardial stromal cell-derived factor 1 expression and beneficial effects of intravenously administered bone marrow stem cells in rats with experimental myocardial infarction. *Basic Res Cardiol, 100*: 217-223.

Majka M, Janowska-Wieczorek A, Ratajczak J, Ehrenman K, Pietrzkowski Z, Kowalska MA, Gewirtz AM, Emerson SG, and Ratajczak MZ (2001). Numerous growth factors, cytokines, and chemokines are secreted by human CD34(+) cells, myeloblasts, erythroblasts, and megakaryoblasts and regulate normal hematopoiesis in an autocrine/paracrine manner. *Blood, 97*: 3075-3085.

Martin-Rendon E, Brunskill SJ, Hyde CJ, Stanworth SJ, Mathur A, and Watt SM (2008). Autologous bone marrow stem cells to treat acute myocardial infarction: a systematic review. *European heart journal, 29*: 1807-1818.

Massa M, Rosti V, Ferrario M, Campanelli R, Ramajoli I, Rosso R, De Ferrari G, Ferlini M, Goffredo L, Bertoletti A, Klersy C, Pecci A, Moratti R, and Tavazzi L (2005). Increased circulating hematopoietic and endothelial progenitor cells in the early phase of acute myocardial infarction. *Blood, 105*: 199-206.

McMurray J, and Pfeffer M (2005). Heart failure. *Lancet, 365*: 1877-1889.

McQuibban GA, Butler GS, Gong JH, Bendall L, Power C, Clark-Lewis I, and Overall CM (2001). Matrix metalloproteinase activity inactivates the CXC chemokine stromal cell-derived factor-1. *The Journal of biological chemistry, 276*: 43503-43508.

McQuibban GA, Gong JH, Wong JP, Wallace JL, Clark-Lewis I, and Overall CM (2002). Matrix metalloproteinase processing of monocyte chemoattractant proteins generates CC chemokine receptor antagonists with anti-inflammatory properties in vivo. *Blood, 100*: 1160-1167.

Meyer GP, Wollert KC, Lotz J, Pirr J, Rager U, Lippolt P, Hahn A, Fichtner S, Schaefer A, Arseniev L, Ganser A, and Drexler H (2009). Intracoronary bone marrow cell transfer after myocardial infarction: 5-year follow-up from the randomized-controlled BOOST trial. *European heart journal, 30*: 2978-2984.

Mooney DJ, and Vandenburgh H (2008). Cell delivery mechanisms for tissue repair. *Cell Stem Cell, 2*: 205-213.

Müller-Ehmsen J, Scheid C, Grundmann F, Hirsch I, Turan G, Tossios P, Mehlhorn U, and Schwinger R (2005). The mobilization of CD34 positive mononuclear cells after myocardial infarction is abolished by revascularization of the culprit vessel. *Int J Cardiol, 103*: 7-11.

Neipp M, Zorina T, Domenick MA, Exner BG, and Ildstad ST (1998). Effect of FLT3 ligand and granulocyte colony-stimulating factor on expansion and mobilization of facilitating cells and hematopoietic stem cells in mice: kinetics and repopulating potential. *Blood, 92*: 3177-3188.

Ohtsuka M, Takano H, Zou Y, Toko H, Akazawa H, Qin Y, Suzuki M, Hasegawa H, Nakaya H, and Komuro I (2004). Cytokine therapy prevents left ventricular remodeling and dysfunction after myocardial infarction through neovascularization. *The FASEB journal : official publication of the Federation of American Societies for Experimental Biology, 18*: 851-853.

Paczkowska E, Larysz B, Rzeuski R, Karbicka A, Jałowiński R, Kornacewicz-Jach Z, Ratajczak M, and Machaliński B (2005). Human hematopoietic stem/progenitor-enriched CD34(+) cells are mobilized into peripheral blood during stress related to ischemic stroke or acute myocardial infarction. *Eur J Haematol, 75*: 461-467.

Peterson JT, Li H, Dillon L, and Bryant JW (2000). Evolution of matrix metalloprotease and tissue inhibitor expression during heart failure progression in the infarcted rat. *Cardiovascular research, 46*: 307-315.

Quaini F, Urbanek K, Beltrami AP, Finato N, Beltrami CA, Nadal-Ginard B, Kajstura J, Leri A, and Anversa P (2002). Chimerism of the transplanted heart. *N Engl J Med, 346*: 5-15.

Ratajczak MZ, Lee H, Wysoczynski M, Wan W, Marlicz W, Laughlin MJ, Kucia M, Janowska-Wieczorek A, and Ratajczak J (2010). Novel insight into stem cell mobilization-plasma sphingosine-1-phosphate is a major chemoattractant that directs the egress of hematopoietic stem progenitor cells from the bone marrow and its level in peripheral blood increases during mobilization due to activation of complement cascade/membrane attack complex. *Leukemia, 24*: 976-985.

Ratajczak MZ, Zuba-Surma EK, Shin DM, Ratajczak J, and Kucia M (2008). Very small embryonic-like (VSEL) stem cells in adult organs and their potential role in rejuvenation of tissues and longevity. *Exp Gerontol, 43*: 1009-1017.

Roger VL, Weston SA, Redfield MM, Hellermann-Homan JP, Killian J, Yawn BP, and SJ. J (2004). Trends in heart failure incidence and survival in a community-based population. *JAMA, 292*: 344-350.

Roger VL, Go AS, Lloyd-Jones DM, Adams RJ, Berry JD, Brown TM, Carnethon MR, Dai S, de Simone G, Ford ES, Fox CS, Fullerton HJ, Gillespie C, Greenlund KJ, Hailpern SM, Heit JA, Ho PM, Howard VJ, Kissela BM, Kittner SJ, Lackland DT, Lichtman JH, Lisabeth LD, Makuc DM, Marcus GM, Marelli A, Matchar DB, McDermott MM, Meigs JB, Moy CS, Mozaffarian D, Mussolino ME, Nichol G, Paynter NP, Rosamond WD, Sorlie PD, Stafford RS, Turan TN, Turner MB, Wong ND, and Wylie-Rosett J (2011). Heart disease and stroke statistics--2011 update: a report from the American Heart Association. *Circulation, 123*: e18-e209.

Sahoo S, Klychko E, Thorne T, Misener S, Schultz KM, Millay M, Ito A, Liu T, Kamide C, Agarwal H, Perlman H, Qin G, Kishore R, and Losordo DW (2011). Exosomes From Human CD34+ Stem Cells Mediate Their Proangiogenic Paracrine Activity. *Circ Res*.

Schachinger V, Assmus B, Erbs S, Elsasser A, Haberbosch W, Hambrecht R, Yu J, Corti R, Mathey DG, Hamm CW, Tonn T, Dimmeler S, and Zeiher AM (2009). Intracoronary infusion of bone marrow-derived mononuclear cells abrogates adverse left ventricular remodelling post-acute myocardial infarction: insights from the reinfusion of enriched progenitor cells and infarct remodelling in acute myocardial infarction (REPAIR-AMI) trial. *Eur J Heart Fail, 11*: 973-979.

Shimoji K, Yuasa S, Onizuka T, Hattori F, Tanaka T, Hara M, Ohno Y, Chen H, Egasgira T, Seki T, Yae K, Koshimizu U, Ogawa S, and Fukuda K (2010). G-CSF promotes the proliferation of developing cardiomyocytes in vivo and in derivation from ESCs and iPSCs. *Cell Stem Cell, 6*: 227-237.

Shintani S, Murohara T, Ikeda H, Ueno T, Honma T, Katoh A, Sasaki K, Shimada T, Oike Y, and Imaizumi T (2001). Mobilization of endothelial progenitor cells in patients with acute myocardial infarction. *Circulation, 103*: 2776-2779.

Stamm C, Kleine HD, Westphal B, Petzsch M, Kittner C, Nienaber CA, Freund M, and Steinhoff G (2004). CABG and bone marrow stem cell transplantation after myocardial infarction. *Thorac Cardiovasc Surg, 52*: 152-158.

Strauer BE, Yousef M, and Schannwell CM (2010). The acute and long-term effects of intracoronary Stem cell Transplantation in 191 patients with chronic heARt failure: the STAR-heart study. *Eur J Heart Fail, 12*: 721-729.

Takahashi K, and Yamanaka S (2006). Induction of pluripotent stem cells from mouse embryonic and adult fibroblast cultures by defined factors. *Cell, 126*: 663-676.

Tendera M, Wojakowski W, Ruzyllo W, Chojnowska L, Kepka C, Tracz W, Musialek P, Piwowarska W, Nessler J, Buszman P, Grajek S, Breborowicz P, Majka M, and Ratajczak MZ (2009). Intracoronary infusion of bone marrow-derived selected CD34+CXCR4+ cells and non-selected mononuclear cells in patients with acute STEMI and reduced left ventricular ejection fraction: results of randomized, multicentre Myocardial Regeneration by Intracoronary Infusion of Selected Population of Stem Cells in Acute Myocardial Infarction (REGENT) Trial. *European heart journal, 30*: 1313-1321.

Tse HF, Thambar S, Kwong YL, Rowlings P, Bellamy G, McCrohon J, Thomas P, Bastian B, Chan JK, Lo G, Ho CL, Chan WS, Kwong RY, Parker A, Hauser TH, Chan J, Fong DY, and Lau CP (2007). Prospective randomized trial of direct endomyocardial implantation of bone marrow cells for treatment of severe coronary artery diseases (PROTECT-CAD trial). *European heart journal, 28*: 2998-3005.

Valgimigli M, Rigolin G, Fucili A, Porta M, Soukhomovskaia O, Malagutti P, Bugli A, Bragotti L, Francolini G, Mauro E, Castoldi G, and Ferrari R (2004). CD34+ and endothelial progenitor cells in patients with various degrees of congestive heart failure. *Circulation, 110*: 1209-1212.

Walter DH, Rochwalsky U, Reinhold J, Seeger F, Aicher A, Urbich C, Spyridopoulos I, Chun J, Brinkmann V, Keul P, Levkau B, Zeiher AM, Dimmeler S, and Haendeler J (2007). Sphingosine-1-phosphate stimulates the functional capacity of progenitor cells by activation of the CXCR4-dependent signaling pathway via the S1P3 receptor. *Arterioscler Thromb Vasc Biol, 27*: 275-282.

Wang Y, Haider H, Ahmad N, Zhang D, and Ashraf M (2006). Evidence for ischemia induced host-derived bone marrow cell mobilization into cardiac allografts. *Journal of molecular and cellular cardiology, 41*: 478-487.

Wojakowski W, Tendera M, Kucia M, Zuba-Surma E, Paczkowska E, Ciosek J, Halasa M, Krol M, Kazmierski M, Buszman P, Ochala A, Ratajczak J, Machalinski B, and Ratajczak MZ (2009). Mobilization of bone marrow-derived Oct-4+ SSEA-4+ very small embryonic-like stem cells in patients with acute myocardial infarction. *J Am Coll Cardiol, 53*: 1-9.

Wojakowski W, Tendera M, Michałowska A, Majka M, Kucia M, Maślankiewicz K, Wyderka R, Ochała A, and Ratajczak M (2004). Mobilization of CD34/CXCR4+, CD34/CD117+, c-met+ stem cells, and mononuclear cells expressing early cardiac, muscle, and endothelial markers into peripheral blood in patients with acute myocardial infarction. *Circulation, 110*: 3213-3220.

Wojakowski W, Tendera M, Zebzda A, Michalowska A, Majka M, Kucia M, Maslankiewicz K, Wyderka R, Król M, Ochala A, Kozakiewicz K, and Ratajczak M (2006). Mobilization of CD34(+), CD117(+), CXCR4(+), c-met(+) stem cells is correlated with left ventricular ejection fraction and plasma NT-proBNP levels in patients with acute myocardial infarction. *Eur Heart J, 27*: 283-289.

Yousef M, Schannwell CM, Kostering M, Zeus T, Brehm M, and Strauer BE (2009). The BALANCE Study: clinical benefit and long-term outcome after intracoronary autologous bone marrow cell transplantation in patients with acute myocardial infarction. *Journal of the American College of Cardiology, 53*: 2262-2269.

Zaruba MM, Theiss HD, Vallaster M, Mehl U, Brunner S, David R, Fischer R, Krieg L, Hirsch E, Huber B, Nathan P, Israel L, Imhof A, Herbach N, Assmann G, Wanke R, Mueller-Hoecker J, Steinbeck G, and Franz WM (2009). Synergy between CD26/DPP-IV inhibition and G-CSF improves cardiac function after acute myocardial infarction. *Cell Stem Cell, 4*: 313-323.

Zohlnhöfer D, Ott I, Mehilli J, Schömig K, Michalk F, Ibrahim T, Meisetschläger G, von Wedel J, Bollwein H, Seyfarth M, Dirschinger J, Schmitt C, Schwaiger M, Kastrati A, Schömig A, and Investigators. R- (2006). Stem cell mobilization by granulocyte colony-stimulating factor in patients with acute myocardial infarction: a randomized controlled trial. *JAMA, 295*: 1003-1010.

Zuba-Surma EK, Guo Y, Taher H, Sanganalmath SK, Hunt G, Vincent RJ, Kucia M, Abdel-Latif A, Tang XL, Ratajczak MZ, Dawn B, and Bolli R (2011). Transplantation of expanded bone marrow-derived very small embryonic-like stem cells (VSEL-SCs) improves left ventricular function and remodelling after myocardial infarction. *Journal of cellular and molecular medicine, 15*: 1319-1328.

Zuba-Surma EK, Kucia M, Abdel-Latif A, Dawn B, Hall B, Singh R, Lillard JW, Jr., and Ratajczak MZ (2008). Morphological characterization of very small embryonic-like stem cells (VSELs) by ImageStream system analysis. *J Cell Mol Med, 12*: 292-303.

20

Gene Therapy of Hematopoietic and Immune Systems: Current State and Perspectives

Maria Savvateeva[1], Fedor Rozov[1,2] and Alexander Belyavsky[1]
[1]*Engelhardt Institute of Molecular Biology, Russian Academy of Sciences*
[2]*University of Oslo, Centre for Medical Studies Russia, Moscow*
Russian Federation

1. Introduction

Hematopoietic stem cells (HSCs) present arguably the best entry point for gene therapy of hematopoietic and immune systems since genetically modified HSCs are long-lived and would eventually transfer the therapeutic constructs to all their descendants. However, gene therapy via HSCs, although conceptually simple, has proven to be a technically formidable problem that has yet to be solved successfully. Despite overtly positive results obtained in gene therapy experiments performed with mouse and larger animal models, these achievements did not translate into clinically acceptable outcomes for non-human primates and human patients, with exception of a few specific disease instances where a therapeutic gene brought about significant survival advantages to transduced cells (Cavazzana-Calvo et al., 2000, Schmidt et al, 2003). Major differences between outcomes of conceptually similar experiments in mice and primates underscore the notion that the fundamental principles governing functioning of hematopoietic system in small short-lived vs. larger long-lived animals differ significantly. Low degree of chimerism obtained in experiments with primates and humans is likely a result of intrinsically low efficiency of viral transduction of long-term repopulating (LTR) HSCs coupled with subsequent massive silencing of integrated constructs (Ellis, 2005; Horn et al, 2002). One may hypothesize that this situation reflects a better protection of hematopoietic system from external influences, in particular invasion of foreign genetic material, in longer-living animals.

However, our deepening knowledge of molecular mechanisms underlying functioning of HSCs within the organism provides hints as to what strategies may lead to the development of the efficient gene therapy via HSCs; some of these strategies are discussed below.

2. Improvements of vectors and ex vivo HSC transduction protocols

Numerous studies indicate that lentiviral vectors that are capable of transducing non-dividing cells may represent a more promising tool for introduction of genetic material into HSCs compared to retroviral vectors (Uchida et al, 1998, Case et al., 1999). This may be attributed to a largely quiescent nature of LTR HSCs, especially in larger animals (Cheshier et al., 1999, Shepherd et al., 2007). Since even lentiviral vectors transduce more efficiently dividing cells than quiescent ones (Trobridge et al., 2004), the current transduction protocols relied until recently on the use of culture conditions that induced entry of HSCs into cell cycle but incidentally failed to maintain their stem cell status (Bunting et al., 1999). This situation seems to have been ameliorated after introduction of transduction protocols that rely on the use of serum-free media that lack factors inducing SC differentiation (Mostoslavsky et al., 2005) and novel growth factors that better preserve cell stemness (Zhang C et al., 2008). It remains yet to see whether these improvements are sufficient to significantly increase the efficiency of HSC gene therapy in clinical settings.

3. Selection of genetically modified HSCs in vivo: Negative selection

As current efficiency of transduction of human LTR HSCs with viral vectors appears to be quite low and there are no clinically proven protocols for expansion of these cells ex vivo, the most promising solution at hand to this problem is an in vivo selection of modified cells after their transduction and re-transplantation back to a patient. Conceptually, one might distinguish negative and positive in vivo selection strategies. The first one can be defined as a strategy that is aimed at elimination of stem and progenitor cells that do not bear integrated functional constructs. Positive selection implies a strategy that does not target the construct-negative stem cells but rather provides selective survival and growth advantage to the cells that bear the inserted construct. The negative selection gains presently much of attention and seems to be better poised for a clinical advancement in the near future. Arguably, the most promising and advanced variant of negative selection is based on the use of O6-MGMT as a selection marker and various alkylating compounds as selection agents (Davis et al., 2000, Ragg et al., 2000). Using this approach and multiple rounds of selection in vivo, overall peripheral blood chimerism has been driven in mice and larger animal models to levels higher than 75%. However, the clinical applicability of this technique is as yet unclear, as recent experiments performed by two research teams with non-human primates using MGMT-mediated selection produced rather conflicting results. One team demonstrated successful implementation of this strategy in monkeys, although with selection efficiencies and chimerism rates highly variable between individual animals (Beard et al., 2010), whereas another team reported a rather negligible increase in chimerism rates upon selection in vivo (Larochelle et al., 2009).

Various implementations of negative selection strategy are listed in the Table 1.

4. Selection of genetically modified HSCs in vivo: Positive selection

Ongoing studies of the mechanisms controlling HSC self-maintenance and commitment continue to identify novel factors that bring about HSC expansion in vivo when over-expressed. A less than exhaustive set of these factors is listed in the Table 2. Arguably, the most extensively studied gene with such properties is the homeobox transcription factor HoxB4. Forced expression of HoxB4 in murine HSCs induces remarkable ex vivo and in vivo cell expansion without compromising their differentiation or inducing leukemic transformation (Sauvageau et al., 1995, Antonchuk et al., 2002). Similar effects were obtained using recombinant TAT-HOXB4 protein (Krosl et al., 2003). In some reports, HoxB4 and negative selection marker MGMT were used together to further increase percentage of modified HSCs (Chinnasamy et al., 2005). However, attempts to use HoxB4 for positive selection of HSCs in larger animals were much less successful, with a major expansion of short-term repopulating cells only (Zhang X et al., 2006). Besides, a significant number of leukemia occurrences apparently related to unregulated expression of HoxB4 were observed in these animals (Zhang X et al., 2008).

Slective marker	Selecting agent	Mode of action	References
O6-MGMT	BCNU, TMZ, other alkylating agents	MGMT protein functions to repair alkylated DNA caused by chemotherapeutic agents like BCNU or TMZ	Sawai et al, 2001; Zielske et al, 2003
Thymidylate synthase	5-fluorouracil (5-FU) 5-fluorodeoxy-uridine (5-FUdR)	Drug-resistant TS can protect bone marrow cells from 5-fluorouracil (5-FU) and related fluoropyrimidines that induce cessation of DNA and RNA synthesis, and subsequent cell death.	Bielas et al, 2009
Tyr22DHFR	Methotrexate	MTX acts on highly proliferative cells, blocking DNA synthesis through competitive inhibition of DHFR. Drug resistant dihydrofolate reductase such as Tyr22 (Tyr22DHFR) has the potential to selectively increase engraftment of gene-modified human hematopoietic cells	Gori et al, 2010
Multidrug resistance gene-1 (MDR)	Taxol, Paclitaxel	Overexpression of the multidrug resistance gene MDR1 in bone marrow cells results in protection from hematopoietic toxicity from chemotherapy drugs that are substrates for the MDR1 drug efflux pump	Cowan et al, 1999

Table 1. Strategies for negative selection of genetically modified HSC

Some other members of the HOX family, either alone or fused with specific cellular partners, are also able to induce expansion of hematopoietic progenitors in mice. Of particular importance is a fusion gene NUP98-HoxA10, which has a remarkable ability of multi-log expansion of murine repopulating cells ex vivo, exceeding that of HoxB4 (Ohta et al., 2007; Watts et al., 2011).

Recently, the powerful effect of overexpression of early acting transcription factor SALL4 on ex vivo expansion of human hematopoietic cells capable of long-term repopulation of NOD/SCID mice was demonstrated (Aguila et al., 2011). Significant ex vivo expansion could be also achieved using recombinant TAT-SALL4B protein.

There are at least a dozen of other genes that, when overexpressed, induce significant expansion of HSCs in mice in vivo. One of the most interesting groups of such factors are epigenetic regulators. Of particular interest is Bmi1, a member of Polycomb group, which is involved in regulation of mantenance of various adult stem cell types. Inactivation of Bmi1 leads to defect in HSC self-renewal (Park et al., 2003), whereas its enforced expression results a striking ex vivo expansion of multipotential progenitors and marked augmentation of HSC repopulating capacity in vivo (Iwama et al., 2004). In addition, enforced expression of Bmi1 in human CD34-positive cells leads to the ex vivo expansion of NOD/SCID repopulating cells (Rizo et al., 2008). Another Polycomb group gene that potentially could be used for positive selection is Ezh2; upon overexpression, it prevents HSC exhaustion (Kamminga et al., 2006). Forced expression of yet another epigenetic regulator, histone demethylase Fbxl10/Jhdm1b in HSCs abolishes exhaustion of the LTR HSCs following serial transplantation. This property of Ezh2 and Fbxl10/Jhdm1b makes them especially appropriate for schemes combining positive and negative selection since the latter one places very significant stress on hematopoietic system.

Another group of genes that might be used for positive selection are those that are frequently activated in predominant hematopoietic cell clones arising after retro- or lentiviral transduction, and are likely therefore to act as factors inducing in vivo expansion of these clones. The most prominent among such genes are MDS1/Evi-1 (Sellers et al., 2010; Métais & Dunbar, 2008), PRDM16 (Du et al., 2005; Ott et al., 2006) HMGA2 (Wang et al., 2010; Cavazzana-Calvo et al., 2010) and LMO2 (McCormack et al., 2003; McCormack et al., 2010). As a note of caution, forced expression of these genes may produce undesired effects; for example, expression of Evi-1 was reported to be associated with chromosomal instability (Stein et al., 2010).

In addition to protein factors, micro RNAs also have effect on HSC function and population size. In particular, miR-125a and miR-125b were shown to increase number of HSCs in vivo or enhance their repopulation capacity (Guo et al., 2010; Ooi et al., 2010).

Having focused on genes that expand stem cell population, one should not overlook another group of genes that exert an opposite effect, namely negative influence on HSC pool size. Thanks to RNA interference technology, suppression of gene expression in various cell types nowadays is nearly as simple as overexpression. If gene knockout or knockdown results in expansion of stem cell population, this property may potentially be used for positive selection. Among genes of interest in this respect are C/EBP alpha, Lnk and Nur77, to name a few. C/EBP alpha-deficient hematopoietic stem cells (HSCs) are hyperproliferative, have increased expression of Bmi-1 and enhanced competitive repopulating activity (Zhang et al. 2004; Heath et al., 2004). Inactivation of Lnk, inhibitory adaptor protein, leads to an expanded HSC pool with enhanced self-renewal (Bersenev et al., 2008). Mice with inactivation of both Nor-1 and Nur77 have abnormal expansion of HSCs and myeloid progenitors and develop lethal acute myeloid leukemia (AML).

Regardless of what gene is being used for positive selection, it is clear that its constitutive expression would eliminate one or more of the negative growth controls imposed on HSCs by organism, and thus increase risks of neoplastic transformation. Therefore, any clinically acceptable protocol for gene therapy using positive selection of transduced HSCs should be based on transient, tightly regulated gene expression. Given that positive selection, if correctly implemented, promises to provide significant advantages over negative selection schemes, further research into creation of robustly regulated expression systems for positive selection in HSCs seem to be fully warranted.

Gene	Observed effects	References
HOXB4	Overexpression of HoxB4 induces significant ex vivo and in vivo expansion of murine long-term repopulating HSCs.	Antonchuk et al., 2002; Sauvageau et al., 1995
NUP98-HOXA10	Enforced expression of NUP98-HOXA10 fusion protein results in significant expansion of murine repopulating cells ex vivo exceeding that of HoxB4.	Ohta et al., 2007; Watts et al., 2011
NF-Ya	Murine HSCs overexpressing NF-Ya demonstrate strongly increased in vivo repopulation.	Zhu et al., 2005
Bmi1	Enforced expression of Bmi1 leads to striking ex vivo expansion of multipotential progenitors and marked augmentation of HSC repopulating capacity in vivo.	Iwama et al., 2004; Rizo et al., 2008
Ezh2	Overexpression prevents exhaustion of long-term repopulating HSCs.	Kamminga et al., 2006
Fbxl10/ Jhdm1b	Same as above.	Konuma et al., 2011
Jab1	Mice with Jab1 overexpression have expanded HSC pool and develop a myeloproliferative disease.	Mori et al., 2008
HMGA2	Frequently found in the vicinity of integrated constructs in gene therapy trials; HMGA2-expressing cells have growth advantage in competitive repopulation and serial transplantation.	Cavazzana-Calvo et al., 2010; Ikeda et al., 2011; Wang et al., 2010
Evi-1	Frequently found in the vicinity of integrated constructs in gene therapy trials.	Métais & Dunbar, 2008; Sellers et al., 2010
PRDM16	Frequently found in the vicinity of integrated constructs in gene therapy trials.	Du et al., 2005; Ott et al., 2006
Sall4	Enforced expression results in ex vivo expansion of long-term NOD/SCID repopulating cells.	Aguila et al., 2011
MicroRNAs miR-125a, miR-125b	Forced expression of miR-125a was capable of increasing the number of HSCs cells several-fold. Overexpression of miR-125b enhances HSC function, as judged by serial transplantation.	Guo et al., 2010; Ooi et al., 2010
Lnk	Mice with Lnk inactivation have an expanded HSC pool with enhanced self-renewal.	Bersenev et al., 2008
Nur77/NR4A1 & Nor-1/NR4A3	Mice with inactivation of both Nor-1 and Nur77 have abnormal expansion of HSCs and myeloid progenitors and develop lethal acute myeloid leukemia.	Mullican et al., 2007
C/EBPα	C/EBP alpha-deficient HSCs are hyperproliferative and have enhanced competitive repopulating activity.	Heath et al., 2004; Zhang P et al. 2004;
Latexin	Mouse strains expressing lower latexin levels have increased numbers of HSCs.	Liang et al., 2007

Table 2. Genes affecting in vivo expansion of HSCs

5. Expansion and selection of genetically modified HSCs ex vivo

Although much hope is currently invested into various schemes aimed at in vivo selection of gene-modified HSCs, a substantially simpler and arguably more elegant solution may be achieved if protocols for long-term culture and robust ex vivo expansion of HSCs could be developed. Very significant expansion of HSCs that occurs during embryonic development indicates that this might be eventually possible.

Over the last two decades, quite a few HSC culture protocols have been developed. The earlier established conditions involved cultivation in the presence of serum and cocktail of "classical" cytokines including SCF, IL3, IL6, FLT3L and TPO. Since bovine serum apparently contains factors that induce differentiation and/or apoptosis of HSCs, recent, more advanced protocols have been developed, which use defined, serum-free conditions that offer better reproducibility and minimize rapid loss of long-term repopulating HSCs during ex vivo culture and transduction with lenti- and retroviral vectors (Mostoslavsky et al., 2005).

In addition to classical cytokines, a number of new growth factors that have pronounced effect on HSC maintenance and expansion were identified in the last years. Among the most important are FGF1 (de Haan et al., 2003), IGFBP2 (Huynh et al., 2008), and several members of angiopoeitin-like family, in particular Angptl3 and 5 (Zhang et al., 2006).

Several major signaling pathways figuring prominently during embryonic development, in particular during specification of hematopoietic lineage, were shown to be important for adult HSC biology. Among those, Notch and Wnt pathways are currently considered as of the most immediate interest as far as HSC-niche interactions and ex vivo expansion are concerned. Stem and progenitor pool-enhancing properties of Notch signaling were demonstrated initially using constitutive Notch1 signaling in murine hematopoietic cells, which produced immortalized, cytokine-dependent stem cell-like cells (Varnum-Finney et al., 2000), and constitutive Notch4 signaling in human cord blood cells, which resulted in significant increase in cells repopulating immunodeficient mice (Vercauteren & Sutherland, 2004). Later on, culture of human CD34+ precursors with the immobilized Notch ligand Delta1 and cytokines was shown to result in a substantial increase in NOD/SCID-repopulating cells (Delaney et al., 2010); similar results were obtained for mouse cells with immobilized Jagged1 ligand (Toda et al., 2011).

As for Wnt signaling, initial studies indicated that overexpression of activated beta-catenin expanded the pool of HSCs in long-term cultures as judged by both phenotype and function. Wnt3a protein induced self-renewal of haematopoietic stem cells, whereas ectopic expression of inhibitors of the Wnt signalling pathway led to suppression of HSC growth in vitro and reduced reconstitution in vivo (Reya et al., 2003; Willert et al., 2003). Later publications demonstrated, though, that inactivation of the beta-catenin gene in bone marrow progenitors does not impair their ability to self-renew and reconstitute all hematopoietic lineages (Cobas et al., 2004), whereas activation of beta-catenin enforced cell cycle entry of hematopoietic stem cells, thus leading to exhaustion of the long-term stem cell pool (Sheller et al., 2006). Some recent studies demonstrate that it is the non-canonical Wnt signaling promoted by Wnt5a rather than the canonical one, that supports maintenance of competitive repopulating murine HSCs in culture (Buckley et al., 2011; Nemeth et al., 2007).

Yet another line of evidence indicates that activation of beta-catenin in the niche components rather than in HSCs may produce support of LTR cells ex vivo (Nemeth et al., 2009). Currently, there is little doubt that Wnt signaling plays important role in HSC biology, but the issue is apparently more complex than was implied by initial publications and remains highly controversial.

Other embryonic signaling pathways also might be exploited in HSC culture. Morphogens of the hedgehog family, namely Sonic and Indian hedgehogs, are able to support ex vivo expansion of human NOD/SCID repopulating cells (Bhardwaj et al., 2001; Kobune et al., 2004), despite the fact that in vivo Hedgehog signaling seems to not be necessary for adult murine hematopoietic stem cell function (Hofmann et al., 2009). BMP4, a member of BMP superfamily, is a critical component of the hematopoietic niche that regulates both HSC number and function (Goldman et al., 2009), and is able to expand NOD/SCID-repopulating cells in culture (Hutton et al., 2006).

In addition to the use of secreted proteins to for ex vivo HSC culture, one apparent trend of the last years is the application of low-molecular weight chemicals, in particular agonists or inhibitors of particular intracellular signaling pathways, for ex vivo culture. Thus, specific inhibitor of p38 kinase induces self-renewal and ex vivo expansion of HSCs as shown by the in vitro cobblestone area forming cell assay and serial transplantation (Wang et al., 2011). GSK-3β inhibitors, which stimulate Wnt signaling, were shown to promote engraftment of cultured HSCs (Ko et al., 2011; Trowbridge et al., 2006). Of significant clinical interest is the finding that ex vivo treatment with stabilized prostaglandin E2 enhances frequency of both hematopoietic progenitors and long-term repopulating HSCs present as analyzed by competitive transplantation (North et al., 2007). According to other data, only the short-term repopulating HSCs are expanded by this treatment, though (Frisch et al., 2009).

The initial studies demonstrating substantial degree of expansion of HSCs ex vivo relied the use of stromal cells as feeder layers (Moore et al., 1997). Based on the substantial progress in identification of HSC niches in bone marrow, there is currently a revival of interest in development of protocols for co-culture of HSC with stromal cell layers (Chou & Lodish, 2010; De Toni et al., 2011). These stromal cells produce a range of factors that significantly improve the maintenance and expansion of HSCs in culture, most likely by mimicking more or less successfully niche conditions. Very prominent components of the HSC niche are cell surface proteins, in particular cell adhesion molecules. The importance of cell-cell interactions was highlighted by the study by Wagner et al., 2007, indicating that maintenance of primitive hematopoietic progenitors by stromal lines is associated with expression of cell adhesion proteins rather than with secretory profiles of these lines. In particular, N-cadherin was shown to be an important component of the osteoblastic HSC niche (Zhang et al., 2003). However, importance of N-cadherin for HSC-niche interactions was later questioned (Kiel et al., 2007), thus rising substantial controversy. In an elegant in vitro study Lutolf et al. (2009) have shown that N-cadherin, as well as Wnt3a, are the only proteins among those tested that were capable of supporting self-renewal divisions of HSCs in vitro. N-cadherin expression was also shown to be important for maintenance of long-term repopulating cells in culture (Hosokawa et al., 2010). Ability of stromal cell line FMS/PA6-P to support primitive murine hematopoietic cells was found to depend critically on N-CAM expression (Wang et al., 2005). Yet another cell adhesion protein, namely mKirre, plays a prominent role in hematopoietic supportive capacity of OP9 stromal cells (Ueno et al., 2003).

Quite promising developments occur currently in the field of 3-D culture (Yuan et al., 2011; Tan et al., 2010; Miyoshi at el., 2011). Despite a relative paucity of data related to the 3-D culture of HSCs, available publications demonstrate significant advantages of this technique and indicate that in combination with correctly chosen or gene-modified stromal cell layers, 3-D culturing may eventually lead to creation of artificial niche that will be able to support substantial expansion of human HSCs ex vivo.

A question of paramount importance for the field is whether specific combinations of soluble factors will be able to attain a bone fide ex vivo expansion of HSCs, or this goal can only be achieved if specific cell surface proteins produced by the niche cells are also employed in the process, or perhaps the only way to the eventual success is the use of supporting stromal cell layers for ex vivo culture? As a number of molecules that contribute to the maintenance of HSCs in vitro and in vivo continues to rise, and there is a steady improvement in techniques for culturing HSCs, chances are that within a matter of a few years, key combination(s) of specific factors and modes of their application that can produce robust self-renewal and expansion of human HSC ex vivo will be identified. Table 3 provides a list, albeit incomplete, of factors and chemicals that, in addition to "classical" cytokines, are being used for maintenance and expansion of HSCs ex vivo.

6. Pre-conditioning and transplantation regimens

A common practice in the field of HSC gene therapy is a transduction of HSCs using viral vectors in the ex vivo setting. The advantages of this strategy include elimination of non-target transduction events, higher transduction efficiency and better control over the overall process. However, the opposite side of the coin in this case is the necessity for transduced cells to compete with the bone marrow-resident ones, which is likely to lower significantly the degree of chimerism after gene therapy. For efficient repopulation of hematopoietic system with gene-modified HSCs, extensive myeloablative treatments eliminating resident HSCs are usually performed. However, since these treatments are of generalized character and connected with substantial risks of morbidity and mortality, especially for elderly patients, they should preferably be avoided whenever possible. A combination of nonmyeloablative pre-conditioning of the recipient animals with in vivo selection strategy can be used to achieve substantial degrees of chimerism (Davis et al., 2000, Zielske et al., 2003). Additional ways to develop more appropriate pretreatment conditions involve the use of molecules that disrupt key signaling pathways within HSCs or niche components thus inducing HSC loss, as was shown for the case of inactivation of c-kit or mpl signaling by neutralizing antibodies (Czechowicz et al., 2007; Yoshihara et al., 2007), and for combined poly(I:C)/5-fluorouracil (5-FU) treatment (Sato et al., 2009). The other approach for nonmyeloablative HSC transplantation is based on disruption of HSC-niche interactions thus aiding in the stem cell mobilization (Chen et al., 2006). This alternative might grow into clinically relevant technique if the efficiency of current protocols for mobilization of HSCs is further improved. The more HSCs are mobilized into circulation and used for viral transduction, the higher is ratio of transduced vs. resident stem cells and better chances to achieve significant engraftment and chimerism of gene-modified cells without resorting to drastic myeloablative regimens. Although current combinations of mobilizing agents (Ramirez et al, 2009) demonstrate much higher mobilization rates than the initially used G-CSF, there is still a long way to go before this strategy may equal or surpass myeloablative pre-conditioning in its efficiency.

Factor	Observed effects	References
FGF1	FGF1 under serum-free conditions stimulates expansion of serially transplantable, long-term repopulating HSCs.	de Haan et al., 2003
Angptl2, 3 and 5	Proteins of angiopoeitin-like family provide 20- to 30-fold net expansion of long-term HSCs according to reconstitution analysis.	Zhang C et al., 2006
IGFBP2	IGFBP2 enhances ex vivo expansion of mouse HSCs.	Huynh et al., 2008
IL32	IL-32 significantly induces the proliferation of HSCs in culture.	Moldenhauer et al., 2011
Delta 1, Jagged1 (Notch ligands)	Culturing murine or human cells with surface-immobilized Notch ligands resulted in expansion of primitive hematopoietic population.	Delaney et al., 2010; Toda et al., 2011;
Wnt3a, Wnt10b (Wnt canonical pathway)	Wnt3a protein induces self-renewal of haematopoietic stem cells. Wnt10b enhances growth of hematopoietic precursors.	Willert et al., 2003; Congdon et al., 2010
Wnt5a (Wnt non-canonical pathway)	Wnt5a inhibits canonical Wnt signaling and supports maintenance of competitive repopulating murine HSCs in culture.	Nemeth et. al, 2007; Buckley et al., 2011
Shh, Ihh	Sonic hedgehog and Indian hedgehog support ex vivo expansion of human NOD/SCID repopulating cells.	Bhardwaj et al., 2001; Kobune et al., 2004
Bmp4	BMP4 expands NOD/SCID-repopulating cells in culture.	Hutton et al., 2006
TAT-HOXB4 fusion protein	TAT-HOXB4 protein produces significant ex vivo expansion of murine HSCs.	Krosl et al., 2003
TAT-NF-Ya fusion protein	TAT-NF-Ya protein treatment produces several-fold increase in the percentage of human cells repopulating immunodeficient mice.	Domashenko et al., 2010
TAT-SALL4B fusion protein	TAT-SALL4B fusion protein rapidly expands long-term NOD/SCID repopulating cells.	Aguila et al, 2011
Prostaglandin E2	Ex vivo incubation with PGE2 increases the frequency of long-term repopulating HSCs as measured by competitive transplantation.	North et al., 2007
SB203580	SB203580, specific p38 inhibitor, leads to increase in HSC self-renewal and ex vivo expansion.	Wang et al., 2011
StemRegenin 1	SR1, aryl hydrocarbon receptor antagonist, provides substantial increase in cells engrafting into immunodeficient mice.	Boitano et al., 2010
zVADfmk, zLLYfmk	Cord blood CD34+ cells cultured in presence of zVADfmk or zLLYfmk (inhibitors of caspases and calpains, respectively) have a higher ability for engraftment in NOD/SCID mice.	Imai et al., 2010; Sangeetha et al, 2010;
GSK-3 inhibitors	Pretreatment with GSK-3 inhibitors (BIO or CHIR-911) promotes engraftment and repopulation of ex vivo-expanded HSCs.	Ko et al., 2011; Trowbridge et al., 2006
Rapamycin	HSCs cultured in vitro in the presence of mTOR inhibitor rapamycin demonstrate enhanced engraftment.	Rohrabaugh et al., 2011
Copper helators	Copper chelator tetraethylenepentamine increases long-term ex vivo expansion and engraftment capabilities of blood progenitors.	Peled et al., 2004
N-cadherin	N-cadherin expression on stromal cells is important for maintenance of long-term repopulating cells in culture.	Hosokawa et al., 2010
N-CAM	N-CAM expression on stromal cells supports primitive murine hematopoietic cells.	Wang et al., 2005
mKirre	mKirre is responsible for hematopoietic supportive capacity of OP9 stromal cells.	Ueno et al., 2003

Table 3. Proteins and compounds affecting ex vivo maintenance and expansion of HSCs ("classical" cytokines not listed)

There are reports indicating that the engraftment of gene-modified stem cells might be significantly improved by their direct intra-bone transplantation (Mazurier et al., 2003). As irradiation commonly used for preconditioning also damages hematopoietic niche, in particular mesenchymal stem cells, HSC co-transplantation with MSCs was tested and showed promising results (Masuda et al., 2009).

Even a more radical departure from the accepted strategies for HSCs would be in situ transduction of HSCs using systemic or intra-bone delivery of viral vectors (McCauslin et al., 2003, Pan, 2009). Currently, this is a rather hypothetical approach due to serious safety concerns connected with potential off-target modifications of non-hematopoetic cells. However, this strategy alleviates the need for hazardous pre-conditioning treatments and will become a viable alternative with further development of modified viral envelops (Zhang X & Roth, 2010) that target vectors specifically to hematopoietic stem and progenitor cells while minimizing off-target events.

7. Safety: Vector genotoxicity, transposon vectors and other issues

The genotoxicity issue is currently the most immediate and direct safety concern related to the gene therapy using HSCs. Several otherwise successful gene therapy trials of severe combined immunodeficiency using retroviral vectors have resulted in occurrence of leukemia in a significant percentage of patients. Substantial efforts were thus devoted to elucidation of integration patterns and clonal population structure in the hematopoietic compartment after viral transduction, both in experimental models and in clinical trials. The obtained results, although not unanimous, demonstrate nevertheless a frequent occurrence of oligoclonal hematopoiesis after gene therapy, with viral integration sites tending to concentrate in the vicinity of a limited number of genes preferentially involved in growth and proliferation control such as above mentioned Evi-1, PRDM16 or HMGA2. Although upregulation of these genes rarely led to overt neoplastic transformation, it is nevertheless clear that the patients with oligoclonal hematopoiesis are at substantial risk of acquiring leukemias at some future time point.

Various strategies are being currently developed to minimize the risk of neoplastic transformations of HSCs after viral transduction. The most promising approaches include using lentiviral instead of retroviral vectors, and insulators to shield cellular oncogenes from activation by strong viral promoters (Puthenveetil et al., 2004). Insulators, however, tends to significantly reduce viral titers (Nielsen et al., 2009), relatively inefficient (Uchida et al., 2011) and do not provide guarantee against insertional activation of potential oncogenes such as HMGA2 (Cavazzana-Calvo et al., 2010). Another approach is to use promoters specific for differentiated cells that are expected to produce negligible activation of oncogenes in stem cells. However, such promoters tend to provide comparably lower expression levels, and although this might be improved by addition of strong enhancers (Gruh et al., 2008), it is far from certain that such combinations would not activate nearby cellular promoters.

Transposon vectors offer an exciting alternative to retro- and lentiviral vectors. The transposon-based gene delivery combines advantages of integrating viral vectors with those of plasmid vectors. Permanent genomic integration of transposon vectors provides long-term expression, whereas there are significantly fewer constraints on vector design and use of various function elements like insulators. Transposon systems are inherently less immunogenic than viral delivery systems, whereas their cargo capacity generally exceeds that of retro- and lentiviral vectors (Zayed *et al.*, 2004). Initial experiments with transposons were plagued by low efficiency of integration, but continuous improvements in molecular design of transposases have significantly increased the efficiency of integration process (Mátés et al., 2009). Currently, transposons based on Sleeping Beauty (SB) system represent the most advanced version of this technology (reviewed by Ivics & Izsvák, 2011), although other system such as piggyBac are also being perfected (Yusa et al., 2011) and may offer some advantages, such as larger cargo capacity, over the SB system (Lacoste et al., 2009).

Although stable SB transposon-mediated gene transfer into hematopoietic cells was reported (Xue et al., 2009), efficient vector delivery to HSCs remains poorly resolved issue, which is currently being addressed by using electroporation or hybrid lentiviral-transposon vectors (Staunstrup et al., 2009). Although certain undesired effects such as SB transposase cytotoxicity were observed, it seems that they might be minimized by controllable mRNA delivery (Galla et al., 2011). Compared to lenti- and retroviral vectors that show preferential integration near active genes, SB transposon vectors demonstrate nearly random integration profiles (Moldt et al., 2011), although this property might not be shared by other transposon systems (Huang et al., 2010).

Another serious safety concern is a direct consequence of a current low efficiency of transduction of LTR HSCs, which necessitates the use of myeloablative pre-conditioning and negative selection strategies to eliminate competing endogenous HSCs and increase chimerism levels. Negative selection strategies using in particular alkylating drugs place a significant stress upon hematopoietic system. However, as demonstrated by Xie et al., 2010, repetitive hematopoietic stress by busulfan administration in a nonhuman primate may rapidly lead to reduction of polyclonality and eventually to cytopenia. In addition, potential long term mutagenic effects of alkylating agents are largely unknown, thus adding more uncertainty as to correct assessment of risks and benefits of this strategy. Apparently, in order to tackle efficiently the problem of low transduction efficiency, it is not sufficient to rely on the use of negative selection only, but is also important to achieve substantial improvements in ex vivo stem cell culturing, expansion and transduction efficiency. Promising approaches also involve use of positive ex vivo and in vivo selection and in situ transduction strategies.

8. Novel technologies

In the recent few years, a group of new exciting and very powerful technologies, namely cell reprogramming using specific combinations of transcription factors and/or micro RNAs appeared (Takahashi & Yamanaka, 2006; Miyoshi et al., 2011). Much hope is invested into development of strategies aiming at derivation of patient-specific induced pluripotent (iPS) cells similar to embryonic stem (ES) cells, with their subsequent differentiation into hematopoetic cells capable of long-term hematopoiesis. In addition to this indirect reprogramming strategy, methods for direct reprogramming that bypass derivation of iPS cells are also being elaborated. There is one report stating that ectopic expression of Oct4 transcription factor in human fibroblasts is sufficient to convert them into hematopoietic cells with in vivo engraftment capacity (Szabo et al., 2010). However, whether the published technique may result in production of bona fide hematopietic stem cells capable of long-term reconstitution, remains to be seen. It should be noted that such a goal has not yet been achieved for ES or iPS cells. If efficient reprogramming into HSCs were possible, the perspectives would look staggering. First of all, since starting primary cell populations such as mesenchymal stem/progenitor cells can be propagated for many generations and are amenable for selection of efficient vector integration events, it will be possible to obtain cell populations in which the majority of reprogrammed HCS-like cells bear functioning transgenes, thus increasing efficiency of gene therapy many-fold. Besides, if this technology were able to generate ex vivo significantly more reprogrammed cells with HSC properties than is possible to obtain from a patient, this would establish basis for a radically increase in a level of chimerism after transplantation, thus further improving the efficiency of gene therapy. Of course, the safety issues, in particular potential epigenetic and genome instability of reprogrammed cells that might result in neoplastic transformations, must be addressed especially carefully in this case.

9. Conclusion

Current protocols of gene therapy of hematopoietic and immune system, despite significant efforts by numerous teams worldwide, demonstrate as yet a relatively modest clinical efficiency. However, there are sufficient reasons to assume that many rather inconspicuous yet significant recent technical developments are preparing the field for a decisive breakthrough in the near future. In addition, new cutting- edge technologies such as direct cell reprogramming are entering the scene and may eventually present a radically different and a more efficient solution of the problem. Given all these considerations, the future of gene therapy of blood and immune system diseases looks definitely bright.

10. Acknowledgment

This work was supported by the Russian Foundation for Basic Research Grants 09-04-01312 to F.R. and 11-04-01814-a to A.B, and a grant of the RAS Program of Molecular Cellular Biology to A.B.

11. References

Aguila, J.R.; Liao, W. ; Yang, J., Avila, C.; Hagag, N.; Senzel, L. & Ma, Y. (2011). SALL4 is a robust stimulator for the expansion of hematopoietic stem cells. *Blood,* Vol.118, No.3, (July 2011), pp. 576-585, ISSN 0006-4971

Antonchuk, J.; Sauvageau, G. & Humphries, R.K. (2002). HOXB4-induced expansion of adult hematopoietic stem cells ex vivo. *Cell,* Vol.109, No.1, (April 2002), pp. 39–45, ISSN 0092-8674

Beard, B.C.; Trobridge, G.D.; Ironside, C.; McCune, J.S.; Adair, J.E. & Kiem, H.P. (2010). Efficient and stable MGMT-mediated selection of long-term repopulating stem cells in nonhuman primates. *The Journal of Clinical Investigation,* Vol.120, No.7, (July 2010), pp. 2345-2354, ISSN 0021-9738

Bersenev, A.; Wu, C.; Balcerek, J. & Tong, W. (2008). Lnk controls mouse hematopoietic stem cell self-renewal and quiescence through direct interactions with JAK2. *The Journal of Clinical Investigation,* Vol.118, No.8, (August 2008), pp. 2832-2844, ISSN 0021-9738

Bhardwaj, G.; Murdoch, B.; Wu, D.; Baker, D.P.; Williams, K.P.; Chadwick, K.; Ling, L.E.; Karanu, F.N. & Bhatia, M. (2001). Sonic hedgehog induces the proliferation of primitive human hematopoietic cells via BMP regulation. *Nature Immunology,* Vol.2, No.2, (February 2001), pp. 172-180, ISSN 1529-2908

Bielas, H.; Schmitt, M., Icreverzi, A.; Ericson,N. & Loeb, L. (2009). Molecularly evolved thymidylate synthase inhibits 5-fluorodeoxyuridine toxicity in human hematopoietic cells. *Human Gene Therapy,* Vol.20, No.12, (December 2009), pp. 703-707, ISSN 1043-0342

Bowman, J.E.; Reese, J.S.; Lingas, K.T. & Gerson, S.L. (2003). Myeloablation is not required to select and maintain expression of the drug-resistance gene, mutant MGMT, in primary and secondary recipients. *Molecular Therapy,* Vol. 8, No.1, (July 2003), pp. 42-50, ISSN 1525-0016

Buckley, S.M.; Ulloa-Montoya, F.; Abts, D.; Oostendorp, R.A.; Dzierzak, E.; Ekker, S.C. & Verfaillie, C.M. (2011). Maintenance of HSC by Wnt5a secreting AGM-derived stromal cell line. *Experimental Hematology,* (January 2011), Vol.39, No.1, pp. 114-123.e1-5, ISSN 0301-472X

Bunting, K.D.; Galipeau, J.; Topham, D.; Benaim, E. & Sorrentino, B.P. (1999). Effects of retroviral-mediated MDR1 expression on hematopoietic stem cell self-renewal and differentiation in culture. *Annals of the New York Academy of Sciences,* Vol.872, (April 1999), pp. 125-141, ISSN 0077-8923

Case, S.S.; Price, M.A.; Jordan, C.T.; Yu, X.J.; Wang, L.; Bauer, G.; Haas, D.L.; Xu, D.; Stripecke, R.; Naldini, L.; Kohn, D.B. & Crooks, G.M. (1999). Stable transduction of quiescent CD34+CD38- human hematopoietic cells by HIV-1-based lentiviral vectors. *Proceedings of the National Academy of Sciences of the United States of America,* Vol.96, No.6, (March 1999), pp. 2988–2993, ISSN 0027-8424

Cavazzana-Calvo, M.; Hacein-Bey, S.; de Saint Basile, G.; Gross, F.; Yvon, E.; Nusbaum, P.; Selz, F.; Hue, C.; Certain, S.; Casanova, J.L.; Bousso, P.; Deist, F.L. & Fischer, A. (2000). Gene therapy of human severe combined immunodeficiency (SCID)-X1 disease. *Science,* Vol.288, No.5466, (April 2000), pp. 669-672, ISSN 0036-8075

Cavazzana-Calvo, M.; Payen, E.; Negre, O.; Wang, G.; Hehir, K.; Fusil, F.; Down, J.; Denaro, M.; Brady, T.; Westerman, K.; Cavallesco, R.; Gillet-Legrand, B.; Caccavelli, L.; Sgarra, R.; Maouche-Chrétien, L.; Bernaudin, F.; Girot, R.; Dorazio, R.; Mulder, G.J.; Polack, A.; Bank, A.; Soulier, J.; Larghero, J.; Kabbara, N.; Dalle, B.; Gourmel, B.; Socie, G.; Chrétien, S.; Cartier, N.; Aubourg, P.; Fischer, A.; Cornetta, K.; Galacteros, F.; Beuzard, Y.; Gluckman, E.; Bushman, F.; Hacein-Bey-Abina, S. & Leboulch, P. (2010). Transfusion independence and HMGA2 activation after gene therapy of human β-thalassaemia. *Nature,* Vol. 467, No.7313, (September 2010), pp. 318-322, ISSN 0028-0836

Chen, J.; Larochelle, A.; Fricker, S.; Bridger, G.; Dunbar, C.E. & Abkowitz J.L. (2006). Mobilization as a preparative regimen for hematopoietic stem cell transplantation. *Blood,* Vol.107, No.9, (May 2006), pp. 3764-3771, ISSN 0006-4971

Cheshier, S.H.; Morrison, S.J.; Liao, X. & Weissman, I.L. (1999). In vivo proliferation and cell cycle kinetics of long-term self-renewing haematopoietic stem cells. *Proceedings of the National Academy of Sciences of the United States of America,* Vol.96, No.6, (March 1999), pp. 3120–3125, ISSN 0027-8424

Chinnasamy, D.; Milsom, M.D.; Shaffer, J.; Neuenfeldt, J.; Shaaban, A.F.; Margison, G.P.; Fairbairn, L.J. & Chinnasamy, N. (2006). Multicistronic lentiviral vectors containing the FMDV 2A cleavage factor demonstrate robust expression of encoded genes at limiting MOI. *Virology Journal,* Vol.3, (March 2006), pp. 14, ISSN 1743-422X

Chou, S. & Lodish, H.F. (2010). Fetal liver hepatic progenitors are supportive stromal cells for hematopoietic stem cells. *Proceedings of the National Academy of Sciences of the United States of America,* Vol.107, No.17, (April 2010), pp. 7799-7804, ISSN 0027-8424

Cobas, M.; Wilson, A.; Ernst, B.; Mancini, S.J.; MacDonald, H.R.; Kemler, R. & Radtke, F. (2004). Beta-catenin is dispensable for hematopoiesis and lymphopoiesis. *The Journal of Experimental Medicine,* Vol.199, No.2, (January 2004), pp. 221-229, ISSN 0022-1007

Cowan, K.H.; Moscow, J.A.; Huang, H.; Zujewski, J.A.; O'Shaughnessy, J.; Sorrentino, B.; Hines, K.; Carter, C.; Schneider, E.; Cusack, G.; Noone, M.; Dunbar, C.; Steinberg, S.; Wilson, W.; Goldspiel, B.; Read, E.J.; Leitman, S.F.; McDonagh, K.; Chow, C.; Abati, A.; Chiang, Y.; Chang, Y.N.; Gottesman, M.M.; Pastan, I. & Nienhuis, A. (1999). Paclitaxel chemotherapy after autologous stem-cell transplantation and engraftment of hematopoietic cells transduced with a retrovirus containing the multidrug resistance complementary DNA (MDR1) in metastatic breast cancer patients. *Clinical Cancer Research,* Vol.5, No.7, (July 1999), pp. 1619-1628, ISSN 1078-0432

Crcareva, A.; Saito, T.; Kunisato, A.; Kumano, K.; Suzuki, T.; Sakata-Yanagimoto, M.; Kawazu, M.; Stojanovic, A.; Kurokawa, M.; Ogawa, S.; Hirai, H. & Chiba, S. (2005). Hematopoietic stem cells expanded by fibroblast growth factor-1 are excellent targets for retrovirus-mediated gene delivery. *Experimental Hematology,* Vol.33, No.12, (December 2005), pp. 1459-1469, ISSN 0301-472X

Czechowicz, A.; Kraft, D.; Weissman, I.L. & Bhattacharya, D. (2007). Efficient transplantation via antibody-based clearance of hematopoietic stem cell niches. *Science,* Vol.318, No. 5854, (November 2007), pp. 1296-1299, ISSN 0036-8075

Davis, B.M.; Koç, O.N. & Gerson, S.L. (2000). Limiting number of G156A O6-methylguanine DNA methyltransferase-transduced marrow progenitors repopulate nonmyeloablated mice after drug selection. *Blood,* Vol. 95, No.10, (May 2000) pp. 3078–3084, ISSN 0006-4971

de Barros, A.P.; Takiya, C.M.; Garzoni, L.R.; Leal-Ferreira, M.L.; Dutra, H.S.; Chiarini, L.B.; Meirelles, M.N.; Borojevic, R. & Rossi, M.I. (2010). Osteoblasts and bone marrow mesenchymal stromal cells control hematopoietic stem cell migration and proliferation in 3D in vitro model. *PLoS One,* Vol.5, No.2, (February 2010), pp. e9093, ISSN 1932-6203

de Haan, G.; Weersing, E.; Dontje, B.; van Os, R.; Bystrykh, L.V.; Vellenga, E. & Miller, G. (2003). In vitro generation of long-term repopulating hematopoietic stem cells by fibroblast growth factor-1. *Developmental Cell,* Vol.4, No.2, (February 2003), pp. 241-251, ISSN 1534-5807

Delaney, C.; Heimfeld, S.; Brashem-Stein, C.; Voorhies, H.; Manger, R.L. & Bernstein, I.D. (2010). Notch-mediated expansion of human cord blood progenitor cells capable of rapid myeloid reconstitution. *Nature Medicine,* Vol.16, No.2, (February 2010), pp. 232-236, ISSN 1078-8956

De Toni, F.; Poglio, S.; Youcef, A.B.; Cousin, B.; Pflumio, F.; Bourin, P.; Casteilla, L. & Laharrague, P. (2011). Human Adipose-Derived Stromal Cells Efficiently Support Hematopoiesis In Vitro and In Vivo: A Key Step for Therapeutic Studies. *Stem Cells and Development,* (April 2011), advance online publication, ISSN 1547-3287

Domashenko, A.D.; Danet-Desnoyers, G.; Aron, A.; Carroll, M.P. & Emerson, S.G. (2010). TAT-mediated transduction of NF-Ya peptide induces the ex vivo proliferation and engraftment potential of human hematopoietic progenitor cells. *Blood,* Vol.116, No.15, (October 2010), pp. 2676-2683, ISSN 0006-4971

Du, Y.; Jenkins, N.A. & Copeland, N.G. (2005). Insertional mutagenesis identifies genes that promote the immortalization of primary bone marrow progenitor cells. *Blood,* Vol.106, No.12, (December 2005), pp. 3932-3939, ISSN 0006-4971

Ellis, J. (2005). Silencing and variegation of gammaretrovirus and lentivirus vectors. *Human Gene Therapy,* Vol.16, No.11, (November 2005), pp. 1241-1246, ISSN 1043-0342

Frisch, B.J.; Porter, R.L.; Gigliotti, B.J.; Olm-Shipman, A.J.; Weber, J.M.; O'Keefe, R.J.; Jordan, C.T. & Calvi, L.M. (2009). In vivo prostaglandin E2 treatment alters the bone marrow microenvironment and preferentially expands short-term hematopoietic stem cells. *Blood,* Vol.114, No.19, (November 2009), pp. 4054-4063, ISSN 0006-4971

Galla, M.; Schambach, A.; Falk, C.S.; Maetzig, T.; Kuehle, J.; Lange, K.; Zychlinski, D.; Heinz, N.; Brugman, M.H.; Göhring, G.; Izsvák, Z.; Ivics, Z. & Baum, C. (2011). Avoiding cytotoxicity of transposases by dose-controlled mRNA delivery. *Nucleic Acids Research,* Vol.39, No.16, (September 2011), pp. 7147-7160, ISSN 0305-1048

Goldman, D.C.; Bailey, A.S.; Pfaffle, D.L.; Al Masri, A.; Christian, J.L. & Fleming, W.H. (2009). BMP4 regulates the hematopoietic stem cell niche. *Blood,* Vol.114, No.20, (November 2009), pp. 4393-4401, ISSN 0006-4971

Gori, J.L.; McIvor, R. & Kaufman, D. (2010). Methotrexate supports in vivo selection of human embryonic stem cell derived-hematopoietic cells expressing dihydrofolate reductase. *Bioengineered Bugs,* Vol.1, No.6, (November 2010), pp. 434-436, ISSN 1949-1018

Guo, S.; Lu, J.; Schlanger, R.; Zhang, H.; Wang, J.Y.; Fox, M.C.; Purton, L.E.; Fleming, H.H.; Cobb, B.; Merkenschlager, M.; Golub, T.R. & Scadden, D.T. (2010). MicroRNA miR-125a controls hematopoietic stem cell number. *Proceedings of the National Academy of Sciences of the United States of America,* Vol.107, No.32, (August 2010), pp. 14229-14234, ISSN 0027-8424

Gruh, I.; Wunderlich, S.; Winkler, M.; Schwanke, K.; Heinke, J.; Blömer, U.; Ruhparwar, A.; Rohde, B.; Li, R.K.; Haverich, A. & Martin, U. (2008). Human CMV immediate-early enhancer: a useful tool to enhance cell-type-specific expression from lentiviral vectors. *The Journal of Gene Medicine,* Vol.10, No.1, (January 2008), pp. 21-32, ISSN 1099-498X

Heath, V.; Suh, H.C.; Holman, M.; Renn, K.; Gooya, J.M; Parkin, S.; Klarmann, K.D.; Ortiz, M.; Johnson, P. & Keller, J. (2004). C/EBPalpha deficiency results in hyperproliferation of hematopoietic progenitor cells and disrupts macrophage development in vitro and in vivo. *Blood,* Vol.104, No.6, (September 2004), pp. 1639-1647, ISSN 0006-4971

Hofmann, I.; Stover, E.H.; Cullen, D.E.; Mao, J.; Morgan, K.J.; Lee, B.H.; Kharas, M.G.; Miller, P.G.; Cornejo, M.G.; Okabe, R.; Armstrong, S.A.; Ghilardi, N.; Gould, S.; de Sauvage, F.J.; McMahon, A.P. & Gilliland, D.G. (2009). Hedgehog signaling is dispensable for adult murine hematopoietic stem cell function and hematopoiesis. *Cell Stem Cell,* Vol.4, No.6, (June 2009), pp. 559-567, ISSN 1934-5909

Horn, P.A.; Morris, J.C.; Bukovsky, A.A.; Andrews, R.G.; Naldini, L.; Kurre, P. & Kiem, H.P. (2002). Lentivirus-mediated gene transfer into hematopoietic repopulating cells in baboons. *Gene Therapy,* Vol.9, No.21, (November 2002), pp. 1464–1471, ISSN 0969-7128

Hosokawa, K.; Arai, F.; Yoshihara, H.; Iwasaki, H.; Nakamura, Y.; Gomei, Y. & Suda, T. (2010). Knockdown of N-cadherin suppresses the long-term engraftment of hematopoietic stem cells. *Blood,* Vol.116, No.4, (July 2010), pp. 554-563, ISSN 0006-4971

Huang, X.; Guo, H.; Tammana, S.; Jung, Y.C.; Mellgren, E.; Bassi, P.; Cao, Q.; Tu, Z.J.; Kim, Y.C.; Ekker, S.C.; Wu, X.; Wang, S.M. & Zhou, X. (2010). Gene transfer efficiency and genome-wide integration profiling of Sleeping Beauty, Tol2, and piggyBac transposons in human primary T cells. *Molecular Therapy,* Vol.18, No.10, (October 2010), pp. 1803-1813, ISSN 1525-0016

Hutton, J.F.; Rozenkov, V.; Khor, F.S.; D'Andrea, R.J. & Lewis, I.D. (2006). Bone morphogenetic protein 4 contributes to the maintenance of primitive cord blood hematopoietic progenitors in an ex vivo stroma-noncontact co-culture system. *Stem Cells and Development,* Vol.15, No.6, (December 2006), pp. 805-813, ISSN 1547-3287

Huynh, H.; Iizuka, S.; Kaba, M.; Kirak, O.; Zheng, J.; Lodish, H.F. & Zhang, C.C. (2008). Insulin-like growth factor-binding protein 2 secreted by a tumorigenic cell line supports ex vivo expansion of mouse hematopoietic stem cells. *Stem Cells,* Vol.26, No.6, (June 2008), pp. 1628-1635, ISSN 1066-5099

Ikeda, K.; Mason, P.J. & Bessler M. (2011). 3'UTR-truncated Hmga2 cDNA causes MPN-like hematopoiesis by conferring a clonal growth advantage at the level of HSC in mice. *Blood,* Vol.117, No.22, (June 2011), pp. 5860-5869, ISSN 0006-4971

Imai, Y.; Adachi, Y.; Shi, M.; Shima, C.; Yanai, S.; Okigaki, M.; Yamashima, T.; Kaneko, K. & Ikehara, S. (2010). Caspase inhibitor ZVAD-fmk facilitates engraftment of donor hematopoietic stem cells in intra-bone marrow-bone marrow transplantation. *Stem Cells and Development,* Vol.19, No.4, (April 2010), pp. 461-468, ISSN 1547-3287

Ivics, Z. & Izsvák, Z. (2011). Non-viral Gene Delivery with the Sleeping Beauty Transposon System. *Human Gene Therapy,* (August 2011), advance online publication, ISSN 1043-0342

Iwama, A.; Oguro, H.; Negishi, M.; Kato, Y.; Morita, Y.; Tsukui, H.; Ema, H.; Kamijo, T.; Katoh-Fukui, Y.; Koseki, H.; van Lohuizen, M. & Nakauchi, H. (2004). Enhanced self-renewal of hematopoietic stem cells mediated by the polycomb gene product Bmi-1. *Immunity,* Vol.21, No.6, (December 2004), pp. 843-851, ISSN 1074-7613

Khoury, M.; Drake, A.; Chen, Q.; Dong, D.; Leskov, I.; Fragoso, M.F.; Li, Y.; Iliopoulou, B.P.; Hwang, W.; Lodish, H.F. & Chen, J. (2011). Mesenchymal stem cells secreting angiopoietin-like-5 support efficient expansion of human hematopoietic stem cells without compromising their repopulating potential. *Stem Cells and Development,* Vol.20, No.8, (August 2011), pp. 1371-1381, ISSN 1547-3287

Kiel, M.J.; Radice, G.L. & Morrison, S.J. (2007). Lack of evidence that hematopoietic stem cells depend on N-cadherin-mediated adhesion to osteoblasts for their maintenance. *Cell Stem Cell,* Vol.1, No.2, (August 2007), pp. 204-217, ISSN 1934-5909

King, K.Y.; Baldridge, M.T.; Weksberg, D.C.; Chambers, S.M.; Lukov, G.L.; Wu, S.; Boles, N.C.; Jung, S.Y.; Qin, J.; Liu, D.; Songyang, Z.; Eissa, N.T.; Taylor, G.A. & Goodell, MA. (2011). Irgm1 protects hematopoietic stem cells by negative regulation of IFN signaling. *Blood,* Vol.118, No. 6, (August 2011), pp. 1525-33, ISSN 0006-4971

Ko, K.H.; Holmes, T.; Palladinetti, P.; Song, E.; Nordon, R.; O'Brien, T.A. & Dolnikov, A. (2011). GSK-3β inhibition promotes engraftment of ex vivo-expanded hematopoietic stem cells and modulates gene expression. *Stem Cells,* Vol.29, No.1, (January 2011), pp. 108-118, ISSN 1066-5099

Kobune, M.; Ito, Y.; Kawano, Y.; Sasaki, K.; Uchida, H.; Nakamura, K.; Dehari, H.; Chiba, H.; Takimoto, R.; Matsunaga, T.; Terui, T.; Kato, J.; Niitsu, Y. & Hamada, H. (2004). Indian hedgehog gene transfer augments hematopoietic support of human stromal cells including NOD/SCID-beta2m-/- repopulating cells. *Blood,* Vol.104, No.4, (August 2004), pp. 1002-1009, ISSN: 0006-4971

Konuma, T.; Nakamura, S.; Miyagi, S.; Negishi, M.; Chiba, T.; Oguro, H.; Yuan, J.; Mochizuki-Kashio, M.; Ichikawa, H.; Miyoshi, H.; Vidal, M. & Iwama, A. (2011). Forced expression of the histone demethylase Fbxl10 maintains self-renewing hematopoietic stem cells. *Experimental Hematology,* Vol.39, No.6, (June 2011), pp. 697-709.e5, ISSN 0301-472X

Krosl, J.; Austin, P.; Beslu, N.; Kroon, E.; Humphries, R.K. & Sauvageau, G. (2003). In vitro expansion of hematopoietic stem cells by recombinant TAT-HOXB4 protein. *Nature Medicine,* Vol.9, No.11, (November 2003), pp. 1428-1432, ISSN 1078-8956

Lacoste, A.; Berenshteyn, F. & Brivanlou, A.H. (2009). An efficient and reversible transposable system for gene delivery and lineage-specific differentiation in human embryonic stem cells. *Cell Stem Cell,* Vol.5, No.3, (September 2009), pp. 332-342, ISSN 1934-5909

Larochelle, A.; Choi, U.; Shou, Y.; Naumann, N.; Loktionova, N.A.; Clevenger, J.R.; Krouse, A.; Metzger, M.; Donahue, R.E.; Kang, E.; Stewart, C.; Persons, D.; Malech, H.L.; Dunbar, C.E. & Sorrentino, B.P. (2009). In vivo selection of hematopoietic progenitor cells and temozolomide dose intensification in rhesus macaques through lentiviral transduction with a drug resistance gene. *The Journal of Clinical Investigation,* Vol.119, No.7, (July 2009), pp. 1952-1963, ISSN 0021-9738

Lutolf, M.P.; Doyonnas, R.; Havenstrite, K.; Koleckar, K. & Blau, H.M. (2009). Perturbation of single hematopoietic stem cell fates in artificial niches. *Integrative biology,* Vol.1, No.1, (January 2009), pp. 59-69, ISSN 1757-9694

Masuda, S.; Ageyama, N.; Shibata, H.; Obara, Y.; Ikeda, T.; Takeuchi, K.; Ueda, Y.; Ozawa, K. & Hanazono, Y. (2009). Cotransplantation with MSCs improves engraftment of HSCs after autologous intra-bone marrow transplantation in nonhuman primates. *Experimental Hematology,* Vol.37, No.10, (October 2009), pp. 1250-1257.e1, ISSN 0301-472X

Mátés, L.; Chuah, M.K.; Belay, E.; Jerchow, B.; Manoj, N.; Acosta-Sanchez, A.; Grzela, D.P.; Schmitt, A.; Becker, K.; Matrai, J.; Ma, L.; Samara-Kuko, E.; Gysemans, C.; Pryputniewicz, D.; Miskey, C.; Fletcher, B.; VandenDriessche, T.; Ivics, Z. & Izsvák, Z. (2009). Molecular evolution of a novel hyperactive Sleeping Beauty transposase enables robust stable gene transfer in vertebrates. *Nature Genetics,* Vol.41, No.6, (June 2009), pp. 753-761, ISSN 1061-4036

Mazurier, F.; Doedens, M.; Gan, O.I. & Dick, J.E. (2003). Rapid myeloerythroid repopulation after intrafemoral transplantation of NOD-SCID mice reveals a new class of human stem cells. *Nature Medicine*, Vol.9, No.7, (July 2003), pp. 959-963, ISSN 1078-8956

McCauslin, C.S.; Wine, J.; Cheng, L.; Klarmann, K.D.; Candotti, F.; Clausen, P.A.; Spence, S.E. & Keller, J.R. (2003). In vivo retroviral gene transfer by direct intrafemoral injection results in correction of the SCID phenotype in Jak3 knock-out animals. *Blood*, Vol.102, No.3, (August 2003), pp. 843-848, ISSN 0006-4971

McCormack, M.P.; Forster, A.; Drynan, L.; Pannell, R. & Rabbitts, T.H. (2003). The LMO2 T-cell oncogene is activated via chromosomal translocations or retroviral insertion during gene therapy but has no mandatory role in normal T-cell development. *Molecular and Cellular Biology*, Vol.23, No.24, (December 2003), pp. 9003-9013, ISSN 0270-7306

McCormack, M.P.; Young, L.F.; Vasudevan, S.; de Graaf, C.A.; Codrington, R.; Rabbitts, T.H.; Jane, S.M. & Curtis, D.J. (2010). The Lmo2 oncogene initiates leukemia in mice by inducing thymocyte self-renewal. *Science*, Vol.327, No.5967, (February 2010), pp. 879-883, ISSN 0036-8075

Métais, J.Y. & Dunbar, C.E. (2008). The MDS1-EVI1 gene complex as a retrovirus integration site: impact on behavior of hematopoietic cells and implications for gene therapy. *Molecular Therapy*, Vol.16, No.3, (March 2008), pp. 439-449, ISSN 1525-0016

Milsom M.D.; Woolford L.B.; Margison G.P.; Humphries R.K. & Fairbairn L.J. (2004). Enhanced in vivo selection of bone marrow cells by retroviral-mediated coexpression of mutant O6-methylguanine-DNA-methyltransferase and HOXB4. *Molecular Therapy*, Vol.10, No.5, (November 2004), pp. 862-873, ISSN 1525-0016

Miyoshi, H.; Murao, M.; Ohshima, N. & Tun T. (2011). Three-dimensional culture of mouse bone marrow cells within a porous polymer scaffold: effects of oxygen concentration and stromal layer on expansion of haematopoietic progenitor cells. *Journal of Tissue Engineering and Regenerative Medicine*, Vol.5, No.2, (February 2011), pp. 112-118, ISSN 1932-6254

Miyoshi, N.; Ishii, H.; Nagano, H.; Haraguchi, N.; Dewi, D.L.; Kano, Y.; Nishikawa, S.; Tanemura, M.; Mimori, K.; Tanaka, F.; Saito, T.; Nishimura, J.; Takemasa, I.; Mizushima, T.; Ikeda, M.; Yamamoto, H.; Sekimoto, M.; Doki, Y. & Mori, M. (2011). Reprogramming of mouse and human cells to pluripotency using mature microRNAs. *Cell Stem Cell*, Vol.8, No.6, (June 2011), pp. 633-638, ISSN 1934-5909

Moldenhauer, A.; Futschik, M.; Lu, H.; Helmig, M.; Götze, P.; Bal, G.; Zenke, M.; Han, W. & Salama, A. (2011). Interleukin 32 promotes hematopoietic progenitor expansion and attenuates bone marrow cytotoxicity. *European Journal of Immunology*, Vol.41, No.6, (June 2011), pp. 1774-1786, ISSN: 0014-2980

Moldt, B.; Miskey., C.; Staunstrup, N.H.; Gogol-Döring, A.; Bak, R.O.; Sharma, N.; Mátés, L.; Izsvák, Z.; Chen, W.; Ivics, Z. & Mikkelsen, J.G. (2011). Comparative Genomic Integration Profiling of Sleeping Beauty Transposons Mobilized With High Efficacy From Integrase-defective Lentiviral Vectors in Primary Human Cells. *Molecular Therapy*, Vol.19, No.8, (August 2011), pp. 1499-1510, ISSN 1043-0342

Moore, K.A.; Ema, H. & Lemischka, I.R. (1997). In vitro maintenance of highly purified, transplantable hematopoietic stem cells. *Blood*, Vol.89, No.12, pp. 4337-4347, ISSN 0006-4971

Mori, M.; Yoneda-Kato, N.; Yoshida, A. & Kato, J.Y. (2008). Stable form of JAB1 enhances proliferation and maintenance of hematopoietic progenitors. *Journal of Biological Chemistry,* Vol.283, No.43, (October 2008), pp. 29011-29021, ISSN 0021-9258

Mostoslavsky, G.; Kotton, D.N.; Fabian, A.J.; Gray, J.T.; Lee, J.S. & Mulligan, R.C. (2005). Efficiency of transduction of highly purified murine hematopoietic stem cells by lentiviral and oncoretroviral vectors under conditions of minimal in vitro manipulation. *Molecular Therapy,* Vol.11, No.6, (June 2005), pp. 932-940, ISSN 1525-0016

Mullican, S.E.; Zhang, S.; Konopleva, M.; Ruvolo, V.; Andreeff, M.; Milbrandt, J. & Conneely, O.M. (2007). Abrogation of nuclear receptors Nr4a3 and Nr4a1 leads to development of acute myeloid leukemia. *Nature Medicine,* Vol.13, No.6, (June 2007), pp. 730-735, ISSN 1078-8956

Neff, T.; Beard, B.C.; Peterson, L.J.; Anandakumar, P.; Thompson, J. & Kiem, H.P. (2005). Polyclonal chemoprotection against temozolomide in a large-animal model of drug resistance gene therapy. *Blood,* Vol.105, No.3, (February 2005), pp. 997-1002, ISSN 0006-4971

Nemeth, M.J.; Topol, L.; Anderson, S.M.; Yang, Y. & Bodine, D.M. (2007). Wnt5a inhibits canonical Wnt signaling in hematopoietic stem cells and enhances repopulation. *Proceedings of the National Academy of Sciences of the United States of America,* Vol.104, No.39, (September 2007), pp. 15436-15441, ISSN 0027-8424

Nemeth, M.J.; Mak, K.K.; Yang, Y. & Bodine, D.M. (2009). beta-Catenin expression in the bone marrow microenvironment is required for long-term maintenance of primitive hematopoietic cells. *Stem Cells,* Vol.27, No.5, (May 2009), pp. 1109-1119, ISSN 1066-5099

Nielsen, T.T.; Jakobsson, J.; Rosenqvist, N. & Lundberg, C. (2009). Incorporating double copies of a chromatin insulator into lentiviral vectors results in less viral integrants. *BMC Biotechnology,* Vol.9, (February 2009), pp. 9-13, ISSN 1472-6750

North, T.E.; Goessling, W.; Walkley, C.R.; Lengerke, C.; Kopani, K.R.; Lord, A.M.; Weber, G.J.; Bowman, T.V.; Jang, I.H.; Grosser, T.; Fitzgerald, G.A.; Daley, G.Q.; Orkin, S.H. & Zon, L.I. (2007). Prostaglandin E2 regulates vertebrate haematopoietic stem cell homeostasis. *Nature,* Vol.447, No.7147, (June 2007), pp. 1007-1011, ISSN 0028-0836

Ohta, H.; Sekulovic, S.; Bakovic, S.; Eaves, C.J.; Pineault, N.; Gasparetto, M.; Smith, C.; Sauvageau, G. & Humphries, R.K. (2007). Near-maximal expansions of hematopoietic stem cells in culture using NUP98-HOX fusions. *Experimental Hematology,* Vol.35, No.5, (May 2007), pp. 817-830, ISSN 0301-472X

Ott, M.G.; Schmidt, M.; Schwarzwaelder, K.; Stein, S.; Siler, U.; Koehl, U.; Glimm, H.; Kühlcke, K.; Schilz, A.; Kunkel, H.; Naundorf, S.; Brinkmann, A.; Deichmann, A.; Fischer, M.; Ball, C.; Pilz, I.; Dunbar, C.; Du, Y.; Jenkins, N.A.; Copeland, N.G.; Lüthi, U.; Hassan, M.; Thrasher, A.J.; Hoelzer, D.; von Kalle, C.; Seger, R. & Grez, M. (2006). Correction of X-linked chronic granulomatous disease by gene therapy, augmented by insertional activation of MDS1-EVI1, PRDM16 or SETBP1. *Nature Medicine,* Vol.12, No.4, (April 2006), pp. 401-409, ISSN 1078-8956

Pan, D. (2009). In situ (in vivo) gene transfer into murine bone marrow stem cells. *Methods in Molecular Biology,* Vol.506, pp. 159-169, ISSN 1064-3745

Park, I.K.; Qian, D.; Kiel, M.; Becker, M.W.; Pihalja, M.; Weissman, I.L.; Morrison, S.J. & Clarke, M.F. (2003). Bmi-1 is required for maintenance of adult self-renewing haematopoietic stem cells. *Nature,* Vol.423, No.6937, (May 2003), pp. 302-305, ISSN 0028-0836

Peled, T.; Landau, E.; Mandel, J.; Glukhman, E.; Goudsmid, N.R.; Nagler, A. & Fibach, E. (2004). Linear polyamine copper chelator tetraethylenepentamine augments long-term ex vivo expansion of cord blood-derived CD34+ cells and increases their engraftment potential in NOD/SCID mice. *Experimental Hematology,* Vol.32, No.6, (June 2004), pp. 547-555, ISSN 0301-472X

Persons, D.A.; Allay, E.R.; Sawai, N.; Hargrove, P.W.; Brent, T.P.; Hanawa, H.; Nienhuis, A.W. & Sorrentino, B.P. (2003). Successful treatment of murine beta-thalassemia using in vivo selection of genetically modified, drug-resistant hematopoietic stem cells. *Blood,* Vol.102, No.2, (July 2003), pp. 506-513, ISSN 0006-4971

Podda, S.; Ward, M.; Himelstein, A.; Richardson, C.; de la Flor-Weiss, E.; Smith, L.; Gottesman, M.; Pastan, I. & Bank, A. (1992). Transfer and expression of the human multiple drug resistance gene into live mice. *Proceedings of the National Academy of Sciences of the United States of America,* Vol. 89, No.20, (October 1992), pp. 9676–9680, ISSN 0027-8424

Puthenveetil, G.; Scholes, J.; Carbonell, D.; Qureshi, N.; Xia, P.; Zeng, L.; Li, S.; Yu, Y.; Hiti, A.L.; Yee, J.K. & Malik, P. (2004). Successful correction of the human beta thalassemia major phenotype using a lentiviral vector. *Blood,* Vol.104, No.12, (December 2004), pp. 3445-3453, ISSN 0006-4971

Ragg, S.; Xu-Welliver, M.; Bailey, J.; D'Souza, M.; Cooper, R.; Chandra, S.; Seshadri, R.; Pegg, A.E. & Williams, D.A. (2000). Direct reversal of DNA damage by mutant methyltransferase protein protects mice against dose intensified chemotherapy and leads to in vivo selection of hematopoietic stem cells. *Cancer Research,* Vol.60, No.18, (September 2000), pp. 5187–5195, ISSN 0008-5472

Ramirez, P.; Rettig, M.P.; Uy, G.L.; Deych, E.; Holt, M.S.; Ritchey, J.K. & DiPersio, J.F. (2009). BIO5192, a small molecule inhibitor of VLA-4, mobilizes hematopoietic stem and progenitor cells. *Blood,* Vol.114, No.7, (August 2009), pp. 1340-1343, ISSN 0006-4971

Reya, T.; Duncan, A.W.; Ailles, L.; Domen, J.; Scherer, D.C.; Willert, K.; Hintz, L.; Nusse, R. & Weissman I.L. (2003). A role for Wnt signalling in self-renewal of haematopoietic stem cells. *Nature,* Vol.423, No.6938, (May 2003), pp. 409-414, ISSN 0028-0836

Richard, E.; Robert, E.; Cario-Andreé, M.; Ged, C.; Géronimi, F.; Gerson, S.L.; de Verneuil, H. & Moreau-Gaudry, F. (2004). Hematopoietic stem cell gene therapy of murine protoporphyria by methylguanine-DNA methyltransferase- mediated in vivo drug selection. *Gene Therapy,* Vol.11, No.22, (November 2004), pp. 1638-1647, ISSN 0969-7128

Rizo, A.; Dontje, B.; Vellenga, E.; de Haan, G. & Schuringa, J.J. (2008). Long-term maintenance of human hematopoietic stem/progenitor cells by expression of BMI1. *Blood,* Vol.111, No.5, (March 2008), pp. 2621-2630, ISSN 0006-4971

Rohrabaugh, S.L.; Campbell, T.B.; Hangoc, G. & Broxmeyer, H.E. (2011). Ex vivo rapamycin treatment of human cord blood CD34(+) cells enhances their engraftment of NSG mice. *Blood Cells, Molecules, & Diseases,* Vol.46, No.4, (April 2011), pp. 318-320, ISSN 1079-9796

Sangeetha, V.M.; Kale, V.P. & Limaye, LS. (2010). Expansion of cord blood CD34 cells in presence of zVADfmk and zLLYfmk improved their in vitro functionality and in vivo engraftment in NOD/SCID mouse. *PLoS One,* Vol.5, No.8, (August 2010), pp. e12221, ISSN 1932-6203

Sato, T.; Onai, N.; Yoshihara, H.; Arai, F.; Suda, T. & Ohteki, T. (2009). Interferon regulatory factor-2 protects quiescent hematopoietic stem cells from type I interferon-dependent exhaustion. *Nature Medicine,* Vol.15, No.6, (June 2009), pp. 696-700, ISSN 1078-8956

Sauvageau, G.; Thorsteinsdottir, U.; Eaves, C.J.; Lawrence, H.J.; Largman, C.; Lansdorp, P.M. & Humphries, R.K. (1995). Overexpression of HOXB4 in hematopoietic cells causes the selective expansion of more primitive populations in vitro and in vivo. *Genes & Development,* Vol.9, No.14, (July 1995), pp. 1753–1765, ISSN. 0890-9369

Sawai, N.; Zhou, S.; Vanin, E.; Houghton, P.; Brent, T. & Sorrentino, B. (2001). Protection and in Vivo Selection of Hematopoietic Stem Cells Using Temozolomide, O6-Benzylguanine, and an Alkyltransferase-Expressing Retroviral Vector. *Molecular Therapy,* Vol.3, No.1, (January 2001), pp. 78–87, ISSN 1525-0016

Scheller, M.; Huelsken, J.; Rosenbauer, F.; Taketo, M.M.; Birchmeier, W.; Tenen, D.G. & Leutz, A. (2006). Hematopoietic stem cell and multilineage defects generated by constitutive beta-catenin activation. *Nature Immunology,* Vol.7, No.10, (October 2006), pp. 1037-1047, ISSN 1529-2908

Schmidt, M.; Carbonaro, D.A.; Speckmann, C.; Wissler, M.; Bohnsack, J.; Elder, M.; Aronow, B.J.; Nolta, J.A.; Kohn, D.B. & von Kalle, C. (2003). Clonality analysis after retroviral-mediated gene transfer to CD34+ cells from the cord blood of ADA-deficient SCID neonates. *Nature Medicine,* Vol.9, No.4, (April 2003), pp. 463–468, ISSN 1078-8956

Sellers, S.; Gomes, T.J.; Larochelle, A.; Lopez, R.; Adler, R.; Krouse, A.; Donahue, R.E.; Childs, R.W. & Dunbar, C.E. (2010). Ex vivo expansion of retrovirally transduced primate CD34+ cells results in overrepresentation of clones with MDS1/EVI1 insertion sites in the myeloid lineage after transplantation. *Molecular Therapy,* Vol.18, No.9, (September 2010), pp. 1633-1639, ISSN 1525-0016

Shepherd, B.E.; Kiem, H.P.; Lansdorp, P.M.; Dunbar, C.E.; Aubert, G.; LaRochelle, A.; Seggewiss, R.; Guttorp, P. & Abkowitz, J.L. (2007). Hematopoietic stem-cell behavior in nonhuman primates. *Blood,* Vol.110, No.6, (September 2007) pp. 1806-1813, ISSN 0006-4971

Sorrentino, B.P.; Brandt, S.J.; Bodine, D.; Gottesman, M.; Pastan, I.; Cline, A. & Nienhuis A.W. (1992). Selection of drug-resistant bone marrow cells in vivo after retroviral transfer of human MDR1. *Science,* Vol.257, No.5066, (July 1992), pp. 99-103, ISSN 0036-8075

Staunstrup, N.H.; Moldt, B.; Mátés, L.; Villesen, P.; Jakobsen, M.; Ivics, Z.; Izsvák, Z. & Mikkelsen, J.G. (2009). Hybrid lentivirus-transposon vectors with a random integration profile in human cells. *Molecular Therapy,* Vol.17, No.7, (July 2009), pp. 1205-1214, ISSN 1525-0016

Stein, S.; Ott, M.G.; Schultze-Strasser, S.; Jauch, A.; Burwinkel, B.; Kinner, A.; Schmidt, M.; Krämer, A.; Schwäble, J.; Glimm, H.; Koehl, U.; Preiss, C.; Ball, C.; Martin, H.; Göhring, G.; Schwarzwaelder, K.; Hofmann, W.K.; Karakaya, K.; Tchatchou, S.; Yang, R.; Reinecke, P.; Kühlcke, K.; Schlegelberger, B.; Thrasher, A.J.; Hoelzer, D.; Seger, R.; von Kalle, C. & Grez, M. (2010). Genomic instability and myelodysplasia with monosomy 7 consequent to EVI1 activation after gene therapy for chronic granulomatous disease. *Nature Medicine,* Vol.16, No.2, (February 2010), pp. 198-204, ISSN 1078-8956

Szabo, E.; Rampalli, S.; Risueño, R.M.; Schnerch, A.; Mitchell, R.; Fiebig-Comyn, A.; Levadoux-Martin, M. & Bhatia, M. (2010). Direct conversion of human fibroblasts to multilineage blood progenitors. *Nature,* Vol.468, No.7323, (November 2010), pp. 521-526, ISSN 0028-0836

Takahashi, K. & Yamanaka, S. (2006). Induction of pluripotent stem cells from mouse embryonic and adult fibroblast cultures by defined factors. *Cell,* Vol.126, No.4, (August 2006), pp. 663-676, ISSN 0092-8674

Tan, J.; Liu, T.; Hou, L.; Meng, W.; Wang, Y.; Zhi, W. & Deng, L. (2010). Maintenance and expansion of hematopoietic stem/progenitor cells in biomimetic osteoblast niche. *Cytotechnology,* Vol.62, No.5, (October 2010), pp. 439-448, ISSN 0920-9069

Toda, H.; Yamamoto, M.; Kohara, H. & Tabata, Y. (2011). Orientation-regulated immobilization of Jagged1 on glass substrates for ex vivo proliferation of a bone marrow cell population containing hematopoietic stem cells. *Biomaterials,* Vol.32, No.29, (October 2011), pp. 6920-6928, ISSN 0142-9612

Trobridge, G. & Russell, D.W. (2004). Cell cycle requirements for transduction by foamy virus vectors compared to those of oncovirus and lentivirus vectors. Journal of Virology, Vol.78, No.5, (March 2004), pp. 2327–2335, ISSN 0022-538X

Trobridge, G.D.; Wu, R.A.; Beard, B.C.; Chiu, S.Y.; Muñoz, N.M.; von Laer, D.; Rossi, J.J. & Kiem, H.P. (2009). Protection of stem cell-derived lymphocytes in a primate AIDS gene therapy model after in vivo selection. *PLoS ONE,* Vol.4, No.11, (November 2009), pp. e7693, ISSN 1932-6203

Trowbridge, J.J.; Xenocostas, A.; Moon, R.T. & Bhatia, M. (2006). Glycogen synthase kinase-3 is an in vivo regulator of hematopoietic stem cell repopulation. *Nature Medicine,* Vol.12, No.1, (January 2006), pp. 89-98, ISSN 1078-8956

Uchida, N.; Sutton, R.E.; Friera, A.M.; He, D.; Reitsma, M.J.; Chang, W.C.; Veres, G.; Scollay, R. & Weissman, I.L. (1998). HIV, but not murine leukemia virus, vectors mediate high efficiency gene transfer into freshly isolated G0/G1 human hematopoietic stem cells. *Proceedings of the National Academy of Sciences of the United States of America,* Vol.95, No.20, (September 1998), pp. 11939–11944, ISSN 0027-8424

Uchida, N.; Washington, K.N.; Lap, C.J.; Hsieh, M.M. & Tisdale, J.F. (2011). Chicken HS4 insulators have minimal barrier function among progeny of human hematopoietic cells transduced with an HIV1-based lentiviral vector. *Molecular Therapy,* Vol.19, No.1, (January 2011), pp. 133-139, ISSN 1525-0016

Ueno, H.; Sakita-Ishikawa, M.; Morikawa, Y.; Nakano, T.; Kitamura, T. & Saito, M. (2003). A stromal cell-derived membrane protein that supports hematopoietic stem cells. *Nature Immunology,* Vol.4, No.5, (May 2003), pp. 457-463, ISSN 1529-2908

VandenDriessche, T.; Ivics, Z.; Izsvák, Z. & Chuah, M.K. (2009). Emerging potential of transposons for gene therapy and generation of induced pluripotent stem cells. *Blood,* Vol.114, No.8, (August 2009), pp. 1461-1468, ISSN 0006-4971

Varnum-Finney, B.; Xu, L.; Brashem-Stein, C.; Nourigat, C.; Flowers, D.; Bakkour, S.; Pear, W.S. & Bernstein, I.D. (2000). Pluripotent, cytokine-dependent, hematopoietic stem cells are immortalized by constitutive Notch1 signaling. *Nature Medicine,* Vol.6, No.11, (November 2000), pp. 1278-1281, ISSN 1078-8956

Vercauteren, S.M. & Sutherland, H.J. (2004). Constitutively active Notch4 promotes early human hematopoietic progenitor cell maintenance while inhibiting differentiation and causes lymphoid abnormalities in vivo. *Blood,* Vol.104, No.8, (October 2004), pp. 2315-2322, ISSN: 0006-4971

Wagner, W.; Roderburg, C.; Wein, F.; Diehlmann, A.; Frankhauser, M.; Schubert, R.; Eckstein, V. & Ho, A.D. (2007). Molecular and secretory profiles of human mesenchymal stromal cells and their abilities to maintain primitive hematopoietic progenitors. *Stem Cells,* Vol.25, No.10, (October 2007), pp. 2638-2647, ISSN 1066-5099

Wang, G.P.; Berry, C.C.; Malani, N.; Leboulch, P.; Fischer, A.; Hacein-Bey-Abina, S.; Cavazzana-Calvo, M. & Bushman, F.D. (2010). Dynamics of gene-modified progenitor cells analyzed by tracking retroviral integration sites in a human SCID-X1 gene therapy trial. *Blood,* Vol.115, No.22, (June 2010), pp. 4356-4366, ISSN 0006-4971

Wang, X.; Hisha, H.; Taketani, S.; Inaba, M.; Li, Q.; Cui, W.; Song, C.; Fan, T.; Cui, Y.; Guo, K.; Yang, G.; Fan, H.; Lian, Z.; Gershwin, M.E. & Ikehara, S. (2005). Neural cell adhesion molecule contributes to hemopoiesis-supporting capacity of stromal cell lines. *Stem Cells,* Vol.23, No.9, (October 2005), pp. 1389-1399, ISSN 1066-5099

Wang, Y.; Kellner, J.; Liu, L. & Zhou, D. (2011). Inhibition of p38 Mitogen-Activated Protein Kinase Promotes Ex Vivo Hematopoietic Stem Cell Expansion. *Stem Cells and Development,* Vol.20, No.7, (July 2011), pp. 1143-1152, ISSN 1547-3287

Watts, K.L.; Zhang, X.; Beard, B.C.; Chiu, S.Y.; Trobridge, G.D.; Humphries, R.K. & Kiem, H.P. (2011). Differential Effects of HOXB4 and NUP98-HOXA10hd on Hematopoietic Repopulating Cells in a Nonhuman Primate Model. *Human Gene Therapy,* (September 2011), advance online publication, ISSN 1525-0016

Willert, K.; Brown, J.D.; Danenberg, E.; Duncan, A.W.; Weissman, I.L.; Reya, T.; Yates, J.R. 3rd & Nusse, R. (2003). Wnt proteins are lipid-modified and can act as stem cell growth factors. *Nature,* Vol.423, No.6938, (May 2003), pp. 448-452, ISSN 0028-0836

Xie, J.; Larochelle, A.; Maric, I.; Faulhaber, M.; Donahue, R.E. & Dunbar, C.E. (2010). Repetitive busulfan administration after hematopoietic stem cell gene therapy associated with a dominant HDAC7 clone in a nonhuman primate. *Human Gene Therapy,* Vol.21, No.6, (June 2010), pp. 695-703, ISSN 1525-0016

Xue, X.; Huang, X.; Nodland, S.E.; Mátés, L.; Ma, L.; Izsvák, Z.; Ivics, Z.; LeBien, T.W.; McIvor, R.S.; Wagner, J.E. & Zhou, X. (2009). Stable gene transfer and expression in cord blood-derived CD34+ hematopoietic stem and progenitor cells by a hyperactive Sleeping Beauty transposon system. *Blood,* Vol.114, No.7, (August 2009), pp. 1319-1330, ISSN 0006-4971

Yoshihara, H.; Arai, F.; Hosokawa, K.; Hagiwara, T.; Takubo, K.; Nakamura, Y.; Gomei, Y.; Iwasaki, H.; Matsuoka, S.; Miyamoto, K.; Miyazaki, H.; Takahashi, T. & Suda, T. (2007). Thrombopoietin/MPL signaling regulates hematopoietic stem cell quiescence and interaction with the osteoblastic niche. *Cell Stem Cell,* Vol.1, No.6, (December 2007), pp. 685-697, ISSN 1934-5909

Yuan, Y.; Tse, K.T.; Sin, F.W.; Xue, B.; Fan, H.H.; Xie, Y. & Xie, Y. (2011). Ex vivo amplification of human hematopoietic stem and progenitor cells in an alginate three-dimensional culture system. *International Journal of Laboratory Hematology,* Vol.33, No.5, (October 2011), pp. 516-525, ISSN 1751-5521

Yusa, K.; Zhou, L.; Li, M.A.; Bradley, A. & Craig, N.L. (2011). A hyperactive piggyBac transposase for mammalian applications. *Proceedings of the National Academy of Sciences of the United States of America,* Vol.108, No.4, (January 2011), pp. 1531-1536, ISSN 0027-8424

Zayed, H.; Izsvák, Z.; Walisko, O. & Ivics, Z. (2004). Development of hyperactive sleeping beauty transposon vectors by mutational analysis. *Molecular Therapy,* Vol. 9, No.2, (February 2004), pp. 292-304, ISSN 1525-0016.

Zhang, C.C.; Kaba, M.; Ge, G.; Xie, K.; Tong, W.; Hug, C. & Lodish, H.F. (2006). Angiopoietin-like proteins stimulate ex vivo expansion of hematopoietic stem cells. *Nature Medicine,* Vol.12, No.2, (February 2006), pp. 240-245, ISSN 1078-8956

Zhang, J.; Niu, C.; Ye, L.; Huang, H.; He, X.; Tong, W.G.; Ross, J.; Haug, J.; Johnson, T.; Feng, J.Q.; Harris, S.; Wiedemann, L.M.; Mishina, Y. & Li, L. (2003). Identification of the haematopoietic stem cell niche and control of the niche size. *Nature,* Vol.425, No.6960, (October 2003), pp. 836-841, ISSN 0028-0836

Zhang, P.; Iwasaki-Arai, J.; Iwasaki, H.; Fenyus, M.L.; Dayaram, T.; Owens, B.M.; Shigematsu, H.; Levantini, E.; Huettner, C.S.; Lekstrom-Himes, J.A.; Akashi, K. & Tenen, D.G. (2004). Enhancement of hematopoietic stem cell repopulating capacity and self-renewal in the absence of the transcription factor C/EBP alpha. *Immunity,* Vol.21, No.6, (December 2004), pp. 853-863, ISSN 1074-7613

Zhang, X. & Roth, M.J. (2010). Antibody-directed lentiviral gene transduction in early immature hematopoietic progenitor cells. *The Journal of Gene Medicine,* Vol.12, No.12, (December 2010), pp. 945-955, ISSN 1099-498X

Zhang, X.B.; Beard, B.C.; Beebe, K.; Storer, B.; Humphries, R.K. & Kiem, H.P. (2006). Differential effects of HOXB4 on nonhuman primate short- and long-term repopulating cells. *PLoS Medicine,* Vol.3, No.5, (May 2006), pp. e173, ISSN 1549-1277

Zhang, X.B.; Beard, B.C.; Trobridge, G.D.; Wood, B.L.; Sale, G.E.; Sud, R.; Humphries, R.K. & Kiem H.P. (2008). High incidence of leukemia in large animals after stem cell gene therapy with a HOXB4-expressing retroviral vector. *The Journal of Clinical Investigation,* Vol.118, No.4, (April 2008), pp. 1502–1510, ISSN 0021-9738

Zhu, J.; Zhang, Y.; Joe, G.J.; Pompetti, R. & Emerson, S.G. (2005). NF-Ya activates multiple hematopoietic stem cell (HSC) regulatory genes and promotes HSC self-renewal. *Proceedings of the National Academy of Sciences of the United States of America,* Vol.102, No.33, (August 2005), pp. 11728-11733, ISSN 0027-8424

Zielske, S.P.; Reese, J.S.; Lingas, K.T.; Donze, J.R. & Gerson, S.L. (2003). In vivo selection of MGMT(P140K) lentivirus–transduced human NOD/SCID repopulating cells without pretransplant irradiation conditioning. *The Journal of Clinical Investigation,* Vol.12, No.10, (November 2003), pp. 1561-1570, ISSN 0021-9738